Modern Birkhäuser Classics

Many of the original research and survey monographs in pure and applied mathematics published by Birkhäuser in recent decades have been groundbreaking and have come to be regarded as foundational to the subject. Through the MBC Series, a select number of these modern classics, entirely uncorrected, are being re-released in paperback (and as eBooks) to ensure that these treasures remain accessible to new generations of students, scholars, and researchers.

Hendrik Van Maldeghem

Generalized Polygons

Reprint of the 1998 Edition

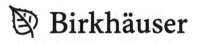 Birkhäuser

Hendrik Van Maldeghem
Department of Mathematics
Ghent University
Krijgslaan 281, S22
9000 Gent
Belgium

ISBN 978-3-0348-0270-3   e-ISBN 978-3-0348-0271-0
DOI 10.1007/978-3-0348-0271-0
Springer Basel Dordrecht Heidelberg London New York

Library of Congress Control Number: 2011941498

Mathematics Subject Classification (2010): 51E12, 51-02, 51E24, 51A50, 51A20, 51H15

© Springer Basel AG 1998
Reprint of the 1st edition 1998 by Birkhäuser Verlag, Switzerland
Originally published as volume 93 in the Monographs in Mathematics series

Printed on acid-free paper

Springer Basel AG is part of Springer Science + Business Media
(www.birkhauser-science.com)

To

Kristien

&

Lieselot, Steven and Jonas

Contents

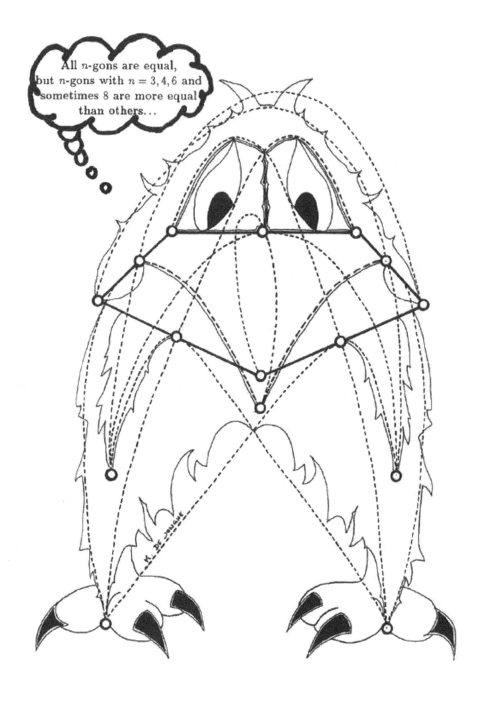

Preface

This book is intended to be an introduction to the fascinating theory of generalized polygons for both the graduate student and the specialized researcher in the field. It gathers together a lot of basic properties (some of which are usually referred to in research papers as belonging to folklore) and very recent and sometimes deep results. I have chosen a fairly strict geometrical approach, which requires some knowledge of basic projective geometry. Yet, it enables one to prove some typically group-theoretical results such as the determination of the automorphism groups of certain Moufang polygons. As such, some basic group-theoretical knowledge is required of the reader.

The notion of a generalized polygon is a relatively recent one. But it is one of the most important concepts in incidence geometry. Generalized polygons are the building bricks of Tits buildings. They are the prototypes and precursors of more general geometries such as partial geometries, partial quadrangles, semi-partial geometries, near polygons, Moore geometries, etc. The main examples of generalized polygons are the natural geometries associated with groups of Lie type of relative rank 2. This is where group theory comes in and we come to the historical *raison d'être* of generalized polygons.

In 1959 Jacques Tits discovered the simple groups of type $^3D_4$ by classifying the trialities with at least one absolute point of a D_4-geometry. The method was predominantly geometric, and so not surprisingly the corresponding geometries (the twisted triality hexagons) came into play. Generalized hexagons were born. In an appendix to his paper on trialities, Tits introduced for the first time the notion of a generalized polygon, remarking that generalized quadrangles are the geometries belonging to the classical groups of type B_2 and C_2. Of course, this birth date is only official, because these geometries were already implicitly present in some of Tits' earlier papers. Also, generalized quadrangles have already been around for a while as quadrics of Witt index 2, or as line systems corresponding to symplectic polarities in three-dimensional projective space over a field (and as such, some characterizations of the symplectic quadrangles already existed). But the explicit idea of studying geometries like generalized polygons is due to Tits. By the way, the class of generalized polygons includes the class of projective planes (and these had already been studied in depth).

*   *   *   *   *

The starting point of this book was a remark made by Bill Kantor in the (Como) Springer Lecture Notes **1181** of 1986 (page 82): "It would be tempting to write a book on this subject". At that time, though, I was not thinking at all about polygons. But I was some five years later. In 1991, during a visit to Braunschweig, Theo Grundhöfer brought up this very line of Kantor's again. I felt attracted and we decided to write a book on polygons together. The actual start was delayed for a while. From 1992 on I occasionally worked on this book, and by 1995 I had written a fair amount. Due to various circumstances, Theo had to drop out. I would never have started the project on my own, but it was already in such an advanced state that I could not give up. And I kept recieving valuable help from Theo. So I decided to go on *alone* and the result is this book. The word *alone* is not to be taken too literally. Indeed, there are some people whom I would like to thank explicitly for their very kind help.

Of course I am indebted to Theo Grundhöfer to start with. Half of the book was written while I was staying in Würzburg, happily enjoying a grant from the DAAD, and the warm hospitality of Theo and his associates, especially Linus Kramer.

Then there is Jacques Tits, who gave some magical lectures at the Collège de France on Moufang polygons. I learned a lot there and I used it in many places in this book. But I am especially grateful to Tits for the time he took to listen to me and answer my questions. He also read part of a preliminary manuscript and needless to say his remarks were both very valuable and stimulating.

Francis Buekenhout, Frank De Clerck, Michael Joswig, Bill Kantor, Norbert Knarr, Linus Kramer, Bernhard Mülherr, Allen Offer, Stanley Payne, Anja Steinbach, Jef Thas, Richard Weiss (in alphabetical order) all read (large) parts of the manuscript and provided very helpful remarks.

I am also greatly indebted to my students Leen Brouns and Eline Govaert for detecting numerous typos and other mistakes while they were reading a preliminary draft of this book for their master thesis. In this connection, I also have to thank the students attending my course in "Buildings and the Geometry of Diagrams" at the University of Ghent in the fall of 1996. My course was based on the manuscript of this book, and while preparing for their exam they discovered some more deficiencies. I would like especially to mention Anne-Marie Acou, Bram Desmet, Francis Gardeyn, Inge Oosterlinck, Nick Sabbe and Valery Vermeulen.

$$* \quad * \quad * \quad * \quad *$$

As I already mentioned, I have chosen a geometrical approach to the theory of generalized polygons. Of course, not everything can be proved with synthetic geometry. Therefore, I have also chosen to work with coordinates, as this is in my opinion the middle way between geometry and algebra. On the one hand it is a very elementary technique, but on the other hand strong enough, for instance, to understand the geometry of the Ree–Tits octagons, or to establish the structure of some automorphism groups. Since the emphasis is on geometry, I have ignored a lot of other features of polygons; this is motivated by the fact that the present book is already voluminous enough. So you will find for instance neither the construction of hexagons with Cayley algebras due to Schellekens, nor the explicit description of all Jordan algebras giving rise to Moufang hexagons. Even worse: there is no rigorous existence proof for the Moufang octagons in this book. But coordinatization and commutation relations provide an elementary description and that is enough for the purpose of this book. In addition, I have given geometric evidence of the existence by proving that a polarity with at least one absolute point in a building of type F_4 gives rise to a generalized octagon.

Since the main goal of this book is to develop a geometric theory for polygons, I have dedicated the first chapter to a lot of elementary geometric properties, every one of which is known, but few of which are in fact explicitly written down somewhere. This should provide a good reference for this material. I have reserved no space for developing matrix techniques, and so some results in that direction are not proved here (the bulk of the theorem of Feit and Higman — which requires such matrix techniques — is proved in an appendix). A consequence of the geometric approach is the fact that I occasionally use notions and results from the theory of buildings. It is not necessary to have a full understanding of that theory, but it helps if one knows some of the major highlights. For instance, I use polar spaces and D_4-geometries to introduce some examples of Moufang hexagons in Chapter 2, but I have tried to make the exposition as self-contained as possible. Of course, projective geometry is a necessary prerequisite and I expect the reader to be familiar with the general theory of projective spaces over skew fields. I do, however, occasionally review some important elementary aspects of projective geometry, and also of group theory. In Chapter 3, the coordinatization of polygons is introduced and some additional examples are explicitly given. The first three chapters can be considered as the basis of the whole book.

From there on, one can read every chapter on its own. Of course there are links between the chapters, but these are *horizontal* rather than *vertical*. For instance, Chapter 4 basically investigates the properties of the automorphism groups of some classical and mixed polygons. In particular, it is proved that those polygons satisfy the *Moufang condition*. Chapter 5 treats the converse: given a Moufang polygon, can we say that it is necessarily a classical or mixed one? However, the classification of Moufang polygons is not proved in this work, as this will be the major theme of a forthcoming book of Tits and Weiss. Chapter 6 brings together the conditions under which we may conclude that a certain polygon is of a certain type (namely, a well-defined subclass of the class of Moufang polygons). So it helps if one knows about Moufang polygons, but this is not required in order to understand the arguments in this chapter. Chapter 7 deals with polarities, ovoids and spreads. We focus on the Suzuki–Tits ovoids and the Ree–Tits unitals. We devote a few words to the recently discovered Moufang quadrangles. In Chapter 8 we gather some projective properties of polygons, such as the determination of some projectivity groups and some little projective groups, and the classification of some projective embeddings. Finally, Chapter 9 is an overview of the topological counterpart of most of what has been done in the other chapters, and it is mainly expository, unlike Chapters 1 through 8, where I have tried to prove most things within a reasonable restriction of space. Some appendices conclude the book. For instance, the only explicit result concerning projective planes is in Appendix B: it is a proof by Tits of the classification of non-associative alternative division rings, independent of the characteristic. Some links between the Moufang quadrangles and the theory of algebraic groups are explained in Appendix C via Tits diagrams. I close the book with ten open problems.

Let me also point out that this book is meant to be complementary to the monograph of Payne and Thas, *Finite Generalized Quadrangles*. That is why I have neither included nor used much of the machinery developed in *op. cit.* That is also the reason why I sometimes skip proofs concerning quadrangles. But it is not necessary to have read *op. cit.* in order to understand this book. The arguments that I use are self-contained and the proofs I omit are not necessary to the rest. So primarily this book is concerned with hexagons and octagons, but it is actually fascinating to see how one can unify some things for generalized polygons, and that is my main goal.

More generalities, motivation and history are contained in the introduction to each chapter. I would like to mention one more point. There are a lot of theorems on generalized n-gons that indicate that the more interesting values of n are $3, 4, 6$, and also 8. The illustration preceding this preface serves to stress this fact. The 15 little circles together with their 15 adjoining lines form a generalized 4-gon. Drawn like that, it inspired Karen De Jonghe to turn it into a little work of art. Thanks, Karen.

$$* \quad * \quad * \quad * \quad *$$

I close this preface with some technical remarks. The notions that I define are written in **boldface**, while an undefined (but in general obvious, or irrelevant, or to be defined) notion that I use for the first time is written in *italics*. The strings $A := B$ or $B =: A$ define the new symbol A as the old one B. In the statements of the lemmas, propositions, theorems and corollaries, and sometimes in the text, I have written the words **if ... then** and **if and only if** in boldface, or, if logically subordinate to some other **if ... then** or **if and only if**, then I have underlined them. This contributes to the clarity. Some more logical structure: the end of a proof, or of a statement without proof, is marked by a little box. But when it concerns a lemma within another proof, then I use the abbreviation "QED". On top of that, the word lemma is in this case put in a box, as are other titles that contribute to the logical structure of the proof, such as "STEP I", etc. Also, I usually denote a map exponentially (but sometimes for clarity, I break my own rules), which means that I write a composition of maps from left to right.

I have included a few illustrations that might help the reader through an argument. Unfortunately, most proofs require either too many diagrams, or too complicated ones. Therefore, I would advise the reader to draw his/her own diagrams.

I have not included an author index. Instead, I have listed in the Bibliography the pages where every reference is mentioned in the book, including unpublished references.

Finally, I am very grateful to the FWO, the Fund for Scientific Research — Flanders (Belgium) for their financial support during the ten years of my scientific career.

Chapter 1

Basic Concepts and Results

1.1 Introduction

This first chapter is devoted mainly to definitions, some examples and a few basic properties. Though generalized polygons are graph-theoretically an (a posteriori) immediate generalization of *projective planes*, the definition requires some preliminaries such as distance in geometries. The preparation of the definition of a generalized polygon is the goal of the first section.

In Section 1.2, we give a few equivalent definitions of generalized polygons. Roughly, we distinguish historically between the original definition by TITS [1959], the one emerging from the more general concept of a *building* (also due to TITS [1974] and which we take as the principal definition), and a graph-theoretic definition given in TITS [1977] (see also KANTOR [1986a]). We specialize to the definition of generalized quadrangles (i.e., generalized 4-gons) and give some first examples of small generalized quadrangles and hexagons, which are related to small Chevalley groups. The constructions are not the standard ones and reflect some sporadic isomorphisms between small simple groups.

One of the most fundamental properties of generalized polygons is the existence of an *order*, i.e., for every generalized polygon Γ, there exists a pair (s, t) of cardinal numbers such that every line is incident with $s+1$ points and every point is incident with $t+1$ lines. This is introduced by considering *perspectivities* and *projectivities* in polygons. The theorem of FEIT & HIGMAN [1964] restricts, amongst other things, the possibility of orders for finite generalized polygons drastically. This result is very important and few other results in that direction are known. We offer a proof in Appendix A, since the methods are completely algebraic. We take a look at small possible orders and comment on existence and uniqueness questions.

We also prove a few results concerning subpolygons, especially with respect to *full* or *ideal* subpolygons. This kind of subpolygons does not occur in the theory of projective planes (generalized 3-gons), but only arises in generalized $2n$-gons.

1

Finally, we introduce in a general setting the various notions of *regularity conditions*. These are very important tools in characterization theorems and they basically reflect the permanent presence of a projective space somewhere . . . .

1.2 Geometries

1.2.1 Definitions. A **geometry (of rank** 2) is a triple $\Gamma = (\mathcal{P}, \mathcal{L}, \mathbf{I})$, where $\mathcal{P}, \mathcal{L}$ are disjoint non-empty sets and $\mathbf{I} \subseteq \mathcal{P} \times \mathcal{L}$ is a relation, the **incidence relation**. Often these objects are called incidence structures or incidence geometries of rank 2 (see BUEKENHOUT [1995], in particular DE CLERCK & VAN MALDEGHEM [1995]). We will often omit the specification "of rank 2". The elements of $\mathcal{P}$ are the **points**, usually denoted by $p, q, x, y, \ldots$, and elements $L, M, K, \ldots$ of $\mathcal{L}$ are called **lines** (or blocks). The sets $\{p, L\}$ with $p \in \mathcal{P}, L \in \mathcal{L}, p \, \mathbf{I} \, L$ are the **(maximal) flags** of Γ. Usually, we will talk about flags instead of maximal flags, although technically, a flag of a geometry is any set (including the empty set) of pairwise incident elements (see BUEKENHOUT [1995]). **Adjacent** flags are distinct flags which have an element in common. The standard notation for the set of flags is $\mathcal{F}$. An **antiflag** is a set $\{p, L\}$, where p and L are *not* incident.

The **double** of Γ is the geometry $2\Gamma = (\mathcal{P} \cup \mathcal{L}, \mathcal{F}, \in)$, where $\in$ denotes set-theoretic inclusion.

The **dual** of $\Gamma = (\mathcal{P}, \mathcal{L}, \mathbf{I})$ is the geometry $\Gamma^D = (\mathcal{L}, \mathcal{P}, \mathbf{I}^D)$, with $u \, \mathbf{I}^D w \iff w \, \mathbf{I} \, u$. So Γ^D is obtained from Γ by interchanging the roles of $\mathcal{P}$ and $\mathcal{L}$. Whenever convenient, we treat $\mathbf{I}$ as a symmetric relation, that is, we do not distinguish between $\mathbf{I}$ and $\mathbf{I}^D$.

For a point $p \in \mathcal{P}$, the set $\Gamma(p) = \{L \in \mathcal{L} \mid p \, \mathbf{I} \, L\}$ of all lines through p is the **pencil** of p, and for $L \in \mathcal{L}$ the **point row** of the line L is defined by $\Gamma(L) = \{p \in \mathcal{P} \mid p \, \mathbf{I} \, L\}$; it is the set of points incident with L. Pencils and point rows are in fact the *residues* of points and lines of Γ (see again BUEKENHOUT [1995]).

Two points p, q are **collinear**, in symbols $p \perp q$, if they are incident with at least one common line, i.e., $p, q \in \Gamma(L)$ for some line L. If L is unique, then we write $L = pq$ and call it the line **joining** p and q. The set $p^\perp$ is defined to be the set of all points collinear with p. In symbols:

$$ p^\perp = \{q \in \mathcal{P} \mid p \perp q\} = \bigcup_{p \, \mathbf{I} \, L} \Gamma(L). $$

Furthermore, for any set $X \subseteq \mathcal{P}$ of points, we define

$$ X^\perp = \bigcap_{x \in X} x^\perp. $$

Hence $X^\perp$ is the set of points collinear with every element of X. We call $X^\perp$ the **perp** of X. Dually, lines L, M are **confluent** or **concurrent** if they share at least one point, i.e., if $L, M \in \Gamma(p)$ for some point p, and we write

$$L^\perp = \bigcup_{p \,\mathbf{I}\, L} \Gamma(p).$$

As for a set of points, one introduces the notation $Y^\perp$ for a set Y of lines. If Z is a set of points or a set of lines then we write $Z^{\perp\perp}$ for $(Z^\perp)^\perp$.

If the lines L and M are concurrent and if there is a unique point p incident with both L and M, then we write $p = L \cap M$ and call p the **intersection point** of L, M. Furthermore, we use standard phrases such as "*a point lies on a line*", or "*a line contains a point*", or "*a line goes through a point*", or "*two lines share a point*" (as we already did above) and their meaning should be clear without explanation.

A **subgeometry** of $\Gamma = (\mathcal{P}, \mathcal{L}, \mathbf{I})$ is a geometry $\Gamma' = (\mathcal{P}', \mathcal{L}', \mathbf{I}')$ with $\mathcal{P}' \subseteq \mathcal{P}, \mathcal{L}' \subseteq \mathcal{L}$ and $\mathbf{I}' = \mathbf{I} \cap (\mathcal{P}' \times \mathcal{L}')$. When considering geometries Γ satisfying special axioms, we are usually interested only in those subgeometries which satisfy the same axioms.

A geometry Γ is called **thick** if all point rows and all line pencils have cardinalities at least 3; thus thickness means there are no "short" lines (i.e., lines with two points) and no "thin" points (points incident with exactly two lines). An element which is incident with at least three elements is sometimes called **thick** itself. If all point rows $\Gamma(L)$, $L \in \mathcal{L}$, have the same (finite or infinite) cardinality $s + 1$, and if all pencils $\Gamma(p)$, $p \in \mathcal{P}$, have the same cardinality $t + 1$, then $\Gamma = (\mathcal{P}, \mathcal{L}, \mathbf{I})$ is said to have **order** (s, t). Thus the existence of an order is a combinatorial homogeneity condition. If $s = 2$, then Γ is a **slim** geometry. If $s = t$, then Γ is said to have **order** s.

A geometry $\Gamma = (\mathcal{P}, \mathcal{L}, \mathbf{I})$ is said to be **finite** if $\mathcal{P}$ and $\mathcal{L}$ are finite sets.

1.2.2 Incidence graphs and distances

Let $\Gamma = (\mathcal{P}, \mathcal{L}, \mathbf{I})$ be a geometry. The **incidence graph** of Γ is the graph with the vertex set $V = \mathcal{P} \cup \mathcal{L}$ and the flags of Γ as edges. Thus the adjacency relation $*$ of this graph is given by

$$x * y \quad \Longleftrightarrow \quad x \,\mathbf{I}\, y \text{ or } y \,\mathbf{I}\, x$$

for $x, y \in V$; this relation $*$ is just the incidence relation $\mathbf{I}$ made symmetric. Every incidence graph is a bipartite graph, and conversely, every bipartite graph is the incidence graph of two mutually dual geometries (in the obvious way). We denote by $\delta(x, y)$ the **distance** of $x, y \in V$ in the incidence graph. This means that $\delta(x, y) = m \in \mathbb{N}_0$ if there exists a **path** of length m from x to y (or **with extremities** x, y), i.e., a sequence $(x = x_0, x_1, \ldots, x_m = y)$ with $x_0 * x_1 * \ldots * x_m$, but no shorter path; furthermore $\delta(x, y) = \infty$ means that there is no path from x to y. A path from x to itself is called **closed**. If $\delta(x, y) \neq \infty$ for all $x, y \in V$, then (the incidence graph of) Γ is said to be **connected**. A path $(x_0, x_1, \ldots, x_m)$ will be called **non-stammering** if $x_{i-1} \neq x_{i+1}$, for all $i = 1, 2, \ldots, m - 1$. Otherwise, the path is **stammering**. A non-stammering closed path will be called a **circuit**. A non-stammering path of length i will be called an i-**path**. This terminology may not be entirely standard. The term "non-stammering" is motivated by the following observation: TITS [1974] uses the word "non-stammering" for a gallery of consecutively distinct chambers in a building. Translated to our situation, this would be a sequence of adjacent flags. Now, every path $(x_0, x_1, \ldots, x_m)$ defines a sequence $(\{x_0, x_1\}, \{x_1, x_2\}, \ldots, \{x_{m-1}, x_m\})$ of flags which are consecutively adjacent or equal. If this sequence is non-stammering in the sense of *op. cit.*, i.e., if two consecutive flags are always distinct, then we have called the path itself non-stammering.

Concerning notation for paths, we will sometimes write things like $\gamma = (x = a_0, \ldots, a_n = y)$, which means that the path γ starts with the element x and ends with the element y, and it is of length n. This is made clear by putting x equal to a_0 and y to a_n; the other elements are a_i for $0 < i < n$.

We define $\Gamma_m(x)$ as the set of objects at distance m from x, i.e., $\Gamma_m(x) = \{y \in V \mid \delta(x, y) = m\}$. Similarly, we define $\Gamma_{\leq m}(x) = \{y \in V \mid \delta(x, y) \leq m\}$ for any $x \in V$. Note that $\Gamma(x) = \Gamma_1(x)$ and $x^\perp = \Gamma_2(x) \cup \{x\}$.

Obviously the triangle inequality holds for the distance function δ. Because the incidence graph is bipartite, the condition $\delta(x, y) = 1$ implies $\delta(x, z) = \delta(y, z) \pm 1$ for all $z \in V$ (here $\infty \pm 1 = \infty$).

Since the relation $*$ is just the same as the incidence relation $\mathbf{I}$ made symmetric, and since we agreed on treating $\mathbf{I}$ as a symmetric relation, we will also use $\mathbf{I}$ from now on for adjacency in the incidence graph of Γ.

1.2.3 Isomorphisms

Let $\Gamma = (\mathcal{P}, \mathcal{L}, \mathbf{I})$ and $\Gamma' = (\mathcal{P}', \mathcal{L}', \mathbf{I}')$ be two geometries. An **isomorphism**, or a **collineation**, of Γ onto Γ' is a pair of bijections $\alpha : \mathcal{P} \to \mathcal{P}'$, $\beta : \mathcal{L} \to \mathcal{L}'$ preserving incidence, or non-incidence, i.e., such that $p \, \mathbf{I} \, L \iff p^\alpha \, \mathbf{I}' \, L^\beta$ for all $p \in \mathcal{P}, L \in \mathcal{L}$. The existence of such an isomorphism is expressed by saying that Γ and Γ' are **isomorphic**, in symbols $\Gamma \cong \Gamma'$.

We shall only deal with geometries Γ and Γ' where each line L is determined uniquely by (and hence may be identified with) its point row $\Gamma(L)$. For geometries of this type, the map β is determined uniquely by α, via $\Gamma'(L^\beta) = \Gamma'(L)^\alpha = \{p^\alpha \mid p \, \mathbf{I} \, L\}$. An isomorphism may then be defined simply to be a bijection $\alpha : \mathcal{P} \to \mathcal{P}'$ which induces a bijection $\mathcal{L} \to \mathcal{L}'$.

An **anti-isomorphism** or **duality** of Γ onto Γ' is a collineation of Γ onto the dual Γ'^D of Γ'. A **correlation** of Γ is an anti-isomorphism of Γ onto itself. A **polarity** is a correlation of order 2.

All isomorphisms of a geometry Γ onto itself form a group Aut Γ, the **automorphism group** (or **collineation group**) of Γ. All isomorphisms and dualities of Γ onto itself form a group, the **correlation group** of Γ, which contains Aut Γ as a subgroup of index at most 2. Usually elements of an automorphism group of a geometry are called **automorphisms**.

1.2.4 Ordinary polygons

Let $n \geq 1$ be a natural number. An **ordinary polygon** is the geometry arising in the obvious way from a (regular) polygon in the real Euclidean plane. In fact, we may define an ordinary n-gon for $n \in \mathbb{N} \setminus \{0\}$ as the unique connected geometry of order 1 (if $n > 1$) or 0 (if $n = 1$) with n points and n lines. For each point p and for each line L of an ordinary polygon, with $p\,\mathbf{I}\,L$, there is a unique circuit $(p, L, \ldots, p)$ of length $2n$. Hence an ordinary 1-gon is just a flag, and an ordinary 2-gon has two points, which are incident with each of the two lines.

We have excluded the possibility $n = \infty$. This means that we will not consider generalized ∞-gons in the sequel.

1.3 Generalized polygons

1.3.1 Definitions. Let $n \geq 1$ be a natural number. A **weak generalized n-gon** is a geometry $\Gamma = (\mathcal{P}, \mathcal{L}, \mathbf{I})$ such that the following two axioms are satisfied:

(*i*) Γ contains no ordinary k-gon (as a subgeometry), for $2 \leq k < n$.

(*ii*) Any two elements $x, y \in \mathcal{P} \cup \mathcal{L}$ are contained in some ordinary n-gon (again as a subgeometry) in Γ, a so-called **apartment**.

A **generalized n-gon** is a weak generalized n-gon Γ which satisfies also the following axiom.

(*iii*) There exists an ordinary $(n + 1)$-gon (as a subgeometry) in Γ.

Instead of 4-gons, 5-gons, 6-gons, 8-gons, n-gons we shall also speak of quadrangles, pentagons, hexagons, octagons, and polygons.

Note that the dual of a (weak) generalized n-gon is again a (weak) generalized n-gon. Therefore, every statement has a dual form which is usually a new statement for generalized polygons. We will generally not write down explicitly this dual form and assume that it has been given. Note also that (*ii*) implies that $\delta(v, w) \leq n$ for all $v, w \in \mathcal{P} \cup \mathcal{L}$. Obviously, a geometry can be a weak generalized n-gon for at most one value of n. Two elements of a weak generalized n-gon Γ are said to be **opposite** if they have the maximal distance n. Two flags are **opposite** if each element of one flag is opposite an element of the other flag. If for two points x, y there exists a unique point collinear with both, then we denote that point by $x \bowtie y$. Dual notation holds for lines. For two elements u, w of a weak generalized n-gon, the number $n - \delta(u, w)$ is called the **codistance (between u and w)**. For any point x of Γ, we denote by $x^{\perp\!\perp}$ the set of points of Γ not opposite x; for any set of points X, we denote by $X^{\perp\!\perp}$ the intersection of all sets $x^{\perp\!\perp}$ for $x \in X$; finally $(X^{\perp\!\perp})^{\perp\!\perp}$ will be written as $X^{\perp\!\perp\perp\!\perp}$. Dual notation holds for (sets of) lines.

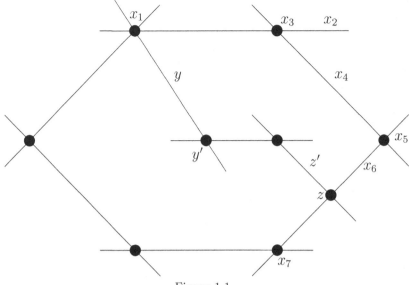

Figure 1.1.

An **ordered apartment** in a generalized n-gon Γ is any circuit of length $2n$ in Γ starting with a point (and hence ending with the same point). An **ordered $(n+1)$-gon** is likewise any circuit of length $2n + 2$ starting with a point. Let $j = n, n + 1$. An ordered (ordinary) j-gon γ is determined by the corresponding ordinary j-gon γ' and a distinguished flag $\{p, L\}$ belonging to γ' by considering the point p as the first element of γ, and L as the second. We say that γ **is ordered by** $\{p, L\}$, or that $\{p, L\}$ **puts an order on** γ'. An n-path will be called a **half apartment (bounded by its extremities)**.

The concept of a generalized polygon is due to TITS [1959]; his original definition is essentially the characterization of Lemma 1.3.5 below.

The following lemma says that for any weak generalized polygon Γ, the additional axiom (iii) holds precisely if Γ is thick. We shall see in Corollary 1.5.3 that every generalized n-gon has an order (s, t) (with $s, t \geq 2$).

1.3.2 Lemma. *A weak generalized n-gon Γ is a generalized n-gon* **if and only if** Γ *is thick.*

Proof. Assume that Γ is thick, and let $(x_0, x_1, \ldots, x_{2n} = x_0)$ be the points and lines of an ordinary n-gon, arranged in a circuit of length $2n$ (see Figure 1.1, where the situation is drawn for $n = 6$). Then $\delta(x_0, x_n) = n$ and $\delta(x_1, x_n) = n - 1 = \delta(x_{2n-1}, x_n)$. As Γ is thick, we can choose $y \in \Gamma(x_1) \setminus \{x_0, x_2\}$ and $z \in \Gamma(x_n) \setminus \{x_{n-1}, x_{n+1}\}$. Then $\delta(y, z) \equiv \delta(y, x_1) + \delta(x_1, x_n) + \delta(x_n, z) \equiv n +$

1 mod 2, Hence $\delta(y, z) \neq n$ and y is not opposite z. Considering any ordinary n-gon containing y and z, we obtain a path $(y, y' \ldots, z', z)$ of length $k < n$. Together with $(y, x_1, x_2, \ldots, x_n, z)$, this forms a closed path γ of length $n+1+k \leq 2n$. Hence $k = n - 1$ and γ is a circuit. This implies that $y' \neq x_1$ and $z' \neq x_n$. Consequently $(x_1, y, y', \ldots, z', z, x_n, x_{n+1}, \ldots, x_{2n}, x_1)$ is a circuit of length $2(n + 1)$. If it would contain less than $2(n + 1)$ distinct elements, then an ordinary ℓ-gon arises, with $\ell < n$ (precisely because the circuit is a non-stammering path). Hence we have an ordinary $(n + 1)$-gon.

Conversely, let $(x_0, x_1, \ldots, x_{2n+2} = x_0)$ be an (ordered) $(n+1)$-gon. We obviously have $\delta(x_0, x_{n+1}) \equiv n + 1 \pmod 2$, hence $\delta(x_0, x_{n+1}) < n$. Let $x_0, x', \ldots, x_{n+1}$ be a shortest path from x_0 to x_{n+1}. If $x' \in \{x_1, x_{2n+1}\}$, then we have an ordinary k-gon with $k < n$, a contradiction. Thus $\Gamma(x_0)$ contains the three distinct elements x_1, x_{2n+1}, x'. Of course, this argument shows that $|\Gamma(x_i)| \geq 3$ for $0 \leq i \leq 2n + 1$.

Now let x be an arbitrary point or line of Γ. Then $\delta(x, x_i) = n$ for some $i \in \{0, 1, \ldots, 2n + 1\}$ (otherwise, we have $\delta(x, x_i) = \delta(x, x_{i+2}) \leq n - 2$ for some i, hence an ordinary k-gon with $k < n$, a contradiction). The map $\theta : \Gamma(x) \to \Gamma(x_i)$ mapping $w \in \Gamma(x)$ to the unique element of $\Gamma(x_i)$ not opposite w is clearly a bijection (in fact, θ is the projectivity $[x; x_i]$ defined in Section 1.5). Thus $|\Gamma(x)| = |\Gamma(x_i)| \geq 3$. □

In the first paragraph of the proof above, we have shown the following result.

1.3.3 Lemma. *Let Γ be a generalized n-gon. Then every path in Γ of length at most $n + 1$ is contained in an ordinary $(n + 1)$-gon.* □

1.3.4 Remark. Let V_0 be a subgraph of the incidence graph $(V, *)$ of a weak generalized n-gon. Then every non-stammering or shortest path in V_0 of length at most n is also a shortest path in $(V, *)$, by 1.3.1(i).

The following two lemmas are graph-theoretic variants of the definition of (weak) generalized polygons.

1.3.5 Lemma. *A geometry $\Gamma = (\mathcal{P}, \mathcal{L}, \mathbf{I})$ is a weak generalized n-gon **if and only if** the following axioms hold for the distance δ in the incidence graph of Γ:*

(i) *If $x, y \in \mathcal{P} \cup \mathcal{L}$ and $\delta(x, y) = k < n$, then there is a unique path of length k joining x to y.*

(ii) *For every $x \in \mathcal{P} \cup \mathcal{L}$, we have $n = \max\{\delta(x, y) : y \in \mathcal{P} \cup \mathcal{L}\}$. (In particular, the incidence graph of Γ has diameter n.)*

Proof. For $k \geq 2$, the uniqueness of paths of length $k < n$ is equivalent to the non-existence of ordinary ℓ-gons with $\ell \leq k$; hence to 1.3.1(i). Furthermore, axiom (ii) follows from Definition 1.3.1. Thus it suffices to derive 1.3.1(ii) from the axioms

(i), (ii) above. We show that every path of length $k \leq n$ is contained in an ordinary n-gon.

By (ii), the incidence graph of Γ has no end-points, that is, every element of $\mathcal{P} \cup \mathcal{L}$ is incident with at least two elements. For otherwise

$$\max\{\delta(x_1, y) : y \in \mathcal{P} \cup \mathcal{L}\} = \max\{\delta(x_2, y) : y \in \mathcal{P} \cup \mathcal{L}\} + 1$$

for a flag $\{x_1, x_2\}$ with x_1 an end-point. Thus every path of length $k < n$ is contained in a path of length $k + 1$. Hence it suffices to consider a path $x_0, \ldots, x_n$ of length n. There exists an element $x_1' \in \Gamma(x_0) \setminus \{x_1\}$. The distance $\delta(x_1', x_n)$ differs from $\delta(x_0, x_n) = n$ by 1, hence $\delta(x_1', x_n) = n - 1$. Now a path of length $n - 1$ from x_1' to x_n completes the path $(x_0, \ldots, x_n)$ to an ordinary n-gon. $\square$

The **girth** of a graph is the length of a minimal circuit (and if there are no circuits, then the girth is put equal to ∞).

1.3.6 Lemma. *A geometry* $\Gamma = (\mathcal{P}, \mathcal{L}, \mathbf{I})$ *is a (weak) generalized n-gon* **if and only if** *the incidence graph of* Γ *is a connected graph of diameter n and girth $2n$, such that each vertex lies on at least three (at least two) edges.*

Proof. By Lemma 1.3.2 we need to prove only the statement about weak generalized n-gons. If Γ is a weak generalized n-gon, then the incidence graph of Γ has all these properties, see Definition 1.3.1.

Conversely, suppose that the incidence graph enjoys all these properties. Then 1.3.5(i) holds, and we have $\delta(x, y) \leq n$ for $x, y \in V = \mathcal{P} \cup \mathcal{L}$. It suffices to verify 1.3.5(ii). Proceeding indirectly, we consider an element $x \in V$ with $\delta(x, y) \leq n - 1$ for all $y \in V$. Fix $y \in V$ such that $\delta(x, y)$ is maximal, say equal to m. Let y', y'' be two distinct elements of $\Gamma_1(y)$ (they exist by assumption). The maximality of m implies $\delta(x, y') = \delta(x, y'') = m - 1$ (remembering that the incidence graph of Γ is bipartite). Consideration of the elements x, y, y', y'' leads to a circuit of length $\leq 2m$ which is smaller than $2n$, a contradiction. $\square$

Since the girth g of the incidence graph of a rank 2 geometry is always even, we sometimes call $g/2$ the **gonality** of both the geometry and its incidence graph. Then a generalized polygon is a thick rank 2 geometry whose incidence graph has a diameter equal to the gonality.

1.3.7 Digression: "Generalized Polygons" versus "Buildings"

In this subsection, we investigate the relation of the notion of "generalized polygon" with the more general notion of "building". We have a double reason for this. First, it is important to know that generalized polygons *are* buildings, and, second, in this book we will use a few times some results from the theory of buildings, so here is a very short introduction. For more information, there are two excellent introductory books about buildings available (RONAN [1989] and BROWN [1989]).

A **simplicial complex** (S, X) is a set S together with a set X of subsets of S, such that the union of X is S and such that every subset of every element of X itself belongs to X. The elements of X are called the **simplices**; the maximal elements of X ("maximal" in the set-theoretic sense) are called the **chambers** of (S, X) (note that the empty set is a simplex). If we remove one element from a chamber, then we talk about a **panel**. **Adjacent** chambers are chambers which meet in a panel. A **chamber complex** is a simplicial complex (S, X) such that all chambers are finite and have the same cardinality (which is then called the **rank**), and such that every two chambers can be joined by a sequence of consecutively adjacent chambers, a so-called (non-stammering) **gallery**. A chamber complex is called **thick** if every panel is in at least three chambers; it is called **thin** if every panel is in exactly two chambers. A **chamber subcomplex** (S', X') of a chamber complex (S, X) is a chamber complex with $S' \subseteq S$ and $X' = \{A \in X : A \subseteq S'\}$, and we will assume that the rank of (S, X) is equal to the rank of (S', X'). Isomorphisms between chamber complices are defined in the usual way.

In order to avoid some unnecessary technicalities, we take a somewhat more restricted definition of building than the usual one. Indeed, we will from the beginning assume that we have an underlying geometry. So we define a **chamber geometry** as a geometry $\Gamma = (\mathcal{V}_1, \mathcal{V}_2, \ldots, \mathcal{V}_j; \mathbf{I})$ of rank j (thus Γ has j different kinds of varieties and $\mathbf{I}$ is an incidence relation between the elements such that no two elements belonging to the same $\mathcal{V}_i$, $1 \leq i \leq j$, can be incident) such that the simplicial complex $(\mathcal{V}, X)$, where $\mathcal{V} = \mathcal{V}_1 \cup \mathcal{V}_2 \cdots \cup \mathcal{V}_j$ and $S \subseteq \mathcal{V}$ belongs to X if and only if every two distinct elements of S are incident, is a chamber complex. **Chambers, thickness, chamber subgeometries, galleries,** etc. are defined in the obvious way (translating the corresponding notions for chamber complices).

A **building** $(\mathcal{V}, X)$ is a thick chamber geometry $(\mathcal{V}_1, \mathcal{V}_2, \ldots, \mathcal{V}_j; \mathbf{I})$, $\mathcal{V} = \mathcal{V}_1 \cup \cdots \cup \mathcal{V}_j$, together with a set $\mathcal{A}$ of thin chamber subgeometries (called **apartments**), such that every two chambers are contained in some element of $\mathcal{A}$, and such that for every two elements $\Sigma, \Sigma' \in \mathcal{A}$ and every two simplices F, F' contained in both Σ and Σ', there exists an isomorphism $\Sigma \to \Sigma'$ which fixes all elements of both F and F'. This definition is taken from TITS [1974], up to the assumption that we already have a geometry.

It is remarkable that with these simple definitions, a very rich theory can be developed. For instance, all buildings with finite apartments and rank at least 3 can be classified, and this is the main theme of *op. cit.* Every Chevalley group, and more generally, every group of Lie type, acts in a natural way on a building, permuting the pairs (Σ, C) transitively (where Σ is an apartment, and C a chamber contained in Σ). Buildings with finite apartments are called **spherical**.

Lots of elementary properties can be mentioned and/or proved at this stage. But we choose not to go too far into detail. We mention that every apartment of a given building is isomorphic to a certain *Coxeter complex*, i.e., a thin chamber complex related to a Coxeter group. When this Coxeter group is a finite dihedral group, then this Coxeter complex (S, X) is of rank 2; the elements of S can be identified with the points and lines of an ordinary polygon, the chambers in X can be viewed as the flags of this polygon. Conversely, in the spherical rank 2 case, the apartments are Coxeter complices related to dihedral groups. We have the following result.

1.3.8 Theorem (Tits [1974]).

(i) Let $(\mathcal{V}, X)$, $\mathcal{V} = \mathcal{V}_1 \cup \mathcal{V}_2$, be a spherical building of rank 2. Then $\Gamma = (\mathcal{V}_1, \mathcal{V}_2, \mathbf{I})$ is a generalized polygon.

(ii) Conversely, if $\Gamma = (\mathcal{P}, \mathcal{L}, \mathbf{I})$ is a generalized polygon with set of flags $\mathcal{F}$, then the structure $(\mathcal{P} \cup \mathcal{L}, \emptyset \cup \{\{v\} : v \in \mathcal{P} \cup \mathcal{L}\} \cup \mathcal{F})$ is a chamber geometry of rank 2. Declaring the thin chamber geometry corresponding to any ordinary subpolygon an apartment, we obtain a spherical building of rank 2.

Thus the pairs of mutual dual generalized polygons can be viewed as the spherical buildings of rank 2.

Proof. We only prove (i), the proof of (ii) being straightforward. Suppose that the Coxeter complex is related to the dihedral group D_{2n}, i.e., every apartment Σ of (S, X) is isomorphic to the chamber geometry of an ordinary n-gon Γ^*. Since every two elements of $\mathcal{V}$ are contained in at least one apartment, we already have that the diameter of Γ is at most equal to n. It suffices to show that the gonality is equal to n. Suppose the gonality is equal to $m < n$ and let Σ' be an ordinary m-gon in Γ. Let x and y be two elements of Σ' at distance $j < m$ from each other (the distance is measured here in Σ', but a smaller value in Γ would contradict the gonality being equal to m, hence, equivalently, the distance is measured in Γ; by the same token there is a *unique* shortest path γ between x and y). We claim that every apartment Σ containing x and y contains every element of γ. Indeed, we use induction on j, the claim for $j = 1$ being trivial. So let $j > 1$. Then let $\gamma = (x, \dots, y', y)$. Let Σ'' be an apartment containing x and $\{y, y'\}$ (the latter is a chamber). There is an isomorphism $\alpha : \Sigma \to \Sigma''$ fixing x and y, hence if $\gamma^\alpha \neq \gamma$, then there arises an ordinary i-gon with $i < j$. The claim follows.

Now let u and v be opposite elements of Σ'. Let u' and v' also be opposite elements of Σ' with $u \mathbf{I} u'$ and $v \mathbf{I} v'$. Consider an apartment Σ containing $\{u, u'\}$ and $\{v, v'\}$. Then our previous claim applied to v, u' (or u, v') implies that all elements of Σ' belong to Σ, contradicting $m \neq n$. Hence the gonality is equal to n. The theorem is proved.   □

Finally, we remark that every building is assigned a *diagram*, which is basically the same as the corresponding Coxeter diagram. If the building is spherical, then these diagrams have names (such as A_n, B_n, etc., where n is the rank of the building), and by definition, the name of the diagram is the type of the building. See Appendix C for some names and types.

1.3.9 Trees and twin trees as generalized ∞-gons

We have already remarked that we do not consider the case $n = \infty$ in this book. In fact, the case $n = \infty$ corresponds to a building of type $\tilde{A}_1$, i.e., a building in which the apartments are related to the infinite dihedral group. In such buildings, which could be called *generalized ∞-gons*, there is no longer a notion of opposite elements. There is no large geometric theory available for buildings of type $\tilde{A}_1$, though they are very important objects. Geometrically, the notion of a *twin building* of type $\tilde{A}_1$, also called a *twin tree*, stands closer to the notion of generalized polygon, and there is also a richer geometric theory; see e.g. RONAN & TITS [1994]. But we will neither define twin trees nor consider them in this book.

We now look at the existence question and some examples.

1.3.10 On the existence question of generalized n-gons

There exists just one weak generalized 1-gon, viz. the ordinary 1-gon (by 1.3.1(ii)). The weak generalized 2-gons are precisely the geometries $(\mathcal{P}, \mathcal{L}, \mathbf{I})$, with $|\mathcal{P}|, |\mathcal{L}| \geq 2$, with trivial incidence relation $\mathbf{I} = \mathcal{P} \times \mathcal{L}$; their incidence graphs are the complete bipartite graphs. Thus generalized n-gons are interesting geometries only for $n \geq 3$. Except when explicitly mentioned otherwise, we assume from now on that $n \geq 3$.

For convenience, the symbol n (or occasionally $2n$) will always denote the diameter of a generalized polygon and a generalized polygon will always be assumed to be a generalized n-gon (thus with $n \geq 3$).

The generalized 3-gons are precisely the projective planes. This important observation can be easily proved: axiom 1.3.1(ii) implies that two distinct points (lines) are incident with at least one line (point), axiom 1.3.1(i) says that they are uniquely determined, and 1.3.1(iii) is the usual non-degeneracy axiom. We emphasize that generalized polygons can (and should) be regarded as natural generalizations of projective planes.

Generalized quadrangles are close relatives of projective planes. We refer to Subsection 1.4.2 for examples of generalized quadrangles.

Each ordinary n-gon is also a weak generalized n-gon of order $(1, 1)$. Furthermore, the double 2Γ of a weak generalized n-gon Γ is a weak generalized $2n$-gon, which is not thick (each line of 2Γ carries precisely two points). But one can do better: there exist (thick) generalized n-gons for every n. Examples are provided by a free construction process due to TITS [1977]; see Subsection 1.3.13 below. Let us just mention here that the double of a generalized digon (or 2-gon) is called a **dual grid**. Of course, the dual of a dual grid is called a **grid**. Hence a grid is a weak generalized quadrangle in which every point is incident with exactly two lines. Note that we do not require here that every line carries the same number of points (unlike PAYNE & THAS [1984]).

On the other hand, there are several results which indicate that the more interesting generalized n-gons have $n = 3, 4, 6$, or 8, see Theorem 1.7.1. Indeed, it seems that for $n = 5, 7$ or $n \geq 9$, no generalized n-gon has been constructed yet without invoking some free construction process like that of TITS [1977] at some stage.

Here we prove a lemma which is a very special case of Theorem 1.7.1 below due to FEIT & HIGMAN [1964]. The proof we offer is due to COOLSAET (unpublished). It only works in this special case.

1.3.11 Lemma. *There exists no generalized pentagon of order* $(2,2)$.

Proof. Let $(\mathcal{P}, \mathcal{L}, \mathbf{I})$ be a generalized pentagon of order $(2,2)$, and consider the geometry $\Gamma = (\mathcal{P}, \{\Gamma_2(p) \mid p \in \mathcal{P}\}, \in)$, with $\Gamma_2(p)$ as defined in Subsection 1.2.2. Any two distinct lines intersect in a unique point, and any two distinct points p, q are contained in a unique line of Γ. Thus Γ is a projective plane, which is finite (of order 5).

The mappings $p \mapsto \Gamma_2(p)$ and $\Gamma_2(p) \mapsto \Gamma_2(p)^{\perp} = \{p\}$ yield a polarity π of Γ. By an algebraic result of Baer, every polarity of a finite projective plane has absolute points; see HUGHES & PIPER [1973], Lemma 12.3, page 240, or DEMBOWSKI [1968](3.3.2), or Theorem 7.2.8 on page 311 in Chapter 7 (alternatively, Γ is the projective plane over the field $\mathbf{GF}(5)$, and all polarities of this plane are known). In fact, COXETER [1987] gives a geometric proof of this result for planes of order 5. But π obviously has no absolute points, a contradiction. □

1.3.12 The smallest generalized hexagon

Now we give a construction, due to TITS [1959], of a generalized hexagon of order $(2,2)$. According to COHEN & TITS [1985] (see also TITS [1959](11.5)), there exists (up to isomorphism and duality) only one generalized hexagon of order $(2,2)$.

Consider in $\mathbf{PG}(2,9)$ a Hermitian curve $\mathcal{H}$, i.e., the set of absolute or self-conjugate points with respect to a unitary polarity, see Subsection 7.5.3 for more details. A **polar triangle** is a set of three non-collinear points $\{x_1, x_2, x_3\}$ such that $x_i x_j$ is conjugate to x_k (with respect to the unitary polarity), for all $\{i, j, k\} = \{1, 2, 3\}$. Define a geometry Γ as follows. The points of Γ are the points off $\mathcal{H}$; the lines of Γ are the polar triangles and incidence is containment. We show that Γ is a generalized hexagon of order $(2,2)$. It is convenient to call the lines of Γ blocks, in order to distinguish them from the lines in $\mathbf{PG}(2,9)$.

Let x be any point of Γ and let L be any block of Γ. We claim that $\delta(x, L) \leq 5$. So we may assume that $x \notin L$ and also that x does not lie on any of the lines $x_i x_j$ of $\mathbf{PG}(2,9)$, $i, j \in \{1, 2, 3\}$, $i \neq j$, where $L = \{x_1, x_2, x_3\}$ (for otherwise either $\delta(x, L) = 1$ or $\delta(x, L) = 3$). Under these assumptions, we show that $\delta(x, L) = 5$. Let L_x be the polar line of x with respect to the unitary polarity. Suppose that L_x meets $x_i x_j$ in $\mathbf{PG}(2,9)$ in the point y_{ij}, $i, j \in \{1, 2, 3\}$, $i < j$, and suppose that all three points y_{12}, y_{13}, y_{23} lie on $\mathcal{H}$. Applying the unitary polarity, we see that the lines $x_i y_{jk}$, $\{i, j, k\} = \{1, 2, 3\}$, $j < k$, meet in the point x. Hence the points x_i, $i = 1, 2, 3$, together with x and the points y_{jk}, $1 \leq j < k \leq 3$, form a subplane of order 2, a contradiction. So without loss of generality we may assume that y_{12} does not belong to $\mathcal{H}$ and hence is a point of Γ. It is now clear that y_{12} is collinear in Γ with both x and x_3. The claim follows.

Now consider any block L. There are three points incident with L in Γ. Since there are three blocks through every point (for there are six points off $\mathcal{H}$ on every line in $\mathbf{PG}(2,9)$ not tangent to $\mathcal{H}$), the number of points at distance 3 from L is equal to 12 **if and only if** Γ does not contain triangles through L. Similarly, the number of points at distance 5 of L equals 48 **if and only if** Γ contains neither pentagons nor quadrangles nor triangles through L. But there are no other points, since the distance to L is at most 5. And the number of points of Γ is exactly equal to $(9^2 + 9 + 1) - (3^3 + 1) = 63 = 3 + 12 + 48$. Hence, since L was arbitrary, we have shown that Γ does not contain m-gons for $3 \leq m \leq 5$. Now consider a point x and a block L at distance 5 (this is possible by the previous counting). Since there are no pentagons, there is a unique block M through x at distance 4 from L. The two other blocks containing x have distance 6 to L, and similarly, the points on these blocks distinct from x lie at distance 5 from L again. So there are cycles of length 12 and hence the girth of the incidence graph of Γ is 12. Since a point and a block lie at distance at most 5 from each other, the diameter is 6. So Γ is a generalized hexagon by virtue of Lemma 1.3.6.

1.3.13 Free polygons

Free constructions of projective planes were introduced by HALL [1943]. The same idea has been used for many other geometries since then. In particular, TITS [1977] considered free polygons. His eventual aim was to construct generalized n-gons with a large automorphism group (see Subsection 4.7.1).

Following TITS [1977], we call a **partial n-gon**, $3 \leq n$, a connected geometry of rank 2 with (finite) gonality $g \geq n$, and we will also assume that it is *not* a weak generalized n-gon. Clearly, every partial n-gon is also a partial m-gon for $3 \leq m \leq n$. Now fix n and let $\Gamma^{(0)}$ be a partial n-gon. We define a geometry $\Gamma^{(i)}$, $i \in \mathbb{N}$, by induction as follows. The geometry $\Gamma^{(i)}$ arises from $\Gamma^{(i-1)}$ by adding a completely new (non-stammering) path of length $n-1$ between every two elements of $\Gamma^{(i-1)}$ which lie at distance $n+1$ from each other (and the path respects the axioms of a geometry, i.e., points are incident with lines but not with points, etc.). It is clear that, for $i \geq 1$, $\Gamma^{(i)}$ is a partial n-gon with gonality n. We show that the union Γ of the family $\{\Gamma^{(i)} : i \in \mathbb{N}\}$ is a weak generalized n-gon, and, in particular, if $\Gamma^{(0)}$ contains some ordinary $(n+1)$-gon, then Γ is a generalized n-gon.

We already remarked that the gonality of $\Gamma^{(i)}$, $i \geq 1$, is equal to n. Hence this will also be the case for Γ, since a finite circuit in Γ must already be contained in some $\Gamma^{(i)}$ (because this is the case for every point and line belonging to that circuit).

Also, if two elements have distance $d \geq n+1$ in $\Gamma^{(i)}$, then they have distance $d-2$ in $\Gamma^{(i+1)}$. This implies that the diameter of Γ is equal to n. (It cannot be less. Indeed, considering a circuit γ of length $< 2n$ in $\Gamma^{(i)}$ that does not exist in $\Gamma^{(i-1)}$, and noting that elements of $\Gamma^{(i)}$ that are not in $\Gamma^{(i-1)}$ are incident with exactly two elements, we readily see that γ contains a path of length $n-1$ which was added to $\Gamma^{(i-1)}$ to connect elements at distance less than $n+1$, a contradiction.)

Lemma 1.5.10 implies now that Γ is a weak generalized n-gon.

If $\Gamma^{(0)}$ contains some ordinary $(n+1)$-gon, then so does Γ and the result follows from the definition.

Note that every automorphism of $\Gamma^{(i)}$ extends uniquely to an automorphism of $\Gamma^{(i+1)}$. So the automorphism group of Γ contains the automorphism group of $\Gamma^{(0)}$.

More results on free constructions of generalized polygons are contained in FUNK & STRAMBACH [1991], [1995]. We will come back to free constructions only in Subsection 4.7.1 on page 160.

We end this section with a characterization of isomorphisms in the case $n > 3$.

1.3.14 Lemma. *Let* $\Gamma = (\mathcal{P}, \mathcal{L}, \mathbf{I})$ *and* $\Gamma' = (\mathcal{P}', \mathcal{L}', \mathbf{I}')$ *be two generalized n-gons, with $n \geq 4$. Let θ be a bijective map from $\mathcal{P}$ to $\mathcal{P}'$ preserving collinearity. Then θ defines a unique isomorphism from Γ to Γ'.*

Proof. Let L be a line of Γ. Let $p, q \in \Gamma_1(L)$. Then we define L^θ as the unique line in Γ' joining p^θ and q^θ. This is well defined because if $r\,\mathbf{I}\,L$, then r^θ is collinear with both p^θ and q^θ, and hence by the non-existence of ordinary triangles in Γ', this implies $r^\theta\,\mathbf{I}'L^\theta$. We first show that θ is injective over $\mathcal{L}$.

The points of an apartment Σ of Γ are mapped either onto the points of an apartment of Γ', or onto the points of a stammering closed path. In the latter case, all points must be sent to $\Gamma'(M)$, for some line M of Γ', otherwise we violate the injectivity of θ. But then by considering suitable apartments through any point p (containing elements of Σ), one similarly sees that $p^\theta\,\mathbf{I}'M$, contradicting the bijectivity of θ. Hence the points of Σ are mapped onto the points of an apartment of Γ'. This already implies that θ is injective over $\mathcal{L}$ (because any two lines can be put into an apartment). Now we show the surjectivity.

We have to show that if the points p^θ and q^θ are collinear in Γ', then the points p and q are collinear in Γ (and hence $(pq)^\theta = p^\theta q^\theta$). Assume, by way of contradiction, that p and q have distance $2j > 2$ in Γ, then there is a non-stammering path of length $2j$ in Γ' between p^θ and q^θ (this follows from the injectivity of θ on the points and lines). Since p^θ and q^θ are collinear, there is a circuit of length not more than $2 + 2j < 2n$ in Γ', a contradiction. Consequently θ is surjective onto $\mathcal{L}$. A similar argument now shows that θ^{-1} preserves incidence and the lemma is proved.   □

1.4 Generalized quadrangles

There is an extensive literature on (finite) generalized quadrangles; see PAYNE & THAS [1984]. Here we just rephrase the definition of generalized quadrangles and describe a few simple constructions. Recall that an **antiflag** is a non-incident point–line pair.

1.4.1 Lemma. *A geometry* $\Gamma = (\mathcal{P}, \mathcal{L}, \mathbf{I})$ *is a weak generalized quadrangle* **if and only if** *the following axioms* (i) *and* (ii), *or* (i) *and* $(ii)'$ *hold:*

(i) *Let* $\{p, L\}$ *be an antiflag. Then there exists a unique flag* $\{q, M\} \in \mathcal{F}$ *such that* $p \, \mathbf{I} \, M \, \mathbf{I} \, q \, \mathbf{I} \, L$.

(ii) *Every point lies on at least two but not on all lines, and dually, every line carries at least two but not all points.*

(ii)' *Every point is on at least two lines, and any two points are contained in at most one line. Also, every line carries at least two points.*

A geometry Γ *is a generalized quadrangle* **if and only if** Γ *is thick, axiom* (i) *holds, and* Γ *contains some antiflag.*

Proof. These axioms hold if Γ is a (weak) generalized quadrangle; see Definitions 1.3.1 and Lemma 1.3.2.

Conversely, assume that a geometry Γ satisfies (i) and (ii), and consider two points p, p' of Γ. If some line through p does not contain p', then (i) implies that p and p' have at most one joining line. If $\Gamma(p) \subseteq \Gamma(p')$, then (ii) gives a line L which avoids p', hence also p; the line M given by (i) contains p, hence also p'. Existence of another line $M' \neq M$ through p and p' gives a contradiction to axiom (i) (applied to $M \cap L$ and M' if these are not incident). We have shown that Γ does not contain ordinary 2-gons. By (i), it does not contain ordinary 3-gons either. This shows 1.3.1(i). The other axiom 1.3.1(ii) is now an easy consequence of (i) and 1.3.1(i). Hence Γ is a weak generalized quadrangle. Similar arguments apply if (i) and $(ii)'$ hold.

If a thick geometry Γ contains an antiflag, then obviously (ii) holds. $\qquad\square$

Also the definition of a generalized n-gon for $n \neq 4$ can be phrased as above in Lemma 1.4.1. We leave this to the interested reader.

1.4.2 Some generalized quadrangles

Denote by $\mathcal{P}$ the set of all 15 transpositions in the symmetric group $\mathbf{S}_6$ of degree 6, and by $\mathcal{L}$ the set of all 15 fixed-point-free involutions in $\mathbf{S}_6$. Define $\sigma \, \mathbf{I} \, \tau \Leftrightarrow \sigma \tau = \tau \sigma$ for $\sigma \in \mathcal{P}, \tau \in \mathcal{L}$. Then $(\mathcal{P}, \mathcal{L}, \mathbf{I})$ is a generalized quadrangle of order $(2, 2)$, as easily shown. In fact, this is the unique generalized quadrangle of order $(2, 2)$, and it is isomorphic to the symplectic quadrangle $\mathsf{W}(2)$ defined in Subsection 2.3.17

on page 63; compare PAYNE & THAS [1984](5.2.3). Note that $\mathbf{S}_6 \cong \mathbf{Sp}_4(2)$; the outer automorphisms of $\mathbf{S}_6$ are dualities of the generalized quadrangle. Another, geometric, construction is as follows. Consider in $\mathbf{PG}(2,4)$ a set $\mathcal{O}$ of six points, no three of which are collinear. Let $\mathcal{P}$ be the set of points off $\mathcal{O}$ and let $\mathcal{L}$ be the set of lines meeting $\mathcal{O}$ in exactly two points. Incidence is the restriction of the incidence relation in $\mathbf{PG}(2,4)$.

The set $\mathcal{O}$ of the previous paragraph is in fact a *hyperoval* in $\mathbf{PG}(2,4)$; see Subsection 7.5.2 on page 320.

The illustration preceding the preface shows $\mathsf{W}(2)$ (the 15 little circles are the points, and the 15 lines adjoining these points are the lines of $\mathsf{W}(2)$).

The 27 lines on a general complex cubic surface give rise to a generalized quadrangle of order $(2,4)$; compare *op. cit.* (see also GRUNDHÖFER, JOSWIG & STROPPEL [1994], Appendix). It is anti-isomorphic to the orthogonal quadrangle $\mathsf{Q}(5,2)$; see Subsection 2.3.12 on page 61.

Consider a unitary polarity in $\mathbf{PG}(3,4)$ with corresponding Hermitian variety $\mathcal{H}$ (see Remark 7.5.5 on page 322). Define the following geometry Γ — very similar to the generalized hexagon of order $(2,2)$ defined in Subsection 1.3.12 above. The points of Γ are the points off $\mathcal{H}$; the lines are the polar quadrangles (i.e., the 4-sets $\{x_1, x_2, x_3, x_4\}$ of non-coplanar points off $\mathcal{H}$ such that the plane $x_i x_j x_k$ is conjugate to the point x_ℓ, for $\{i, j, k, \ell\} = \{1, 2, 3, 4\}$). As above, one can check that Γ is a generalized quadrangle of order $(3,3)$. It is unique up to isomorphism and duality. It is in fact isomorphic to the classical symplectic quadrangle $\mathsf{W}(3)$ (see Subsection 2.3.17 again).

See also Chapter 2 for generalized quadrangles constructed from forms (the so-called *classical examples*) and Chapter 3 for more non-classical examples, in particular Subsection 3.7.

1.5   Projections and projectivities

Let $\Gamma = (\mathcal{P}, \mathcal{L}, \mathbf{I})$ be a weak generalized n-gon, and let $b \in V = \mathcal{P} \cup \mathcal{L}$. For every $x \in \Gamma_{\leq n-1}(b)$ there is a unique path $(x, \ldots, x', b)$ of length $\delta(x, b)$,; see Lemma 1.3.5. We say that the element $x' \in V$ is the **projection** $\mathrm{proj}_b x$ **of** x **onto** b. Thus the projection proj_b is a map from $\Gamma_{\leq n-1}(b)$ into $\Gamma(b)$.

Now we consider opposite elements $a, b \in V$, i.e., elements with $\delta(a, b) = n$. Then $\Gamma(a) \subseteq \Gamma_{n-1}(b)$, hence the projection considered above restricts to a mapping

$$[a; b] : \Gamma(a) \to \Gamma(b).$$

This mapping $[a; b]$ is characterized by $[a; b](x) = x' \Leftrightarrow \delta(x', b) = 1$ and $\delta(x, x') = n - 2$, for $x \in \Gamma(a)$. We call $[a; b]$ the **perspectivity** from a to b. Clearly $[a; b]$ is a bijection, with inverse $[b; a]$. Given a sequence $a_0, \ldots, a_k$ of elements of V with $\delta(a_i, a_{i+1}) = n$ for $0 \leq i < k$ we define

$$[a_0; a_1; \ldots; a_k] := [a_0; a_1][a_1; a_2] \ldots [a_{k-1}; a_k] : \Gamma(a_0) \to \Gamma(a_k),$$

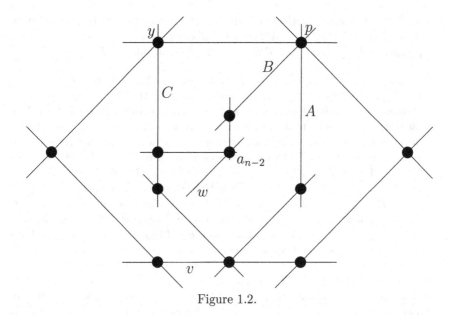

Figure 1.2.

and we call this bijection a **projectivity**. For $a_0 = a_k = a$, such a projectivity is a bijection of $\Gamma(a)$ onto itself. The set of all projectivities of $\Gamma(a)$ onto itself form a permutation group $\Pi(a)$, which is called the **group of projectivities** of a, or the **projectivity group** of a. Note that in the literature the notation $[a, b]$ is used for $[a; b]$; we prefer $[a; b]$ to avoid confusion with coordinates of lines later on (see Chapter 3).

These definitions generalize well-known concepts of projective geometry. For the definition of projections in buildings see TITS [1974](3.19) and RONAN [1989], page 33. The following lemma is implicitly in TITS [1974] and has passed into folklore. Our proof is based on KNARR [1988].

1.5.1 Lemma.   *Let $\Gamma = (\mathcal{P}, \mathcal{L}, \mathbf{I})$ be a (thick) generalized n-gon.*

(i) *There is a projectivity $\Gamma(p) \to \Gamma(q)$ for any two points $p, q \in \mathcal{P}$. If n is odd, then there exists a projectivity $\Gamma(v) \to \Gamma(w)$ for any two elements $v, w \in \mathcal{P} \cup \mathcal{L}$.*

(ii) *The group $\Pi(p)$ acts doubly transitively on $\Gamma(p)$, and all groups $\Pi(p)$ with $p \in \mathcal{P}$ are equivalent as permutation groups.*

Proof.

(i) Since Γ is connected, it suffices to consider the special case where $\delta(p, q) = 2$. By Lemma 1.3.3, p and q are contained in an ordinary $(n + 1)$-gon γ. In γ

we find an element r opposite both p and q. Thus $[p; r; q]$ is a projectivity from $\Gamma(p)$ to $\Gamma(q)$. If n is odd, then it suffices to show that there exists some perspectivity $[x; L]$ for $x \in \mathcal{P}$ and $L \in \mathcal{L}$. We can take $\{x, L\} = \{p, r\}$.

(ii) Conjugation by a projectivity as in (i) shows that all groups $\Pi(p)$ with $p \in \mathcal{P}$ are equivalent as permutation groups. For the double transitivity, it suffices to show that for all lines $L \in \Gamma(p)$ the stabilizer $\Pi(p)_L$ is transitive on $\Gamma(p) \setminus \{L\}$ (note that Γ is thick). Choose $y \in \Gamma(L) \setminus \{p\}$, and choose an element $v \in \mathcal{P} \cup \mathcal{L}$ opposite both p and y in an ordinary $(n + 1)$-gon containing p and y (see Figure 1.2, where we have drawn the situation for $n = 5$). Let $A, B \in \Gamma(p) \setminus \{L\}$, and define $C = A^{[p;v;y]}$. Considering any apartment through B and C, we see that there is a non-stammering path $\gamma' = (B = a_0, a_1, a_2, \dots, a_{2n-4} = C)$ of length $2n - 4$ joining the lines B and C, with $a_{n-2} \neq L$ (this condition is superfluous for $n \neq 4$), and we have $\delta(B, a_{n-2}) = \delta(C, a_{n-2}) = n - 2$. Choose $w \in \Gamma(a_{n-2}) \setminus \{a_{n-3}, a_{n-1}\}$. Then $[p; v; y; w; p]$ fixes L and maps A to B.   □

Both assertions of Lemma 1.5.1 are false for arbitrary weak generalized polygons. As to the transitivity assertion, it is well known that the groups $\Pi(p)$ of projectivities of projective planes ($n = 3$) are always triply transitive, but this is not generally true in generalized n-gons. For examples, and for more information on (groups of) projectivities, see Section 8.4 and KNARR [1988] (see also GRUNDHÖFER, JOSWIG & STROPPEL [1994]).

It is worth remarking here that KNARR [1988] conjectures that for finite generalized polygons Γ which are not classical polygons (one of the examples given in Chapter 2), the groups $\Pi(p)$ always contain the alternating group of $\Gamma(p)$. This is true for projective planes, possibly apart from projective planes of order 23, which might or might not exist (see GRUNDHÖFER [1988]). For finite generalized quadrangles, not much is known, apart from some characterizations in BROUNS, TENT & VAN MALDEGHEM [19**]; in particular the group of projectivities of a line L of any non-classical quadrangle (see Chapter 2) of order (s, t), with $s = t^2$, contains a non-trivial element fixing three points on L. For finite hexagons and octagons, nothing is known.

A direct consequence of Lemma 1.5.1 is the following statement.

1.5.2 Corollary. *Let* $\Gamma = (\mathcal{P}, \mathcal{L}, \mathbf{I})$ *be a generalized n-gon. Let* $v, w \in \mathcal{P} \cup \mathcal{L}$ *be arbitrary, but with the restriction that, if n is even, both v, w are points, or both v, w are lines. Let* $v_i \, \mathbf{I} \, v$ *and* $w_i \, \mathbf{I} \, w$, $i = 1, 2$, *be arbitrary,* $v_1 \neq v_2$, $w_1 \neq w_2$. *Then there exists a projectivity* $\Gamma(v) \to \Gamma(w)$ *mapping the ordered pair* (v_1, v_2) *to* (w_1, w_2).

Proof. It suffices to consider any projectivity from $\Gamma(v)$ to $\Gamma(w)$ and then apply Lemma 1.5.1(ii).   □

1.5.3 Corollary. *Every generalized n-gon has an order (s,t) (with $s,t \geq 2$); if n is odd, then $s = t$.*

Proof. Noting that the perspectivities are bijective, the existence of an order follows from Lemma 1.5.1(i) and its dual. If n is odd, then there is a perspectivity from a point row to a line pencil and hence $s = t$. □

In the next lemma, the symbol $\lfloor i \rfloor$, $i \in \mathbb{Q}$, means the largest integer smaller than or equal to i.

1.5.4 Lemma. *Let $\Gamma = (\mathcal{P}, \mathcal{L}, \mathbf{I})$ be a finite generalized n-gon of order (s,t) with $n \geq 3$, and let $x \in \mathcal{P}$. Then*

(i) **if n is even, then**

$$|\mathcal{P}| = (s+1)\frac{(st)^{\frac{n}{2}} - 1}{st - 1}, \qquad |\mathcal{L}| = (t+1)\frac{(st)^{\frac{n}{2}} - 1}{st - 1};$$

(ii) **if n is odd, then**

$$|\mathcal{P}| = |\mathcal{L}| = \frac{s^n - 1}{s - 1};$$

(iii) *we have* $|\Gamma_1(x)| = t + 1$,
$|\Gamma_2(x)| = (t+1)s$,
$|\Gamma_3(x)| = (t+1)st \ (n \geq 3)$,
and, in general,

$$|\Gamma_i(x)| = (t+1)s^{\lfloor\frac{i}{2}\rfloor} t^{\lfloor\frac{i-1}{2}\rfloor}, \quad 1 \leq i \leq n-1,$$

and

$$|\Gamma_n(x)| = s^{\frac{n}{2}} t^{\frac{n}{2}-1}, \quad n \text{ even},$$
$$|\Gamma_n(x)| = s^{n-1}, \quad n \text{ odd}.$$

Proof. It suffices to prove the formula for $|\Gamma_i(x)|$, because $|\mathcal{P}|$, or $|\mathcal{L}|$, is the sum of all terms $|\Gamma_i(x)|, 0 \leq i \leq n$, with i even, or odd, respectively.

If $i = \delta(x, z) \leq n - 1$, then z is incident with a unique element y of $\Gamma_{i-1}(x)$. Thus double counting of all flags $\{y, z\}$ with $\delta(x, y) = i - 1$, $\delta(x, z) = i$ gives $|\Gamma_i(x)| = |\Gamma_{i-1}(x)|s$ if i is even (i.e., z is a point), and $|\Gamma_i(x)| = |\Gamma_{i-1}(x)|t$ if i is odd (i.e., z is a line).

If $\delta(x, z) = n$, then $\Gamma_1(z) \subseteq \Gamma_{n-1}(x)$. Then double counting of all flags $\{y, z\}$ with $\delta(x, y) = n - 1$, $\delta(x, z) = n$ gives the formula $(t + 1)|\Gamma_n(x)| = |\Gamma_{n-1}(x)|s$ if n is even, and $(s + 1)|\Gamma_n(x)| = |\Gamma_{n-1}(x)|t$ if n is odd (then $s = t$). The result now easily follows. □

Throughout this book, we will make freely use of the above lemma without referring explicitly to it. In fact, we will only need the various values for $n = 3, 4, 6, 8$ and we write some of them down separately in the next result.

1.5.5 Corollary. *Let* $\Gamma = (\mathcal{P}, \mathcal{L}, \mathbf{I})$ *be a finite (weak) generalized n-gon of order* (s,t), *with* $n \in \{3, 4, 6, 8\}$. *Then we have*

$$|\mathcal{P}| = \begin{cases} s^2 + s + 1 & \text{if } n = 3, \\ (1+s)(1+st) & \text{if } n = 4, \\ (1+s)(1+st+s^2t^2) & \text{if } n = 6, \\ (1+s)(1+st)(1+s^2t^2) & \text{if } n = 8. \end{cases}$$

Dually,

$$|\mathcal{L}| = \begin{cases} s^2 + s + 1 & \text{if } n = 3, \\ (1+t)(1+st) & \text{if } n = 4, \\ (1+t)(1+st+s^2t^2) & \text{if } n = 6, \\ (1+t)(1+st)(1+s^2t^2) & \text{if } n = 8. \end{cases}$$

$\square$

We end this section with a few useful, though elementary, lemmas.

The non-existence of ordinary k-gons with $2 \leq k < n$ stipulated in 1.3.1(*i*) gives the following.

1.5.6 Lemma. *Let* $\Gamma = (\mathcal{P}, \mathcal{L}, \mathbf{I})$ *be a weak generalized n-gon, and let* $x, y, z \in \mathcal{P} \cup \mathcal{L}$. *If* $\delta(x,y) + \delta(y,z) \leq n$ *and* $\mathrm{proj}_y x \neq \mathrm{proj}_y z$, **then** $\delta(x,z) = \delta(x,y) + \delta(y,z)$.
$\square$

1.5.7 Lemma. *Let* $\Gamma = (\mathcal{P}, \mathcal{L}, \mathbf{I})$ *be a weak generalized n-gon, and let* $x, y, z \in \mathcal{P} \cup \mathcal{L}$. *If* $\delta(x,y) + \delta(y,z) + \delta(z,x) \leq 2n$, $\mathrm{proj}_a b \neq \mathrm{proj}_a c$, *for* $\{a,b,c\} = \{x,y,z\}$, *with* $\delta(a,b) < n$, *for all* $\{a,b\} \subseteq \{x,y,z\}$, **then** $\delta(x,y) + \delta(y,z) + \delta(z,x) = 2n$ *and* x, y, z *are contained in a unique apartment of* Γ.

Proof. This follows immediately from the previous lemma considering the element u opposite x and contained in the path connecting y with z. From the assumptions it indeed follows that u is unique, and there are two different paths of length n joining x and u: one containing the path from x to y; the other one containing the path connecting x with z. It readily follows that the union of these two paths of length n is an apartment. The lemma follows.
$\square$

1.5.8 Lemma. *Let* $\Gamma = (\mathcal{P}, \mathcal{L}, \mathbf{I})$ *be a generalized n-gon, and let* $x, y \in \mathcal{P}$. *Then there exists* $w \in \mathcal{P} \cup \mathcal{L}$ *opposite both* x *and* y.

Proof. Consider the middle element u of any minimal path $\gamma = (x, \dots, y)$. Put $k = \delta(x,u) = \delta(u,y)$. We have $k \leq \frac{n}{2}$. Consider an element u' incident with u and not belonging to γ. Note that u' exists by the thickness of Γ. By Lemma 1.5.6 above, u' is at distance $k+1$ from both x and y. Now consider any element w of Γ at distance $n-k$ of u and such that $\mathrm{proj}_u w = u'$. The existence of w is ensured by considering any apartment through u and u'. But Lemma 1.5.6 now implies that $\delta(x,w) = \delta(x,u) + \delta(u,w) = k + (n-k) = n$. So x is opposite w. Similarly, y is opposite w.
$\square$

1.5.9 Lemma. *Let x be a point or a line of a weak generalized n-gon Γ, and let Σ be an apartment in Γ. Then some element of Σ is opposite x. If x does not belong to Σ, then at least two elements of Σ are opposite x.*

Proof. Pick y in Σ such that $\delta(x,y)$ is minimal. Then $\delta(x,y) + \delta(y,z) = n$ for some z in Σ, and if $x \neq y$, then there exist at least two such elements z in Σ. By minimality of $\delta(x,y)$ we have $\mathrm{proj}_y x \neq \mathrm{proj}_y z$, hence 1.5.6 gives the assertion. $\qquad\square$

By a similar argument, we can also do a little better now than Lemma 1.3.6, at least for weak polygons.

1.5.10 Lemma. *A geometry $\Gamma = (\mathcal{P}, \mathcal{L}, \mathbf{I})$ is a weak generalized n-gon* **if and only if** *the incidence graph of Γ is a graph of diameter n and girth $2n$.*

Proof. By Lemma 1.3.6, we have to show that the incidence graph of Γ is connected (which follows immediately from the assumption that the diameter is finite) and that every element is incident with at least two other elements. Suppose $w \in \mathcal{P} \cup \mathcal{L}$ is incident with a unique element w' (every element is incident with *some* element by connectedness). Since the diameter is n, every element lies at distance no greater than n from w, hence at distance no greater than $n - 1$ from w'. But there is some circuit of length $2n$ by the assumption on the girth. As in the previous lemma, it is easily shown that at least one element of that circuit is at distance n from w', a contradiction. $\qquad\square$

1.6 Structure of weak generalized polygons

Here we explain how weak generalized polygons can be described in terms of (thick) generalized polygons. For this, we need to define the multiple $m\Gamma$ of a geometry Γ for any natural number $m \geq 1$.

1.6.1 Definition. Let $m \geq 1$ be an integer, and let $\Gamma = (\mathcal{P}, \mathcal{L}, \mathbf{I})$ be a geometry. We define a graph on the set

$$W = \mathcal{P} \cup \mathcal{L} \cup (\mathbf{I} \times \{1, 2, \ldots, m-1\})$$

by specifying the adjacency $*$ as follows:

$$p * (p, L, 1) * (p, L, 2) * \ldots * (p, L, m-1) * L$$

for every incident pair $(p, L) \in \mathbf{I} \subseteq \mathcal{P} \times \mathcal{L}$. Thus the graph $(W, *)$ arises from the incidence graph of Γ by replacing every path of length 1 by a path of length m. The graph $(W, *)$ is bipartite and hence determines a geometry $m\Gamma$ up to duality; by insisting that the points of Γ should also be points of $m\Gamma$, the geometry $m\Gamma$ is determined uniquely.

Thus, if m is odd, then $m\Gamma$ has the point set $\mathcal{P} \cup \mathbf{I} \times \{2, 4, \ldots, m-1\}$ and the line set $\mathcal{L} \cup \mathbf{I} \times \{1, 3, \ldots, m-2\}$, and if m is even, then $m\Gamma$ has the point set $\mathcal{P} \cup (\mathbf{I} \times \{2, 4, \ldots, m-2\}) \cup \mathcal{L}$ and the line set $\mathbf{I} \times \{1, 3, \ldots, m-1\}$; the incidence of $m\Gamma$ is as indicated above.

If Γ is a weak generalized n-gon, where we now include the case $n = 2$, then $m\Gamma$ is easily seen to be a weak generalized nm-gon. If $m > 1$, then $m\Gamma$ is not thick; in fact, if Γ has order (s, t), then 2Γ has order $(1, t)$ if $s = t$; for $s \neq t$, $m \geq 2$, and for $st > 1$, $m > 2$, the geometry $m\Gamma$ does not have an order. The following result says that all weak generalized polygons arise in this or similar fashion from (thick) generalized polygons. This is due to TITS [1976a] (see also YANUSHKA [1981]). We will sometimes refer to it as the **Structure Theorem (for weak polygons)**.

1.6.2 Theorem (Tits [1976a]). *Let* $\Gamma = (\mathcal{P}, \mathcal{L}, \mathbf{I})$ *be a weak generalized n-gon with* $n \geq 3$. *Recall that an element* $x \in \mathcal{P} \cup \mathcal{L}$ *is thick if* $|\Gamma(x)| \geq 3$. *Then we have precisely one of the following cases:*

(i) *Γ is an ordinary n-gon (with order $(1,1)$, without thick elements).*

(ii) *Γ is obtained from an ordinary n-gon $x_0 \mathbf{I} x_1 \mathbf{I} \ldots \mathbf{I} x_{2n} = x_0$ by inserting an arbitrary non-zero number of (mutually disjoint) paths of lenght n from x_0 to x_n. Then x_0, x_n are the only thick elements of Γ, and Γ does not have an order.*

(iii) *There exists a divisor d of n with $d < n$ and a (thick) generalized $\frac{n}{d}$-gon Γ' (with $\frac{n}{d}$ possibly equal to 2, but not to 1) such that $\Gamma \cong d\Gamma'$ or $\Gamma \cong (d\Gamma')^D$, the dual of $d\Gamma'$. There is also a bijection from the set of points and lines of Γ' onto the set of thick elements of Γ which maps elements at distance i onto elements at distance id.*

Proof. If Γ does not contain any thick element, then obviously (by connectedness), Γ is an ordinary n-gon. So we may assume that there is at least one thick element, say, x. By Lemma 1.3.5(ii), there exists at least one element y opposite x. By considering the projectivity $[x; y]$ (see Section 1.5), we also see that y is thick.

Now let d be the minimal distance between any two distinct thick elements of Γ. Since there are at least two thick elements by the preceding paragraph, the number d is well defined and less than or equal to n. First suppose that $d = n$. Let x and y be two thick elements of Γ for which $\delta(x, y) = n$. We claim that there are no other thick elements. Indeed, let z be any thick element, $x \neq z \neq y$. Let y_n be any element in $\Gamma(x)$. There is a unique path $\gamma_y = (y = y_1, y_2, \ldots, y_n)$ from y to y_n. Let $\gamma_z = (z = z_1, z_2, \ldots, z_n = y_n)$ be the path from z to y_n. Let i be minimal with respect to the property $y_i = z_i$. Then i is well defined since $z_n = y_n$; hence $i \leq n$. Also, $i > 1$ since $y \neq z$. But y_i is a thick element since it is incident with the distinct elements y_{i-1}, z_{i-1} and y_{i+1} (putting $x = y_{n+1}$ if necessary). Since $\delta(x, y_i) = n - i + 1 > 0$, we have by assumption $n - i + 1 = n$, hence $i = 1$, a contradiction. The claim is proved. And we easily see that (ii) holds.

So we may assume $d < n$. We note that, as above, every element opposite a thick element is also thick. From this, we infer that $2d \leq n$. Indeed, let x, y be thick elements such that $\delta(x, y) = d$. Let Σ be an apartment containing x and y. Let z be the element of Σ opposite y. Then z is thick and $\delta(x, z) = n - d$. By the minimality of d, we must have $d \leq n - d$.

Now let x and y be two thick elements for which $\delta(x, y) = d$. We claim that every element, say z, at distance d from x is a thick element. By the thickness of x, there exists an element w at distance $n - d$ from x such that $\mathrm{proj}_x y \neq \mathrm{proj}_x w \neq \mathrm{proj}_x z$. By Lemma 1.5.6 we know that $\delta(y, w) = \delta(z, w) = \delta(y, x) + \delta(x, w) = n$. Hence y and w are opposite and w is thick. But w and z are also opposite and our claim follows. We now easily see that all elements at distance id (with i any non-negative integer such that $id \leq n$) from x are thick. Now we claim that for any thick element u of Γ the number $d' = \delta(x, u)$ is a multiple of d. Indeed, let $r < d$ be the remainder after dividing d' by d. By considering a path from x to u, we see that there is a thick element at distance r from u. The minimality of d implies $r = 0$ and hence d divides d'. Hence the claim. In particular, d divides n.

So we have shown that the set of thick elements of Γ is equal to the set of elements v of Γ such that $\delta(v, x)$ is a multiple of d. If $d = 1$, this implies that Γ is thick and (iii) holds with $\Gamma' = \Gamma$. Hence we may assume from now on that $d \geq 2$. Obviously, we may take for x any thick element.

Next we claim that the relation "$\delta(x, y)$ *is divisible by* $2d$" is an equivalence relation on the set of all thick elements of Γ (and obviously there are exactly two equivalence classes). Indeed, otherwise we find a closed (not necessarily non-stammering) path $\gamma = (x_0, x_1, \dots)$ of length $(2k + 1)d$ such that x_i is thick precisely if i is divisible by d. Pick such a path of minimal length (i.e., with minimal k). Then the elements x_i are mutually distinct (except that $x_0 = x_{(2k+1)d}$). We have $\delta(x_0, x_{kd}) = kd = \delta(x_0, x_{(k+1)d})$, by minimality of k. Indeed, if $\delta(x_0, x_{kd}) = \ell d$, $\ell < k$, then $\ell \equiv k$ mod 2 gives us a circuit of length no more than $(\ell + k + 1)d < (2k+1)d$. If $\ell \equiv (k+1)$ mod 2, then we obtain a circuit of length no more than $(k + l)d < (2k + 1)d$. Consider an apartment Σ which contains x_0 and x_{kd+1}. Since x_{kd+1} is *not* thick (remembering $d \geq 2$), we infer that x_{kd} and $x_{(k+1)d}$ also belong to Σ. Hence $kd \neq n$ and so γ coincides with Σ. But then $(2k + 1)d = 2n$, contradicting the fact that d divides n. Hence our claim.

Now we define a geometry Γ' as follows: the points are the elements of one equivalence class of thick elements; the lines are the elements of the other equivalence class; a point is incident with a line if the corresponding elements are at distance d from each other. It is clear that $\Gamma \cong d\Gamma'$ or $\Gamma \cong (d\Gamma')^D$. There is an obvious bijection between the set of paths of length $i \leq n/d$ in Γ' and the set of paths of length id in Γ which start and end with a thick element. This readily implies that the gonality and the diameter of Γ' is equal to n/d. Noting that every element of Γ' is thick, the complete result now follows from Lemma 1.3.6. $\qquad \square$

1.7   Finite and semi-finite generalized polygons

The theorem of FEIT & HIGMAN [1964] says that finite generalized n-gons exist only for a few values of n. Moreover, it imposes restrictions on the orders (s, t) of finite polygons.

1.7.1 Theorem (Feit & Higman [1964]). *Let Γ be a weak generalized n-gon of order (s, t) with $n \geq 3$.* **If** Γ *is finite,* **then** *one of the following holds:*

(i) $s = t = 1$, *and Γ is an ordinary n-gon;*

(ii) $n = 3$, $s = t > 1$, *and Γ is a projective plane;*

(iii) $n = 4$ *and the number*
$$\frac{st(1 + st)}{s + t}$$
is an integer;

(iv) $n = 6$, *and if $s, t > 1$, then st is a perfect square. In that case, we put $u = \sqrt{st}$ and $w = s + t$. The number*
$$\frac{u^2(1 + w + u^2)(1 \pm u + u^2)}{2(w \pm u)}$$
is an integer for both choices of signs (either two times $+$ or two times $-$);

(v) $n = 8$, *and if $s, t > 1$, then $2st$ is a perfect square; in particular $s \neq t$. If we put $u = \sqrt{\frac{st}{2}}$ and $w = s + t$, then the number*
$$\frac{u^2(1 + w + 2u^2)(1 + 2u^2)(1 \pm 2u + 2u^2)}{2(w \pm 2u)}$$
is an integer for both choices of signs (either two times $+$ or two times $-$);

(vi) $n = 12$ *and $s = 1$ or $t = 1$.*   □

The proof of part of this result is given in Appendix A. A complete proof is in e.g. HIGMAN [1975].

1.7.2 Theorem. *Let Γ be a finite generalized n-gon of order (s, t), $n \geq 4$. Then one of the following holds.*

(i) **(Higman [1975])** $n = 4$ *and $s \leq t^2$; dually $t \leq s^2$;*

(ii) **(Haemers & Roos [1981])** $n = 6$ *and $s \leq t^3$; dually $t \leq s^3$;*

(iii) **(Higman [1975])** $n = 8$ *and $s \leq t^2$; dually $t \leq s^2$.*   □

Note that CAMERON [1975] provides a very simple and geometric proof of the inequality in Theorem 1.7.2(i), which also provides information on the case $t = s^2$ (see below). See also PAYNE & THAS [1984] for a discussion on the history of that result. Unlike many other results on quadrangles that can be found in *op. cit.*, we will reproduce CAMERON's proof below because it illustrates a technique which we will need later on for a result on octagons. Anyway, CAMERON's is a combinatorial proof that can compete with eigenvalue techniques. No other part of Theorem 1.7.2 has yet been proved in a combinatorial or geometrical fashion.

A very important corollary to the previous results is the following fact.

1.7.3 Corollary (Feit & Higman [1964]). *Finite generalized n-gons exist only for* $n \in \{3, 4, 6, 8\}$. □

Let us now get back to the inequalities of Theorem 1.7.2. For generalized quadrangles, there is a sufficient and necessary geometric condition for equality; for generalized hexagons and octagons, there are only necessary geometric conditions known; it is conjectured that the necessary conditions under (ii) below are also sufficient for $t = s^3$.

By *mutual position of some elements of* Γ, we mean the subgraph of the incidence graph of Γ obtained by considering all shortest paths connecting any two of these elements. The proofs of (ii) and (iii) below will be omitted; they require some deeper results on graphs (matrix techniques, Krein conditions).

1.7.4 Theorem. *Let* Γ *be a finite generalized n-gon of order* (s, t).

(i) **(Bose & Shrikhande [1972], Cameron [1975])** If $n = 4$, then $t = s^2$ **if and only if** *for some (and hence for all) pair(s) of opposite points* x, y, *the number of points collinear with* x, y *and some other point* z *opposite both* x *and* y *is a constant (hence does not depend on the choice of* z; *the constant is* $s + 1$).

(ii) **(Haemers [1979])** If $n = 6$, then $t = s^3$ *implies that, if* $0 \le i, j, k \le 6$, <u>then</u> *the number* $p_{i,j,k}(x, y, L)$ *of points at distance* i *from a point* x, *at distance* j *from a point* y *and at distance* k *from a line* L, *only depends on the mutual position of the points* x, y *and the line* L *in* Γ.

(iii) **(Coolsaet (unpublished))** If $n = 8$, then $t = s^2$ *implies that, if* $0 \le i, j, k \le 8$, *and* x, y, z *are points of* Γ *such that* $\{y, z\} \subseteq \Gamma_{\le 4}(x)$, <u>then</u> *the number* $q_{i,j,k}(x, y, z)$ *of points at distance* i *from* x, *at distance* j *from* y *and at distance* k *from* z *only depends on the mutual position of* x, y, z *in* Γ.

Proof. We only prove (i), and at the same time Theorem 1.7.2(i).

We fix two opposite points x and y of the generalized quadrangle Γ. For each point z opposite both x and y, we denote by t_z the number $|\{x, y, z\}^\perp|$. We count in two different ways the pairs (z, u), where $z \notin x^\perp \cup y^\perp$ and $u \in \{x, y, z\}^\perp$. We obtain

$$\sum_z t_z = (t+1)(t-1)s.$$

Indeed, there are $t + 1$ points in $\{x, y\}^\perp$; there are $t - 1$ lines not incident with x or y through each such point; and there are s points opposite both x and y on each such line. Next, we count the number of triples (z, u, v), where z and u are as before and $v \neq u$ also belongs to $\{x, y, z\}^\perp$. Similarly as above, we obtain

$$\sum_z t_z(t_z - 1) = (t+1)t(t-1).$$

From this one calculates $\sum_z t_z^2 = (t+1)(t-1)(s+t)$. Denote by μ the mean value of t_z and put d equal to the number of points of Γ opposite both x and y, i.e., after a short computation, $d = s^2t - st - s + t$. Then d times the *variance* is equal to

$$d\sum_z(\mu - t_z)^2 = d\sum_z t_z^2 - \left(\sum_z t_z\right)^2 = (t+1)t(t-1)(s-1)(s^2 - t).$$

This number has to be non-negative, hence $s^2 \geq t$ and, dually, $t^2 \geq s$. If $t = s^2$, then the variance is zero and hence t_z is constant. Conversely, if t_z is constant, then of course, the variance is zero and $t = s^2$. Hence the result, noting that, if $t = s^2$, the constant is equal to $s + 1$. $\square$

The method in the previous proof (calculating the variance) will be called the **variance trick**.

1.7.5 Definition.   A finite generalized n-gon of order (s, t) such that $s = t^2$ or $t = s^2$ (if $n = 4, 8$), or such that $s = t^3$ or $t = s^3$ (for $n = 6$) is called an **extremal polygon**. If $s = t$, then we call the (finite) polygon a **square polygon**.

1.7.6 Example. A lot of information is provided by Theorem 1.7.4(ii). It is actually easy to compute all numbers $p_{i,j,k}(x, y, L)$ and the reader might calculate some of them as an excercise. We will give here one example, namely, of a case which we will use later on (see the proof of Theorem 6.3.8 on page 256). Suppose $\delta(x, y) = 4$ and $\delta(x \bowtie y, L) = 3$, while $\delta(x, L) = \delta(y, L) = 5$. We compute $p_{4,4,3}(x, y, L)$. There are $s^4 = st$ points in $\Gamma_3(L) \cap \Gamma_4(y)$ not collinear with $x \bowtie y$, and there are $s - 1$ such points collinear with $x \bowtie y$. Hence

$$\sum_{z \in X} p_{4,4,3}(z, y, L) = s^4 + s(s-1),$$

where X is the set of points of the line joining x and $x \bowtie y$ distinct from $x \bowtie y$. Since there are s such points, and since $p_{4,4,3}(z, y, L)$ is constant, we deduce that $p_{4,4,3}(x, y, L) = s^3 + s - 1$.

1.7.7 Remarks on orders of projective planes

Apart from the above results, not too much is known about the orders of finite generalized polygons. All finite projective planes presently known have prime power order.

Of course, there is the celebrated result of BRUCK & RYSER [1949] which states that, if the order s of a projective plane is congruent to 1 or 2 modulo 4, then s can be written as the sum of two perfect squares, or, equivalently, the square-free part of n contains no prime divisors equal to 3 modulo 4. This for instance rules out the existence of planes of order 6. It is also a well-known computer result that a projective plane of order 10 does not exist; see LAM, THIEL & SWIERCZ [1989] and LAM [1991]. But there ends our knowledge about the existence of projective planes with certain orders.

For finite generalized n-gons with $n \geq 4$, no analogue of the BRUCK & RYSER theorem is known.

1.7.8 Semi-finite polygons

Clearly, a generalized polygon of order (s, t) is finite **if and only if** s and t are finite. A question that can be posed here is whether there exist generalized $2n$-gons with $2n \geq 4$ having an order (s, t) such that $s > 1$ is finite and t is infinite. Such a polygon is called **semi-finite of semi-order** s. The above question was posed by TITS (unpublished) many years ago. If $2n = 4$ and $s \leq 4$, then t is finite by results of CAMERON [1981] ($s = 2$), KANTOR (unpublished) ($s = 3$) and CHERLIN [19**] ($s = 4$). The latter makes use of model theory. For $t = 3$, there is also a proof by BROUWER [1991].

Let us reproduce first CAMERON's result here. We will provide two proofs, the second of which prepares the proof of the analogous result for $s = 3$. More exactly, we show:

1.7.9 Theorem (Cameron [1981]).
If Γ is a generalized quadrangle of order $(2, t)$, $t \geq 2$ (and possibly infinite), then $t = 2$ or $t = 4$. Hence there do not exist semi-finite generalized quadrangles of semi-order 2. Moreover, the generalized quadrangles of order $(2, 2)$ and $(2, 4)$, respectively, are unique (up to isomorphism).

Proof. *(First proof)* We may suppose that $t > 2$, so there are at least four lines through each point. Let p be any point and let L_1, L_2, L_3, L_4 be four different lines through p (see Figure 1.3). Let x_0 be a point opposite p; let, with an inductive definition, M_i be the unique line through x_{i-1} confluent with L_i; and let x_i be the unique point different from x_{i-1} on M_i and opposite p, $i = 1, 2, 3, 4$. For $i = 0, 1, 2$, we have that the projection of x_i onto L_4 is different from the projection of x_{i+1} onto L_4, hence the projection of x_0 onto L_4 differs from the projection of x_3 onto L_4. So x_0 is not collinear with $M_4 \cap L_4$. A similar reasoning shows that the projection of x_0 onto L_3 coincides with the projection of x_2 onto L_3, hence x_0 is

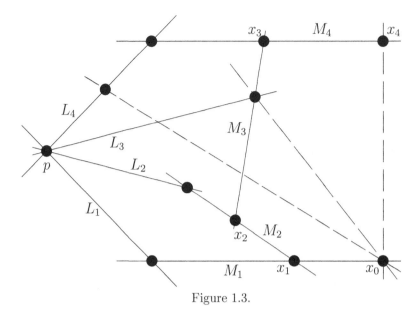

Figure 1.3.

collinear with $M_3 \cap L_3$, and consequently x_0 is not collinear with x_3. So x_0 is not collinear with the two points x_3 and $M_4 \cap L_4$ on the lines M_4, hence it is collinear with the third point x_4 on that line. As above, we have that the projection of x_0 onto L_3 differs from the projection of x_4 onto L_3, because the latter differs from the projection of x_3 (or equivalently x_2) onto L_3. So x_0x_4 does not meet L_3. Similarly, the projection of x_0 onto L_2 differs from the projection of x_2 onto L_2, which equals the projection of x_4 onto L_2 (because they both differ from the projection of x_3 onto L_2). Hence x_0x_4 does not meet L_2. But clearly, x_0x_4 meets neither L_1 nor L_4, hence there must be a line L_5 different from L_i, $i = 1, 2, 3, 4$, incident with p and concurrent with x_0x_4. This shows $t \geq 4$. If $t > 4$, then consider a further line L_6 through p. The projection of x_i onto L_6 differs from the projection of x_{i+1} onto L_6, $i = 0, 1, 2, 3$, hence the projection of x_0 onto L_6 equals the projection of x_4 onto L_6, implying $L_6 = L_5$. This proves the first part of the theorem.

(*Second proof*) Let L_1 and L_2 be two opposite lines and let $\Gamma(L_i) = \{p_{i1}, p_{i2}, p_{i3}\}$, $i = 1, 2$, where we have chosen the notation in such a way that $p_{1j} \perp p_{2j}$, for all $j \in \{1, 2, 3\}$. Let p_{3j} be the third point on the line $M_j = p_{1j}p_{2j}$, $j = 1, 2, 3$. The line L_3 incident with p_{31} and confluent with M_2 must of course contain p_{32}. The projections of the two collinear points p_{13} and p_{23} onto L_3 differ from both p_{31} and p_{32}, hence they are collinear with the same point on L_3. This can clearly only happen when M_3 is concurrent with L_3. Hence we have a grid Γ' with lines L_i and M_j, $i, j \in \{1, 2, 3\}$. Let us denote by P the set of (nine) points of that grid, and let us denote by Q the set of points of Γ not on that grid. Note that we have shown that any two opposite lines are contained in a grid of order $(2,1)$. Let x be any point of Q, then the set of points of P collinear with x defines a unique permutation σ_x of $\{1,2,3\}$ as follows: if $p_{ij} \in P$ is collinear with x, then $i^{\sigma_x} = j$. From now on (and also in the proof of the next theorem), we view a permutation σ also as the set of pairs (i, i^σ). Clearly, if x and y are collinear points of Q, and if σ_x and σ_y are not disjoint, then the permutations σ_x and σ_y share exactly one element and $\sigma_x \sigma_y^{-1}$ is an involution, which is completely determined by the ordered pair (i,j) for which $j = i^{\sigma_x} = i^{\sigma_y}$. Indeed, the number of fixed elements in $\{1,2,3\}$ of $\sigma_x \sigma_y^{-1}$ is equal to the number of points of P collinear to both x and y, but that is clearly equal to 1.

Now let p be any element of Q. Without loss of generality, we may assume that σ_p is the identity. For all $i \in \{1,2,3\}$, the unique point x_i of Q on pp_{ii}, $x_i \neq p$, satisfies $\sigma_{x_i} = \{(i,i),(j,k),(k,j)\}$, with $\{i,j,k\} = \{1,2,3\}$. The unique point y_{ijk} of Q on $x_i p_{jk}$, $y_{ijk} \neq x_i$, corresponds to the permutation $\{(j,k),(k,i),(i,j)\}$, again for all $\{i,j,k\} = \{1,2,3\}$. Now consider the grid containing the lines M_k and M_i', with $\Gamma(M_i') = \{p, x_i, p_{ii}\}$. The three points neither on M_k nor on M_i', but incident with a line meeting both M_k and M_i', are x_k, y_{ijk} and p_{ij}. Hence these three points are on a common line, which implies $y_{kij} = y_{ijk}$. It is now easy to see that the set

$$P \cup \{p, x_1, x_2, x_3, y_{123}, y_{321}\}$$

of 15 points of Q defines a subquadrangle of order $(2,2)$ of Γ. In particular, we conclude that the number of points x of Q such that σ_x is some fixed permutation σ, is independent of σ. Let that number be ℓ. If $\ell = 1$, then Γ has order $(2,2)$ (and is unique, as follows from our construction). If $\ell = 2$, then Γ has order $(2,3)$, which violates the divisibility condition in Theorem $1.7.1(iii)$. Alternatively, we have a full subquadrangle of order $(2,2)$, and this violates Theorem $1.8.7(ii)$. If $\ell = 3$, then we have a generalized quadrangle of order $(2,4)$. Now let $\ell > 3$. Let x and y be two points of Q so that σ_x and σ_y are disjoint, and let $y' \in Q$ be so that $\sigma_y = \sigma_{y'}$. Let p be any point of P collinear with both y and y'. Let z and z' be the unique points of Q on py and py', respectively (and different from y and y', respectively). We clearly have $\sigma_z = \sigma_{z'}$, but σ_z is not disjoint with σ_x, for otherwise $\sigma_z \sigma_x^{-1}$ has order 3, and hence it is an even permutation, as is $\sigma_x \sigma_y^{-1}$, and $\sigma_y \sigma_z^{-1}$ is a transposition, which is a contradiction. Put $\sigma_x \cap \sigma_z = \sigma_x \cap \sigma_{z'} = \{(i,j)\}$. If x is collinear with z, then z is the unique third point on the line xp_{ij}, hence x is not collinear with z'. But x is also not collinear with p, which forces x to be collinear with y'. Hence x is collinear with all but at most one point y such that σ_y is a given permutation disjoint from σ_x. Of course, in the subquadrangle determined by P and x and constructed above, x is opposite each such y, hence we may conclude that, for given $x \in Q$ and any given permutation σ of $\{1,2,3\}$ such that σ_x and σ are disjoint, the point x is opposite a unique point $y \in Q$ with $\sigma_y = \sigma$. Now consider a point x and two disjoint permutations σ and σ' which are also disjoint from σ_x. Since $\ell > 3$, there is a point y with $\sigma_y = \sigma$ collinear with x. Let R be the set of points of Q with associated permutation σ'. We know $|R| = \ell$. There is a unique element of R opposite x, there is a unique element of R opposite y, hence there are at least two elements of R collinear with both x and y. But there is only room for one more point on the line xy, so $\ell > 3$ is impossible.

The preceding argument (about x being opposed to precisely one element y of Q with σ_y given and disjoint from σ_x) easily leads to the conclusion that for $\ell = 3$ the quadrangle is unique.   □

Also semi-order 3 is ruled out for quadrangles, by the following theorem due to Kantor (except for the uniqueness when $t = 9$, which is due to DIXMIER & ZARA [19**] and CAMERON (unpublished)). The proof we present is essentially due to BROUWER [1991], but we have refined his arguments so that we can also prove the inequality $t \leq s^2$ in this case, and so that we can derive uniqueness of the generalized quadrangle of order $(3, 9)$, given Theorem 6.7.1 on page 278.

1.7.10 Theorem (Kantor, unpublished). *Let Γ be a generalized quadrangle of order $(3, t)$. Then $t \leq 9$. Hence there do not exist semi-finite generalized quadrangles of semi-order $s = 3$. Moreover, if $t = 9$, then Γ is unique.*

Proof. Suppose Γ is a semi-finite generalized quadrangle of semi-order 3. We follow closely the second proof of the previous theorem. Consider two opposite lines L_1 and L_2 and let $\Gamma(L_i) = \{p_{i1}, p_{i2}, p_{i3}, p_{i4}\}$, $i = 1, 2$. We can choose the notation such that p_{1j} is collinear with p_{2j}, $j = 1, 2, 3, 4$, and we denote the line $p_{1j}p_{2j}$ by M_j. Let p_{31} and p_{41} be the other points on M_1. We show first that L_1 and L_2 are contained in a unique grid of order $(3, 1)$. Suppose that a line L through p_{31} meets exactly two of the lines M_2, M_3, M_4, say, M_2 and M_3. Then besides the intersections with M_1, M_2, M_3, the line L is also incident with the two distinct projections of p_{14} and p_{24} onto L, contradicting $s = 3$. Suppose now that a line L through p_{31} meets exactly one of the lines M_2, M_3, M_4. Then it is readily seen that every line meeting M_i and M_j, $j \neq i$, $i, j \in \{1, 2, 3, 4\}$, does not meet M_k for $i \neq k \neq j$. So there are points p_{3j} and p_{4j} on M_j, $j \in \{1, 2, 3, 4\}$, such that p_{ij} is collinear with $p_{i'j'}$, $i, i' \in \{3, 4\}$, $j, j' \in \{1, 2, 3, 4\}$, **if and only if** $i \neq i'$. Now let M be any line through p_{11} with $L_1 \neq M \neq M_1$, and such that M does not meet $p_{3j}p_{4j'}$, for all $j, j' \in \{2, 3, 4\}$, $j \neq j'$. The existence of M is only ensured if $t > 7$, so we suppose from now on $t > 7$. Let x, y and z be the projections of p_{32}, p_{42} and p_{22}, respectively, onto M, then clearly $\Gamma(M) = \{p_{11}, x, y, z\}$. Without loss of generality we may suppose that p_{23} is collinear with y and p_{24} with x. Since p_{43} is collinear with both p_{23} and p_{32}, it is collinear with z, and hence p_{33} is collinear with x. But p_{34} is collinear with p_{42}, p_{43} and p_{24}, hence the projection of p_{34} onto M differs from, respectively, y, z and x. Since obviously $p_{11} \neq \mathrm{proj}_M p_{34}$ we obtain a contradiction.

Thus we have shown that there are two lines L_3 and L_4, both different from both L_1 and L_2, meeting all of M_1, M_2, M_3 and M_4. We put $p_{ij} = L_i \cap M_j$, $i, j \in \{1, 2, 3, 4\}$ (this extends our previous notation). Let P be the set of all points p_{ij} just defined, and let Q be the set of remaining points of Γ. As in the previous proof, we associate to every element x of Q a permutation σ_x of $\{1,2,3,4\}$ as follows: if $p_{ij} \in P$ is collinear with x, then $i^{\sigma_x} = j$, and we denote $(i,j) \in \sigma_x$. This time, if x and y are collinear points of Q, and if σ_x and σ_y are not disjoint, then the latter share exactly one element and $\sigma_x\sigma_y^{-1}$ has order 3, hence it is an even permutation; this is similar to the argument in the second proof of the previous theorem.

Let p be any element of Q. Without loss of generality, we may assume that σ_p is the identity. Similar to the previous proof, we can consider the points x_{ijk} of Q on pp_{mm} such that x_{ijk} has corresponding permutation $\{(m,m),(i,j),(j,k),(k,i)\}$. And then we can look at the points of Q which are incident with a line through such an x_{ijk} and through an element of P. Again one can use the fact that any two opposite lines are contained in a unique grid of order $(3,1)$ to conclude that we obtain in this way a set of $|P| + (2 \times 12) = 40$ points forming a subquadrangle Γ' of order $(3,3)$. The permutation associated to any point of Γ' in Q is an even permutation and for every such even permutation σ, there are exactly two points x with $\sigma_x = \sigma$. Suppose there are ℓ such subquadrangles (all points of some of them induce even permutations, all points of the others induce odd permutations). Then $t = 1 + 2\ell$ (this can be seen either by looking at the number of lines through a point of P, or by counting the points, which gives $16 + 24\ell = (1+s)(1+st) = 4 + 12t$; this also holds in the infinite case), hence, since we assume $t > 7$, we have $\ell > 3$. If $\ell = 4$, then we have $t = 9$ and the previous arguments show that the conditions of the dual of Theorem 6.7.1 (see page 278) are satisfied, and the uniqueness follows. So suppose that $\ell \geq 5$. Then there are at least three subquadrangles of order $(3,3)$ containing P such that the permutation associated to any element of Q of these quadrangles has fixed parity. So, without loss of generality, we may assume that there are three different such subquadrangles with all associated permutations even. Considering two fixed disjoint even permutations σ and σ', we see as in the second proof of the previous theorem that every point x with $\sigma_x = \sigma$ is opposite exactly two points y with $\sigma_y = \sigma'$, and these two points lie in the same subquadrangle of order $(3,3)$ together with x and P. In particular, there exist collinear points x and y of Q such that $\sigma_x = \sigma$ and $\sigma_y = \sigma'$. Now consider a third even permutation σ'' disjoint form both σ and σ'. Let R be the set of points with associated permutation equal to σ'', then $|R| \geq 6$. Since there are exactly two elements of R opposite x, and exactly two elements of R opposite y, there are at least two elements of R collinear with both x and y. But the line xy contains at most one point z of R (indeed, two points with the same associated permutation can never be collinear). The final contradiction is obtained. $\square$

If both s and t are infinite, then one cannot expect any relation between s and t. Indeed, for generalized quadrangles constructed from forms on vector spaces of infinite dimension (compare Chapter 2) one parameter depends only on the field, but the other parameter depends also on the dimension of the vector space. Hence s and t need not be of the same infinite cardinality.

For topological generalized polygons, the topological dimension yields an analogue of the order of a finite generalized polygon; see Chapter 9.

1.7.11 Classification results for small orders

Without proof we mention the following results (all results on quadrangles can be found, including proofs, in PAYNE & THAS [1984]).

All projective planes of order no greater than 10 are classified. For the orders $2, 3, 4, 5, 7$ and 8, there is just one projective plane in each case (the classical one, compare Chapter 2), see e.g. STEVENSON [1972], page 71. For order 9 there are exactly four isomorphism classes of projective planes, see LAM, KOLESOVA & THIEL [1991] (another computer result). For the orders 6 and 10, there are no projective planes at all, as already mentioned above.

A generalized quadrangle of order (s, t) with $s = 2$ has necessarily $t = 2$ or $t = 4$, and in both cases there is exactly one generalized quadrangle (respectively $\mathsf{W}(2)$ and $\mathsf{Q}(5, 2)$; see Subsection 2.3.12 and Subsection 2.3.17) by Theorem 1.7.9. If $s = 3$, then Theorem 1.7.1(iii) and Theorem 1.7.2(i) imply that $t = 3, 5, 6$ or $t = 9$. For $t = 5$ there exists a unique quadrangle, of $*$-Tits type (see Subsection 3.7.2), as there does for $t = 9$, $\mathsf{Q}(5, 3)$ (compare Theorem 1.7.10); for $t = 3$ there is a unique one up to duality (namely $\mathsf{W}(3)$ which is the dual of $\mathsf{Q}(4, 3)$) and for $t = 6$ none exists. Finally, there is a unique quadrangle of order $(4, 4)$ (the symplectic $\mathsf{W}(4)$).

A finite generalized hexagon of order (s, t) with $s = 2$ has necessarily $t = 2$ or $t = 8$. This follows from Theorem 1.7.1(iv) and Theorem 1.7.2(ii). In the first case there are exactly two mutually dual examples, namely $\mathsf{H}(2)$ and its dual $\mathsf{H}(2)^D$; in the second case there is a unique hexagon: the dual twisted triality hexagon $\mathsf{T}(2, 8)$ (see COHEN & TITS [1985]). Nothing on uniqueness is known for other orders.

For generalized octagons, nothing is known concerning uniqueness and non-existence. The smallest example has order $(2, 4)$ (or dually $(4, 2)$). It is believed to be unique, but that is not yet proved. Actually, all finite known examples (the Ree–Tits octagons, see Section 2.5) have, up to duality, order $(2^{2e+1}, 2^{4e+2})$, $e \in \mathbb{N}$.

1.7.12 Another result of Brouwer's

BROUWER [1993] considers the following problem, which he solves using an eigenvalue matrix technique. Let Γ be a generalized n-gon and let p and F be a point and a flag of Γ, respectively. Consider the subgeometry Γ' of Γ induced by the points and lines of Γ which lie at distance at least $n - 1$ from p. Consider also the subgeometry Γ'' of Γ induced by all flags opposite F. The question can be posed: are these geometries connected? BROUWER [1993] answers this question for very general — though finite — Γ. His result is essentially the following.

Let us call the geometry Γ' the **opposite-point geometry of Γ (with respect to p)**; and Γ'' the **opposite-flag geometry of Γ (with respect to F)**.

1.7.13 Theorem (Brouwer [1993]). *Let Γ be a finite generalized n-gon of order (s, t). Then the opposite-point geometry of Γ with respect to any point is connected if $(n, s, t) \neq (6, 2, 2)$ and $(n, s, t) \neq (8, 2, 4)$. The opposite-flag geometry of Γ with*

respect to any flag is connected **if** $(n, st) \neq (4, 4)$, $(6, 4)$, $(6, 9)$ *and* $(8, 8)$. *For all values of* (n, s, t) *mentioned as possible exceptions, there do exist examples (namely classical polygons) for which the corresponding opposite-point or opposite-flag geometries are not connected.* □

1.7.14 Remark. Every opposite-point geometry of any semi-finite polygon of semi-order 2 must be disconnected (using an argument similar to the first proof of Theorem 1.7.9); this was also noted by KANTOR (unpublished).

All opposite-flag geometries of all infinite Moufang polygons (see Chapter 5) are connected (see ABRAMENKO & VAN MALDEGHEM [19**]). Also, for all $n \geq 5$, there exist n-gons with disconnected opposite-flag geometries (by ABRAMENKO [1996]). For $n = 4$, all infinite generalized n-gons have connected opposite-flag geometries, as proved by CUYPERS (unpublished). A proof is contained in BROUWER [1993]. We present a different proof.

1.7.15 Theorem. *Every opposite-flag geometry of any generalized quadrangle* Γ *of order* (s, t), $st > 4$, *is connected.*

Proof. Let $\{p, L\}$ be a flag of Γ. Without loss of generality, we may assume that $s > 2$. Given any line M opposite L, it clearly suffices to show that every line N opposite both L and M is connected to M in the opposite-flag geometry Γ'' of Γ with respect to $\{p, L\}$. It is easily seen that each connected component of Γ'' contains a line with prescribed projection (different from L) onto p. Hence we may assume that $\text{proj}_p M = \text{proj}_p N =: X$. Put $a = X \cap M$ and $b = X \cap N$. If there is a point $x \in \Gamma(M) \setminus \{a\}$ with $\text{proj}_x N \neq \text{proj}_x L$, then the result follows. If no such point exists, then we consider a line $N' \neq N$ of Γ'' incident with any point $y \neq b$ of N. Now the $s - 1$ points x on M belonging to Γ'' with $x \neq \text{proj}_M y$ all satisfy $\text{proj}_x N' \in \Gamma''$. Since for exactly one of them we have $\text{proj}_{N'} x \perp p$, there are $s - 2 > 0$ paths in Γ'' connecting M with N', and hence with N. □

As an interesting exercise, I leave it to the reader to prove that all opposite-point geometries of any generalized pentagon are connected.

1.8 Subpolygons

A (weak) sub-n-gon Γ' of a geometry Γ is a subgeometry (see Definitions 1.2.1) which is a (weak) generalized n-gon. If Γ is a weak generalized m-gon, then $n \geq m$. The weak subpolygon Γ' is called **ideal** if every pencil of Γ' coincides with the corresponding pencil of Γ. Dually, a weak subpolygon Γ' is called **full** if every point row of Γ' is a point row of Γ. It will be very useful to have these different names for *full* and *ideal* subpolygons.

We now prove some general statements about (weak), mainly full and ideal, subpolygons.

1.8.1 Proposition. *A sub-n-gon Γ' of a generalized n-gon Γ is ideal* **if and only if** *at least one pencil of Γ' coincides with the corresponding pencil in Γ.*

Proof. We claim that distances are the same whether measured in Γ' or Γ. Indeed, any two elements of Γ' can be put in an apartment of Γ' and this must also be an apartment of Γ. The claim follows (compare also Remark 1.3.4 on page 7). Hence every perspectivity in Γ' is also a perspectivity in Γ. It follows from Lemma 1.5.1(i) that Γ' is an ideal subpolygon. $\square$

1.8.2 Proposition. *A full and ideal weak sub-m-gon Γ' of a generalized n-gon Γ, $m \geq n$, coincides with Γ itself.*

Proof. Clearly we have $\Gamma'_2(x) = \Gamma_2(x)$, for all points x of Γ'. By connectedness, Γ' coincides with Γ. $\square$

1.8.3 Corollary. **If** *a sub-n-gon Γ' of a generalized n-gon Γ has at least one pencil in common with Γ, and* **if** *n is odd,* **then** *Γ' coincides with Γ.*

Proof. By Proposition 1.8.1, Γ' is an ideal subpolygon. Now note that, if n is odd, then every perspectivity maps a pencil on a point row; so we see that Γ' is also a full subpolygon. Proposition 1.8.2 concludes the proof. $\square$

1.8.4 Proposition. *The intersection of two weak sub-n-gons of a generalized n-gon is a weak sub-n-gon* **if and only if** *it contains an apartment.*

Proof. It suffices to show that, if the intersection $\Gamma^* = \Gamma' \cap \Gamma''$ of two weak sub-n-gons Γ' and Γ'' of a generalized n-gon Γ contains an apartment Σ, then it is a weak sub-n-gon.

Clearly, the girth of the incidence graph of Γ^* is equal to $2n$ (since Γ^* contains Σ). Now let u and w be two arbitrary elements of Γ^*. If $\delta(u, w) < n$ (in Γ), then the unique path joining u to w belongs to both Γ' and Γ'', hence to Γ^*, and so the distance from u to w in Γ^* is smaller than n. Suppose now that u and w are opposite. By Lemma 1.5.9, we can pick an element v in Σ opposed to w. The projectivity $[v; w]$ maps each element of $\Gamma^*(v)$ onto an element of $\Gamma^*(w)$, hence there exists at least one element w' in Γ^* incident with w. By the previous argument, the distance from v to u in Γ^* is $n-1$, hence the distance from u to w in Γ^* does not exceed n. So the diameter of the incidence graph of Γ^* equals n. By Lemma 1.5.10, the assertion follows. $\square$

1.8.5 Corollary. *Let Γ' be a sub-n-gon of a generalized n-gon Γ. Then Γ' is uniquely determined by an apartment, a pencil $\Gamma'(p)$ and a point row $\Gamma'(L)$ of Γ' with $p\,\mathbf{I}\,L$. In particular, let p and L be a point and a line, respectively, of a weak sub-n-gon Γ'' of Γ, with $p\,\mathbf{I}\,L$. If $\Gamma''(p) = \Gamma(p)$ and $\Gamma''(L) = \Gamma(L)$,* **then** *Γ'' coincides with Γ.*

Proof. Let Γ'' be a second sub-n-gon containing an apartment Σ of Γ', the pencil $\Gamma'(p)$ of Γ' and the point row $\Gamma'(L)$ of Γ'. The intersection $\Gamma' \cap \Gamma''$ is a weak sub-n-gon of Γ containing two thick elements at mutual distance 1. By the Structure Theorem (Theorem 1.6.2), $\Gamma' \cap \Gamma''$ is thick. It is full and ideal in both Γ' and Γ'' by Proposition 1.8.1. Proposition 1.8.2 now implies that $\Gamma' = \Gamma' \cap \Gamma'' = \Gamma''$.

As for the second assertion, the Structure Theorem implies that Γ'' is thick. The corollary follows. □

1.8.6 Corollary. *Let Γ' and Γ'' be weak ideal sub-n-gons of a generalized n-gon Γ. Suppose that neither Γ' nor Γ'' is thick. **If** Γ' and Γ'' contain a common opposite pair of points, **then** Γ' coincides with Γ''.*

Proof. Since all points of both Γ' and Γ'' are thick, the Structure Theorem implies that Γ' and Γ'' are isomorphic to the double of two generalized $\frac{n}{2}$-gons, Γ_0' and Γ_0'', respectively. The intersection of Γ' and Γ'' contains two opposite points by assumption, therefore it contains an apartment (remembering that Γ' and Γ'' are ideal weak subpolygons). So the intersection Γ^* is a weak sub-n-gon of both Γ' and Γ''. But, as before, Γ^* is the double of a thick generalized $\frac{n}{2}$-gon Γ_0^*. The latter corresponds to an ideal and full subpolygon of both Γ_0' and Γ_0'' (by Proposition 1.8.1); Proposition 1.8.2 now implies that $\Gamma_0' = \Gamma_0^* = \Gamma_0''$, hence Γ' and Γ'' coincide. □

Not too much is known about sub-m-gons of generalized n-gons for $n \neq m$. Examples exist because one can start a free construction of a generalized n-gon with a given generalized m-gon, $m > n$, viewed as a partial n-gon. From now on, however, we define a **subpolygon of a generalized polygon** as a sub-n-gon of a generalized n-gon.

In the finite case, there are some heavy restrictions on the orders of subpolygons, in particular on ideal and full subpolygons. A **proper** subpolygon Γ' of a polygon Γ is a subpolygon with $\Gamma' \neq \Gamma$.

1.8.7 Proposition (Thas [1972b], [1974], [1976], [1979]). *Let Γ' be an ideal weak proper sub-n-gon of order (s', t) of a finite generalized n-gon Γ of order (s, t). Then one of the following cases occurs.*

  (i) $n = 3$ and $\Gamma' = \Gamma$;

  (ii) $n = 4$ and $s \geq s't$;

  (iii) $n = 6$ and $s \geq s'^2 t$;

  (iv) $n = 8$ and $s \geq s'^2 t$.

Proof. By Theorem 1.7.1, we have $n \in \{3, 4, 6, 8\}$.

For $n = 3$, the result follows from Corollary 1.8.3 above, noting that Γ' is thick (Theorem 1.7.1(ii) implies that $s' = t$).

Let $n = 4$. The number of points of Γ not belonging to Γ' and incident with a line of Γ' is equal to $(1 + t)(1 + s't)(s - s')$. Hence if we add that number to the number of points of Γ', the resulting number should not exceed the total number of points. After a short calculation, this gives the inequality (using Lemma 1.5.4 or Corollary 1.5.5)

$$(s - s')t(s - s't) \geq 0.$$

Since $s - s' > 0$ (Γ' is a proper subpolygon), the result follows.

Let $n = 6$. The number of lines of Γ not belonging to Γ', but confluent with some (and hence a unique!) line of Γ' is equal to

$$(t + 1)(1 + s't + s'^2t^2)(s - s')t.$$

The same argument as above for $n = 4$ gives us

$$(t + 1)t^2(s - s')(s - s'^2t) \geq 0,$$

from which the result easily follows.

Finally, let $n = 8$. This time we need the *variance trick*; see the proof of Theorem 1.7.4(i). We consider the set W of lines of Γ which lie at distance at least 4 from each line of Γ'. For $L \in W$, we let t_L be equal to the number of lines of Γ' at distance 4 from L. Applying the variance trick to the numbers t_L (and leaving the details of the computation to the reader), we obtain consecutively

$$\left\{ \begin{array}{rcl} \sum t_L & = & st^2(s - s')(1 + t)(1 + s't)(1 + s'^2t^2), \\ \sum t_L(t_L - 1) & = & s'^3t^4(s - s')(1 + t)(1 + s't)(1 + s'^2t^2), \\ |W| & = & t^2(s - s')(t + 1)(s + s^2t + ss't - s'^3t^2). \end{array} \right.$$

Hence

$$|W| \sum t_L^2 - \left(\sum t_L \right)^2 =$$
$$t^5(s - s')^2(t + 1)^2(s^2 + s'^4t^2)(1 + s't)(1 + s'^2t^2)(s - s'^2t) \geq 0.$$

The proposition is proved.   □

When combined with the inequalities of Theorem 1.7.2, we obtain some strong — and very useful — conditions on the existence of (towers of) subpolygons in finite polygons.

1.8.8 Theorem. *Let Γ be a finite generalized $2m$-gon, $2 \leq m \leq 4$, of order (s, t). If $m = 4$, then Γ has neither full nor ideal proper suboctagons. If $m = 2, 3$ and Γ has a proper ideal weak subpolygon Γ' of order (s', t), then we have*

(i) $s' \leq t \leq s$;

(ii) *if* $s' = t$, *then* $s = t^m$;

(iii) *if* $s = t$, *then* $s' = 1$;

(iv) *if* Γ' *contains a proper ideal weak subpolygon* Γ'' *of order* (s'', t), *then* $s = t^m$, $s' = t$ *and* $s'' = 1$.

Proof. First let $m = 4$. By Proposition 1.8.7, we know $s \geq s'^2 t$. By Theorem 1.7.2(*iii*), $s'^2 \geq t$, so $s \geq t^2$, implying, by Theorem 1.7.2(*iii*) again, $s = t^2$. Plugging this into our first inequality, we see that $t \geq s'^2$, hence $t = s'^2$. It follows that $s = s'^4$ and hence $st = s'^6 = (s'^3)^2$, contradicting the fact that $2st$ must be a perfect square by Theorem 1.7.1(*v*).

Now let $m = 2, 3$. By Proposition 1.8.7, we know $s \geq s'^{m-1} t$. Since $s' \geq 1$, this already implies $s \geq t$. Since $t^m \geq s$, this also implies that $t^m \geq s'^{m-1} t$, hence $t \geq s'$, proving (*i*). If $s' = t$, then $s \geq t^m$, hence $s = t^m$ by Theorem 1.7.2. This shows (*ii*). If $s = t$, then $1 \geq s'^{m-1}$, hence $s' = 1$ and (*iii*) is proved.

We now prove (*iv*). By (*i*) applied to Γ'' inside Γ', we have $s'' \leq t \leq s'$. Combined with (*i*) as above, this gives us $t = s'$. Hence $s = t^m$ by (*ii*) and $s'' = 1$ by (*iii*). $\qquad\square$

1.8.9 Definition. Let Γ be a generalized polygon. If Γ has no proper ideal subpolygons, then we call Γ **point-minimal**. Dually, if Γ has no proper full subpolygons, then we say that Γ is a **line-minimal** polygon. If a generalized polygon is point-minimal and line-minimal, then we say that it is **minimal**.

Combining (*i*) and (*iii*) of Theorem 1.8.8, we immediately see:

1.8.10 Corollary. *A full subpolygon of a finite generalized polygon* Γ, *where* Γ *has order* (s, t) *with* $s \geq t$, *coincides with* Γ, *i.e., every finite polygon of order* (s, t) *with* $s \geq t$ *is line-minimal. Hence every finite generalized polygon is, up to duality, a line-minimal polygon.* $\qquad\square$

Finally, we note:

1.8.11 Corollary (Thas [1974], [1976]).

(i) *If* Γ', *of order* (s', t), *is a proper ideal subquadrangle of the generalized quadrangle* Γ *of order* (s, t) *and* $s = s't$, *then every point of* Γ *not in* Γ' *lies on a unique line of* Γ' *and every line of* Γ *not belonging to* Γ' *is confluent with exactly* $1 + s$ *lines of* Γ'.

(ii) *If* Γ', *of order* (s', t), *is a proper ideal subhexagon of the generalized hexagon* Γ *of order* (s, t) *and* $s = s'^2 t$, *then every line of* Γ *not in* Γ' *meets a unique line of* Γ' *and every point of* Γ *not incident with a line of* Γ' *is at distance 3 from exactly* $1 + t$ *lines of* Γ'.

Proof. We show (ii). The proof of (i) is completely similar.

From the proof of Proposition 1.8.7(ii), we infer that $s = s'^2 t$ **if and only if** there are no other lines than those of Γ' and those meeting a line of Γ'. The assertion is now clear.   □

1.8.12 Remark. THAS [1974] (see also PAYNE [1973]) shows that for any weak subquadrangle Γ' of order (s', t') of a quadrangle Γ of order (s, t) the inequality $s \geq s't'$ (if $s \neq s'$), and dually $t \geq s't'$ (if $t \neq t'$), holds. Moreover, if $s = s't'$, then each point of Γ not incident (in Γ) with any line of Γ' is collinear with exactly $1 + s'$ points of Γ'. For any weak subhexagon Γ' of order (s', t') of a hexagon Γ of order (s, t), THAS [1976] shows the inequality $st \geq s'^2 t'^2$. Moreover, if $s > s'$ and $st = s'^2 t'^2$, and if x is a point of Γ which is not collinear with any point of Γ', then there are exactly $1 + t'$ lines of Γ' in $\Gamma_3(p)$. Also a more sophisticated inequality exists for generalized octagons, but we will not use it, and it has, up to the special case above, never been used to my knowledge. But see THAS [1979], where it is proved that the order (s', t') of any suboctagon Γ' of a generalized octagon Γ of order (s, t), $t \neq t'$, satisfies

$$s^3 t^2 (s+1) + st(s^2 + s'^4 t'^3 - ss'(t'+1)(1 + s't' + s'^2 t'^2)) + s'^4 t'^3 (s+1) \geq 0.$$

1.9 Regularity

1.9.1 Definitions. Let u be a point or a line of a weak generalized n-gon Γ and let $2 \leq i \leq \frac{n}{2}$. The i-**derivation** of Γ at u is the geometry

$$\Gamma_{i,u} = (\Gamma_i(u), \{\Gamma_i(u) \cap \Gamma_{n-i}(w) : w \in \Gamma_n(u)\}, \in).$$

This is the geometry induced on $\Gamma_i(u)$ by the **distance-i-traces** $\Gamma_i(u) \cap \Gamma_{n-i}(w)$, where w is opposite u (and we sometimes call such a distance-i-trace a **block** of the geometry $\Gamma_{i,u}$). We denote $\Gamma_i(u) \cap \Gamma_{n-i}(w)$ by $u^w_{[i]}$. Furthermore, it is convenient to write u^w instead of $u^w_{[2]}$. Similarly, a distance-2-trace u^w is sometimes called simply a **trace** (**with respect to** u or **in** $u^\perp$).

It is easy to see that any two elements (points or lines of Γ) of $\Gamma_i(u)$ which are at distance $2i$ from each other are contained in some distance-i-trace, hence a block of $\Gamma_{i,u}$. Indeed, those two elements are together with u contained in some apartment Σ (remembering $2 \leq i \leq \frac{n}{2}$ and using Lemma 1.5.6) and it suffices to consider $u^w_{[i]}$ for the element w of Σ opposite u. We say that u is **distance-i-regular** if the i-derivation $\Gamma_{i,u}$ at u is a so-called **partial linear space**, that is, if distinct distance-i-traces $u^w_{[i]}$, w opposite u, have at most one element in common.

The reader can check for himself that these definitions are trivial whenever we would allow $i = 1$ or $i > \frac{n}{2}$.

We say that an element u of Γ is **regular** if it is distance-i-regular for all $2 \leq i \leq \frac{n}{2}$. If all points, respectively lines of Γ are (distance-i-)regular, then we say that Γ is **point-(distance-i-) regular**, respectively **line-(distance-i-)regular**.

One can easily see that for generalized quadrangles, the notion of a regular point coincides with the classical notion of regular point given in PAYNE & THAS [1984]. Regularity is a highly non-trivial condition. Some polygons are point-regular, other polygons do not contain any regular point or line. All point-regular generalized hexagons are classified, see Theorem 6.3.2, page 243. On the other hand, such a classification for generalized quadrangles does not (yet) exist, although no non-classical examples are known. For generalized n-gons with $n > 4$ and $n \neq 6$, the situation is very simple: no such point-regular polygon exists; see Corollary 6.4.8, page 263.

1.9.2 Definitions. Let Γ be a generalized polygon and let p be some point of Γ. Suppose that p is a distance-2-regular point. We define the following geometry $\Gamma_p^{\triangle} = (\mathcal{P}_p^{\triangle}, \mathcal{L}_p^{\triangle}, \mathbf{I}_p)$, which we call the **perp-geometry in** p (of Γ). The points of $\Gamma_p^{\triangle}$ are the elements of $p^{\perp}$; the lines of $\Gamma_p^{\triangle}$ are the ordinary lines through p together with the traces p^x, x opposite p. Incidence is defined as the natural one. The triangle $\triangle$ in the notation is explained by the following lemma.

1.9.3 Proposition. *The perp-geometry $\Gamma_p^{\triangle}$ in a distance-2-regular point p of a generalized polygon Γ is a projective plane* **if and only if** *for every trace X in $p^{\perp}$ and every element v at codistance 1 from all points of X, the restriction to X of the projection onto v is surjective (in which case it is automatically bijective).*

Proof. Let X be a trace in $p^{\perp}$ and let v be an element of Γ at codistance 1 from every point of X. Suppose that the projection of X onto v is not injective. We seek a contradiction. At least two points x, y of X are projected onto the same element w incident with v and so $X = p^w$. Let L be the projection of v onto p (v and p have odd codistance!) and let $z = \text{proj}_L(v)$. There is a (non-stammering) path $(p, L, z, \ldots, v, w)$ of length n, hence $\delta(w, z) = n - 2$ and $z \in p^w$. But $\delta(v, z) = 3$, contradicting the fact that v is at codistance 1 from *all* elements of X. We conclude that the projection of X onto v is injective.

Let u be any element incident with v and opposite p. If u is the projection of some point x of X onto v, then $x \in X \cap p^u$. Conversely, if $X \cap p^u$ contains some point x, then u is the projection of x onto v. The result now follows rather easily. $\square$

1.9.4 Definition. If the perp-geometry in a distance-2-regular point is a projective plane, then we also call this plane the **perp-plane** and p is called a **projective point**.

The previous proposition immediately implies:

1.9.5 Corollary. *Let Γ be a generalized polygon of order (s, t) containing a distance-2-regular point p. Then $s \geq t$, and* **if** *Γ is finite,* **then** *p is projective* <u>*if and only if $s = t$.*</u> $\square$

As a preliminary study, we now take a closer look at the cases $n = 4, 5, 6$.

1.9.6 Regular points in generalized quadrangles

For generalized quadrangles it is obvious that the notions of "distance-2-regular point" and "regular point" coincide. So let p be a regular point in a generalized quadrangle. We define the following geometry Γ_p^∇. The points of Γ_p^∇ are the perps $x^\perp$ of points x collinear with p. The lines of Γ_p^∇ are the ordinary lines through p together with all **spans** $\{p, z\}^{\perp\perp}$, z opposite p. Incidence is defined as the natural one. We call this geometry the **span-geometry in** p of Γ. It can be easily checked that the following map θ is well defined and that it defines an isomorphism from $\Gamma_p^\triangle$ to Γ_p^∇:

$$\theta \; : \; \begin{array}{ll} x \; \mapsto \; x^\perp, & x \in p^\perp; \\ p^z \; \mapsto \; \{p, z\}^{\perp\perp}, & z \in \Gamma_4(x). \end{array}$$

Hence, the perp-geometry and the span-geometry in a regular point p of a generalized quadrangle are isomorphic. Therefore, we also refer to a regular point as a **span-regular point**. If p is projective, then both the perp-geometry and the span-geometry are projective planes. In that case, the span-geometry is also called the **span-plane**.

1.9.7 Regular points in generalized pentagons

Again, it is easily seen that the notions of "distance-2-regular point" and "regular point" in a generalized pentagon coincide.

Suppose a generalized pentagon Γ has two collinear regular points p and q (see Figure 1.4). Let L be some line opposite p such that $q \in p^L$. Let a and b be two other points of p^L and let the projection of L onto a and b be, respectively, L_a and L_b. Let x be incident with L and collinear with q. Finally, let M be any line through q such that $qx \neq M \neq pq$. The traces q^{L_a} and q^{L_b} contain both p and x, hence, by the regularity of q, they coincide. So there are lines M_a and M_b meeting, respectively, L_a and L_b and which are both incident with the same point y on M. Note that $M_a \neq M_b$. We now see that $\{a, q\} \subseteq p^L \cap p^{M_a}$, hence $p^L = p^{M_a}$. This implies that $\delta(M_a, b) = 3$. But then we have a quadrangle $(M_a, \mathrm{proj}_{M_a}(b), \mathrm{proj}_b(M_a), b, L_b, L_b \cap M_b, M_b, y, M_a)$. Hence *no generalized pentagon can contain two collinear regular points*.

1.9.8 Distance-2-regular points in generalized hexagons

Let p be a distance-2-regular point in a generalized hexagon Γ. Our aim is to define a *span-geometry* in p, as for generalized quadrangles above. It turns out, however, that we need a little stronger condition on p. We say that p is **span-regular** if it is distance-2-regular and if for each point x collinear with p, and each pair of points a, b opposite x, the conditions $p \in x^a \cap x^b$ and $|x^a \cap x^b| \geq 2$ imply that $x^a = x^b$. Note that this is automatic if x is a distance-2-regular point!

We have the following lemma.

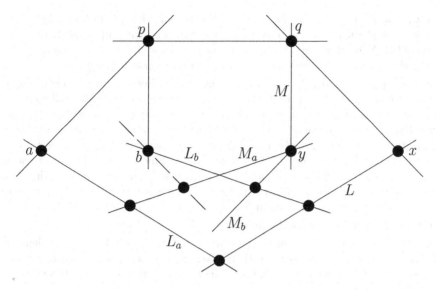

Figure 1.4.

1.9.9 Lemma. *Let p be a span-regular point in a generalized hexagon Γ, and let $x \in \Gamma_2(p)$. Let a, b be two points opposite x with $p \in x^a \cup x^b$.* **If** $|x^a \cap x^b| \geq 2$, **then** $x^a = x^b$.

Proof. Without loss of generality, we may assume that $p \in x^a \setminus x^b$ with $|x^a \cap x^b| \geq 2$. It suffices to find a contradiction. Let c be any point in $x^a \cap x^b$. Let $L = \mathrm{proj}_b(c)$. Put $b' = \mathrm{proj}_L(p)$. Then $b \neq b'$ and $\{c, p\} \subseteq x^{b'}$. But now $p \in x^a \cap x^{b'}$ and $|x^a \cap x^{b'}| \geq 2$, hence by the span-regularity, $x^a = x^{b'}$. But if $c' \in x^a \cap x^b$, with $c \neq c' \neq p$, then, as $b \neq b'$, $\delta(c', b') = \delta(c', b) + \delta(b, b') = 6$, a contradiction. The lemma is proved. □

Suppose now that p is a span-regular point of a generalized hexagon Γ. Let q be opposite p. Our first aim is to show that there exists a unique weak non-thick ideal subhexagon $\Gamma(p, q)$ through p and q. Consider the following set S of points. A point x of Γ belongs to S **if and only if** $x \in \Gamma_2(z) \cap \Gamma_4(y)$ with either $y \in p^q$, $z \in q^p$, y and z not collinear, or $y \in q^p$, $z \in p^q$ and y and z not collinear. The set of points of $\Gamma(p, q)$ is then S; a line of $\Gamma(p, q)$ is any line of Γ containing at least two points of S. Incidence is natural.

1.9.10 Lemma. *Let p be a span-regular point in a generalized hexagon Γ. With the above notation, the geometry $\Gamma(p, q)$ is the unique weak non-thick ideal subhexagon of Γ containing p, q.*

Proof. We show that $\Gamma(p, q)$ is a weak non-thick ideal subhexagon.

Fix a point $y \in p^q$. An arbitrary point of $\mathcal{S}$ collinear with y is obtained by taking a point z of q^p such that z and y are not collinear, and then all elements of y^z are points of $\mathcal{S}$. Note, however, that y^z contains p and $q \bowtie y$, hence y^z is independent of the choice of z and is just the trace in $y^\perp$ through p and $q \bowtie y$.

Now fix a point $z \in q^p$. An arbitrary point of $\mathcal{S}$ collinear with z is obtained by taking a point y of p^q such that z and y are not collinear, and then all elements of z^y are points of $\mathcal{S}$. If y' is a second element of p^q not collinear with z, then we show that $z^y = z^{y'}$. Clearly every line L through z contains a point of z^y. Let L be arbitrary but such that it is incident with z and such that it lies at codistance 1 from p. Let a be the point of z^y incident with L. Since p^a contains $p \bowtie z$ and y, we have $y' \in p^a$, so $\delta(a, y') = 4$ and $a \in z^{y'}$. Since L was arbitrary but incident with z (and at codistance 1 from p; if $\delta(p, L) = 3$, then $p \bowtie z \in z^y \cap z^{y'}$), we infer $z^y = z^{y'}$. Hence the set of points in $\mathcal{S}$ collinear with z is just a trace z^y, y as above.

We refer to the preceding paragraphs as *the first part of the proof*.

We now show that any two distinct elements a, b of the set $\Gamma^+(p, q)$ of elements of $\mathcal{S}$ at distance 0 or 4 from p are themselves at distance 4 from each other, and moreover we show that $a \bowtie b$ belongs to the set $\Gamma^-(p, q) = \Gamma^+(q, p)$ of elements of $\mathcal{S}$ at distance 0 or 4 from q.

This is obvious if a or b coincides with p or if $a \bowtie p = b \bowtie p$ (assuming $a \neq p \neq b$). So we assume that $b' = b \bowtie p$ is distinct from $a' = a \bowtie p$. By the first part of the proof, we may also assume that a and b are opposite q, and also, b lies at distance 4 from $a'' = a' \bowtie q$. Putting $b'' = b' \bowtie q$, we see that, by the span-regularity of p, $a'^{b''} = a'^b$ (since both sets contain p and a''). Hence $\delta(a, b) = 4$ (because $a \in a'^{b''}$ by the first part of the proof). Put $c = a \bowtie b$. We have to show $\delta(q, c) = 4$. Suppose by way of contradiction that q and c are opposite. Let c' be the projection of q onto ac and let L be the projection of c' onto q. Also, put $x = \text{proj}_L(p)$. Since $x \in q^p$, we know $\delta(x, a) = 4$, hence $x = q \bowtie c'$. Notice that this implies $L \neq qb''$. But then, again by the first part of this proof, $\delta(x, b) = 4$, which is impossible if $c \neq c'$. Consequently $c = c'$ is at distance 4 from q and since it belongs to $x^{a'}$, it also belongs to $\Gamma^-(p, q)$ by definition.

Next, we show that every two distinct elements a, b of $\Gamma^-(p, q)$ are at distance 4 from each other, and moreover $a \bowtie b$ belongs to the set $\Gamma^+(p, q)$.

Again, we may assume that $a \neq q \neq b$, that $a' = (a \bowtie q) \neq (b \bowtie q) = b'$, and that $\delta(a, p) = \delta(b, p) = 6$ (see Figure 1.5). Let c be the projection of b onto aa' and put $x = b \bowtie c$. Put $a'' = p \bowtie a'$. By the first part of the proof $\delta(a'', b) = 4$, hence x is at distance 4 from the two points a' and $b \bowtie a''$ of $a''^{b'} \subseteq \mathcal{S}$. By Lemma 1.9.9, $p \in a''^x$, in particular $\delta(p, x) = 4$. But now $p \bowtie x \in p^b \supseteq \{a'', p \bowtie b'\} \subseteq p^a$. Distance-2-regularity in p implies $\delta(a, p \bowtie x) = 4$ and so $c = a$. Hence we have shown $\delta(a, b) = 4$. We also have shown that $p \bowtie x$ belongs to $\mathcal{S}$, hence $x \in (p \bowtie x)^{b'}$ belongs to $\Gamma^+(p, q)$.

So one can see that $\mathcal{S} = \Gamma^+(p, q) \cup \Gamma^-(p, q)$ and also that the geometry

$$(\Gamma^+(p, q), \Gamma^-(p, q), \perp)$$

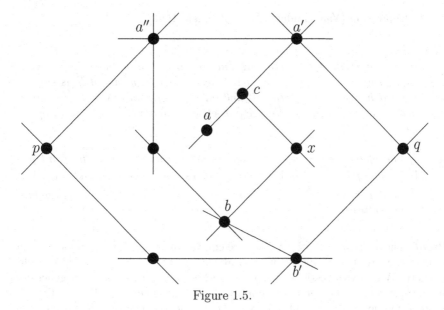

Figure 1.5.

is a projective plane. Consequently $\Gamma(p,q)$ is an ideal weak subhexagon which is the double of that projective plane.

The uniqueness of $\Gamma(p,q)$ is obvious; see also Corollary 1.8.6. □

1.9.11 Remark. Note that the weak subhexagon $\Gamma(p,q)$ is completely determined by two of its points y and z with y opposite z; see Corollary 1.8.6.

From now on, we will use $\Gamma^+(p,q)$ and $\Gamma^-(p,q)$ as a standard notation.

1.9.12 Span-geometries in span-regular points of a generalized hexagon

We define the following geometry $\Gamma_p^{\square}$, called the **span-geometry** in p of Γ. The points are of two types. The points of type (I) are the perps $x^\perp$, for any $x \in p^\perp$. The points of type (II) are the sets $\Gamma^+(p,q)$, for any point q opposite p. The lines of $\Gamma_p^{\square}$ are also of two types. The first type, type (A), contains all ordinary lines through p; the second, type (B), are the traces containing p. Incidence is the natural inclusion.

In this case, $\Gamma_p^{\square}$ is not isomorphic to $\Gamma_p^{\triangle}$, since not every pair of points in $\Gamma_p^{\square}$ is joined by a line, e.g. the point $p^\perp$ is not joined to any point of type II. Also the dual of $\Gamma_p^{\square}$ is not isomorphic to $\Gamma_p^{\triangle}$, by a similar argument. The quadrangle $\square$ in the notation is explained by the following result.

1.9.13 Proposition (Van Maldeghem & Bloemen [1993]).

(i) *The span-geometry $\Gamma_p^{\square}$ in a span-regular point p of a generalized hexagon Γ is a generalized quadrangle **if and only if** the projection of any trace X containing p onto a line L at codistance 1 from all points of X is surjective (in which case it is bijective) **if and only if** every trace $X = y^z$ (y collinear with p and z opposite y) through p meets every other trace $Y = y^q$ (q opposite y).*

(ii) *If one of the equivalent conditions in (i) is satisfied, then the point $p^{\perp}$ of $\Gamma_p^{\square}$ is a regular point in $\Gamma_p^{\square}$ and the corresponding perp-geometry $(\Gamma_p^{\square})_{p^{\perp}}^{\triangle}$ is naturally isomorphic to $\Gamma_p^{\triangle}$. Hence, $p^{\perp}$ is projective in $\Gamma_p^{\square}$ if and only if p is projective in Γ.*

Proof. The last two conditions of (i) are equivalent by a similar argument to that used in the proof of Proposition 1.9.3 above. Also it is clear that $\Gamma_p^{\square}$ is a thick geometry. We now investigate when it satisfies the main axiom of a generalized quadrangle, i.e., when for every non-incident point–line pair (P, l) of $\Gamma_p^{\square}$ there exists a unique incident point–line pair (Q, m) such that Q is incident with l and m incident with P. As the reader has probably already noticed, we denote points of $\Gamma_p^{\square}$ by capital letters $P, Q, \ldots$ and lines of $\Gamma_p^{\square}$ by lower-case letters such as $l, m, \ldots$. This should make the distinction between the two geometries Γ and $\Gamma_p^{\square}$ more clear. We now suppose that P and l are given as above, and we seek Q and m (also as above). There are four possibilities.

1. P *is of type* (I) *and* l *is of type* (A).
   Since P and l are not incident, we must have $P = x^{\perp}$, for some $x \in \Gamma_2(p)$, and $l = L$ is an ordinary line through p not incident (in Γ) with x. We can put $Q = p^{\perp}$ and $m = px$. The pair (Q, m) is unique since any other point on l, say $y^{\perp}$, $y \neq p$, is not collinear with P, and since every other line through P, say x^z, z opposite x and $\delta(p, z) = 4$, does not meet l.

2. P *is of type* (I) *and* l *is of type* (B).
   So let $P = x^{\perp}$, $x \in p^{\perp}$, and let $l = y^z$, $y \in \Gamma_2(p)$, and $z \in \Gamma_4(p) \cap \Gamma_6(y)$. Note that $x \neq y$, otherwise P and l are incident. If x is on py, then we can put $Q = y^{\perp}$ and $m = xy$ and this is unique by a similar argument to that above.
   Suppose now that x is not incident with py (in Γ). Clearly Q must have type (II). So we may put $Q = \Gamma^+(p, q)$, for some point q opposite p in Γ. Obviously $\Gamma(p, q)$, as a weak non-thick ideal subhexagon, must contain p, x and all elements of y^z. This completely determines Q in a well-defined and unique way. Indeed, we can take $q \in \Gamma_4(x) \cap \Gamma_2(z')$, with $z' \in y^z$, $z' \neq p$. The line m is then the trace $x^{z'}$.

3. *P is of type* (II) *and l has type* (A).
So let $P = \Gamma^+(p, q)$, for some point q opposite p, and let $l = L$ be some line through p in Γ. There is a unique point x of $\Gamma^-(p, q)$ incident with L. It is easy to see that Q must be $x^\perp$ and that m must be the trace consisting of all points of $\Gamma^+(p, q)$ in $\Gamma_2(x)$.

4. *P is of type* (II) *and l has type* (B).
Let $P = \Gamma^+(p, q)$ be as above, and let $l = y^z$, $y \in \Gamma_2(p)$, $\delta(p, z) = 4$ and z opposite y. If y is a point of $\Gamma^-(p, q)$, then Q must be equal to $y^\perp$. Indeed, if Q were equal to some $\Gamma^+(p, q')$, then $\Gamma(p, q)$ must intersect $\Gamma(p, q')$ in some trace x^u through p (with x not on yp) and it must contain y; but then it coincides with $\Gamma(p, q)$ itself since $\Gamma(p, q)$ is uniquely defined by x^u and y. Hence m is the trace consisting of all points of $\Gamma^+(p, q)$ collinear with y.
So suppose y is not a point of $\Gamma^-(p, q)$. Then q is opposite y and note that q can be considered as a *general* point opposite both p and y. Clearly Q is of type (II), if it exists. We seek a necessary and sufficient condition for the existence of Q. Put $Q = \Gamma^+(p, q')$, for some point q' yet to be determined. Then $\Gamma^+(p, q')$ contains y^z and some point u of $\Gamma^+(p, q)$ at distance 4 from p which we may assume to be collinear with q (since we can always consider another point on the unique trace through p and u). Hence $y^z = y^u$. Consider the unique point w collinear with y and at distance 3 from uq. Then w is opposite every element of q^p except for u. Hence u is the only point of $\Gamma^+(p, q)$ which is collinear with q and such that $y^z = y^u$. This shows the uniqueness of Q, if it exists, and the uniqueness of m (which is the unique trace through p and u). Now the existence of Q is equivalent to the existence of a point u of $\Gamma^+(p, q)$ collinear with q and at distance 4 from every element of y^z, but this is equivalent to the existence of a point w of y^z at distance 3 from some line uq (with u as above). This in turn is clearly equivalent to the existence of a point w of y^z at distance 4 from q, which is equivalent to saying that y^q has a non-trivial intersection with y^z.
Noting that q is essentially arbitrary (since the previous conditions must be satisfied for all P), and also noting that every trace $y^{z'}$ with z' opposite y, but z' not opposite p, meets y^z in p, (i) follows.

Now suppose that $\Gamma_p^\square$ is a generalized quadrangle. Clearly the points of $\Gamma_p^\square$ collinear with $p^\perp$ and at the same time collinear with some point $\Gamma(p, q)$, q opposite p, all are of type (I) and they can be written as $y^\perp$ with $y \in p^q$. This shows (ii) and the proposition is proved. □

1.9.14 Definition. A span-regular point p in a generalized hexagon Γ such that the span-geometry $\Gamma_p^\square$ is a generalized quadrangle, is called a **polar point**. The span-geometry is in this case also called the **span-quadrangle**.

As an immediate corollary to the previous proposition, we have:

1.9.15 Corollary. *Let p be a span-regular point in a finite generalized hexagon Γ of order (s,t). Then $s \geq t$ and p is a polar point* **if and only if** *$s = t$. Hence a span-regular point of a finite generalized hexagon is a polar point* **if and only if** *it is a projective point.*   □

The perp- and span-geometries, which are in the literature sometimes called *derived geometries*, will be important in some characterization theorems (see Chapter 6) and in the theory of self-dual and self-polar generalized quadrangles and hexagons (see Chapter 7). Note that, if every point collinear to some fixed point p of a generalized hexagon Γ, including p, is projective, then p is a polar point.

1.9.16 Distance-3-regular points in generalized hexagons

Let p be a distance-3-regular point of a generalized hexagon Γ. Every pair of lines (L, M) such that L and M are opposite, but both at distance 3 from p, defines a unique distance-3-trace, which we will denote by $R(L, M)$ and also call the **(line-)regulus** defined by L and M. The set of points q opposite p such that $p_{[3]}^q = R(L, M)$ is called a **(point-)regulus** and denoted by $R(p, q)$. The reguli $R(L, M)$ and $R(p, q)$ are called **complementary**. They determine each other completely. If every point of Γ is distance-3-regular, then also every line of Γ is distance-3-regular and conversely. Each of these equivalent conditions was called *the regulus condition* by RONAN [1980a], [1980b].

More generally, all points of a generalized $2m$-gon are distance-m-regular **if and only if** all lines are distance-m-regular. We will later on prove that distance-m-regularity in generalized $2m$-gons is implied by point-distance-$\frac{m+1}{2}$-regularity. The most important case of this is $m = 3$, and this special case is due to RONAN [1980a].

1.9.17 Proposition (Ronan [1980a]). *If a generalized hexagon is point-distance-2-regular, then it is also distance-3-regular.*

Proof. Let L and M be opposite lines in Γ and let p, q and r be mutually opposite points all at distance 3 from both L and M. Let N be a line at distance 3 from both p and q. Clearly p^q and p^r have at least two points in common (on L and M) and so r lies at distance 4 from the point $x = \mathrm{proj}_N(p)$. Similarly, r lies at distance 4 from $y = \mathrm{proj}_N(q)$. Clearly $\delta(x, y) = 2$ and so we obtain a cycle $(r, \ldots, x, N, y, \ldots, r)$ of length 10 unless $\delta(r, N) = 3$. The proposition is proved.   □

For generalized hexagons, distance-3-regularity will play an important role in some characterization theorems (see Chapter 6). Also, we refer to Section 6.4 for more results on regularity in generalized polygons.

To close this section, we mention the following result.

1.9.18 Proposition. *If x is a distance-j-regular point of a generalized n-gon Γ, $2 \leq j \leq \frac{n}{2}$, and Γ' is a subpolygon of Γ containing x,* **then** *x is a distance-j-regular point of Γ'.*

Proof. If two distinct distance-j-traces meet in Γ' in at least two points, then, considered as traces in Γ, they are still distinct and they meet in at least two points of Γ as well. Hence the result. $\qquad\square$

This result is not true for projective points or polar points; see for instance Corollary 2.3.20 on page 65.

1.9.19 Remark. We have developed the theory of regularity and span-regularity mainly for points. Of course, this can be dualized. In particular, the geometries $\Gamma_L^\triangle$, $\Gamma_L^\triangledown$, $\Gamma_L^\square$ for a line L of Γ are defined in a dual way. That means that (to fix the duality class) the points on L are *points* of these geometries. We will need to use these "dual" perp-geometries and span-geometries in Section 7.8.

Chapter 2

Classical Polygons

2.1 Introduction

The notion of a *generalized polygon* arose from the classification of *trialities* of the geometry of the quadric $\mathbf{Q}(7, \mathbb{K})$ in $\mathbf{PG}(7, \mathbb{K})$, $\mathbb{K}$ any field (see Subsection 2.4.2 below) in TITS [1959]. But already in TITS [1954], [1955] generalized polygons arise as a geometrical interpretation of complex Lie groups. This generalized the well-known connection between the *classical* complex Lie groups and the complex projective spaces and quadrics. It was noticed by TITS [1959] that an axiomatization of these geometries in the relative rank 2 case gave rise to geometries (generalized polygons) corresponding to the exceptional groups of Lie type G_2 (*Dickson's groups*). Also the geometries related to *twisted versions* of groups of Lie type are covered; in particular the Ree groups of type $^2F_4$ produce generalized octagons (see TITS [1960]). The projective planes and generalized quadrangles corresponding to classical groups were called *classical polygons* (see KANTOR [1986a]). Soon the term *classical* became a synonym of *corresponding to a group of Lie type* for some authors, but others stuck to the corresponding group notion, i.e., classical polygons correspond to classical groups. When TITS [1974], [1976a] introduced his ideas on the *Moufang condition* and *Moufang polygons*, the term *classical* was sometimes indistinguishable from the term *Moufang*. The situation at this moment is that the class of *Moufang polygons* is a well-defined class of generalized polygons, but the proof of Tits' (and Weiss') enumeration has not yet been completely published; see Chapter 5 for more details. On the other hand, the class of *classical polygons* is not so well defined, but once a point of view is taken, all classical polygons can usually be enumerated. For instance, in PAYNE & THAS [1984], a finite classical quadrangle is one that can be fully embedded in a finite projective space. All of them are known (see Chapter 8, in particular Theorem 8.5.16).

We will take the following point of view. Roughly, the generalized polygons which will be defined in this chapter will be called *classical*. This term is motivated by the fact that these polygons are studied much more than others and hence serve as the *classical objects* (in the sense of *standard examples*). One exception will be the Ree–Tits octagons, which we will only explicitly introduce in Chapter 3 and which we also want to call *classical*; another exception is the projective plane related to a non-associative alternative division ring, which we introduce in this chapter, but which will not be called *classical* for historical reasons. In this way, all finite Moufang polygons will be called *classical*. In addition to these classical examples, there are other examples of *Moufang polygons*, and we will define some in Chapter 3, the notable examples being the *mixed quadrangles* and *mixed hexagons*. Finally, when discussing the Moufang property in Chapter 5, we will meet other polygons, and we will give them names, too. The main thing is that we will have a name for large classes of polygons sharing some important common group-theoretic (and geometric) properties. This enables one to state propositions in an elegant way. Note that the classes that we will define will not necessarily be disjoint. For instance, we will see that some symplectic quadrangles are both classical and mixed.

It is worth remarking that, by definition, the dual of a classical polygon is always itself a classical polygon, unlike in PAYNE & THAS [1984] and unlike the other names that we will assign to subclasses of classical polygons. So if it matters what the points are, then we call a classical polygon by its specific name, such as orthogonal quadrangle, twisted triality hexagon. If it doesn't matter, then we simply say "classical polygon".

Concerning notation, we have chosen to follow PAYNE & THAS [1984] for the finite generalized quadrangles, and one class of hexagons. This implies that we do not use the group notation used by KANTOR [1986a]. A table with "translations" (Table 2.1) can be found at the end of this chapter.

2.2 Classical and alternative projective planes

Since we want to emphasize generalized n-gons for $n > 3$, we will be brief in this section. The **classical projective planes** are in fact the *Desarguesian planes*. Other, related, planes are the *alternative planes*. Among the Desarguesian planes, one has the more restricted class of *Pappian* planes.

Desarguesian projective planes

2.2.1 Construction

Let V be a three-dimensional right vector space over a skew field $\mathbb{K}$ (i.e., the scalars are written on the right of the vectors). We define $\Gamma = \mathbf{PG}(2, \mathbb{K}) = (\mathcal{P}, \mathcal{L}, I)$ as follows. The points of Γ are the 1-spaces of V; the lines of Γ are the 2-spaces of V; incidence is symmetrized inclusion (here, an i-space is just an i-dimensional subspace, $i = 1, 2$; we will use this terminology later for arbitrary i in n-dimensional vector spaces, $n \geq i$).

Another way of seeing $\mathbf{PG}(2,\mathbb{K})$ is as follows. The elements of $\mathcal{P}$ are the triples $(x,y,z) \in \mathbb{K} \times \mathbb{K} \times \mathbb{K}$ up to a right non-zero scalar, $(x,y,z) \neq (0,0,0)$; the elements of $\mathcal{L}$ are the triples $[u,v,w]$ up to a left non-zero scalar, $[u,v,w] \neq [0,0,0]$; the point represented by (x,y,z) is incident with the line represented by $[u,v,w]$ if and only if $ux + vy + wz = 0$.

A third way of seeing this geometry is the following. The points are of three types: the pairs $(x,y) \in \mathbb{K} \times \mathbb{K}$, the elements (m), $m \in \mathbb{K}$ and a symbol (∞). The lines are dually also of three types: the pairs $[m,k] \in \mathbb{K} \times \mathbb{K}$, the elements $[x]$, $x \in \mathbb{K}$ and the symbol $[\infty]$. Incidence is defined as follows: the point (∞) is incident with $[\infty]$ and $[x]$, for all $x \in \mathbb{K}$; the point (m), $m \in \mathbb{K}$, is incident with $[\infty]$ and $[m,k]$ for all $k \in \mathbb{K}$; the point (x,y), $x,y \in \mathbb{K}$, is incident with $[x]$, for all $x \in \mathbb{K}$, and with $[m,k]$ **if and only if** $mx + y = k$.

It is an easy exercise to show that these three ways produce isomorphic projective planes (in particular, the second description arises from the first by introducing coordinates in V). We call them the **Desarguesian planes (over $\mathbb{K}$)**. They have the following well-known characterizing property (see HILBERT [1899], BAER [1942]):

2.2.2 Theorem. *A projective plane Γ is Desarguesian* **if and only if** *the following holds: for all triples (a_1, a_2, a_3) and (b_1, b_2, b_3) of pairwise distinct points, with $a_i \neq b_i$ and $a_i b_i \neq a_j b_j$, $i,j \in \{1,2,3\}$, $i \neq j$, the lines $a_1 b_1$, $a_2 b_2$ and $a_3 b_3$ are concurrent if and only if the points $a_1 a_2 \cap b_1 b_2$, $a_2 a_3 \cap b_2 b_3$ and $a_1 a_3 \cap b_1 b_3$ are collinear.* $\square$

The configuration formed by the ten points a_i, b_i, $a_i a_j \cap b_i b_j$, $a_1 b_1 \cap b_1 b_2$, $i,j \in \{1,2,3\}$, $i \neq j$, and the ten lines $a_i b_i$, $a_i a_j$, $b_i b_j$ and the line containing the points $a_1 a_2 \cap b_1 b_2$ and $a_2 a_3 \cap b_2 b_3$, $i,j \in \{1,2,3\}$, $i \neq j$, in a Desarguesian projective plane (but also in any plane whenever the two conditions of the previous theorem are satisfied for these points and lines) is usually called a **Desargues configuration**.

If the skew field $\mathbb{K}$ is commutative, i.e., if $\mathbb{K}$ is a field, then we sometimes call $\mathbf{PG}(2,\mathbb{K})$ a **Pappian plane**. All Pappian planes share the following well-known characterizing property (see e.g. HUGHES & PIPER [1973]):

2.2.3 Theorem. *A projective plane Γ is Pappian* **if and only if** *for all triples (a_1, a_2, a_3) and (b_1, b_2, b_3) of distinct collinear points, with $a_i \notin b_1 b_2$ and $b_i \notin a_1 a_2$, $i = 1,2,3$, the points $a_1 b_2 \cap b_1 a_2$, $a_2 b_3 \cap b_2 a_3$ and $a_1 b_3 \cap b_1 a_3$ are collinear.* $\square$

The configuration induced by the points and lines in the statement of the previous theorem is, similarly as above with the Desargues configuration, usually called a **Pappus configuration**.

Desarguesian planes are also called **classical planes**.

Alternative projective planes

2.2.4 Construction

Let $\mathbb{D}$ be an **alternative field**, also called an **alternative division ring**, i.e., addition in $\mathbb{D}$ defines a commutative group; multiplication in $\mathbb{D}^\times$, where $\mathbb{D}^\times$ denotes $\mathbb{D}$ without the neutral element 0 with respect to addition, has a neutral element 1; there is a two-sided inverse x^{-1} for every element $x \neq 0$; both distributive laws hold; and $(yx)x^{-1} = y$, $y = x^{-1}(xy)$ for all $x, y \in \mathbb{D}$. We define the following geometry Γ. The points of Γ are of three types: the pairs $(x, y) \in \mathbb{D} \times \mathbb{D}$, the elements (m), $m \in \mathbb{D}$ and a symbol (∞). The lines are dually also of three types: the pairs $[m, k] \in \mathbb{D} \times \mathbb{D}$, the elements $[x]$, $x \in \mathbb{D}$ and the symbol $[\infty]$. Incidence is defined as follows: the point (∞) is incident with $[\infty]$ and $[x]$, for all $x \in \mathbb{D}$; the point (m), $m \in \mathbb{D}$, is incident with $[\infty]$ and $[m, k]$ for all $k \in \mathbb{D}$; the point (x, y), $x, y \in \mathbb{D}$, is incident with $[x]$, for all $x \in \mathbb{D}$, and with $[m, k]$ **if and only if** $mx + y = k$.

This is completely similar to the third construction of a Desarguesian plane above. Hence, if $\mathbb{D}$ is a skew field, then Γ is the Desarguesian plane over $\mathbb{D}$. If $\mathbb{D}$ is not a skew field, then we call Γ an **alternative plane (over $\mathbb{D}$)**. Alternative planes and Desarguesian planes share the following characterizing property (see again e.g. HUGHES & PIPER [1973]):

2.2.5 Theorem. *A projective plane is alternative or classical **if and only if** for every point p and every line L incident with p, for all triples of pairwise distinct points (a_1, a_2, a_3) and (b_1, b_2, b_3) with $a_i \neq b_i$ and $a_i b_i \neq a_j b_j$, $i, j \in \{1, 2, 3\}$, $i \neq j$, such that p is incident with $a_k b_k$, $k = 1, 2$, and $a_1 a_2 \cap b_1 b_2$ and $a_2 a_3 \cap b_2 b_3$ are incident with L, we have that p is incident with $a_3 b_3$ if and only if L is incident with $a_1 a_3 \cap b_1 b_3$.* $\qquad\square$

The configuration induced by the elements in the previous theorem is a special case of the Desargues configuration; it is often called the **little Desargues configuration** (see HUGHES & PIPER [1973]).

The reason why we mention these characterizations is the following. From the description of alternative planes, one might get the impression that one flag, namely $\{(\infty), [\infty]\}$, plays a special role. But that is not true. It only plays a special role in the construction. The characterization theorems make this clear because no special flag is hypothesized there.

We classify non-associative alternative fields in Appendix B.

2.3 Classical generalized quadrangles

In this section, we define the classical generalized quadrangles. Our definition is based on Chapter 10 of BRUHAT & TITS [1972], modified slightly by TITS [1995]; see also Chapter 8 of TITS [1974].

2.3.1 σ-quadratic forms

Let $\mathbb{K}$ be a skew field and σ an anti-automorphism of order at most 2. This implies in particular that $\mathbb{K}$ is a field if σ is the identity because in that case $ab = (ab)^\sigma = b^\sigma a^\sigma = ba$, for all $a, b \in \mathbb{K}$. Let V be a — not necessarily finite-dimensional — right vector space over $\mathbb{K}$ and let $g : V \times V \to \mathbb{K}$ be a $(\sigma, 1)$-linear form, i.e., for all $v_1, v_2, w_1, w_2 \in V$ and all $a_1, a_2, b_1, b_2 \in \mathbb{K}$, we have

$$g(v_1 a_1 + v_2 a_2, w_1 b_1 + w_2 b_2) =$$
$$a_1^\sigma g(v_1, w_1) b_1 + a_1^\sigma g(v_1, w_2) b_2 + a_2^\sigma g(v_2, w_1) b_1 + a_2^\sigma g(v_2, w_2) b_2.$$

We define $f : V \times V \to \mathbb{K}$ as follows:

$$f(x, y) = g(x, y) + g(y, x)^\sigma.$$

It is clear that f is also $(\sigma, 1)$-linear and moreover f satisfies $f(x, y)^\sigma = f(y, x)$, for all $x, y \in V$. Therefore we say that f is a $(\sigma$-$)$**Hermitian form**. Denote $\mathbb{K}_\sigma := \{t^\sigma - t : t \in \mathbb{K}\}$. We define $q : V \to \mathbb{K}/\mathbb{K}_\sigma$ as

$$q(x) = g(x, x) + \mathbb{K}_\sigma,$$

for all $x \in V$. We call q a σ-**quadratic form (over $\mathbb{K}$)**. Let W be a subspace of V. We say that q is **anisotropic over** W if $q(w) = 0$ **if and only if** $w = 0$, for all $w \in W$ (where we have written the zero vector as 0, and the element $0 + \mathbb{K}_\sigma$ also as 0; the context always makes it clear which zero is meant by "0"). It is **non-degenerate** if it is anisotropic over the subspace $\{v \in V : f(v, w) = 0, \text{ for all } w \in V\}$. From now on we assume that q is non-degenerate.

Note that, if $q(v) = 0$, then $q(vk) = 0$, for all $k \in \mathbb{K}$. Indeed, $q(v) = 0$ is equivalent to $g(v, v) \in \mathbb{K}_\sigma$, so put $g(v, v) = t^\sigma - t$. But then $g(vk, vk) = k^\sigma g(v, v) k = (k^\sigma t k)^\sigma - (k^\sigma t k)$. Hence the inverse image $q^{-1}(0)$ is a union of 1-spaces. We say that q **has Witt index** 2, if $q^{-1}(0)$ contains 2-spaces, but no higher-dimensional subspaces.

For a non-degenerate σ-quadratic form q over $\mathbb{K}$ with Witt index 2, we define the following geometry $\Gamma = \mathsf{Q}(V, q)$. The points of Γ are the 1-spaces in $q^{-1}(0)$; the lines are the 2-spaces in $q^{-1}(0)$; and incidence is symmetrized inclusion.

Before showing that $\mathsf{Q}(V, q)$ is a weak generalized quadrangle, we start with a lemma.

2.3.2 Lemma. *With the above notation, we have $\mathbb{K} \neq \mathbb{K}_\sigma$.*

Proof. The lemma is clear if σ is the identity. If σ is not the identity, then we remark that every element x of $\mathbb{K}_\sigma$ satisfies the relation $x^\sigma = -x$. Hence if $\mathbb{K} = \mathbb{K}_\sigma$, then $x^\sigma = -x$ for all $x \in \mathbb{K}$. If the characteristic of $\mathbb{K}$ is equal to 2, then this implies that $\sigma = 1$; if the characteristic of $\mathbb{K}$ is not equal to 2, then this implies that $1^\sigma = -1 \neq 1$, a contradiction as well. $\qquad\square$

2.3.3 Theorem (Bruhat & Tits [1972]). *The geometry* $Q(V, q)$ *defined above is a weak generalized quadrangle.*

Proof. We first show that, given a point p and a line L of $Q(V, q)$, with p not incident with L, there exists a unique point p' incident with L and collinear with p. Since the line pp' will then be well defined (because two distinct 1-spaces in V define a unique 2-space), this implies that there is also a unique line pp' incident with p and concurrent with L. Let v be a vector on the 1-space corresponding to p, and let a, b be two non-proportional vectors in the 2-space corresponding to L. To show that there is at least one such a point p', we may assume that p is not collinear with the points of $Q(V, q)$ represented by a and b.

Now we investigate what it means algebraically for two vectors v and a to represent collinear points. By definition, this means that $q(vk_1 + ak_2) = 0$, for all $k_1, k_2 \in \mathbb{K}$. Since $q(v) = q(a) = 0$, this is equivalent to

$$
\begin{aligned}
g(vk_1 + ak_2, vk_1 + ak_2) &= k_1^\sigma g(v, a)k_2 + k_2^\sigma g(a, v)k_1 \\
&= k_1^\sigma f(v, a)k_2 + (k_1^\sigma g(a, v)^\sigma k_2)^\sigma - k_1^\sigma g(a, v)^\sigma k_2 \in \mathbb{K}_\sigma,
\end{aligned}
$$

hence to $k_1^\sigma f(v, a)k_2 \in \mathbb{K}_\sigma$, for all $k_1, k_2 \in \mathbb{K}$. If $f(v, a) \neq 0$, then this implies $\mathbb{K} = \mathbb{K}_\sigma$, contradicting Lemma 2.3.2. Hence $f(v, a) = 0$. Conversely, it follows from $f(v, a) = 0$ that the points $v\mathbb{K}$ and $a\mathbb{K}$ of $Q(V, q)$ are collinear (use the same equality above).

Hence we may assume that $f(v, a) \neq 0 \neq f(v, b)$, and we know that $f(a, b) = 0$. We are looking for a scalar $k \in \mathbb{K}$ such that $f(v, a + bk) = 0$. Clearly

$$
k = -f(v, b)^{-1} f(v, a)
$$

satisfies that condition. Hence we have shown that there is at least one point p' on L collinear with p. Suppose there are two such points. Without loss of generality, we may take the points $a\mathbb{K}$ and $b\mathbb{K}$. Hence $f(v, a) = f(v, b) = 0$. Note that $f(v, v) = 0$; indeed, $q(v) = 0$ by definition, so $g(v, v) \in \mathbb{K}_\sigma$, hence we can put $g(v, v) = t^\sigma - t$. But $f(v, v) = g(v, v) + g(v, v)^\sigma = t^\sigma - t + t - t^\sigma = 0$. So by the linearity $f(vk_1 + ak_2 + bk_3, v\ell_1 + a\ell_2 + b\ell_3) = 0$, for all $k_i, \ell_i \in \mathbb{K}$, $i \in \{1, 2, 3\}$. This readily implies that the subspace of V generated by v, a, b is contained in $q^{-1}(0)$. Since the Witt index equals 2, we necessarily have $p \mathbf{I} L$, a contradiction.

So we have shown condition (i) of Lemma 1.4.1 (page 15). We now show condition $(ii)'$ of that same lemma.

It is clear that every line in $Q(V, q)$ contains $|\mathbb{K}| + 1$ points, hence all lines are thick. Also, it is clear that two distinct points of $Q(V, q)$ are contained in at most one line, since two different 1-spaces of V are contained in exactly one 2-space of V.

All that is left to show is that every point is incident with at least two lines. To that end, we first claim that for every point p in $Q(V, q)$, there is at least one point p' of $Q(V, q)$ not collinear with p in $Q(V, q)$. Indeed, suppose $p = v\mathbb{K}$. Let w be any vector of V and suppose that $w\mathbb{K}$ is not a point of $Q(V,q)$. Note that $w\mathbb{K}$ certainly exists since we may otherwise assume that every 1-space of V is a point of $Q(V,q)$ collinear with p. This would imply $f(v,w) = 0$, for every vector w, and since $q(v) = 0$, this contradicts the non-degeneracy of q. Note that this argument implies also that we may assume that $f(v,w) \neq 0$. Now the vector $v + wk$, $k \in \mathbb{K}$, defines a point of $Q(V,q)$ **if and only if** $q(v+wk) = 0$. This condition is equivalent to

$$k^\sigma g(w,w)k + g(v,w)k + k^\sigma g(w,v) \in \mathbb{K}_\sigma.$$

Noting that $k^\sigma g(w,v) - g(w,v)^\sigma k \in \mathbb{K}_\sigma$, we deduce that, by dividing on the right by k,

$$k = -g(w,w)^{-\sigma} f(w,v)$$

is a non-zero solution, since $f(v,w) \neq 0$. Hence the claim.

Now let p be any point of $Q(V,q)$. Since the Witt index is equal to 2, there is at least one line L in $Q(V,q)$. By the first part of the proof, we may assume that $L \, \mathbf{I} \, p$. Let p' be a point of $Q(V,q)$ not collinear with p, then there is a line M through p' meeting L in some point $p'' \neq p$. There is some point p''' not collinear with p'' in $Q(V,q)$. Hence there is a line M' through p''' meeting M in a point distinct from p''. But now there is a line L' through p meeting M', and clearly $L' \neq L$ (otherwise we violate condition (i) of Lemma 1.4.1 that we showed above). So we have at least two lines L, L' in $Q(V,q)$ through p.

The theorem now follows from Lemma 1.4.1. □

Next, we look for a standard equation for q.

2.3.4 Proposition. *Let q be a non-degenerate σ-quadratic form over $\mathbb{K}$ on the vector space V. Then there exist four vectors e_i, $i \in \{-2, -1, 1, 2\}$, a direct sum decomposition*

$$V = e_{-2}\mathbb{K} \bigoplus e_{-1}\mathbb{K} \bigoplus V_0 \bigoplus e_1\mathbb{K} \bigoplus e_2\mathbb{K}$$

and a non-degenerate anisotropic σ-quadratic form $q_0 : V_0 \to \mathbb{K}/\mathbb{K}_\sigma$ such that for all $v = e_{-2}x_{-2} + e_{-1}x_{-1} + v_0 + e_1x_1 + e_2x_2$, with $x_i \in \mathbb{K}$, $i \in \{-2, -1, 1, 2\}$ and $v_0 \in V_0$,

$$q(v) = x^\sigma_{-2}x_2 + x^\sigma_{-1}x_1 + q_0(v_0).$$

Proof. We already know that $Q(V,q)$ is a weak quadrangle. Let $e_{-2}\mathbb{K}$, $e_{-1}\mathbb{K}$, $e_1\mathbb{K}$ and $e_2\mathbb{K}$ be four points of $Q(V,q)$ such that $e_i\mathbb{K}$ and $e_j\mathbb{K}$ are opposite **if and only if** $i+j = 0$. It is readily seen that these four 1-spaces cannot be contained in a 3-space (otherwise there arise triangles in $Q(V,q)$). Hence the sum $e_{-2}\mathbb{K}+e_{-1}\mathbb{K}+e_1\mathbb{K}+e_2\mathbb{K}$ is direct. Now note that by the previous proof, we have

$$f(e_i, e_j) = 0, \qquad\qquad i+j \neq 0.$$

Upon replacing e_i by a multiple, we may assume that $f(e_i, e_{-i}) = 1$, for all $i \in \{-2, -1, 1, 2\}$. Now we define the subspace

$$V_0 := \{v \in V : f(v, e_i) = 0, \text{ for all } i = -2, -1, 1, 2\}.$$

For any $v \in V$, it is an elementary exercise to check that the vector

$$v - e_{-2}f(e_2, v) - e_{-1}f(e_1, v) - e_1 f(e_{-1}, v) - e_2 f(e_{-2}, v)$$

belongs to V_0. Also, no non-zero vector v^* generated by e_{-2}, e_{-1}, e_1 and e_2 belongs to V_0. Indeed, such a vector v^* would satisfy by assumption $f(v^*, e_i) = 0$, for all $i \in \{-2, -1, 1, 2\}$, which implies, putting $v^* = e_{-2}a_{-2} + e_{-1}a_{-1} + e_1 a_1 + e_2 a_2$, $a_i \in \mathbb{K}$, that all a_i are equal to 0, $i \in \{-2, -1, 1, 2\}$, a contradiction. Therefore, the following sum is indeed direct:

$$V = e_{-2}\mathbb{K} \bigoplus e_{-1}\mathbb{K} \bigoplus V_0 \bigoplus e_1 \mathbb{K} \bigoplus e_2 \mathbb{K}.$$

Now let $v = e_{-2}x_{-2} + e_{-1}x_{-1} + v_0 + e_1 x_1 + e_2 x_2$, with $x_i \in \mathbb{K}$ and $v_0 \in V_0$, $i \in \{-2, -1, 1, 2\}$. Then one calculates easily that, using $a + \mathbb{K}_\sigma = a^\sigma + \mathbb{K}_\sigma$,

$$
\begin{aligned}
q(v) &= x_{-2}^\sigma f(e_{-2}, e_2)x_2 + x_{-1}^\sigma f(e_{-1}, e_1)x_1 + q(v_0) \\
&= x_{-2}^\sigma x_2 + x_{-1}^\sigma x_1 + q(v_0).
\end{aligned}
$$

Now let $q_0 : V_0 \to \mathbb{K}/\mathbb{K}_\sigma$ be the restriction of q to V_0. Suppose $q_0(v) = 0$, for some non-zero vector $v \in V_0$. Since we also have $f(v, e_i) = 0$, the point $v\mathbb{K}$ of $\mathsf{Q}(V, q)$ is collinear with $e_i\mathbb{K}$, for all $i \in \{-2, -1, 1, 2\}$, a contradiction.

The proposition is proved. $\square$

To determine the order of $\mathsf{Q}(V, q)$, we still need to know how many lines there are through any point. The next proposition gives the answer.

2.3.5 Proposition. *Let* $\mathsf{Q}(V, q)$ *be a classical weak quadrangle and suppose* q *has the standard equation of Proposition 2.3.4. Define the set*

$$\widehat{X} = \{(v, k) \in V_0 \times \mathbb{K} : k \in -q_0(v)\}.$$

Then $\mathsf{Q}(V, q)$ *contains exactly* $|\widehat{X}| + 1$ *lines through each point.*

Proof. Without loss of generality, we may consider the point $e_1\mathbb{K}$. We already know that all points of $\mathsf{Q}(V, q)$ collinear with $e_1\mathbb{K}$ are represented by vectors v such that $f(v, e_1) = 0$. Also, the number of lines through $e_1\mathbb{K}$ is equal to the number of points of $\mathsf{Q}(V, q)$ collinear with both $e_1\mathbb{K}$ and $e_{-1}\mathbb{K}$. Such points have representatives satisfying in addition $f(v, e_{-1}) = 0$. It is now readily seen that $v \in e_{-2}\mathbb{K} + V_0 + e_2\mathbb{K}$, so we can put (with obvious notation) $v = e_{-2}x_{-2} + v_0 + e_2 x_2$. If $x_{-2} = 0$, then $0 = q(v) = q_0(v_0)$, so $v_0 = 0$ (since q_0 is anisotropic) and $v = e_2$. This already gives one point of $\mathsf{Q}(V, q)$ collinear with both $e_1\mathbb{K}$ and $e_{-1}\mathbb{K}$.

Suppose now $x_{-2} \neq 0$, then we may take $x_{-2} = 1$. We have $0 = q(v) = x_2 + q_0(v_0)$. Hence $x_2 \in -q_0(v_0)$. Conversely, if $x_2 \in -q_0(v_0)$, then the vector $v = e_{-2} + v_0 + e_2 x_2$ defines a point of $\mathsf{Q}(V, q)$ collinear with both $e_1\mathbb{K}$ and $e_{-1}\mathbb{K}$.

The proposition now follows easily. $\square$

There is an interesting corollary to Proposition 2.3.5.

2.3.6 Corollary. *The weak quadrangle* Q(V, q) *is a generalized quadrangle* **if and only if** *V has dimension at least 5 or* σ *is not the identity.*

Proof. According to Proposition 2.3.5, the weak quadrangle Q(V, q) is not thick if and only if the set $\widehat{X}$ has only one element, i.e., if $0 \in V$ is the only element of V_0 and if $0 \in \mathbb{K}$ is the only element of $\mathbb{K}_\sigma$. Hence Q(V, q) is non-thick **if and only if** $V_0 = \{0\}$ and $\sigma = 1$. The corollary follows. □

The points and lines of the quadrangle Q(V, q) can be seen as living in the projective space **PG**(V) associated to V in the standard way. Therefore we sometimes refer to that representation of Q(V, q) as a **standard embedding** of Q(V, q).

For a discussion of the regular points and lines of the classical quadrangles, we refer to Proposition 3.4.8 on page 106. Also, in the subsections following Subsection 3.4.7, we prove that some classical quadrangles with $\sigma = 1$ are anti-isomorphic to certain other classical quadrangles with $\sigma \neq 1$.

2.3.7 Definitions. Members of the special class of classical quadrangles with $\sigma = 1$ are called **orthogonal quadrangles**. The rest is called **Hermitian quadrangles**. The duals of the classical quadrangles are also called **classical**.

Sometimes we also denote the orthogonal quadrangle Q(V, q) by Q($d - 1, \mathbb{K}, q$), where V is d-dimensional over $\mathbb{K}$. And the Hermitian quadrangle Q(V, q) with anti-automorphism σ is also sometimes denoted by H($d - 1, \mathbb{K}, q, \sigma$), where V is d-dimensional over $\mathbb{K}$.

In the finite case, a σ-quadratic form q of Witt index 2 is, up to isomorphism and up to a scalar factor, determined by the dimension, the field and the kind (= orthogonal or Hermitian). Hence we delete the q and the σ from the notation. For $\mathbb{K} \cong \mathbf{GF}(s)$, we then use Q($d, s$) for the orthogonal quadrangle in d-dimensional projective space over $\mathbf{GF}(s)$ (and $d = 4, 5$, see below), and H(d, s) is the Hermitian quadrangle in d-dimensional projective space over $\mathbf{GF}(s)$ with corresponding involutory field automorphism $x \mapsto x^{\sqrt{s}}$ (and here, $d = 3, 4$, see also below).

2.3.8 Quadrics as orthogonal quadrangles

Let V be a right $(d+1)$-dimensional vector space over a (commutative) field $\mathbb{K}$ and let **PG**(V) denote the corresponding d-dimensional projective space. Let Q be a quadric in **PG**(V) of **Witt index** 2, i.e., Q contains lines but no planes of **PG**(V). Then the points and lines of Q are, with the natural incidence relation, the points and lines of a generalized quadrangle Γ. We show that *there is a 1-quadratic form* $q : V \rightarrow \mathbb{K}$ *such that* Γ *is isomorphic to* Q($d, \mathbb{K}, q$).

Let Q have equation

$$\sum_{i \leq j = 0}^{d} a_{ij} x_i x_j = 0,$$

with $a_{ij} \in \mathbb{K}$, with respect to a basis in $\mathbf{PG}(V)$, or equivalently, in V. Put

$$g((x_0, x_1, \ldots, x_d), (y_0, y_1, \ldots, y_d)) = \sum_{i \leq j = 0}^{d} a_{ij} x_i y_j.$$

This is clearly a bilinear form. The corresponding 1-quadratic form

$$q((x_0, x_1, \ldots, x_d)) = \sum_{i \leq j = 0}^{d} a_{ij} x_i x_j$$

is zero precisely on the 1-spaces of V which correspond to points of Q (noting that $\mathbb{K}_\sigma = \{0\}$ here).

Conversely, it is easily seen that any orthogonal quadrangle $\mathsf{Q}(d, \mathbb{K}, q)$ arises from a quadric with equation $q(v) = 0$. Hence the class of orthogonal quadrangles coincides with the class of quadrics of Witt index 2 in projective space.

2.3.9 Hermitian varieties as Hermitian quadrangles

Let V be a vector space over the skew field $\mathbb{K}$ and suppose that $f : V \times V \to \mathbb{K}$ is a σ-Hermitian form, i.e., f is $(\sigma, 1)$-linear and $f(v, w) = f(w, v)^\sigma$, with σ non-trivial. The corresponding Hermitian variety $\mathcal{H}$ in $\mathbf{PG}(V)$ is a generalized quadrangle $\Gamma(\mathcal{H})$ **if and only if** $\mathcal{H}$ contains lines but no planes.

Recall that the points of $\mathcal{H}$ correspond to the 1-spaces of V with representatives v such that $f(v, v) = 0$.

Suppose that the characteristic of $\mathbb{K}$ is not equal to 2. We choose a basis $(e_i)_{i \in J}$ in V and we put an arbitrary order on J. We define

$$\begin{aligned} g(e_i, e_j) &= f(e_i, e_j) & \text{if } i < j, \\ g(e_i, e_i) &= \tfrac{1}{2} f(e_i, e_i) \\ g(e_i, e_j) &= 0 & \text{if } i > j. \end{aligned}$$

One can check easily that the associated σ-quadratic form q of the thus defined $(\sigma, 1)$-linear form reads

$$q(v) = \frac{1}{2} f(v, v) + \mathbb{K}_\sigma.$$

It is clear that, if $f(v, v) = 0$, then also $q(v) = 0$. Suppose now $q(v) = 0$. This means that, since $2^\sigma = 2$, $f(v, v) \in \mathbb{K}_\sigma$. Write $f(v, v) = k^\sigma - k$. Since $f(x, x) = f(x, x)^\sigma$, we have $k^\sigma = k$, hence $f(v, v) = 0$. Consequently $\mathsf{Q}(V, q)$ defines a quadrangle which is isomorphic to the quadrangle $\Gamma(\mathcal{H})$ defined above.

Suppose now that the characteristic of $\mathbb{K}$ is 2. We choose a basis $(e_i)_{i \in J}$ in V such that $f(e_i, e_i) = 0$, for all $i \in J$ (this can easily be done since we may assume that the points of $\mathcal{H}$ generate $\mathbf{PG}(V)$). Now we can define $g : V \times V \to \mathbb{K}$ as above (deleting the factor $\frac{1}{2}$ of course). One checks that $f(x, y) = g(x, y) + g(y, x)^\sigma$. Define

$$q : V \to \mathbb{K}/\mathbb{K}_\sigma : v \mapsto g(v, v) + \mathbb{K}_\sigma.$$

If k is a representative of $q(v)$ in $\mathbb{K}$, then $k = g(v,v) + l^\sigma + l$. Hence

$$f(v,v) = g(v,v) + g(v,v)^\sigma = g(v,v) + l^\sigma + l + g(v,v)^\sigma + l + l^\sigma = k + k^\sigma.$$

In general, however (i.e., when $\mathbb{K}$ is not commutative; see below for the commutative case), the quadrangle $\mathsf{Q}(V,q)$ is only a subquadrangle of $\Gamma(\mathcal{H})$. Indeed, if $q(v) = 0$, then 0 is a representative of $q(v)$, and hence by the above, $f(v,v) = 0 + 0^\sigma = 0$. To obtain $\Gamma(\mathcal{H})$, we have to consider another σ-quadratic form. Define $V_* = \mathbb{K}^{(\sigma)}/\mathbb{K}_\sigma$, with $\mathbb{K}^{(\sigma)}$ the set of fixed points of σ in $\mathbb{K}$ (and note that, if $k \in \mathbb{K}_\sigma$, then $k^\sigma = k$ so that this expression makes sense). We turn V_* into a (not necessarily finite-dimensional) right vector space over $\mathbb{K}$ by defining $v \cdot k = k^\sigma v k$, for all $v \in V_*$ and $k \in \mathbb{K}$. We choose a basis $(e_i^*)_{i \in J_*}$ in V_*, choose a representative g_i in $\mathbb{K}^{(\sigma)}$ for every e_i^* and define $g_*(e_i^*, e_j^*) = 0$ for $i \neq j$, and $g_*(e_i^*, e_i^*) = g_i$. It is readily checked that this determines a $(\sigma, 1)$-linear form g_* with $g_*(v_*, v_*)$ a representative of v_* in $\mathbb{K}^{(\sigma)}$. We put $V' = V \bigoplus V_*$ and define, with the obvious notation,

$$q'(v + v_*) = g_*(v_*, v_*) + q(v) = v_* + q(v).$$

So q' is by definition a σ-quadratic form. We claim that the canonical projection onto V of $q'(0)^{-1}$ is bijective and coincides with $\mathcal{H}$.

Indeed, first suppose that $q'(v + v_*) = 0$. Then any representative k of $q(v)$ belongs to $\mathbb{K}^{(\sigma)}$ and hence $f(v,v) = k^\sigma + k = k + k = 0$ (see above). Now suppose that $f(v,v) = 0$. Then putting $v_* = q(v)$, we obtain $q'(v + v_*) = q(v) + v_* = 0$. Clearly if $q'(v + v_*') = 0$, then $q(v) = v_*'$ and hence $v_* = v_*'$. The claim is proved.

Hence we have shown that every Hermitian variety $\mathcal{H}$ containing lines but no planes gives rise to a classical quadrangle $\Gamma(\mathcal{H})$. The representation of $\Gamma(\mathcal{H})$ as the Hermitian variety $\mathcal{H}$ in a projective space will also be called a **standard embedding of** $\Gamma(\mathcal{H})$.

Finally, we would like to introduce the following notation. Since the σ-quadratic form q does not play any role in $\mathsf{H}(3, \mathbb{K}, q, \sigma)$ (simply look at the standard equation for q), we may denote that quadrangle by $\mathsf{H}(3, \mathbb{K}, \sigma)$. In particular, inequivalent σ-quadratic forms have inequivalent associated anti-automorphisms σ.

2.3.10 D_ℓ-quadrangles

In general, the **Witt index** of a σ-quadratic form q defined on some vector space V is the dimension of the subspaces of highest dimension in $q^{-1}(0)$. For a fixed dimension and a fixed field, a lot of cases can occur. For instance, there may be several non-equivalent σ-quadratic forms (of different Witt index, for different σ). The corresponding geometries are *(classical) polar spaces (of rank r)*, where r is the Witt index of the σ-quadratic form. So the classical generalized quadrangles are in fact (up to duality) classical polar spaces of rank 2. Polar spaces (we will not need the precise definition of such geometries) can be viewed as spherical buildings and thus they are assigned a diagram and a type (see Subsection 1.3.7 on page 8). For one particular such type and diagram (namely, D_ℓ), the polar space is completely and uniquely determined by the field and the dimension of the vector space (or alternatively, the rank of the polar space), and this gives rise to an important subclass of classical quadrangles, as we will now explain. However, we take the projective point of view.

Consider the projective space $\mathbf{PG}(2\ell - 1, \mathbb{K})$ and let Q be a quadric in that space, i.e., the null set of a (homogeneous) quadratic equation in the coordinates in $\mathbf{PG}(2\ell - 1, \mathbb{K})$. We say the Q is **non-degenerate** if no point of Q is collinear (on Q, i.e., the joining line has all its points on Q) with all other points of Q. The **Witt index** of Q is said to be k if the (projective) dimension of the projective subspace of highest dimension contained in Q is equal to $k - 1$. For arbitrary ℓ and $\mathbb{K}$, there always exists a quadric Q_ℓ of Witt index ℓ (the "split case", or, in French, "forme déployée") in $\mathbf{PG}(2\ell - 1, \mathbb{K})$ and it is projectively unique (which means that one such quadric can always be transformed into any other by a collineation of the projective space, and this collineation can be chosen to come from a linear map in the underlying vector space). The standard equation is given by $X_0 X_1 + X_2 X_3 + \cdots + X_{2l-2} X_{2l-1} = 0$ and one can see that for instance the subspace with equations $X_0 = X_2 = \cdots = X_{2l-2} = 0$ of projective dimension $\ell - 1$ is contained in Q. The corresponding polar space is of type D_ℓ. Note that Q is never a generalized quadrangle; weak quadrangles appear for $\ell = 2$. Now let Q' be any non-degenerate quadric in $\mathbf{PG}(2\ell - 1, \mathbb{K})$ with equation $F(X_1, \ldots, X_{2\ell}) = 0$. Then by extending the field $\mathbb{K}$ to its quadratic closure (or algebraic closure) $\overline{\mathbb{K}}$ and considering the equation $F(X_1, \ldots, X_{2\ell}) = 0$ over $\overline{\mathbb{K}}$, we obtain a polar space of type D_ℓ over $\overline{\mathbb{K}}$. If $\ell > 1$ and Q' has Witt index 2, then we say that the corresponding generalized quadrangle is **of type** (D_ℓ), or a D_ℓ-**quadrangle**. If the characteristic of $\mathbb{K}$ is not equal to 2, then every orthogonal quadrangle in odd-dimensional projective space over $\mathbb{K}$ is a D_ℓ-quadrangle; if $\mathbb{K}$ has characteristic 2, then the quadric over $\overline{\mathbb{K}}$ might be degenerate and if it is, we do not have a D_ℓ-quadrangle.

Some quadrangles of type (D_ℓ) will turn up as ideal subquadrangles of the so-called *exceptional Moufang quadrangles*; see Chapter 5.

Classical quadrangles over special fields

2.3.11 Commutative fields

We already know that, if $\sigma = 1$, then $Q(V, q)$, with q a σ-quadratic form as above, is isomorphic to the quadrangle arising from a quadric of Witt index 2 in $\mathbf{PG}(V)$. Suppose now that $\mathbb{K}$ is commutative and $\sigma \neq 1$. Let q be a σ-quadratic form and f the associated σ-Hermitian form. If the characteristic of $\mathbb{K}$ is not equal to 2, then

$$q^{-1}(0) = \{v \in V : f(v, v) = 0\}.$$

This can be shown as in Subsection 2.3.9 above. If the characteristic of $\mathbb{K}$ is equal to 2, then clearly $\mathbb{K}_\sigma = \mathbb{K}^{(\sigma)}$. Indeed, $\mathbb{K}_\sigma \subseteq \mathbb{K}^{(\sigma)}$ because $(k^\sigma + k)^\sigma = k + k^\sigma$; $\mathbb{K}^{(\sigma)} \subseteq \mathbb{K}_\sigma$ because $l(k^\sigma + k) = (lk)^\sigma + (lk)$ for all $l \in \mathbb{K}^{(\sigma)}$, and hence, since

$\mathbb{K}_\sigma$ is non-trivial ($\sigma \neq 1$), we deduce $\mathbb{K}^{(\sigma)} \cdot \mathbb{K}_\sigma \subseteq \mathbb{K}_\sigma$. Let g be the $(\sigma, 1)$-linear form associated with q, i.e., $g(x, y) + g(y, x)^\sigma = f(x, y)$. If $f(v, v) = 0$, then $g(v, v) = g(v, v)^\sigma \in \mathbb{K}^{(\sigma)} = \mathbb{K}_\sigma$, hence $q(v) = 0$. Conversely, if $q(v) = 0$, then $g(v, v) \in \mathbb{K}_\sigma = \mathbb{K}^{(\sigma)}$, so $f(v, v) = 0$.

This shows that in the commutative case *any classical quadrangle arises from a quadric or a Hermitian variety containing lines but no planes, in some projective space.*

Also remark that in **PG**$(4, \mathbb{K})$, all non-degenerate quadrics are projectively equivalent (this follows from Proposition 2.3.4). Hence we denote such an orthogonal quadrangle by $\mathsf{Q}(4, \mathbb{K})$, without referring to the (unique) 1-quadratic form.

2.3.12 Finite fields

Since finite skew fields are fields, all finite classical quadrangles arise from quadrics or Hermitian varieties. Let $\mathbb{K} = \mathbf{GF}(s)$. It is well known (see e.g. ARTIN [1957], page 144, or O'MEARA [1971], page 157) that for d odd there are exactly two isomorphism classes of (non-degenerate) quadrics in **PG**(d, s): members of one class have Witt index $\frac{d-1}{2}$, members of the other class have Witt index $\frac{d+1}{2}$ (and are of type $(D_{\frac{d+1}{2}})$). This is essentially due to the fact that every element of $\mathbf{GF}(s)$ can be written as a sum of two squares. Hence the dimension determines the quadrangle and the cases are: $d = 3$ (weak non-thick quadrangle) and $d = 5$. The classical quadrangle corresponding to the latter is a D_3-quadrangle and is denoted by $\mathsf{Q}(5, s)$, since the quadratic form is — up to isomorphism and a factor — determined by the dimension. If d is even, there is a unique isomorphism class of non-degenerate quadrics in **PG**(d, s) and they contain maximal projective subspaces of dimension $\frac{d-2}{2}$. So only $d = 4$ produces generalized quadrangles, and we denote such a quadrangle by $\mathsf{Q}(4, s)$.

The situation for Hermitian varieties is even simpler: for every dimension d, there is — up to isomorphism — just one example over $\mathbf{GF}(s)$ (with s a perfect square; the involutory field automorphism $x \mapsto x^{\sqrt{s}}$ is uniquely determined) and it has maximal projective subspaces of dimension $\frac{d-1}{2}$ (for d odd) or $\frac{d-2}{2}$ (for d even), see SCHARLAU [1985], page 39. So only $d = 3$ and $d = 4$ give us quadrangles and we denote them, respectively, by $\mathsf{H}(3, s)$ and $\mathsf{H}(4, s)$, in conformity with previous notation.

2.3.13 Algebraically closed fields

Not surprisingly, in any algebraically closed field $\mathbb{K}$ of characteristic not equal to 2, or, more generally, in any quadratically closed field of characteristic not equal to 2, a quadric of Witt index 2 has the standard equation

$$X_0^2 + X_1^2 + \cdots + X_\ell^2 = 0,$$

with $\ell = 3, 4$ (see O'MEARA [1971], Section 61B), but the case $\ell = 3$ corresponds to a weak non-thick quadrangle (see Corollary 2.3.6 above). So there is a unique orthogonal quadrangle over $\mathbb{K}$, namely $\mathsf{Q}(4, \mathbb{K})$.

2.3.14 The classical (skew) fields $\mathbb{R}, \mathbb{C}$ and $\mathbb{H}$

SYLVESTER's theorem implies that for any ordered field $\mathbb{K}$ where each positive element is a square, coordinates can be chosen in such a way that any quadric of Witt index 2 in $\mathbf{PG}(\ell, \mathbb{K})$ is given by the equation

$$-X_0^2 - X_1^2 + X_2^2 + X_3^2 + \cdots + X_\ell^2 = 0,$$

with $\ell \geq 4$ (see for instance ARTIN [1957], page 149 or O'MEARA [1971], Section 61A). Hence every orthogonal quadrangle over $\mathbb{R}$ is uniquely determined by the dimension ℓ. Therefore we can denote this orthogonal quadrangle by $\mathsf{Q}(\ell, \mathbb{R})$.

Of course, there are no Hermitian quadrangles over $\mathbb{R}$ since there are no non-trivial field automorphisms in $\mathbb{R}$.

Every involutory field automorphism in $\mathbb{C}$ is conjugate (in $\operatorname{Aut}\mathbb{C}$) to the standard conjugation map $a + ib \mapsto (a + ib)^* = a - ib$, where $i = \sqrt{-1}$ and $a, b \in \mathbb{R}$. It follows that, unlike the finite case, for every dimension $\ell \geq 3$, there is a unique Hermitian quadrangle over $\mathbb{C}$ in $\mathbf{PG}(\ell, \mathbb{C})$ and we briefly denote it by $\mathsf{H}(\ell, \mathbb{C})$. The corresponding Hermitian form is equivalent to

$$f(x, y) = -x_0^* y_0 - x_1^* y_1 + \sum_{r=2}^{\ell} x_r^* y_r,$$

where $x = (x_0, x_1, \ldots, x_\ell)$ and similarly for y.

Since $\mathbb{C}$ is algebraically closed, it follows from the previous subsection that there is only one orthogonal quadrangle over $\mathbb{C}$, namely $\mathsf{Q}(4, \mathbb{C})$.

Every involutory anti-automorphism of $\mathbb{H}$, the standard quaternions $\mathbb{R} + i\mathbb{R} + j\mathbb{R} + k\mathbb{R}$ over $\mathbb{R}$, is conjugate (in $\operatorname{Aut}\mathbb{H}$) to **either** the standard conjugation $a + ib + jc + kd \mapsto (a + ib + jc + kd)^* = a - ib - jc - kd$, **or** the skew conjugation $a + ib + jc + kd \mapsto (a + ib + jc + kd)^\diamond = a - ib + jc + kd$, where $a, b, c, d \in \mathbb{R}$. If σ is the standard conjugation, then for every dimension $\ell \geq 3$, there exists a unique σ-Hermitian form

$$f(x, y) = -x_0^* y_0 - x_1^* y_1 + \sum_{r=2}^{\ell} x_r^* y_r,$$

where $x = (x_0, x_1, \ldots, x_\ell)$ and similarly for y. The corresponding quadrangle only depends on ℓ and hence can be denoted by $\mathsf{H}(\ell, \mathbb{H}, \mathbb{R})$, where the presence of $\mathbb{R}$ replaces the notation σ in that $\mathbb{R}$ is the field of fixed elements of σ. This quadrangle will sometimes be referred to as a **real quaternion Hermitian quadrangle**.

If σ is the skew conjugation, then every σ-Hermitian form in $\mathbf{PG}(\ell, \mathbb{H})$ can be written as

$$f(x, y) = \sum_{r=0}^{\ell} x_r^\diamond y_r,$$

where again $x = (x_0, x_1, \ldots, x_\ell)$ and similarly for y. It is readily checked that the Witt index is equal to 2 **if and only if** $\ell = 3$ or $\ell = 4$. In this case we obtain the

Hermitian quadrangles $H(3, \mathbb{H}, \mathbb{C})$ and $H(4, \mathbb{H}, \mathbb{C})$, using similar notation to that above, and they will be called the **complex quaternion Hermitian quadrangles**, because the centre $\mathbb{R}$ together with the set $\mathbb{H}_\sigma$ generate a complex subfield of $\mathbb{H}$.

Certain generalized quadrangles discussed in this subsection are dual to others; for a complete account on this matter, see Subsection 9.6.4 on page 417 in Chapter 9.

2.3.15 Local fields

Quadrics in $\mathbf{PG}(\ell, \mathbb{K})$ of Witt index 2, and with $\mathbb{K}$ a finite extension of $\mathbb{Q}_p$ (the field of p-adic numbers) or $\mathbf{GF}(q)((t))$ (the field of Laurent series over $\mathbf{GF}(q)$), exist only for $\ell \in \{3, 4, 5\}$ and they can all be classified; see O'MEARA [1971], Section 63C (compare also SCHARLAU [1985], pages 91, 185, 217).

2.3.16 Number fields

Here, the situation is much more complicated. We simply refer to O'MEARA [1971], Section 66 and LAM [1973], Chapter 6. We will not need those results in the rest of the book.

The symplectic quadrangle

We are now going to define a very important class of classical quadrangles separately and in a way different from that above. We will indicate the proof of the fact that the quadrangle is classical (and postpone a detailed proof to the next chapter; see Proposition 3.4.13 on page 109). The construction below has also some nice applications to the theory of projective spaces (for instance, to the construction of *ovoids*; see Subsection 7.6.25 on page 340), and it is similar to the construction of a class of classical hexagons.

2.3.17 Symplectic polarities and their quadrangles

Let $\mathbb{K}$ be any (commutative) field and consider in $\mathbf{PG}(3, \mathbb{K})$, with respect to some chosen basis, the symplectic polarity τ which maps the point (y_0, y_1, y_2, y_3) to the plane with equation $y_1 X_0 - y_0 X_1 + y_3 X_2 - y_2 X_3 = 0$. A line L of $\mathbf{PG}(3, \mathbb{K})$ is called **totally isotropic** if $L^\tau = L$. We define the following geometry $\mathsf{W}(\mathbb{K})$. The points of $\mathsf{W}(\mathbb{K})$ are the points of $\mathbf{PG}(3, \mathbb{K})$; the lines of $\mathsf{W}(\mathbb{K})$ are the totally isotropic lines of τ. We show that $\mathsf{W}(\mathbb{K})$ is a generalized quadrangle. Note that every point of $\mathbf{PG}(3, \mathbb{K})$ is incident with its image (that is what a symplectic polarity is all about).

If L is a line of $\mathbf{PG}(3, \mathbb{K})$ such that $L^\tau = L$, and p is a point on L, then, since $p \mathbf{I} L$ in $\mathbf{PG}(3, \mathbb{K})$, also $L \mathbf{I} p^\tau$. Conversely, if L is a line of $\mathbf{PG}(3, \mathbb{K})$ through some point p and incident with p^τ, then for any other point $x \mathbf{I} L$, x^τ is incident with x (since every point is incident with its image) and with p (since $x \mathbf{I} p^\tau$), hence $L^\tau = (xp)^\tau = xp = L$. We have shown that a line L is totally isotropic if and only if for every point p on L, it is incident with p^τ if and only if this property holds for at least one point p on L.

Now let p be a point of $\mathsf{W}(\mathbb{K})$ not incident with some line L of $\mathsf{W}(\mathbb{K})$. Then there is a unique line M of $\mathsf{W}(\mathbb{K})$ through p meeting L. Indeed, M must be incident with p, it must be contained in p^τ and it must meet L, so M joins p with $L \cap p^\tau$; the latter is a singleton since L is not contained in p^τ (otherwise $p\,\mathbf{I}\,L$, a contradiction). Furthermore, the points $(1,0,0,0)$, $(0,1,0,0)$, $(0,0,1,0)$ and $(0,0,0,1)$ form an ordinary 4-gon. This shows that $\mathsf{W}(\mathbb{K})$ is a weak generalized quadrangle.

Noting that each line contains $|\mathbb{K}|+1$ points and each point is incident with $|\mathbb{K}|+1$ lines, we conclude that $\mathsf{W}(\mathbb{K})$ is a (thick) generalized quadrangle, the **symplectic quadrangle (over** $\mathbb{K}$**)**. The polarity τ defines an anti-symmetric bilinear form q as follows:

$$q((x_0, x_1, x_2, x_3), (y_0, y_1, y_2, y_3)) = x_0 y_1 - x_1 y_0 + x_2 y_3 - x_3 y_2.$$

By previous remarks, it is clear that two points x and y of $\mathbf{PG}(3, \mathbb{K})$, and hence of $\mathsf{W}(\mathbb{K})$, are collinear in $\mathsf{W}(\mathbb{K})$ **if and only if** $q(x,y) = 0$. We call the form q the bilinear form **associated with** $\mathsf{W}(\mathbb{K})$. It defines $\mathsf{W}(\mathbb{K})$ completely.

As for $\mathsf{Q}(V, q)$ above, we will sometimes refer to the representation of $\mathsf{W}(\mathbb{K})$ in $\mathbf{PG}(3, \mathbb{K})$ just described as the **standard embedding**.

2.3.18 Grassmann coordinates

We now introduce *Grassmann coordinates* (in the special case $d = 3$: *Plücker coordinates*) for the lines of a projective space $\mathbf{PG}(d, \mathbb{K})$ over a field $\mathbb{K}$. Choose a basis and coordinates and let L be a line of $\mathbf{PG}(d, \mathbb{K})$, $d \geq 2$. Consider two arbitrary points x and y on L with respective coordinates $(x_0, x_1, \ldots, x_d)$ and $(y_0, y_1, \ldots, y_d)$. Then one can easily verify that the $\binom{d+1}{2}$-tuple $(p_{ij})_{0 \leq i < j \leq d}$, where

$$p_{ij} = \begin{vmatrix} x_i & x_j \\ y_i & y_j \end{vmatrix} = x_i y_j - x_j y_i,$$

is, up to a non-zero scalar multiple, independent of the points x and y on L. Hence the line L defines a unique point $p_L = (p_{ij})_{0 \leq i < j \leq d}$ of $\mathbf{PG}(\binom{d+1}{2}) - 1, \mathbb{K})$. The coordinates of the point p_L are the **Grassmann coordinates** of L. For $d = 3$, all these points constitute a quadric, the so-called **Klein quadric** (see e.g. Chapter 12 of TAYLOR [1992]), and the Grassmann coordinates are then called **Plücker coordinates**.

Now, from the expression of the bilinear form associated with τ, one immediately sees that the Grassmann coordinates of a line of $\mathsf{W}(\mathbb{K})$ satisfy $p_{01} + p_{23} = 0$, and conversely, every line whose Grassmann coordinates satisfy $p_{01} + p_{23} = 0$ is a line of $\mathsf{W}(\mathbb{K})$. This is thus another way to describe $\mathsf{W}(\mathbb{K})$.

This description has the advantage of making apparent the isomorphism of $\mathsf{W}(\mathbb{K})$ and the dual of $\mathsf{Q}(4, \mathbb{K})$. Indeed, the quadrangle $\mathsf{Q}(4, \mathbb{K})$ is nothing other than a quadric Q of Witt index 2 in $\mathbf{PG}(4, \mathbb{K})$. The relation $p_{01} + p_{23} = 0$ determines a hyperplane in $\mathbf{PG}(5, \mathbb{K})$ which meets the Klein quadric exactly in a non-degenerate quadric containing lines (the pencils of $\mathsf{W}(\mathbb{K})$) but no planes (because of the non-degeneracy). Since there is essentially only one such quadric, it must be isomorphic to Q.

Recall from Definition 1.9.4 (see page 39) that a *projective point* in a generalized quadrangle is a regular point for which the perp-geometry is a projective plane.

2.3.19 Theorem. *All points of the symplectic quadrangle* $\mathsf{W}(\mathbb{K})$ *over any field* $\mathbb{K}$ *are projective.*

Proof. Let τ be a symplectic polarity in $\mathbf{PG}(3, \mathbb{K})$ corresponding to the symplectic quadrangle $\mathsf{W}(\mathbb{K})$. All traces are of the form $x^\tau \cap y^\tau$ for x and y two non-collinear points in $\mathsf{W}(\mathbb{K})$, viewed as points of $\mathbf{PG}(3, \mathbb{K})$. So every trace is a line of $\mathbf{PG}(3, \mathbb{K})$ and hence determined by any two of its points. Therefore, every point is regular. It is now easily seen that the perp-geometry in a point p is nothing other than the projective plane p^τ. $\qquad\square$

2.3.20 Corollary. *No proper full or ideal subquadrangle* Γ *of a symplectic quadrangle* $\mathsf{W}(\mathbb{K})$ *can be isomorphic to a symplectic quadrangle.*

Proof. Using Proposition 1.9.18 on page 46, we see that the perp-geometry in a point x of Γ is a proper subgeometry of the corresponding perp-geometry of $\mathsf{W}(\mathbb{K})$, except that all lines through x in both perp-geometries are the same (if Γ is an ideal subquadrangle), or all points on some line through x are the same (if Γ is a full subquadrangle). So they cannot be both projective planes by Corollary 1.8.3 (see page 34). $\qquad\square$

Later on, we will define certain subquadrangles of the symplectic quadrangles over a field of characteristic 2, the so-called *mixed quadrangles*; see Subsection 3.4.2 on page 100. Some of these will be full or ideal proper subquadrangles of $\mathsf{W}(\mathbb{K})$.

2.4 Classical generalized hexagons

From the point of view of group theory, there are no such things as *classical hexagons*, except maybe for an example of order $(2, 2)$, because no classical group is naturally associated with a generalized hexagon. The exception noted is related to the group $\mathbf{PSU}_3(3)$. This group is (sporadically) isomorphic to the exceptional group of type G_2 over $\mathbf{GF}(2)$ and a construction of a generalized hexagon of order $(2, 2)$ related to the group $\mathbf{PSU}_3(3)$ is given in Subsection 1.3.12. However, we will introduce a class of classical hexagons, the name "classical" being motivated by the fact that they naturally live on classical objects like quadrics (in particular, we will see that they all live on the quadric (of type D_4) in seven-dimensional space (hence containing projective 3-spaces), whereas other important examples are related to exceptional groups of type E_6 and E_8 (see Appendix C); these only exist in the infinite case). We will also see that there is a great similarity between the symplectic quadrangle and one of the classes of classical hexagons, namely, the so-called *split Cayley hexagons*. This is an extra motivation for the name *classical hexagons*.

We start with an important definition in the theory of buildings and, in particular, the theory of polar spaces.

2.4.1 Definition. Let $\Gamma = (\mathcal{P}, \mathcal{L}, \mathbf{I})$ be a geometry of rank 2. Then we say that Γ satisfies the **Buekenhout–Shult one-or-all axiom** if for every point $p \in \mathcal{P}$ and every line $L \in \mathcal{L}$ not incident with p, either all points of L are collinear with p, or exactly one point on L is collinear with p.

If the gonality of Γ is at least 3, then together with some non-degeneracy conditions, the Buekenhout–Shult one-or-all axiom characterizes the class of all polar spaces and thus provides a definition for these objects. For more details, see BUEKENHOUT & SHULT [1974].

2.4.2 The quadric $\mathbf{Q}(7, \mathbb{K})$

For the definition of the classical hexagons, we will need some understanding of the geometry of the quadric $\mathbf{Q}(7, \mathbb{K})$ of type D_4 over a field $\mathbb{K}$ in $\mathbf{PG}(7, \mathbb{K})$. It is by definition the quadric containing projective 3-spaces. Recall that there is essentially only one such for any field $\mathbb{K}$. A standard equation is given by $X_0 X_1 + X_2 X_3 + X_4 X_5 + X_6 X_7 = 0$. We will give some general properties below. Our goal is to understand geometrically how *triality* produces hexagons. The classification of trialities will not be carried out, but we will explicitly describe the examples giving rise to hexagons.

The quadric $\mathbf{Q}(7, \mathbb{K})$ has as characteristic property that every plane contained in it is itself contained in exactly two projective three-dimensional subspaces of $\mathbf{Q}(7, \mathbb{K})$. The set of three-dimensional subspaces on $\mathbf{Q}(7, \mathbb{K})$ can be subdivided in two subsets in the following way. Two 3-subspaces belong to the same subset **if and only if** their intersection is a projective space of odd dimension (empty, a line or a 3-space). Each subset is called a **set of generators**. It follows that there is a unique element of each set of generators through a plane of $\mathbf{Q}(7, \mathbb{K})$. The D_4-**geometry** $\Omega(\mathbb{K})$ **attached to** $\mathbf{Q}(7, \mathbb{K})$ is defined as follows. There are four different types of elements. The 0-*points* are the points of $\mathbf{Q}(7, \mathbb{K})$; the *lines* are the lines of $\mathbf{Q}(7, \mathbb{K})$, and we denote this set by $\mathcal{L}$; the 1-*points* are the elements of one set of generators; the 2-*points* are the elements of the other set of generators. We denote the set of i-points by $\mathcal{P}^{(i)}$, $i = 0, 1, 2$. *Incidence* is symmetrized containment for i-points and lines, $i = 0, 1, 2$; also for 0-points and j-points, $j = 1, 2$; and a 1-point is incident with a 2-point if the corresponding 3-spaces meet in a plane of $\mathbf{Q}(7, \mathbb{K})$. The key property is that every permutation of the set $\{\mathcal{P}^{(0)}, \mathcal{P}^{(1)}, \mathcal{P}^{(2)}\}$ defines a geometry which is isomorphic to $\Omega(\mathbb{K})$. For $i = 0, 1, 2$, we call two i-points p and q **collinear** when they are incident with a common line.

Let p be a point, and let L be a line on $\mathbf{Q}(7, \mathbb{K})$. Then, since $\mathbf{Q}(7, \mathbb{K})$ is a quadric, either there is exactly one point on L collinear on $\mathbf{Q}(7, \mathbb{K})$ with p, or p and L are contained in a plane of $\mathbf{Q}(7, \mathbb{K})$. In the latter case, all points of L are collinear on $\mathbf{Q}(7, \mathbb{K})$ with p. Thus the geometry of points and lines on $\mathbf{Q}(7, \mathbb{K})$ satisfies the **Buekenhout–Shult one-or-all axiom** (and indeed $\mathbf{Q}(7, \mathbb{K})$ is a polar space, one of type D_4). An immediate consequence is that for a point p and a projective 3-space S (plane π) either p is collinear with all points of a plane of S (line of π), or p is contained in S (p and π generate a projective 3-space of $\mathbf{Q}(7, \mathbb{K})$ or p belongs to π).

2.4.3 Definition. Let $\Omega(\mathbb{K})$ be the geometry defined from $\mathbf{Q}(7, \mathbb{K})$ as above. A **triality** of $\Omega(\mathbb{K})$ is a map

$$\theta : \mathcal{L} \to \mathcal{L}, \mathcal{P}^{(0)} \to \mathcal{P}^{(1)}, \mathcal{P}^{(1)} \to \mathcal{P}^{(2)}, \mathcal{P}^{(2)} \to \mathcal{P}^{(0)}$$

preserving incidence in $\Omega(\mathbb{K})$ and such that θ^3 is the identity.

Let a triality θ be given. An **absolute i-point** p is an element of $\mathcal{P}^{(i)}$ which is incident with p^θ, $i = 0, 1, 2$. An **absolute line** is a line which is fixed by θ. There is some similarity with polarities in three-dimensional projective space, as pointed out by TITS [1959], and we will come back to that matter in Subsection 2.4.18.

2.4.4 Theorem (Tits [1959]). *Let θ be a triality of $\Omega(\mathbb{K})$. Suppose that one of the following hypotheses is satisfied:*

(i) *there exists at least one absolute i-point, for some $i \in \{0, 1, 2\}$, and every absolute i-point is incident with at least two absolute lines;*

(ii) *there exists a cycle $(L_0, L_1, \ldots, L_d)$, $d > 2$, of absolute lines (with L_i concurrent with $L_{i+1} \neq L_i$; subscripts modulo d).*

Then for every $i \in \{0, 1, 2\}$, the geometry $\Gamma^{(i)}$ with point set $\mathcal{P}^{(i)}_{\mathrm{abs}}$ the set of absolute i-points, with line set $\mathcal{L}_{\mathrm{abs}}$ the set of absolute lines and with the natural incidence, is a weak generalized hexagon with thick lines (and hence with some order). Also, the isomorphism class of this geometry is independent of $i \in \{0, 1, 2\}$.

Proof. We prove this in several steps. Without loss of generality, we take $i = 0$ and we briefly talk about *points* instead of 0-points. Also, we use the symbol $\in$ to denote incidence between a point and some other element, i.e., we consider $\Omega(\mathbb{K})$ as $\mathbf{Q}(7, \mathbb{K})$. In particular, we will also talk about *planes*, and these are the planes of $\mathbf{Q}(7, \mathbb{K})$. There is no loss of generality in doing so. In fact, a plane can abstractly be viewed as a pair of incident 1- and 2-points. Note that everything we prove for θ also holds for θ^2, since θ^2 is also a triality with obviously the same absolute lines and the same absolute i-points, for all $i \in \{0, 1, 2\}$.

| **Lemma 1** | *Every point p of any absolute line L is an absolute point.*

Proof. Indeed, $p \in L = L^\theta \subseteq p^\theta$ and similarly for θ^2. $\qquad$ QED

Lemma 2. *Whenever two distinct absolute points p and q are such that $p \in q^\theta$, then p is collinear with q and pq is an absolute line.*

Proof. Since q is absolute, $q \in q^\theta$ and so there is a unique line pq incident with both p and q. The line pq is in q^θ, so $(pq)^\theta$ belongs to q^{θ^2}. Also, $(pq)^\theta$ is the intersection (viewed in $\mathbf{Q}(7, \mathbb{K})$) of p^θ and q^θ, both of which contain p. Therefore, p is incident with $(pq)^\theta$ and hence p also belongs to q^{θ^2}.

Also, $p \in q^\theta$ implies $q \in p^{\theta^2}$. Hence similarly as we showed $p \in q^{\theta^2}$, this implies that $q \in p^\theta$. Interchanging the roles of p and q, we infer from the previous paragraph that $q \in (pq)^\theta$. Therefore $pq = (pq)^\theta$.   QED

Lemma 3. **If** *an absolute point p is collinear with an absolute point q, but $p \notin q^\theta$ and hence $q \notin p^\theta$,* **then** *there is an absolute point x such that px and qx are absolute lines.*

Proof. We first claim that the planes $\pi_p := p^\theta \cap p^{\theta^2}$ and $\pi_q := q^\theta \cap q^{\theta^2}$ meet in a unique point. Suppose first that they meet in at least two points, say x and y. Then p is collinear with x, y, q and there are two possibilities. First, q belongs to xy. In that case $q \in p^\theta$, contrary to our assumptions. Second, x, y, q forms a triangle. Then x, y, q, p are contained in a 3-space containing π_q. But there are only two 3-spaces containing x, y, q and these are q^θ and q^{θ^2}. These give, respectively, $p \in q^\theta$ and $q \in p^\theta$, a contradiction. So we have shown that π_p and π_q meet in at most one point.

Suppose now that π_p and π_q are disjoint. We consider $L := (pq)^\theta$. This line is contained in $p^\theta \cap q^\theta$ and hence it meets the planes π_p and π_q, necessarily in unique distinct points u and v, respectively, for otherwise π_p and π_q share a common point. Similarly L^θ meets π_p and π_q in unique distinct points u' and v', respectively. So q^θ is generated by π_q and u; q^{θ^2} is generated by π_q and u', with u and u' collinear (because they belong to π_p). But that implies that u' is collinear with all points of a plane π_q of q^θ plus an extra point u, in contradiction with the consequences of the Buekenhout–Shult one-or-all axiom (see above), remembering that $q^\theta \neq q^{\theta^2}$. Hence this situation cannot occur and our first claim is proved.

Our next claim is that whenever a line L is incident with p and contained in π_p, with π_p as above, then L^θ is also incident with p and contained in π_p. Indeed, as L is incident with π_p, it is incident with p^θ and with p^{θ^2}. So L^θ is incident with p^{θ^2}, with p and with p^θ. Hence with π_p as well (on $\mathbf{Q}(7, \mathbb{K})$).

Now let x be the unique point in the intersection of π_p and π_q. If we denote by $\mathbf{I}$ the incidence relation in $\Omega(\mathbb{K})$, then from $p\mathbf{I}px\mathbf{I}x\mathbf{I}xq\mathbf{I}q$ follows $p\mathbf{I}(px)^\theta\mathbf{I}x^\theta\mathbf{I}(xq)^\theta\mathbf{I}q$. Therefore x^θ contains both p and q and hence the line pq. Now $(pq)^\theta$ is the intersection of p^θ and q^θ. Since this intersection also contains x, we see that $x \mathbf{I} (pq)^\theta$. It follows that $x^{\theta^2} \mathbf{I} (pq)^\theta \mathbf{I} x$. Hence $x \mathbf{I} x^{\theta^2}$ and applying θ, we conclude $x^\theta \mathbf{I} x$. Thus, x is an absolute point and the lines px and qx are absolute lines by Lemma 2. Therefore, Lemma 3 is proved.   QED

Lemma 4. *The diameter of the incidence graph $(G, *)$ of $\Gamma^{(0)}$ is less than or equal to 6.*

Proof. If δ denotes distance (as usual), then we have to prove that $\delta(v, w) \le 6$, for all points and lines v, w of $\Gamma^{(0)}$. If v is a point and w is a line, then by the Buekenhout–Shult one-or-all axiom, there is at least one point x on w collinear in $\Omega(\mathbb{K})$ with v. By Lemma 1, x is an absolute point at distance no more than 4 from v by Lemma 2 and Lemma 3. Hence $\delta(v, w) \le 5$. If v, w are both points or both lines, then by considering an element z incident with w, we see that $\delta(v, w) \le \delta(v, z) + \delta(z, w) \le 5 + 1 = 6$. Hence the diameter of $(G, *)$ is 6 or less. QED

Lemma 5. *The gonality of $(G, *)$ is larger than 3.*

Proof. Suppose the lines L, M, N form a triangle, i.e., L, M, N are absolute lines and $\{x\} = L \cap M \ne \{y\} = M \cap N$. The 3-space x^θ is the unique element of the set of generators corresponding to 1-points containing the lines L and M. Similarly, y^θ contains M and N, so $y^\theta = x^\theta$ implying $x = y$. The lemma is proved. QED

Lemma 6. *The gonality of $(G, *)$ is larger than 4.*

Proof. Suppose the lines L, M, N, P form a quadrilateral. If they are contained in a plane of $\mathbf{Q}(7, \mathbb{K})$, then any three of them form a triangle, contradicting Lemma 5. If $\{x\} = L \cap M$, then $L \cup M \subseteq x^\theta \cap x^{\theta^2}$, hence the four "vertices" of the quadrilateral are two by two collinear (in $\mathbf{Q}(7, \mathbb{K})$). Hence they are contained in a 3-space U of $\mathbf{Q}(7, \mathbb{K})$. Since U contains the plane $\langle L, M \rangle$, and since there are only two 3-spaces through that plane, we must have $U = x^\theta$ or $U = x^{\theta^2}$. We may assume without loss of generality $U = x^\theta$. But then also $y^\theta = U$ with $\{y\} = N \cap P$, a contradiction. QED

Lemma 7. *The gonality of $(G, *)$ is larger than 5.*

Proof. Suppose we have a pentagon L, M, N, P, Q of lines. Let x again be the intersection of L and M. Let y and z be the intersection point of, respectively, N and P, and of P and Q. As above, M and N lie in a plane of $\mathbf{Q}(7, \mathbb{K})$, hence x is collinear with y and similarly with z. So by the Buekenhout–Shult one-or-all axiom, x is collinear with all points of the space generated by N, P, Q. We conclude that the pentagon lies entirely in a 3-space U (since a 2-space is ruled out by Lemma 5). Without loss of generality, we may again assume that $x^\theta = U$; compare Lemma 6. But then also $U = y^\theta = z^\theta$, a contradiction. QED

Lemma 8. *The gonality of $(G, *)$ is equal to 6.*

Proof. This is obvious if we assume that there is a circuit in $\Gamma^{(0)}$ (since the diameter is at most 6, we can always reduce that circuit to one of length 12). So we may assume that each absolute point is incident with at least two absolute lines. Also, by assumption, there exists an absolute line L. By Lemma 1, each point on L is absolute. Let $x \mathbf{I} L$. By assumption, there exists an absolute line $M \mathbf{I} x$ with $M \ne L$. Let $y \mathbf{I} M$ with $y \ne x$. Then again, y is absolute. Let N be an absolute line through y distinct from M, and let $z \neq y$ be incident with N. Let P be an absolute line through z distinct from N, and let $u \neq z$ be incident with P. Finally, let $Q \neq P$ be an absolute line through u. By the Buekenhout–Shult one-or-all axiom, there is a point v on Q collinear in $\Omega(\mathbb{K})$ with x. By Lemma 3, we have $\delta(x, v) \leq 4$. Since the gonality of $(G, *)$ is at least 6, we see that $\delta(v, x) = 4$ and the sequence (x, y, z, u, v, w), where w is collinear with both x and v in $\Gamma^{(0)}$, is an ordinary hexagon. So the gonality of $(G, *)$ is equal to 6 and its diameter is also equal to 6.   QED

By Lemma 1.5.10 (see page 21), $\Gamma^{(0)}$ is a weak generalized hexagon with thick lines. Applying triality, the last assertion follows and the proof of the theorem is complete.   □

2.4.5 Remark. Note that the proof above implies that two points x and y of $\Gamma^{(0)}$ are opposite **if and only if** they are not collinear in $\Omega(\mathbb{K})$ (or equivalently in $\mathbf{Q}(7, \mathbb{K})$).

2.4.6 Trilinear forms

To give an explicit example of a triality, we should have a convenient description of $\mathbf{Q}(7, \mathbb{K})$, i.e., a description of $\Omega(\mathbb{K})$ in which the i-points play the same role as the j-points for $i, j \in \{0, 1, 2\}$. This is possible by introducing a *trilinear form*, see CARTAN [1938]. We follow TITS [1959] for the notation.

The points of $\mathbf{Q}(7, \mathbb{K})$ can be viewed as 8-tuples $(x_0, x_1, \ldots, x_7)$, up to a scalar multiple, with elements in $\mathbb{K}$ and satisfying the relation

$$x_0 x_4 + x_1 x_5 + x_2 x_6 + x_3 x_7 = 0.$$

The philosophy of trilinear forms is that since 1-points and 2-points play the same role as 0-points, it must be possible to label the 1-points and 2-points in the same way as the 0-points and to introduce an algebraic operation that tells one when two elements are incident. In fact, it is possible to do even better: let $J = \{0, 1, \ldots, 7\}$ and let V be an eight-dimensional vector space over $\mathbb{K}$; then there exists a trilinear form $\mathcal{T} : V \times V \times V \to \mathbb{K}$ such that a pair of points (x, y) of $\mathbf{Q}(7, \mathbb{K})$ represents an incident (0-point, 1-point)-pair in $\Omega(\mathbb{K})$ **if and only if** the linear form $\mathcal{T}(x, y, z')$ is identical zero in z'; and similarly for any cyclic permutation of the letters x, y, z. This trilinear form has the following explicit description:

$$\mathcal{T}(x, y, z) = \begin{vmatrix} x_0 & x_1 & x_2 \\ y_0 & y_1 & y_2 \\ z_0 & z_1 & z_2 \end{vmatrix} + \begin{vmatrix} x_4 & x_5 & x_6 \\ y_4 & y_5 & y_6 \\ z_4 & z_5 & z_6 \end{vmatrix}$$
$$+ x_3(z_0 y_4 + z_1 y_5 + z_2 y_6) + x_7(y_0 z_4 + y_1 z_5 + y_2 z_6)$$
$$+ y_3(x_0 z_4 + x_1 z_5 + x_2 z_6) + y_7(z_0 x_4 + z_1 x_5 + z_2 x_6)$$
$$+ z_3(y_0 x_4 + y_1 x_5 + y_2 x_6) + z_7(x_0 y_4 + x_1 y_5 + x_2 y_6)$$
$$- x_3 y_3 z_3 - x_7 y_7 z_7.$$

For example, in order to find the equation of the 3-space on $\mathbf{Q}(7,\mathbb{K})$ which corresponds to the 1-point $y = (1,0,\ldots,0)$, we simply plug in the value for y in $\mathcal{T}(x,y,z)$ and require that the coefficients of all z_i, $i \in J$, vanish. This gives us $x_1 = x_2 = x_4 = x_7 = 0$.

2.4.7 Trialities that produce generalized hexagons

Now we give the formulae for all trialities which produce (thick) generalized hexagons. Let $\mathcal{T}$ be the trilinear form as introduced above. Since a line of Ω is determined by two i-points, for all $i \in \{0,1,2\}$, it is readily seen that every permutation θ of $\mathcal{P}^{(0)} \cup \mathcal{P}^{(1)} \cup \mathcal{P}^{(2)}$ preserving incidence (and well defined on the types of points) induces a not necessarily type-preserving automorphism of $\Omega(\mathbb{K})$.

Let σ be an automorphism of $\mathbb{K}$ of order 1 or 3. Then the map

$$\tau_\sigma : \mathcal{P}^{(i)} \to \mathcal{P}^{(i+1)} : (x_j)_{j \in J} \mapsto (x_j^\sigma)_{j \in J}, \qquad i = 0,1,2 \bmod 3,$$

clearly preserves incidence in $\Omega(\mathbb{K})$ (because the trilinear form $\mathcal{T}$ is preserved). Moreover, the order of τ_σ is clearly 3. Hence τ_σ is a triality. We call τ_σ a triality of **type** $(\mathbf{I}_\sigma)$, closely following TITS [1959]. There are other types, but we will not need them. We review them briefly in Subsection 2.4.18.

2.4.8 Theorem (Tits [1959]). *The geometry* $\Gamma^{(i)} = (\mathcal{P}_{\mathrm{abs}}^{(i)}, \mathcal{L}_{\mathrm{abs}}, \mathbf{I})$ *arising from the triality* τ_σ *is a generalized hexagon of order* $(|\mathbb{K}|, |\mathbb{K}^{(\sigma)}|)$, *where* $\mathbb{K}^{(\sigma)}$ *is the subfield of* $\mathbb{K}$ *consisting of those elements fixed by* σ. *Replacing* σ *by* σ^{-1} *produces an isomorphic hexagon.*

Proof. According to Theorem 2.4.4, it suffices to show that there is an ordinary hexagon in $\Gamma^{(0)}$, and that there is an absolute point incident with exactly $|\mathbb{K}^{(\sigma)}|+1$ absolute lines.

Let e_i be the 0-point with coordinates x_j, $j \in J$, all zero except x_i, which can be chosen to be equal to 1. Clearly

$$\mathcal{T}(e_i, e_i^{\tau_\sigma}, z) \equiv 0$$

if and only if $i \neq 3,7$. The 0-points incident with $e_0^{\tau_\sigma}$ are those whose coordinates satisfy $x_1 = x_2 = x_4 = x_7 = 0$ (this is the example at the end of Subsection 2.4.6); so clearly the absolute points e_5 and e_6 are incident with $e_0^{\tau_\sigma}$ and hence the lines e_5e_0 and e_0e_6 are absolute lines. Similarly, the lines e_6e_1, e_1e_4, e_4e_2 and e_2e_5 are absolute lines. These six lines in total now clearly form an ordinary hexagon.

Using the trilinear form $\mathcal{T}$ again, it takes an elementary calculation to see that the 0-points $(x_j)_{j \in J}$ incident with both $e_0^{\tau_\sigma}$ and $e_0^{\tau_\sigma^2}$ are precisely the points satisfying $x_1 = x_2 = x_3 = x_4 = x_7 = 0$ (and these indeed form a plane π in $\mathbf{PG}(7,\mathbb{K})$; this plane is denoted by π_{e_0} in the proof of Theorem 2.4.4). Every absolute line incident with e_0 lies in π; moreover, by the proof of Theorem 2.4.4, every absolute point p in π, $p \neq e_0$, gives rise to an absolute line $e_0 p$. Now consider the 0-point p with coordinates $(0,0,0,0,0,k,1,0)$, $k \in \mathbb{K}$. Its image under τ_σ is the 3-space in $\mathbf{PG}(7, \mathbb{K})$ with equations

$$\left\{ \begin{array}{rcl} 0 & = & x_2 + k^\sigma x_1, \\ 0 & = & x_5 - k^\sigma x_6, \\ 0 & = & x_3, \\ 0 & = & x_4. \end{array} \right.$$

This space contains p **if and only if** $k = k^\sigma$. Noting that also the point with coordinates $(0,0,0,0,0,1,0,0)$ of π is absolute, we see that the order of $\Gamma^{(i)}$ is equal to $(|\mathbb{K}|, |\mathbb{K}^{(\sigma)}|)$.

If we replace σ by σ^{-1}, then we interchange $\mathcal{P}^{(1)}$ and $\mathcal{P}^{(2)}$. The theorem now follows directly. $\square$

2.4.9 Definitions. Taking $\sigma = 1$, we see that over every field $\mathbb{K}$ there exists a triality that produces a generalized hexagon. We call this hexagon **classical**, and, more specifically, we speak of the **split Cayley hexagon (over $\mathbb{K}$)**, denoted by $\mathsf{H}(\mathbb{K})$. The reason for that name is that this hexagon can also be constructed using a split Cayley algebra over $\mathbb{K}$, see for instance SCHELLEKENS [1962a], [1962b] (and moreover, the corresponding simple algebraic group is also split). The hexagon $\mathsf{H}(\mathbb{K})$ deserves the name "classical" in more than one way: on top of the reasons already mentioned (lying on a classical polar space), it is the most important hexagon, it is the main example, and in fact, the *only* example for many fields $\mathbb{K}$. The dual of $\mathsf{H}(\mathbb{K})$ is also a **classical hexagon** and denoted $\mathsf{H}(\mathbb{K})^D$. In the finite case, the split Cayley hexagon over the Galois field $\mathbf{GF}(q)$ is denoted by $\mathsf{H}(q)$.

We call a generalized hexagon arising from a triality as in Theorem 2.4.8 with $\sigma \neq 1$ also a **classical hexagon**, or more specifically, a **twisted triality hexagon**, and we denote it by $\mathsf{T}(\mathbb{K}, \mathbb{K}^{(\sigma)}, \sigma)$. Note that $\mathbb{K}$ is a Galois extension of degree 3 of $\mathbb{K}^{(\sigma)}$. The dual of $\mathsf{T}(\mathbb{K}, \mathbb{K}^{(\sigma)}, \sigma)$ is denoted by $\mathsf{T}(\mathbb{K}^{(\sigma)}, \mathbb{K}, \sigma)$ and is also called **classical**. In the finite case, the field automorphism σ is — up to inverse — determined by the field $\mathbf{GF}(q^3)$ and hence we can unambiguously denote the unique twisted triality hexagon over the field $\mathbf{GF}(q^3)$ by $\mathsf{T}(q^3, q)$, and its dual by $\mathsf{T}(q, q^3)$. Note that we do not follow THAS [1995] (who writes $\mathsf{H}(q^3, q)$ for $\mathsf{T}(q^3, q)$ and has no special notation for the dual) in this notation in order to avoid confusion with the Hermitian quadrangles, in particular with $\mathsf{H}(4, 64)$.

The representation of $\mathsf{T}(\mathbb{K}, \mathbb{K}^{(\sigma)}, \sigma)$ on $\mathsf{Q}(7, \mathbb{K})$ as above is sometimes referred to as the **standard embedding** of $\mathsf{T}(\mathbb{K}, \mathbb{K}^{(\sigma)}, \sigma)$.

We now show a property of $\mathsf{H}(\mathbb{K})$ that will allow us to construct $\mathsf{H}(\mathbb{K})$ in a more direct way on a quadric in projective 6-space.

2.4.10 Theorem (Tits [1959]). *The points and lines of* $\mathsf{H}(\mathbb{K})$, *considered as the geometry* $\Gamma^{(0)}$ *of absolute points and lines of the triality* τ_σ *with* $\sigma = 1$, *all lie in* *the hyperplane of* $\mathbf{PG}(7, \mathbb{K})$ *with equation* $x_3 + x_7 = 0$. *Conversely, every point of* $\mathbf{Q}(7, \mathbb{K})$ *in that hyperplane belongs to* $\mathsf{H}(\mathbb{K})$.

Proof. The necessary and sufficient condition for a 0-point p with coordinates $(x_j)_{j \in J}$ to be an absolute point is that $\mathcal{T}((x_j)_{j \in J}, (x_j)_{j \in J}, z)$ vanishes. This is equivalent to the following condition (as is easily computed by looking at the coefficients of the z_i, $i \in J$, and taking subscripts modulo 8):

$$\begin{cases} 0 & = & x_{i+4}(x_3 + x_7), & i \neq 3, 7, \\ 0 & = & x_0 x_4 + x_1 x_5 + x_2 x_6 - x_i^2 & i = 3, 7. \end{cases}$$

The result now follows readily.   □

2.4.11 Proposition. *The twisted triality hexagon* $\mathsf{T}(\mathbb{K}, \mathbb{K}^{(\sigma)}, \sigma)$ *has an ideal subhexagon isomorphic to* $\mathsf{H}(\mathbb{K}^{(\sigma)})$.

Proof. This follows by restricting coordinates in $\Omega(\mathbb{K})$ to $\mathbb{K}^{(\sigma)}$.   □

2.4.12 Remark. There is another class of hexagons closely related to the twisted triality hexagons; we will define this class in Subsection 3.5.8 (see page 114).

The dual of the twisted triality hexagons are called in the literature the *hexagons related to the groups of type* $^3D_4$ and, in the finite case, denoted by $^3D_4(q)$ or $^3D_4(q^3)$ (cf. KANTOR [1986a]).

Split Cayley hexagons

2.4.13 Tits' description of $\mathsf{H}(\mathbb{K})$

Recall that the absolute points of a triality of type ($\mathbf{I}_{\mathrm{id}}$) are exactly the points of the intersection of a hyperplane of $\mathbf{PG}(7, \mathbb{K})$ with $\mathbf{Q}(7, \mathbb{K})$. Considering coordinates as above, this hyperplane has equation $X_3 + X_7 = 0$. Hence substituting X_7 for X_3 and deleting X_7 (which amounts to the same as deleting X_7 and substituting $-X_3$ for X_3; this substitution is for historical reasons), we can identify the point set of $\Gamma = \mathsf{H}(\mathbb{K})$ with the point set of the "parabolic" quadric $\mathbf{Q}(6, \mathbb{K})$ in $\mathbf{PG}(6, \mathbb{K})$ with equation

$$X_0 X_4 + X_1 X_5 + X_2 X_6 = X_3^2.$$

A tedious explicit computation (which we will not perform) shows that the Grassmann coordinates of the lines of $\mathsf{H}(\mathbb{K})$ satisfy the following six linear equations:

$$\begin{array}{lll} p_{12} = p_{34}, & p_{54} = p_{32}, & p_{20} = p_{35}, \\ p_{65} = p_{30}, & p_{01} = p_{36}, & p_{46} = p_{31}, \end{array}$$

and conversely, every line on $\mathbf{Q}(6, \mathbb{K})$ whose Grassmann coordinates satisfy these equations is a line of Γ. This gives a complete and explicit description of $\mathsf{H}(\mathbb{K})$ on the quadric $\mathbf{Q}(6, \mathbb{K})$. It is due to TITS [1959]. By the way, one can deduce all the above equations from the first one by consecutively applying the following rule: if $p_{ij} = p_{3k}$ is in the list, then so are $p_{(i\pm4)k} = p_{3j}$ and $p_{k(j\pm4)} = p_{3i}$, where in ±4 one should choose the appropriate sign in order to obtain a number between 0 and 7.

We sometimes refer to this representation of the split Cayley hexagons as a **standard embedding**.

2.4.14 (Perfect) Symplectic hexagons

Now assume that the characteristic of $\mathbb{K}$ is 2. We first recall some properties of the quadric $\mathbf{Q}(6, \mathbb{K})$. Let $\mathbf{Q}(6, \mathbb{K})$ have equation

$$X_0 X_4 + X_1 X_5 + X_2 X_6 = X_3^2.$$

Consider the point k with coordinates $(0, 0, 0, 1, 0, 0, 0)$. Let L be any line of $\mathbf{PG}(6, \mathbb{K})$ through k and suppose that L contains the point with coordinates $(x_0, x_1, x_2, 0, x_4, x_5, x_6)$, $x_i \in \mathbb{K}$, $i = 0, 1, 2, 4, 5, 6$. A point $(x_0, x_1, x_2, \ell, x_4, x_5, x_6)$ of L (with $\ell \in \mathbb{K}$) is contained in $\mathrm{Q}(6, \mathbb{K})$ if and only if

$$\ell^2 = x_0 x_4 + x_1 x_5 + x_2 x_6.$$

Since the characteristic of $\mathbb{K}$ is 2, L meets $\mathbf{Q}(6, \mathbb{K})$ in at most one point. Note that, if $\mathbb{K}$ is perfect (and hence every element of $\mathbb{K}$ is a square in $\mathbb{K}$), then L meets $\mathbf{Q}(6, \mathbb{K})$ always in exactly one point. In any case, k is called the **nucleus** of $\mathbf{Q}(6, \mathbb{K})$. Hence we may project $\mathbf{Q}(6, \mathbb{K})$ from k onto the hyperplane H with equation $X_3 = 0$. Let $p(x_0, x_1, x_2, x_3, x_4, x_5, x_6)$ be any point of $\mathbf{Q}(6, \mathbb{K})$. The set of points of $\mathbf{Q}(6, \mathbb{K})$ collinear with p on $\mathbf{Q}(6, \mathbb{K})$ is given by the equations:

$$\begin{cases} 0 & = & x_0 X_4 + x_4 X_0 + x_1 X_5 + x_5 X_1 + x_2 X_6 + x_6 X_2, \\ X_3^2 & = & X_0 X_4 + X_1 X_5 + X_2 X_6. \end{cases}$$

Hence the coordinates $(X_0, X_1, X_2, 0, X_4, X_5, X_6)$ of the projection of these points from k onto the hyperplane H satisfy the equation

$$0 = x_0 X_4 + x_4 X_0 + x_1 X_5 + x_5 X_1 + x_2 X_6 + x_6 X_2,$$

which is the equation of a hyperplane H_p of H, and clearly the correspondence $p \mapsto H_p$ uniquely defines a symplectic polarity ρ in H. This implies that the lines of $\mathbf{Q}(6, \mathbb{K})$ are projected onto totally isotropic lines for ρ. If $\mathbb{K}$ is perfect, then one can now easily calculate that all totally isotropic lines for ρ in H are obtained in this way. Hence for $\mathbb{K}$ perfect, the geometry of $\mathbf{Q}(6, \mathbb{K})$ is isomorphic to the geometry of the symplectic space $\mathbf{W}(5, \mathbb{K})$ (which is the geometry of totally isotropic subspaces for a symplectic polarity in $\mathbf{PG}(5, \mathbb{K})$).

Considering the standard embedding of $\mathsf{H}(\mathbb{K})$ in $\mathbf{Q}(6, \mathbb{K})$, we now see that, if $\mathbb{K}$ has characteristic 2, we can represent $\mathsf{H}(\mathbb{K})$ inside the symplectic space $\mathbf{W}(5, \mathbb{K})$. This means that the points of $\mathsf{H}(\mathbb{K})$ are some points of $\mathbf{PG}(5,\mathbb{K})$, and the lines of $\mathsf{H}(\mathbb{K})$ are some lines of $\mathbf{PG}(5,\mathbb{K})$ which are moreover totally isotropic with respect to some symplectic polarity. Therefore we sometimes call $\mathsf{H}(\mathbb{K})$ a **symplectic hexagon**. The above representation is also called a **standard embedding** of $\mathsf{H}(\mathbb{K})$. Hence these geometries have two standard embeddings, and we should always make it clear which one we mean. This will usually be achieved by the choice of the name *symplectic* or *split Cayley*, and it is clear which embedding we associate with each of these names.

If $\mathbb{K}$ is perfect, then the points of the standard embedding of the symplectic hexagon $\mathsf{H}(\mathbb{K})$ are *all* points of $\mathbf{PG}(5,\mathbb{K})$. In this case, we sometimes call $\mathsf{H}(\mathbb{K})$ a **perfect symplectic hexagon**.

Since the absolute lines of a triality θ incident with an absolute point p all lie in the plane $p^\theta \cap p^{\theta^2}$, we have the following property:

2.4.15 Theorem (Ronan [1980a]). *All points of any split Cayley or twisted triality hexagon are distance-2-regular.*

Proof. By the remark preceding the theorem, we know that, for two opposite points p and q (where p and q are not collinear in $\Omega(\mathbb{K})$), the set p^q is contained in the set of points of the plane $p^\theta \cap p^{\theta^2}$ collinear in $\Omega(\mathbb{K})$ with q, which forms a line. So p^q is contained in a line of $\Omega(\mathbb{K})$ and therefore it is determined by any two of its points. $\square$

More exactly, for the split Cayley hexagons we can be more specific.

2.4.16 Theorem. *All points of the split Cayley hexagon $\mathsf{H}(\mathbb{K})$ over any field $\mathbb{K}$ are polar points.*

Proof. Let $\mathsf{H}(\mathbb{K})$ be represented on the quadric $\mathbf{Q}(6,\mathbb{K})$. As in the proof of Theorem 2.3.19, one can see easily that the perp-geometry in a point p is exactly the projective plane on $\mathbf{Q}(6,\mathbb{K})$ containing the lines of $\mathsf{H}(\mathbb{K})$ through p. It is also readily seen that for opposite points p and q, the set $\Gamma^+(p,q)$ (see Remark 1.9.11 on page 43) "is" a projective plane of $\mathbf{Q}(6,\mathbb{K})$ through p, and every plane of $\mathbf{Q}(6,\mathbb{K})$ through p and not containing lines of $\mathsf{H}(\mathbb{K})$ arises in such a way. It follows that the span-geometry at a point p is equal to the geometry of planes and lines on $\mathbf{Q}(6,\mathbb{K})$ through p and this is known to be a generalized quadrangle, namely one isomorphic to $\mathsf{W}(\mathbb{K})$ (the points of $\mathsf{W}(\mathbb{K})$ corresponding to the planes through p and the lines of $\mathsf{W}(\mathbb{K})$ to the lines through p). $\square$

As in Corollary 2.3.20, one can now also prove:

2.4.17 Corollary. *No proper full or ideal subhexagon Γ of a split Cayley hexagon $\mathsf{H}(\mathbb{K})$ can be isomorphic to a split Cayley hexagon.* $\square$

Later on, we will define full and ideal subhexagons of the split Cayley hexagons over a field of characteristic 3, the so-called *mixed hexagons*; see Subsection 3.5.3 on page 112.

2.4.18 Polarities of $\mathbf{PG}(3, \mathbb{K})$ versus trialities of $\Omega(\mathbb{K})$

We now come back to the similarity with polarities in projective 3-space $\mathbf{PG}(3, \mathbb{K})$. There are essentially four kinds of polarities in $\mathbf{PG}(3, \mathbb{K})$ having absolute points and lines (in $\mathbf{PG}(3, \mathbb{K})$ itself). One can distinguish the types by looking at the set of absolute points. In $\Omega(\mathbb{K})$, there are four kinds of trialities having absolute points (and they automatically have absolute lines). We give a brief survey of results due to TITS [1959].

For a given polarity θ in $\mathbf{PG}(3, \mathbb{K})$, or for a given triality θ in $\Omega(\mathbb{K})$, and for a given absolute point p, we denote the one-dimensional projective space formed by the lines incident with p and with p^θ, or incident with p, p^θ and with p^{θ^2}, by $\mathbf{PG}(1, \mathbb{K})^{(p)}$.

1. **Pseudo-polarities** in $\mathbf{PG}(3, \mathbb{K})$ are polarities for which the set of absolute points is a proper subspace π. These only exist in characteristic 2. Suppose π is a plane. Then the set of absolute lines is the pencil of lines in π through the image of π. Hence the geometry of absolute points and lines can be considered as a *degenerate generalized quadrangle*: "degenerate", because it does not contain a proper cycle (or quadrilateral in this case); "quadrangle", because the diameter is equal to 4, as for generalized quadrangles. For a given absolute point p, the collineation induced on $\mathbf{PG}(1, \mathbb{K})^{(p)}$ is either the identity or an involution with one fixed point (an *elation*), according as $p = \pi^\theta$ or $p \neq \pi^\theta$.

   In $\Omega(\mathbb{K})$, there is a similar phenomenon, namely, in characteristic 3, there are trialities which have absolute points and lines, but there is no ordinary hexagon contained in $\Gamma^{(0)}$. More exactly, if we denote such a triality by θ, then all absolute points are contained in a four-dimensional space $\mathbf{PG}(4, \mathbb{K})$ which meets $\mathbf{Q}(7, \mathbb{K})$ in a degenerate quadric Q which is the projection from some line D of $\mathbf{Q}(7, \mathbb{K})$ of a non-degenerate conic lying in a plane of $\mathbf{PG}(4, \mathbb{K})$ skew to D. All points of Q are absolute. For each point p on L, every line incident with p, with p^θ and with p^{θ^2} is absolute (the plane $p^\theta \cap p^{\theta^2}$ lies on Q and contains D); for every other point p on Q, there is a unique absolute line incident with p, namely, the line pp', where p' lies on D such that $p \in p'^\theta$. Hence the geometry of absolute points and lines can be considered here as a *degenerate generalized hexagon*: there are no proper cycles and the diameter is equal to 6. For a given absolute point p, the collineation induced on $\mathbf{PG}(1, \mathbb{K})^{(p)}$ is either the identity or a collineation of order 3 with one fixed point (an *elation*), according as $p \mathbf{I} D$ or $p \in Q \setminus D$.

   Note that there exist fields (necessarily of characteristic 2) such that, in $\mathbf{PG}(3, \mathbb{K})$, there are pseudo-polarities for which the set of absolute points is a line, a point or the empty set; $\mathbb{K}$ cannot be perfect in this case. For trialities in $\Omega(\mathbb{K}')$ (where $\mathbb{K}'$ has charactistic 3), this has no analogue (so the perfectness of $\mathbb{K}'$ does not change the above described situation, provided the set of absolute points is non-empty).

2. **Orthogonal polarities** in $\mathbf{PG}(3, \mathbb{K})$ are polarities for which the set of absolute points is a non-degenerate ruled quadric $\mathbf{Q}(3, \mathbb{K})$ (i.e., a quadric containing lines, hence a quadric of type D_2). The set of absolute lines is precisely the union of the two sets of generators of $\mathbf{Q}(3, \mathbb{K})$. This is clearly a weak generalized quadrangle (denoted by $\mathsf{Q}(3, \mathbb{K})$) which is the dual of the double of a generalized digon. For a given absolute point p, the collineation induced on $\mathbf{PG}(1, \mathbb{K})^{(p)}$ is an involution with two fixed points (a *homology*). These polarities do not exist in characteristic 2, but they do exist over every field of characteristic $\neq 2$.

   In $\Omega(\mathbb{K})$, there is again a similar phenomenon. Indeed, for some fields $\mathbb{K}$ (see below for examples), there exists a triality θ such that each absolute point is incident with exactly two absolute lines. The geometry $\Gamma^{(0)}$ is in this case a weak generalized hexagon which is the dual of the double of a (thick) projective plane (over $\mathbb{K}$). For a given absolute point p, the collineation induced on $\mathbf{PG}(1, \mathbb{K})^{(p)}$ has order 3 and has two fixed points (a *homology*). These trialities exist for all fields admitting such a homology in the corresponding projective 1-space; in particular, the characteristic of the field is not 3. Examples are provided by the finite fields $\mathbf{GF}(q)$ with $q \equiv 1$ modulo 3.

3. **Symplectic polarities** in $\mathbf{PG}(3, \mathbb{K})$ are polarities θ with the property that for every point p, every line incident with p and with p^θ is an absolute line. This is enough to conclude that every point is an absolute point. So the set of absolute points forms a linear subspace of $\mathbf{PG}(3, \mathbb{K})$, namely, $\mathbf{PG}(3, \mathbb{K})$ itself. The Grassmann coordinates of the absolute lines satisfy a linear equation (see Subsection 2.3.18 above). For a given absolute point p, the collineation induced on $\mathbf{PG}(1, \mathbb{K})^{(p)}$ is always the identity. These polarities exist for all fields $\mathbb{K}$.

   In $\Omega(\mathbb{K})$, there is a similar phenomenon. Indeed, the trialities of type $(\mathbf{I}_{\mathrm{id}})$ above have the property that for every absolute point p, all lines incident with p, p^θ and p^{θ^2} are absolute lines. Moreover, the set of absolute points is the set of points lying in a linear subspace of $\mathbf{PG}(7, \mathbb{K})$, namely a hyperplane (see Subsection 2.4.13 above). Also, the Grassmann coordinates of the absolute lines satisfy a system of linear equations. For a given absolute point p, the collineation induced on $\mathbf{PG}(1, \mathbb{K})^{(p)}$ is always the identity. Finally, these trialities exist for all fields $\mathbb{K}$.

More similarities between the class of symplectic quadrangles and the class of split Cayley hexagons can be found in Chapter 3 (they both contain "mixed subpolygons"), Chapter 4 (their members admit collineations with analogous properties), Chapter 5 (they behave similarly with respect to point-

minimality and line-minimality), Chapter 6 (they share a number of combinatorial, geometric and algebraic characterizations), Chapter 7 (they contain
self-dual and self-polar members and the "ovoids" arising this way have similar properties) and Chapter 9 (many similarities already mentioned have
topological analogues).

4. **Unitary** or **Hermitian polarities** in $\mathbf{PG}(3, \mathbb{K})$ are in bijective correspondence
with the involutory field automorphisms of $\mathbb{K}$. For a given such involution
σ and corresponding polarity θ, the set of absolute lines through an absolute point p is parametrized by the subfield of $\mathbb{K}$ of fixed elements under σ,
together with one extra element ∞. For a given absolute point p, the collineation induced on $\mathbf{PG}(1, \mathbb{K})^{(p)}$ is the involution arising from the semi-linear
map with identity matrix and the non-trivial involutory field automorphism
σ. These polarities exist for all fields admitting such an involution.

In $\Omega(\mathbb{K})$, the trialities of type $(\mathbf{I}_\sigma)$, $\sigma \neq 1$, have similar properties. They
are in bijective correspondence with the class of field automorphisms σ of
order 3 in $\mathbb{K}$, and $\mathbb{K}^{(\sigma)} \cup \{\infty\}$ parametrizes the set of absolute lines through
an absolute point. For a given absolute point p, the collineation induced
on $\mathbf{PG}(1, \mathbb{K})^{(p)}$ arises from the semi-linear map with identity matrix and
associated field automorphism σ. These trialities exist for all fields admitting
such an automorphism.

Note that all points of the quadrangles arising from polarities in $\mathbf{PG}(3, \mathbb{K})$ as above
are distance-2-regular, just like the points of all hexagons arising from trialities in
$\Omega(\mathbb{K})$; see Proposition 2.4.15.

2.5   Classical generalized octagons

There is at present no elementary geometric construction known of the classical
octagons. In the literature, one is usually referred to the construction of this geometry using the *Tits system* (or (B, N)-*pair*; for these notions see Section 4.7) in
the Chevalley groups of type $^2F_4$ (the so-called Ree groups of characterictic 2),
which was also the original construction by TITS [1960] (see also TITS [1983]).
The construction which we would like to give uses *metasymplectic spaces*, i.e., the
point–line geometries arising from buildings of type F_4. So in fact, we should first
construct such spaces. Of course, this is beyond the scope of this book. Hence-
forth, a rigorous existence proof of the classical octagons will not be considered
in this book. But for those readers who are more or less familiar with elementary
properties of metasymplectic spaces — which are nevertheless more popular than
the classical octagons — we include the proof of the following theorem. The result
is well known, but to the best of my knowledge no proof exists in print. The result
was first announced by TITS [1960]; see also SARLI [1986]. Note, however, that I
will need some elementary results concerning *polarities* of Moufang quadrangles,

in particular about *ovoids* (this is unavoidable since buildings of type F_4 contain Moufang quadrangles as *residues*, and a polarity of such a building induces a polarity in some of these quadrangles). I therefore advise the reader to read first the relevant parts of Chapter 7, if necessary.

2.5.1 Definition. For a field $\mathbb{K}$ of positive characteristic p, we will call an endomorphism $\sigma : \mathbb{K} \to \mathbb{K}$ such that $x^{\sigma^2} = x^p$ for all $x \in \mathbb{K}$, a **Tits endomorphism**. Note that the endomorphism $x \mapsto x^p$ itself is called the **Frobenius endomorphism**. So a Tits endomorphism is a square root of the Frobenius endomorphism (but in general not unique; see the introduction of Section 7.6, page 322). We denote the field of squares of a field $\mathbb{K}$ of characteristic 2 by $\mathbb{K}^2$.

2.5.2 Theorem (Tits (unpublished)). *Let $\mathcal{M}$ be a metasymplectic space over some field $\mathbb{K}$, i.e., the planes of $\mathcal{M}$ are planes over $\mathbb{K}$. Suppose $\mathcal{M}$ is self-polar and let θ be a polarity. Then $\mathbb{K}$ has characteristic 2, it admits a Tits endomorphism σ and the geometry $\mathsf{O}(\mathbb{K}, \sigma)$ whose points and lines are the absolute points and lines, respectively, with natural incidence relation, is a generalized octagon with $|\mathbb{K}| + 1$ points per line and $|\mathbb{K}|^2 + 1$ lines per point.* □

Proof. We first define metasymplectic spaces axiomatically, then list some properties that we will use (these properties can be read off the diagram in most cases), and then proceed to the proof of the theorem.

Definition and properties of metasymplectic spaces

A **metasymplectic space** $\mathcal{M}$ is a building (see Subsection 1.3.7) with four types of elements, usually called *points, lines, planes* and *hyperlines*, together with a binary reflexive and symmetric incidence relation satisfying the axioms (M1) to (M4) stated below. The term building already implies that the incidence graph is connected, that no element has two or more different types and that every flag (a **flag** is a set of mutual incident elements) is contained in a chamber, i.e., a flag consisting of just four elements, one of each type.

The **residue** of a flag F is the geometry of elements distinct from those belonging to F and incident with all elements of F, subject to the incidence relation inherited from $\mathcal{M}$. The **type** of a flag is the set of types of its elements.

(M1) The residue of any flag of type {point, line} or {plane, hyperline} is a projective plane.

(M2) The residue of any flag of type {point, plane}, {line, hyperline} or {line, plane} is a generalized digon.

(M3) The residue of any flag of type {point, hyperline} is a generalized quadrangle.

(M4) If we call a **shadow** the set of points incident with a given element, then the intersection of any two shadows is again a shadow or empty. Furthermore, two distinct elements have distinct shadows.

The first three axioms tell you exactly that $\mathcal{M}$ belongs to the diagram F_4. The last axiom is the **intersection property**; see TITS [1981]. In fact, we have taken this definition in terms of points, lines, planes and hyperlines from TITS [1981].

From axiom (M4) it follows immediately that every element is determined by the set of points incident with it. So we may identify every element with the set of points incident with it. This way it makes sense to talk about intersections of hyperlines, planes, etc., and to use set-theoretic symbols as $\in$, $\subseteq$, ...

A metasymplectic space $\mathcal{M}$ now has the following properties. The proofs use only standard *diagram arguments*, i.e., using the axioms (M1), (M2) and (M3) combined with incidence properties of the generalized polygons corresponding to the various residues, or standard *apartment arguments* (see the first three chapters of TITS [1974]), i.e., the mutual position of two elements can be seen in an *apartment* of the corresponding building. Note that (M6) follows directly from (M1) and (M4).

(M5) *Let x and y be two points of $\mathcal{M}$. Then one of the following situations occurs:*

(0) $x = y$.

(1) *There is a unique line incident with both x and y. In this case, we call x and y* **collinear** *and we denote the unique line by xy.*

(2) *There is a unique hyperline incident with both x and y. In this case there is no line incident with both x and y, and we call x and y* **cohyperlinear**. *We denote the unique hyperline by $x \Diamond y$.*

(3) *There is a unique point z collinear with both x and y. In this case we call x and y* **almost opposite** *and we denote z by $x \bowtie y$.*

(4) *There is no point collinear with both x and y. In this case we call x and y* **opposite**.

(M6) *The intersection of two hyperlines is either empty, or a point, or a plane.*

(M7) *Let x be a point and h a hyperline of $\mathcal{M}$. Then one of the following situations occurs:*

(0) $x \in h$.

(1) *There is a unique line L in h such that x is collinear with all points of L. Every point y of h which is collinear with all points of L is cohyperco-linear with x and $x \Diamond y$ contains L. Every other point z of h (i.e., every point z of h collinear with a unique point z' of L) is almost opposite x and $x \bowtie z = z' \in L$.*

(2) *There is a unique point u of h cohyperlinear with x. We have $h \cap (x \Diamond u) = \{u\}$. All points v of h collinear with u are almost opposite x and $x \bowtie v \notin h$. All points w of h cohyperlinear with u are opposite x.*

(M8) *Two elements are incident* **if and only if** *the shadow of one of these elements is contained in the shadow of the other.*

(M9) *The points, lines and planes incident with a hyperline form a polar space; in particular, the Buekenhout–Shult one-or-all axiom (see* Definition 2.4.1*) and its consequences hold.*

(M10) *There is a principle of <u>duality</u>: if we replace point, line, plane, hyperline by, respectively, hyperline, plane, line, point in the above definitions and statements, then we obtain (mostly new) true properties.*

This principle of duality is a necessary condition for the existence of polarities. A sufficient condition is that the quadrangles which appear as a residue of a flag of type {point, hyperline} are Suzuki quadrangles (see Subsection 3.4.6 on page 103). We will not prove this. A proof would require the construction of such metasymplectic spaces and this would lead us into the theory of either Chevalley — or more generally, algebraic — groups, or Tits buildings (cf. TITS [1974], RONAN & TITS [1987] or RONAN [1989]). In Lemma 2 below, we give evidence for the fact that the Suzuki quadrangles are needed in order to have a polarity.

| Proof of the theorem |

Suppose now $\mathcal{M}$ is a metasymplectic space and θ is a polarity of $\mathcal{M}$, i.e., θ permutes the elements of $\mathcal{M}$ in such a way that points are mapped to hyperlines and vice versa, and lines are mapped to planes and vice versa; moreover θ preserves the incidence relation and θ^2 is the identity. An element of $\mathcal{M}$ is called **absolute** if it is incident with its image under θ. We claim that there is at least one absolute element. Indeed, let x be any point of $\mathcal{M}$. We apply (M7) to $h = x^\theta$. If $x \in x^\theta$, then x is an absolute point and the claim is proved. Suppose that there is a unique line L in x^θ all points of which are collinear with x. It is not so difficult to see that L^θ is equal to the plane generated by x and L, and hence L is an absolute line. Hence, by (M7), we may assume that there is a unique point $y \in x^\theta$ cohyperlinear with x. But then y^θ is the unique hyperline through x meeting x^θ in a point and so y is an absolute point. The claim follows.

We put Γ equal to the geometry with point set the set of absolute points, with line set the set of absolute lines and with natural incidence relation. Our goal is to show that Γ is a generalized octagon. We will do so in a sequence of lemmas.

| **Lemma 1.** | *Every point of an absolute line is absolute.*

Proof. Let L be an absolute line and $p \in L$. Applying θ, we obtain $L^\theta \subseteq p^\theta$ (using (M8)). Since L is absolute, $L \subseteq L^\theta$. Hence $p \in L \subseteq L^\theta \subseteq p^\theta$ and so p is absolute.   QED

It is well known that metasymplectic spaces satisfy a so-called *Moufang condition* (see TITS [1974], [1976a]). This implies that all generalized polygons appearing as residues are Moufang polygons. So we may assume that the projective planes arising as residues of flags of type {plane, hyperline} are defined over an alternative field $\mathbb{K}$, which is also known to parametrize one of the two kinds of root groups of the generalized quadrangles arising as residues of flags of type {point, hyperline}. We denote by $\mathbf{PGL}_2(\mathbb{K})$ the group induced on a line L of $\mathbf{PG}(2, \mathbb{K})$ by the stabilizer of L in the group of automorphisms of $\mathbf{PG}(2, \mathbb{K})$ generated by all *elations* (hence in the so-called *little projective group* of $\mathbf{PG}(2, \mathbb{K})$; see Definitions 4.4.4 on page 143).

Lemma 2. *There is at least one absolute point. For every absolute point p, the residue $Q(p, p^\theta)$ of the flag $\{p, p^\theta\}$ is a Suzuki quadrangle and the absolute lines through p form an ovoid in $Q(p, p^\theta)$. Hence there are $|\mathbb{K}|^2 + 1$ absolute lines incident with p. Also, $\mathbb{K}$ has characteristic 2 and admits a Tits endomorphism.*

Proof. Since there is at least one absolute element, there is at least one absolute point. Indeed, if there is no absolute point, then there must be an absolute line (noting that an element a is absolute **if and only if** a^θ is absolute). But Lemma 1 implies that all points on that line are absolute. Hence the claim.

So let p be an absolute point. The polarity θ clearly induces a polarity in $Q(p, p^\theta)$. By Proposition 7.2.5 (see page 308), the absolute points with respect to that polarity form an ovoid of $Q(p, p^\theta)$. But it is easily seen that these absolute points are in fact absolute lines of $\mathcal{M}$ with respect to θ (if the lines of $\mathcal{M}$ incident with p and p^θ are called points of $Q(p, p^\theta)$). Since $Q(p, p^\theta)$ is a self-polar Moufang quadrangle, it must be either a mixed quadrangle $\mathsf{Q}(\mathbb{L}, \mathbb{L}'; L, L')$ (for appropriate $\mathbb{L}, \mathbb{L}', L$ and L') or a Moufang quadrangle of type $(BC - CB)_2$, by Theorem 7.3.2 on page 312. The group induced on a residue of a flag of type {point, plane, hyperline} contains $\mathbf{PGL}_2(\mathbb{K})$ (looking in a residue of type {plane, hyperline}). This implies that all non-trivial root elations of $Q(p, p^\theta)$ are conjugate. Hence $Q(p, p^\theta)$ cannot be a Moufang quadrangle of type $(BC - CB)_2$, because in such a quadrangle, elements of $[U_1, U_3] \subseteq U_2$ (with the notation of Subsection 5.5.5) are never conjugate to elements of $U_2 \setminus [U_1, U_3]$ (and the latter is non-empty!). Hence $Q(p, p^\theta) \cong \mathsf{Q}(\mathbb{L}, \mathbb{L}'; L, L')$. Now the group induced on a residue of a flag of type {point, plane, hyperline} is also contained in $\mathbf{PGL}_2(\mathbb{L})$ (looking in a residue of a flag of type {point, hyperline}). Hence, looking at the stabilizer of two points in such a residue, we readily deduce that $\mathbb{K}$ is a (commutative) field and that $\mathbb{K} = \mathbb{L} = L$, and consequently also $\mathbb{L}' = L'$, with $\mathbb{L}'$ the image of $\mathbb{L}$ under a Tits endomorphism (this follows from Theorem 7.3.2 on page 312). So $Q(p, p^\theta)$ is a Suzuki quadrangle (see Subsection 3.4.6 on page 103). The rest of the lemma follows from Proposition 7.2.3 (page 307). QED

The next lemma shows that θ has properties entirely different from polarities in projective spaces.

Lemma 3. *If a line of $\mathcal{M}$ contains two absolute points, then it is an absolute line. Also, no plane contains more than one absolute line.*

Proof. Let p and q be two distinct absolute points incident with a line M. Applying θ we see that p^θ and q^θ share a plane π. If p is collinear with all points of π, then $p \in q^\theta$ by (M7). If p is collinear with all points of a line L of π, then again by (M7), $q \in L$ and hence again $q \in p^\theta$. By (M9), there are no other possibilities. So we have shown that $q \in p^\theta$. Thus, $M \subseteq p^\theta$, implying $p \in M^\theta$. Similarly $q \in M^\theta$. Axiom (M4) implies that $M \subseteq M^\theta$.

Let the plane π contain two absolute lines L and M. Let L and M meet in p. It is easily seen that $L, M \subseteq p^\theta$. But this contradicts the fact that L and M represent non-collinear points (of an ovoid) in $Q(p, p^\theta)$.   QED

Lemma 4. *Let p and q be cohyperlinear absolute points. Then there exists a unique absolute point x collinear with both p and q.*

Proof. Suppose first that $p \in q^\theta$. Then $q \in p^\theta$ and hence $p^\theta = (p \Diamond q) = q^\theta$, implying $p = q$, a contradiction.

Suppose now that the hyperlines p^θ and $p \Diamond q$ have only p in common. By (M7), q is opposite all points of p^θ which are cohyperlinear with p. Since p and q are contained in a unique hyperline, the hyperlines p^θ and q^θ meet in a unique point z. By (M7), we must have $z = p$ and hence $p \Diamond q = q^\theta$. Hence $p \in q^\theta$, a contradiction again.

So we may assume that $p \Diamond q$ meets p^θ in a plane π. This plane π is a line of the quadrangle $Q(p, p^\theta)$ and it is therefore incident with a unique absolute point of $Q(p, p^\theta)$ with respect to the polarity induced by θ in $Q(p, p^\theta)$. This point represents an absolute line L of $\mathcal{M}$ incident with p. Since q is not collinear with p, the set of points of π collinear with q is a line N distinct from L. Let $\{x\} = L \cap N$. Then x is an absolute point (since it lies on the absolute line L) collinear with both p and q. Since q is absolute, the previous lemma implies that qx is an absolute line.

If y is another absolute point of $\mathcal{M}$ collinear with both p and q, then $p \in py\,\mathbf{I}\,(py)^\theta$ implies $py\,\mathbf{I}\,p^\theta$. Similarly $qy\,\mathbf{I}\,q^\theta$, hence $y \in p^\theta \cap q^\theta$ and consequently $y = x$.   QED

The previous proof also shows that, if x, y, z are three absolute points with x and y collinear, and y and z collinear, then x and z are cohyperlinear and $z \notin x^\theta$. Indeed, both x and z are contained in y^θ.

This in turn now implies:

Lemma 5. Γ *does not contain a proper pentagon.*

Proof. Let x, y, z, u, v be the consecutive collinear vertices of a proper pentagon (with x and v collinear). The point x is collinear with $v \in u^\theta$; it is cohyperlinear with $z \in u^\theta$, hence by (M7) either v and z are collinear (a contradiction), or $x \in u^\theta$. The latter contradicts our previous remark.   QED

Lemma 6. *Let p and q be two almost opposite absolute points. Then there exist unique collinear absolute points x and y such that x is collinear with p and y with q.*

Proof. Since p and q are not contained in a common hyperline, the point $p \bowtie q$ is not absolute and the hyperlines p^θ and q^θ are disjoint. Let $h^\theta = p \bowtie q$. Since both p and q are collinear with $p \bowtie q$, the hyperline h meets both p^θ and q^θ in a plane, say, π_p and π_q, respectively. Using the Buekenhout–Shult one-or-all axiom in p^θ, one easily sees that there is at least one plane π in p^θ containing p and sharing a line L with π_p, $p \notin L$ (L is uniquely determined if $p \notin \pi_p$). It follows that there is a unique absolute line L_p through p meeting L and hence there is some absolute point x collinear with p and lying in π_p. Similarly, there is some absolute point y collinear with q and lying in π_q. Since the planes π_p and π_q are disjoint, the points x and y do not coincide. Hence they are either collinear — in which case the lemma is proved, up to uniqueness of x and y — or cohyperlinear. In the latter case, there is a unique absolute point z collinear with both x and y by Lemma 4. Since $x, y \in z^\theta$, we see that $z \in h$. By uniqueness of the hyperline through x and y, $z^\theta = h$. Applying θ, we deduce $z = p \bowtie q$, contradicting the fact that $p \bowtie q$ is not absolute.

There remains to show that x and y are unique. Suppose x' and y' are collinear absolute points collinear with, respectively, p and q. The point $q \notin p^\theta$ is cohyperlinear with both x and x' of p^θ. If $x \neq x'$, then (M7) implies that q is collinear with all points of a line L in p^θ and that the unique hyperline through q and $x(x')$, namely $y^\theta (y'^\theta)$, contains L. Thus y^θ would meet p^θ in a line, hence a plane and so, applying θ, p and y would be collinear, a contradiction.   QED

An almost identical argument as in the last part of the previous proof can be used for the following lemma.

Lemma 7. Γ *does not contain proper heptagons.*

Proof. Let $(p_1, p_2, \dots, p_7, p_8 = p_1)$ be a heptagon with p_i collinear with p_{i+1}, for all i modulo 7. The point p_1 is collinear with the point p_2 belonging to p_3^θ. Hence by (M7), it is not opposite $p_4 \in p_3^\theta$. Since Γ does not contain proper pentagons by Lemma 5, p_1 and p_4 are almost opposite (cohyperlinear would imply a pentagon by Lemma 4). Also, p_1 is cohyperlinear with $p_6 \in p_5^\theta$ and almost opposite $p_5 \in p_5^\theta$. Since p_4 and p_6 are not collinear, (M7) implies that the unique hyperline through p_1 and p_6, namely p_7^θ, meets p_5^θ in at least a line, hence a plane. Therefore p_5 and p_7 are collinear, a contradiction.   QED

Lemma 6 says in fact that there are no proper ordinary hexagons in Γ. Indeed, if p, x, y, q are consecutively collinear absolute points, then p is collinear with $x \in y^\theta \ni q$, and so we deduce as in the beginning of the proof of Lemma 7 that p and q are almost opposite. This brings us back to the situation of Lemma 6.

Lemma 8. *The gonality of Γ is equal to* 8.

Proof. In view of previous lemmas stating that there are no proper j-gons for $j \leq 7$ in Γ, we only have to exhibit a proper ordinary octagon in Γ. Let p_0 be an absolute point. Let L_0 be an absolute line through p_0 (existing by Lemma 2). Let $p_1 \neq p_0$ be on L_0. Let $L_1 \neq L_0$ be an absolute line through p_1. Continuing thus, we obtain a sequence

$$p_0 \, \mathbf{I} \, L_0 \, \mathbf{I} \, p_1 \, \mathbf{I} \, L_1 \, \mathbf{I} \, \dots \, \mathbf{I} \, p_4 \, \mathbf{I} \, L_4,$$

with p_4 opposite p_0; otherwise there arises an ordinary proper j-gon in Γ with $j \leq 8$. Let h be any hyperline containing L_4 (for instance one can take $h = p_4^\theta$). Since p_0 is opposite $p_4 \in h$, there is a unique point $x \in h$ cohyperlinear with p_0. All points in h collinear with x are almost opposite p_0 and there is a unique such point p_5 on L_4 (unique because otherwise, by the Buekenhout–Shult one-or-all axiom, all points of L_4 are almost opposite p_0, including p_4, a contradiction). By Lemma 6, there are collinear absolute points p_6 and p_7 collinear with, respectively, p_5 and p_0. We have established an octagon $p_0 \perp p_1 \perp \ldots \perp p_7 \perp p_0$.   QED

To finish the proof of Theorem 2.5.2 we only have to prove that the diameter of Γ is equal to 8. Note that we have shown above that, whenever a line L of $\mathcal{M}$ contains a point opposite some point p (in $\mathcal{M}$), then there is a (unique) point x on L almost opposite p. In fact, this follows readily from (M7)(2).

Lemma 9. *The diameter of Γ is equal to 8.*

Proof. This will follow as soon as we have shown that a point and a line are always at distance $j \leq 7$ from each other in the incidence graph of Γ. This will follow if we show that, for every line L and every point p, there is a point x on L at distance at most 6 from p. By Lemmas 3, 4 and 6, we may assume that all points of L are opposite p. But then we remarked above that there must be a unique point on L almost opposite x, a contradiction.   QED

The theorem is proved.   □

2.5.3 Definitions. We call the generalized octagon $\mathsf{O}(\mathbb{K}, \sigma)$ as in the statement of the theorem the **Ree–Tits octagon**, for obvious reasons. Both the Ree–Tits octagons and their duals will be called **classical**, but recall that the term *Ree–Tits octagon* is, as a matter of convenience, reserved for $\mathsf{O}(\mathbb{K}, \sigma)$ itself. In the finite case, it follows from the previous theorem that the field $\mathbf{GF}(q)$ has even order and $q = 2^{2e+1}$. In that case, the Tits endomorphism σ is an automorphism and is determined by $\mathbf{GF}(q)$. The corresponding Ree–Tits octagon is therefore denoted by $\mathsf{O}(q)$. We will give an explicit description of $\mathsf{O}(\mathbb{K}, \sigma)$ in the next chapter, but without proof, in view of the remarks we made in this section.

A special case occurs when $\mathbb{K}$ is a perfect field. Indeed, the corresponding Ree–Tits octagons have nicer geometric properties which also characterize them; see Subsection 6.9 on page 298. In that case, we call $\mathsf{O}(\mathbb{K}, \sigma)$ a **perfect Ree–Tits octagon**.

Commenting on things to come, we note that the proof of the characterization Theorem 6.9.3 (page 300) consists of reconstructing the metasymplectic space $\mathcal{M}$ for a given octagon satisfying the given axioms. The proof of the fact that there is essentially one polarity with at least one absolute element for a given metasymplectic space over a field $\mathbb{K}$ of characteristic 2 and a given Tits endomorphism of $\mathbb{K}$ can be deduced from various results of TITS [1974], [1962b], [1964]. An explicit proof is contained in VAN MALDEGHEM [1998].

2.6   Table of notation for some classical polygons

We summarize part of the notation and terminology that we have introduced in this chapter in Table 2.1. For a generalized n-gon, we display n, Γ, the corresponding (simple) group G, its name and a reference. We denote by $\mathbb{K}$ any field, $\mathbb{L}$ any skew field, q any prime power, ρ (or τ) an appropriate 1-quadratic (or σ-quadratic) form on a $(d+1)$-dimensional vector space over $\mathbb{K}$ (respectively $\mathbb{L}$) (and where σ is an anti-automorphism of order 2 of $\mathbb{L}$), $\mathbb{F}$ denotes a cubic Galois extension of $\mathbb{K}$ (if that exists), and σ' is a non-trivial element of the corresponding Galois group. Also, we denote by char$\mathbb{K}$ the characteristic of the field $\mathbb{K}$. In the infinite case, it is impossible to give all orthogonal and Hermitian quadrangles. Therefore, we have restricted ourselves to mentioning only the finite cases.

n	Γ	G	Name	Remarks	Sub-section
3	$\mathbf{PG}(2,\mathbb{K})$	$\mathbf{PSL}_3(\mathbb{K})$	Pappian plane		2.2.3
3	$\mathbf{PG}(2,\mathbb{L})$	$\mathbf{PSL}_3(\mathbb{L})$	Desarguesian plane		2.2.1
4	$\mathsf{W}(\mathbb{K})$	$\mathbf{PSp}_4(\mathbb{K})$	Symplectic quadrangle		2.3.17
4	$\mathsf{Q}(d,\mathbb{K},\rho)$		Orthogonal quadrangle		2.3.7
4	$\mathsf{H}(d,\mathbb{L},\tau,\sigma)$		Hermitian quadrangle		2.3.7
4	$\mathsf{Q}(4,q)$	$\mathbf{PSO}_5(q)$	Orthogonal quadrangle		2.3.12
4	$\mathsf{Q}(5,q)$	$\mathbf{PSO}_6^-(q)$	Orthogonal quadrangle		2.3.12
4	$\mathsf{H}(3,q^2)$	$\mathbf{PSU}_4(q)$	Hermitian quadrangle		2.3.12
4	$\mathsf{H}(4,q^2)$	$\mathbf{PSU}_5(q)$	Hermitian quadrangle		2.3.12
6	$\mathsf{H}(\mathbb{K})$	$\mathbf{G}_2(\mathbb{K})$	Split Cayley hexagon		2.4.9
6	$\mathsf{H}(\mathbb{K})$	$\mathbf{G}_2(\mathbb{K})$	Symplectic hexagon	char$\mathbb{K}=2$	2.4.14
6	$\mathsf{H}(\mathbb{K})$	$\mathbf{G}_2(\mathbb{K})$	Perfect symplectic hexagon	char$\mathbb{K}=2$ $\mathbb{K}$ perfect	2.4.14
6	$\mathsf{T}(\mathbb{F},\mathbb{K},\sigma')$	${}^3\mathbf{D}_4(\mathbb{F},\sigma')$	Twisted triality hexagon		2.4.9
8	$\mathsf{O}(\mathbb{K},\theta)$	${}^2\mathbf{F}_4(\mathbb{K},\theta)$	Ree–Tits octagon	char$\mathbb{K}=2$; $\theta^2=2$	2.5.3

Table 2.1. Some classical generalized polygons with their simple groups and name.

Chapter 3

Coordinatization and Further Examples

3.1 Introduction

The coordinatization theory of projective planes is one of the most powerful tools for constructing non-Desarguesian examples, but it is also very helpful in proving numerous results in the classical planes. For generalized polygons, coordinates have helped in proving results in both general and classical polygons. No generalized polygon, apart from many projective planes, was first constructed via coordinatization, but some have otherwise no elementary description. The prototype of this is the class of Ree–Tits octagons. One very interesting feature of coordinates in Desarguesian projective planes is the fact that one can produce "homogeneous" coordinates. This means that all points and all lines play the same role initially, and we can write down explicitly collineations moving any point to any other, using matrices for example. This is no longer true for the other classical generalized polygons. Of course one can take the coordinates of the ambient projective space (using standard embeddings); but these coordinates are not intrinsic. So homogeneous questions are best handled without (intrinsic) coordinates, while problems related to a fixed flag will usually have a solution using the coordinatization theory. Examples will be given throughout this book. However, we will never have to use explicitly coordinatization of generalized $(2n + 1)$-gons. So, to fix ideas, we establish the theory only for $2n$-gons, assuming the theory is well known for projective planes (see e.g. HUGHES & PIPER [1973], Chapter V), the only interesting "odd" case that we will explicitly need.

This chapter is structured as follows. We introduce the general coordinatization theory in Section 3.2. Then we give a description of some additional examples of polygons strongly related to the classical ones. In Section 3.7, we construct some non-classical generalized quadrangles, for instance the flock quadrangles, and we show the relation of one class of these with the classical generalized hexagon $H(\mathbb{K})$ (this result is due to KANTOR [1980] in the finite case). We also give the construction due to BADER & LUNARDON [1993] of $H(q)$, $q \equiv 2 \bmod 3$, extended to all fields of characteristic different from 3 in which raising to the third power is a bijection. As a curiosity, we construct an infinite generalized quadrangle with a projective point in which the perp-geometry is a non-Desarguesian projective plane, and an infinite hexagon with a polar point in which the perp-geometry and span-geometry are non-classical.

3.2   General coordinatization theory

3.2.1 The general setting

As remarked in the introduction above, we may assume that we are given a generalized $2n$-gon $\Gamma = (\mathcal{P}, \mathcal{L}, I)$ of order (s, t). We choose an ordered apartment

$$\Omega = (x_0, L_0, x_1, L_2, x_3, \dots, L_{2n-2}, x_{2n-1}, L_{2n-1}, x_{2n-2}, \dots, L_3, x_2, L_1, x_0),$$

with $x_0, x_1, \dots, x_{2n-1} \in \mathcal{P}$ and $L_0, L_1, \dots, L_{2n-1} \in \mathcal{L}$. For convenience, we put $x_{-1} := x_0$, $L_{-1} := L_0$, $x_{2n} := x_{2n-1}$ and $L_{2n} := L_{2n-1}$. We will call the ordered apartment Ω the **hat-rack** of the coordinatization.

Let R_1 and R_2 be two sets satisfying the following:

(R1) There is a *zero* element $0 \in R_1 \cap R_2$.

(R2) Neither R_1 nor R_2 contains the symbol ∞.

(R3) The cardinality of R_1 as a set is s and the cardinality of R_2 as a set is t.

By (R3), it is possible to fix bijections π_i and λ_i, $i = 0, 1, \dots, 2n - 1$,

$$\pi_i : R_1 \to \Gamma(L_i) \setminus \{x_{i-1}\},$$

$$\lambda_i : R_2 \to \Gamma(x_i) \setminus \{L_{i-1}\}.$$

We require these bijections to have the following two properties.

(B1) For $0 \leq i \leq 2n - 1$, one has $0^{\pi_i} = x_{i+1}$ and $0^{\lambda_i} = L_{i+1}$.

(B2) The projection of the point a^{π_i} onto the line L_{2n-i-1} is exactly the point $a^{\pi_{2n-i-1}}$, $a \in R_1$, and, dually, the projection of the line k^{λ_i} onto the point x_{2n-i-1} is exactly the line $k^{\lambda_{2n-i-1}}$, $k \in R_2$, $i = 0, 1, \dots, 2n - 1$.

In fact, condition (B2) means that the bijections π_i and λ_i for $i = n, n+1, \dots, 2n - 1$ are determined by π_{i-n} and λ_{i-n}. As a next condition, one would like to relate the bijections π_i, $i = 1, 2, \dots, n-1$ to the bijection π_0 and, dually, λ_i, $i = 1, 2, \dots, n - 1$, to λ_0. There might be various ways to do so, and we select one by introducing the following condition (N).

(N) There are special elements $1 \in R_1$ and $1^* \in R_2$ with the property that for all $a \in R_1$, the projection of the point a^{π_0} onto the line $(1^*)^{\lambda_{2n-1}}$ coincides with the projection of the point a^{π_1} onto that line, and that, dually, for all $k \in R_2$, the projection of the line k^{λ_0} onto the point $1^{\pi_{2n-1}}$ coincides with the projection of the line k^{λ_1} onto that point. This also defines 1^{π_0}, 1^{π_1} and $1^{\pi_{2n-2}}$, and dually $(1^*)^{\lambda_0}$, $(1^*)^{\lambda_1}$ and $(1^*)^{\lambda_{2n-2}}$. We now relate π_1 and π_j, j even and $j \leq 2n-4$, and also λ_0 and λ_j, j odd and $j \leq 2n-3$, by induction on j. First let j be even and put $2i = j$. Suppose, to fix the ideas, that i is even. Then the element of Ω in the middle of L_1 and L_j is the point x_{i-1} (for i odd, this is the line L_{i-1}). By induction, there is a line $L' = (1^*)^{\lambda_{i-1}}$ incident with x_{i-1} and we consider the line M at distance $2n - i - 2$ from L' and at distance $i + 1$ from x_{2n-i}. The line M is opposite both L_1 and L_j. We require that the projection of the point a^{π_1} onto the line M coincides with the projection of a^{π_j} onto M. Similarly for j odd.

Obviously, there is a dual way to the procedure explained in (N). Also, it is in fact not necessary to assume condition (N) to introduce coordinates and to work with them. So in the future, we will always explicitly say when we assume condition (N). Also, for a given coordinatization, the elements 1 and 1^* are not necessarily unique with respect to property (N). One can check this on many examples that will be given later on (usually, for generalized quadrangles when not dealing with characteristic 2, the element -1 (or -1^*) produces identical coordinates).

Let us also remark that there is a simple explanation for distinguishing the 1s in R_1 and R_2, but identifying the 0s in these sets. The reason is that the 0s are always there, while it will sometimes happen that we only introduce a 1 for one set, e.g., for R_1 (and then only part of condition (N) will be satisfied). This happens for instance when we coordinatize the classical quadrangles: in full generality, the coordinatization will also work for the weak non-thick quadrangles arising from a ruled quadric in three-dimensional projective space.

The procedure explained in (N) is not necessarily self-dual. For the explicitly given examples later on, especially the Moufang quadrangles and hexagons, we would like to have a self-dual procedure. This can be obtained as follows. For quadrangles, one can check that (N) *is always* self-dual; for hexagons, we remark that we do not need induction to relate π_2 with π_1, hence we may also relate λ_2 with λ_1 in a dual way. This relates π_i with π_1 and λ_i with λ_1, for every $i \in \{0,1,2,3,4,5\}$. Also for octagons, we do not need the induction in (N) and so we may perform a similar self-dual normalization by relating π_2 and π_4 to π_1 as in (N), and doing the dual for λ_2 and λ_4. For n-gons with $n \geq 10$, the induction in (N) *is* needed. We will refer to this self-dual normalization process for quadrangles, hexagons and octagons as a **symmetric normalization**.

It is convenient to abbreviate the sequence $0, 0, 0, \ldots, 0$ of i zeros, $i \in \mathbb{N}$, as 0^i. We also make the general notational assumption of denoting elements of R_1 by letters a, b, c (together with subscripts and/or superscripts) and elements of R_2 by letters

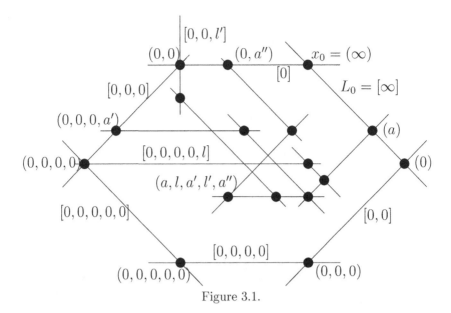

Figure 3.1.

k, l, m (same remark). As a result, we use sentences such as *"for all a"*, and *"there is an element k"* instead of "for all $a \in R_1$", and "there is an element $k \in R_2$".

We now proceed with the actual coordinatization. To each point and each line will be associated its **coordinates**, i.e., an ordered i-tuple of elements in $R_1 \cup R_2$, $i \in \{0, 1, \dots, 2n-1\}$. This association will be called a **labelling**. The coordinates of points will be written in parentheses and those of lines in brackets. For clarity's sake, the 0-tuples ( ) and [ ] will be written as (∞) and $[\infty]$. It will turn out that each i-tuple with $i \in \{0, 1, \dots, 2n-1\}$ and consisting alternately of elements of R_1 and R_2 corresponds to exactly one element of $\mathcal{P} \cup \mathcal{L}$.

3.2.2 The labelling of the elements

We start by labelling the point x_0 by (∞) and the line L_0 by $[\infty]$ (see Figure 3.1, where we have drawn the labelling for a general point in a generalized hexagon). Let y_0 be any point opposite x_0. Then there is a unique path

$$(y_0, M_1, y_1, M_2, y_2, \dots, M_{n-1}, y_{n-1}, L_0)$$

of length $2n - 1$ connecting y_0 with L_0. It is easily verified that M_i is opposite L_{2i-1} and y_j is opposite x_{2j}, $0 < i < n$, $0 \le j < n$. Let the pre-image of the projection of M_i onto x_{2i} under λ_{2i} be l_i, $i = 1, 2, \dots, n-1$, and let similarly the pre-image of the projection of y_j onto L_{2j+1} under π_{2j+1} be a_j, $j = 0, 1, \dots, n-1$. Then we label the line M_i, $1 \le i \le n - 1$, by

$$[a_{n-1}, l_{n-1}, a_{n-2}, l_{n-2}, \dots, a_i, l_i],$$

and the point y_j, $0 \leq j \leq n-1$ by

$$(a_{n-1}, l_{n-1}, a_{n-2}, l_{n-2}, \dots, l_{i+1}, a_i).$$

If we define the **distance** $\delta(x, \{p, L\})$ **from a point** x **to a flag** $(p, L) \in \mathcal{P} \times \mathcal{L}$ to be the smallest of the numbers $\delta(x, p)$ and $\delta(x, L)$, then we already have given coordinates to all points (lines) of Γ at odd (even) distance from the flag (x_0, L_0). In a completely similar dual way, we give coordinates to all other elements. Note that in this manner the point a^{π_i} has the coordinates $(0^i, a)$ and, dually, the line k^{λ_i} has been given the coordinates $[0^i, k]$, $i = 0, 1, \dots, 2n-2$. The point $a^{\pi_{2n-1}}$ and the line $k^{\lambda_{2n-1}}$ have the coordinates $(a, 0^{2n-2})$ and $[k, 0^{2n-2}]$. This follows from condition (B2).

We now make a few observations.

3.2.3 Some properties

(O1) *For all $i \in \mathbb{N}$, $1 \leq i \leq n-1$ and i odd, the coordinatization induces a bijection from the set of points of Γ at distance i from (x_0, L_0) to the set $R_1 \times R_2 \times \cdots \times R_2 \times R_1$ (i factors).*

(O2) *For all $i \in \mathbb{N}$, $0 \leq i \leq n-1$ and i even, the coordinatization induces a bijection from the set of points of Γ at distance i from (x_0, L_0) to the set $R_2 \times R_1 \times \cdots \times R_2 \times R_1$ (i factors).*

(O3) *For all $i \in \mathbb{N}$, $1 \leq i \leq n-1$ and i odd, the coordinatization induces a bijection from the set of lines of Γ at distance i from (x_0, L_0) to the set $R_2 \times R_1 \times \cdots \times R_1 \times R_2$ (i factors).*

(O4) *For all $i \in \mathbb{N}$, $0 \leq i \leq n-1$ and i even, the coordinatization induces a bijection from the set of lines of Γ at distance i from (x_0, L_0) to the set $R_1 \times R_2 \times \cdots \times R_1 \times R_2$ (i factors).*

In particular, the last coordinate of any point different from x_0 is an element of R_1 and the last coordinate of any line different from L_0 is an element of R_2.

As an example, we show the surjectivity for (O3). So let $[k_1, b_2, k_3, \dots, b_{i-1}, k_i]$ be an appropriate i-tuple. The point (k_1, b_2) is the projection of $b_2^{\pi_{2n-2}}$ onto $[k_1] = k_1^{\lambda_0}$. The line $[k_1, b_2, k_3]$ is the projection of $k_3^{\lambda_{2n-3}}$ onto (k_1, b_2). It is now easy to see that $[k_1, b_2, k_3, \dots, b_{i-1}, k_i]$ is the projection of $k_i^{\lambda_{2n-i}}$ onto $(k_1, b_2, k_3, \dots, b_{i-1})$ (which is a point by induction).

(O5) *If the number of coordinates of a point p differs by at least 2 from the number of coordinates of a line L, then p and L are not incident.*

(O6) *If the number i_p of coordinates of a point p differs by exactly 1 from the number i_L of coordinates of a line L, then p is incident with L **if and only if** p and L share the same first i coordinates, where i is the smallest among i_p and i_L.*

(O7) *If the number of coordinates of a point p is the same as the number of coordinates of a line L and if this number is not $2n - 1$, then p is incident with L **if and only if** $p = (\infty)$ and $L = [\infty]$.*

The proofs of all these observations are easy exercises.

In order to have a complete description of Γ in terms of coordinates, we should find a rule to decide whether a point and a line having both $2n - 1$ coordinates are incident or not. This will be done by introducing $(2n - 2)$ $2n$-ary operations Ψ_i, $i = 1, 2, \ldots, 2n - 2$, as follows:

$$\Psi_i \; : \quad \overbrace{R_2 \times R_1 \times \cdots \times R_2 \times R_1}^{2n \text{ factors}} \quad \rightarrow \quad R_{i+1 \bmod 2} \qquad (*)$$
$$(k, a_1, l_2, a_3, \ldots, l_{2n-2}, a_{2n-1}) \quad \mapsto \quad \Psi_i(k, a_1, l_2, \ldots),$$

where $\Psi_i(k, a_1, l_2, \ldots, a_{2n-1})$ is the $(2n - i)$th coordinate of the projection of $[k]$ onto the point $(a_1, l_2, a_3, \ldots, l_{2n-2}, a_{2n-1})$.

Dually, one can also define $(2n - 2)$ $2n$-ary operations Φ_i, $i = 1, 2, \ldots, 2n - 2$, as follows:

$$\Phi_i \; : \quad \overbrace{R_1 \times R_2 \times \cdots \times R_1 \times R_2}^{2n \text{ factors}} \quad \rightarrow \quad R_{i \bmod 2}$$
$$(a, k_1, b_2, k_3, \ldots, b_{2n-2}, k_{2n-1}) \quad \mapsto \quad \Phi_i(a, k_1, b_2, \ldots),$$

where $\Phi_i(a, k_1, b_2, \ldots, k_{2n-1})$ is the $(2n - i)$th coordinate of the projection of (a) onto the line $[k_1, b_2, k_3, \ldots, b_{2n-2}, k_{2n-1}]$.

From these definitions, it immediately follows that

(O8) *A point p with coordinates $(a_1, l_2, a_3, \ldots, l_{2n-2}, a_{2n-1})$ is incident with a line L having coordinates $[k_1, b_2, k_3, \ldots, b_{2n-2}, k_{2n-1}]$ **if and only if***

$$\begin{cases} \Psi_i(k_1, a_1, l_2, \ldots, a_{2n-1}) = k_{2n-i}, & i = 1, 3, \ldots, 2n - 5, 2n - 3, \\ \Psi_i(k_1, a_1, l_2, \ldots, a_{2n-1}) = b_{2n-i}, & i = 2, 4, \ldots, 2n - 4, 2n - 2, \end{cases}$$

if and only if

$$\begin{cases} \Phi_i(a_1, k_1, b_2, \ldots, k_{2n-1}) = a_{2n-i}, & i = 1, 3, \ldots, 2n - 5, 2n - 3, \\ \Phi_i(a_1, k_1, b_2, \ldots, k_{2n-1}) = l_{2n-i}, & i = 2, 4, \ldots, 2n - 4, 2n - 2. \end{cases}$$

It is now also clear that the coordinates together with the $2n$-ary operations Ψ_i, $i = 1, 2, \ldots, 2n - 2$, completely determine the geometry Γ, as well as the coordinates together with the operations Φ_i, $i = 1, 2, \ldots, 2n - 2$.

The labelling of the points and lines implies the following properties of the operations Ψ_i, $i = 1, 2, \ldots, 2n - 2$.

(O9) *For all* $i = 1, 2, \ldots, n-1$, *for all* $a_1, a_3, \ldots, a_{2i-1}$ *and for all* $l_2, l_4, \ldots, l_{2i}$, *one has*

$$l_{2i} = \Psi_{2i-1}(0, a_1, l_2, a_3, \ldots, a_{2i-1}, l_{2i}, 0^{2n-2i-1}).$$

(O10) *For all* $i = 1, 2, \ldots, n-1$, *for all* $a_1, a_3, \ldots, a_{2i+1}$ *and for all* $l_2, l_4, \ldots, l_{2i}$, *one has*

$$a_{2i+1} = \Psi_{2i}(0, a_1, l_2, a_3, \ldots, l_{2i}, a_{2i+1}, 0^{2n-2i-2}).$$

(O11) *For all* $i = 1, 2, \ldots, n-1$, *for all* $k, l_{2i}, l_{2i+2}, \ldots, l_{2n-2}$ *and for all* $a_{2i+1}, \ldots$, a_{2n-3}, a_{2n-1}, *one has*

$$l_{2i} = \Psi_{2i-1}(k, 0^{2i-1}, l_{2i}, a_{2i+1}, l_{2i+2}, \ldots, l_{2n-2}, a_{2n-1}).$$

(O12) *For all* $i = 1, 2, \ldots, n-1$, *for all* $k, l_{2i}, l_{2i+2}, \ldots, l_{2n-2}$ *and for all* $a_{2i-1}, \ldots$, a_{2n-3}, a_{2n-1}, *one has*

$$a_{2i-1} = \Psi_{2i-2}(k, 0^{2i-2}, a_{2i-1}, l_{2i}, a_{2i+1}, \ldots, l_{2n-2}, a_{2n-1}).$$

All these equalities directly follow from the labelling above, e.g. (O9) says that the projection of the line $l_{2i}^{\lambda_{2n-2i}}$ onto the point $(a_1, l_2, a_3, \ldots, l_{2i-2}, a_{2i-1})$ is exactly the line $[a_1, l_2, \ldots, a_{2i-1}, l_{2i}]$. Of course, the properties (O9) to (O12) can also be stated dually in terms of the Φ_i, $i = 1, 2, \ldots, 2n - 2$. We leave it to the reader to do so explicitly.

3.2.4 2n-gonal 2n-ary rings

Suppose now we are given a $2n$-tuple $\mathcal{R} = (R_1, R_2, \Psi_1, \Psi_2, \ldots, \Psi_{2n-2})$, where R_1 and R_2 are two sets satisfying conditions (R1) and (R2) above, and where Ψ_i, $i = 1, 2, \ldots, 2n - 2$ is a $2n$-ary operation as in (*), satisfying moreover (O9) to (O12). Then we can define in the obvious way a geometry $\Gamma(\mathcal{R})$ in which points and lines are certain i-tuples of elements of $R_1 \cup R_2$ satisfying the observations (O1) up to (O4). Incidence is defined as in observations (O5) to (O8). If this geometry happens to be a generalized $2n$-gon, then we call $(R_1, R_2, \Psi_1, \ldots, \Psi_{2n-2})$ a 2n-**gonal** 2n-**ary ring**. If moreover condition (N), or its self-dual variant (for $2n \leq 8$) is satisfied, then we say that the $2n$-gonal $2n$-ary ring is **normalized**, or **symmetrically normalized**. We can coordinatize $\Gamma(\mathcal{R})$ using R_1 and R_2 and such that any element coincides with its coordinates (by (O9) to (O12)). The resulting operations $\Psi(\mathcal{R})_i$, $i = 1, 2, \ldots, 2n - 2$, coincide with the original Ψ_i. So any $2n$-gonal $2n$-ary ring defines in a unique and explicit way a generalized $2n$-gon. For $2n = 4$, one can still write down a reasonable number of algebraic conditions on $(R_1, R_2, \Psi_1, \Psi_2)$ which are equivalent to being a 4-gonal 4-ary ring, usually called here a quadratic quaternary ring. If $2n > 4$, then this becomes rather messy and tedious. Since we will define certain generalized quadrangles using the coordinatization, we will write down and prove these algebraic conditions below.

We can do the same with $(R_1, R_2, \Phi_1, \Phi_2, \dots, \Phi_{2n-2})$. The corresponding notion will here be a **((symmetrically) normalized) dual $2n$-gonal $2n$-ary ring**.

For $n = 2$ and $n = 3$, a $2n$-gonal $2n$-ary ring is also called a **quadratic quaternary ring** and a **hexagonal sexternary ring**, respectively. In general, for polygons rather than for $2n$-gons, one could also speak of a **polygonal domain** (no $2n$ is hypothesized).

3.2.5 Remark. As for projective planes, the polygonal domain of a generalized polygon (even normalized or symmetrically normalized) is in general not uniquely determined: it depends on the hat-rack and on the choice of the 1 and 1^*. However, if the $2n$-gon admits a collineation group acting transitively on ordered ordinary $(2n+1)$-gons, then all normalized or symmetrically normalized polygonal domains are isomorphic (with the obvious definition of isomorphic polygonal domains). This can be seen through the proof of Corollary 4.5.13 on page 151, which also gives some examples of polygons satisfying this condition. In general, we could call polygonal domains **isotopic** if they coordinatize isomorphic polygons. Unlike the situation for classical projective planes, non-isomorphic symmetrically normalized isotopic polygonal domains exist for classical polygons. The situation becomes worse if we delete the symmetric normalization. Then a lot of non-isomorphic polygonal domains are possible, even for classical projective planes (imagining a similar (not necessarily normalized) coordinatization for $2n+1$-gons, in particular for projective planes). But this will not bother us much, though we will have to deal with it (see for instance Subsection 3.4.6 (page 103) and Proposition 3.4.9 (page 107)). Note that for projective planes, a normalized polygonal domain is sometimes called a **planar ternary field** or **planar ternary ring**, and we refer to HUGHES & PIPER [1973], Chapter V, for properties.

3.3 Generalized quadrangles and hexagons

3.3.1 Theorem. *Let R_1 and R_2 be two sets both containing at least two elements and sharing some element 0. Suppose $\infty \notin R_1 \cup R_2$. Let*

$$\Psi_1 : R_2 \times R_1 \times R_2 \times R_1 \to R_2,$$

$$\Psi_2 : R_2 \times R_1 \times R_2 \times R_1 \to R_1$$

*be two quaternary operations. Then $\mathcal{R} = (R_1, R_2, \Psi_1, \Psi_2)$ is a quadratic quaternary ring **if and only if** the following properties hold.*

(O) $\Psi_1(0, a, l, 0) = l = \Psi_1(k, 0, l, a')$, *for all* a, a', k, l,
     $\Psi_2(0, a, l, a') = a' = \Psi_2(k, 0, 0, a')$, *for all* a, a', k, l.

(A) *(i) For all a, b, k, k' there exist unique a^*, l^* such that* $\begin{cases} \Psi_1(k, a, l^*, a^*) = k' \\ \Psi_2(k, a, l^*, a^*) = b; \end{cases}$
     (ii) for all a, b, k, l, there exists a unique a^* such that $\Psi_2(k, a, l, a^*) = b$.

(B) (*i*) *For all a, b, k, k', l, $k \neq l$, there exist unique l^*, a^*, b^* such that*

$$\begin{cases} k' &=& \Psi_1(k, a^*, l^*, b^*), \\ b &=& \Psi_2(k, a^*, l^*, b^*), \\ a &=& \Psi_2(l, a^*, l^*, b^*); \end{cases}$$

(*ii*) *for all a, a', b, k, l, $a \neq b$, there exist unique k^*, b^* such that*

$$\begin{cases} \Psi_1(k^*, a, l, a') &=& \Psi_1(k^*, b, k, b^*), \\ \Psi_2(k^*, a, l, a') &=& \Psi_2(k^*, b, k, b^*). \end{cases}$$

(C) *For all a, a', b, k, k', l, the system*

$$\begin{cases} \Psi_1(k, a^*, l^*, b^*) &=& k', \\ \Psi_1(k^*, a^*, l^*, b^*) &=& \Psi_1(k^*, a, l, a'), \\ \Psi_2(k, a^*, l^*, b^*) &=& b, \\ \Psi_2(k^*, a^*, l^*, b^*) &=& \Psi_2(k^*, a, l, a'), \end{cases}$$

   has

   (*i*) *exactly one solution (a^*, b^*, k^*, l^*) if we have $b \neq \Psi_2(k, a, l, a')$ and also $k' \neq \Psi_1(k, a, l, a'')$, where a'' is defined by $\Psi_2(k, a, l, a'') = b$ (see (A)(ii));*

   (*ii*) *no solution in (a^*, b^*, k^*, l^*) if* **either** *$b = \Psi_2(k, a, l, a')$ and $k' \neq \Psi_1(k, a, l, a'')$, where a'' is defined by $\Psi_2(k, a, l, a'') = b$ (see (A)(ii)), or $b \neq \Psi_2(k, a, l, a')$ and $k' = \Psi_1(k, a, l, a'')$, where a'' is defined by $\Psi_2(k, a, l, a'') = b$.*

Proof. Conditions (R1) and (R2) are hypothesized. Conditions (O9) to (O12) are equivalent to (O). So we have to prove that $\Gamma = \Gamma(\mathcal{R})$ is a generalized quadrangle if and only if (A), (B) and (C) hold (given (O)). Let us write down explicitly the set $\mathcal{P}$ of points, the set $\mathcal{L}$ of lines and the incidence relation in Γ.

The points of Γ are all elements of the form (∞), (a), (k, b) and (a, l, a'). The lines of Γ are the elements of the form $[\infty]$, $[k]$, $[a, l]$ and $[k, b, k']$. Incidence is given by the chain

$$(a, l, a')\, \mathbf{I}\, [a, l]\, \mathbf{I}\, (a)\, \mathbf{I}\, [\infty]\, \mathbf{I}\, (\infty)\, \mathbf{I}\, [k]\, \mathbf{I}\, (k, b)\, \mathbf{I}\, [k, b, k']$$

and

$$(a, l, a')\, \mathbf{I}\, [k, b, k']$$

$$\Updownarrow$$

$$\begin{cases} \Psi_1(k, a, l, a') = k' \\ \Psi_2(k, a, l, a') = b. \end{cases}$$

We have to show that conditions (A), (B) and (C) are equivalent to the main axiom for generalized quadrangle, i.e., for every non-incident pair $(p, L) \in \mathcal{P} \times \mathcal{L}$, there exists a unique pair $(q, M) \in \mathcal{P} \times \mathcal{L}$ such that $p\, \mathbf{I}\, M\, \mathbf{I}\, q\, \mathbf{I}\, L$.

Condition (A)(*i*) is equivalent to that axiom for $p = (a)$ and $L = [k, b, k']$.
Condition (A)(*ii*) is equivalent to that axiom for $p = (k, b)$ and $L = [a, l]$.
Condition (B)(*i*) is equivalent to that axiom for $p = (l, a)$ and $L = [k, b, k']$, $k \neq l$.
Condition (B)(*ii*) is equivalent to that axiom for $p = (a, l, a')$ and $L = [b, k]$, $a \neq b$.
Condition (C) is equivalent to that axiom for $p = (a, l, a')$ and $L = [k, b, k']$.
All other cases are trivially satisfied.   □

Note that HANSSENS & VAN MALDEGHEM [1989] study the relation between the quadratic quaternary ring of a generalized quadrangle and the presence of certain collineations in Γ. This gives some different conditions on a ring for it to be a quadratic quaternary ring.

A theorem like Theorem 3.3.1 for e.g. generalized hexagons is possible, but there are so many possibilities to consider (cf. the above proof) that it becomes too messy in the general case. However, if the hexagon has a regular point and a regular line incident with that regular point, then the conditions become reasonable (see Subsection 7.8.8 on page 358). Also, if the generalized hexagon has a lot of automorphisms, then the conditions become reasonable.

We now study the relation of a (span-)regular point with the algebraic operations.

3.3.2 Theorem (Hanssens & Van Maldeghem [1988]). *Let* Γ *be a generalized quadrangle coordinatized by a quadratic quaternary ring* $\mathcal{R} = (R_1, R_2, \Psi_1, \Psi_2)$. *Let* $\mathcal{R}' = (R_1, R_2, \Phi_1, \Phi_2)$ *be the dual quadratic quaternary ring. Then the point* (∞) *is regular* **if and only if** *the operation* Ψ_2 *is independent of its third argument, i.e.,* $\Psi_2(k, a, l, a') = \Psi_2(k, a, l', a')$, *for all* a, a', k, l, l', **if and only if** *the operation* Φ_1 *is independent of its last argument, i.e.,* $\Phi_1(a, k, b, k') = \Phi_1(a, k, b, k'')$ *for all* a, b, k, k', k''. **If** *moreover* (∞) *is projective (see* Definition 1.9.4*),* **then** $T(a, k, b) = \Phi_1(a, k, b, 0)$ *defines a coordinatizing planar ternary field of the perpplane* $\Gamma_{(\infty)}$, *if* $\mathcal{R}$ *is normalized and if we identify* R_1 *with* R_2 *by the bijection* $k \mapsto \Phi_1(1, k, 0, 0)$.

Proof. Consider the points (∞) and (a, l, a'). We have

$$\{(\infty), (a, l, a')\}^{\perp} = \{(k, \Psi_2(k, a, l, a')) : k \in R_2\} \cup \{(a)\}$$

and

$$\{(0, a'), (a)\}^{\perp} = \{(a, l^*, a') : l^* \in R_2\} \cup \{(\infty)\}.$$

So the point (∞) is regular **if and only if** the points $(k, \Psi_2(k, a, l, a'))$ and (a, l^*, a') are collinear for all a, a', k, l, l^*. The latter happens **if and only if** $\Psi_2(k, a, l, a') = \Psi_2(k, a, l^*, a')$ for all a, a', k, l, l^*, and also **if and only if**

$$\begin{cases} \Phi_1(a, k, \Psi_2(k, a, l, a'), k') = a \\ \Phi_2(a, k, \Psi_2(k, a, l, a'), k') = l^* \end{cases}$$

for all a, a', k, l, l^* and some k' depending on the free variables. Now we fix a, a', k and l. By the dual of 3.3.1(A)(ii) above, we can get any value in R_2 for k', varying l^*. The result follows.

Now suppose that (∞) is projective. We first show the claim that $k \mapsto \Phi_1(1, k, 0, 0)$ is bijective. Let $a \in R_1$ be arbitrary. The traces $\{(\infty), (1, 0, a)\}^\perp$, $\{(0, 0, 0), (\infty)\}^\perp$ are distinct lines of $\Gamma_{(\infty)}^\triangle$ and hence they meet in a unique point (k, b). By the coordinatization process, $b = 0$ and hence the point $(1, \Phi_2(1, k, 0, 0), a)$ is collinear with $(k, 0)$ (by the regularity of (∞)) and so since $(1, \Phi_2(1, k, 0, 0), \Phi_1(1, k, 0, 0))$ is collinear with $(k, 0)$ (it is incident with $[k, 0, 0]$), we must have $a = \Phi_1(1, k, 0, 0)$ (otherwise there arises a triangle). Suppose now $\Phi_1(1, k, 0, 0) = \Phi_1(1, k', 0, 0) =: a$; then the traces $\{(\infty), (1, 0, a)\}^\perp$ and $\{(\infty), (0, 0, 0)\}^\perp$ have the points $(k, 0)$ and $(k', 0)$ in common. Since they are not equal (indeed, the first one contains (1) and the second one (0)), we must have $k = k'$. This proves our claim.

The points of $\Gamma_{(\infty)}^\triangle$ are labelled (∞), (a) and (k, b), as in Γ itself. We now label the trace $\{(\infty), (a, l, a')\}^\perp$ by $[[a, a']]$. By the regularity of (∞), this is independent of l. The lines through (∞) keep their labelling $[k]$ and $[\infty]$. We have

$$(k, b) \ \mathbf{I}_{(\infty)} \ [k] \ \mathbf{I}_{(\infty)} \ (\infty) \ \mathbf{I}_{(\infty)} \ [\infty] \ \mathbf{I}_{(\infty)} \ (a) \ \mathbf{I}_{(\infty)} \ [[a, a']],$$

for all a, a', b, k. Furthermore, (k, b) is incident with $[[a, a']]$ **if and only if** $(a, 0, a')$ is collinear with (k, b). This is the case **if and only if** there exists k' such that $\Phi_1(a, k, b, k') = a'$. But this is equivalent to $T(a, k, b) = a'$. All properties of a planar ternary field are now satisfied. This proves the theorem. $\qquad\square$

3.3.3 Theorem (Van Maldeghem & Bloemen [1993]). *Let Γ be a generalized hexagon coordinatized by a hexagonal sexternary ring $\mathcal{R} = (R_1, R_2, \Psi_1, \Psi_2, \Psi_3, \Psi_4)$, and let $\mathcal{R}' = (R_1, R_2, \Phi_1, \Phi_2, \Phi_3, \Phi_4)$ be the dual hexagonal sexternary ring. Then*

(1) *the point (∞) is distance-2-regular **if and only if** the operation Ψ_4 is independent of its third, fourth and fifth argument, i.e., $\Psi_4(k, a, l, a', l', a'') = \Psi_4(k, a, l'', a''', l''', a'')$, for all $a, a', a'', a''', k, l, l', l'', l'''$, **if and only if** the operation Φ_1 is independent of its three last arguments, i.e., for all $a, b, b', b'_*, k, k', k'', k''_*$, we have $\Phi_1(a, k, b, k', b', k'') = \Phi_1(a, k, b, k''_*, b'_*, k''_*)$. **If** (∞) is projective, **then** the ternary operation $T(a, k, b) := \Phi_1(a, k, b, 0, 0, 0)$ defines a coordinatizing planar ternary field of the perp-plane $\Gamma_{(\infty)}^\triangle$, if $\mathcal{R}$ is (symmetrically) normalized and if we identify R_1 with R_2 by the bijection $k \mapsto \Phi_1(1, k, 0, 0, 0, 0)$.*

(2) *the point (∞) is span-regular **if and only if** it is distance-2-regular and the operation Ψ_2 is independent of both its third and its fifth argument, i.e., $\Psi_2(k, a, l, a', l', a'') = \Psi_2(k, a, l'', a', l''', a'')$, for all $a, a', a'', k, l, l', l'', l'''$, **if and only if** it is distance-2-regular and the operation Φ_3 is independent of both its fourth and its sixth argument, i.e., $\Phi_3(a, k, b, k', b', k'') =$*

$\Phi_3(a, k, b, k'_*, b', k''_*)$ *for all* $a, b, b', k, k', k'', k'_*, k''_*$. **If** *moreover* (∞) *is a polar point,* **then**

$$\begin{cases} Q_1(a, k, b, b') = \Phi_1(a, k, b, 0, 0, 0) \\ Q_2(a, k, b, b') = \Phi_3(a, k, b, 0, b', 0) \end{cases}$$

defines a (not necessarily normalized) dual coordinatizing quadratic quaternary ring (R_1, R_1, Q_1, Q_2) *of the span-quadrangle* $\Gamma^{\square}_{(\infty)}$, *if* $\mathcal{R}$ *is (symmetrically) normalized and if we identify* R_1 *with* R_2 *by the bijection* $k \mapsto \Phi_1(1, k, 0, 0, 0, 0)$.

Proof. Consider the points $(0, a'')$ and (a). A generic point p at distance 4 from both these points and at distance 6 from (∞) has coordinates (a, l, a', l', a'') (with l, a', l' arbitrary). The projection of p onto $[k]$ is the point $(k, \Psi_4(k, a, l, a', l', a''))$. If (∞) is distance-2-regular, then clearly $\Psi_4(k, a, l, a', l', a'')$ must be independent of l, a' and l'.

Conversely, suppose that $\Psi_4(k, a, l, a', l', a'')$ is independent of l, a' and l'. Let (k_0, b_0) and (k_1, b_1) be two points of $(\infty)^\perp$, with $k_0 \neq k_1$, and assume without loss of generality $k_0 \neq 0$. Let p and p' be two points with $(k_0, b_0), (k_1, b_1) \in (\infty)^p, (\infty)^{p'}$. We show that $(\infty)^p = (\infty)^{p'}$. Let (a) and $(0, a'')$ be the projections of p' onto $[\infty]$ and $[0]$, respectively. Let L_0 be the projection of (k_0, b_0) onto p. Let x be the projection of $(0, a'')$ onto L_0; let L be the projection of $(0, a'')$ onto x; and let y be the projection of (a) onto L. Then y is a point at distance 4 from both $(0, a'')$ and (a) and hence it has the same first and last coordinate as p' (namely a and a'', respectively). Since $\Psi_4(k, a, l, a', l', a'')$ is independent of l, a', l', we have $(\infty)^y = (\infty)^{p'}$. Hence y is at distance 4 from (k_0, b_0), but clearly this can only happen when $x = y$, in which case $\delta(x, (k_1, b_1)) = 4$ implies $x = y = p$. Hence (∞) is a distance-2-regular point.

Similarly, one shows that the point (∞) is distance-2-regular **if and only if** the operation $\Phi_1(a, k, b, k', b', k'')$ is independent of k', b' and k''.

Suppose now that (∞) is projective and that $\mathcal{R}$ is (symmetrically) normalized. The mapping $k \mapsto \Phi_1(1, k, 0, 0, 0, 0)$ is a bijection from R_2 to R_1 and $T(a, k, b) = \Phi_1(a, k, b, 0, 0, 0)$ defines a coordinatizing planar ternary field of the projective plane $\Gamma^{\triangle}_{(\infty)}$. Both of these statements are shown in a completely similar way to their counterparts for generalized quadrangles; see Theorem 3.3.2. This shows the first part of the theorem.

From now on suppose that (∞) is distance-2-regular. We claim that there is a unique trace containing (∞) and any point (a, l, a') **if and only if** Φ_3 is independent of its last argument and we label these equivalent conditions by $(*)$ for further reference. A generic point at distance 4 from (∞) and (a, l, a') has coordinates (k, b, k', b'), where k, b and k' are arbitrary and such that there exists k'' with (b, k'') a solution of

$$\begin{cases} \Phi_3(a, k, b, k', b', k'') = a', \\ \Phi_4(a, k, b, k', b', k'') = l. \end{cases}$$

Now note that there is a unique trace through (∞) and (a, l, a') **if and only if** every such trace contains (a, l^*, a'), for any $l^* \in R_2$. Indeed, the point $(0, 0, 0, a')$ is at distance 4 from all these points and from (∞). So let $l^* \in R_2$ be arbitrary; then (a, l^*, a') is at distance 4 from (k, b, k', b') **if and only** if there exists $k''' \in R_2$ such that

$$\begin{cases} \Phi_3(a, k, b, k', b', k''') = a', \\ \Phi_4(a, k, b, k', b', k''') = l^*. \end{cases}$$

But the mapping $l^* \mapsto k'''$ obtained by the last equation is a bijection (this is easily seen, cf. (A)(ii) of Theorem 3.3.1). Hence the claim.

Now, given that $\Phi_3(a, k, b, k', b', k'')$ is independent of k'', one similarly shows that it is independent of k' **if and only if** the following condition $(**)$ is satisfied: for every (fixed) point (k, b, k', b') and every (variable) point (a, l, a') at distance 4 from (k, b, k', b'), the trace $(k, b)^{(a, l, a')}$ is independent of (a, l, a'). Note that, in this case, these traces consist of (∞) and all points (k, b, k'_1, b'), where k, b and b' are fixed and k'_1 varies.

It is clear that $(**)$ follows from the span-regularity of (∞). We now show that this condition also *implies* that (∞) is span-regular (still under the assumption that $\Phi_3(a, k, b, k', b', k'')$ is independent of k''). To that end, let $p = (k, b, k', b')$, $p_1 = (k, b, k'_1, b')$ and $p_0 = (k_0, b_0, k'_0, b'_0)$ with $\delta(p, p_0) = 4$ and $k \neq k_0$. We have to show that $\delta(p_0, p_1) = 4$. Let $x = p \bowtie p_0$; let p'_1 be the projection of p_0 onto $[k, b, k'_1]$; let $x' = p_0 \bowtie p'_1$. Both x and x' are at distance 4 from both (k, b) and (k_0, b_0), hence since (∞) is distance-2-regular, they are at distance 4 from a common point z on $[\infty]$. Let $y = x \bowtie z$ and put $y' = x' \bowtie z$; then y is at distance 4 from both p and p_1 by condition $(**)$. Now both p_0 and p_1 are at distance 4 from both (∞) and y, hence they determine the same trace through (∞) and y in $z^{\perp}$ by condition $(*)$. Since $y' \in z^{\perp}$ and $\delta(p_0, y') = 4$, y' is on that trace. Hence $\delta(p_1, y') = 4$. Unless $p_1 = p'_1$, we have a pentagon $p_1 \perp p'_1 \perp x' \perp y' \perp (y' \bowtie p_1) \perp p_1$. This shows that (∞) is span-regular.

If (∞) is a polar point, then, as in the proof of Theorem 3.3.2, one shows by direct and obvious coordinatization that

$$\begin{cases} Q_1(a, k, b, b') = \Phi_1(a, k, b, 0, 0, 0) \\ Q_2(a, k, b, b') = \Phi_3(a, k, b, 0, b', 0) \end{cases}$$

defines a not necessarily (symmetrically) normalized dual quadratic quaternary ring (R_1, R_1, Q_1, Q_2) coordinatizing $\Gamma^{\square}_{(\infty)}$, if $\mathcal{R}$ is normalized and if we identify R_1 with R_2 by the bijection $k \mapsto \Phi_1(1, k, 0, 0, 0, 0)$. $\qquad\qquad\square$

3.4 The classical and mixed quadrangles

In this and the following sections, we write down coordinatizing rings for some of the classical generalized polygons. For the finite classical generalized quadrangles, see also HANSSENS & VAN MALDEGHEM [1987].

The coordinatization itself is always an easy exercise. The computation of the rings is a simple straightforward calculation once the coordinates are introduced. As everything is determined by the choice of the hat-rack Ω, the sets R_1 and R_2, and the bijections π_i and λ_i $(i = 0, 1, \ldots, n-1$ for not normalized rings; $i = 0$ for (symmetrically) normalized rings, $n = 2$ or $n = 3$), we will restrict ourselves by giving only this information.

3.4.1 The symplectic quadrangles

Although we will coordinatize all classical quadrangles, we first handle the symplectic one separately, and differently. One of the reasons is that we would like to be able to go from the three-dimensional space to the quadrangle and back; see for instance Section 7.6 of Chapter 7.

Let $\mathbb{K}$ be any field and let $\mathsf{W}(\mathbb{K})$ be the corresponding symplectic quadrangle (see Subsection 2.3.17 on page 63). For the associated bilinear form in $\mathbf{PG}(3, \mathbb{K})$ we can take

$$x_0 y_1 - x_1 y_0 + x_2 y_3 - x_3 y_2.$$

We put $R_1 = R_2 = \mathbb{K}$ and define (an arrow means "is labelled by")

$$
\begin{aligned}
(1, 0, 0, 0) &\rightarrow (\infty), \\
(a, 0, 1, 0) &\rightarrow (a), \\
(0, 0, 0, 1) &\rightarrow (0, 0), \\
(0, 1, 0, 0) &\rightarrow (0, 0, 0),
\end{aligned}
$$

while we define for the lines

$$
\begin{aligned}
X_1 = X_3 = 0 &\rightarrow [\infty], \\
X_1 = X_2 - kX_3 = 0 &\rightarrow [k], \\
X_0 = X_3 = 0 &\rightarrow [0, 0], \\
X_0 = X_2 = 0 &\rightarrow [0, 0, 0],
\end{aligned}
$$

with $a, k \in \mathbb{K}$. This defines the following normalized quadratic quaternary ring (and also its dual):

$$
\begin{cases}
\Psi_1(k, a, l, a') = a^2 k + l - 2aa', \\
\Psi_2(k, a, l, a') = -ak + a',
\end{cases}
\qquad
\begin{cases}
\Phi_1(a, k, b, k') = ak + b, \\
\Phi_2(a, k, b, k') = a^2 k + k' + 2ab.
\end{cases}
$$

The connection between the coordinates in $\mathbf{PG}(3, \mathbb{K})$ and those in $\mathsf{W}(\mathbb{K})$ is listed in Table 3.1. For the sake of clarity, we denote there the line in $\mathbf{PG}(3, \mathbb{K})$ defined by two points p and q by $\langle p, q \rangle$.

3.4.2 The mixed quadrangles

Let $\mathbb{K}$ be a field of characteristic 2, $\mathbb{K}'$ a subfield of $\mathbb{K}$ containing all squares, let L be a subspace of $\mathbb{K}$ considered as a vector space over $\mathbb{K}'$ and let L' be a subspace of $\mathbb{K}'$ considered as a vector space over $\mathbb{K}^2$ (the field of all squares of

POINTS	
Coordinates in $W(\mathbb{K})$	Coordinates in $\mathbf{PG}(3, \mathbb{K})$
(∞)	$(1, 0, 0, 0)$
(a)	$(a, 0, 1, 0)$
(k, b)	$(-b, 0, k, 1)$
(a, l, a')	$(l - aa', 1, -a', -a)$

LINES	
Coordinates in $W(\mathbb{K})$	Coordinates in $\mathbf{PG}(3, \mathbb{K})$
$[\infty]$	$\langle (1, 0, 0, 0), (0, 0, 1, 0) \rangle$
$[k]$	$\langle (1, 0, 0, 0), (0, 0, k, 1) \rangle$
$[a, l]$	$\langle (a, 0, 1, 0), (l, 1, 0, -a) \rangle$
$[k, b, k']$	$\langle (-b, 0, k, 1), (k', 1, -b, 0) \rangle$

Table 3.1. Coordinatization of $W(\mathbb{K})$.

$\mathbb{K}$). We suppose that $1 \in L \cap L'$, also that $L^{-1} = L$ and $L'^{-1} = L'$, and that L and L' generate, respectively, $\mathbb{K}$ and $\mathbb{K}'$, where $\mathbb{K}$ and $\mathbb{K}'$ are viewed as rings. The latter conditions are only necessary for a standard unique description; see the next proposition. The **mixed quadrangle** $Q(\mathbb{K}, \mathbb{K}'; L, L')$, which was first defined by TITS [19**], is the following subquadrangle of the symplectic quadrangle $W(\mathbb{K})$. In the coordinatizing ring for $W(\mathbb{K})$, we restrict $\mathbb{K}$ viewed as R_1 to L and $\mathbb{K}$ viewed as R_2 to L'. We obtain the normalized (ordinary and dual) quadratic quaternary rings (L, L', Ψ_1, Ψ_2) and (L, L', Φ_1, Φ_2), where

$$\begin{cases} \Psi_1(k, a, l, a') = a^2 k + l, \\ \Psi_2(k, a, l, a') = ak + a', \end{cases} \qquad \begin{cases} \Phi_1(a, k, b, k') = ak + b, \\ \Phi_2(a, k, b, k') = a^2 k + k'. \end{cases}$$

Of course, if the field $\mathbb{K}$ is perfect (for instance finite), there are no mixed quadrangles over $\mathbb{K}$ different from the symplectic one.

3.4.3 Proposition (Tits [19]).** *Let, for $i \in \{1, 2\}$, $\mathbb{K}_i$ be a field of characteristic 2, let $\mathbb{K}_i'$ be a subfield containing all squares of $\mathbb{K}_i$, let L_i and L_i' be subspaces of, respectively, $\mathbb{K}_i$ and $\mathbb{K}_i'$ considered as a vector space over, respectively, $\mathbb{K}_i'$ and $\mathbb{K}_i^2$ and suppose $1 \in L_i \cap L_i'$, $L_i^{-1} = L_i$, $L_i'^{-1} = L_i'$. The quadrangles $Q(\mathbb{K}_1, \mathbb{K}_1'; L_1, L_1')$ and $Q(\mathbb{K}_2, \mathbb{K}_2'; L_2, L_2')$ are isomorphic **if and only if** there exist a field isomorphism $\theta : \mathbb{K}_1 \to \mathbb{K}_2$ and elements $k \in \mathbb{K}_2$, $k' \in \mathbb{K}_2'$ such that $\mathbb{K}_2' = \mathbb{K}_1'^{\theta}$, $L_2 = k \cdot L_1^{\theta}$ and $L_2' = k' \cdot L_1'^{\theta}$.*

Proof. It is straightforward to see that, if the last condition is satisfied, then the given quadrangles are isomorphic. Now suppose that the two quadrangles are isomorphic. Since they are subquadrangles of the symplectic quadrangle $W(\mathbb{K}_i)$ (which is the same as $Q(\mathbb{K}_i, \mathbb{K}_i; \mathbb{K}_i, \mathbb{K}_i)$) for $i = 1, 2$, respectively, they are Moufang quadrangles (see Lemma 5.2.2 on page 174; cf. Theorem 4.5.5 on page 146).

Hence they satisfy the Tits condition by Theorem 5.2.9 on page 177. So we may assume that the isomorphism between the two quadrangles identifies the respective points (∞), (0), $(0,0)$ and $(0,0,0)$ and the respective lines $[\infty]$, $[0]$, $[0,0]$ and $[0,0,0]$. Therefore we may conceive $Q(\mathbb{K}_2, \mathbb{K}_2'; L_2, L_2')$ as a recoordinatization of $Q(\mathbb{K}_1, \mathbb{K}_1'; L_1, L_1')$ with respect to the same hat-rack (identifying elements which correspond under the isomorphism). Hence there are maps $\varphi : L_1 \to L_2$ and $\varphi' : L_1' \to L_2'$ such that the points (a) and (a^φ) coincide, for all $a \in L_1$, and such that the lines $[k]$ and $[k^{\varphi'}]$ coincide, for all $k \in L_1'$. Define X and X' by $X^\varphi = 1$ and $X'^{\varphi'} = 1$. Then the normalization implies that $(0,b)$ and $(0, (bX'^{-1})^\varphi)$ are identical, for all $b \in L_1$. Similarly, the lines $[0,l]$ and $[0, (lX^{-2})^{\varphi'}]$ coincide, for all $l \in L_1'$. It now easily follows that the points (a,l,a') and $(a^\varphi, (lX^{-2})^{\varphi'}, (a'X'^{-1})^\varphi)$ coincide, and also the lines $[k,b,k']$ and $[k^{\varphi'}, (bX'^{-1})^\varphi, (k'X^{-2})^{\varphi'}]$ are the same. Since also the incidence relation is the same, one obtains immediately:

$$\begin{cases} (a^2kX^{-2} + lX^{-2})^{\varphi'} &= (a^\varphi)^2 k^{\varphi'} + (lX^{-2})^{\varphi'}, \\ (akX'^{-1} + a'X'^{-1})^\varphi &= a^\varphi k^{\varphi'} + (a'X'^{-1})^\varphi, \end{cases}$$

for all $a, a', b \in L_1$ and all $k, k', l \in L_1'$.

One readily deduces that both φ and φ' are additive. The second equation is then equivalent to $(akX'^{-1})^\varphi = a^\varphi k^{\varphi'}$. Putting $a = X$, we see that $(kXX'^{-1})^\varphi = k^{\varphi'}$. Plugging this in into the last equation, we obtain, by putting $a = x$ and $kX'^{-1} = y$,

$$(xy)^\varphi = x^\varphi \cdot (yX)^\varphi \qquad (3.1)$$

for all $x \in L_1$ and $y \in L_1'$. Now define $\theta : L_1 \to L_1' : x \mapsto (xX)^\varphi$. It follows that $(xy)^\theta = (xyX)^\varphi = ((xX)y)^\varphi = (xX)^\varphi \cdot (yX)^\varphi = x^\theta y^\theta$, for all $x \in L_1$ and all $y \in L_1'$, by using Equation (3.1). Since L_i' generates $\mathbb{K}_i'$ (as a ring), and since $\mathbb{K}_i'$ is a subset of L_i, $i \in \{1,2\}$, this implies that the restriction of θ to $\mathbb{K}_1'$ is a field isomorphism from $\mathbb{K}_1'$ to $\mathbb{K}_2'$. The same arguments applied to L_i^2 (which generates $\mathbb{K}_i^2$ as a ring), $i \in \{1,2\}$, show that θ maps $\mathbb{K}_1^2$ to $\mathbb{K}_2^2$ (as an isomorphism). If we define $a^\theta = b$, $a \in \mathbb{K}_1$, $b \in \mathbb{K}_2$, **if and only if** $(a^2)^\theta = b^2$, then we see that θ is extended to an isomorphism between $\mathbb{K}_1$ and $\mathbb{K}_2$ which maps $\mathbb{K}_1'$ to $\mathbb{K}_2'$ and such that for any $x \in L_1$ the equality $x^\varphi = (xX^{-1})^\theta = X^{-\theta} x^\theta$ holds; hence, putting $k = X^{-\theta}$, we see that $L_2 = k \cdot L_1^\theta$. Similarly, one shows that $L_2' = k' \cdot L_1'^\theta$ for $k' = X'^{-\theta}$.

The proposition is proved. $\square$

The dual pairs among the mixed quadrangles can be traced with the following result.

3.4.4 Proposition (Tits [19]).** *The dual of the mixed Moufang quadrangle* $Q(\mathbb{K}, \mathbb{K}'; L, L')$ *is isomorphic to the mixed quadrangle* $Q(\mathbb{K}', \mathbb{K}^2; L', L^2)$.

Proof. We coordinatize $Q(\mathbb{K}, \mathbb{K}'; L, L')$ as above and then dualize. Denote this dual quadrangle by Γ. Then in Γ, the quadratic quaternary ring (L', L, Ψ_1, Ψ_2) is as follows:

$$\left\{ \begin{array}{l} \Psi_1(k,a,l,a') = ak + l, \\ \Psi_2(k,a,l,a') = ak^2 + a'. \end{array} \right.$$

If we rename $[k]$ as $[k^2]$, then we obtain a quadratic quaternary ring $(L', L^2, \Psi_1, \Psi_2)$ with

$$\left\{ \begin{array}{l} \Psi_1(k^2,a,l^2,a') = (ak + l)^2, \\ \Psi_2(k^2,a,l^2,a') = ak^2 + a', \end{array} \right.$$

or, in other words,

$$\left\{ \begin{array}{l} \Psi_1(k,a,l,a') = a^2 k + l, \\ \Psi_2(k,a,l,a') = ak + a', \end{array} \right.$$

and it follows from Subsection 3.4.2 above that this coordinatizes the mixed quadrangle $Q(\mathbb{K}', \mathbb{K}^2; L', L^2)$. $\qquad\Box$

This implies the following result:

3.4.5 Corollary (Tits [19**]). *All points and lines of a mixed quadrangle are regular.*

Proof. Since any mixed quadrangle is a subquadrangle of a symplectic quadrangle, and every point of a symplectic quadrangle is regular, we deduce from Proposition 1.9.18 on page 46 that every point of a mixed quadrangle is regular. The result now follows from Proposition 3.4.4. $\qquad\Box$

3.4.6 The Suzuki quadrangles

Let us focus on one special case of these mixed quadrangles, which we shall need when we discuss the Ree–Tits octagons. Suppose $\mathbb{K}$ (with characteristic 2) admits a Tits endomorphism σ (see Definition 2.5.1 on page 79). Put $L' = \mathbb{K}' = \mathbb{K}^\sigma$ and $L = \mathbb{K}$. The corresponding mixed quadrangle $Q(\mathbb{K}, \mathbb{K}^\sigma; \mathbb{K}, \mathbb{K}^\sigma)$ is called a **Suzuki quadrangle** $W(\mathbb{K}, \sigma)$ (because we will later on see that the Suzuki groups and Suzuki–Tits ovoids can be defined directly using these quadrangles; see Section 7.6). We now recoordinatize $W(\mathbb{K}, \sigma)$. Clearly, $R_1 = \mathbb{K}$ is isomorphic to $R_2 = \mathbb{K}^\sigma$, and so we can redefine R_2 as $\mathbb{K}$ using the injective endomorphism σ. In the coordinates, we have to substitute $k^{\sigma^{-1}}$ for k, etc. In the quadratic operations, we have to do the inverse substitution, i.e., write k^σ whenever we see k, etc. Hence we obtain the normalized (ordinary and dual) quadratic quaternary rings $(\mathbb{K}, \mathbb{K}, \Psi_1, \Psi_2)$ and $(\mathbb{K}, \mathbb{K}, \Phi_1, \Phi_2)$, where

$$\left\{ \begin{array}{l} \Psi_1(k,a,l,a') = a^\sigma k + l, \\ \Psi_2(k,a,l,a') = ak^\sigma + a', \end{array} \right. \qquad \left\{ \begin{array}{l} \Phi_1(a,k,b,k') = ak^\sigma + b, \\ \Phi_2(a,k,b,k') = a^\sigma k + k'. \end{array} \right.$$

Obviously, $\Psi_i \equiv \Phi_i$, $i = 1, 2$, so the Suzuki quadrangles are self-dual; in fact they are self-polar (for definitions see Subsection 7.2.1 on page 306) and a polarity is given by swapping the parentheses with the square brackets. By Proposition 7.2.5

on page 308, the set of absolute points is an ovoid of $\mathsf{W}(\mathbb{K}, \sigma)$, and the set of absolute lines a spread. For more on this see Section 7.6, starting on page 322.

Note that, if σ is an automorphism, then $\mathsf{W}(\mathbb{K}, \sigma)$ is nothing else than the usual symplectic quadrangle $\mathsf{W}(\mathbb{K})$. Indeed, $\mathsf{W}(\mathbb{K})$ is obviously isomorphic to the mixed quadrangle $\mathsf{Q}(\mathbb{K}, \mathbb{K}; \mathbb{K}, \mathbb{K})$, and if σ is onto, then the identity map from $\mathbb{K}$ to $\mathbb{K}$ maps $\mathbb{K}$ onto $\mathbb{K}^\sigma$. The claim now follows from Proposition 3.4.3.

The classical quadrangles

3.4.7 Coordinatization

We now consider a general classical quadrangle. We use the notation of Subsection 2.3.1 on page 53. So let V be a right vector space over some skew field $\mathbb{K}$, and let $g : V \times V \to \mathbb{K}$ be a $(\sigma, 1)$-linear form for some anti-automorphism σ of $\mathbb{K}$ whose square is the identity. Put again

$$\begin{cases} \mathbb{K}_\sigma &= \{t^\sigma - t : t \in \mathbb{K}\}, \\ q &: \quad V \to \mathbb{K}/\mathbb{K}_\sigma : x \mapsto g(x, x) + \mathbb{K}_\sigma, \\ f &: \quad V \times V \to \mathbb{K} : (x, y) \mapsto g(x, y) + g(y, x)^\sigma \end{cases}$$

and suppose that q is non-degenerate and has Witt index 2; cf. Subsection 2.3.1. We know that we can write V as

$$V = e_{-2}\mathbb{K} \bigoplus e_{-1}\mathbb{K} \bigoplus V_0 \bigoplus e_1\mathbb{K} \bigoplus e_2\mathbb{K},$$

such that

$$q(x_{-2}, x_{-1}, x_0, x_1, x_2) = x_{-2}^\sigma x_2 + x_{-1}^\sigma x_1 + q_0(x_0),$$

with $x_i \in e_i\mathbb{K}$, $i = -2, -1, 1, 2$ and $x_0 \in V_0$, and where q_0 is a non-degenerate anisotropic σ-quadratic form (so $q_0^{-1}(0) = 0 \in V_0$).

Let $R_1 = \mathbb{K}$ and put $R_2 = \{(k_0, k_1) \in V_0 \times \mathbb{K} : k_1 \in -q_0(k_0)\}$. We define an addition in R_2, and a scalar multiplication as follows. For $(k_0, k_1), (l_0, l_1) \in R_2$ and $a \in \mathbb{K}$, we put:

$$\begin{aligned} (k_0, k_1) \oplus (l_0, l_1) &= (k_0 + l_0, k_1 + l_1 - f(k_0, l_0)), \\ a \otimes (k_0, k_1) &= (k_0 a, a^\sigma k_1 a). \end{aligned}$$

It is straightforward to check that all these operations are well defined. Also $R_2, \oplus$ is a group, not necessarily commutative, and the inverse of an element (k_0, k_1) is $\ominus(k_0, k_1) := (-k_0, -k_1 - f(k_0, k_0))$. We denote that group briefly by $\widehat{X}$. Now we coordinatize $\mathsf{Q}(V, q)$ as follows. Let us represent by a 5-tuple $(x_{-2}, x_{-1}, x_0, x_1, x_2) \in \mathbb{K} \times \mathbb{K} \times V_0 \times \mathbb{K} \times \mathbb{K}$ (defined up to a right multiple $k \in \mathbb{K}$) the point of the projective space $\mathbf{PG}(V)$ represented by the vector $e_{-2}x_{-2} + e_{-1}x_{-1} + x_0 + e_1x_1 + e_2x_2$. Then we put

$$\begin{aligned} (1, 0, 0, 0, 0) &\mapsto (\infty), \\ (a, 0, 0, 1, 0) &\mapsto (a), \\ (b, 1, 0, 0, 0) &\mapsto (0, b), \\ (0, 0, 0, 0, 1) &\mapsto (0, 0, 0), \end{aligned}$$

POINTS	
Coordinates in $Q(V, q)$	Points in $\mathbf{PG}(V)$
(∞)	$(1, 0, 0, 0, 0)$
(a)	$(a, 0, 0, 1, 0)$
$((k_0, k_1), b)$	$(-b, 1, k_0, k_1, 0)$
$(a, (l_0, l_1), a')$	$(l_1 + aa'^\sigma, -a^\sigma, l_0, a'^\sigma, 1)$

LINES	
Coordinates in $Q(V, q)$	Lines in $\mathbf{PG}(V)$
$[\infty]$	$\langle (1, 0, 0, 0, 0), (0, 0, 0, 1, 0) \rangle$
$[(k_0, k_1)]$	$\langle (1, 0, 0, 0, 0), (0, 1, k_0, k_1, 0) \rangle$
$[a, (l_0, l_1)]$	$\langle (a, 0, 0, 1, 0), (l_1, -a^\sigma, l_0, 0, 1) \rangle$
$[(k_0, k_1), b, (k'_0, k'_1)]$	$\langle (-b, 1, k_0, k_1, 0), (k'_1, 0, k'_0, b^\sigma - f(k_0, k'_0), 1) \rangle$

Table 3.2. Coordinatization of $Q(V, q)$.

for $a, b \in \mathbb{K}$, and, with our convention above of writing lines in a projective space,

$$\langle (1, 0, 0, 0, 0), (0, 1, k_0, k_1, 0) \rangle \mapsto [(k_0, k_1)]$$

(the lines $[\infty]$, $[0, 0]$ and $[0, 0, 0]$ follow from the labelling of the points of the hat-rack above). Note that we are not able to define a normalized quadratic quaternary ring in this general description because this description also gives the weak quadrangles obtained from a ruled quadric in projective 3-space. If we assume that the dimension of V_1 is at least 1, then we may normalize by selecting the element $1 \in \mathbb{K}$ as the unit element of R_1, and, as unit element of R_2, we may take any pair (x_0, x_1) with $x_1 \neq 0$ (this choice can always be made since q_0 is anisotropic). It is not always possible to choose $x_1 = 1$, therefore we will not consider this and we will normalize the quadratic quaternary ring only "half-way", i.e., we choose the coordinates of the lines through (0) such that $[(k_0, k_1)]$ and $[0, (k_0, k_1)]$ have the same projection onto $(1, 0, 0)$.

This completely defines the coordinatization and we write in Table 3.2 the coordinates of all points and lines.

The corresponding quadratic quaternary ring $(R_1, R_2, \Psi_1, \Psi_2)$ is as follows:

$$\begin{cases} \Psi_1((k_0, k_1), a, (l_0, l_1), a') &= (l_0, l_1) \oplus (a^\sigma \otimes (k_0, k_1)) \oplus (0, aa'^\sigma - a'a^\sigma), \\ \Psi_2((k_0, k_1), a, (l_0, l_1), a') &= a' - ak_1 + f(l_0, k_0). \end{cases}$$

In the calculations, one first finds

$$\Psi_1((k_0, k_1), a, (l_0, l_1), a') = \\ (l_0 + k_0 a^\sigma, l_1 - ak_1^\sigma a^\sigma - f(l_0 + k_0 a^\sigma, k_0) a^\sigma + aa'^\sigma - a'a^\sigma),$$

which implies, by the half-way normalization, putting $a' = l_1 = 0$, $l_0 = 0$ (zero vector) and $a = 1$, that

$$k_1 + k_1^\sigma + f(k_0, k_0) = 0,$$

for all $(k_0, k_1) \in R_2$. This can also be proved directly, but the computations boil
down to the same thing (BRUHAT & TITS [1972],(10.1.1)(5)).

We now have:

3.4.8 Proposition.

(i) *Every point of the classical quadrangle* $\mathsf{Q}(V, q)$ *is regular* **if and only if** $f|(V_0 \times V_0) \equiv 0$ *or* $V_0 = \{0\}$.

(ii) *Every line of the classical quadrangle* $\mathsf{Q}(V, q)$ *is regular* **if and only if** $\sigma = 1$ *(i.e.,* **if and only if** *the quadrangle is an orthogonal one).*

(iii) **If** *all points and lines of the classical quadrangle* $\mathsf{Q}(V, q)$ *are regular,* **then** *the characteristic of* $\mathbb{K}$ *is equal to* 2 *and* $\mathsf{Q}(V, q)$ *is isomorphic to the mixed quadrangle* $\mathsf{Q}(\mathbb{K}, \mathbb{K}'; \mathbb{K}, L')$, *where* L' *is the vector space over* $\mathbb{K}^2$ *consisting of all elements* $q_0(v_0)$, $v_0 \in V_0$ *(and where we scale* q_0 *such that* $1 \in L'$), *and* $\mathbb{K}'$ *is the subfield of* $\mathbb{K}$ *generated by* L' *(as a ring).*

Proof. Part (ii) follow immediately from Theorem 3.3.2 and the fact that (∞) was basically chosen arbitrarily by Proposition 2.3.4. Part (i) would follow similarly if we knew that $f|(V_0 \times V_0)$ being identical zero is independent of the coordinates in $\mathbf{PG}(V)$; or if we knew that the automorphism group of $\mathsf{Q}(V, q)$ acts transitively on the point set. But the latter follows independently from Theorem 4.5.2 (page 144), whence (i). We now show (iii).

By parts (i) and (ii) we know that $\sigma = 1$ and that $f|(V_0 \times V_0) \equiv 0$ or $V_0 = 0$. If $V_0 = 0$, then $\mathsf{Q}(V, q)$ is a non-thick weak quadrangle by Corollary 2.3.6. Hence we may assume that f, restricted to $V_0 \times V_0$, is identically 0. But we know that $2k_1 + f(k_0, k_0) = 0$, for all $(k_0, k_1) \in R_2$. Hence the characteristic of $\mathbb{K}$ must be equal to 2. Also we have, for $v, w \in V_0$, that $q_0(v + w) = q_0(v) + q_0(w)$ and $q_0(vk) = k^2 q_0(v)$. If we rescale q_0 such that $1 \in q_0(V_0)$, then this implies that $L' := q_0(V_0)$ is a vector space over $\mathbb{K}^2$ and that it generates some subfield $\mathbb{K}'$ as a ring. Also, every image $q_0(v)$ completely determines v (since q_0 is anisotropic). Hence we may identify R_2 with L' and the simplified quaternary operations show the result.   □

The previous result allows one sometimes to conclude that two given classical quadrangles (or duals) are not isomorphic. The next result shows an opposite example. It is a special, sporadic and important case of an isomorphism between an orthogonal quadrangle and the dual of a Hermitian one. Note that any anisotropic quadratic form in two variables over a field $\mathbb{K}'$ defines a Galois extension $\mathbb{K}$ of $\mathbb{K}'$ by putting one variable equal to 1 (leaving a quadratic irreducible polynomial), unless the characteristic of $\mathbb{K}'$ is equal to 2 and the quadratic form $q_0(x, y) = Ax^2 + Bxy + Cy^2$ has $B = 0$.

3.4.9 Proposition. *Let* $\mathbb{L}$ *be a quadratic Galois extension of the field* $\mathbb{K}$ *and let* σ *be the non-trivial element of the Galois group* $\mathrm{Gal}(\mathbb{L}/\mathbb{K})$. *The Hermitian quadrangle* $\mathsf{H}(3, \mathbb{L}, q, \sigma)$ *over the field* $\mathbb{L}$, *with* $V_0 = \{0\}$ *and corresponding (anti-) automorphism* σ, *is isomorphic to the dual of the orthogonal* D_3-*quadrangle* $\mathsf{Q}(5, \mathbb{K}, q')$ *over the field* $\mathbb{K}$, *where the dimension of* V_0' *equals* 2 *and where the quadratic form* q_0' *in two variables defines the Galois extension* $\mathbb{L}$ *of* $\mathbb{K}$.

Proof. We have to distinguish between characteristic 2 and the rest. The case of characteristic 2 is actually easier, so we will restrict ourselves to the rest. Indeed: this case requires a recoordinatization at the end, whereas in characteristic 2 this is not necessary. But the rest of the proof is quite the same.

So we assume that the characteristic of the fields $\mathbb{L}$ and $\mathbb{K}$ is different from 2. Let $\mathbb{L} = \mathbb{K}(\alpha)$, where $\alpha^2 = u \in \mathbb{K}$. Hence we can write every element k of $\mathbb{L}$ as a pair (k_0, k_1) with $k = k_0 + \alpha k_1$. It follows that $\sigma : (k_0, k_1) \mapsto (k_0, -k_1)$. The quadrangle $\mathsf{Q}(5, \mathbb{K}, q')$ (we add to the standard notation a prime), where $q_0' : (k_0, k_1) \mapsto k_0^2 - u k_1^2$, has the following coordinatizing operations (see above, but we have identified R_2' with V_0', after all, $-q_0'(v)$ is well defined in $\mathbb{K}$ since $\mathbb{K}_{\sigma'} = \{0\}$ because $\sigma' = 1$):

$$\begin{cases} \Psi_1'((k_0, k_1), a, (l_0, l_1), a') &= (l_0, l_1) + a(k_0, k_1), \\ \Psi_2'((k_0, k_1), a, (l_0, l_1), a') &= a' + a(k_0^2 - u k_1^2) + 2l_0 k_0 - 2u l_1 k_1, \end{cases}$$

where $(k_0, k_1), (l_0, l_1) \in V_0'$. Indeed, since $\sigma' = 1$, f is completely determined by q and one calculates that $f((k_0, k_1), (l_0, l_1)) = 2k_0 l_0 - 2u k_1 l_1$, for all (k_0, k_1) and (l_0, l_1) in V_0'.

In the Hermitian quadrangle $\mathsf{H}(3, \mathbb{L}, q, \sigma)$, with $q_0 = 0$, we have $V_0 = 0$ and hence R_2 can be identified with $\mathbb{L}_\sigma$. It is readily seen that $\mathbb{L}_\sigma = \alpha \mathbb{K}$. So we redefine R_2 as $\mathbb{K}$ where the new coordinate is k if the old one was αk. After a short calculation, one finds (remembering $f|V_0 \equiv 0$ and writing every element a of $\mathbb{L}$ as $a = (a_0, a_1)$, where $a_0, a_1 \in \mathbb{K}$ and $a = a_0 + \alpha a_1$):

$$\begin{cases} \alpha \Psi_1(k, (a_0, a_1), l, (a_0', a_1')) &= \alpha l + (a_0^2 - u a_1^2)\alpha k + 2\alpha a_1 a_0' - 2\alpha a_1' a_0, \\ \Psi_2(k, (a_0, a_1), l, (a_0', a_1')) &= (a_0', a_1') - (a_0, a_1)\alpha k. \end{cases}$$

If we recoordinatize replacing a_0 by a_1, ua_1 by a_0, ul by l and uk' by k', leaving the other coordinates unchanged, then we obtain the dual of $(R_1', R_2'; \Psi_1', \Psi_2')$ above. $\square$

An alternative proof consists in applying the Klein correspondence. One writes down the conditions on the Plücker coordinates for a line in $\mathbf{PG}(3, \mathbb{L})$ to belong to $\mathsf{H}(3, \mathbb{L}, q, \sigma)$; the image under the Klein correspondence is then seen to define precisely the orthogonal quadrangle $\mathsf{Q}(5, \mathbb{K}, q')$.

3.4.10 Remark. Noting that for a commutative field $\mathbb{L}$, the fixed point set of an involution is *always* a field $\mathbb{K}$ such that $\mathbb{L}$ is a quadratic (Galois) extension of $\mathbb{K}$, we see that every Hermitian quadrangle $\mathsf{H}(3, \mathbb{L}, q, \sigma)$, for $\mathbb{L}$ commutative, is anti-isomorphic to an orthogonal quadrangle $\mathsf{Q}(5, \mathbb{K}, q')$. We now show conversely that an orthogonal quadrangle $\mathsf{Q}(5, \mathbb{K}, q')$ is anti-isomorphic to a Hermitian quadrangle $\mathsf{H}(3, \mathbb{L}, q, \sigma)$ for some quadratic extension $\mathbb{L}$ of $\mathbb{K}$ **if and only if** $\mathsf{Q}(5, \mathbb{K}, q')$ is not a mixed quadrangle (or, equivalently, $\mathsf{Q}(5, \mathbb{K}, q')$ does not contain regular points). Indeed, by Proposition 3.4.8, the latter is equivalent to $f'|(V_0' \times V_0') \not\equiv 0$ (with obvious notation). Hence for the restriction g_0' of the corresponding 1-quadratic form to $V_0' \times V_0'$ we have $g_0'((x, y), (x', y')) \neq g_0'((x', y'), (x, y))$. This means that we can write

$$g_0'((x, y), (x', y')) = Axx' + Bxy' + B'x'y + Cyy',$$

with $B \neq B'$. So $q_0' = q'|(V_0 \times V_0)$ can be written as $q_0'((x, y)) = Ax^2 + (B + B')xy + Cy^2$, with $B + B' \neq 0$. We can now put $\mathbb{L}$ equal to the Galois extension of $\mathbb{K}$ defined by the irreducible quadric polynomial $Ax^2 + (B + B')x + C$. The result now follows from the previous proposition.

Similarly, one has the following anti-isomorphism (the proof is completely similar to the one above, therefore we omit it). Note that every quaternion skew field has a unique involution σ' for which the set of fixed points is precisely the centre, and such that both $xx^{\sigma'}$ and $x + x^{\sigma'}$ belong to the centre for every element x. This involution is usually called the **standard involution**.

3.4.11 Proposition. *Let $\mathbb{L}$ be a skew field which is a quaternion algebra over its centre $\mathbb{K}$; in particular, $\mathbb{L}$ is four-dimensional over $\mathbb{K}$. Let σ be a "skew conjugation", i.e., an involutory anti-automorphism of $\mathbb{L}$ whose set of fixed points contains $\mathbb{K}$ and such that $\mathbb{K}_\sigma$ is one-dimensional over $\mathbb{K}$. The Hermitian quadrangle $\mathsf{H}(3, \mathbb{L}, q, \sigma)$ over the skew field $\mathbb{L}$, with $V_0 = \{0\}$ and corresponding anti-automorphism σ, is dual to the orthogonal D_4-quadrangle $\mathsf{Q}(7, \mathbb{K}, q')$ over the field $\mathbb{K}$, where the dimension of V_0' equals 4 and where the quadratic form q_0' in four variables describes the norm form $z \mapsto zz^{\sigma'} \in \mathbb{K}$ in $\mathbb{L}$ with respect to the standard involution σ'.* $\quad\square$

It is not a surprise that the standard involution σ' shows up in the previous statement, because every involution of $\mathbb{L}$ is of the form $x \mapsto ex^{\sigma'}e^{-1}$, for some $e \in \mathbb{L}$. If $\mathbb{L}_\sigma$ is one-dimensional over $\mathbb{K}$, then it is easy to check that every element $e \in \mathbb{L}_\sigma$ does the job. So let $e = t^\sigma - t \neq 0$, for some $t \in \mathbb{L}$; then any "punctured" pencil in $\mathsf{H}(3, \mathbb{L}, q, \sigma)$ is parametrized by $R_2 = e\mathbb{K}$ ("punctured" here just means "with one element removed"). So we have to put R_1' (which parametrizes any "punctured" point row of $\mathsf{Q}(7, \mathbb{K}, q')$) equal to $R_2 e^{-1}$. In the expression of Ψ_1 (which becomes Φ_2' after multiplying on the right with e^{-1}), we have $a^\sigma \otimes (0, k_1) = (0, ak_1a^\sigma)$. But $k_1 = ek$ with $k \in \mathbb{K}$, hence in Φ_2', this expression becomes (deleting the first zero) $k(aea^\sigma e^{-1}) = kaa^{\sigma'}$, and the standard involution shows up.

As an important example, consider the quaternion division algebra $\mathbb{H}$ over the reals $\mathbb{R}$. Here, an involution σ as above is given by

$$a + ib + jc + kd \mapsto a - ib + jc + kd,$$

where $a, b, c, d \in \mathbb{R}$, and $\mathbb{H} = \mathbb{R} + i\mathbb{R} + j\mathbb{R} + k\mathbb{R}$. The space $\mathbb{H}_\sigma$ is equal to $i\mathbb{R}$ and we can take $e = i$, e as above. Clearly $x \mapsto -ix^\sigma i$ is the standard involution in $\mathbb{H}$.

3.4.12 Remark. Let $\mathbb{L}$ be a quaternion skew field with skew conjugation σ and standard conjugation σ'. It is clear from Table 3.2 that the points of $\mathsf{H}(3, \mathbb{L}, \sigma)$ are precisely the points of $\mathbf{PG}(3, \mathbb{L})$ satisfying the equation

$$X_{-2}^\sigma X_2 + X_2^\sigma X_{-2} + X_{-1}^\sigma X_1 + X_1^\sigma X_{-1} = 0.$$

This is in fact true for an arbitrary Hermitian quadrangle $\mathsf{H}(3, \mathbb{L}, \sigma)$. By giving a point $(x_{-2}, x_{-1}, x_1, x_2)$ the new coordinates $(x_{-2}, x_{-1}, kx_1, k_2)$, with $k \in \mathbb{L}_\sigma$, we obtain the new equation

$$X_{-2}^{\sigma'} X_2 - X_2^{\sigma'} X_{-2} + X_{-1}^{\sigma'} X_1 - X_1^{\sigma'} X_{-1} = 0.$$

Since in the commutative case, the latter is the standard form of a Hermitian variety in $\mathbf{PG}(3, \mathbb{L})$ with associated involution σ', it is reasonable to call the quadrangle $\mathsf{H}(3, \mathbb{L}, \sigma)$ the **standard quaternion quadrangle (over** $\mathbb{L}$) despite the fact that it is defined using the non-standard involution σ. Hence the standard quaternion quadrangle over $\mathbb{H}$ is the real quaternion quadrangle $\mathsf{H}(3, \mathbb{H}, \mathbb{R})$. We may formulate the previous proposition as: *every standard Hermitian quadrangle is anti-isomorphic to an orthogonal quadrangle.*

Finally, we show that the symplectic quadrangle over $\mathbb{K}$ is isomorphic to the dual of $\mathsf{Q}(4, \mathbb{K})$, and that these quadrangles are self-dual **if and only if** $\mathbb{K}$ is perfect of characteristic 2. Note that we have already shown part of the next proposition in Subsection 2.3.18 (page 64).

3.4.13 Proposition. *The orthogonal quadrangle* $\mathsf{Q}(4, \mathbb{K})$ *is isomorphic to the dual of* $\mathsf{W}(\mathbb{K})$. *Also,* $\mathsf{W}(\mathbb{K})$ *is self-dual* **if and only if** $\mathbb{K}$ *has characteristic 2 and is perfect.*

Proof. It is easily seen that the 1-quadratic form q_0 corresponding to $\mathsf{Q}(4, \mathbb{K})$ (standard notation) is proportional to squaring. Hence by rescaling the other base vectors, we may represent q as

$$q : \mathbb{K} \times \mathbb{K} \times \mathbb{K} \times \mathbb{K} \times \mathbb{K} \to \mathbb{K} : (x_{-2}, x_{-1}, x_0, x_1, x_2) \mapsto x_{-2}x_2 + x_{-1}x_1 + x_0^2.$$

Furthermore, we can identify $(k_0, k_1) \in R_2$ with k_0 since $k_1 = -q_0(k_1) = -k_0^2$. Also, one computes easily that $f(l_0, k_0) = 2l_0 k_0$. The coordinatizing operations are now as follows:

$$\begin{cases} \Psi_1(k, a, l, a') &= l + ak, \\ \Psi_2(k, a, l, a') &= a' + ak^2 + 2lk, \end{cases}$$

which are exactly the operations of the dual normalized quadratic quaternary ring of the symplectic quadrangle over $\mathbb{K}$. This shows the first assertion.

For the second assertion, we note that, if $W(\mathbb{K})$ is self-dual, then every point and every line is regular, hence $\mathbb{K}$ has characteristic 2 by Proposition 3.4.8(iii). Since now $W(\mathbb{K})$ can be written as $Q(\mathbb{K}, \mathbb{K}; \mathbb{K}, \mathbb{K})$, and since by Proposition 3.4.4 its dual is isomorphic to $Q(\mathbb{K}, \mathbb{K}^2; \mathbb{K}, \mathbb{K}^2)$, Proposition 3.4.3 implies that there is an isomorphism $\theta : \mathbb{K} \to \mathbb{K}$ mapping $\mathbb{K}$ onto $\mathbb{K}^2$. Therefore $\mathbb{K}$ is perfect. Conversely, if $\mathbb{K}$ is perfect and of characteristic 2, then $W(\mathbb{K}) = Q(\mathbb{K}, \mathbb{K}; \mathbb{K}, \mathbb{K})$ is isomorphic to $W(\mathbb{K})^D = Q(\mathbb{K}, \mathbb{K}^2; \mathbb{K}, \mathbb{K}^2)$ by considering the identity map $\mathbb{K} \to \mathbb{K}$. $\square$

The cases considered in the previous three theorems are the only anti-isomorphisms that exist between orthogonal and Hermitian quadrangles, or between symplectic and orthogonal quadrangles. This follows from the results in BOREL & TITS [1973]. As for other (anti-)isomophisms, there are other possibilities for describing the symplectic quadrangle over a non-perfect field of characteristic 2 as an orthogonal quadrangle, depending on the dimension of $\mathbb{K}$ over $\mathbb{K}^2$. If this dimension is $d < \infty$, then one has for instance an orthogonal quadrangle in $\mathbf{PG}(d + 3, \mathbb{K})$ isomorphic to $W(\mathbb{K})$.

3.5   The classical hexagons

Although the split Cayley hexagons and the twisted triality hexagons could be coordinatized at the same time, it is convenient to treat the split Cayley hexagon separately for similar reasons to those which led us to treat the symplectic quadrangle separately. After all, the split Cayley hexagon has a representation in projective 6-space unlike the twisted triality hexagons.

3.5.1 The split Cayley hexagons

Let $\mathbb{K}$ be any field and let $H(\mathbb{K})$ be the split Cayley hexagon over $\mathbb{K}$ (see Definitions 2.4.9 on page 72). We use the explicit description of $H(\mathbb{K})$ in six-dimensional projective space given in Subsection 2.4.13 on page 73. So let $H(\mathbb{K})$ be embedded in the quadric $Q(6, \mathbb{K})$ with equation

$$X_0 X_4 + X_1 X_5 + X_2 X_6 = X_3^2$$

in $\mathbf{PG}(6, \mathbb{K})$. Recall that the points of $H(\mathbb{K})$ are all points of $Q(6, \mathbb{K})$ and that the lines of $H(\mathbb{K})$ are those lines on $Q(6, \mathbb{K})$ whose Grassmannian coordinates satisfy

$$p_{12} = p_{34}, \qquad p_{20} = p_{35}, \qquad p_{01} = p_{36},$$
$$p_{03} = p_{56}, \qquad p_{13} = p_{64}, \qquad p_{23} = p_{45}.$$

When introducing coordinates for $H(\mathbb{K})$, there will be no possible confusion for points in $\mathbf{PG}(6, q)$ have seven coordinates there. A line through the points p and q is denoted by $\langle p, q \rangle$, as above.

POINTS	
Coordinates in $\mathsf{H}(\mathbb{K})$	Coordinates in $\mathbf{PG}(6, \mathbb{K})$
(∞)	$(1, 0, 0, 0, 0, 0, 0)$
(a)	$(a, 0, 0, 0, 0, 0, 1)$
(k, b)	$(b, 0, 0, 0, 0, 1, -k)$
(a, l, a')	$(-l - aa', 1, 0, -a, 0, a^2, -a')$
(k, b, k', b')	$(k' + bb', k, 1, b, 0, b', b^2 - b'k)$
(a, l, a', l', a'')	$(-al' + a'^2 + a''l + aa'a'', -a'', -a, -a' + aa'',$
	$1, l + 2aa' - a^2 a'', -l' + a'a'')$

LINES	
Coordinates in $\mathsf{H}(\mathbb{K})$	Coordinates in $\mathbf{PG}(6, \mathbb{K})$
$[\infty]$	$\langle (1, 0, 0, 0, 0, 0, 0), (0, 0, 0, 0, 0, 0, 1) \rangle$
$[k]$	$\langle (1, 0, 0, 0, 0, 0, 0), (0, 0, 0, 0, 0, 1, -k) \rangle$
$[a, l]$	$\langle (a, 0, 0, 0, 0, 0, 1), (-l, 1, 0, -a, 0, a^2, 0) \rangle$
$[k, b, k']$	$\langle (b, 0, 0, 0, 0, 1, -k), (k', k, 1, b, 0, 0, b^2) \rangle$
$[a, l, a', l']$	$\langle (-l - aa', 1, 0, -a, 0, a^2, -a'),$
	$(-al' + a'^2, 0, -a, -a', 1, l + 2aa', -l') \rangle$
$[k, b, k', b', k'']$	$\langle (k' + bb', k, 1, b, 0, b', b^2 - b'k),$
	$(b'^2 + k''b, -b, 0, -b', 1, k'', -kk'' - k' - 2bb') \rangle$

Table 3.3. Coordinatization of $\mathsf{H}(\mathbb{K})$.

We put $R_1 = R_2 = \mathbb{K}$ and label the following points as indicated:

$$
\begin{aligned}
(1, 0, 0, 0, 0, 0, 0) &\rightarrow (\infty), \\
(a, 0, 0, 0, 0, 0, 1) &\rightarrow (a), \\
(0, 0, 0, 0, 0, 1, 0) &\rightarrow (0, 0), \\
(0, 1, 0, 0, 0, 0, 0) &\rightarrow (0, 0, 0), \\
(0, 0, 1, 0, 0, 0, 0) &\rightarrow (0, 0, 0, 0), \\
(0, 0, 0, 0, 1, 0, 0) &\rightarrow (0, 0, 0, 0, 0);
\end{aligned}
$$

and the lines:

$$
\begin{aligned}
X_1 = X_2 = X_3 = X_4 = X_5 = 0 &\rightarrow [\infty], \\
X_1 = X_2 = X_3 = X_4 = X_6 + kX_5 = 0 &\rightarrow [k], \\
X_0 = X_2 = X_3 = X_4 = X_5 = 0 &\rightarrow [0, 0], \\
X_1 = X_3 = X_4 = X_6 = X_0 = 0 &\rightarrow [0, 0, 0], \\
X_0 = X_2 = X_3 = X_5 = X_6 = 0 &\rightarrow [0, 0, 0, 0], \\
X_0 = X_1 = X_3 = X_5 = X_6 = 0 &\rightarrow [0, 0, 0, 0, 0].
\end{aligned}
$$

This determines the coordinates of every point and line in $\mathsf{H}(\mathbb{K})$. The results of these calculations are listed in Table 3.3. The corresponding symmetrically normalized (ordinary and dual) hexagonal sexternary rings are as follows.

$$\begin{cases} \Psi_1(k,a,l,a',l',a'') = a^3 k + l - 3a''a^2 + 3aa', \\ \Psi_2(k,a,l,a',l',a'') = a^2 k + a' - 2aa'', \\ \Psi_3(k,a,l,a',l',a'') = a^3 k^2 + l' - kl - 3a^2 a'' k - 3a'a'' + 3aa''^2, \\ \Psi_4(k,a,l,a',l',a'') = -ak + a'', \end{cases}$$

and

$$\begin{cases} \Phi_1(a,k,b,k',b',k'') = ak + b, \\ \Phi_2(a,k,b,k',b',k'') = a^3 k^2 + k' + kk'' + 3a^2 kb + 3bb' + 3ab^2, \\ \Phi_3(a,k,b,k',b',k'') = a^2 k + b' + 2ab, \\ \Phi_4(a,k,b,k',b',k'') = -a^3 k + k'' - 3ba^2 - 3ab'. \end{cases}$$

The Pappian projective planes, the symplectic quadrangles and the split Cayley hexagons share a number of interesting properties and characterizations. One of them is that they are defined over any field, in particular over any algebraically closed field. So in a certain sense they are the prototypes of the polygons, the universal examples (not in a strictly mathematical sense). They are also related with each other via *derivation* (the *perp-geometries* and *span-geometries*); see the next theorem. Therefore, Linus Kramer suggested that I call the members of these three classes of polygons the **Pappian polygons**, and that is exactly what I will do.

Now we prove algebraically Theorem 2.3.19 (see page 65) and Theorem 2.4.16 (on page 75) and be more precise in view of the remarks just made.

3.5.2 Theorem. *Let* Γ *be a Pappian generalized n-gon, $n \in \{3,4,6\}$, defined over the field* $\mathbb{K}$ *(so either* $\Gamma = \mathbf{PG}(2,\mathbb{K})$ *or* $\Gamma = \mathsf{W}(\mathbb{K})$ *or* $\Gamma = \mathsf{H}(\mathbb{K})$*). All points of* Γ *are projective; all corresponding perp-planes are Pappian projective planes isomorphic to* $\mathbf{PG}(2,\mathbb{K})$*. Also, if $n = 6$, then all points are polar and all span-quadrangles are isomorphic to the Pappian quadrangle* $\mathsf{W}(\mathbb{K})$*; if $n = 4$, then all points are projective and the span-planes are isomorphic to the Pappian projective plane* $\mathbf{PG}(2,\mathbb{K})$*. Hence the functor "taking span-geometries in a span-regular point of a generalized polygon" maps* $\mathsf{H}(\mathbb{K})$ *and any of its points to* $\mathsf{W}(\mathbb{K})$*, and* $\mathsf{W}(\mathbb{K})$ *and any of its points to* $\mathbf{PG}(2,\mathbb{K})$*.*

Proof. All these results follow straightforwardly by considering the coordinatizing rings in Theorem 3.3.2 on page 96 and Theorem 3.3.3 on page 97, noting that by Subsection 1.9.6 the perp-plane and span-plane in a projective point of a generalized quadrangle are isomorphic. □

3.5.3 The mixed hexagons

Let $\mathbb{K}$ be a field of characteristic 3 and let $\mathbb{K}'$ be a subfield such that $\mathbb{K}^3 \leq \mathbb{K}' \leq \mathbb{K}$. If we restrict in the coordinatization of $\mathsf{H}(\mathbb{K})$ the set R_2 to $\mathbb{K}'$, then all operations are still well defined, and we obtain a subhexagon $\mathsf{H}(\mathbb{K},\mathbb{K}')$, which we call a **mixed hexagon**. The mixed hexagons were discovered by TITS [1974](10.3.2) in connection with "mixed groups", hence the terminology. The symmetrically normalized (ordinary and dual) hexagonal sexternary rings $(\mathbb{K}, \mathbb{K}', \Psi_1, \ldots, \Psi_4)$ and $(\mathbb{K}, \mathbb{K}', \Phi_1, \ldots, \Phi_4)$ are as follows:

$$\left\{ \begin{array}{l} \Psi_1(k, a, l, a', l', a'') = a^3 k + l, \\ \Psi_2(k, a, l, a', l', a'') = a^2 k + a' + aa'', \\ \Psi_3(k, a, l, a', l', a'') = a^3 k^2 + l' - kl, \\ \Psi_4(k, a, l, a', l', a'') = -ak + a'', \end{array} \right.$$

and

$$\left\{ \begin{array}{l} \Phi_1(a, k, b, k', b', k'') = ak + b, \\ \Phi_2(a, k, b, k', b', k'') = a^3 k^2 + k' + kk'', \\ \Phi_3(a, k, b, k', b', k'') = a^2 k + b' - ab, \\ \Phi_4(a, k, b, k', b', k'') = -a^3 k + k''. \end{array} \right.$$

The following results are similar to their counterparts for mixed quadrangles.

3.5.4 Proposition. *Let $\mathbb{K}_i$ be a field of characteristic 3 and let $\mathbb{K}_i'$ be a subfield such that $\mathbb{K}_i^3 \leq \mathbb{K}_i' \leq \mathbb{K}_i$, $i \in \{1, 2\}$. Then the mixed hexagons $\mathsf{H}(\mathbb{K}_1, \mathbb{K}_1')$ and $\mathsf{H}(\mathbb{K}_2, \mathbb{K}_2')$ are isomorphic **if and only if** there exists an isomorphism $\theta : \mathbb{K}_1 \to \mathbb{K}_2$ which maps $\mathbb{K}_1'$ to $\mathbb{K}_2'$.*

Proof. Clearly, if such an isomorphism between $\mathbb{K}_1$ and $\mathbb{K}_2$ exists, then the two mixed hexagons are isomorphic.

Now suppose that the two mixed hexagons $\mathsf{H}(\mathbb{K}_1, \mathbb{K}_1')$ and $\mathsf{H}(\mathbb{K}_2, \mathbb{K}_2')$ are isomorphic. As in the proof of Proposition 3.4.3, we may assume that the isomorphism, say ϕ, maps (∞) to (∞), $(0, \ldots, 0)$ to $(0, \ldots, 0)$, etc. Also we will show independently in the next chapter (see Proposition 4.5.11 on page 150) that the automorphism group of any mixed hexagon stabilizing an apartment Σ acts transitively on the pairs (p, L), where p is a point on a line M of Σ, but does not belong to Σ, and L is a line, not in Σ, through a (fixed) point M on Σ. Hence we may assume that the isomorphism ϕ maps the point (1) to (1) and the line $[1]$ to $[1]$. For all $x \in \mathbb{K}_1$, we denote by x^θ the coordinate of $(x)^\phi$; for all $k \in \mathbb{K}_1'$, we denote by $k^{\theta'}$ the coordinate of the line $[k]^\phi$. Clearly $0^\theta = 0$, $1^\theta = 1$, $0^{\theta'} = 0$ and $1^{\theta'} = 1$. Since the hexagonal sexternary rings corresponding to $\mathsf{H}(\mathbb{K}_i, \mathbb{K}_i')$, $i \in \{1, 2\}$ are normalized, we derive from the map Φ_1 of the dual hexagonal sexternary ring that

$$a^\theta k^{\theta'} + b^\theta = (ak + b)^\theta,$$

for all $a, b \in \mathbb{K}_1$ and $k \in \mathbb{K}_1'$. Putting $k = 1$, we see that θ is additive. Putting $b = 0$ and $a = 1$, we see that θ and θ' are identical on $\mathbb{K}_1'$. Putting $b = 0$, we see that θ' is an isomorphism from $\mathbb{K}_1'$ to $\mathbb{K}_2'$. From Φ_2, we similarly deduce that $(a^\theta)^3 = (a^3)^\theta = (a^3)^{\theta'}$. Since $\mathbb{K}_1^3 \subseteq \mathbb{K}_1'$, this implies that θ is an isomorphism from $\mathbb{K}_1$ to $\mathbb{K}_2$.

This completes the proof of the proposition. $\square$

The proof of the next result is completely similar to the proof of Proposition 3.4.4. We leave it to the reader as an easy exercise.

3.5.5 Proposition. *The dual of the mixed hexagon* $\mathsf{H}(\mathbb{K}, \mathbb{K}')$ *is the mixed hexagon* $\mathsf{H}(\mathbb{K}', \mathbb{K}^3)$. $\qquad\qquad\qquad\qquad\qquad\qquad\qquad\qquad\qquad\qquad\qquad\qquad\square$

And also the following corollary has a similar proof to that of its counterpart for mixed quadrangles.

3.5.6 Corollary. *All points and lines of a mixed hexagon are span-regular.* $\qquad\square$

Note also:

3.5.7 Corollary. *All points of a split Cayley hexagon* $\mathsf{H}(\mathbb{K})$ *are regular. All lines are regular* **if and only if** $\mathsf{H}(\mathbb{K})$ *is a mixed hexagon* **if and only if** $\mathbb{K}$ *has characteristic* 3. *The split Cayley hexagon* $\mathsf{H}(\mathbb{K})$ *is self-dual* **if and only if** $\mathrm{char}\,\mathbb{K} = 3$ *and* $\mathbb{K}$ *is perfect.*

Proof. The first two assertions follow immediately from the transitivity of the automorphism group of $\mathsf{H}(\mathbb{K})$ (see Theorem 4.5.6 on page 147), the shape of the corresponding coordinatizing rings and Theorem 3.3.3. The last assertion follows now from Proposition 3.5.4 above. $\qquad\qquad\qquad\qquad\qquad\qquad\qquad\qquad\qquad\square$

3.5.8 The twisted triality hexagons $\mathsf{T}(\mathbb{L}, \mathbb{K}, \sigma)$

In this subsection, we consider the twisted triality hexagon $\mathsf{T}(\mathbb{L}, \mathbb{K}, \sigma)$, where σ is a field automorphism of order 3 of the Galois extension $\mathbb{L}$ of degree 3 of the (commutative) field $\mathbb{K}$, and which fixes $\mathbb{K}$ pointwise; see Subsection 2.4.9 on page 72. Close relatives are generalized hexagons related to the groups of type ${}^6D_4$ (see below).

The coordinatization we present is deduced from the representation of the twisted triality hexagon as "absolute geometry" of a triality. In DE SMET & VAN MALDEGHEM [1993b] $\mathsf{T}(\mathbb{L}, \mathbb{K}, \sigma)$ is coordinatized using KANTOR's description [1986a] of $\mathsf{T}(\mathbb{L}, \mathbb{K}, \sigma)$ (in the finite case). The result is, however, exactly the same, by the choice of our coordinates. For practical use, Kantor's description is essentially equivalent to the description with coordinates as we have already remarked in Chapter 2.

Let $\mathbb{K}$ be any field and let $\mathbb{L}$ be a Galois extension of $\mathbb{K}$ of degree 3. Let σ be a non-trivial element of order 3 of the Galois group of $\mathbb{L}/\mathbb{K}$. Let $\mathsf{T}(\mathbb{L}, \mathbb{K}, \sigma)$ be the twisted triality hexagon as defined in Subsection 2.4.7 on page 71. This hexagon lies on the quadric $\mathbf{Q}(7, \mathbb{L})$ in $\mathbf{PG}(7, \mathbb{L})$.

We put $R_1 = \mathbb{L}$, $R_2 = \mathbb{K}$ and we label the following points as indicated below (for clarity's sake, we put a semicolon in the middle of the coordinate-tuple of a point in $\mathbf{PG}(7, \mathbb{L})$:

$$
\begin{aligned}
(1,0,0,0;0,0,0,0) &\rightarrow (\infty), \\
(a,0,0,0;0,0,1,0) &\rightarrow (a), \\
(0,0,0,0;0,1,0,0) &\rightarrow (0,0), \\
(0,1,0,0;0,0,0,0) &\rightarrow (0,0,0), \\
(0,0,1,0;0,0,0,0) &\rightarrow (0,0,0,0), \\
(0,0,0,0;1,0,0,0) &\rightarrow (0,0,0,0,0);
\end{aligned}
$$

and the lines:

$$
\begin{aligned}
X_1 = X_2 = X_3 = X_4 = X_5 = X_7 = 0 &\rightarrow [\infty], \\
X_1 = X_2 = X_3 = X_4 = X_7 = X_6 + kX_5 = 0 &\rightarrow [k], \\
X_0 = X_2 = X_3 = X_4 = X_5 = X_7 = 0 &\rightarrow [0,0], \\
X_1 = X_3 = X_4 = X_6 = X_7 = X_0 = 0 &\rightarrow [0,0,0], \\
X_0 = X_2 = X_3 = X_5 = X_6 = X_7 = 0 &\rightarrow [0,0,0,0], \\
X_0 = X_1 = X_3 = X_5 = X_6 = X_7 = 0 &\rightarrow [0,0,0,0,0].
\end{aligned}
$$

This determines, by the general rules, the coordinates of every point and line in $\mathsf{T}(\mathbb{L}, \mathbb{K}, \sigma)$. We briefly indicate how the calculations were carried out. To find a point p on a line L and at distance 4 from a given point x (with $\delta(x, L) = 5$), we find the unique point p on $\mathbf{Q}(7, \mathbb{L})$ collinear (in $\mathbf{Q}(7, \mathbb{L})$) with x. To find a line L through a point p and at distance 4 from a given line M (with $\delta(p, M) = 5$), we proceed as follows. First we find the point x on M at distance 4 from p. Then, with the notation of Subsection 2.4.7 again, the planes $x^{\tau_\sigma} \cap x^{\tau_\sigma^2}$ and $p^{\tau_\sigma} \cap p^{\tau_\sigma^2}$ meet in a unique point y (by part (2) of the proof of Theorem 2.4.4 on page 67) and, by the same reference, y is a point on L.

The results of the calculations are listed in Table 3.4. The corresponding symmetrically normalized (ordinary and dual) hexagonal sexternary rings are as follows:

$$
\left\{
\begin{aligned}
\Psi_1(k,a,l,a',l',a'') &= kN(a) + l - \mathrm{Tr}(a''a^{\sigma + \sigma^2}) + \mathrm{Tr}(a'a), \\
\Psi_2(k,a,l,a',l',a'') &= ka^{\sigma + \sigma^2} + a' - a''^\sigma a^{\sigma^2} - a''^{\sigma^2} a^\sigma, \\
\Psi_3(k,a,l,a',l',a'') &= k^2 N(a) + l' + \mathrm{Tr}(a''^{\sigma + \sigma^2} a) - k\,\mathrm{Tr}(a''a^{\sigma + \sigma^2}) - \\
&\qquad \mathrm{Tr}(a'a'') - kl, \\
\Psi_4(k,a,l,a',l',a'') &= -ka + a'',
\end{aligned}
\right.
$$

and

$$
\left\{
\begin{aligned}
\Phi_1(a,k,b,k',b',k'') &= ka + b, \\
\Phi_2(a,k,b,k',b',k'') &= k^2 N(a) + k' + \mathrm{Tr}(b^{\sigma + \sigma^2} a) + k\,\mathrm{Tr}(ba^{\sigma + \sigma^2}) + \\
&\qquad \mathrm{Tr}(bb') + kk'', \\
\Phi_3(a,k,b,k',b',k'') &= ka^{\sigma + \sigma^2} + b' + a^\sigma b^{\sigma^2} + a^{\sigma^2} b^\sigma, \\
\Phi_4(a,k,b,k',b',k'') &= -kN(a) + k'' - \mathrm{Tr}(ba^{\sigma + \sigma^2}) - \mathrm{Tr}(ab'),
\end{aligned}
\right.
$$

where $\mathrm{Tr}(x) = x + x^\sigma + x^{\sigma^2}$ and $N(x) = x^{1 + \sigma + \sigma^2}$. Restricting the elements a, a', a'', b, b' to $\mathbb{K}$ (noting $\mathrm{Tr}(x) = 3x$ and $N(x) = x^3$ for $x \in \mathbb{K}$), we see that we get exactly the coordinatization of $\mathsf{H}(\mathbb{K})$ above.

POINTS	
Coordinates in $\mathsf{T}(\mathbb{L}, \mathbb{K}, \sigma)$	Coordinates in $\mathbf{PG}(7, \mathbb{L})$
(∞)	$(1, 0, 0, 0; 0, 0, 0, 0)$
(a)	$(a, 0, 0, 0; 0, 0, 1, 0)$
(k, b)	$(b, 0, 0, 0; 0, 1, -k, 0)$
(a, l, a')	$(-l - aa', 1, 0, a^{\sigma}; 0, a^{\sigma+\sigma^2}, -a', -a^{\sigma^2})$
(k, b, k', b')	$(k' + bb', k, 1, -b^{\sigma}; 0, b', b^{\sigma+\sigma^2} - b'k, b^{\sigma^2})$
(a, l, a', l', a'')	$(-al' + a'^{\sigma+\sigma^2} + a''l + aa'a'', -a'', -a, a'^{\sigma^2} - a^{\sigma}a'';$ $\quad 1, l + (aa')^{\sigma} + (aa')^{\sigma^2} - a^{\sigma+\sigma^2}a'', -l' + a'a'', a^{\sigma^2}a'' - a'^{\sigma})$

LINES	
Coordinates in $\mathsf{T}(\mathbb{L}, \mathbb{K}, \sigma)$	Coordinates in $\mathbf{PG}(7, \mathbb{L})$
$[\infty]$	$\langle(1, 0, 0, 0; 0, 0, 0, 0), (0, 0, 0, 0; 0, 0, 1, 0)\rangle$
$[k]$	$\langle(1, 0, 0, 0; 0, 0, 0, 0), (0, 0, 0, 0; 0, 1, -k, 0)\rangle$
$[a, l]$	$\langle(a, 0, 0, 0; 0, 0, 1, 0), (-l, 1, 0, a^{\sigma}; 0, a^{\sigma+\sigma^2}, 0, -a^{\sigma^2})\rangle$
$[k, b, k']$	$\langle(b, 0, 0, 0; 0, 1, -k, 0), (k', k, 1, -b^{\sigma}; 0, 0, b^{\sigma+\sigma^2}, b^{\sigma^2})\rangle$
$[a, l, a', l']$	$\langle(-l - aa', 1, 0, a^{\sigma}; 0, a^{\sigma+\sigma^2}, -a', -a^{\sigma^2}),$ $(a'^{\sigma+\sigma^2} - al', 0, -a, a'^{\sigma^2}; 1, l + (aa')^{\sigma} + (aa')^{\sigma^2}, -l', -a'^{\sigma})\rangle$
$[k, b, k', b', k'']$	$\langle(k' + bb', k, 1, -b^{\sigma}; 0, b', b^{\sigma+\sigma^2} - b'k, b^{\sigma^2}),$ $(b'^{\sigma+\sigma^2} + k''b, -b, 0, b'^{\sigma^2};$ $\quad 1, k'', -kk'' - k' - (bb')^{\sigma} - (bb')^{\sigma^2}, -b'^{\sigma})\rangle$

Table 3.4. Coordinatization of $\mathsf{T}(\mathbb{L}, \mathbb{K}, \sigma)$.

3.5.9 Hexagons of type $^6D_4$

We keep the same notation as in the previous subsection, with the exception that we now suppose that $\mathbb{L}$ is a cubic extension of $\mathbb{K}$ which is *not* a Galois extension, but such that there is a quadratic extension $\mathbb{M}$ of $\mathbb{L}$ which *is* a Galois extension of $\mathbb{K}$ of degree 6. Let σ be an element of order 3 of the Galois group $\mathrm{Gal}(\mathbb{M}/\mathbb{K})$. It is clear that σ fixes an element of $\mathbb{L}$ **if and only if** that element belongs to $\mathbb{K}$. Hence for all $x \in \mathbb{L}$ we have that $\mathrm{N}(x)$ as defined above still belongs to $\mathbb{K}$. Also, $x^{\sigma+\sigma^2} = \frac{\mathrm{N}(x)}{x}$ belongs to $\mathbb{L}$ and hence $x^{\sigma}y^{\sigma^2} + x^{\sigma^2}y^{\sigma} = (x+y)^{\sigma+\sigma^2} - x^{\sigma+\sigma^2} - y^{\sigma+\sigma^2}$ belongs to $\mathbb{L}$, for all $x, y \in \mathbb{L}$. This implies that in this case, all operations Ψ_i and Φ_i, $i = 1, 2, 3, 4$, are still well defined and we obtain $\mathsf{T}(\mathbb{L}, \mathbb{K}, \sigma)$, a **hexagon of type** $^6D_4$. We call this hexagon and also its dual **classical**.

3.5.10 Example. The previous situation occurs when one has a Galois extension $\mathbb{M}$ of degree 6 of a field $\mathbb{K}$ such that the Galois group $\mathrm{Gal}(\mathbb{M}/\mathbb{K})$ is permutation isomorphic to the symmetric group $\mathbf{S}_3$ acting on three letters. Then one can take for $\mathbb{L}$ the fixed field of an involution of $\mathrm{Gal}(\mathbb{M}/\mathbb{K})$. For example, let $\mathbb{K}$ be the field $\mathbb{Q}$ of rational numbers, then one might take for $\mathbb{L}$ the field $\mathbb{Q}(\sqrt[3]{2})$ of rational numbers with the real third root of 2 adjoint. To obtain the field $\mathbb{M}$ (the so-called *Galois closure*), one further adjoins a non-trivial cubic root of unity. Note that $\mathbb{L}$ is a separable extension of $\mathbb{K}$, but it is not a normal extension.

As in Corollary 3.5.7, one can easily show the following result.

3.5.11 Corollary. *All points of a twisted triality hexagon or a hexagon of type* $^6D_4$ *are span-regular. No line of such a hexagon is distance-2-regular.*

3.6   The Ree–Tits octagons

3.6.1   Coordinatization and commutation relations

In order to coordinatize the Ree–Tits octagons we use the description of TITS [1983] via the commutation relations of *root* groups. These commutation relations can be found in Subsection 5.5.20 on page 225, and the reader might first want to go through some parts of Chapter 5. We restrict ourselves here to briefly mentioning how one derives a coordinatization from the commutation relations, as in JOSWIG & VAN MALDEGHEM [1995]. For the definition of root groups, see Subsection 5.2.1, page 174. In fact, this procedure can be carried out for any Moufang polygon, once the commutation relations are available. Since a Moufang n-gon has $n \leq 8$ (see Theorem 5.3.3 on page 179), the explanation below for the Ree–Tits octagons is general enough. We note that it provides an algorithmic way to go from the commutation relations to the polygonal domain (as a test for the calculations by hand, this was done by JOSWIG & VAN MALDEGHEM [1995]).

Let $\mathbb{K}$ be a field of characteristic 2 admitting a Tits endomorphism σ. Recall from Section 2.5 (page 78) that the Ree–Tits generalized octagon $\mathsf{O}(\mathbb{K}, \sigma)$ is unambiguously associated with $\mathbb{K}$ and σ. We choose an apartment Ω in $\mathsf{O}(\mathbb{K}, \sigma)$ and label its elements (∞), $[\infty]$, (0), $[0,0]$, $(0,0,0)$, etc. For each 6-path ϕ in Ω, there is a root group U_ϕ. According to TITS [1983], half of these root groups, say those containing point-elations (see Definitions 4.4.1 on page 140), are parametrized by the elements of the field $\mathbb{K}$ (and the root group is in fact isomorphic to $(\mathbb{K}, +)$); the other half are parametrized by the pairs $(k_0, k_1) \in \mathbb{K} \times \mathbb{K}$ with operation law $(k_0, k_1) \oplus (l_0, l_1) = (k_0 + l_0, k_1 + l_1 + l_0 k_0^\sigma)$. Following TITS [1983] we denote this group by $\mathbb{K}_\sigma^{(2)}$. It is isomorphic to the regular normal subgroup of a point stabilizer in the Suzuki group $\mathbf{Sz}(\mathbb{K}, \sigma)$ (see Subsection 7.6.5 on page 325).

We put $R_1 = \mathbb{K}$ and $R_2 = \mathbb{K}_\sigma^{(2)}$.

Every 6-path in Ω is determined by its middle element v; we denote the 6-path by ϕ_v and the corresponding root group by U_v. The image of the point (0) (line $[0]$) under the action of an element of the root group $U_{[0^3]}$ $(U_{(0^3)})$ parametrized by $a \in \mathbb{K}$ $((k_0, k_1) \in \mathbb{K}_\sigma^{(2)})$ is given the coordinate (a) $([(k_0, k_1)])$. We will, however, abbreviate $(0,0)$ by 0, to be consistent. The element of $U_{[0^3]}$ $(U_{(0^3)})$ mapping (0) to

(a) ([0] to [(k_0, k_1)]) will be denoted by $u_{[0^3]}(a)$ ($u_{(0^3)}(k_0, k_1)$). In the same fashion, we give the coordinates $(0^2, a')$ ($(0^4, a'')$, $(0^6, a''')$) to the image of the point (0^3) ((0^5), (0^7)) under the action of an element of the root group $U_{[0]}$ ($U_{[\infty]}$, $U_{[0^2]}$) parametrized by a' (a'', a'''). Dually, we coordinatize lines. As above, we define $u_{[0]}(a')$, $u_{[\infty]}(a'')$, $u_{[0^2]}(a''')$ and their duals. This is enough to obtain coordinates for all elements of $\mathsf{O}(\mathbb{K}, \sigma)$.

In order to have a complete description of $\mathsf{O}(\mathbb{K}, \sigma)$, we must find the octagonal octanary ring. Let $x = (a, l, a', l', a'', l'', a''')$ and $L = [k, b, k', b', k'', b'', k''']$ be a point and a line, respectively, where $l = (l_0, l_1)$, $l' = (l'_0, l'_1)$, ..., $k = (k_0, k_1)$, etc. We now further abbreviate $u_a = u_{[0^3]}(a)$, $u_l = u_{(0^2)}(l)$, $u_{a'} = u_{[0]}(a')$, etc. Similarly we define (in a dual way) the root elations u_k, u_b, $u_{k'}$, etc. Put

$$u_x = u_{a'''} u_{l''} u_{a''} u_{l'} u_{a'} u_l u_a \quad \text{and} \quad u_L = u_{k'''} u_{b''} u_{k''} u_{b'} u_{k'} u_b u_k.$$

Then it is clear that $x = (0^7)^{u_x}$ and $L = [0^7]^{u_L}$. Also remark that the automorphism u_x does not change the first coordinate of any line. This implies that $x \mathbf{I} L$ if and only if $(0^7) \mathbf{I} L^{u_x^{-1}}$ (the latter must then be $[k, 0^6]$ by the previous remark, hence:) **if and only if** $L^{u_x^{-1}} = [0^7]^{u_k}$ **if and only if** $[0^7]^{u_L u_x^{-1} u_k^{-1}} = [0^7]$. Since $[0^7]^{u_a} = [0^7]$, the latter is equivalent to $[0^7]^{u_a u_L u_x^{-1} u_k^{-1}} = [0^7]$. Dually, $x \mathbf{I} L$ **if and only if** $(0^7)^{u_k u_x u_L^{-1} u_a^{-1}} = (0^7)$, which can be rewritten as $(0^7)^{u_a u_L u_x^{-1} u_k^{-1}} = (0^7)$. Since by Lemma 5.2.4 (page 175) the group generated by all root elations fixing the flag $F = ((\infty), [\infty])$ acts regularly on the set of flags opposite F (recall from Definitions 1.3.1 that opposite flags are flags whose respective points and lines are opposite) we have that $x \mathbf{I} L$ if and only if $u_a u_L = u_k u_x$. Using the commutation relations displayed in Subsection 5.5.20 on page 225, one computes this condition and obtains the octagonal octanary ring.

3.6.2 The octagonal octanary ring

For $k = (k_0, k_1)$, set $\mathrm{Tr}(k) = k_0^{\sigma+1} + k_1$ (the *trace* of k) and set $\mathrm{N}(k) = k_0^{\sigma+2} + k_0 k_1 + k_1^\sigma$ (the *norm* of k). Define a multiplication $a \otimes k = a \otimes (k_0, k_1) = (ak_0, a^{\sigma+1} k_1)$ for $a \in \mathbb{K}$ and $k \in \mathbb{K}_\sigma^{(2)}$. Also write $(k_0, k_1)^\sigma$ for (k_0^σ, k_1^σ). Then we have

$$
\begin{aligned}
\Psi_1(k, a, l, \ldots, a''') &= (l_0, l_1) \oplus a \otimes (k_0, k_1) \oplus (0, al'_0 + a^\sigma l''_0) \\[4pt]
\Psi_2(k, a, l, \ldots, a''') &= a' + a^{\sigma+1} \mathrm{N}(k) + k_0(al'_0 + a^\sigma l''_0 + \mathrm{Tr}(l)) \\
&\quad + a^\sigma(a''' + l_0 k_1) + al''_0 + l_0 l'_0 \\[4pt]
\Psi_3(k, a, l, \ldots, a''') &= a^\sigma \otimes (k_1, \mathrm{Tr}(k)\,\mathrm{N}(k)) \oplus k_0 \otimes (l_0, l_1)^\sigma \\
&\quad \oplus (0, \mathrm{Tr}(k)\,\mathrm{N}(l) + a^{\sigma+1} l_0 \mathrm{N}(k)^\sigma \\
&\quad + \mathrm{Tr}(k)(aa' + a^\sigma l_0 l''_0 + a^{\sigma+1} a''') \\
&\quad + \mathrm{Tr}(l)(k_1^\sigma a + a''') + k_1^\sigma a^{\sigma+1} l''_0 + k_0^{\sigma+1} a^2 l''^\sigma_0 \\
&\quad + k_0(a' + al''^\sigma_0 + k_1 a^\sigma l_0 + a^\sigma a''')^\sigma \\
&\quad + k_0^\sigma l_0(a' + al''^\sigma_0 + k_1 a^\sigma l_0 + a^\sigma a''')
\end{aligned}
$$

$$+a(l_1'' + a''''^{\sigma} l_0 + a''' l_0')$$
$$+l_0''(a' + a^{\sigma} a''') + a'' l_0 + l_0 l_0' l_0'')$$
$$\oplus (l_0', l_1')$$

$$\Psi_4(k, a, l, \dots, a''') = a'' + a^{\sigma+1} \mathrm{N}(k)^{\sigma} + a(k_0 l_0'' + l_0 k_1 + a''')^{\sigma}$$
$$+ \mathrm{Tr}(k)(l_1 + a^{\sigma} l_0'') + k_0^{\sigma}(a' + a^{\sigma} a''') + l_0' l_0'' + l_0^{\sigma} a'''$$

$$\Psi_5(k, a, l, \dots, a''') = (l_0'', l_1'') \oplus a \otimes (\mathrm{Tr}(k), k_0 \mathrm{N}(k)^{\sigma}) \oplus l_0 \otimes (k_0, k_1)^{\sigma}$$
$$\oplus (0, \mathrm{N}(k)(a^{\sigma} l_0'' + l_1) + k_0(a'' + l_0' l_0'' + aa''''^{\sigma} + l_0^{\sigma} a''')$$
$$+ k_1(k_1 l_0 a^{\sigma} + a' + a l_0''^{\sigma} + a^{\sigma} a''')$$
$$+ k_0 k_1^{\sigma} a l_0^{\sigma} + a''''^{\sigma} l_0 + a''' l_0')$$

$$\Psi_6(k, a, l, \dots, a''') = a''' + a \mathrm{N}(k) + l_0 k_1 + k_0 l_0''$$

This provides a very explicit description of the Ree–Tits octagons. We are now in a position to show the following result by TITS [1983].

3.6.3 Theorem (Tits [1983]). *The Ree–Tits octagons contain full weak suboctagons which are anti-isomorphic to the double of Suzuki quadrangles.*

Proof. If we replace R_2 by $\{0\}$ and keep R_1, then all octanary operations Ψ_1 up to Ψ_6 are still well defined, and they become either trivial or

$$\Psi_2(0, a, 0, a', 0, a'', 0, a''') = a' + a^{\sigma} a''' \qquad \text{(S1)}$$
$$\Psi_4(0, a, 0, a', 0, a'', 0, a''') = a'' + aa''''^{\sigma} \qquad \text{(S2)}$$
$$\Psi_6(0, a, 0, a', 0, a'', 0, a''') = a''' \qquad \text{(S3)}$$

This coordinatizes a non-thick full suboctagon Γ which is the double of a thick generalized quadrangle Γ^* as follows. The points of Γ^* are the lines of Γ having $0, 3, 4$ or 7 coordinates; the lines of Γ^* are the lines of Γ having $1, 2, 5$ or 6 coordinates; incidence is defined by the existence of a common point. We now coordinatize Γ^*. As there are no longer any elements of R_2 involved, we drop here locally our general notational assumption about $k, l \dots \in R_2$ and replace it by $k, l, \dots \in R_2^*$, where the superscript "$*$" means "with respect to Γ^*". Now put $R_1^* = R_2^* = \mathbb{K}$. Put $(\infty)^* = [\infty]$, $[\infty]^* = [0]$, $(a)^* = [0, a, 0]$, $[k]^* = [k, 0]$, $(0, b)^* = [0^2, b, 0]$, $[0, l]^* = [0^3, l, 0]$, $(0^3)^* = (0^7)$ and $[0^3]^* = [0^6]$. This determines a coordinatization of Γ^* in a unique way. It is easy to check that we have the following correspondence:

$(\infty)^*$	$=$	$[\infty]$	$[\infty]^*$	$=$	$[0]$
$(a)^*$	$=$	$[0, a, 0]$	$[k]^*$	$=$	$[k, 0]$
$(k, b)^*$	$=$	$[k, 0, b, 0]$	$[a, l]^*$	$=$	$[0, a, 0, l, 0]$
$(a, l, a')^*$	$=$	$[0, a, 0, l, 0, a', 0]$	$[k, b, k']^*$	$=$	$[k, 0, b, 0, k', 0]$

The quaternary operations expressing incidence between points and lines with three coordinates can be deduced immediately from (S1), (S2) and (S3) above and they show that Γ^* is a generalized quadrangle with quaternary operations

$$\begin{cases} \Psi_1^*(k,a,l,a') & = & a^\sigma k + l \\ \Psi_2^*(k,a,l,a') & = & k^\sigma a + a' \end{cases}$$

The theorem follows from the description of the Suzuki quadrangles in Subsection 3.4.6.   □

In Sections 6.9 and 8.2, we show some more properties of the Ree–Tits octagons derived from this coordinatization.

3.7 Some non-classical quadrangles

3.7.1 The semi-classical quadrangles of Tits type

The following construction of quadrangles is due to TITS, but appeared first in DEMBOWSKI [1968], page 304.

An **ovoid** $\mathcal{O}$ in a projective space of dimension $d \geq 2$ is a set of points having the following three properties:

(Ov1) Each line meets $\mathcal{O}$ in at most two points.

(Ov2) Through each point x of $\mathcal{O}$, there is a unique hyperplane H_x tangent to $\mathcal{O}$ (i.e., having exactly one point in common with $\mathcal{O}$).

(Ov3) If a line through $x \in \mathcal{O}$ is tangent to $\mathcal{O}$, then it is contained in the tangent hyperplane H_x.

An ovoid in a projective plane is usually called an **oval**; in the literature, one can also find the name *ovaloid* for an ovoid. The notion of an ovoid is due to TITS [1962c].

Let $\mathcal{O}$ be an ovoid in $\mathbf{PG}(d,\mathbb{K})$, for some (skew) field $\mathbb{K}$. Let $H = \mathbf{PG}(d,\mathbb{K})$ be embedded as a hyperplane in $\mathbf{PG}(d+1,\mathbb{K})$. Define the following geometry Γ. The points of Γ have three types:

[P1] the points of $\mathbf{PG}(d+1,\mathbb{K})$ not lying in H;

[P2] the hyperplanes of $\mathbf{PG}(d+1,\mathbb{K})$ meeting $\mathcal{O}$ in a single point (and hence containing a hyperplane in $\mathbf{PG}(d,\mathbb{K})$ tangent to $\mathcal{O}$);

[P3] the ovoid $\mathcal{O}$.

The lines have two types:

[L1] the lines of $\mathbf{PG}(d+1,\mathbb{K})$ which do not lie in H and which meet $\mathcal{O}$ in a necessarily unique point,

[L2] the points of $\mathcal{O}$.

Incidence is containment or reversed containment. It is easy to check that Γ is a generalized quadrangle, called of **Tits type**. We denote it usually by $\mathsf{T}_d(\mathcal{O})$. Lots of examples of ovoids are known (see e.g. DEMBOWSKI [1968], or various places in BUEKENHOUT [1995]). If $\mathcal{O}$ is a quadric, then $\mathsf{T}_d(\mathcal{O})$ is a classical generalized quadrangle. This follows from the fact that, if one projects a quadric Q from a point x on the quadric onto a hyperplane H not containing x, then the points of Q not collinear with x on Q are mapped bijectively onto an affine subspace A of H (the hyperplane A_∞ at infinity being the intersection of H with the tangent space H_x of Q at x); the points of Q collinear with x on Q are mapped onto a quadric Q' in A_∞ and each such point can be identified with the projection of its tangent hyperplane; and x itself can be identified with the quadric Q' (one can easily check that this defines an isomorphism from any orthogonal quadrangle to a certain quadrangle of Tits type; see also PAYNE & THAS [1984](3.2.2; 3.2.4) for the finite case). Conversely, if $\mathsf{T}_d(\mathcal{O})$ is a classical quadrangle, then $\mathcal{O}$ is a quadric. This is an excercise in Galois geometry: using the regularity of the lines (indeed, one can easily check that the lines of type [L2] are regular; hence all lines are regular), one deduces that all plane sections of $\mathcal{O}$ are conics. Standard arguments apply now to conclude that $\mathcal{O}$ is a quadric.

There is one other interesting example, which we would like to call **semi-classical of Tits type**. Let $\mathbb{K}$ be a field of characterictic 2 admitting a Tits endomorphism σ (see Definition 2.5.1 on page 79). Let $d = 3$ and consider the points of **PG**$(3, \mathbb{K})$ with coordinates $(0, 0, 0, 1)$ and $(1, a, b, a^{\sigma+2} + b^\sigma + ab)$, $a, b \in \mathbb{K}$. We shall prove in Chapter 7 that these points indeed form an ovoid $\mathcal{O}_{ST}(\mathbb{K})$ in **PG**$(3, \mathbb{K})$. This ovoid $\mathcal{O}$ is called the **Suzuki–Tits ovoid** and was discovered by TITS [1962a]; see also Section 7.6, in particular Proposition 7.6.14 on page 329.

In general, it is easy to see that coordinates can be chosen such that an ovoid $\mathcal{O}$ in **PG**$(d, \mathbb{K})$ consists of the point $(0, 0, \ldots, 0, 1)$ and the points

$$(1, x_1, x_2, \ldots, x_{d-1}, f(x_1, x_2, \ldots, x_{d-1})),$$

for some function f, with $f(x_1, x_2, \ldots, x_{d-1}) = 0$ if and only if $x_1 = x_2 = \cdots = x_{d-1} = 0$. In **PG**$(d, \mathbb{K})$, we consider $X_0, X_1, \ldots, X_d$-coordinates, while for **PG**$(d + 1, \mathbb{K})$, we add an X_{-1}-coordinate, writing in front of the X_0-coordinate. We coordinatize $\mathsf{T}_d(\mathcal{O})$ as follows. We put $R_1 = \mathbb{K}$ and $R_2 = \mathbb{K}^{d-1}$, and we abbreviate $0 = (0, 0, \ldots, 0)$. The unique point of type [P3] gets the coordinate (∞). The line of type [L2] corresponding to the point $(0, 0, \ldots, 0, 1)$ of $\mathcal{O}$ gets the coordinate $[\infty]$. The point $(1, x_1, x_2, \ldots, x_{d-1}, f(x_1, x_2, \ldots, x_{d-1}))$ of $\mathcal{O}$ has, as a line of type [L2] of $\mathsf{T}_d(\mathcal{O})$, coordinates $[(x_1, x_2, \ldots, x_{d-1})]$. The hyperplane with equation $X_0 + aX_{-1} = 0$ meets $\mathcal{O}$ in the unique point $(0, 0, \ldots, 0, 1)$ and receives, as a point of $\mathsf{T}_d(\mathcal{O})$ incident with $[\infty]$, the coordinate (a). Similarly, the hyperplane with equation $X_d + bX_{-1} = 0$ is given the coordinates $(0, b)$, $b \in \mathbb{K}$ (without bothering about normalization). Finally, the line of **PG**$(d + 1, \mathbb{K})$ containing the points $(0, 0, \ldots, 0, 1)$ and $(1, 0, x_1, x_2, \ldots, x_{d-1}, 0)$ is denoted with coordinates $[0, (x_1, x_2, \ldots, x_{d-1})]$. One can now compute the coordinates of all points and lines of $\mathsf{T}_d(\mathcal{O})$ according to the general rules, obtaining for the point $(a, (l_1, l_2, \ldots, l_{d-1}), a')$ of $\mathsf{T}_d(\mathcal{O})$ the point of $\mathbf{PG}(d+1, \mathbb{K})$ with coordinates $(1, -a, l_1, l_2, \ldots, l_{d-1}, -a')$. Now let

$$g(x_1, x_1', x_2, x_2', \ldots, x_{d-1}, x_{d-1}') \in \mathbb{K}, x_i, x_i' \in \mathbb{K}, i \in \{1, 2, \ldots, d-1\},$$

be defined as the unique element $g \in \mathbb{K}$ such that the line joining the points

$$(0, 1, x_1, x_2, \ldots, x_{d-1}, f(x_1, x_2, \ldots, x_{d-1})) \text{ and } (0, 0, x_1', x_2', \ldots, x_{d-1}', g)$$

is a tangent to $\mathcal{O}$ in $\mathbf{PG}(d, \mathbb{K})$, if $(x_1', x_2', \ldots, x_{d-1}') \neq (0, 0, \ldots, 0)$, and we put $g = 0$ if $(x_1', x_2', \ldots, x_{d-1}') = (0, 0, \ldots, 0)$. By condition [Ov3], and since $(0, 0, \ldots, 0, 1)$ belongs to $\mathcal{O}$, the element g is well defined. Then the line of $\mathbf{PG}(d+1, \mathbb{K})$ through the points

$$(1, 0, k_1', \ldots, x_{d-1}', g(k_1, k_1', \ldots) - b) \text{ and } (0, 1, k_1, \ldots, k_{d-1}, f(k_1, \ldots))$$

is the line $[(k_1, \ldots, k_{d-1}), b, (k_1', \ldots, k_{d-1}')]$. Hence a short calculation shows that, putting $x \in R_2$ equal to $(x_1, x_2, \ldots, x_{d-1})$,

$$(a, l, a') \text{ is incident with } [k, b, k']$$

$$\Longleftrightarrow$$

$$\begin{cases} k_i' &= ak_i + l_i, \ i \in \{1, 2, \ldots, d-1\}, \\ a' &= a \cdot f(k_1, k_2, \ldots, k_{d-1}) - g(k_1, k_1', \ldots, k_{d-1}, k_{d-1}') + b, \end{cases}$$

for all $a, a', b \in R_1$ and $k, k', l \in R_2$ (where we consider coordinates of points in $\mathbf{PG}(d+1, \mathbb{K})$ determined up to a left multiple).

If $\mathcal{O}_{ST}(\mathbb{K})$ is the Suzuki–Tits ovoid as defined above, then

$$f(k_1, k_2) = k_1^{\sigma+2} + k_2^{\sigma} + k_1 k_2.$$

Now we compute $g = g(k_1, k_1', k_2, k_2')$. The element g is such that the point

$$(0, 1, k_1 - \ell k_1', k_2 - \ell k_2', f(k_1, k_2) - \ell g)$$

belongs to $\mathcal{O}_{ST}(\mathbb{K})$ **if and only if** $\ell = 0$. We obtain that the equation

$$k_1^{\sigma} k_1'^2 \ell^2 + k_1^2 k_1'^{\sigma} \ell^{\sigma} + k_1'^{\sigma+2} \ell^{\sigma+2} + k_2'^{\sigma} \ell^{\sigma} + k_1 k_2' \ell + k_2 k_1' \ell + k_1' k_2' \ell^2 + g\ell = 0$$

must have a unique solution in ℓ, namely $\ell = 0$. We rewrite this equation as follows:

$$f(k_1'\ell, (k_2' + k_1' k_1^{\sigma})\ell) = (k_1 k_2' + k_2 k_1' + g)\ell.$$

It is now obvious, since $f(a, b) = 0$ **if and only if** $a = b = 0$, that $g = k_1 k_2' + k_2 k_1'$. Hence a quadratic quaternary ring for the semi-classical quadrangle $\mathsf{T}_3(\mathcal{O}_{ST}(\mathbb{K}))$ of Tits type can be written as follows:

$$\begin{cases} \Phi_1(a, k, b, k') &= a \cdot (k_1^{\sigma+2} + k_1 k_2 + k_2^{\sigma}) + k_1 k_2' + k_2 k_1' + b, \\ \Phi_2(a, k, b, k') &= a \cdot (k_1, k_2) + (l_1, l_2). \end{cases}$$

This and every quadrangle of Tits type is an *elation generalized quadrangle*. We will define that notion in Subsection 3.7.3 below. Moreover, the *elation group* is always abelian, so it is a translation generalized quadrangle; see Subsection 4.9.8 on page 171. Indeed, the elation group is, with our above notation, the full group of translations of $\mathbf{PG}(d+1, \mathbb{K})$ with axis $\mathbf{PG}(d, \mathbb{K})$.

3.7.2 The quadrangles of *-Tits type and of Payne type

There are a few other quadrangles that can be mentioned here. We give a general construction due to PAYNE [1971] (in the finite case), but omit the (fairly straightforward) proof. Let Γ be a generalized quadrangle having a projective point p. We define a new quadrangle $^p\Gamma$ as follows. The points of $^p\Gamma$ are the points of Γ opposite p. The lines of $^p\Gamma$ are the lines of Γ not through p and the sets $\{p, x\}^{\perp\perp}$ (the spans) in Γ for any x opposite p in Γ. The incidence relation is the natural one inherited from Γ. Then one can check that $^p\Gamma$ is indeed a generalized quadrangle. We call it **of Payne type**. If Γ is finite of order (s, s), then $^p\Gamma$ is finite of order $(s-1, s+1)$. If Γ is isomorphic to a symplectic quadrangle $\mathsf{W}(\mathbb{K})$ naturally embedded in $\mathbf{PG}(3, \mathbb{K})$ with associated symplectic polarity θ, then we obtain the quadrangles discovered by AHRENS & SZEKERES [1969]. The points of $^p\Gamma$ in this case are the points of $\mathbf{PG}(3, \mathbb{K})$ not in the plane p^θ; hence these form an affine space $\mathbf{AG}(3, \mathbb{K})$. The lines of $^p\Gamma$ which are not lines of Γ are the lines of $\mathbf{AG}(3, \mathbb{K})$ which are incident in $\mathbf{PG}(3, \mathbb{K})$ with p. This gives us a natural embedding of $^p\mathsf{W}(\mathbb{K})$ in an affine 3-space. It is proved by GRUNDHÖFER, JOSWIG & STROPPEL [1994] that, whenever $|\mathbb{K}| > 4$, then Aut $^p\mathsf{W}(\mathbb{K})$ is induced by $\mathbf{AGL}_3(\mathbb{K})$. They also show that all groups of projectivities of $^p\mathsf{W}(\mathbb{K})$ contain the finitary alternating groups in their standard permutation representation.

Suppose that Γ is isomorphic to $\mathsf{T}_2(\mathcal{O})$ for some oval $\mathcal{O}$ in the projective plane $\mathbf{PG}(2, \mathbb{K})$ over some perfect field $\mathbb{K}$ of characteristic 2. Suppose that $\mathcal{O}$ has a **nucleus** c (i.e., every tangent to $\mathcal{O}$ contains c and every line through c is a tangent; if $\mathbb{K}$ is finite there is always a unique nucleus). If p denotes the unique point of type [P3], then $^p\Gamma$ may be constructed in a different, more direct way, as follows. Let $\mathbf{PG}(2, \mathbb{K})$ be embedded as a hyperplane in $\mathbf{PG}(3, \mathbb{K})$ and let O be the union of $\mathcal{O}$ and its nucleus c. Then the points of $\mathsf{T}_2^*(O)$ are the points of $\mathbf{PG}(3, \mathbb{K})$ not in the hyperplane $\mathbf{PG}(2, \mathbb{K})$; the lines of $^p\Gamma$ are the lines in $\mathbf{PG}(3, \mathbb{K})$ not in $\mathbf{PG}(2, \mathbb{K})$ and meeting O in a unique point; incidence is that inherited from $\mathbf{PG}(3, \mathbb{K})$. Using the definition of $\mathsf{T}_2(\mathcal{O})$, one can see that $^p\mathsf{T}_2(\mathcal{O})$ is isomorphic to $\mathsf{T}_2^*(O)$. We say that the quadrangle $\mathsf{T}_2^*(O)$ is of *-**Tits type**. Such a set O is usually called a **hyperoval**; see also Subsection 7.5.2.

Two special cases are worth considering separately here. If $|\mathbb{K}| = 4$, then any hyperoval O has six elements and admits the symmetric group $\mathbf{S}_6$ as its automorphism group inside $\mathbf{PG}(2, 4)$ (see also Subsection 7.5.2); hence the quadrangle $\mathsf{T}_2^*(O)$ admits in this case a large automorphism group; in fact a *distance transitive automorphism group* (see Section 4.8.1). Also, $\mathsf{T}_2^*(O)$ is in this case isomorphic to the unique generalized quadrangle of order $(3, 5)$.

If $|\mathbb{K}| = 16$, then there is a hyperoval O' admitting a transitive group (this is not true if O is a conic and $|\mathbb{K}| > 4$), the so-called *Lunelli–Sce hyperoval* (see LUNELLI & SCE [1958]). In this case the quadrangle $\mathsf{T}_2^*(O')$ admits a flag-transitive group (this is easily seen) and has order $(15, 17)$.

3.7.3 Definition of elation generalized quadrangles

Let $\Gamma = (\mathcal{P}, \mathcal{L}, I)$ be a generalized quadrangle and let p be some point of Γ. If there exists a group of automorphisms of Γ fixing every line through p and acting regularly on the set of points opposite p, then we call Γ an **elation generalized quadrangle** with **elation point** p. All classical quadrangles are elation generalized quadrangles with all points elation points. There is a general group-theoretical method to construct finite elation generalized quadrangles, due to KANTOR [1980]. Details of this construction can be found in PAYNE & THAS [1984]; see also Section 4.9. A special case of this construction is equivalent to the construction of *flocks of a quadratic cone*, as was shown by THAS [1987]. It is this class of quadrangles that we will define here. For obvious reasons, we call them **flock quadrangles**. Our point of view will be a little bit different from Kantor's because we also want to include the infinite case. The theory of coordinatization seems to be very suitable for this purpose and allows one to skip the group-theoretical background. At the same time, we obtain a natural algebraic definition of a *derived flock* in the general case.

3.7.4 Flock quadrangles

Let $\mathbb{K}$ be any field and let $x : \mathbb{K} \to \mathbb{K}$, $y : \mathbb{K} \to \mathbb{K}$ and $z : \mathbb{K} \to \mathbb{K}$ be three arbitrary maps with the only restriction that they all map 0 to 0. As an exception to our general rule, we will denote the image of an element $k \in \mathbb{K}$ under the map x by x_k, and similarly for y and z. We define the following (dual) quaternary ring $\mathcal{R}_{x,y,z} = (\mathbb{K} \times \mathbb{K}, \mathbb{K}, \Phi_1, \Phi_2)$,

$$\begin{cases} \Phi_1(a, k, b, k') = (b_0 + 2a_0 x_k + a_1 y_k, b_1 + a_0 y_k + 2a_1 z_k), \\ \Phi_2(a, k, b, k') = k' + a_0^2 x_k + a_0 a_1 y_k + a_1^2 z_k + a_0 b_0 + a_1 b_1, \end{cases}$$

where $a = (a_0, a_1)$, $b = (b_0, b_1)$, etc. For a field $\mathbb{K}$ with even characteristic, we put $\mathcal{C}_1(\mathbb{K})$ equal to the set of elements $k \in \mathbb{K}$ such that the equation $X^2 + X + k = 0$ has no solutions in $\mathbb{K}$. Furthermore, for a quadratic form $aX^2 + bX + c$, we call the **discriminant** the field element $b^2 - 4ac$ in the odd characteristic case (including characteristic 0), or acb^{-2} in the even characteristic case. Also, we call a field $\mathbb{K}$ of characteristic 2 (distinct from 2) **full** if $\mathcal{C}_1(\mathbb{K}) \cap (\mathcal{C}_1(\mathbb{K}) + \mathcal{C}_1(\mathbb{K}))$ is empty (if the product of two non-squares is a square). We have the following theorem.

3.7.5 Theorem. *The quaternary ring $\mathcal{R}_{x,y,z}$ defined above is a (not necessarily normalized) dual quadratic quaternary ring* **if and only if** *the following two conditions are satisfied:*

(i) *for every $r \in \mathbb{K} \cup \{\infty\}$, the mapping $\mathbb{K} \to \mathbb{K} : k \mapsto (x_k)r^2 + (y_k)r + (z_k)$ is a bijection (for $r = \infty$, this means that $k \mapsto x_k$ is a bijection);*

(ii) for all $r \in \mathbb{K} \cup \{\infty\}, l \in \mathbb{K}$, the mapping

$$\mathbb{K} \setminus \{l\} \to \mathbb{K} \setminus \{0\} : k \mapsto \frac{(x_k - x_l)r^2 + (y_k - y_l)r + (z_k - z_l)}{(y_k - y_l)^2 - 4(x_k - x_l)(z_k - z_l)}$$

is a bijection.

If $\mathbb{K}$ is full and the characteristic of $\mathbb{K}$ is not 2, then this is equivalent to the following three conditions:

(a) for every $k, l \in \mathbb{K}$, $k \neq l$, the element $(y_k - y_l)^2 - 4(x_k - x_l)(z_k - z_l)$ is not a square in $\mathbb{K}$;

(b) for every $r \in \mathbb{K}$, the mapping $\mathbb{K} \to \mathbb{K} : k \mapsto x_k r^2 + y_k r + z_k$ is surjective;

(c) for all $r, l \in \mathbb{K}$, the mapping

$$\mathbb{K} \setminus \{l\} \to \mathbb{K} \setminus \{0\} : k \mapsto \frac{(x_k - x_l)r^2 + (y_k - y_l)r + (z_k - z_l)}{(y_k - y_l)^2 - 4(x_k - x_l)(z_k - z_l)}$$

is surjective.

If $\mathbb{K}$ is full and the characteristic of $\mathbb{K}$ is 2, then conditions (i) and (ii) are equivalent to conditions (a'), (b) and (c), where (a') is:

(a') for every $k, l \in \mathbb{K}$, $k \neq l$, we have $y_k \neq y_l$, and the element $(x_k + x_l)(z_k + z_l)(y_k + y_l)^{-2}$ belongs to $C_1(\mathbb{K})$.

In the finite case (a), (b) and (c) hold if and only if (a) holds, and, similarly, (a'), (b) and (c) hold if and only if (a') holds.

Proof. We have to prove that conditions (i) and (ii) are equivalent to (the dual of) conditions (O), (A), (B) and (C) of Theorem 3.3.1. Conditions (O) and (A) are trivially satisfied. After some calculations, one finds that condition (B)(i) is satisfied whenever the equation

$$m = (a_0 - b_0)^2 x_{k*} + (a_0 - b_0)(a_1 - b_1)y_{k*} + (a_1 - b_1)^2 z_{k*},$$
$$\text{with} \quad m = k - l - (b_0 - a_0)a_0' - (b_1 - a_1)a_1',$$

has a unique solution k^* for every $a, b \in \mathbb{K} \times \mathbb{K}$, $a \neq b$, and every $k, l \in \mathbb{K}$. This is clearly equivalent to condition (i). Similarly, one can check that (B)(ii) is satisfied whenever the system of equations

$$\begin{cases} 2(x_l - x_k)a_0^* + (y_l - y_k)a_1^* &= b_0 - a_0, \\ (y_l - y_k)a_0^* + 2(z_l - z_k)a_1^* &= b_1 - a_1, \end{cases}$$

has a unique solution $a^* \in \mathbb{K} \times \mathbb{K}$ for all $k, l \in \mathbb{K}$, $k \neq l$ and all $a, b \in \mathbb{K} \times \mathbb{K}$. Clearly, this follows from the following condition:

(i') for every $k, l \in \mathbb{K}$, $k \neq l$, the element $(y_k - y_l)^2 - 4(x_k - x_l)(z_k - z_l)$ differs from 0.

One can see that (i') is an easy consequence of (i). Condition (C) requires some more calculations, but assuming (i) and (i'), one finds that it is equivalent to (ii). The calculations are all straightforward and can be found in detail in DE CLERCK & VAN MALDEGHEM [1994].

It is also straightforward to see that the injectivity of the map in (i) implies conditions (a) and (a'), and hence (a), (a'), (b) and (c) follow easily from (i) and (ii).

Conversely, each of the conditions (a) and (a') implies the injectivity of the mapping in (i) and so (i) follows from (a) (or (a')) and (b). It remains to show that the map in (c) is injective, given $\mathbb{K}$ is full. Therefore we have to show that the discriminant $D(k,l,m)$ of the quadratic form

$$\left(\frac{x_k - x_l}{D(k,l)} - \frac{x_m - x_l}{D(m,l)}\right) r^2 + \left(\frac{y_k - y_l}{D(k,l)} - \frac{y_m - y_l}{D(m,l)}\right) r + \left(\frac{z_k - z_l}{D(k,l)} - \frac{z_m - z_l}{D(m,l)}\right),$$

with

$$D(i,l) = (y_i - y_l)^2 - 4(x_i - x_l)(z_i - z_l), \qquad i \in \{k,m\},$$

is a non-square if the characteristic of $\mathbb{K}$ is not equal to 2, and that it belongs to $\mathcal{C}_1(\mathbb{K})$ if the characteristic of $\mathbb{K}$ is equal to 2. If the characteristic is not equal to 2, then one computes

$$D(k,l,m) = \frac{D(k,m)}{D(k,l)D(m,l)},$$

which is a non-square since $D(k,l)D(m,l)$ is a square as $\mathbb{K}$ is full.

Suppose now that the characteristic of $\mathbb{K}$ is 2. A similar computation shows that in this case

$$D(k,l,m) = \frac{(x_k + x_l)(z_k + z_l)}{(y_k + y_l)^2} + \frac{(x_l + x_m)(z_l + z_m)}{(y_l + y_m)^2} + \frac{(x_m + x_k)(z_m + z_k)}{(y_m + y_k)^2}.$$

Since $\mathbb{K}$ is full, $D(k,l,m) \in \mathcal{C}_1(\mathbb{K})$.

Suppose now that $\mathbb{K}$ is finite. Then it is certainly full. We have already seen that the injectivity of the mappings in (i) and (ii) is a consequence of (a) or (a'). But in the finite case an injective map from a set to itself is also surjective. Hence the theorem. $\qquad\qquad\qquad\square$

3.7.6 Remark. Condition (i) is equivalent to the condition that the set of planes with equation $x_k X_0 + z_k X_1 + y_k X_2 + X_3 = 0$ in $\mathbf{PG}(3, \mathbb{K})$ partitions the quadratic cone with equation $X_0 X_1 = X_2^2$ into disjoint conics, $k \in \mathbb{K}$. This is called a **flock** of $\mathcal{C}$ and the triple (x, y, z) is called a **$\mathbb{K}$-clan**. For every $l \in \mathbb{K} \cup \{\infty\}$, we define the triple $({}^l X, {}^l Y, {}^l Z)$ of maps as

$${}^l X_k = \frac{x_k - x_l}{D(k,l)}, \qquad {}^l Y_k = \frac{y_k - y_l}{D(k,l)}, \qquad {}^l Z_k = \frac{z_k - z_l}{D(k,l)},$$

for $k \neq l$ (and $D(k, l)$ as before) and ${}^l X_l = {}^l Y_l = {}^l Z_l = 0$, where we put $x_\infty = y_\infty = z_\infty = 0$ and $D(k, \infty) = 1$. Hence we see that the maps x, y, z define a quadratic quaternary ring $\mathcal{R}_{x,y,z}$ and hence a generalized quadrangle if and only if for every $l \in \mathbb{K} \cup \{\infty\}$ the triple $({}^l X, {}^l Y, {}^l Z)$ is a $\mathbb{K}$-clan. In that case, $({}^l X, {}^l Y, {}^l Z)$ defines a flock, which we call a **derivation** of the one defined by (x, y, z). If follows from Theorem 3.7.5 that all derivations of a finite flock are well defined and hence a finite flock is equivalent to a flock quadrangle. This was first observed by THAS [1987]. A geometric relation between these objects has been established by KNARR [1992] in the odd characteristic case. In the finite case, there is a whole theory about these flocks, which are also related to translation planes, but this goes beyond the scope of this book. We refer the reader to THAS [1987], [1995]. We restrict ourselves here to three things: first we present a planar ternary field for the projective plane associated with a certain flock (via the Klein correspondence); secondly we prove that flock quadrangles are elation quadrangles and we mention some examples; finally we present an alternative construction due to BADER & LUNARDON [1993] (based on Knarr's construction mentioned above) of the finite split Cayley hexagons $\mathsf{H}(q)$ for $q \equiv 2 \bmod 3$ (see Subsection 3.7.14 below).

3.7.7 Flock projective planes

Let $\mathbb{K}$ be a field and $x, y, z : \mathbb{K} \to \mathbb{K}$ three arbitrary maps with the restriction that they all map 0 to 0 and that x is bijective. We define the following ternary ring $F_{x,y,z} = (\mathbb{K} \times \mathbb{K}, T)$:

$$T((k, t), (k_1, t_1), (k_2, t_2)) = (k', t'),$$

where k' and t' are defined by

$$k' = k_2 + kk_1 - z_t x_{t_1}, \qquad x_{t'} = x_{t_2} + k_1 x_t + k x_{t_1} + y_t x_{t_1}.$$

We have the following result.

3.7.8 Theorem. *The ternary ring $F_{x,y,z}$ is a planar ternary field (i.e., it coordinatizes a projective plane)* **if and only if** *(x, y, z) is a $\mathbb{K}$-clan. In this case, it is a left quasi-field and the corresponding projective plane is a translation plane.*

Proof. One can easily verify that the equation

$$T(m, a, b) = T(m, a', b'), \quad a \neq a',$$

in m has a unique solution for all $a, a', b, b', a \neq a'$, **if and only if** condition (*i*) of Theorem 3.7.5 is satisfied. Similarly, the system of equations

$$\begin{cases} T(m, a, b) &= k, \\ T(m', a, b) &= k', \end{cases} \quad m \neq m',$$

in a and b has a unique solution for all $m, m', k, k', m \neq m'$, **if and only if** condition (*i'*) of the proof of Theorem 3.7.5 is satisfied.

The second part of the theorem follows from an easy calculation. $\qquad\square$

Note that the Desarguesian projective plane over $\mathbb{K}$ is a subplane of the projective plane associated with $F_{x,y,z}$. This can be seen by viewing $\mathbb{K}$ as $\mathbb{K} \times \{0\} \subseteq \mathbb{K} \times \mathbb{K}$. Throughout the rest of this section, we denote $a.b = a_0 b_0 + a_1 b_1$, for $a, b \in \mathbb{K} \times \mathbb{K}$.

3.7.9 Theorem. *Every flock quadrangle is an elation generalized quadrangle.*

Proof. We show that the flock quadrangle Γ coordinatized by the dual quadratic quaternary ring $(\mathbb{K} \times \mathbb{K}, \mathbb{K}, \Phi_1, \Phi_2)$ and defined by the functions x, y and z is an elation generalized quadrangle with elation point (∞). Therefore, we define the map

$$\theta : \Gamma \to \Gamma : \begin{aligned} (a, l, a') &\mapsto (a + A, l + L - a.A', a' + A') \\ [k, b, k'] &\mapsto [k, b + A' - (2x_k A_0 + y_k A_1, y_k A_0 + 2z_k A_1), \\ &\qquad k' + b.A + A'.A - (x_k A_0^2 + y_k A_0 A_1 + z_k A_1^2)], \end{aligned}$$

where the image of a point or a line with zero, one or two coordinates is obtained by restriction. Clearly, θ preserves the incidence relation (defined by the quaternary operations Φ_1 and Φ_2). The set of all such collineations θ is a group which apparently acts regularly on the set of points opposite (∞) and which fixes all lines through (∞). This proves the assertion. $\square$

3.7.10 Examples. Usually the functions x, y and z of the previous paragraphs are written in a matrix

$$A_k = \begin{pmatrix} x_k & y_k \\ 0 & z_k \end{pmatrix}.$$

We give some known examples, presenting them without the original finiteness restriction. We again put $D(k, l) = (y_k - y_l)^2 - 4(x_k - x_l)(z_k - z_l)$ and $f(k, r) = x_k r^2 + y_k r + z_k$. Note also that $\mathbb{R}$, the field of real numbers, is full.

1. Let $X^2 + bX + c$ be an irreducible polynomial over $\mathbb{K}$. Then put

$$A_k = \begin{pmatrix} k & bk \\ 0 & ck \end{pmatrix}.$$

   This flock quadrangle is in fact anti-isomorphic to the orthogonal quadrangle defined by the quadric $\mathcal{Q}$ in $\mathbf{PG}(5, \mathbb{K})$ with equation

   $$cX_0^2 - bX_0 X_1 + X_1^2 = (4c - b^2)(X_2 X_3 + X_4 X_5).$$

   Indeed, using standard notation, we can take the 1-quadratic form

   $$q_0((x_0, x_1)) = \frac{c}{4c - b^2} x_0^2 - \frac{b}{4c - b^2} x_0 x_1 + \frac{1}{4c - b^2} x_1^2,$$

while for the associated bilinear form g_0 (restricted to V_0, here two-dimensional) we may choose

$$g_0((x_0, x_1), (y_0, y_1)) = \frac{c}{4c - b^2} x_0 y_0 - \frac{b}{4c - b^2} x_0 y_1 + \frac{1}{4c - b^2} x_1 y_1.$$

Hence the corresponding Hermitian form, restricted to V_0 is given by

$$f_0((x_0, x_1), (y_0, y_1)) = \frac{2c}{4c - b^2} x_0 y_0 - \frac{b}{4c - b^2}(x_0 y_1 + x_1 y_0) + \frac{2}{4c - b^2} x_1 y_1.$$

The corresponding quadratic quaternary operations are (restricting R_2 to V_0 and writing the elements of V_0 as pairs (k_0, k_1)):

$$\begin{cases} \Psi_1((k_0, k_1), a, (l_0, l_1), a') &= (l_0, l_1) + a(k_0, k_1), \\ \Psi_2((k_0, k_1), a, (l_0, l_1), a') &= a' + aq_0((k_0, k_1)) + f_0((k_0, k_1), (l_0, l_1)). \end{cases}$$

Substituting $k_0 = 2\bar{k}_0 + b\bar{k}_1$ and $k_1 = b\bar{k}_0 + 2c\bar{k}_1$, we obtain after an easy calculation

$$\begin{cases} \Psi_1((\bar{k}_0, \bar{k}_1), a, (l_0, l_1), a') &= (l_0, l_1) + a(2\bar{k}_0 + b\bar{k}_1, b\bar{k}_0 + 2c\bar{k}_1), \\ \Psi_2((k_0, k_1), a, (l_0, l_1), a') &= a' + a(\bar{k}_0^2 + b\bar{k}_0\bar{k}_1 + c\bar{k}_1^2) + \bar{k}_0 l_0 + \bar{k}_1 l_1, \end{cases}$$

which is exactly the dual of the dual quaternary operations of the flock quadrangle defined by the matrices A_k above. The flock is called the **linear flock**.

Note that all planes of the above linear flock contain the line with equations $X_0 + cX_1 + bX_2 = X_3 = 0$. This explains the name "linear".

2. Let $\mathbb{K}$ be a field of characteristic different from 3 in which raising to the third power is a bijection, e.g. $\mathbb{K} \cong \mathbf{GF}(q)$, $q \equiv 2 \bmod 3$, or $\mathbb{K} \cong \mathbb{R}$. Then put

$$A_k = \begin{pmatrix} k & 3k^2 \\ 0 & 3k^3 \end{pmatrix}.$$

Here, $D(k, l) = -3(k - l)^4$ and $f(k, r) = \frac{1}{9}[(3k + r)^3 - r^3]$. We will come back to this very interesting example in the next subsection.

These examples (in the finite case) are due to KANTOR [1980]. In the literature, they are usually denoted by $K(q)$ for a finite field $\mathbf{GF}(q)$. To avoid confusion with the notation for a general field $\mathbb{K}$, we will adopt the notation $\mathsf{Ka}(\mathbb{K})$ here.

3. Let $m \in \mathbb{K}$ and let σ be a field automorphism of $\mathbb{K}$. Put

$$A_k = \begin{pmatrix} k & 0 \\ 0 & -mk^\sigma \end{pmatrix}.$$

For each field $\mathbb{K}$, one can seek conditions on m and σ such that the conditions (i) and (ii) of Theorem 3.7.5 are satisfied. We will call the corresponding quadrangles **semi-classical flock quadrangles**, or **semi-classical of Kantor type** (because, in the finite case, the corresponding flock is characterized by the property that all planes go through a common point (see THAS [1987]); this is a little weaker than for the linear flocks where all planes go through a common line (see above)). For a finite field, one must have that m is a non-square. Of course, when σ is trivial, then we are back at example 1, so we get nothing new for $\mathbb{K} \cong \mathbb{R}$. Let us give another example. Let k be any field and put $\mathbb{K} = k((t))$, the field of all Laurent series over k in the variable t. Let σ be an automorphism of $\mathbb{K}$ preserving the *natural valuation* on $\mathbb{K}$ (here the order of the series, i.e., the exponent of the smallest non-vanishing power of t; for the definition of the notion of a valuation on a field, see Subsection 9.7.3, page 421). Put $m = t$. Then the completeness of $\mathbb{K}$ with respect to the valuation ensures us that condition (ii) of Theorem 3.7.5 is satisfied. In fact, we can generalize this example to any field with a complete non-trivial valuation (with values in $\mathbb{Z}$), taking for m an element with odd valuation (and σ an automorphism of the field preserving the valuation). These examples are due to KANTOR [1986b]. The infinite examples were constructed by VAN MALDEGHEM [1989a] in connection with affine buildings of type $\tilde{C}_2$ and are generalized quadrangles with valuation (see Section 9.7 on page 419).

Other examples are given in THAS [1995] (finite case), DE CLERCK & VAN MALDEGHEM [1994] and BADER & PAYNE [19**] (infinite case).

3.7.11 Quadrangles arising from hexagons

In this subsection, we give another construction, due to KANTOR [1980], of the flock quadrangles under example 2 in Subsection 3.7.10. We take a somewhat more general point of view by allowing infinite hexagons. We approach the problem via coordinatization.

Consider the split Cayley hexagon $\mathsf{H}(\mathbb{K})$ over some field $\mathbb{K}$ of characteristic different from 3 and fix a line L of $\mathsf{H}(\mathbb{K})$. Let us, following KANTOR [1980], define a geometry $\Gamma(L)$ as follows. The points of $\Gamma(L)$ are the line L, the points at distance 3 from L and the lines opposite L; the lines are the points incident with L and the lines at distance 4 from L; two elements of $\Gamma(L)$ are incident if they lie at distance 1 or 2 from each other in Γ. We now give a description of $\Gamma(L)$ in terms of the coordinates, putting $L = [\infty]$.

First, we remark that $\Gamma(L)$ does not contain triangles through its point L. This makes it possible to coordinatize $\Gamma(L)$ as if it were a generalized quadrangle, choosing $R_1 = \mathbb{K} \times \mathbb{K}$ and $R_2 = \mathbb{K}$ and denoting $(0,0)$ by 0, as before. We denote the coordinates in $\Gamma(L)$ with a subscript L and make the following labelling (following the general rules about coordinatization of a generalized quadrangle):

$$
\begin{aligned}
[\infty] &\to (\infty)_L, & (\infty) &\to [\infty]_L, \\
(-a_1, -a_0) &\to ((a_0, a_1))_L, & (3k) &\to [k]_L, \\
(0, 9b_1, 3b_0) &\to (0, (b_0, b_1))_L, & [0, 0, 9l] &\to [0, l]_L.
\end{aligned}
$$

The factors 3 and 9 and the minus signs are not essential and can be deleted. We have introduced them to facilitate later identification of $\Gamma(L)$. This labelling defines coordinates for all elements of $\Gamma(L)$ and after a few calculations, one finds:

$$
\begin{aligned}
(3k, 9b_1 - 27kb_0, 3b_0) &\rightarrow (k, (b_0, b_1))_L, \\
[-a_1, -a_0, 9l] &\rightarrow [(a_0, a_1), l]_L, \\
[-a_1, -a_0, 9l, 3a_0', 9a_1'] &\rightarrow ((a_0, a_1), l, (a_0', a_1'))_L, \\
[3k, 9b_1 - 27kb_0, 3b_0, 9k'] &\rightarrow [k, (b_0, b_1), k']_L.
\end{aligned}
$$

Incidence between points and lines not both having three coordinates is the usual one. Using the hexagonal sexternary ring for $\mathsf{H}(\mathbb{K})$, one computes the condition under which a point $(a, l, a')_L$ (with $a = (a_0, a_1)$, etc. as before) is incident with a line $[k, b, k']_L$. This condition is

$$
\begin{cases}
(a_0', a_1') = (b_0 + 2a_0 k + 3a_1 k^2, b_1 + 3a_0 k^2 + 6a_1 k^3), \\
l = k' + a_0^2 k + 3a_0 a_1 k^2 + 3a_1^2 k^3 + a_0 b_0 + a_1 b_1.
\end{cases}
$$

The right-hand sides are exactly the quaternary operations of the ring $\mathcal{R}_{x,y,z}$ with x, y, z the maps corresponding to example 2 in Subsection 3.7.10. Hence we conclude:

3.7.12 Theorem. *The geometry $\Gamma(L)$ derived from the split Cayley hexagon as defined above is a generalized quadrangle **if and only if** $\mathbb{K}$ is a field of characteristic different from 3 in which raising to the third power is a bijection in $\mathbb{K}$. In this case, $\Gamma(L)$ is isomorphic to the flock quadrangle $\mathsf{Ka}(\mathbb{K})$.*

Proof. To prove this theorem, we only have to show that $\Gamma(L)$ is not a generalized quadrangle for fields $\mathbb{K}$ of characteristic 3, and for fields $\mathbb{K}$ where $x \mapsto x^3$ is not bijective. But this can easily be done using the same method as above. We only needed the characteristic to be different from 3 in order to be able to introduce the factors 3 and 9. Deleting them gives us a quaternary ring which can never be quadratic. We leave the details of the computations to the reader. □

3.7.13 Remarks. (1) The non-existence of triangles in $\Gamma(L)$ implies the non-existence of four apartments in $\mathsf{H}(\mathbb{K})$ sharing in pairs a chain of length 4 starting with a line (and one of these "starting lines" is L); see Figure 3.2. Let us call this configuration a **Kantor configuration** based at L. Requiring that no Kantor configuration based at L exists is just expressing that lines opposite L, viewed as points of $\Gamma(L)$, cannot form a triangle. But apparently, no other triangles can arise by the properties of a generalized hexagon. Hence the non-existence of a Kantor configuration based at L is equivalent to the non-existence of triangles in $\Gamma(L)$. In the finite case, this is equivalent to $\Gamma(L)$ being a generalized quadrangle (because it has the right order). This equivalence is true for every finite generalized hexagon for which $s = t$. However, it is conjectured by KANTOR (unpublished) that the absence of any Kantor configuration in a finite generalized hexagon Γ of order (s, s)

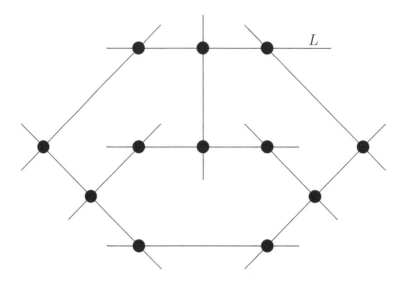

Figure 3.2. The Kantor configuration based at the line L.

implies that Γ is classical. For $\mathsf{H}(\mathbb{K})$, with $\mathbb{K}$ an infinite field not of characteristic 3, the absence of any Kantor configuration based at L is equivalent to raising to the third power being injective in $\mathbb{K}$. Examples of this arise in Laurent series fields, e.g. over finite fields of order $q \equiv 2 \bmod 3$.

(2) A similar construction method for quadrangles from finite octagons of order (s, s^2) has been worked out by LÖWE [1992]. The condition on the octagon is, however, never satisfied in the classical octagons, as was shown by JOSWIG & VAN MALDEGHEM [1995].

3.7.14 The Bader–Lunardon construction of $\mathsf{H}(q)$, $q \equiv 2 \bmod 3$

The BADER & LUNARDON [1993] construction is based on KNARR's [1992] construction of generalized quadrangles from flocks. Since we do not want to go into details (see THAS [1995] for an excellent survey of the matter) concerning this, we just write down the construction of BADER & LUNARDON without referring to the related construction of KNARR. As usual, we give the construction in general, whereas the original one is only given in the finite case.

Let $\mathbb{K}$ be a field of characteristic different from 3 in which raising to the third power is a bijection. Let $\mathbf{W}(5, \mathbb{K})$ be the symplectic space in $\mathbf{PG}(5, \mathbb{K})$ defined by the form

$$X_0 Y_5 - X_5 Y_0 + X_1 Y_3 - X_3 Y_1 + X_2 Y_4 - X_4 Y_2$$

and let $\mathbf{W}(3, \mathbb{K})$ be the symplectic space induced by $\mathbf{W}(5, \mathbb{K})$ in the 3-space $X_0 = X_5 = 0$. Let p be the point $(0, 0, 0, 0, 0, 1)$ and define for each $t \in \mathbb{K} \cup \{\infty\}$ a point

p_t, and a line L_t as follows:

$$p_t = (0, 1, t, t^3, -3t^2, 0),$$

$$p_\infty = (0, 0, 0, 1, 0, 0),$$

$$L_t = \langle (0, 1, 0, -2t^3, 3t^2, 0), (0, 0, 1, 3t^2, -6t, 0) \rangle,$$

$$L_\infty = \langle (0, 0, 0, 1, 0, 0), (0, 0, 0, 0, 1, 0) \rangle.$$

Define the following geometry Γ. The points of Γ are of four types:

(i) the point p;

(ii) the points on the lines pp_t different from p;

(iii) the totally singular planes of $\mathbf{W}(5, \mathbb{K})$ not contained in the polar hyperplane of p with respect to $\mathbf{W}(5, \mathbb{K})$ (which has equation $X_0 = 0$) and meeting some plane spanned by p and L_t in a line;

(iv) the points of $\mathbf{PG}(5, \mathbb{K})$ not lying in the hyperplane $X_0 = 0$.

The lines are of three different types:

(a) the lines spanned by p and p_t;

(b) the lines of the planes spanned by p and L_t not going through p;

(c) the totally singular lines of $\mathbf{W}(5, \mathbb{K})$ not contained in the hyperplane $X_0 = 0$, and meeting this hyperplane in a point of some line pp_t.

The incidence relation is inherited from $\mathbf{PG}(5, \mathbb{K})$. Then Γ is isomorphic to the dual of $\mathsf{H}(\mathbb{K})$. The set $\mathcal{C} = \{p_t : t \in \mathbb{K} \cup \{\infty\}\}$ is a twisted cubic. The polar plane with respect to $\mathbf{W}(3, \mathbb{K})$ of any point of $\mathcal{C}$ is an osculating plane at that point of $\mathcal{C}$. The line L_t is the tangent line at p_t to $\mathcal{C}$. For more details, we refer to BADER & LUNARDON [1993]. Note also that LUNARDON [1993] constructs in a similar way some of the finite twisted triality hexagons.

3.8 Other generalized polygons

3.8.1 Curious generalized quadrangles and hexagons

We end this chapter by constructing an infinite generalized quadrangle with a projective regular point for which the perp-plane is not classical. We do that by giving the quadratic quaternary ring.

Let $R_1 = R_2 = \mathbb{K}((t)) \times \mathbb{K}((t))$, where $\mathbb{K}$ is any field of characteristic 2 or 3. Let r be any element of $\mathbb{K}((t))$ having natural valuation 1 and such that the coefficient of t in r is equal to 1. Define the field automorphism

$$\theta : \mathbb{K}((t)) \to \mathbb{K}((t)) : f(t) \mapsto f(r).$$

This is well defined. Next, we define the multiplication $a \odot b = (a_0, a_1) \odot (b_0, b_1) = (a_0 b_0 + t a_1^\theta b_1^\theta, a_0 b_1 + a_1 b_0)$ on $\mathbb{K}((t)) \times \mathbb{K}((t))$. If we denote by $a \cdot b$ the multiplication $a \odot b$ for $\theta = 1$ (i.e., $r = t$ and $a \cdot b$ can be viewed as a multiplication of $\mathbb{K}((t))(\sqrt{t}) = \mathbb{K}((\sqrt{t})))$, then we can define

$$\begin{cases} \Psi_1(k, a, l, a') = a \cdot a \cdot k + l, \\ \Psi_2(k, a, l, a') = a \odot k + a', \end{cases}$$

and this defines a quadratic quaternary ring $\mathcal{R} = (R_1, R_2, \Psi_1, \Psi_2)$, if the characteristic of $\mathbb{K}$ is equal to 2. The perp-geometry $\Gamma(\mathcal{R})_{(\infty)}$ is a projective plane coordinatized by the planar ternary field $T(a, k, b) = a \odot k + b$ (see Theorem 3.3.2). If $r \neq t$, this is a proper division ring plane in the sense of HUGHES & PIPER [1973]. This example is taken from VAN MALDEGHEM [1989a]. It arises in the theory of affine buildings of rank 3, and in particular, of type $\tilde{C}_2$, and it is a generalized quadrangle with valuation (see Section 9.7 on page 419).

It is clear how one constructs a similar example in characteristic 3 defining an infinite generalized hexagon with a polar point, a non-classical perp-plane and a non-classical span-geometry. We give the coordinatizing ring. Let $\mathbb{K}$ have characteristic 3 and define $R_1 = R_2 = \mathbb{K} \times \mathbb{K}$ again, with the operations $\cdot$ and $\odot$ as above. The hexagonal sexternary ring is defined as:

$$\begin{cases} \Psi_1(k, a, l, a', l', a'') = a \cdot a \cdot a \cdot k + l, \\ \Psi_2(k, a, l, a', l', a'') = a \cdot (a \odot k) + a' + a \cdot a'', \\ \Psi_3(k, a, l, a', l', a'') = a \cdot a \cdot a \cdot k \cdot k + l' - k \cdot l, \\ \Psi_4(k, a, l, a', l', a'') = -a \odot k + a''. \end{cases}$$

3.8.2 Conclusion

So it seems that in the infinite case, one can twist slightly the defining algebraic relations of a Moufang polygon to obtain a new polygon which has typically non-Moufang properties. Tricks like that have also successfully been applied in other cases, like the compact connected case for quadrangles; see Subsection 9.3.4 (page 410).

In conclusion, we may say that for every $n \geq 3$, there are lots of generalized n-gons (via free constructions; see Subsection 1.3.13 on page 13). For $n = 3, 4, 6$, other nice examples can be constructed explicitly. No explicit construction of a non-Moufang generalized n-gon is known to me for $n \neq 3, 4, 6$ without invoking a free construction process at some stage. In the finite case, a lot of generalized n-gons are known for $n = 3, 4$, but for $n = 6, 8$, the only known examples are the Moufang n-gons. Needless to say that one of the most important problems in the theory of generalized polygons is to find an example of a non-Moufang finite generalized hexagon or octagon or to show there is none; see Problem 1 in Appendix E.

Chapter 4

Homomorphisms and Automorphism Groups

4.1 Introduction

This chapter deals with morphisms of polygons and is, together with Chapter 5, the most group-related chapter in this book.

Most of the morphisms that one ever meets are isomorphisms, especially automorphisms. We have already discussed isomorphisms between some classical polygons, so the emphasis of this chapter will be on collineations and collineation groups. But we start with a result by PASINI [1983] stating that every epimorphism between finite generalized polygons is an isomorphism. Then there are two ways to go. The first way is to study *local properties* of collineations, i.e., properties of certain special collineations that can act on a generalized polygon; in particular, we will study the fixed point structure of such automorphisms. Also, one can try to see what kind of collineations live in the automorphism group of the classical polygons that we have defined so far.

The second thing to study are *global properties*, i.e., answering questions like "how does the automorphism group of a polygon look"? For the classical polygons, the class of groups with a so-called *Tits system of rank* 2 is essentially equivalent to the class of generalized polygons having a group acting transitively on the set of ordered apartments (we will call this the *Tits condition*; in the literature, the corresponding automorphism groups are sometimes called *strongly transitive*) and so we study those groups a little. Using the classification of the finite simple groups, one can show that for finite polygons, the Tits condition implies that the polygon is classical. It is an open problem to find a proof which does not use this classification and which works for n-gons, $n \neq 3$. The most far-reaching results in this direction can be found in Chapter 6; see e.g. Theorem 6.8.9. In fact, using the classification of finite simple groups, one can classify all *point-distance transitive* polygons. We will state these results, but their proofs are beyond the scope of this book.

Furthermore, some non-classical polygons (different from projective planes) can be described merely in terms of an automorphism group. The prototypes are probably the Kantor systems that give rise to elation generalized quadrangles (see Section 3.7.3 on page 124). Since this theory is developed in PAYNE & THAS [1984] (see also THAS [1995]), we will not consider it here in full detail, another reason being that no relevant analogue has yet been found useful for dealing with (non-classical) hexagons or octagons (on which the emphasis of this book lies). So we only give a brief introduction.

We shall also calculate the order of the automorphism groups of the finite classical generalized polygons.

4.2 A theorem of Pasini on epimorphisms

An **epimorphism** of some generalized n-gon $\Gamma = (\mathcal{P}, \mathcal{L}, \mathbf{I})$ onto some generalized n-gon $\Gamma' = (\mathcal{P}', \mathcal{L}', \mathbf{I}')$ is a pair of surjections $\mathcal{P} \to \mathcal{P}'$, $\mathcal{L} \to \mathcal{L}'$ preserving incidence (and hence diminishing distances). The following result essentially says that an epimorphism between generalized n-gons either is an isomorphism or has some infinite "fibres". This is due to PASINI [1983] in the general case; for $n = 3$, see SKORNJAKOV [1957], HUGHES [1960] and MORTIMER [1975]. However, the proof below is inspired by BÖDI & KRAMER [1995]. In fact, we slightly simplify their proof and keep to purely geometric arguments.

4.2.1 Theorem (Skornjakov [1957], Pasini [1983]). *Let θ be an epimorphism from a generalized n-gon Γ onto a generalized n-gon Γ'. Then either θ is an isomorphism, or there exists an element v in Γ' with infinite pre-image (i.e., with $|v^{\theta^{-1}}|$ infinite). In the latter case, the pre-image of every element is infinite; more exactly, for any flag $\{u, w\}$ of Γ, there are infinitely many elements v incident with u and such that $v^\theta = w^\theta$. In particular, if θ is not an isomorphism, then for the order (s, t) of Γ, we have that both s and t are infinite, and hence Γ itself is infinite.*

Proof. Note that θ diminishes distances. This implies in particular that, if $\delta(u^\theta, w^\theta) \in \{n, n-1\}$, then $\delta(u, w) = \delta(u^\theta, w^\theta)$.

Suppose that θ is not an isomorphism. For any ordered flag (u, w) of Γ, let $\eta_{u,w}$ be the cardinality of the set $\{v \mathbf{I} u : v^\theta = w^\theta\}$.

First we claim that for a line L of Γ and a point x on L, the number $\eta_{L,x}$ is independent of both L and x. Let u^θ be opposite L^θ in Γ' and let w^θ be the projection of x^θ onto u^θ. Then L and u are opposite. Let p be incident with L, with $p^\theta = x^\theta$. Let v be the projection of p onto u. Then

$$n - 1 = \delta(p, u) = \delta(p, v) + \delta(v, u) \geq \delta(x^\theta, v^\theta) + \delta(v^\theta, u^\theta) \geq \delta(x^\theta, u^\theta) = n - 1,$$

hence $\delta(x^\theta, v^\theta) = n - 2$. Consequently $v^\theta = w^\theta$ and $\{p \mathbf{I} L : p^\theta = x^\theta\}^{[L;u]} \subseteq \{v \mathbf{I} u : v^\theta = w^\theta\}$. Similarly (or by symmetry), the reverse inclusion $\supseteq$ holds. The claim

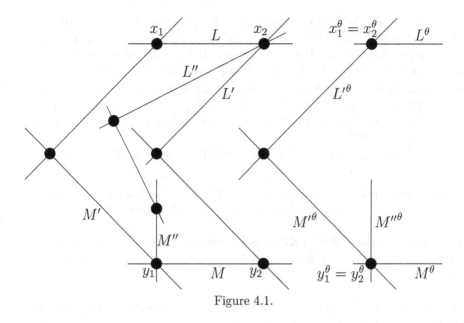

Figure 4.1.

now follows easily from Lemma 1.5.1(i) (see page 17). Note that the dual claim also holds and that for odd n, the number $\eta_{u,w}$ is independent of the ordered flag (u,w). This follows also from Lemma 1.5.1(i).

We put $\eta_{L,x}$ equal to η and $\eta_{x,L}$ equal to η'. So if n is odd, then $\eta = \eta'$.

Now we claim that there exists at least one element u in Γ incident with two distinct elements w, w' such that $w^\theta = w'^\theta$; or in other words, at least one of η, η' is greater than 1. Indeed, since θ is not an isomorphism, there exist distinct elements u_1 and u_2 of Γ such that $u_1^\theta = u_2^\theta$. Let γ be a minimal path from u_1 to u_2. Clearly γ^θ must stammer and our claim follows easily.

Now we show that, if $\eta > 1$, then η' is infinite. So by assumption there exists a line L of Γ incident with two distinct points x_1, x_2 such that $x_1^\theta = x_2^\theta$. To fix ideas, suppose that n is even. The proof for odd n is completely the same, up to the names "point" and "line" for some elements. Let M^θ be opposite L^θ (see Figure 4.1, where we have drawn the situation for $n = 6$). Then L is opposite M. Let y_i be the projection of x_i onto M, $i = 1, 2$. Also, let L' (M') be the projection of M onto x_2 (of L onto y_1). Let (s', t'), s', t' possibly infinite, be the order of Γ'. Let M'' be any line incident with y_1 such that $M''^\theta \neq M'^\theta$. According to our first claim, there are $\eta' t'$ choices for M''. Consider the projection L'' of M'' onto x_2 (the latter is opposite y_1 so that the projection map is bijective from $\Gamma_1(y_1)$ to $\Gamma_1(x_2)$). Since $\delta(x_2^\theta, y_1^\theta) = n - 2$, and since $M'^\theta = \text{proj}_{y_1^\theta} x_2^\theta$, we infer from Lemma 1.5.6 that $\delta(x_2^\theta, M''^\theta) = n - 1$. So the only element of Γ' incident with

x_2^θ and not opposite M''^θ is the projection of the latter onto the former, and that is L'^θ. Since θ diminishes distances, this implies that $L''^\theta = L'^\theta$. But there are only η' choices for such L''. Hence, by the injectivity of the projection map, we conclude that $\eta't' \leq \eta'$. But $t' \geq 2$; this is only possible if η' is infinite.

Now we can finish our proof. We have shown that at least one of η, η' is greater than 1. Without loss of generality we may assume $\eta > 1$. But the previous paragraph implies that η' is infinite, in particular, $\eta' > 1$. The dual argument now implies that η is infinite.

Of course, for n odd, the last paragraph is superfluous since in this case $\eta = \eta'$.

The theorem is proved. □

A **proper epimorphism** is an epimorphism which is not an isomorphism. An immediate consequence of the previous theorem is the following result.

4.2.2 Corollary (Skornjakov [1957], Pasini [1983]).   *There does not exist any proper epimorphism from a finite generalized n-gon onto a finite generalized n-gon.* □

From the proof of Theorem 4.2.1 also follows a local characterization of isomorphisms.

4.2.3 Corollary (Bödi & Kramer [1995]).   *An epimorphism θ between generalized polygons is an isomorphism **if and only if** the restriction of θ to at least one point row or to at least one line pencil is bijective.* □

4.2.4 Remark.   The thickness assumption in both Theorem 4.2.1 and Corollary 4.2.2 cannot be dispensed with. Indeed, consider any generalized n-gon Γ coordinatized as in Chapter 3. Then the map sending an element with i coordinates (counting (∞) and $[\infty]$ as elements with zero coordinates) to the corresponding element with i coordinates all of which are 0, is a proper epimorphism (a *retraction onto an apartment* in TITS' building language [1974]) from Γ onto the hat-rack of the coordinatization, viewed as an ordinary n-gon.

4.2.5 Examples.   Examples of proper epimorphisms onto finite n-gons exist for all $n \in \{3,4,6,8\}$. For projective planes, generalized quadrangles and generalized hexagons, it suffices to consider the Pappian examples over a locally finite local field and define an epimorphism in the obvious way onto the Pappian polygon over the residue field by extending the epimorphism between the two fields mentioned to the projective space in which the Pappian polygon has a standard embedding (i.e., a projective space of dimension 2, 3 and 6, respectively). See also Theorem 9.7.4 on page 422. For generalized octagons, one can do the same in a projective space of dimension 25 with a Ree–Tits octagon over the field of Laurent series over a finite field $\mathbf{GF}(2^{2e+1})$ with variable X, where the exponents of X are in $\mathbb{Z}(\sqrt{2})$.

4.2.6 Remark. In VAN MALDEGHEM [1990a], it is proved without introducing a variation of the free construction of polygons (but by using some theory of affine buildings) that for $n = 3, 4, 6$, there exists a generalized n-gon (of order $|\mathbb{R}|$) admitting an epimorphism onto every finite generalized n-gon.

From now on, we concentrate on automorphisms. We first collect some notation and elementary known results in group theory that we will need.

4.3 Notation and results from group theory

Let G be any (non-trivial) group, with identity (or neutral) element 1. Every element g of G defines an **inner automorphism** of G by **conjugation** as $x \mapsto x^g = g^{-1}xg$. A subgroup N which is stabilized under every inner automorphism of G is a **normal subgroup**; in symbols, $N \trianglelefteq G$. The **quotient group** G/N is the group of all cosets $xN = \{xn : n \in N\}$ with multiplication law $xN \cdot yN = xyN$ (and this is well-defined exactly because N is a normal subgroup). A group without proper normal subgroups (*proper* means: distinct from $\{1\}$ and G itself) is a **simple** group. An **almost simple group** with **socle** H is a subgroup of $\operatorname{Aut} H$ containing the simple non-abelian group H. An example of a normal subgroup is the **derived group** or **commutator subgroup** $G' = [G, G]$, generated by all **commutators** $[x, y] = x^{-1}y^{-1}xy$, $x, y \in G$. If we define inductively the dth **normal derivative** $G^{(d)}$ of G by $G^{(0)} = G$ and $G^{(d)} = [G^{(d-1)}, G^{(d-1)}]$, then we say that G is **soluble** if $G^{(d)} = \{1\}$ for some d. It is easy to prove that, if $N \trianglelefteq G$, then G is soluble **if and only if** both N and G/N are soluble. If we set $[A, B] = \langle [a, b] : a \in A, b \in B \rangle$ (where $\langle S \rangle$, $S \subseteq G$, denotes the subgroup of G generated by S, i.e., the intersection of all subgroups of G containing S), then we can define by induction the dth **central derivative** $G^{[d]}$ as $[G^{[d-1]}, G]$, where $G^{[1]} = G^{(1)} = [G, G]$. A **nilpotent** group G is a group for which $G^{[d]} = \{1\}$, for some $d \in \mathbb{N}$. If $G^{[d]} = \{1\}$ and $G^{[d-1]} \neq \{1\}$, then we say that G is nilpotent **of class** d. The nilpotent groups of class 1 are precisely the non-trivial abelian groups. Note that every nilpotent group is soluble.

Let $N \trianglelefteq G$, and suppose that H is a subgroup of G (written $H \leq G$) with the property that $H \cap N = \{1\}$ and $\langle H, N \rangle = G$. Then $H \cong G/N$ and we denote $G = N : H$. We also say that G is the **semi-direct product of N and H**.

If for subsets A, B of G we have $[A, B] = \{1\}$, then we say that A **centralizes** B. The **centre** $Z(G)$ of G is defined by $Z(G) = \{x \in G : [x, G] = \{1\}\}$, and it is clearly a normal subgroup of G. For $g \in G$, we put $A^g = \{x^g : x \in A\}$. If $A^g = A$, for $g \in G$, then we say that g **normalizes** A. If all elements of B normalize A, then we likewise say that B **normalizes** A.

A subgroup H of G, $H \neq G$, is a **maximal subgroup** if $\langle H, g \rangle = G$, for every element $g \in G \setminus H$. If G acts on a set X, then we say that G acts **primitively** if G does not stabilize any non-trivial partition of X (the trivial partitions of X are the partitions $\{X\}$ and $\{\{x\} : x \in X\}$). We denote the action of G on X exponentially.

Then, if $G_x = \{g \in: x^g = x\}$, $x \in X$, then we have that G_x is a maximal subgroup of G precisely if G acts primitively on X. Also, we say that G acts **semi-regularly** on X if $|G_x| = 1$, for all $x \in X$. If moreover G acts **transitively** (i.e., there exist $g \in G$ with $x^g = y$, for all $x, y \in X$), then G is said to act **regularly** on X. An element $g \in G$ acts **freely** on X if $\langle g \rangle$ acts semi-regularly on X.

4.4 Root elations and generalized homologies

4.4.1 Definitions. Let Γ be a generalized n-gon. If a collineation g of Γ fixes all elements incident with at least one element of a given path γ of length $n - 2$, then we call g a **root elation**, **γ-elation**, or briefly an **elation**. When n is even, then there are two kinds of $(n - 2)$-paths, namely, a path of length $n - 2$ can start and end with a point, or with a line. In the first case, we sometimes call a corresponding elation a **point-elation**; in the other case logically a **line-elation**. For n even, we define a **central elation** (also called a **central collineation**) g as a collineation fixing all elements at distance no more than $\frac{n}{2}$ of a certain point p, which is then called the **centre (of g)**. Dually, an **axial elation** or **axial collineation** g is a collineation fixing everything at distance no more than $\frac{n}{2}$ from a certain line, called the **axis (of g)**. We will see later on that every classical generalized n-gon (with n still even) admits central or axial elations (see Corollary 5.4.7 on page 206), but not necessarily both. Let v and w be two opposite elements of Γ (and now n is no longer necessarily even); then a $\{v, w\}$-**homology**, or briefly a **(generalized) homology** is a collineation fixing every element incident with v or incident with w.

The next theorem gathers some conditions under which we may conclude that the fixed point structure of a collineation is some weak generalized subpolygon of certain type.

4.4.2 Theorem.

  (*i*) *Let g be a collineation of a generalized polygon Γ. Then the subgeometry Γ' of Γ of points and lines fixed by g is a weak subpolygon of Γ **if and only if** g fixes some apartment pointwise.*

  (*ii*) *Let g be a collineation of a generalized n-gon Γ. **Then** the subgeometry Γ' of Γ of points and lines fixed by g is isomorphic or dual to the multiple of some generalized n''-gon Γ'', for $n'' \geq 2$ a multiple of some positive divisor d of n **if** g fixes some apartment Σ and two elements v, w not belonging to Σ, but incident with respective (unique) elements v_Σ, w_Σ of Σ such that the greatest common divisor of $\delta(v_\Sigma, w_\Sigma)$ and n is equal to $\frac{n}{d}$ (**if** in particular $\delta(v_\Sigma, w_\Sigma)$ and n are relatively prime, **then** Γ' is a subpolygon; this happens for instance in the case where v_Σ and w_Σ are incident).*

(*iii*) *Let g be a collineation of a generalized n-gon Γ, with n even. Then the subgeometry Γ' of Γ of points and lines fixed by g is a full weak subpolygon* **if and only if** *g fixes an apartment Σ and all points on certain lines L_1 and L_2 of Σ with $\frac{\delta(L_1,L_2)}{2}$ relatively prime to $\frac{n}{2}$.*

(*iv*) *Let g be a collineation of a generalized n-gon Γ with n even. Then the subgeometry Γ' of Γ of points and lines fixed by g is a full subpolygon* **if and only if** *g fixes an apartment Σ, all points on a certain line L of Σ and at least three lines through a point p of Γ with $\delta(p, L)$ relatively prime to n (this happens in particular when p is incident with L).*

(*v*) *Let g be a collineation of a generalized n-gon Γ with n even. Then g is the identity* **if and only if** *g fixes an apartment Σ, all points on a certain line L of Σ and all lines through a certain point p of Γ with $\delta(p, L)$ relatively prime to n (this happens in particular when p is incident with L).*

(*vi*) *Let g be a collineation of a generalized n-gon Γ with n odd. Then g is the identity* **if and only if** *g fixes an apartment Σ, all elements incident with a certain element v_1 of Σ and at least three elements incident with a certain element v_2 of Γ, with $\delta(v_1, v_2)$ relatively prime to n (this happens in particular when v_1 is incident with v_2).*

Proof. This theorem states necessary and/or sufficient conditions for an automorphism g to have as fixed point structure a weak subpolygon satisfying some additional properties. In each case, however, the necessary condition is obvious. So it remains to consider the respective sufficient conditions.

(*i*) Since Γ' is a subgeometry of Γ, the girth of the incidence graph of Γ' cannot be smaller than the girth of the incidence graph of Γ. But as Γ' contains an apartment, these girths are equal.

Now suppose that two elements v and w are at distance i from each other in Γ' (i is possibly infinite). If v and w are not opposite in Γ, and if they are at distance j from each other in Γ, then the path of length j connecting v and w is unique. Since g fixes v and w, it must also fix that path, hence $i = j$. Now suppose that v and w are opposite in Γ. Since Γ' contains an apartment, there is at least one element u in Γ', $u \neq w$, not opposite w. By the previous argument, the projection w' of u onto w belongs to Γ' and is not opposite v. Hence w' has the same distance to v in Γ as in Γ'. Therefore w has the same distance to v in Γ as in Γ'. So also the diameter of Γ' is equal to the diameter of Γ. The result now follows from Lemma 1.5.10 (see page 21).

(*ii*) We already know that g fixes a weak sub-n-gon Γ' pointwise. By the Structure Theorem of weak polygons (see Theorem 1.6.2 on page 22), Γ' is isomorphic or dual to the multiple of a generalized n''-gon Γ'' and two thick elements

v_Σ and w_Σ of Γ' lie at distance $k\frac{n}{n''}$ from each other, for some integer k, $0 \leq k \leq n''$. Since $\frac{n}{n''}$ divides both $\delta(v_\Sigma, w_\Sigma)$ and n, the greatest common divisor of these two numbers, which is equal to $\frac{n}{d}$, is a multiple of $\frac{n}{n''}$. Therefore n'' is a multiple of d. This shows the first part of (ii). If in particular $\delta(v_\Sigma, w_\Sigma)$ and n are relatively prime, then $d = n$ and hence $n'' = n$. This completes the proof of (ii).

(iii) The given condition immediately implies that the greatest common divisor of $\delta(L_1, L_2)$ and n is equal to $2 = \frac{n}{n/2}$, hence by (ii), Γ' is either a subpolygon, or isomorphic or dual to the double of a generalized $\frac{n}{2}$-gon Γ''. If Γ' is a subpolygon, then we use (the dual of) Proposition 1.8.1 (see page 34) to conclude that Γ' is full; suppose now that Γ' is isomorphic or dual to the double of a generalized $\frac{n}{2}$-gon Γ''. Note that, since Γ' has some thick lines, it is dual to the double of Γ''. If $\frac{n}{2}$ is odd, then, by Lemma 1.5.1(i) on page 17, there exists a projectivity $\Gamma''(v) \to \Gamma''(w)$ for every two elements v and w of Γ''. Now every projectivity in Γ'' induces a projectivity in Γ' and hence in Γ. Since $\Gamma(L_1) = \Gamma'(L_1)$, it follows that Γ' is full. If $\frac{n}{2}$ is even, then the distance from L_1 to L_2 in Γ' is not divisible by 4 (this follows immediately from the assumptions), hence they represent distinct types of elements in Γ''. A similar argument to that used before completes the proof.

(iv) By (ii), g fixes a subpolygon Γ' pointwise (noting that we can always replace Σ by an apartment containing both L and p). Using Lemma 1.8.1 again, one deduces that Γ' is full (as in the previous paragraph).

(v) By (iv) and its dual, g fixes a subpolygon Γ' pointwise and Γ' is full and ideal, hence it coincides with Γ by Proposition 1.8.2, therefore g fixes all points and lines of Γ and thus is the identity.

(vi) By (ii), g fixes a subpolygon Γ'. Since n is odd, there is a projectivity $\Gamma(w_1) \to \Gamma(w_2)$ for every pair of elements w_1, w_2 of Γ'. Since every such projectivity is also a projectivity in Γ, and since for at least one element v_1 of Γ we have $\Gamma(v_1) = \Gamma'(v_1)$, the subpolygon Γ' is full and ideal, hence it coincides with Γ.

The proof of the theorem is complete. $\square$

It is known that for projective planes, elations and homologies act freely on a set of points incident with a certain well-defined fixed line. There is a similar result for generalized polygons.

4.4.3 Proposition.

(i) *Let* $\gamma = (v_1, v_2 \ldots, v_{n-1})$ *be a path of length* $n - 2$ *in the generalized* n-*gon* Γ. *Let* v_0 *be an element of* Γ *which is incident with* v_1, $v_0 \neq v_2$. *Then the set* G *of all* γ-*elations forms a group which acts semi-regularly on the set* S *of* elements incident with v_0 but distinct from v_1. In particular, every γ-elation acts freely on that set. Equivalently, **if** v_n is incident with v_{n-1}, $v_n \neq v_{n-2}$, **then** G acts semi-regularly on the set $\mathcal{A}$ of apartments containing v_0, γ and v_n. Every γ-elation acts freely on $\mathcal{A}$. Also, G acts transitively on S **if and only if** it acts transitively on $\mathcal{A}$.

(ii) Let v and w be opposite elements in the generalized n-gon Γ. Then the set of all $\{v, w\}$-homologies forms a group G. Let v' be incident with v and put $S' := \Gamma(v') \setminus \{v, \mathrm{proj}_{v'} w\}$. **If** n is odd, **then** G acts semi-regularly on S'.

Proof.

(i) Clearly G is a group. Suppose $g \in G$ fixes an element v of S. There is a unique path γ' of length $n - 1$ from v to v_n and g fixes v and v_n, hence it fixes that path pointwise. Thus g fixes an apartment (consisting of v_0 and the union of γ and γ'). But g fixes $\Gamma(v_1) \cup \Gamma(v_2)$ pointwise and so the result follows from Theorem 4.4.2(v), noting that every element of S uniquely defines an element of $\mathcal{A}$ (as just described) and conversely, every apartment containing v_0, γ, v_n contains an element $v \mathbf{I} v_0$, $v \neq v_1$ (and so $v \in S$).

(ii) Again the fact that G is a group is trivial. Clearly every apartment through v and w is fixed by any element g of G. If g, moreover, fixes an element of S', then by Theorem 4.4.2(vi) it must be the identity, provided n is odd.

The proposition is proved. □

4.4.4 Definitions. With the notation of Proposition 4.4.3(i), we say that γ is a **Moufang path** if the group of all γ-elations acts transitively on S, or equivalently on $\mathcal{A}$. If all paths of length $n - 2$ in Γ are Moufang paths, then we say that Γ is a **Moufang polygon**, or that Γ **satisfies the Moufang condition.** If Γ is a Moufang polygon, then the collineation group generated by all elations is often called the **little projective group** of Γ. Also, for any path γ of length $n - 2$ in a Moufang n-gon, the group of all γ-elations is called a **root group.**

With the notation of Proposition 4.4.3(ii), we say that Γ **is** $\{v, w\}$-**transitive,** if the group of all $\{v, w\}$-homologies acts transitively (but not necessarily regularly) on the set S'. Now put $w' = \mathrm{proj}_{v'} w$. If the group of all $\{v, w\}$-homologies acts transitively on the set $\Gamma_1(w') \setminus \{v', \mathrm{proj}_{w'} w\}$, then we say that Γ is $\{v, w\}$-**quasi-transitive.**

4.4.5 Remark. There do exist $\{v, w\}$-homologies in generalized n-gons, with n even, which do not act freely on the set S' defined in (ii) of the previous proposition. Indeed, consider the coordinatization of the twisted triality hexagon $\mathsf{T}(\mathbb{L}, \mathbb{K}, \sigma)$ as in Subsection 3.5.8 (page 114). Raising every coordinate to the power σ induces a collineation which fixes the hat-rack, all lines through (∞) and $|\mathbb{K}| + 1$ points on

$[\infty]$, namely, all those with their coordinate in $\mathbb{K} \cup \{\infty\}$. A similar example exists for the Hermitian quadrangle $\mathsf{H}(3, \mathbb{K}', \sigma')$, with $\mathbb{K}'$ a (commutative) field.

However, a sufficient condition to rule this out in the finite case is provided by the following result.

4.4.6 Proposition. *Let L and M be opposite lines in a finite generalized n-gon Γ of order (s, t), with n even. Let x be incident with L and put $S := \Gamma(x) \setminus \{L, L'\}$, where L' is the projection of M onto x. If $s \geq t$, then every $\{L, M\}$-homology acts freely on S.*

Proof. Let g be an $\{L, M\}$-homology. If g fixes an element of S, then by Theorem 4.4.2(iv), g fixes a full subpolygon Γ'. As $s \geq t$, Corollary 1.8.10 on page 37 implies that $\Gamma' = \Gamma$. $\qquad\qquad\square$

4.5　Collineations of classical polygons

The classical polygons have large automorphism groups. They all admit the maximal number of root elations and lots of generalized homologies. We will show this for large classes of classical polygons. In the finite case, we will use the results of this section to determine the order of the collineation groups of all classical polygons in the next section.

Root elations

For projective planes, it is obvious that every path of length 1 in a Desarguesian plane is a Moufang path, and it is well known that this is also the case in every alternative plane (where the name "Moufang path" comes from). Hence the theorem:

4.5.1 Theorem. *Every classical or alternative projective plane is a Moufang plane.*

For generalized quadrangles we have the following results, all due to TITS [1976a].

4.5.2 Theorem (Tits [1976a]). *Every classical generalized quadrangle is a Moufang quadrangle.*

Proof. Let
$$V = e_{-2}\mathbb{K} \bigoplus e_{-1}\mathbb{K} \bigoplus V_0 \bigoplus e_1\mathbb{K} \bigoplus e_2\mathbb{K}$$
be a right vector space over the skew field $\mathbb{K}$, let σ be an anti-automorphism of $\mathbb{K}$ with $\sigma^2 = 1$, let $g : V \times V \to \mathbb{K}$ be a $(\sigma, 1)$-linear form and let

$$q : V \to \mathbb{K}/\mathbb{K}_\sigma : (x_{-2}, x_{-1}, x_0, x_1, x_2) \mapsto x_{-2}^\sigma x_2 + x_{-1}^\sigma x_1 + q_0(x_0) + \mathbb{K}_\sigma$$

be the associated non-degenerate σ-quadratic form, where q_0 is a non-degenerate anisotropic σ-quadratic form on V_0, and where $q(x) = g(x, x) + \mathbb{K}_\sigma$. Let $\mathbb{Q}(V, q)$ be the corresponding generalized quadrangle (see Subsection 2.3.1 on page 53). We consider the coordinatization of $\mathbb{Q}(V, q)$ as presented in Section 3.4.7. In particular, we use the operations $\oplus$ and $\otimes$ again, defined as:

$$
\begin{aligned}
(k_0, k_1) \oplus (l_0, l_1) &= (k_0 + l_0, k_1 + l_1 - f(k_0, l_0)), \\
a \otimes (k_0, k_1) &= (k_0 a, a^\sigma k_1 a).
\end{aligned}
$$

First note that the description of $\mathbb{Q}(V, q)$ is symmetric with respect to the hat-rack of the coordinatization. So in order to show that every path of length 2 contained in the hat-rack is a Moufang path, it suffices to show this for two paths, respectively corresponding to point-elations and line-elations. We consider the following map $E(A)$, $A \in \mathbb{K}$ on the coordinates:

$$
E(A) : \quad
\begin{aligned}
(\infty) &\mapsto (\infty), \\
(a) &\mapsto (a), \\
[a, (l_0, l_1)] &\mapsto [a, (l_0, l_1) \oplus (0, Aa^\sigma - aA^\sigma)], \\
(a, (l_0, l_1), a') &\mapsto (a, (l_0, l_1) \oplus (0, Aa^\sigma - aA^\sigma), a' + A), \\
[\infty] &\mapsto [\infty], \\
[(k_0, k_1)] &\mapsto [(k_0, k_1)], \\
((k_0, k_1), b) &\mapsto ((k_0, k_1), b + A), \\
[(k_0, k_1), b, (k_0', k_1')] &\mapsto [(k_0, k_1), b + A, (k_0', k_1')].
\end{aligned}
$$

Clearly $E(A)$ fixes all lines through (∞) and through (0), and it fixes all points on $[\infty]$. It is readily checked that $E(A)$ preserves incidence. Hence $E(A)$ is an $((\infty), [\infty], (0))$-elation for every $A \in \mathbb{K}$ and it maps $(0, 0)$ to $(0, A)$. Since A was arbitrary, this means that the path $((\infty), [\infty], (0))$ is a Moufang path.

Dually, one shows that $([\infty], (\infty), [0])$ is a Moufang path by considering the following map $E((K_0, K_1))$, $(K_0, K_1) \in R_2$:

$$
E((K_0, K_1)) : \quad
\begin{aligned}
(\infty) &\mapsto (\infty), \\
(a) &\mapsto (a), \\
[a, (l_0, l_1)] &\mapsto [a, (K_0, K_1) \oplus (l_0, l_1)], \\
(a, (l_0, l_1), a') &\mapsto (a, (K_0, K_1) \oplus (l_0, l_1), a'), \\
[\infty] &\mapsto [\infty], \\
[(k_0, k_1)] &\mapsto [(k_0, k_1)], \\
((k_0, k_1), b) &\mapsto ((k_0, k_1), b - f(K_0, k_0)), \\
[(k_0, k_1), b, (k_0', k_1')] &\mapsto [(k_0, k_1), b - f(K_0, k_0), (K_0, K_1) \oplus (k_0', k_1')].
\end{aligned}
$$

Now note that the hat-rack was chosen arbitrarily (it depends on the standard form of q). Alternatively, one can use Lemma 5.2.8 on page 177; either way we conclude that every path of length 2 of a classical quadrangle is a Moufang path. The theorem is proved. $\qquad\square$

It follows now for instance, as remarked in the proof of Proposition 3.4.8 on page 106, that for the quadrangle $Q(V,q)$ (using standard notation), $f|(V_0 \times V_0)$ being identical zero is independent of the coordinates in $\mathbf{PG}(V)$.

4.5.3 Remark. In the proof of the previous theorem, it was in fact enough to give the action of the elations on the points and lines opposite (∞) and $[\infty]$, respectively. In general, if a collineation θ of some coordinatized generalized n-gon Γ fixes both (∞) and $[\infty]$, then it suffices to give the action of θ on the points and lines opposite one of (∞) or $[\infty]$ to determine θ completely and to check that θ is a collineation. Indeed, the image of any other element v, where we may assume without loss of generality that $\delta(v,(\infty)) < \delta(v,[\infty])$, is uniquely determined by the image of any element x opposite $[\infty]$ such that $\delta(x,v) + \delta(v,(\infty)) = n-1$. If this is independent of x (which is readily checked), then incidence is for these elements automatically preserved, if it is preserved for all elements opposite (∞) or $[\infty]$.

It is clear that the group of all point-elations in a classical quadrangle $Q(V,q)$ belonging to some fixed Moufang path is isomorphic to the additive group of the underlying skew field $\mathbb{K}$. We also observe that the group of root line-elations belonging to a fixed root is isomorphic to the group $\widehat{X}$ (with the notation of Subsection 3.4.7 on page 104). We use that observation to prove the following result.

4.5.4 Proposition. *A self-dual classical generalized quadrangle is a mixed quadrangle.*

Proof. If $Q(V,q)$ is self-dual, then by the observations just made, $\widehat{X}$ must be abelian, hence, with our standard notation, $f(v,w) = f(w,v)$, for all $v, w \in V_0$. So $f(v,w) = f(v,w)^\sigma$, for all $v, w \in V_0$. Multiplying w by $k \in \mathbb{K}$, this implies that $f(v,wk) = f(v,w)k = k^\sigma f(v,w)$. If $f(v,w) \neq 0$, we can choose v such that $f(v,w) = 1$ and so $k = k^\sigma$ for all $k \in \mathbb{K}$, hence $\sigma = 1$, the quadrangle is orthogonal and therefore every line is regular. Consequently, every point is regular and the result follows from Proposition 3.4.8(iii) on page 106. So we may assume that $\sigma \neq 1$ and consequently that $f(v,w) = 0$, for all $v, w \in V_0$. But then Ψ_2 of the coordinatizing ring (see Subsection 3.4.7 on page 104) is independent of the third argument, implying that (∞) is a regular point (by Theorem 3.3.2). Hence all points and all lines are regular and the result again follows from Proposition 3.4.8(iii). This shows the assertion. $\square$

The next result has a proof similar to the one of Theorem 4.5.2. However, using Lemma 5.2.2 on page 174 (which is proved independently and could as well be proved here), the assertion is obvious.

4.5.5 Theorem (Tits [1976a],[19**]). *Every mixed generalized quadrangle is a Moufang quadrangle.*

Proof. Every mixed quadrangle is a subquadrangle of a symplectic quadrangle. The theorem follows now from Theorem 4.5.2 and Lemma 5.2.2. □

4.5.6 Theorem (Tits [1976a]). *Every classical generalized hexagon is a Moufang hexagon.*

Proof. We only show the result for the twisted triality hexagons and the hexagons of type $^6D_4$. For the split Cayley hexagons, it suffices to put $\mathbb{L} = \mathbb{K}$, $\sigma = 1$ and $\mathrm{Tr} = 3$ in the formulae below.

By Remark 4.5.3, it suffices to give the action of the elations on the points and lines opposite (∞) and $[\infty]$, respectively.

With respect to the coordinatization of the classical hexagon $\mathsf{T}(\mathbb{L}, \mathbb{K}, \sigma)$ given in Section 3.5.8 on page 114, we have the mapping $E(A)$, $A \in \mathbb{L}$, defined as follows:

$$\begin{aligned}
E(A) \quad : \quad (a, l, a', l', a'') \quad &\mapsto \quad (a, l - \mathrm{Tr}(Aa^{\sigma+\sigma^2}), a' + A^\sigma a^{\sigma^2} + A^{\sigma^2} a^\sigma, \\
&\qquad l' + \mathrm{Tr}(aA^{\sigma+\sigma^2}) + \mathrm{Tr}(a'A), a'' + A), \\
[k, b, k', b', k''] \quad &\mapsto \quad [k, b + A, k', b', k''].
\end{aligned}$$

Dually, we have the mapping $E(K)$, $K \in \mathbb{K}$, defined as:

$$\begin{aligned}
E(K) \quad : \quad (a, l, a', l', a'') \quad &\mapsto \quad (a, l + K, a', l', a''), \\
[k, b, k', b', k''] \quad &\mapsto \quad [k, b, k' - kK, b', k'' + K].
\end{aligned}$$

It takes a few elementary calculations to check that these mappings preserve incidence. And obviously, they are γ-elations respectively for $\gamma = ((\infty), [\infty], \ldots, (0,0,0))$ and (dually) $\gamma = ([\infty], (\infty), \ldots, [0,0,0])$. □

The proof of the next result is similar to the proof of Theorem 4.5.5, now combining Theorem 4.5.6 and Lemma 5.2.2.

4.5.7 Theorem (Tits [1976a]). *Every mixed hexagon is a Moufang hexagon.* □

It is a little silly to construct a root elation for the Ree–Tits octagons using coordinates. Indeed, the coordinatization was derived from the commutation relations which already express that every 6-path is a Moufang path. So this would be no proof at all. In fact, we did not prove the existence of the Ree–Tits octagon, and so it seems reasonable also to accept the fact that these octagons have Moufang 6-paths. Proofs of these assertions would either lead us into the Ree groups of characteristic 2, or require uninformative and tiresome calculations with coordinates. Of course, in order to work explicitly with root elations in the Ree–Tits octagons, we could try to find an expression in terms of coordinates, similar to the elations $E(\cdots)$ above. But since we do not need this for the rest of this book, we leave it to the reader. We only remark that the groups of point-elations are isomorphic to the additive group of the underlying field $\mathbb{K}$, and that the groups of line-elations are isomorphic to the group $\mathbb{K}_\sigma^{(2)}$ (see Subsection 3.2), and we state:

4.5.8 Theorem (Tits [1983]). *Every classical generalized octagon is a Moufang octagon.*   □

4.5.9 Remark. In fact, using the coordinatization of the Ree–Tits octagons, one could try to prove the existence of the Ree–Tits octagons (performing tedious calculations). Then one could compute explicitly some root elations and thus show that these octagons satisfy the Moufang condition. In fact, the commutation relations displayed in the next chapter are equivalent to the coordinatization. Sometimes it is better to use these relations to solve a question, sometimes it is better to use the coordinates. The latter has the advantage of being more elementary (but less elegant).

Generalized homologies

We start with the classical quadrangles.

4.5.10 Proposition.

(*i*) **If** Γ *is a classical quadrangle* $Q(V, q)$, *for some vector space* V *and non-degenerate* σ-*quadratic form* q *in* V, *and* **if** x, y *are two opposite points of* $Q(V, q)$, **then** $Q(V, q)$ *is* $\{x, y\}$-*transitive.*

(*ii*) **If** Γ *is a symplectic quadrangle,* **then** *for every pair of opposite elements* v, w, *the quadrangle* Γ *is* $\{v, w\}$-*transitive.*

(*iii*) *The mixed quadrangle* $Q(\mathbb{K}, \mathbb{K}'; L, L')$ *is* $\{x, y\}$-*transitive, for some pair of opposite points* x, y, **if and only if** *it is* $\{x, y\}$-*transitive for all pairs of opposite points* x, y **if and only if** $\mathbb{K} = L$. *Also the dual holds, from which it follows that the mixed quadrangle* $Q(\mathbb{K}, \mathbb{K}'; L, L')$ *is* $\{v, w\}$-*transitive for all pairs* v, w *of opposite elements* **if and only if** $\mathbb{K} = L$ *and* $\mathbb{K}' = L'$.

Proof.

(*i*) Using the coordinatization in Subsection 3.4.7 (page 104), we see that the map h_A, which is completely determined by its action on elements with three coordinates by Remark 4.5.3, is a $\{(\infty), (0, 0, 0)\}$-homology and it maps (1) to the arbitrary point (A), $A \in \mathbb{K} \setminus \{0\}$:

$$
h_A \ : \quad
\begin{aligned}
(a, (l_0, l_1), a') &\mapsto (Aa, A^\sigma \otimes (l_0, l_1), Aa'), \\
[(k_0, k_1), b, (k_0', k_1')] &\mapsto [(k_0, k_1), Ab, A^\sigma \otimes (k_0', k_1')].
\end{aligned}
$$

This proves (*i*).

(*ii*) By dualizing (*i*) (noting that the symplectic quadrangle over $\mathbb{K}$ is nothing other than the dual of an orthogonal quadrangle over $\mathbb{K}$; see Proposition 3.4.13 on page 109), we see that we only have to prove the result for opposite pairs of points x, y. We use the coordinatization of Subsection 3.4.1 (see page 100). The following map h_A, given by its action on the elements with three coordinates, is a $\{(\infty), (0,0,0)\}$-homology and maps (1) to the arbitrary point (A) on $[\infty]$, $A \in \mathbb{K} \setminus \{0\}$:

$$h_A \ : \ \begin{aligned} (a,l,a') &\ \mapsto\ (Aa, A^2l, Aa'), \\ [k,b,k] &\ \mapsto\ [k, Ab, A^2k']. \end{aligned}$$

The fact that h_A preserves incidence can be checked easily. This proves (*ii*).

(*iii*) We consider the coordinatization of the mixed quadrangle $Q(\mathbb{K}, \mathbb{K}'; L, L')$ as given in Subsection 3.4.2 on page 100. Let h_A be an $\{(\infty), (0,0,0)\}$-homology which maps (1) to (A). Then h_A induces a permutation $\mu : L \to L$ by the rule

$$(a^\mu) = (a)^{h_A}, \qquad a \in L.$$

Since h_A fixes $[1, 0, 0]$, the same permutation can be equivalently described as

$$(a^\mu, 0, 0) = (a, 0, 0)^{h_A} \Leftrightarrow (0, a^\mu) = (0, a)^{h_A} \Leftrightarrow (0, 0, a^\mu) = (0, 0, a)^{h_A}.$$

Similarly, there is a map $\nu : L' \to L'$ defined by

$$[0, 0, k^\nu] = [0, 0, k]^{h_A}, \qquad k \in L',$$

and we also have

$$[0, k^\nu] = [0, k]^{h_A}.$$

This implies that h_A acts as follows on elements with three coordinates:

$$h_A \ : \ \begin{aligned} (a,l,a') &\ \mapsto\ (a^\mu, l^\nu, a'^\mu), \\ [k,b,k'] &\ \mapsto\ [k, b^\mu, k'^\nu]. \end{aligned}$$

Since h_A preserves incidence, the maps μ and ν must satisfy the following equations for all $a, a' \in L$ and all $k, l \in L'$:

$$\begin{cases} (ak + a')^\mu &= a^\mu k + a'^\mu, \\ (a^2 k + l)^\nu &= (a^\mu)^2 k + l^\nu. \end{cases}$$

Putting $a = 1$ and $l = 0$ in the second equality, we see that $k^\nu = A^2 k$, for all $k \in L'$ (since $1^\mu = A$ by assumption). Now we set $k = 1$ and $l = 0$ in the same equation to obtain, using $(a^2)^\nu = A^2 a^2$ (since $a^2 \in \mathbb{K}^2 \subseteq L'$), $A^2 a^2 = (a^\mu)^2$. Therefore $a^\mu = Aa$, for all $a \in L$. So $Aa \in L$, for all $a \in L$. Conversely, it follows that if $Aa \in L$ for every $a \in L$, then h_A is well defined (since $A^2 \in \mathbb{K}^2$ and hence $A^2 k$ is always an element of L'). Hence, if all homologies h_A, $A \in L \setminus \{0\}$, exist, then the ring generated by L must be L itself. But L generates $\mathbb{K}$ as a ring, so $\mathbb{K} = L$. Whence the first assertion.

The dual holds by considering $Q(\mathbb{K}', \mathbb{K}^2; L', L^2)$; see Proposition 3.4.4 on page 102.

The proposition is proved.   □

4.5.11 Proposition. *Let Γ be a mixed or classical hexagon. Let x, y be two opposite elements of Γ. Then Γ is $\{x, y\}$-transitive.*

Proof. Suppose x and y are points. Let L be a line through x, and p, p' two points on L distinct from x and from $\mathrm{proj}_L(y)$. Then we have to show that there exists a collineation of Γ fixing all lines through x, fixing all lines through y and mapping p to p'. We coordinatize Γ in such a way that $x = (\infty)$, $y = (0, 0, 0, 0, 0)$, $L = [\infty]$ and $p = (1)$. We use the same notation as in Section 3.5. Then $p' = (A)$ for some $A \in R_1 \setminus \{0\}$. For the split Cayley hexagons and the mixed hexagons, we consider the map

$$
\begin{aligned}
h_1 : \quad & (a, l, a', l', a'') \mapsto (Aa, A^3 l, A^2 a', A^3 l', Aa''), \\
& [k, b, k', b', k''] \mapsto [k, Ab, A^3 k', A^2 b', A^3 k''],
\end{aligned}
$$

which clearly preserves incidence (in view of the hexagonal sexternary ring in Section 3.5.1) and acts as desired.

For the twisted triality hexagons and the hexagons of type $^6 D_4$, we consider the map

$$
\begin{aligned}
h_2 : \quad & (a, l, a', l', a'') & \mapsto \quad & (Aa, \mathrm{N}(A)l, (\mathrm{N}(A)/A)a', \mathrm{N}(A)l', Aa''), \\
& [k, b, k', b', k''] & \mapsto \quad & [k, Ab, \mathrm{N}(A)k', (\mathrm{N}(A)/A)b', \mathrm{N}(A)k''].
\end{aligned}
$$

which also preserves incidence and acts as desired. The proposition will be proved if we show that the split Cayley hexagons, the twisted triality hexagons and the hexagons of type $^6 D_4$ are $\{L, M\}$-transitive for every pair (L, M) of opposite lines. We can coordinatize such that $L = [\infty]$ and $M = [0, 0, 0, 0, 0]$. We have to prove that there is a map fixing all points on $[\infty]$ and $[0, 0, 0, 0, 0]$ and mapping $[1]$ to any line $[K]$, $K \in R_2 \setminus \{0\}$. Consider the following automorphism:

$$
\begin{aligned}
h_3 : \quad & (a, l, a', l', a'') & \mapsto \quad & (a, Kl, Ka', K^2 l', Ka''), \\
& [k, b, k', b', k''] & \mapsto \quad & [Kk, Kb, K^2 k', Kb', Kk''].
\end{aligned}
$$

Reading this inside the respective coordinatizing rings, this gives us the desired generalized homology.

This completes the proof of the proposition.   □

The Ree–Tits octagons only have $\{x, y\}$-transitivity for opposite points x and y. If the underlying field is perfect, then we have $\{L, M\}$-quasi-transitivity for opposite lines L, M (see Definitions 4.4.4).

4.5.12 Proposition. *Let Γ be a Ree–Tits octagon. Let x, y be two opposite points. Then Γ is $\{x, y\}$-transitive. Also, if Γ is perfect, then Γ is $\{L, M\}$-quasi-transitive, for every pair of opposite lines $\{L, M\}$.*

Proof. We coordinatize with respect to the same conditions as in the previous proof with the additional assumption that $N = [0]$ (and with the understanding that $(0, 0, 0, 0, 0)$ is replaced by $(0, 0, 0, 0, 0, 0, 0)$, etc.). We consider the map

$$
\begin{aligned}
h_5 : \quad (a, l, a', \ldots, a''') &\mapsto (Aa, A \otimes l, A^{\sigma+1}a', A^\sigma \otimes l', A^{\sigma+1}a'', A \otimes l'', Aa'''), \\
[k, b, k', \ldots, k'''] &\mapsto [k, Ab, A \otimes k', A^{\sigma+1}b', A^\sigma \otimes k'', A^{\sigma+1}b'', A \otimes k'''],
\end{aligned}
$$

with $A \in R_1 \setminus \{0\}$. Again h_5 acts as desired. If Γ is perfect, then the Tits endomorphism σ is bijective and hence has an inverse σ^{-1}. We consider the map

$$
\begin{aligned}
h_6 : \quad (a, l, a', l', a'', l'', a''') &\mapsto (Aa, A^{\sigma^{-1}} \otimes l, A^\sigma a', A^{\sigma^{-1}} \otimes l', Aa'', \\
&\qquad A^{1-\sigma+\sigma^{-1}} \otimes l'', a'''), \\
[k, b, k', b', k'', b'', k'''] &\mapsto [A^{\sigma^{-1}-1} \otimes k, b, A^{1-\sigma+\sigma^{-1}} \otimes k', Ab', \\
&\qquad A^{\sigma^{-1}} \otimes k'', A^\sigma b'', A^{\sigma^{-1}} \otimes k'''].
\end{aligned}
$$

This completes the proof of the proposition. $\qquad\qquad\qquad\qquad\qquad\quad$ □

There is an immediate corollary to these propositions.

4.5.13 Corollary.

    (i) *A symplectic quadrangle and a mixed quadrangle $Q(\mathbb{K}, \mathbb{K}'; \mathbb{K}, \mathbb{K}')$, in particular a Suzuki quadrangle, have a collineation group acting transitively on the set of ordered (ordinary) pentagons.*

    (ii) *A mixed or classical hexagon has a collineation group acting transitively on the set of ordered (ordinary) heptagons.*

    (iii) *A Ree–Tits octagon has no collineation group acting transitively on the set of all ordered (ordinary) nonagons.*

Proof. We only show (ii) and (iii), the proof of (i) being similar to the proof of (ii).

    (ii) Let Γ be a mixed or classical hexagon. Let $\mathcal{H}$ be any ordinary heptagon in Γ. Put an order on $\mathcal{H}$ by choosing a flag $\{p, L\}$ in $\mathcal{H}$. Let M be opposite in $\mathcal{H}$. Consider the apartment Σ containing the path $(p, L, \ldots, M)$ of $\mathcal{H}$. Let x be the point of $\mathcal{H}$ on M not in Σ and let N be the line $\mathcal{H}$ through p not in Σ. It is easily seen that $\mathcal{H}$ together with the above order is completely and unambiguously defined by Σ, x and N. Using the appropriate homologies guaranteed to exist by Proposition 4.5.11, we see that the full collineation group of Γ acts transitively on the triples (Σ, x, N) described above.

(*iii*) If $\mathsf{O}(\mathbb{K}, \sigma)$ admitted a collineation group acting transitively on the set of all ordered nonagons, then by a similar reasoning to that above, there would be a collineation group fixing an ordered apartment Σ and acting transitively on the lines through one of the points of Σ but not belonging to Σ, and dually. By conjugating elations, this would imply that all non-trivial root elations belonging to the same 6-path have the same order, but one of these groups (corresponding to line-elations) is by Subsection 4.5 isomorphic to $\mathbb{K}_\sigma^{(2)}$ and this has always elements of order 2 and of order 4, a contradiction. $\square$

Later on, we will see in Theorem 6.8.9 (see page 288) that no finite generalized octagon can have a collineation group acting transitively on the set of all ordered nonagons. Moreover, in the finite case, the above properties characterize the corresponding polygons; see again Theorem 6.8.9, and also Theorem 6.8.10 on page 295.

4.6 Collineation groups of finite classical polygons

We consider the following six classes of finite classical polygons: $\mathsf{W}(q)$, $\mathsf{H}(3, q^2)$, $\mathsf{H}(4, q^2)$, $\mathsf{H}(q)$, $\mathsf{T}(q, q^3)$ and $\mathsf{O}(q)$ for appropriate fields $\mathbf{GF}(q)$ in each case. The other finite classical polygons are duals of these, so they can be disregarded concerning their collineation groups. However, some results for $\mathsf{W}(q)$ and $\mathsf{H}(q)$ can as well be stated in general for $\mathsf{W}(\mathbb{K})$ and $\mathsf{H}(\mathbb{K})$ and we will do so.

In this section, we always assume that $q = p^h$ is a prime power (of the prime p). So we will accordingly not use the symbol p for a point. We will also use the notation (a, b) for the greatest common divisor of the positive integers a and b.

All results are well known, and if a reference should be given, then one could say that we are only restating some very special cases of results in TITS [1974].

A **torus** of $G \leq \mathrm{Aut}(\Gamma)$, for a classical polygon Γ, is the pointwise stabilizer in G of an apartment of Γ.

In the statement of the next propositions, we introduce some groups, where we use the notation of Section 4.3. Concerning terminology, we will call elements of $\mathbf{PGL}_d(\mathbb{K})$ **projective linear transformations**, those of $\mathbf{P\Gamma L}_d(\mathbb{K})$ **projective semilinear transformations**, and those of $\mathbf{PSL}_d(\mathbb{K})$ **special projective linear transformations**.

4.6.1 Proposition. *Let* $\mathsf{W}(\mathbb{K})$ *be embedded naturally in* $\mathbf{PG}(3, \mathbb{K})$. *Then the following statements hold:*

(*i*) *The group* $\mathbf{PGSp}_4(\mathbb{K})$ *of all collineations of* $\mathsf{W}(\mathbb{K})$ *induced by* $\mathbf{PGL}_4(\mathbb{K})$ *is generated by all root elations and all generalized homologies.*

(*ii*) *Every collineation of* W(K) *is induced by a projective semi-linear transformation of* **PG**(3, K).

(*iii*) *The full collineation group of* W(K) *is isomorphic to the semi-direct product* **PGSp**$_4$(K) : Aut K.

(*iv*) *All root groups of* W(K) *are isomorphic to the additive group of* K.

(*v*) *Every torus in* **PGSp**$_4$(K) *is generated by generalized homologies and is isomorphic to the direct product of two copies of the multiplicative group of* K.

Note first that if $g \in$ **PGL**$_4$(K) fixes all points of W(K), then g is the identity.

As before, we use the coordinatization of Subsection 3.4.1, especially also Table 3.1 on page 101. A general element of the group generated by all ([∞], (∞), [0])-elations and ((∞), [∞], (0))-elations is given by (and we only give the action on the elements with three coordinates):

$$\begin{aligned} E(A, K) \quad : \quad (a, l, a') \quad &\mapsto \quad (a, l + K + 2Aa, a' + A), \\ [k, b, k] \quad &\mapsto \quad [k, b + A, k' + K] \end{aligned}$$

(this is easily checked). The set of all elements $E(0, K)$ (or $E(A, 0)$) is a group isomorphic to the additive group of K and forms a root group of line-elations (or point-elations). This proves (*iv*). An elementary calculation shows that the special projective linear transformation $L(A, K)$ of **PG**(3, q) with associated matrix

$$\begin{pmatrix} 1 & K & 0 & -A \\ 0 & 1 & 0 & 0 \\ 0 & -A & 1 & 0 \\ 0 & 0 & 0 & 1 \end{pmatrix}$$

induces $E(A, K)$ in W(q), see also Appendix D. Hence the little projective group (see Definitions 4.4.4) is induced by projective linear transformations in **PG**(3, q). It is easily seen (see also Theorem 5.2.9 on page 177) that this group acts transitively on the set of ordered apartments.

Similarly, a general element of the group generated by all generalized homologies that stabilize the hat-rack of the coordinatization can be written as:

$$\begin{aligned} h_{A,K} \quad : \quad (a, l, a') \quad &\mapsto \quad (Aa, A^2 Kl, AKa'), \\ [k, b, k'] \quad &\mapsto \quad [Kk, AKb, A^2 Kk'] \end{aligned}$$

and it is induced in W(q) by the projective linear transformation $L_{A,K}$ with associated matrix

$$\begin{pmatrix} A^2 K & 0 & 0 & 0 \\ 0 & 1 & 0 & 0 \\ 0 & 0 & AK & 0 \\ 0 & 0 & 0 & A \end{pmatrix}.$$

Note that $h_{1,K}$ commutes with $h_{A,1}$ and that the set of all $h_{1,K}$ or of all $h_{A,1}$ forms a group of generalized homologies isomorphic to the multiplicative group of $\mathbb{K}$. Let $\theta \in \mathbf{PGL}_4(\mathbb{K})$ be any collineation of $\mathsf{W}(\mathbb{K})$. Then by the foregoing, we may assume that θ fixes the hat-rack of the coordinatization pointwise, and that it fixes the point (1) and the line $[1]$, hence it fixes the points $(1,0,0)$, $(0,0,1)$ and $(0,1)$ as well. But then its matrix must be the identity, so $\theta = 1$. So we have shown (i) and (v).

Now suppose that θ is any element of $\mathrm{Aut}(\mathsf{W}(\mathbb{K}))$. In order to show that it is induced by a projective semi-linear transformation in $\mathbf{PG}(3, \mathbb{K})$, we may suppose by the foregoing that it again fixes the points (∞), (0), $(0,0)$, $(0,0,0)$, (1), $(1,0,0)$, $(0,1)$ and $(0,0,1)$, and also the line $[1]$. This implies that there exist maps $\mu : \mathbb{K} \to \mathbb{K}$ and $\nu : \mathbb{K} \to \mathbb{K}$ defined by $(a^\mu) = (a)^\theta$ and $[k^\nu] = [k]^\theta$, for all $a, k \in \mathbb{K}$, with the properties $1^\mu = 1^\nu = 1$, $0^\mu = 0^\nu = 0$ and (expressing that θ preserves incidence by looking at Φ_1 and Ψ_1)

$$\begin{cases} (ak + b)^\mu &= a^\mu k^\nu + b^\mu, \\ (a^2 k + l - 2aa')^\nu &= (a^\mu)^2 k^\nu + l^\nu - 2a^\mu a'^\mu \end{cases}$$

for all $a, b, k, l \in \mathbb{K}$. Putting $a = 1$ and $b = 0$ in the first equality, we see that $\mu = \nu$. It then follows from the first equality again that μ preserves addition and multiplication in $\mathbb{K}$, hence $\mu = \nu$ is a field automorphism. It is now obvious that θ is induced by a projective semi-linear transformation, namely, the semi-linear map with corresponding matrix the identity matrix and field automorphism μ. Conversely, for every field automorphism $\mu = \nu$, the above equations are identities and so the stabilizer of a pentagon in $\mathsf{W}(\mathbb{K})$ is isomorphic to the automorphism group of $\mathbb{K}$. This proves (ii) and (iii).

The proposition is completely proved. $\square$

We will see later that the group $\mathbf{PSp}_4(\mathbb{K}) := \mathbf{PGSp}_4(\mathbb{K}) \cap \mathbf{PSL}_4(\mathbb{K})$ is the little projective group of $\mathsf{W}(\mathbb{K})$; see Theorem 8.3.2(i) on page 368.

4.6.2 Proposition. *Let* $\mathsf{W}(q)$ *be embedded naturally in* $\mathbf{PG}(3, q)$*. Every colline-ation of* $\mathsf{W}(q)$ *is induced by a unique projective semi-linear transformation of* $\mathbf{PG}(3, q)$*. The full collineation group of* $\mathsf{W}(q)$ *is isomorphic to the semi-direct product* $\mathbf{PGSp}_4(q) : \mathrm{Aut}\,\mathbf{GF}(q)$*, where* $\mathbf{PGSp}_4(q) = \mathrm{Aut}\,\mathsf{W}(q) \cap \mathbf{PGL}_4(q)$*. Put* $\mathbf{PSp}_4(q) = \mathrm{Aut}\,\mathsf{W}(q) \cap \mathbf{PSL}_4(q)$*; then*

$$| \mathrm{Aut}\,\mathsf{W}(q)| = q^4(q^4 - 1)(q^2 - 1)h,$$

$$|\mathbf{PGSp}_4(q)| = q^4(q^4 - 1)(q^2 - 1)$$

and

$$|\mathbf{PSp}_4(q)| = \frac{q^4(q^4 - 1)(q^2 - 1)}{(q - 1, 2)}.$$

All root groups are isomorphic to the additive group of $\mathbf{GF}(q)$ *and every torus in* $\mathbf{PGSp}_4(q)$ *is generated by generalized homologies and is isomorphic to the direct product of two copies of the multiplicative group of* $\mathbf{GF}(q)$*.*

Proof. From the previous proposition it follows that $\mathbf{PGSp}_4(q)$ acts regularly on the ordered pentagons of $\mathsf{W}(q)$ (in view of the proof of Corollary 4.5.13). This number equals $q^4(q^4 - 1)(q^2 - 1) = |\mathbf{PGSp}_4(q)|$. Also, the number of automorphisms of $\mathbf{GF}(q)$ is equal to h (remember $q = p^h$ with p prime). Hence $|\operatorname{Aut}\mathsf{W}(q)| = h|\mathbf{PGSp}_4(q)|$.

Using the notation of the previous proof, we note that $L_{A,K}$ belongs to $\mathbf{PSL}_4(q)$ if and only if K^2 is a fourth power in $\mathbf{GF}(q)$, or equivalently, K is square in $\mathbf{GF}(q)$. Hence

$$|\mathbf{PSp}_4(q)| = \frac{|\mathbf{PGSp}_4(q)|}{(q-1,2)}.$$

The proposition is proved. □

4.6.3 Proposition. *Let* $\mathsf{H}(3,q^2)$ *be embedded naturally in* $\mathbf{PG}(3,q^2)$. *Every collineation of* $\mathsf{H}(3,q^2)$ *is induced by a projective semi-linear transformation of* $\mathbf{PG}(3,q^2)$. *The full collineation group of* $\mathsf{H}(3,q^2)$ *is isomorphic to the semi-direct product* $\mathbf{PGU}_4(q) : \operatorname{Aut}\mathbf{GF}(q^2)$, *where* $\mathbf{PGU}_4(q) = \operatorname{Aut}(\mathsf{H}(3,q^2)) \cap \mathbf{PGL}_4(q^2)$. *Put* $\mathbf{PSU}_4(q) = \operatorname{Aut}(\mathsf{H}(3,q^2)) \cap \mathbf{PSL}_4(q^2)$; *then*

$$|\operatorname{Aut}(\mathsf{H}(3,q^2))| = q^6(q^4 - 1)(q^3 + 1)(q^2 - 1)2h,$$

$$|\mathbf{PGU}_4(q)| = q^6(q^4 - 1)(q^3 + 1)(q^2 - 1)$$

and

$$|\mathbf{PSU}_4(q)| = \frac{(q+1,2)q^6(q^4 - 1)(q^3 + 1)(q^2 - 1)}{(q+1,4)}.$$

The root groups (see Definitions 4.4.4*) are isomorphic to the additive group of* $\mathbf{GF}(q^2)$ *(point-elations) or* $\mathbf{GF}(q)$ *(line-elations). Every torus of the group* $\mathbf{PGU}_4(q)$ *is generated by generalized homologies and is isomorphic to the direct product of the multiplicative group of* $\mathbf{GF}(q)$ *with the multiplicative group of* $\mathbf{GF}(q^2)$.

Proof. The proof of this proposition is completely similar to that of Propositions 4.6.1 and 4.6.2. Note that $\mathsf{H}(3,q^2)$ is $\{x,y\}$-transitive for all pairs of opposite points x and y by Proposition 4.5.10(*i*) and that it is $\{L,M\}$-transitive for all pairs of opposite lines L and M by the dual of Proposition 4.5.10(*i*) (since the dual of $\mathsf{H}(3,q^2)$ is the classical orthogonal quadrangle $\mathsf{Q}(4,q)$). Using the coordinatization of Subsection 3.4.7, a general element of the group generated by the homologies fixing the hat-rack is given by

$$
\begin{array}{rl}
h_{A,K} : & (a,l,a') \mapsto (Aa, A^{q+1}Kl, AKa'), \\
& [k,b,k'] \mapsto [Kk, AKb, A^{q+1}Kk'],
\end{array}
$$

for $A \in \mathbf{GF}(q^2)^\times$ and $K \in \mathbf{GF}(q)^\times$. This has corresponding matrix $M(A, K)$ in $\mathbf{PG}(3, q^2)$, with

$$M(A, K) = \begin{pmatrix} A^{q+1}K & 0 & 0 & 0 \\ 0 & A^q & 0 & 0 \\ 0 & 0 & A^q K & 0 \\ 0 & 0 & 0 & 1 \end{pmatrix}.$$

The determinant of $M(A, K)$ equals $A^{3q+1}K^2$. Since $K \in \mathbf{GF}(q)$, K^2 is a fourth power in $\mathbf{GF}(q^4)$. So the only condition for $h_{A,K}$ to belong to $\mathbf{PSL}_4(q^2)$ is that A^{3q+1} is a fourth power in $\mathbf{GF}(q^2)$. If A is a square, then clearly A^{3q+1} is a fourth power. If A is not a square in $\mathbf{GF}(q^2)$, then, in view of $4|q^2 - 1$, there must hold $4|2q - 1$, hence $(q + 1, 4) = 2$ and in this case all elements $h_{A,K}$ belong to $\mathbf{PSL}_4(q^2)$. In the other case, i.e., if $(q+1, 4) = 4$, we have to exclude all non-squares $A \in \mathbf{GF}(q^2)^\times$. The proposition follows easily. $\qquad\square$

The generalized quadrangle $\mathsf{H}(4, q^2)$ is not $\{L, M\}$-transitive for any pair of opposite lines L and M. The same situation occurs in the Ree–Tits octagon $\mathsf{O}(q)$. We choose to show the result for the Ree–Tits octagons in more detail, and so the proof of the next proposition is omitted, but it is similar to (and in fact easier than) the proof of Proposition 4.6.10 below.

4.6.4 Proposition. *Let* $\mathsf{H}(4, q^2)$ *be embedded naturally in* $\mathbf{PG}(4, q^2)$. *Every collineation of* $\mathsf{H}(4, q^2)$ *is induced by a projective semi-linear transformation of* $\mathbf{PG}(4, q^2)$. *The full collineation group of* $\mathsf{H}(4, q^2)$ *is isomorphic to the semi-direct product* $\mathbf{PGU}_5(q) : \mathrm{Aut}\,\mathbf{GF}(q^2)$, *where* $\mathbf{PGU}_5(q) = \mathrm{Aut}(\mathsf{H}(4, q^2)) \cap \mathbf{PGL}_5(q^2)$. *Put* $\mathbf{PSU}_5(q) = \mathrm{Aut}(\mathsf{H}(4, q^2)) \cap \mathbf{PSL}_5(q^2)$; *then*

$$|\mathrm{Aut}(\mathsf{H}(4, q^2))| = q^{10}(q^5 + 1)(q^4 - 1)(q^3 + 1)(q^2 - 1)2h,$$

$$|\mathbf{PGU}_5(q)| = q^{10}(q^5 + 1)(q^4 - 1)(q^3 + 1)(q^2 - 1)$$

and

$$|\mathbf{PSU}_5(q)| = \frac{q^{10}(q^5 + 1)(q^4 - 1)(q^3 + 1)(q^2 - 1)}{(q + 1, 5)}.$$

The root groups are isomorphic to the additive group of $\mathbf{GF}(q^2)$ *(point-elations), or to the group consisting of pairs*

$$(k_0, k_1) \in \mathbf{GF}(q^2) \times \{t \in \mathbf{GF}(q^2) : t^q + t = 0\}$$

with operation law

$$(k_0, k_1) \oplus (l_0, l_1) = (k_0 + l_0, k_1 + l_1 + \theta^q k_0 l_0^q - \theta k_0^q l_0),$$

where θ *is any non-zero fixed element of* $\mathbf{GF}(q^2)$ *satisfying* $(1 + \theta)^{q+1} = 1$ *(line-elations).*
Every torus of $\mathbf{PGU}_5(q)$ *is generated by generalized homologies and is isomorphic to the direct product of two copies of the multiplicative group of* $\mathbf{GF}(q^2)$. $\qquad\square$

As a general remark, we can in fact state the following result, the proof of which runs similarly to the above proofs.

4.6.5 Proposition. *All collineations of any classical quadrangle* $Q(V, \tilde{q})$, *where* $\tilde{q}$ *is a non-degenerate* σ-*quadratic form in* V, *are induced by projective semi-linear transformations of* V. $\qquad\square$

We now turn our attention to the hexagons.

4.6.6 Proposition. *Let* $H(\mathbb{K})$ *be embedded naturally in* $\mathbf{PG}(6, \mathbb{K})$. *Then the following statements hold:*

(*i*) *The group* $\mathbf{G}_2(\mathbb{K})$ *of collineations of* $H(\mathbb{K})$ *induced by* $\mathbf{PSL}_7(\mathbb{K})$ *is the same as the group of collineations of* $H(\mathbb{K})$ *induced by* $\mathbf{PGL}_7(\mathbb{K})$.

(*ii*) *The group* $\mathbf{G}_2(\mathbb{K})$ *is generated by all root elations and all generalized homologies.*

(*iii*) *Every collineation of* $H(\mathbb{K})$ *is induced by a unique projective semi-linear transformation of* $\mathbf{PG}(6, \mathbb{K})$.

(*iv*) *The full collineation group of* $H(\mathbb{K})$ *is isomorphic to the semi-direct product* $\mathbf{G}_2(\mathbb{K}) : \mathrm{Aut}\,\mathbb{K}$.

(*iv*) *All root groups of* $H(\mathbb{K})$ *are isomorphic to the additive group of* $\mathbb{K}$.

(*v*) *Every torus in* $\mathbf{G}_2(\mathbb{K})$ *is generated by generalized homologies and is isomorphic to the direct product of two copies of the multiplicative group of* $\mathbb{K}$.

Proof. The proof of this proposition is completely similar to the proof of Proposition 4.6.1. To obtain matrix representations of elements of $\mathrm{Aut}\, H(\mathbb{K})$, one can use the results of Appendix D. In particular, one can see there that the deteminant of a matrix corresponding to any generalized homology is a perfect seventh power in $\mathbb{K}$. Hence every generalized homology is induced by a special projective linear transformation. This is needed to prove (*i*). The reader can now reconstuct easily the whole proof. $\qquad\square$

We will see later on (Theorem 8.3.2(*i*) on page 368), that the group $\mathbf{G}_2(\mathbb{K})$ is the little projective group of $H(\mathbb{K})$.

4.6.7 Proposition. *Let* $H(q)$ *be embedded naturally in* $\mathbf{PG}(6, q)$. *Every collineation of* $H(q)$ *is induced by a projective semi-linear transformation of* $\mathbf{PG}(6, q)$. *The full collineation group of* $H(q)$ *is isomorphic to the semi-direct product* $\mathbf{G}_2(q) : \mathrm{Aut}\,\mathbf{GF}(q)$, *where* $\mathbf{G}_2(q) = \mathrm{Aut}(H(q)) \cap \mathbf{PGL}_7(q)$. *We have* $\mathbf{G}_2(q) = \mathrm{Aut}(H(q)) \cap \mathbf{PSL}_7(q)$ *and*

$$|\mathrm{Aut}(H(q))| = q^6(q^6 - 1)(q^2 - 1)h,$$
$$|\mathbf{G}_2(q)| = q^6(q^6 - 1)(q^2 - 1).$$

All root groups are isomorphic to the additive group of $\mathbf{GF}(q)$ *and every torus in* $\mathbf{G}_2(q)$ *is generated by generalized homologies and is isomorphic to the direct product of two copies of the multiplicative group of* $\mathbf{GF}(q)$.

Proof. Again, this is similar to the proof of Proposition 4.6.2. $\square$

4.6.8 Remark. Both Propositions 4.6.6 and 4.6.7 hold also for the symplectic hexagons in five-dimensional projective space. See Appendix D for explicit forms of elations and generalized homologies.

Completely similarly to Proposition 4.6.7 the following holds (and the proof, which we omit, is as before, using the results of Appendix D):

4.6.9 Proposition. *Let* $\mathsf{T}(q^3, q)$ *be embedded naturally in* $\mathbf{PG}(7, q^3)$. *Every collineation of* $\mathsf{T}(q^3, q)$ *is induced by a projective semi-linear transformation of* $\mathbf{PG}(7, q^3)$. *The full collineation group of* $\mathsf{T}(q^3, q)$ *is isomorphic to the semi-direct product* $^3\mathbf{D}_4(q) : \mathrm{Aut}\,\mathbf{GF}(q)$, *where* $^3\mathbf{D}_4(q) = \mathrm{Aut}(\mathsf{T}(q^3, q)) \cap \mathbf{PGL}_8(q^3)$. *We have* $^3\mathbf{D}_4(q) = \mathrm{Aut}(\mathsf{T}(q^3, q)) \cap \mathbf{PSL}_8(q^3)$ *and*

$$|\mathrm{Aut}(\mathsf{T}(q^3, q))| = q^{12}(q^8 + q^4 + 1)(q^6 - 1)(q^2 - 1)3h,$$
$$|^3\mathbf{D}_4(q)| = q^{12}(q^8 + q^4 + 1)(q^6 - 1)(q^2 - 1).$$

The root groups are isomorphic to the additive group of $\mathbf{GF}(q)$ *or* $\mathbf{GF}(q^3)$ *and every torus in* $^3\mathbf{D}_4(q)$ *is generated by generalized homologies and is isomorphic to the direct product of the multiplicative group of* $\mathbf{GF}(q)$ *with that of* $\mathbf{GF}(q^3)$. $\square$

In the infinite case, there are examples of twisted triality hexagons $\mathsf{T}(\mathbb{L}, \mathbb{K}, \sigma)$ for which

$$\mathrm{Aut}(\mathsf{T}(\mathbb{L}, \mathbb{K}, \sigma)) \cap \mathbf{PGL}_8(\mathbb{L}) \neq \mathrm{Aut}(\mathsf{T}(\mathbb{L}, \mathbb{K}, \sigma)) \cap \mathbf{PSL}_8(\mathbb{L}),$$

see also Remark 8.3.6 on page 372.

Although the Ree–Tits octagon $\mathsf{O}(q)$ has some embedding in a 25-dimensional projective space, we shall not allude to that embedding in the statement of the next result, since we have not described it (it arises from the embedding of a building of type F_4 in that projective space; see e.g. COHEN [1995]). But we do need it in the proof at one point, similarly to the previous proofs.

4.6.10 Proposition. *Let* $\mathsf{O}(q)$ *be the classical Ree–Tits octagon,* $q = 2^{2e+1}$; *then the order of the full collineation group of* $\mathsf{O}(q)$ *is equal to*

$$|\mathrm{Aut}(\mathsf{O}(q))| = q^{12}(q^6 + 1)(q^4 - 1)(q^3 + 1)(q - 1)(2e + 1).$$

Also, the group $^2\mathbf{F}_4(q)$ *generated by all root elations and homologies has order*

$$\frac{|\mathrm{Aut}(\mathsf{O}(q))|}{2e + 1} = q^{12}(q^6 + 1)(q^4 - 1)(q^3 + 1)(q - 1).$$

The root groups are isomorphic to the additive group of $\mathbf{GF}(q)$ *(point-elations)*
and to the group $\mathbf{GF}(q)_{2^{e}+1}^{(2)}$ *(line-elations). Every torus in G is generated by gen-*
eralized homologies and is isomorphic to the direct product of two copies of the
multiplicative group of $\mathbf{GF}(q)$.

Proof. As before, one deduces that G acts transitively on the set of apartments.
Since $\mathbf{GF}(q)$ is perfect, we know by Proposition 4.5.12 that there is a group of
order $(q-1)^2$ generated by homologies (of the form h_5 and h_6 with the notation
of the proof of Proposition 4.5.12). Now we show that any element $g \in \mathrm{Aut}(\mathsf{O}(q))$
can be written as the product of an element of G and a field automorphism. By
the above results, we may assume that g fixes the hat-rack of a coordinatization,
and also the point (1). Conjugating elements of the group $\mathbf{GF}(q)_{2^{e}+1}^{(2)}$ by g, we
see that g must preserve the set of lines $[(0,k_1)]$, $k_1 \in \mathbf{GF}(q)$. Hence we may
also assume that g fixes [1]. But now an exercise similar to the last part of the
proof of Proposition 4.6.2 shows that g is induced by a field automorphism of
$\mathbf{GF}(q)$. From the embedding in 24-dimensional space, we know that such a g is a
projective semi-linear transformation, hence it is in G (which consists of projective
linear transformations) **if and only if** the field automorphism is trivial. Counting
the number of apartments in $\mathsf{O}(q) = (\mathcal{P}, \mathcal{L}, \mathbf{I})$, we otain (with $s = q$ and $t = q^2$)

$$|\mathcal{P}|(t+1)stststst = q^{12}(q+1)(q^3+1)(q^6+1)(q^2+1)$$

and we see that G has the desired order. The result now follows. □

4.7 The Tits condition

Let Γ be a generalized polygon and G a group of automorphisms of Γ. Suppose
that G acts transitively on the set of ordered apartments of Γ. Then we say that
Γ **satisfies the Tits condition**, or that Γ **is a Tits polygon** (with respect to G).

Now let Γ be a Tits n-gon with respect to some group G. Fix an apartment Σ of
Γ and a flag $F = \{p, L\}$ in Σ. Let B be the stabilizer in G of F and let N be
the (not necessarily pointwise) stabilizer of Σ in G. Then we say that (B, N) is a
(saturated) Tits system in G for Γ. This definition is equivalent to what is called
in the literature *irreducible saturated Tits systems of relative rank 2* and which
is usually defined group-theoretically. We will, however, only need the geometric
definition given above. For the introduction of Tits systems in the literature, see
TITS [1962b] (called (B, N)-pairs there) and TITS [1961].

Tits n-gons exist for every $n \geq 3$ by a modification of the free construction process
explained in Subsection 1.3.13 (page 13); see below. Hence a complete classification
of all Tits polygons is out of the question. The construction of a Tits n-gon for
arbitrary n is our first aim in this section. Then we prove a general result on
parabolic subgroups (which we shall need in the next section; parabolic subgroups
are, with the above notation, subgroups containing B). Lastly, we look at the finite

case, mention some related results and say a few words about finite flag-transitive polygons.

4.7.1 A free construction of Tits n-gons

Let us fix $n \geq 3$. The goal of this subsection is to construct a Tits n-gon. This will be achieved by a modification of the free construction in Subsection 1.3.13. We are not interested in the abstract details, but rather in the geometric essence.

Suppose $\Gamma = (\mathcal{P}, \mathcal{L}, \mathbf{I})$ is a partial n-gon containing at least one ordinary n-gon. Let us call an ordinary n-gon in Γ an **apartment**, just as in the case of a generalized n-gon. Also, an **ordered apartment** Σ is an apartment containing a distinguished flag $\{p, L\}$, $p \in \mathcal{P}$, $L \in \mathcal{L}$, whose elements can be regarded as the first two elements of the circuit associated with Σ (see Definitions 1.3.1). Let us fix an ordered apartment Ω in Γ, and we consider the set $\mathcal{A}$ of all ordered apartments of Γ which are not in the orbit of Ω under $\operatorname{Aut}(\Gamma)$. We define subsequently the geometries $\Gamma_{(i)}$, $i \in \mathbb{N}$, with $\Gamma = \Gamma_{(0)}$ ($\Gamma_{(i)}$ will depend on Γ and Ω). For each $\Sigma \in \mathcal{A}$, there is a unique isomorphism θ^Σ mapping Σ to Ω (mapping the respective distinguished flags onto one another). Also, we choose two copies $\Gamma_{(0)}^{(\Sigma,+)}$ and $\Gamma_{(0)}^{(\Sigma,-)}$ of $\Gamma_{(0)}$ with corresponding isomorphisms $\varphi^{(\Sigma,\pm)} : \Gamma_{(0)} \to \Gamma_{(0)}^{(\Sigma,\pm)}$ and put $\Gamma_{(0)} = \Gamma_{(0)}^{(\Omega,\pm)}$ (with $\varphi^{(\Omega,\pm)}$ equal to the identity). We consider the disjoint union of all $\Gamma_{(0)}^{(\Sigma,\pm)}$, where Σ ranges over $\mathcal{A} \cup \{\Omega\}$, and for every element x in $\Gamma_{(0)}$, and every ordered apartment $\Sigma \in \mathcal{A}$ containing x, we identify in that disjoint union the element x with $x^{\theta^\Sigma \varphi^{(\Sigma,+)}}$, and the element $x^{\varphi^{(\Sigma,-)}}$ with x^{θ^Σ}. The result is by definition $\Gamma_{(1)}'$. To obtain $\Gamma_{(1)}$, we remark that any automorphism g of $\Gamma_{(0)}$ can be cannonically extended to the union of all $\Gamma_{(0)}^{(\Sigma,+)}$, because $\mathcal{A}$ is mapped into itself by g. We now choose for every ordered apartment $\Upsilon \neq \Omega$ in Γ not belonging to $\mathcal{A}$ a particular automomorphism g_Υ that maps Ω to Σ. We consider a copy $\Gamma_{(0)}^{(\Upsilon)}$ of the union of $\Gamma_{(0)}$ and all $\Gamma_{(0)}^{(\Sigma,-)}$ (with corresponding isomorphism v^Υ from that union to $\Gamma_{(0)}^{(\Upsilon)}$), $\Sigma \in \mathcal{A}$, consider again the disjoint union of $\Gamma_{(1)}'$ and all $\Gamma_{(0)}^{(\Upsilon)}$, and identify each element x of $\Gamma_{(0)} \subseteq \Gamma_{(1)}$ with $x^{g_\Upsilon v^\Upsilon}$. The result is by definition $\Gamma_{(1)}$. Note that every map θ^Σ now extends to a map

$$\theta_1^\Sigma : \Gamma_{(0)}^{(\Sigma,+)} \cup \Gamma_{(0)} \to \Gamma_{(0)} \cup \Gamma_{(0)}^{(\Sigma,-)} : \quad \begin{aligned} x &\mapsto x^{(\varphi^{(\Sigma,+)})^{-1}}, &\quad x \in \Gamma_{(0)}^{(\Sigma,+)}, \\ x &\mapsto x^{\varphi^{(\Sigma,-)}}, &\quad x \in \Gamma_{(0)}. \end{aligned}$$

Indeed, this follows immediately from the identifications made. For instance, if x belongs to $\Sigma \in \mathcal{A}$, then

$$x^{\theta_1^\Sigma} = (x^{\theta^\Sigma \varphi^{(\Sigma,+)}})^{(\varphi^{(\Sigma,+)})^{-1}} = x^{\theta^\Sigma}.$$

Also, every automorphism g_Υ (defined as above) extends to $\Gamma_{(1)}$ in the obvious way.

To obtain $\Gamma_{(k+1)}$ and $\theta_{k+1}^{\Sigma} : \Gamma_{(k)}^{(\Sigma,+)} \cup \Gamma_{(k)} \rightarrow \Gamma_{(k)} \cup \Gamma_{(k)}^{(\Sigma,-)}$ (for every $\Sigma \in \mathcal{A}$) from $\Gamma_{(k)}$ and all maps $\theta_k^{(\Sigma)}$ inductively, we proceed as follows. For every $\Sigma \in \mathcal{A}$ we consider two copies $\Gamma_{(k)}^{(\Sigma,\pm)}$ of $\Gamma_{(k)}$ with corresponding isomorphisms $\varphi_k^{(\Sigma,\pm)}$: $\Gamma_{(k)} \rightarrow \Gamma_{(k)}^{(\Sigma,\pm)}$ and put $\Gamma_{(k)} = \Gamma_{(k)}^{(\Omega,\pm)}$ (with $\varphi_k^{(\Omega,\pm)}$ equal to the identity). As before, we consider the disjoint union of all $\Gamma_{(k)}^{(\Sigma,\pm)}$, where Σ ranges over $\mathcal{A} \cup \{\Omega\}$, and for every element x in $\Gamma_{(k)}$, and every ordered apartment $\Sigma \in \mathcal{A}$ such that $\Gamma_{(k-1)}^{(\Sigma,+)}$ contains x, we identify in that disjoint union the element x with $x^{\theta_k^{\Sigma} \varphi_k^{(\Sigma,+)}}$, and the element $x^{\varphi_k^{(\Sigma,-)}}$ with $x^{\theta_k^{\Sigma}}$. The result is by definition $\Gamma_{(k+1)}'$. To obtain $\Gamma_{(k+1)}$, we proceed as before, as well as to obtain the maps θ_{k+1}^{Σ}. We have that $\Gamma_{(k)} = \Gamma_{(k)}^{(\Omega,\pm)}$ is a subgeometry of $\Gamma_{(k+1)}$, and so we can take the union $\Gamma_{(\infty)}^{\Omega}$ of all these partial n-gons. It is not so difficult to see that $\Gamma_{(\infty)}^{\Omega}$ is again a partial n-gon. Now, as k approaches ∞, the map θ_k^{Σ} clearly tends to an automorphism θ_{∞} of $\Gamma_{(\infty)}^{\Omega}$, because every element of $\Gamma_{(\infty)}^{\Omega}$ lives in some $\Gamma_{(k)}$, and so it is in both the pre-image and the image of θ_{k+1}^{Σ}. Also, every automorphism g_{Υ} extends canonically to $\Gamma_{(\infty)}^{\Omega}$ (in the obvious way), hence $\text{Aut}(\Gamma_{(\infty)}^{\Omega})$ acts transitively on the class of ordered apartments of $\Gamma \subseteq \Gamma_{(\infty)}^{\Omega}$.

Now, we start again from a partial n-gon $\Gamma^{(0)}$, which is not a weak generalized n-gon. We define the geometries $\Gamma^{(i+1)}$, $i \in \mathbb{N}$, by induction as follows. If i is even, then we again join any two elements of Γ_i at distance $n + 1$ from each other by a path of length $n - 1$, as in Subsection 1.3.13 (page 13). Note that each automorphism of $\Gamma^{(i)}$ extends uniquely to an automorphism of $\Gamma^{(i+1)}$, and that $\Gamma^{(1)}$ contains at least one ordered apartment Ω. If i is odd, then we put $\Gamma^{(i+1)}$ equal to $(\Gamma^{(i)})_{(\infty)}^{\Omega}$. As $\Gamma^{(i)}$ is now always a subgeometry of $\Gamma^{(i+1)}$, we can take the union $\Gamma^{(\infty)}$ and oviously obtain a Tits n-gon (leaving the reader to check the details).

This construction, wrapped in a different package, is due to TITS [1977], who generalized a similar construction carried out for projective planes by KEGEL & SCHLEIERMACHER [1973].

Some general properties of saturated Tits systems

4.7.2 The Weyl group of a Tits system

Let G be an automorphism group of a generalized n-gon Γ and suppose that (B, N) is a saturated Tits system in G for Γ. The group $H := B \cap N$ is the pointwise stabilizer of the apartment Σ of Γ stabilized by N. Hence $H \trianglelefteq N$. Clearly $W := N/H$ can be thought of as a group acting on Σ. From the assumptions on G, it follows easily that W acts flag-transitively on Σ, hence W is the dihedral group of order $2n$. It is called the **Weyl group** of G (with respect to the saturated Tits system (B, N)).

A useful lemma is the following result (see Section 4.3 for definitions).

4.7.3 Lemma. *If* $\Gamma = (\mathcal{P}, \mathcal{L}, \mathbf{I})$ *has the Tits property with respect to the group* G, *then* G *acts primitively on the set of points of* Γ *and, dually, also primitively on the set of lines of* Γ. *Consequently the stabilizer of any point or line of* Γ *in* G *is a maximal subgroup of* G.

Proof. We prove the result for points. Let C be an element of some partition of $\mathcal{P}$ stabilized by G, and assume that C contains at least two points x and y. We must show $C = \mathcal{P}$. By the Tits property, we can fix x and map y to some other point y' collinear with y. Indeed, let Σ be any ordered apartment containing both x and y and ordered by the flag $(x, \text{proj}_x y)$ (if x is not opposite y; otherwise choose any flag in Σ containing x). By the thickness assumption, there is at least one point y' at distance $\delta(x, y)$ from x and collinear with y. If Σ' is any ordered apartment containing x and y', and ordered in the same way as Σ, then we map Σ to Σ'. One now sees that x is fixed and y is mapped to y'. This implies that we may assume that x and y are collinear themselves (by renaming y' as x). But now we can fix y and map x to any point in $\Gamma_2(y)$. So every point collinear to y belongs to C. Since y was arbitrary in C, applying this property a finite number of times, we conclude that C is the whole point set of Γ. $\qquad\square$

4.7.4 Definitions. A subgroup of G containing some conjugate of B is called a **parabolic subgroup** of G. The conjugates of B themselves are also called the **Borel subgroups** of G. These definitions are independent of Σ and F. Indeed, the conjugates of B are precisely the stabilizers of the flags of Γ (and all flags occur since G acts flag-transitively on Γ). The next theorem says exactly what the parabolic subgroups of G are. It is the rank 2 case of a general result due to Tits (cf. RONAN [1989] and TAYLOR [1992]).

4.7.5 Theorem. *Let* P *be a parabolic subgroup of the group* G *which contains a saturated Tits system for the generalized polygon* Γ. *Then either* $P = G$, *or* P *is a Borel subgroup, or* P *is the stabilizer in* G *of a point or of a line of* Γ.

Proof. Without loss of generality we may assume that P properly contains B. Hence there is an element $g \in P$ which does not fix $F = \{p, L\}$. Upon considering the dual polygon, we may assume that $\delta(F, p^g) \geq \delta(F, L^g)$, recalling that $\delta(F, x)$ is the minimum of $\{\delta(p, x), \delta(L, x)\}$. Let x be any point of Γ incident with L^g and such that $\delta(F, x) = \delta(F, p^g)$. By the Tits property, there exists some $b \in B$ mapping $\{p^g, L^g\}$ onto $\{x, L^g\}$. So P contains the stabilizer B^g of the flag F^g and the stabilizer B^{gb} of the flag $F^{gb} = \{x, L^g\}$. We claim that P contains the full stabilizer of L in G. Indeed, let $h \in G$ fix L. First suppose that h does not map x

to p^g. Then by the Tits property there exists an element $h' \in B^g$ mapping x^h back to x. Hence $hh' \in B^{gb} \leq P$ and so, since $h' \in P$, we conclude that $h \in P$. Now suppose that $x^h = p^g$. Then bh^{-1} fixes the flag $\{p^g, L^g\}$, hence $bh^{-1} \in B^g \leq P$, therefore $h \in P$ and our claim follows. So P contains G_{L^g}. The result follows from Lemma 4.7.3. $\qquad\square$

The last results in this subsection essentially state that we can reconstruct Γ from G and B.

4.7.6 Theorem (Tits [1974]). *Let Γ be a Tits polygon with respect to the group G, and let B be a Borel subgroup stabilizing some flag $\{p, L\}$. Then there are exactly two (parabolic) subgroups G_p and G_L properly containing B and distinct from G. Also, the mapping $F \mapsto G_F$ defines a bijection between the set of flags of Γ and the class of Borel subgroups in G. Furthermore, Γ may be described as follows.*
* *The points of Γ are the conjugates of G_p.*
* *The lines of Γ are the conjugates of G_L.*
* *A point and a line are incident **if and only if** the corresponding conjugates of G_p and G_L meet in a Borel subgroup.*

Proof. The first assertion follows immediately from Theorem 4.7.5. The second assertion follows from the fact that any Borel subgroup fixes a unique flag and G acts transitively on the set of flags (this is an immediate consequence of the Tits property). We now show the last assertion. Clearly any conjugate of G_p is the stabilizer of some point of Γ, and, conversely, the stabilizer of any point of Γ is a conjugate of G_p by the Tits property (actually, only the transitivity on points is needed here). Suppose now that $G_x = G_y$ for points x and y. Then the Tits property readily implies that $x = y$ (indeed, we can fix x and map y to any point at distance $\delta(x, y)$ of x). If x and M are incident (x a point, M a line), then $G_x \cap G_M = G_{x,M}$, which is a Borel subgroup. Now suppose that x and M are not incident and that $G_x \cap G_M$ contains some Borel subgroup, which we may assume to be $B = G_{p,L}$ without loss of generality. From the second assertion it follows that B fixes x **if and only if** $x = p$. Dually $M = L$ and hence x is incident with M. $\qquad\square$

An alternative way of reconstructing Γ is given by the following result, which goes back to the 1950s (see DEMBOWSKI [1968](1.2.17)).

4.7.7 Theorem (Tits [1974]). *Let Γ be a Tits polygon with respect to the group G, and let $\{p, L\}$ be some flag in Γ. Then Γ may be described as follows.*
* *The points of Γ are the right cosets $(G_p)g$ of G_p, $g \in G$.*
* *The lines of Γ are the right cosets $(G_L)g$ of G_L, $g \in G$.*
* *A point and a line are incident **if and only if** the corresponding cosets of G_p and G_L are not disjoint.*

Proof. Clearly the map $p^g \mapsto (G_p)g$ is well defined and bijective from the set of points of Γ to the set of right cosets of (G_p) in G. So we can identify the point

p^g with $(G_p)g$. Similarly, we can identify the line L^g with the right coset $(G_L)g$. Now suppose p^g and L^h are incident, $g, h \in G$. Let $f \in G$ be such that f maps $\{p, L\}$ to $\{p^g, L^h\}$ (this exists by the Tits property; in fact only flag-transitivity is needed here). Then $p^g = p^f$, $L^h = L^f$, also $(G_p)g = (G_p)f$ and $(G_L)h = (G_L)f$. Hence $f \in (G_p)g \cap (G_L)h$. Conversely, if $f \in (G_p)g \cap (G_L)h$, for $f, g, h \in G$, then $p^g = p^f$ and $L^h = L^f$. Therefore p^g is incident with L^h.   □

4.7.8 Remark. The Ree–Tits octagons are examples of classical polygons which were first constructed via the algebraic notion of a Tits system by TITS [1960], although already in the same paper, TITS [1960] alludes to the construction of these octagons via the absolute elements of polarities in metasymplectic spaces. This construction was later announced again by SARLI [1986], and eventually proved by SARLI [1988] using group-theoretical arguments. VAN MALDEGHEM [1998] gives a purely geometric proof in the perfect case. Also, there is a general procedure for constructing "twisted buildings", developed by MÜHLHERR (unpublished), and the construction of the Ree–Tits octagons is a special case of this.

We now turn to the finite case, where under some weaker hypotheses one can still classify. We postpone the *topological compact* case to Chapter 9, but it should be noted that also in this case, one can carry out a classification using much weaker hypotheses (in fact, the situation is even more satisfying than in the finite case!).

4.8   Finite point-distance transitive and flag-transitive polygons

Finite point-distance transitive polygons

4.8.1 Definitions and conjectures

The Tits condition for generalized n-gons can be restated as the requirement that G acts transitively on (ordered, non-stammering) paths of length $n + 1$ starting with a point (or equivalently, a line). Generalizing this condition, and inspired by TUTTE [1966], we say that the generalized n-gon Γ is s-**path transitive** (with respect to an automorphism group G of Γ), $1 \leq s \leq n+1$, if G acts transitively on all paths of length s starting with a point. So, $(n+1)$-path transitivity is equivalent to the Tits condition. Two other cases need our special attention. A slightly weaker condition than n-path transitivity is *point-distance transitivity*. A generalized n-gon is called **point-distance transitive** (with respect to an automorphism group G) if for all $2i$, $1 \leq 2i \leq n$, $i \in \mathbb{N}$, the group G acts transitively on the pairs (x, y) of points at distance $2i$ from each other. Also, a 1-path transitive polygon is simply called a **flag-transitive** polygon.

There are two important conjectures related to finite s-path transitive polygons. The first is by TITS [1974], page 221, saying that all finite Tits polygons are classical polygons. The second is by KANTOR [1991] saying that a finite flag-transitive generalized n-gon, $n \geq 4$, is either a classical polygon or isomorphic (or anti-isomorphic) to the unique generalized quadrangle of order $(3, 5)$ (of $*$-Tits type), or isomorphic (or anti-isomorphic) to a generalized quadrangle of $*$-Tits type and of order $(15, 17)$ arising from the Lunelli–Sce oval in $\mathbf{PG}(2, 16)$ (see Section 3.7.2). Without using the classification of the finite simple groups, only the case $n = 3$ of the first conjecture has been solved. Indeed, OSTROM & WAGNER [1959] prove that a finite 2-path transitive projective plane is necessarily classical. Another contribution to Tits' conjecture is the result of YANUSHKA [1976]. This says that every point-distance transitive generalized hexagon of prime order s is a Moufang hexagon. One can also show a similar result for generalized quadrangles (KANTOR, unpublished, VAN MALDEGHEM, unpublished). Of course, from the celebrated result of OSTROM & WAGNER [1959] mentioned above it follows that every finite point-distance transitive projective plane is classical.

4.8.2 Theorem.
(1) (Ostrom & Wagner [1959]). *Every finite point-distance transitive projective plane is classical.*
(2) *Every point-distance transitive generalized quadrangle of prime order s is a classical quadrangle (and hence, up to duality, a symplectic quadrangle).*
(3) (Yanushka [1976]). *Every point-distance transitive generalized hexagon of prime order s is a classical hexagon (and hence, up to duality, a split Cayley hexagon).*
$\square$

Using the classification of finite simple groups, Tits' conjecture can be proved completely. In fact, one can classify all finite point-distance transitive weak polygons. Since this result is especially interesting from a group-theoretical point of view, we will mention the complete result, including the restrictions on the group G. Recall that we denote the double of a generalized polygon Γ by 2Γ (and recall also that the double is in fact the incidence graph), where lines of 2Γ contain exactly two points. We use the notation of CONWAY, CURTIS, NORTON, PARKER & WILSON [1985].

4.8.3 Theorem (Buekenhout & Van Maldeghem [1994]).
Suppose Γ is a point-distance transitive weak finite generalized n-gon, $n \geq 3$, with respect to some automorphism group G of Γ; then either Γ is a grid or a dual grid (and we do not insist on the groups), or (Γ, G) is one of the examples of Table 4.1 (where q denotes an arbitrary prime power, and where G does not contain a correlation unless explicitly mentioned).

As an immediate consequence, one obtains the classification of all finite Tits polygons.

4.8.4 Theorem (Buekenhout & Van Maldeghem [1994]).
Given the classification of finite simple groups, every finite Tits polygon is a classical polygon.
$\square$

n	Γ	G	Restrictions
3	$\mathbf{PG}(2,q)$	$\mathbf{PSL}_3(q) \trianglelefteq G \leq \operatorname{Aut} \mathbf{PSL}_3(q)$	
4	$\mathsf{W}(q)$	$\mathbf{PSp}_4(q) \trianglelefteq G \leq \operatorname{Aut} \mathbf{PSp}_4(q)$	
4	$\mathsf{Q}(4,q)$	$\mathbf{PSO}_5(q) \trianglelefteq G \leq \operatorname{Aut} \mathbf{PSO}_5(q)$	
4	$\mathsf{Q}(5,q)$	$\mathbf{PSO}_6^-(q) \trianglelefteq G \leq \operatorname{Aut} \mathbf{PSO}_6^-(q)$	
4	$\mathsf{H}(3,q^2)$	$\mathbf{PSU}_4(q) \trianglelefteq G \leq \operatorname{Aut} \mathbf{PSU}_4(q)$	
4	$\mathsf{H}(4,q^2)$	$\mathbf{PSU}_5(q) \trianglelefteq G \leq \operatorname{Aut} \mathbf{PSU}_5(q)$	
6	$\mathsf{H}(q)$	$\mathbf{G}_2(q) \trianglelefteq G \leq \operatorname{Aut} \mathbf{G}_2(q)$	
6	$\mathsf{T}(q^3,q)$	$^3\mathbf{D}_4(q) \trianglelefteq G \leq \operatorname{Aut} {}^3\mathbf{D}_4(q)$	
8	$\mathsf{O}(q)$	$^2\mathbf{F}_4(q) \trianglelefteq G \leq \operatorname{Aut} {}^2\mathbf{F}_4(q)$	q odd power of 2
4	$\mathsf{H}(4,q^2)^D$	$\mathbf{PSU}_5(q) \trianglelefteq G \leq \operatorname{Aut} \mathbf{PSU}_5(q)$	
4	$\mathsf{W}(2)$	$\mathbf{A}_6$	
4	$\mathsf{T}_2^*(O)$	$2^6 : 3 : \mathbf{A}_6 \leq G \leq 2^6 : 3 : \mathbf{S}_6$	O a complete oval in $\mathbf{PG}(2,4)$
6	$\mathsf{H}(q)^D$	$\mathbf{G}_2(q) \trianglelefteq G \leq \operatorname{Aut} \mathbf{G}_2(q)$	
6	$\mathsf{T}(q,q^3)$	$^3\mathbf{D}_4(q) \trianglelefteq G \leq \operatorname{Aut} {}^3\mathbf{D}_4(q)$	
6	$\mathsf{H}(2)$	$\mathbf{PSU}_3(3) \cong \mathbf{G}_2(2)'$	
6	$2\,\mathbf{PG}(2,q)$	$\mathbf{PSL}_3(q) : 2 \leq G \leq \operatorname{Aut} \mathbf{PSL}_3(q)$	G contains a correlation
6	$(2\,\mathbf{PG}(2,q))^D$	$\mathbf{PSL}_3(q) : 2 \leq G \leq \operatorname{Aut} \mathbf{PSL}_3(q)$	G contains a correlation
8	$\mathsf{O}(2^e)^D$	$^2\mathbf{F}_4(q) \trianglelefteq G \leq \operatorname{Aut} {}^2\mathbf{F}_4(q)$	e odd
8	$\mathsf{O}(2)$	$^2\mathbf{F}_4(2)' \cong \mathbf{T}$ (the **Tits group**)	
8	$2\,\mathsf{W}(q)$	$\mathbf{PSp}_4(q).2 \leq G \leq \operatorname{Aut} \mathbf{PSp}_4(q)$	q even, G contains a correlation
8	$(2\,\mathsf{W}(q))^D$	$\mathbf{PSp}_4(q).2 \leq G \leq \operatorname{Aut} \mathbf{PSp}_4(q)$	q even, G contains a correlation
8	$2\,\mathsf{W}(2)$	$\mathbf{A}_6 : 2$	G contains a correlation
12	$2\,\mathsf{H}(q)$	$\mathbf{G}_2(q).2 \leq G \leq \operatorname{Aut} \mathbf{G}_2(q)$	q is a power of 3, G contains a correlation
12	$(2\,\mathsf{H}(q))^D$	$\mathbf{G}_2(q).2 \leq G \leq \operatorname{Aut} \mathbf{G}_2(q)$	q is a power of 3, G contains a correlation

Table 4.1. Finite point-distance transitive weak generalized polygons.

Let us mention one step in the proof of Theorem 4.8.3. Namely, it is proved first that a distance transitive group acts primitively on the point set of the corresponding generalized polygon. The proof is the same as for Lemma 4.7.3 and can be taken over unchanged.

4.8.5 Lemma. *If G acts distance-transitively on the generalized polygon Γ, then G acts primitively on the set of points of Γ.* $\qquad\square$

Finite flag-transitive polygons

4.8.6 Classifying finite flag-transitive polygons

Finite flag-transitive projective planes have also been conjectured by many authors to be classical. The most far-reaching result in that direction is a result due to KANTOR [1983], stating that the order s of a non-classical projective plane admitting a flag-transitive group G must satisfy (1) $|G| = (1 + s)(1 + s + s^2)$, (2) s is even and (3) $1 + s + s^2$ is a prime.

For $n > 3$, there are only partial results towards the classification of all finite flag-transitive generalized n-gons. The question remains wide open. Worth mentioning is the fact that BUEKENHOUT & VAN MALDEGHEM [1993] prove that no sporadic group can be the socle of an almost simple group acting point-transitively on a finite generalized hexagon or octagon. In fact, a method is developed for handling any small almost simple group and tested successfully on every group displayed in the main section of CONWAY et al. [1985]. I have also proved in some unpublished notes that the finite Ree groups in characteristic 3 and the finite Suzuki groups cannot act flag-transitively on a generalized hexagon and octagon, respectively. More such results dealing with infinite classes of simple groups seem to be with reach, but we are still far away from even a preliminary result, especially concerning generalized quadrangles.

So the determination of all finite flag-transitive generalized polygons remains one of the most important goals for future researchers in this area.

On the other hand, SEITZ [1973] classified all flag-transitive groups acting on a finite classical polygon (as part of the classification of all flag-transitive subgroups of a finite Chevalley group acting on its natural building). Since we do not want to go into details concerning the structure of these groups, we only mention a weak form of this result; part of it was written up as an unpublished correction to *op. cit.*

4.8.7 Theorem (Seitz [1973]). *Let Γ be a finite Moufang n-gon and G a flag-transitive collineation group of Γ. Then G contains the little projective group of Γ, except in the following cases, and up to duality:*

(i) *$n = 3$, Γ is isomorphic to $\mathbf{PG}(2,2)$ or $\mathbf{PG}(2,8)$ and G is a (Frobenius) group of order $3 \cdot 7 = 21$ or $9 \cdot 67 = 653$ and it acts sharply transitively on the set of flags of Γ.*

(*ii*) $n = 4$, Γ *is isomorphic to* $\mathsf{W}(2)$, $\mathsf{W}(3)$ *or* $\mathsf{H}(3,9)$. *If* $\Gamma \cong \mathsf{W}(2)$, *then* $G \cong \mathbf{A}_6$, *the alternating group on six elements, and this is the derived group of the little projective group (which is the full automorphism group* $\mathbf{S}_6$ *of* Γ; *the group* $\mathbf{A}_6$ *does not contain any root elation); if* $\Gamma \cong \mathsf{W}(3)$, *then* G *has a normal elementary abelian subgroup of order 16 and a complement isomorphic to* $\mathbf{A}_5$, $\mathbf{S}_5$ *(the full symmetric group on five letters) or a (Frobenius) group of order 20; if* $\Gamma \cong \mathsf{H}(3,9)$, *then* G *has a normal subgroup isomorphic to* $\mathbf{PSL}_3(4)$ *and the corresponding factor group has order 2 or 4.*

(*iii*) $n = 6$, Γ *is the split Cayley hexagon* $\mathsf{H}(2)$ *and* G *is the derived group* $\mathbf{PSU}_3(3)$ *of the full automorphism group* $\mathbf{G}_2(2)$ *of* $\mathsf{H}(2)$. *It contains all line-elations, but no point-elation, and it has index 2 in* $\mathbf{G}_2(2) \cong \mathbf{PGU}_3(3)$.

(*iv*) $n = 8$, Γ *is the Ree–Tits octagon of order* $(2,4)$ *and* G *is the Tits group* $\mathbf{T}$, *which is the derived group of the full automorphism group* ${}^2\mathbf{F}_4(2)$ *of* Γ. *It contains all point-elations, all line-elations of order 2 and it has index 2 in* ${}^2\mathbf{F}_4(2)$.

The cases $\Gamma \cong \mathsf{W}(2)$, $\Gamma \cong \mathsf{W}(3)$ and $\Gamma \cong \mathsf{H}(2)$ can be seen geometrically on the models given in Subsections 1.3.12 and 1.4.2.

4.9   Kantor systems

Recall from Subsection 3.7.3 on page 124 that an *elation generalized quadrangle* $(\Gamma^{(p)}, G)$ is a quadrangle Γ for which there is a point p of Γ such that there is a group G of collineations acting regularly on the set of points opposite p. If Γ has finite order (s, t), then clearly G has order $s^2 t$. Let y be a point opposite p and let $\{L_i : i = 0, 1, \ldots, t\}$ be the set of lines of Γ incident with p. Let z_i be the projection of y onto L_i, $i \in \{0, 1, \ldots, t\}$. Let S_i be the stabilizer in G of the line $y z_i$. We have the following properties:

4.9.1 Proposition (Kantor [1980]). *With the above notation, the following hold:*

[Ka1] *the set* $S_i^* := \bigcup \{ S_i g : g \in G \text{ and } S_i g \cap (S_j \setminus \{1\}) = \emptyset, \text{ for all } j \neq i \}$ *is a subgroup of* G;

[Ka2] *the order of* G *is* $s^2 t$; *the order of each* S_i *is* s; *the order of each* S_i^* *is* st;

[Ka3] $S_i S_j \cap S_k = \{1\}$, *for distinct* $i, j, k \in \{0, 1, \ldots, y\}$;

[Ka4] $S_i \cap S_j^* = \{1\}$, *for distinct* $i, j \in \{0, 1, \ldots, t\}$.

Also, for every $i \in \{0, 1, \ldots, t\}$, S_i^* *is the stabilizer in* G *of* z_i.

Proof. It is convenient to prove the assertions in a different order.

[Ka3] If for $g_i \in S_i$ and $g_j \in S_j$ the product $g_i g_j$ belongs to S_k, then there arises a triangle y, y^{g_j}, $y^{g_i g_j}$.

[Ka1], [Ka4] Let T_i be the stabilizer of z_i in G. If $g \in T_i \cap S_j$, $i \neq j$, then there arises a triangle y, y^g, z_i, unless $g = 1$ (and remarking that $g \notin S_i$, otherwise we have a digon y, y^g). Hence $T_i \cap S_j = \{1\}$ for $i \neq j$. Since obviously $S_i \leq T_i$ and $|T_i| = st$, the group T_i is the union of t cosets of S_i which are all disjoint from $S_j \setminus \{1\}$. Hence every $g \in S_j \setminus \{1\}$ defines a coset $S_i g$ which is not contained in T_i. If $S_i g_j = S_i g_k$, for $g_j \in S_j \setminus \{1\}$, $g_k \in S_k \setminus \{1\}$, $g_j \neq g_k$, then $g_j g_k^{-1} \in S_i$, contradicting [Ka3] (even if $j = k$). Hence the $t(s-1)$ non-identity elements of all S_j, $j \neq i$, define $t(s-1)$ cosets of S_i which do not belong to T_i. The remaining t cosets must then form T_i and are therefore the only cosets which are disjoint from all sets $S_j \setminus \{1\}$. Hence $T_i = S_i^*$.

[Ka2] This follows from the regular action of G on the points opposite p and the fact that there are precisely $s^2 t$ points opposite p; s points on $y z_i$ opposite p; st points collinear with z_i and opposite p, $i \in \{0, 1, \dots, t\}$.

The proposition is proved. □

4.9.2 Definition. Now let G be an arbitrary finite group admitting a set $\mathcal{J}$ of $t + 1 > 2$ subgroups S_i, $i \in \{0, 1, \dots, t\}$, which satisfy conditions [Ka1], [Ka2], [Ka3] and [Ka4] of Proposition 4.9.1 (for some positive integer $s > 1$). Then we call $(G, \mathcal{J})$ a **Kantor system**. The motivation to consider Kantor systems is the fact that, conversely, these provide elation generalized quadrangles with respect to the group G.

4.9.3 Theorem (Kantor [1980]). *Let $(G, \mathcal{J})$ be a Kantor system and use the above notation. Define the following geometry $\Gamma(G, \mathcal{J})$. The points of Γ are of three types:*

(i) the elements of G,

(ii) the right cosets $S_i^* g$, $g \in G$, $i \in \{0, 1, \dots, t\}$,

(iii) a symbol (∞).

The lines are of two types:

(a) the right cosets $S_i g$, $g \in G$, $i \in \{0, 1, \dots, t\}$,

(b) the symbols $[S_i]$.

Incidence is defined as follows: the point (∞) *of type* (iii) *is incident with all lines of type* (b); *the point* $S_i^* g$ *is incident with the line* $[S_i]$ *of type* (b) *and with the lines* $S_i h$ *with* $h \in S_i^* g$; *a point* g *of type* (i) *is incident with all lines* $S_i g$ *of type* (a). *The geometry* $\Gamma(G, \mathcal{J})$ *is a generalized quadrangle of order* (s, t). *Also,* $\Gamma(G, \mathcal{J})$ *is an elation generalized quadrangle with elation point* (∞).

Proof. If L is a line and p a point of $\Gamma(G, \mathcal{J})$ not incident with L, then we show that there exists a unique point q incident with L and collinear with p, and a unique line M incident with both p and q. This is easy if p has type (ii) or (iii) or if L has type (b). So suppose $L = S_i h$ has type (a) and $p = g \in G$ has type (i). If the line M exists, then it certainly has type (a). If the point q exists, then it has either type (i) or type (ii). First we claim that the existence of $q = S_j^* g'$ (of type (ii)) and $M = S_k h'$ as above is equivalent to $S_i h \subseteq S_i^* g$. Indeed, from the assumptions it follows readily that $i = j = k$; hence $g \in S_i h' \subseteq S_i^* g' \supseteq S_i h$, implying $S_i^* g \supseteq S_i h$. Conversely, if $S_i h \subseteq S_i^* g$, then $g \in S_i g \subseteq S_i^* g \supseteq S_i h$, hence the claim.

Next we claim that the existence of $q = g' \in G$ (of type (i)) and $M = S_k h'$ as above is equivalent to $S_i h \nsubseteq S_i^* g$. Indeed, the assumption is equivalent to the existence of a number $k \in \{0, 1, \dots, t\} \setminus \{i\}$ such that there exist $g', h' \in G$ with $g' \in S_i h \cap S_k h'$ and $g \in S_k h'$. This is equivalent to the existence of $k \neq i$ such that $S_i h g^{-1} \cap S_k$ is non-empty. It follows immediately from the definition of S_i^* that this means $S_i h g^{-1} \nsubseteq S_i^*$. Hence our claim.

In order to prove that $\Gamma(G, \mathcal{J})$ is a generalized quadrangle, we only need to show that the k from the previous paragraph is unique. If it were not unique, then we could write $h g^{-1}$ in two ways as $h g^{-1} = g_i g_k = g_i' g_\ell'$, $g_m \in S_m$, $m = i, k, \ell$; hence $g_k = g_\ell$ by Condition [Ka3], implying $h g^{-1} \in S_i$, a contradiction.

The natural action of G on the elements of $\Gamma(G, \mathcal{J})$ on the right (with trivial action on (∞) and on the lines of type (b)) clearly turns $\Gamma(G, \mathcal{J})$ into an elation generalized quadrangle. $\qquad\square$

Moreover, we have the following result, which is almost immediate and so we omit the proof.

4.9.4 Theorem (Kantor [1980]). *Let* Γ *be a finite elation generalized quadrangle with respect to the group* G *and with elation point* p. *Let* y *be a point opposite* p *and let* $\{z_i : i = 0, 1, \dots, t\}$ *be the collection of points collinear with* p *and* y. *Let* S_i *be the stabilizer of* $y z_i$. *Put* $\mathcal{J} = \{S_i : i = 0, 1, \dots, t\}$. *Then* $(G, \mathcal{J})$ *is a Kantor system and the corresponding quadrangle* $\Gamma(G, \mathcal{J})$ *is naturally isomorphic with* Γ.
$\qquad\square$

BADER & PAYNE [19**] introduce infinite Kantor systems and prove results similar to those above for infinite generalized quadrangles.

4.9.5 Translation generalized quadrangles

The conditions [Ka1] to [Ka4] are sometimes not so easy to handle. A few sufficient conditions that are easier to check have been found in some special cases (see for instance PAYNE & THAS [1984], Chapters 8 and 10). We mention one of them. If $s = t$ and if all groups S_i are normal subgroups of G, then we only need [Ka2] and [Ka3]. Moreover, in this case, all elements of S_i are root elations, for all $i \in \{0, 1, \ldots, t\}$, and G is abelian. In fact, if G is abelian, then an elation generalized quadrangle with elation point p is called a **translation generalized quadrangle (with translation centre p)**. It is not difficult to show that all lines through p are regular in this case.

4.9.6 Elation generalized hexagons

A similar construction and description holds in some special cases for hexagons, and KANTOR (unpublished) has some results in that direction. But no non-classical examples have been found, which shows once again that the classical hexagons (and certainly the octagons) are very special and rare objects.

Kantor's construction of finite generalized n-gons from groups is treated with proofs in COHEN & COOPERSTEIN [1992].

We restrict ourselves here by proving that, if the group G — which acts regularly on the elements of some generalized n-gon Γ opposite some fixed point p, and which fixes all lines through p — is abelian, then $n = 3$ or $n = 4$. This is due to JOSWIG [1994].

4.9.7 Theorem (Joswig [1994]). *Let Γ be a generalized n-gon, and let $G \leq \mathrm{Aut}(\Gamma)$ fix all lines through some point p. Suppose that G is abelian and that it acts regularly on the set $\Gamma_n(p)$. Then $n \in \{3, 4\}$.*

Proof. To fix ideas, suppose that n is even. Let x be a point opposite p and let L_1 and L_2 be two lines incident with x, $L_1 \neq L_2$. Let $x_i \, \mathbf{I} \, L_i$ with $x \neq x_i$, $i = 1, 2$. Let θ_i be the unique element of G mapping x to x_i, $i = 1, 2$. Put $x' = x^{\theta_1}_2$. Clearly x' is collinear with $x^{\theta_1} = x_1$. But $x' = x_2^{\theta_2^{-1}\theta_1\theta_2} = x_1^{\theta_2}$, hence, similarly, x' is collinear with x_2. It follows that, if $n > 4$, $x = x'$, hence $\theta_1 = \theta_2$ (otherwise we have a quadrangle (x, x_1, x', x_2)), a contradiction. $\qquad\square$

4.9.8 Translation generalized polygons

A generalized polygon Γ with an abelian group G and a point p as in the statement of the previous theorem is called a **translation generalized polygon (with translation point p and translation group G)**. Examples are the classical projective planes, the translation planes (in the "classical" sense), the semi-classical quadrangles of Tits type (see Subsection 3.7.1 on page 122), the dual of the semi-classical quadrangles of Kantor type (see Subsection 3 on page 129) and every Moufang quadrangle with regular lines.

Chapter 5

The Moufang Condition

5.1 Introduction

A lot of examples of generalized polygons arise from *algebraic groups* in one way or another, for instance all classical polygons; see also Appendix D. In this chapter, we aim at a description (but not proof) of a characterization of all these examples. Namely, they are the only polygons satisfying the Moufang condition; see Definitions 4.4.4 on page 143. The main results are due to TITS [1976a], [1976b], [19**], [1979], [1983], [1994a], WEISS [1979] and TITS & WEISS [19**]. FAULKNER [1977] has also made a contribution.

The main idea of the Moufang condition is to generalize the notion of a *Moufang projective plane* to a *Moufang (generalized) polygon*. The classification of all Moufang polygons is one of the major achievements in the theory of generalized polygons. The result was announced by TITS in 1976 (although he missed one class of Moufang quadrangles), but at present (1997) has not yet been completely published, except for projective planes (by combined results of MOUFANG [1933], KLEINFELD [1951] and BRUCK & KLEINFELD [1951]), generalized octagons (see TITS [1983]) and the cases $n \neq 3, 4, 6, 8$ of non-existence (see TITS [1976b], [1979] and WEISS [1979]). There is also an unpublished manuscript of TITS [19**] classifying all Moufang quadrangles with regular points and lines. It should also be mentioned that the Moufang hexagons were explicitly classified in the 1960s by TITS (unpublished), up to the classification of the (exceptional) Jordan algebras that arose; see below. Finally, a book by TITS & WEISS [19**] is in preparation containing the complete proof of the classification of Moufang polygons. In the finite case, TITS [1976a] remarks that the classification of Moufang polygons follows from an entirely group-theoretical result of FONG & SEITZ [1973], [1974]. I will show how this happens in Section 5.7. I will, however, not repeat here the proof of FONG & SEITZ's result (it is about 100 pages of purely group-theoretical arguments, with more than that number of pages of background). Instead, I will outline the proof in the general case, occasionally in some detail, especially where the ar-

guments are geometric. I have based my approach on the lectures given by TITS
[1994b], [1995] in Paris (Collège de France) in January, February and December
1994, and January 1995, and his seminars in 1996 and 1997. I would like to thank
Jacques Tits for permission to reproduce large parts of his beautiful lectures.

5.2 First properties of Moufang polygons

5.2.1 Definitions. We recall some notions from Definitions 4.4.1 (page 140) and
from Definitions 4.4.4 (page 143), at the same time fixing the notation we will
use in the course of this chapter. Let Γ be a generalized n-gon and let us fix
some apartment Σ in Γ. Denote the elements of Σ by x_i, $i \in \mathbb{Z}$, with $x_i \mathbf{I} x_{i+1}$
and $x_i = x_{i+2n}$. Let G be a collineation group of Γ and let us denote by U_i the
pointwise stabilizer in G of $\Gamma_1(x_{i+1}) \cup \Gamma_1(x_{i+2}) \cup \ldots \cup \Gamma_1(x_{i+n-1})$. If some element
u of U_i fixes x_{i-1}, then it fixes Σ and hence by Theorem 4.4.2$(v), (vi)$ on page 140
it is the identity. So U_i acts semi-regularly on the set of half apartments bounded
by $\{x_i, x_{i+n}\}$ not containing x_{i+1}. If that action is transitive (hence regular), then
we call the path $(x_{i+1}, x_{i+2}, \ldots, x_{i+n-1})$ a **Moufang path**. The elements of U_i are
called **root elations** (independent of whether the path is a Moufang path or not). If
all paths of length $n-2$ in Σ are Moufang paths, then we say that Σ **is a Moufang
apartment**. If every apartment of Γ is a Moufang apartment, then we say that Γ
satisfies the Moufang condition, or that Γ **is a Moufang polygon**.

In fact, all the above definitions depend on the group G. To emphasize this, or to
avoid confusion, we will occasionally add "**with respect to** G" to all these notions.

If $i \leq j$, $i, j \in \mathbb{Z}$, then, with the above notation, we denote by $U_{[i,j]}$ the group
generated by $U_i, U_{i+1}, \ldots, U_j$. If $i > j$, then by definition, $U_{[i,j]}$ is the trivial
group. Also, we denote the set of non-trivial elements of U_i by U_i^*.

We will also use the notation and results of Section 4.3.

We now collect a few preliminary consequences of these definitions. Throughout,
Γ, Σ, G, the x_i and the U_i, $i \in \mathbb{Z}$ are as above, except when explicitly mentioned
otherwise.

5.2.2 Lemma. *Every subpolygon* Γ' *of a Moufang polygon* Γ *is itself a Moufang
polygon.*

Proof. We may assume that Σ is contained in Γ'. It suffices to show that, if
$u \in U_i$ and if x_{i-1}^u belongs to Γ', then $\Gamma'^u = \Gamma'$. So suppose $u \in U_i$ for some
$i \in \mathbb{Z}$ and suppose x_{i-1}^u is an element of Γ'. Clearly the polygons Γ' and Γ'^u have
the apartment Σ^u in common. They also share all elements of $\Gamma_1'(x_{i+1})$ and all
elements of $\Gamma_1'(x_{i+2})$. Hence by Corollary 1.8.5 (see page 34), they coincide and
the lemma is proved. $\square$

5.2.3 Lemma. *Suppose that Σ is a Moufang apartment.*

(i) **If** $i - 1 \leq j \leq i + n - 3$, **then** $U_{[i,j]}$ *is the pointwise stabilizer in* G *of* $\{x_j, x_{j+1}, \ldots, x_{n+i-4}\} \cup \Gamma_1(x_{n+i-2}) \cup \Gamma_1(x_{n+i-1})$ *and fixes every element of* $\Gamma_1(x_{j+1}) \cup \Gamma_1(x_{j+2}) \cup \ldots \cup \Gamma_1(x_{n+i-1})$.

(ii) **If** $i + 1 \leq j \leq i + n - 1$, **then** $[U_i, U_j] \leq U_{[i+1,j-1]}$.

(iii) **If** $i \leq j < i + n$, **then** *the product* $U_i U_{i+1} \ldots U_j$ *is a group and every element* u *of this group has a unique decomposition* $u = u_i u_{i+1} \ldots u_j$, *where* $u_k \in U_k$ *for* $i \leq k \leq j$.

Proof. (i) Clearly, $U_{[i,j]}$ fixes $\Gamma_1(x_k)$ for $j < k < i + n$. Conversely, suppose an element $u \in G$ fixes $\{x_j, x_{j+1}, \ldots, x_{n+i-4}\} \cup \Gamma_1(x_{n+i-2}) \cup \Gamma_1(x_{n+i-1})$ pointwise. We show by induction on $j \geq i - 1$ (for i fixed) that $u \in U_{[i,j]}$. If $j = i - 1$, then u fixes Σ and the result follows from Theorem 4.4.2(v), (vi) on page 140. So suppose $i \leq j \leq i + n - 3$. Let $u_j \in U_j$ be such that $x_{j-1}^{u_j} = x_{j-1}^{u}$. This is possible by the fact that Σ is a Moufang apartment. The collineation uu_j^{-1} fixes x_{j-1} and it fixes the set $\{x_j, x_{j+1}, \ldots, x_{n+i-4}\} \cup \Gamma_1(x_{n+i-2}) \cup \Gamma_1(x_{n+i-1})$ pointwise. By the induction hypothesis, $uu_j^{-1} \in U_{[i,j-1]}$, hence $u \in U_{[i,j]}$. Part (i) is proved.

(ii) If $i < j \leq i + n - 1$, then every element of U_i fixes x_k for $j \leq k \leq i + n$, and it fixes every element of $\Gamma_1(x_l)$ for $j \leq l \leq i + n - 1$. Similarly, all elements of U_j fix x_k for $j \leq k \leq i + n$ and they all fix every element of $\Gamma_1(x_{l'})$ for $j + 1 \leq l' \leq i + n$. Hence every element of $[U_i, U_j]$ fixes $\Gamma_1(x_k)$, $j \leq k \leq i + n$, pointwise. By part (i) of this lemma, $[U_i, U_j]$ must then be a subgroup of $U_{[i+1,j-1]}$. Hence (ii).

(iii) We show this assertion by induction on $j \geq i$. For $j = i$, there is nothing to prove. So suppose $j > i$. Let $u = u_i u_{i+1} \ldots u_j$ and let $u' = u'_i u'_{i+1} \ldots u'_j$ with $u_k, u'_k \in U_k$, $i \leq k \leq j$. By part (ii) of this lemma, we can write uu'^{-1} as $u^* u_j (u'_j)^{-1}$ with $u^* \in U_{[i,j-1]}$. By the induction hypothesis, $U_1 U_2 \ldots U_j$ is a group. Suppose that $u = u'$, then $u_j (u'_j)^{-1} = 1$, since otherwise $u_j (u'_j)^{-1}$ acts non-trivially on $\Gamma_1(x_j)$, unlike u^*. The result now follows easily from the induction hypothesis. $\square$

5.2.4 Lemma.

(i) *The group* $U_+ := U_1 U_2 \ldots U_n$ *acts regularly on the set of flags of* Γ *opposite* $\{x_n, x_{n+1}\}$.

(ii) *The group* $U_1 U_2 \ldots U_{n-1}$ *acts regularly on the set of elements of* Γ *opposite* x_n.

Proof. (i) Let $\{y_0, y_1\}$ be a flag opposite $\{x_n, x_{n+1}\}$ (with y_0 opposite x_n) and suppose that we have the path $(y_0, y_1, y_2, \ldots, y_{n-1}, x_n)$, with $y_{n-1} \neq x_{n+1}$ (this is possible by the fact that $\{y_0, y_1\}$ is opposite $\{x_n, x_{n+1}\}$). Using an element of U_n, we can map y_{n-1} to x_{n-1}. Using an element of U_{n-1}, we can map the image of y_{n-2} to x_{n-2}. Proceeding in this way, we can map $\{y_0, y_1\}$ on $\{x_0, x_1\}$ by an element of U_+. This shows that U_+ is transitive on the set of flags opposite $\{x_n, x_{n+1}\}$. Now suppose that $u_1 u_2 \cdots u_n \in U_+$ (with $u_i \in U_i$, $1 \leq i \leq n$) fixes $\{x_0, x_1\}$, hence Σ. Since u_j, $1 \leq j \leq n-1$, fixes x_{n-1}, u_n must also fix x_{n-1}, hence u_n is trivial. Since u_j, $1 \leq j \leq n-2$, fixes x_{n-2}, u_{n-1} must also fix x_{n-2}, hence u_{n-1} is trivial. Proceeding in this way, we find that all u_i, $1 \leq i \leq n$, are trivial and hence U_+ acts semi-regularly, hence regularly on the set of flags opposite $\{x_n, x_{n+1}\}$.

The proof of (ii) is completely similar. $\square$

5.2.5 Lemma. *The group $U_+ = U_1 U_2 \ldots U_n$ is a normal subgroup of the stabilizer in the full collineation group $\operatorname{Aut} \Gamma$ of Γ of the flag $\{x_n, x_{n+1}\}$.*

Proof. The conjugate U_i^g of U_i, with $g \in (\operatorname{Aut} \Gamma)_{\{x_n, x_{n+1}\}}$, $i \in \{1, \ldots, n\}$, is a group of root elations with respect to the apartment Σ^g and hence completely determined by i and $\{x_0, x_1\}^g$. By Lemma 5.2.4, we may choose $g \in U_1 U_2 \ldots U_n$, hence the result. $\square$

5.2.6 Lemma. *Let $u \in U_i^*$ for some $i \in \mathbb{Z}$. Then there exist unique elements $u', u'' \in U_{i+n}$ such that $\mu = u'uu''$ stabilizes Σ. Moreover, $x_j^\mu = x_{2i-j}$ and ${}^\mu U_j = U_{n+2i-j}$. The map $\omega : U_i^* \to U_{i+n}^* : u \mapsto u''$ (defined for all $i \in \mathbb{Z}$), is a bijection and is the inverse of $\omega' : U_i^* \to U_{i+n}^* : u \mapsto u'$ (also defined for all $i \in \mathbb{Z}$). We also have*

$$\omega^{-1}(u) u \omega(u) = u \omega(u) \omega^2(u) = \omega^{-2}(u) \omega^{-1}(u) u.$$

Proof. Let Θ and Θ' be the half apartments $(x_i, x_{i+1}, \ldots, x_{i+n})$ and $(x_{i-n}, x_{i-n+1}, \ldots, x_i)$, respectively. Let $u', u'' \in U_{i+n}$. Then $u'uu''$ stabilizes Σ **if and only if** it interchanges Θ and Θ' (if $u'uu''$ fixed Θ', then $\Theta'^{u'uu''} = \Theta^{uu''} = \Theta'$ which implies $\Theta'^u = \Theta'$, contradicting the non-triviality of u). Now $\Theta^{u'uu''} = \Theta'$ is equivalent to $\Theta^{u'} = \Theta'^{u^{-1}}$ (which defines u' uniquely), and $\Theta'^{u'uu''} = \Theta$ is equivalent to $\Theta'^u = \Theta^{u''^{-1}}$ (and this determines u'' uniquely). Since all of u, u' and u'' fix x_i and x_{i+n}, the second statement follows easily (the collineation $u'uu''$ acts as a "reflection" on Σ).

The bijectivity of the map ω follows immediately from the equality $\Theta'^u = \Theta^{u''^{-1}}$, which can now be taken as the definition of ω. Similarly for ω'. Also, from $\Theta^{u'} = \Theta'^{u^{-1}}$ it follows (by interchanging the roles of Θ and Θ' in the new definition of ω just mentioned) that $\omega(u') = u$, hence ω is the inverse of ω'.

Noting that $u'uu'' = uu''u'^{u'uu''}$, we see that $\omega^2(u) = \omega(u'') = u'^{u'uu''}$ and hence

$$\omega^{-1}(u) u \omega(u) = u \omega(u) \omega^2(u).$$

Similarly $\omega^{-1}(u) u \omega(u) = \omega^{-2}(u) \omega^{-1}(u) u$. This completes the proof of the lemma.
                                    $\square$

For given $u \in U_i^*$, we denote the element μ of the preceding lemma by $\mu(u)$. Also the symbols $\omega(u)$ and $\omega^{-1}(u)$ for $u \in U_i$, $i \in \mathbb{Z}$, will be frequently used. Some consequences of the preceding lemma and this notation are collected in the next result.

5.2.7 Corollary. *For all $u \in U_i$, $i \in \mathbb{Z}$, and all $k \in \mathbb{Z}$, we have*

$$\mu(u) = \mu(\omega^k(u)) = \omega^{k-1}(u)\omega^k(u)\omega^{k+1}(u) = \mu(u^{\mu(u)}).$$

Proof. The first two equalities follow from the previous lemma by an inductive argument. For the last one we note that

$$u^{\mu(u)} = (u\omega(u)\omega^2(u))^{-1} \cdot u \cdot (\omega(u)\omega^2(u)\omega^3(u)) = \omega^3(u)$$

and the result follows. □

5.2.8 Lemma. *If Γ contains one Moufang apartment Σ (with respect to G), then Γ is a Moufang polygon (with respect to G).*

Proof. Let $\gamma = (x_0, x_1', \ldots, x_{n+1}')$ be an arbitrary path of length $n+1$ in Γ starting in x_0. Up to renumbering the elements of Σ, we can suppose that $x_1' \neq x_{2n-1}$. Hence there is a unique $u_n \in U_n$ mapping x_1' to x_1. Now there is a unique $u_{n+1} \in U_{n+1}$ mapping $(x_2')^{u_n}$ to x_2; there is a unique $u_{n+2} \in U_{n+2}$ mapping $(x_3')^{u_n u_{n+1}}$ to x_3. Continuing this way, we see that there is an element $u \in U_{[n,2n]}$ mapping γ onto $(x_0, x_1, \ldots, x_{n+1})$. This implies that G is transitive on the set of all paths of length $n+1$ starting with x_0. This means in particular that G acts transitively on the set of elements opposite x_0. Actually, x_0 was arbitrary in Σ and so G acts transitively on the set of all elements opposite x_i, for all $i \in \mathbb{Z}$. If x is any element of Γ, then by Lemma 1.5.9 on page 21, x is opposite some element of Σ, say x_n. Hence there is a collineation in G mapping x to x_0. So from the previous arguments, it follows that every half apartment can be mapped by an element of G to a half apartment contained in Σ. Hence every path of length $n-2$ is a Moufang path and the result follows. □

Note that the transitivity on paths of length $n+1$ starting at a point (or line) is in fact equivalent to transitivity on ordered apartments. We have called this the **Tits condition**; see also Section 4.7. The proof of Lemma 5.2.8 implies that every Moufang polygon satisfies the Tits condition.

5.2.9 Theorem. *Every Moufang polygon satisfies the Tits condition with respect to the little projective group.*

5.3 Weiss' theorem

5.3.1 Trees

A **tree** is a graph (undirected, without multiple edges and loops) without circuits. A **thick** tree is a tree such that every vertex is adjacent to at least three vertices. A thick tree is necessarily infinite. Examples of trees are the universal covers of the incidence graphs of geometries, in particular of generalized polygons. This goes as follows.

Consider the incidence graph (X, E) of a generalized polygon Γ. So X is the set of all points and lines of Γ and E is the set of edges, i.e., the flags of Γ. But as our general notation suggests, the following construction holds for any graph. Let x_0 be any vertex of (X, E). We first note that every path in (X, E) can be **reduced** to a non-stammering one by deleting (x, y) whenever (x, y, x) is a sub-path. We define a new graph $(\tilde{X}_{x_0}, \tilde{E}_{x_0})$ as follows. The set $\tilde{X}_{x_0}$ consists of all non-stammering paths starting in x_0. Two such paths form an edge if one path is transformed into the other when the last vertex is deleted. If x_1 is another vertex of (X, E), then the graphs $(\tilde{X}_{x_0}, \tilde{E}_{x_0})$ and $(\tilde{X}_{x_1}, \tilde{E}_{x_1})$ are isomorphic and an isomorphism from the first to the second graph is given by the juxtaposition (in front) with any path $(x_1, \dots, x_0)$ (and then reducing). Hence we may denote the graph we have just defined by $(\tilde{X}, \tilde{E})$. But in practice, we always choose a vertex x_0 and put $(\tilde{X}, \tilde{E}) = (\tilde{X}_{x_0}, \tilde{E}_{x_0})$. Note that by taking $x_0 = x_1$ in the foregoing, we obtain an automorphism of $(\tilde{X}, \tilde{E})$. Such an automorphism is called a **deck transformation** and all such deck transformations form a group $\tilde{D}$.

There is a natural epimorphism from $(\tilde{X}, \tilde{E})$ to (X, E) (mapping a path to its end vertex) and we denote that epimorphism by $\eta_{(X,E)}$.

Let θ be an automorphism of (X, E) and denote the automorphism group of the latter by G. Then θ clearly induces an isomorphism from $(\tilde{X}_{x_0}, \tilde{E}_{x_0})$ to $(\tilde{X}_{x_0^\theta}, \tilde{E}_{x_0^\theta})$ (by taking the images of paths starting at x_0) and hence this defines an automorphism of $(\tilde{X}, \tilde{E})$, up to a deck transformation. The group generated by all such automorphisms, together with all deck transformations, is denoted by $\tilde{G}$. The epimorphism $\eta_{(X,E)}$ induces a natural epimorphism from $\tilde{G}$ onto G the kernel of which is exactly the group of deck transformations. So $G \cong \tilde{G}/\tilde{D}$ in a natural way.

5.3.2 Definitions. Let (X, E) be a thick connected tree, G an automorphism group of (X, E) and $n \in \mathbb{N}$, $n \geq 3$. Following WEISS [1995a], [1995b], [1996], [1997], we call (X, E) a (G, n)**-Moufang tree** if for every non-stammering path $(x_0, x_1, \dots, x_n)$ of length n, the following two conditions are satisfied (where $X_i(x)$, $x \in X$, denotes the set of vertices of (X, E) at distance i from x).

(W1) The pointwise stabilizer in G of $X_1(x_1) \cup X_1(x_2), \cup \dots \cup X_1(x_{n-1})$ acts transitively on the set $X_1(x_n) \setminus \{x_{n-1}\}$.

(W2) The pointwise stabilizer in G of $X_1(x_0) \cup X_1(x_1) \cup \{x_3, x_4, \dots, x_n\}$ is trivial.

An element of G stabilizing $X_1(x_1) \cup X_1(x_2), \cup \ldots \cup X_1(x_{n-1})$ pointwise is called an n-**root elation**. An **inversion** is an automorphism of (X, E) which maps each element x to an element at odd distance from x.

The next result will imply that Moufang n-gons only exist for $n = 3, 4, 6$ or 8. However, we show the general version for trees. For the time being, we forget our general notation since we will deal now with trees rather than with polygons.

5.3.3 Theorem (Weiss [1979]). *Let (X, E) be a thick connected tree and $n \in \mathbb{N}$, $n \geq 3$. Let G be an automorphism group of (X, E). If (X, E) is a (G, n)-Moufang tree,* **then** *$n = 3, 4, 6$ or 8. If, moreover, G contains an inversion,* **then** *$n \neq 8$.*

Proof. The proof is based on the lectures given by TITS [1994b] in Paris in January 1994, except for the case $n = 12$, where we follow closely the original proof by WEISS [1979].

The result is shown by a sequence of lemmas.

In what follows, we assume that all paths we consider are non-stammering, except when the contrary is explicitly stated. Also, we denote the pointwise stabilizer of a set $E_1(x_1) \cup E_1(x_2) \cup \ldots \cup E_1(x_m)$, $m \in \mathbb{N}$, by $G(x_1, \ldots, x_m)$. Usually, $(x_1, \ldots, x_m)$ is a path.

Denote by X_O and X_I the set of vertices of X at even and odd distance respectively from some fixed element of X (these sets are well defined and, up to renaming, independent of that fixed element). We know $X = X_O \cup X_I$.

Lemma 1. *Let $(x_0, x_1, \ldots, x_m)$ be a path of length $m \leq n$ and let $i, j \in \mathbb{N}$ with $i + m + j \leq n + 1$. Then $G(x_1, \ldots, x_{m-1})$ (if $m \geq 2$), G_{x_0, x_1} (for $m = 1$) or G_{x_0} (for $m = 0$) acts transitively on the set of paths of the form*

$$(x_{-i}, \ldots, x_{-1}, x_0, x_1, \ldots, x_m, x_{m+1}, \ldots, x_{m+j}).$$

Proof. Similar to the first part of the proof of Lemma 5.2.8. QED

As a corollary, we obtain:

Lemma 2.
(i) For every vertex $x \in X$, G_x acts doubly transitively on $X_1(x)$.
(ii) G_{X_O} acts transitively on X_O and G_{X_I} acts transitively on X_I; if G contains an inversion, then G acts transitively on X.
(iii) G acts transitively on E. □

Lemma 3. *Let $(x_0, x_1, \ldots, x_m)$ be a path of length $m < n$, $m \geq 2$. Put $\gamma = (x_2, \ldots, x_m)$. Then the group $T(x_0, x_1; \gamma)$ of permutations on $X_1(x_0)$ induced by $G(x_1, x_2, \ldots, x_m)$ depends only on x_1 and neither on γ nor on m.*

Proof. By Lemma 1, the group $G(x_0)$ acts transitively on the set of paths of fixed length $m - 2$ starting with (x_0, x_1). Hence $T(x_0, x_1; \gamma)$ depends only on the length of γ.

Now let $n - 1 = m$ and so $\gamma = (x_2, x_3, \ldots, x_{n-1})$. Let $x_n \neq x_{n-2}$ be any vertex adjacent to x_{n-1} and let $x_{n+1} \neq x_{n-1}$ be any vertex adjacent to x_n. We show that $T(x_0, x_1; x_2) \leq T(x_0, x_1; \gamma)$. This will imply Lemma 3, since the inverse inclusion is obvious. We may assume $m > 2$. Then $n > 3$. Let $g \in T(x_0, x_1; x_2)$. By multiplying with an element of $G(x_0, x_1, x_2)$, we may assume by Lemma 1 that g fixes $\gamma \cup \{x_n\}$ pointwise. Let h be an element of $G(x_1, x_2, \ldots, x_{n-1})$ mapping x_{n+1} to x_{n+1}^g. The element $h^{-1}g$ fixes both $X_1(x_1)$ and $X_1(x_2)$ pointwise and it also fixes x_k, $3 \leq k \leq n+1$. By hypothesis (W2) of the theorem, $g = h$ and so Lemma 3 follows.

QED

Hence we can denote the group $T(x_0, x_1; \gamma)$ by $T(x_0, x_1)$. In the classical case, this is the group of translations of the affine line consisting of vertices adjacent to x_0 where x_1 is the point at infinity. For that reason, we will call — in this proof — the elements of $T(x_0, x_1)$ **translations in** $X_1(x_0)$ **with centre** x_1. By (W1), the group $T(x_0, x_1)$ acts regularly on $X_1(x_0) \setminus \{x_1\}$.

From now on we fix a path $(x_i)_{i \in \mathbb{Z}}$ and we put $U_i = G(x_{i+1}, \ldots, x_{i+n-1})$. Also, for $i \leq j$, we let $U_{[i,j]}$ be the group generated by the groups $U_i, \ldots, U_j$ and for $j < i$, $U_{[i,j]}$ is the trivial group.

The following lemma is essentially equivalent to Lemma 5.2.3, except for the cases $j = i + n - 2$ and $j = i + n - 1$ of (i).

Lemma 4.

(i) **If** $i \leq j \leq i + n - 3$, **then** $U_{[i,j]} = G(x_{j+1}, \ldots, x_{i+n-1})$.
  If $j = i + n - 2$, **then** $U_{[i,j]} = \{g \in G(x_{i+n-1}) : g$ *induces a translation in* $X_1(x_{i+n-2})$ *with centre* $x_{i+n-1}\} = \{g \in G(x_{i+n-1}) : g$ *induces a translation in* $X_1(x_{i+n})$ *with centre* $x_{i+n-1}\}$.
  If $j = i + n - 1$, **then** $U_{[i,j]} = \{g \in G_{x_{i+n-1}, x_{i+n}} : g$ *induces a translation in* $X_1(x_{i+n-1})$ *with centre* x_{i+n} *and* g *induces a translation in* $X_1(x_{i+n})$ *with centre* $x_{i+n-1}\}$.

(ii) **If** $i + 1 \leq j \leq i + n - 1$, **then** $[U_i, U_j] \leq U_{[i+1, j-1]}$.

(iii) **If** $i \leq j < i + n$, **then** *the product* $U_i U_{i+1} \ldots U_j$ *is a group and every element* u *of this group has a unique decomposition* $u = u_i u_{i+1} \ldots u_j$, *where* $u_k \in U_k$ *for* $i \leq k \leq j$.

Proof. The cases $j = i + n - 2$ and $j = i + n - 1$ of (i) follow like the other cases by an inductive argument. For instance, let g be in $G_{x_{i+n-1}, x_{i+n}}$ such that g induces a translation in $X_1(x_{i+n-1})$ with centre x_{i+n} and another one in $X_1(x_{i+n})$ with centre x_{i+n-1}. After multiplying with a suitable element of U_{i+n-1}, we may suppose that g fixes $X_1(x_{i+n-1})$ pointwise (since the translation group in $X_1(x_{i+n-1})$ acts regularly on $X_1(x_{i+n-1}) \setminus \{x_{i+n}\}$). But then we are back at the case $j = i+n-2$.

QED

For two adjacent vertices x and y, we denote by $V(x, y)$ the inverse image of $T(x, y)$ in the group $G_{x,y}$. So an element of $G_{x,y}$ belongs to $V(x, y)$ if and only if it induces in $X_1(x)$ and in $X_1(y)$ a translation with centre y and x, respectively.

The next lemma is crucial for the classification of Moufang polygons. For the definition of centre, we refer to Section 4.3.

Lemma 5. *For every $i \in \mathbb{Z}$, the centre $Z(U_{[i+1,i+n]})$ of $U_{[i+1,i+n]}$ is non-trivial. Equivalently, for every pair of adjacent vertices x and y, the centre $Z(V(x, y))$ of $V(x, y)$ is non-trivial.*

Proof. Let m be the smallest positive integer such that there exists $i \in \mathbb{Z}$ with the property that $[U_i, U_{m+i}]$ is non-trivial. To see that this exists, we claim that whenever $u \in U_{m+i}^*$ centralizes U_i (see Section 4.3) and $m \geq 0$, then $m \leq \lfloor \frac{n}{2} \rfloor$. Indeed, if $m > \lfloor \frac{n}{2} \rfloor$, then u does not fix x_{m-1} and hence $(x_0, x_1, \ldots, x_m, x_{m-1}^u, \ldots, x_0^u)$ is a path of length $2m$, which must be fixed by every element of U_i. Since every element of U_i fixes $X_1(x_1) \cup X_1(x_2)$ pointwise, axiom (W2) implies $2m \leq n$. This proves the claim.

Remark that for $i \neq j$, an element $u \in U_j$ normalizes U_i **if and only if** it centralizes U_i. Indeed, if $u_i \in U_i$ and $u^{-1} u_i u \in U_i$, then also $[u_i, u] \in U_i$, but $[u_i, u] \in U_{[i+1,j-1]}$ (assuming $i < j$; if not, then consider $U_{[j+1,i-1]}$). But $U_{[i+1,j-1]} \cap U_j$ is trivial and the remark is proved.

We return to the situation where we have $m \in \mathbb{Z}$, $m > 0$ minimal with respect to the property that there exists $i \in \mathbb{Z}$ such that $[U_i, U_{m+i}]$ is non-trivial. Note that $[U_i, U_{m+i}] \leq U_{[i+1,m+i-1]}$ and by the minimality of m, $U_{[i+1,m+i-1]}$ centralizes both U_i and U_{m+i} (this follows from the fact that, by Lemma 1 and by the fact that $2m \leq n$, $-m$ is the largest negative integer for which there exists $i \in \mathbb{Z}$ such that $[U_i, U_{i-m}]$ is non-trivial). But now since U_i and U_{m+i} centralize $U_{[i+1,m-1]}$, then $[U_i, U_{m+i}]$, which is non-trivial, also centralizes $U_{[i+1,m-1]}$. But since $[U_i, U_{m+i}] \leq U_{[i+1,m-1]}$ and the latter centralizes both U_i and U_{m+i}, we conclude that $[U_i, U_{m+i}]$ centralizes $U_{[i,m+i]}$. So $Z(U_{[i,m+i]})$ is non-trivial.

Now let $j \in \mathbb{Z}$, $0 < j \leq n$, be maximal with the property that there exists $i \in \mathbb{Z}$ such that $Z(U_{[i+1,i+j]})$ is non-trivial. Such a j exists and clearly $j > m \geq 1$, so in fact $j > 1$. We will now show that $j = n$. Suppose on the contrary that $j < n$. Let $k \leq j$ be the smallest positive integer such that $U_{[i+1,i+k]} \cap Z(U_{[i+1,i+j]})$ is non-trivial and let u be a non-trivial element of that intersection. If u centralized U_i, then $Z(U_{[i,(i-1)+(j+1)]})$ would contain a non-trivial element (namely u), contradicting the maximality of j. Hence there is a non-trivial element

$u' = [u_i, u]$, with $u_i \in U_i^*$. Note that by Lemma $4(ii)$, $u' \in U_{[i+1,i+k-1]}$. But since u_i normalizes $U_{[i+1,i+j]}$ (using $j < n$ and Lemma $4(ii)$) and u centralizes $U_{[i+1,i+j]}$, its commutator u' centralizes $U_{[i+1,i+j]}$. This contradicts the minimality of k.

Now by Lemma $4(i)$, $U_{[i+1,i+n]} = V(x_{i+n}, x_{i+n+1})$ and hence by Lemma 5 and Lemma $2(iii)$, $Z(V(x,y))$ is non-trivial for every pair of adjacent vertices. The lemma follows. QED

We briefly denote $Z(x,y) = Z(V(x,y))$. For a positive integer r, we denote $X_{\leq r}(x)$ by $B(x;r)$ and $\{z \in X : \delta(x,z) = \delta(y,z) + 1 \leq r\}$ by $B(x,y;r)$. One easily verifies that

$$B(x;r) = B(x,y;r) \cup B(y,x;r+1) \tag{5.1}$$

for adjacent vertices x and y. We denote by $F(x;r)$ and $F(x,y;r)$ the pointwise stabilizer of $B(x;r)$ and $B(x,y;r)$, respectively.

The next lemma follows immediately from the fact that $V(x,y)$ acts transitively on the set of paths of length $n+1$ starting with (x,y).

Lemma 6. *Let $\{x,y\} \in E$. If $u \in Z(x,y)$ fixes a path of length $r \leq n+1$ starting with (x,y), then $u \in F(x,y;r)$.* QED

We will now show that the elements of $Z(x,y)$ have to fix a lot. The idea of the rest of the proof is then to try to prove that they have to fix even more under the condition $n \neq 3,4,6,8,12$, and this leads to a contradiction.

For the remainder of the proof we put $n' = \lfloor \frac{n-1}{2} \rfloor$.

Lemma 7. *For every edge $\{x,y\}$, we have $Z(x,y) \leq F(x,y;n'+1) \cap F(x;n')$.*

Proof. With our previous notation, we must show that $Z(x_0, x_1)$ fixes $(x_0, x_1, \ldots, x_{n'+1})$. If not, let k be the smallest positive integer such that there exists $z \in Z(x_0, x_1)$ which does not fix x_{k+1}. Now, U_0 fixes $\gamma = (x_0, \ldots, x_n)$, hence every element of U_0 fixes γ^z (since U_0 and $Z(x_0, x_1)$ commute). But the path

$$(x_n, x_{n-1}, \ldots, x_k, x_{k+1}^z, \ldots, x_n^z)$$

has length $2(n-k)$, and since every element of U_0 fixes $X_1(x_{n-1}) \cup X_1(x_{n-2})$ pointwise, hypothesis (W2) implies that $2(n-k) \leq n$. This implies $k \geq n'+1$ as required.

We have shown that $Z(x,y) \leq F(x,y;n'+1)$. For reasons of symmetry, also $Z(x,y) \leq F(y,x;n'+1)$, in particular $Z(x,y) \leq F(x,y;n') \cap F(y,x;n'+1) = F(x;n')$. QED

Lemma 8. *If $Z(x_0, x_1)$ and $Z(x_{n'+1}, x_{n'+2})$ do not centralize each other*, **then** $n \in \{3,4,6\}$.

Proof. Let $z \in [Z(x_0, x_1), Z(x_{n'+1}, x_{n'+2})]$ be non-trivial. Since $Z(x_0, x_1)$ fixes $x_{n'+1}$, it stabilizes $B(x_{n'+1}; n')$. Since $Z(x_{n'+1}, x_{n'+2})$ fixes $B(x_{n'+1}; n')$ pointwise, we have that z fixes $B(x_{n'+1}; n')$ pointwise. Also, z fixes $B(x_1; n')$ pointwise by symmetry. So z fixes the path $(x_{1-n'}, \ldots, x_{2n'+1})$ and all neighbours of $x_{2-n'}$ and $x_{3-n'}$. By condition (W2), $3n' \le n$. If n is odd, $n' = \frac{n-1}{2}$ and so $n \le 3$. If n is even, then $n' = \frac{n-2}{2}$ and so $n \le 6$. $\qquad$ QED

The next lemma will take care of the odd case.

Lemma 9. If $[Z(x_0, x_1), Z(x_{n'+1}, x_{n'+2})]$ *is trivial,* **then either** $Z(x_0, x_1)$ *fixes* $B(x_1; n' + 1)$ *pointwise* **or** $Z(x_{n'+1}, x_{n'+2})$ *fixes* $B(x_{n'+1}; n' + 1)$ *pointwise. In particular,* **if** n *is odd,* **then** *this situation cannot occur and hence by* Lemma 8, $n = 3$.

Proof. If $Z(x_0, x_1)$ fixes $x_{n'+2}$, then by Lemma 6 above, $Z(x_0, x_1)$ fixes $B(x_0, x_1; n'+2)$ pointwise. Since it also fixes $B(x_1, x_0; n'+1)$ by Lemma 7, the result follows from Equation (5.1).

If $Z(x_0, x_1)$ does not fix $x_{n'+2}$, then let z be an element in $Z(x_0, x_1)$ which does not fix $x_{n'+2}$ and let z' be an arbitrary element of $Z(x_{n'+1}, x_{n'+2})$. Since z and z' commute, z' fixes $(x_{n'+2}, x_{n'+1}, x_{n'+2}^z, \ldots, x_{2n'+2}^z)$. Lemmas 6 and 7 complete the proof (using Equation (5.1)).

If n is odd, then we have a non-trivial element fixing, without loss of generality, the path $(x_{-n'}, \ldots, x_{n'+2})$ of length $2n'+2 = n+1$ and fixing all vertices adjacent to $x_{1-n'}$ and $x_{2-n'}$. This contradicts (W2). $\qquad$ QED

Lemma 10. If n *is even,* **then** $n = 4, 6, 8$ *or* 12. **If,** *moreover,* G *contains an inversion,* **then** 8 *and* 12 *cannot occur.*

Proof. By Lemma 9, we may assume without loss of generality that $F(x_0; n' + 1)$ is not trivial. Lemma 2 implies that $F(x_k; n' + 1)$ is non-trivial for all even integers k. If G contains an inversion, then $F(x_k; n' + 1)$ is non-trivial for all integers k. Note that $2n' + 2 = n$.

Let us suppose that for $m \in \mathbb{Z}$ the group $F(x_m; n' + 1)$ is non-trivial, $n' + 1 \le m \le n$. Now we have that $x_{(n'+1)+(n-m)}^z \in B(x_m; n' + 1)$ for all $z \in F(x_0; n' + 1)$ since the distance from $x_{(n'+1)+(n-m)}$ to x_m is at most $(n - m) + (m - (n' + 1)) = n' + 1$. So any element of $[F(x_0; n' + 1), F(x_m; n' + 1)]$ fixes (by symmetry) the path $(x_{2m-3n'-3}, \ldots, x_{3n'+3-m})$. This path has length $3n - 3m$. As before, this implies that $[F(x_0; n' + 1), F(x_m; n' + 1)]$ is trivial whenever $m \le \frac{2n-1}{3}$. But if $[F(x_0; n' + 1), F(x_m; n' + 1)]$ is trivial, then, as before, this means that every element of $F(x_0; n' + 1)$ fixes the path

$$(x_{-n'-1}, \ldots, x_{m-(n'+1)}, x_{m-(n'+1)-1}^f, \ldots, x_{-n'-1}^f)$$

of length $2m$, for every non-trivial element $f \in F(x_m; n' + 1)$ (because f does not fix $x_{m-(n'-1)+1}$ by condition (W2)). As before, this means that $2m \le n$.

Hence we have shown that, whenever $m \leq \frac{2n-1}{3}$, then $m \leq \frac{n}{2}$. If n' is even, then Lemma 9 shows that for every $m \in \mathbb{Z}$ the group $F(x_m; n' + 1)$ is non-trivial and hence in this case there do not exist integers m with $\frac{n}{2} < m \leq \frac{2n-1}{3}$. But this implies $n = 6$. If n' is odd, then there do not exist even integers m satisfying $\frac{n}{2} < m \leq \frac{2n-1}{3}$. This implies that $n = 4, 8$ or 12. If, moreover, G contains an inversion, then there do not exist integers m such that $\frac{n}{2} < m \leq \frac{2n-1}{3}$. This implies that $n = 4$.   QED

We now further analyse and kill the case $n = 12$.

Without loss of generality, we may assume that $F(x_0, 6)$ is non-trivial.

Lemma 11. *Let f and g, respectively, be an arbitrary non-trivial element of $F(x_2, 6)$ and $F(x_{10}, 6)$, respectively, and let $u \in G(x_9, x_{10}, x_{11}, x_{12})$ be such that $(x_9, x_{10})^f = (x_7, x_6)^{u^{-1}}$. Then*

(i) *we have $[f, g] = g^{-fu} \in F(x_6, 6)$;*

(ii) *for every element $h \in F(x_6, 6)$, there exists an element $g \in F(x_{10}, 6)$ such that $[f, g] = h$.*

Proof. First note that $x_9^f \neq x_9$ by condition (W2). Hence an element g as in the lemma exists (this can be deduced easily from (W1); see also Lemma 1). Since g maps $B(x_4, 4)$ into $B(x_4, 4) \subseteq B(x_2, 6)$, the commutator $[f, g]$ fixes all these elements. Similarly, $[f, g]$ fixes all elements at distance no more than 4 from x_8. Hence $[f, g]$ belongs to $G(x_1, x_2, \ldots, x_{11})$. Since $x_{10}^{fu} = x_6$, we have $g^{-fu} \in F(x_6, 6) \subseteq G(x_1, x_2, \ldots, x_{11})$. Now

$$x_{13}^{u^{-1}f^{-1}g^{-1}fu} = (x_{13}^{f^{-1}g^{-1}f})^u = (x_{13}^{f^{-1}g^{-1}f})^g = x_{13}^{[f,g]},$$

remembering $x_{12}^{f^{-1}g^{-1}f} = x_{12}$, and (i) follows from condition (W2).

For h given as in (ii), it suffices to put $g = fuh^{-1}u^{-1}f^{-1}$. The proof of the lemma is complete.   QED

Lemma 12. **If,** *with the same notation as in the previous lemma, we also have* $u \in G(x_9, \ldots, x_{16})$, **then** *there exists $k \in F(x_{14}, 6)$ such that*

$$(x_6, x_7)^k = (x_{10}, x_9)^f = (x_6, x_7)^{u^{-1}}.$$

Proof. By the previous lemma $h = [f, g] = g^{-fu}$ belongs to $F(x_6, 6)$. We claim that $[u, g] = 1$. Indeed, since u maps $B(x_8, 4) \subseteq B(x_{10}, 6)$ into itself, we already have $[u, g] \in G(x_5, x_6, \ldots, x_{15})$. But g stabilizes $X_1(x_{16})$, while u fixes this set pointwise; hence the claim follows from condition (W2). We calculate

$$
\begin{aligned}
[h, u] = [[f, g], u] &= h^{-1}u^{-1}[f, g]u \\
&= h^{-1}u^{-1}f^{-1}g^{-1}fgu \\
&= h^{-1}(fu)^{-1}g^{-1}fug \\
&= h^{-1}g^{-fu}g \\
&= g.
\end{aligned}
$$

By Lemma 11(ii) (applied after performing the transformation $i \mapsto 24 - i$ on the indices of the U_i and the x_i), there exists $k \in F(x_{14}, 6)$ such that $g = [k, h] = [h, k]^{-1}$. Remembering $[u, g] = 1$, one easily calculates

$$
[h, ku] = [h, u] \cdot [h, k] = g \cdot g^{-1} = 1.
$$

Hence $h = h^{ku}$ and so $h \in F(x_6, 6) \cap F(x_6, 6)^{ku}$. Since obviously $h \neq 1$, this can only happen if $x_6^{ku} = x_6$. Since k and u fix x_8, we also have $x_7^{ku} = x_7$. The lemma follows.   QED

We can now finish the proof of the theorem. Indeed, let v be any element of $G(x_2, x_3, \dots, x_{12})$. Note that v acts freely on $X_1(x_{13}) \setminus \{x_{12}\}$. Let f and k be as above (for arbitrary u as in Lemma 12). First we claim that $[v, k] = 1$.

Indeed, since k stabilizes $X_1(x_8)$ and v fixes that set pointwise, the commutator $[v, k]$ fixes it pointwise. Also, k fixes $B(x_{13}, 5)$ pointwise and v fixes x_{13}, so $[v, k] \in G(x_8, x_9, \dots, x_{17})$. Since fk^{-1} maps (x_9, x_{10}) onto (x_7, x_6) by Lemma 12, and since v fixes $X_1(x_9)$ and $X_1(x_{10})$ pointwise, the element $v^{fk^{-1}}$ will fix $X_1(x_6)$ and $X_1(x_7)$ pointwise. Now note that v and f commute since $[v, f]$ belongs to both $F(x_2, 6)$ (because v fixes x_2) and $G(x_8)$ (because f fixes x_8). Therefore $v^{fk^{-1}} = v^{k^{-1}}$. But $[v, k^{-1}] = v^{-1} \cdot v^{k^{-1}} \in G(x_6, x_7)$. Consequently $[v, k^{-1}] \in G(x_6, \dots, x_{17}) = \{1\}$ (using (W2)). The claim follows.

From $[v, k] = 1$, we deduce $k = k^v$ and hence $k \in F(x_{14}, 6) \cap F(x_{14}, 6)^v$. As before this implies that v fixes x_{14} and so, by (W2), we should have $v = 1$. But v was arbitrary, a final contradiction.   □

As a corollary we now have the following result.

5.3.4 Theorem (Tits [1976b],[1979], Weiss [1979]) *Moufang n-gons exist only for* $n = 3, 4, 6$ *or* 8.

Proof. If Γ is a Moufang n-gon and G is the little projective group of Γ, then the universal cover is a $(\tilde{G}, n)$-Moufang tree. The result follows from Theorem 5.3.3.
                                                                                                          □

5.4 Root systems

5.4.1 Polygons belonging to root systems

The root groups of the Moufang polygons arising from algebraic groups, mixed groups or Ree groups all satisfy certain relations that are implied by a *root system*. A **root system** is a set of non-zero vectors in a real Euclidean space satisfying certain conditions. It will suffice for us to consider root systems in the real Euclidean plane $V_2(\mathbb{R})$. With every root group U_i (we use the same notation as in Subsection 5.2.1) one associates a non-zero vector e_i in $V_2(\mathbb{R})$, and possibly with a certain subgroup V_i of U_i, one associates a vector f_i which is a positive multiple of e_i. For a given Moufang n-gon Γ and an associated root system $\mathcal{R}$, we say that Γ **belongs to** $\mathcal{R}$ if

(RS) for every $i, j \in \mathbb{Z}$, $i \le j < i+n$, the decomposition of any element of $[X_i, Y_j] \subseteq U_i U_{i+1} U_{i+2} \ldots U_{j-1} U_j$ as given by Lemma 5.2.3(iii) has no factor in Z_k, $i < k < j$, whenever z_k cannot be written as a linear combination in x_i and y_j with positive non-zero integer coefficients, where $\{(X, x), (Y, y), (Z, z)\} \subseteq \{(U, e), (V, f)\}$ (this includes U_i abelian if V_i is not defined; for $n = 8$, an "integer" means an "algebraic integer" in $\mathbb{Q}(\sqrt{2})$).

For each $n \in \{3, 4, 6, 8\}$, we introduce a certain root system and we will show that any Moufang polygon must belong to one of these root systems. This is a very important first step in the classification of all Moufang polygons and it can be proved using very elementary group theory. This enables one to define algebras over fields but the further determination of these algebras is less elementary and beyond the scope of this book (making an exception for the case $n = 3$, see Appendix B). It involves long and tedious calculations in certain types of algebras that are rather far away from the geometry in this book. So we restrict ourselves to the first step and to the announcement of the final result in each case of n. Also, for reasons of simplicity, our notion of "belonging to a root system" is somewhat weaker than the notion of "possessing a filtration into a root datum" introduced by TITS [1994a]. We can afford this because we do not carry out the full classification.

Note that the root systems we will define arise naturally in the theory of algebraic groups of relative rank 2 (disregarding the octagon case). The commutation relations that follow from root systems with no multiple roots are in the literature referred to as the *Steinberg relations*. In TITS [1962d] root systems with multiple roots are introduced, and the rank 2 case of *op. cit.* amounts precisely to what we describe here.

5.4.2 The root system of type A_2

Define in $V_2(\mathbb{R})$ the following six vectors:

$$e_0(0, 1), \quad e_1(\tfrac{\sqrt{3}}{2}, \tfrac{1}{2}), \quad e_2(\tfrac{\sqrt{3}}{2}, -\tfrac{1}{2}),$$
$$e_3(0, -1), \quad e_4(-\tfrac{\sqrt{3}}{2}, -\tfrac{1}{2}), \quad e_5(-\tfrac{\sqrt{3}}{2}, \tfrac{1}{2}).$$

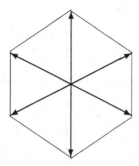

Figure 5.1. The root system of type A_2

We call this set of vectors the **root system of type** A_2; see Figure 5.1. If Γ is a Moufang projective plane, then according to condition (RS), it belongs to that root system **if and only if** every U_i is abelian. Consider, however, the following conditions (which are, by the presence of (3M2), a priori somewhat stronger than belonging to the root system of type A_2):

(3M1) All groups U_i, $i \in \mathbb{Z}$, are abelian.

(3M2) For all $i \in \mathbb{Z}$ we have $[U_{i-1}, U_{i+1}] = U_i$.

These properties can easily be shown directly in a geometric fashion. For instance, it is a standard fact that for a translation plane the group of translations is abelian. So every Moufang plane belongs to the root system of type A_2. We will show this again in a more sophisticated way below using the properties of Lemma 5.2.3(*ii*), (*iii*) and Lemma 5.2.6.

5.4.3 The root systems of type C_2 and BC_2

Define in $V_2(\mathbb{R})$ the following 12 vectors:

$$
\begin{array}{llll}
e_0(1,0), & e_2(0,1), & e_4(-1,0), & e_6(0,-1), \\
e_1(\tfrac{1}{2},\tfrac{1}{2}), & e_3(-\tfrac{1}{2},\tfrac{1}{2}), & e_5(-\tfrac{1}{2},-\tfrac{1}{2}), & e_7(\tfrac{1}{2},-\tfrac{1}{2}), \\
f_1(1,1), & f_3(-1,1), & f_5(-1,1), & f_7(1,-1).
\end{array}
$$

We call the set of vectors $\{e_0, e_1, \ldots, e_7\}$ the **root system of type** C_2, and the set of vectors $\{e_0, e_1, \ldots, e_7, f_1, f_3, f_5, f_7\}$ the **root system of type** BC_2 (see also Figure 5.2). If Γ is a Moufang quadrangle, then it belongs to one of these root systems (the system C_2 being a special case of BC_2) if there exist subgroups $V_{2i+1} \le U_{2i+1}$ (which are trivial for C_2) such that the following conditions are satisfied:

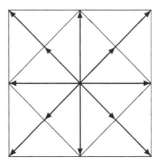

Figure 5.2. The root system of type BC_2

(4M1) All U_{2i}, $i \in \mathbb{Z}$, are abelian.

(4M2) All U_i have (nilpotency) class at most 2, i.e., $[U_i, U_i']$ is trivial for all $i \in \mathbb{Z}$.

(4M3) We have $[U_{2i+1}, U_{2i+1}] \leq V_{2i+1}$, for all $i \in \mathbb{Z}$.

(4M4) The commutators $[V_{2i-1}, U_{2i+1}]$ and $[U_{2i-1}, V_{2i+1}]$ are trivial for all $i \in \mathbb{Z}$.

(4M5) For all $i \in \mathbb{Z}$ we have $[U_{2i}, U_{2i+2}] \leq V_{2i+1}$.

(4M6) For all $i \in \mathbb{Z}$ we have $[U_{2i}, V_{2i+3}] \leq V_{2i+1} U_{2i+2}$; symmetrically $[V_{2i-3}, U_{2i}] \leq U_{2i-2} V_{2i-1}$.

We will show below that every Moufang quadrangle satisfies these properties (possibly up to reindexing the groups U_i). Of course, there is no canonical choice for the notation C_2, BC_2. We could also have used B_2 and CB_2 respectively. When we talk about classification, we will make some conventions about this choice (see page 211).

We will always assume that, if we say that some quadrangle belongs to a root system of type BC_2, the groups V_{2i+1} above are non-trivial, i.e., they do not have size 1 and they do not equal U_{2i+1}. This way, it will be possible for a quadrangle to belong to a root system of either type C_2 or BC_2.

5.4.4 The root system of type G_2

Define in $V_2(\mathbb{R})$ the following 12 vectors:

$$
\begin{array}{llll}
e_1(0,1), & e_2(\tfrac{\sqrt{3}}{6}, \tfrac{1}{2}), & e_3(\tfrac{\sqrt{3}}{2}, \tfrac{1}{2}), & e_4(\tfrac{\sqrt{3}}{3}, 0), \\
e_5(\tfrac{\sqrt{3}}{2}, \tfrac{1}{2}), & e_6(\tfrac{\sqrt{3}}{6}, -\tfrac{1}{2}), & e_7(0, -1), & e_8(-\tfrac{\sqrt{3}}{6}, -\tfrac{1}{2}), \\
e_9(-\tfrac{\sqrt{3}}{2}, -\tfrac{1}{2}), & e_{10}(-\tfrac{\sqrt{3}}{3}, 0), & e_{11}(-\tfrac{\sqrt{3}}{2}, \tfrac{1}{2}), & e_{12}(-\tfrac{\sqrt{3}}{6}, \tfrac{1}{2}).
\end{array}
$$

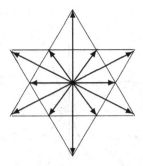

Figure 5.3. The root system of type G_2

We call this set of vectors the **root system of type** G_2 (see Figure 5.3). We will show below that every Moufang hexagon belongs to this root system. Noting that the roots with even index form an A_2-subsystem, this means equivalently that we can choose the indices in such a way that the following conditions are satisfied for all U_i, $i \in \mathbb{Z}$:

(6M1) All U_i, $i \in \mathbb{Z}$, are abelian.

(6M2) The commuators $[U_{2i}, U_{2i+2}]$ and $[U_i, U_{i+3}]$ are trivial for all $i \in \mathbb{Z}$.

(6M3) For all $i \in \mathbb{Z}$ we have $[U_{2i-2}, U_{2i+2}] = U_{2i}$.

5.4.5 The root system of type $^2F_4$

Define in $V_2(\mathbb{R})$ the following 24 vectors:

$$
\begin{array}{lll}
e_0(\sqrt{2}, 0), & f_0(\sqrt{2} + 2, 0), & e_1(\sqrt{2} + 1, 1), \\
e_2(1, 1), & f_2(\sqrt{2} + 1, \sqrt{2} + 1), & e_3(1, \sqrt{2} + 1), \\
e_4(0, \sqrt{2}), & f_4(0, \sqrt{2} + 2), & e_5(-1, \sqrt{2} + 1), \\
e_6(-1, 1), & f_6(-\sqrt{2} - 1, \sqrt{2} + 1), & e_7(-\sqrt{2} - 1, 1), \\
e_8(-\sqrt{2}, 0), & f_8(-\sqrt{2} - 2, 0), & e_9(-\sqrt{2} - 1, -1), \\
e_{10}(-1, -1), & f_{10}(-\sqrt{2} - 1, -\sqrt{2} - 1), & e_{11}(-1, -\sqrt{2} - 1), \\
e_{12}(0, -\sqrt{2}), & f_{12}(0, -\sqrt{2} - 2), & e_{13}(1, -\sqrt{2} - 1), \\
e_{14}(1, -1), & f_{14}(\sqrt{2} + 1, -\sqrt{2} - 1), & e_{15}(\sqrt{2} + 1, -1).
\end{array}
$$

We call this set of vectors the **root system of type** $^2F_4$; see also Figure 5.4. We will show that every Moufang octagon belongs to this root system. This is equivalent to showing that we can choose the indices in such a way that there exist abelian subgroups V_{2i} of U_{2i}, such that the groups U_i and V_{2i}, $i \in \mathbb{Z}$, satisfy Conditions (8M1) to (8M12) below.

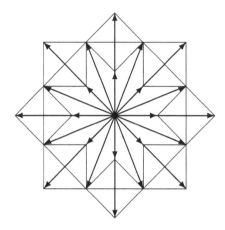

Figure 5.4. The root system of type $^2F_4$

A **shift** or **flip**, of a statement is the statement that we get by performing the transformation $i \mapsto i + 2k$ or $i \mapsto 2k - i$, respectively, to the indices, for arbitrary $k \in \mathbb{Z}$.

(8M1) All U_{2i+1}, $i \in \mathbb{Z}$, are abelian.

(8M2) All U_{2i}, $i \in \mathbb{Z}$, are of class at most 2, i.e., $U'_{2i} = [U_{2i}, U_{2i}]$ centralizes U_{2i}; moreover, $U'_{2i} \leq V_{2i}$.

(8M3) For all $i \in \mathbb{Z}$ the commutator $[U_{2i-2}, U_{2i+2}]$ is trivial.

(8M4) For all $i \in \mathbb{Z}$ the commutators $[U_{2i-1}, U_{2i+1}]$ and $[U_{2i-1}, U_{2i+3}]$ are trivial.

(8M5) For all $i \in \mathbb{Z}$ the commutators $[V_{2i-2}, U_{2i}U_{2i+1}]$ and $[U_{2i-1}U_{2i}, V_{2i+2}]$ are trivial.

(8M6) For all $i \in \mathbb{Z}$ we have $[U_{2i}, U_{2i+5}] \leq V_{2i+2}U_{2i+3}V_{2i+4}$ (and the "flip" of this).

(8M7) For all $i \in \mathbb{Z}$ we have $[U_{2i-2}, U_{2i+1}] \leq V_{2i}$ (and the "flip" of this).

(8M8) For all $i \in \mathbb{Z}$ we have $[V_{2i}, U_{2i+5}] \leq V_{2i+2}$ (and the "flip" of this).

(8M9) For all $i \in \mathbb{Z}$ we have $[V_{2i}, U_{2i+6}] \leq U_{2i+1}V_{2i+2}U_{2i+3}V_{2i+4}$ (and the "flip" of this).

(8M10) For all $i \in \mathbb{Z}$ we have $[V_{2i}, V_{2i+6}] \leq V_{2i+2}U_{2i+3}V_{2i+4}$ (and the "flip" of this).

(8M11) For all $i \in \mathbb{Z}$ we have $[V_{2i}, U_{2i+7}] \leq U_{[2i+1,2i+5]}V_{2i+6}$ (and the "flip" of this).

(8M12) For all $i \in \mathbb{Z}$ we have $[U_{2i-3}, U_{2i+3}] \leq U_{2i-1}U_{2i+1}$.

We now prove the assertion that every Moufang polygon belongs to one of the above root systems. We will use a sequence of lemmas to attain our goal.

5.4.6 Theorem (Tits [1983],[1994a]). *Every Moufang polygon belongs to one of the root systems A_2, C_2, BC_2, G_2 or $^2F_4$.*

Proof. By Theorem 5.3.4, we only have to consider generalized n-gons for $n \in \{3, 4, 6, 8\}$. So let Γ be a Moufang n-gon for some $n \in \{3, 4, 6, 8\}$. We start with a few general lemmas and from time to time, we refer to specific n. Furthermore, for $i \in \mathbb{Z}$, an element of G carrying i as a subscript will automatically be assumed to be inside U_i. Also we will denote by i an arbitrary integer, except when explicilty mentioned.

The proof below is, for the cases $n = 3, 4, 6$, based directly on the lectures given by Tits at the Collège de France in 1994. For $n = 8$, we follow closely TITS [1983], except for a few lemmas, and we also use some arguments appearing already in the proof of Weiss' theorem above.

Every property we prove is also valid for a "shift" or a "flip" of the integers. We will use the "shift" and the "flip" of a property sometimes without further notice.

Note that by Lemma 5.2.3 every element u of the group $[U_i, U_j]$, $i < j < i+n$, has a decomposition $u = u_i \ldots u_j$. We will call the element u_k of that decomposition the **factor of u in U_k**.

| **Lemma 1.** | If $i < j < i+n$, then $(u_iu_{i+1} \ldots u_{j-1})^{u_j} = u_iu'_{i+1} \ldots u'_{j-1}$, for every
$u_k \in U_k$, $i \leq k \leq j$, and some $u'_l \in U_l$, $i < l < j$.

Proof. Indeed, by Lemma 5.2.3(ii), we have $u_k^{u_j} = u_k[u_k, u_j] \in u_kU_{[k+1,j-1]}$, for all $k \in \{i, i+1, \ldots, j-1\}$. QED

As an immediate consequence of Lemma 1 (and its "flip"), we have that the factor in U_i or U_j of the product of two elements of $U_{[i,j]}$, $i < j < i+n$, is equal to the product of the factors in U_i or U_j respectively of the elements.

| **Lemma 2.** | Let $u \in U_i$ and $v \in U_{i+n-1}$. Put

$$[u, v] = u_{i+1}u_{i+2} \ldots u_{i+n-2}.$$

Then $u_{i+n-2} = u^{\mu(v)}$.

Proof. Indeed, the element $u^{\mu(v)}$ belongs to U_{n+i-2} by Lemma 5.2.6 on page 176. On the other hand, we calculate (keeping in mind that U_i commutes with U_{i-1})

$$
\begin{aligned}
u^{\mu(v)} &= \left(u^{\omega^{-1}(v)}\right)^{v\omega(v)} \\
&= (u^v)^{\omega(v)} \\
&= (u[u,v])^{\omega(v)} \\
&= (uu_{i+1}\ldots u_{i+n-2})^{\omega(v)} \\
&= u_i' u_{i+1}' \ldots u_{i+n-3}' u_{i+n-2}
\end{aligned}
$$

(the last equality by Lemma 1 above). Since the latter must belong to U_{i+n-2}, all u_k', $i \le k < i+n-2$, must be trivial and the result follows.   QED

We will use this lemma a few times in the following form (after a "flip"). For given $u_i \in U_i$ and given $u_{i+1} \in U_{i+1}$, there exists $u_{i+n-1} \in U_{i+n-1}$ such that the factor of $[u_i, u_{i+n-1}]$ in U_{i+1} is equal to u_{i+1}.

Proof of the Theorem for $n = 3$.
If $n = 3$, then from Lemma 2 it follows immediately that $[U_{i-1}, U_{i+1}] = U_i$ (this is property (3M2)). Since U_i commutes with both U_{i-1} and U_{i+1}, it commutes with U_i, hence U_i is abelian and this is property (3M1).

We now continue with some lemmas. But we may suppose from now on that $n \ge 4$. In particular, n is even and we put $2n' = n$.
We write $V_i = [U_{i-1}, U_{i+1}]$. It is clear that $V_i \le Z(U_{[i-1,i+1]})$.

Lemma 3.   *The group $[U_{i-1}, V_{i+1}]$ is a subgroup of $Z(U_{[i-1,i+2]})$.*
Proof. Indeed, since V_{i+1} centralizes $U_{[i,i+2]}$ and U_{i-1} normalizes $U_{[i,i+2]}$, it follows that $[U_{i-1}, V_{i+1}]$ centralizes $U_{[i,i+2]}$. But since $[U_{i-1}, V_{i+1}] \le U_i$, the first group also centralizes U_{i-1}. Hence the result.   QED

Lemma 4.   **If $n = 4$, then** *we have $[U_i, U_i] = [U_{i-1}, V_{i+1}]$ and hence $[U_i, U_i] \le Z(U_i)$.*
Proof. Let $v_{i+1} = [u_i, u_{i+2}] \in V_{i+1}$, with u_i and u_{i+2} arbitrary. Furthermore, let u_{i-1} be arbitrary in U_{i-1}. We compute (keeping in mind that "adjacent" root groups commute and that $u_{i+2}^{u_{i-1}} = [u_{i-1}, u_{i+2}^{-1}]u_{i+2} \in U_{[i,i+1]}u_{i+2}$)

$$
\begin{aligned}
v_{i+1}^{u_{i-1}} &= [u_i, u_{i+2}]^{u_{i-1}} \\
&= [u_i^{u_{i-1}}, u_{i+2}^{u_{i-1}}] \\
&= [u_i, u_i' u_{i+1} u_{i+2}] \\
&= [u_i, u_i' u_{i+2}] \\
&= [u_i, u_{i+2}][u_i, u_i']^{u_{i+2}} \\
&= v_{i+1}[u_i, u_i']^{u_{i+2}}.
\end{aligned}
$$

Hence we have $[v_{i+1}, u_{i-1}] = [u_i, u_i']^{u_{i+2}}$. This shows $[U_{i-1}, V_{i+1}] \leq [U_i, U_i]$. On the other hand, given u_i and u_i', we can choose for any u_{i-1} (by a "flip" of Lemma 2) the element u_{i+2} such that $[u_{i-1}, u_{i+2}^{-1}] \in U_{i+1} u_i'$. Hence $u_{i+2} u_{i-1} u_{i+2}^{-1}$ belongs to $U_{i-1} U_{i+1} u_i'$. Noting that u_i centralizes $U_{i-1} U_{i+1}$, we calculate

$$
\begin{aligned}
[u_i, u_i'] &= u_i^{-1} u_i^{u_i'} \\
&= u_i^{-1} u_i^{u_{i+2} u_{i-1} u_{i+2}^{-1}} \\
&= u_i^{-1} (u_{i+2} u_{i-1}^{-1} u_{i+2}^{-1}) u_i (u_{i+2} u_{i-1} u_{i+2}^{-1}) \\
&= u_i^{-1} u_{i+2} u_i u_{i-1}^{-1} u_i^{-1} u_{i+2}^{-1} u_i u_{i+2} u_{i-1} u_{i+2}^{-1} \\
&= u_{i+2} [u_i, u_{i+2}]^{-1} u_{i-1}^{-1} [u_i, u_{i+2}] u_{i-1} u_{i+2}^{-1} \\
&= [[u_i, u_{i+2}], u_{i-1}]^{u_{i+2}^{-1}}.
\end{aligned}
$$

This completes the proof of Lemma 4. QED

Lemma 5. **If** $n = 4$, **then** $[U_{i+1}, U_{i+1}]$ *is trivial whenever* $U_i \cap Z(U_{[i,i+2]})$ *is non-trivial.*

Proof. Let u be a non-trivial element of $U_i \cap Z(U_{[i,i+2]})$. Since U_{i+3} normalizes $U_{[i,i+2]}$, every element of the form $[u, u_{i+3}]$ centralizes $U_{[i,i+2]}$. So every element u_{i+1} of U_{i+1} commutes with $[u, u_{i+3}]$. But given $u_{i+1}' \in U_{i+1}$, there exist u_{i+2}, u_{i+3} such that $[u, u_{i+3}] = u_{i+1}' u_{i+2}$, by Lemma 2. Since u_{i+1} commutes with the latter and with u_{i+2}, it commutes with u_{i+1}'. QED

Lemma 6. **If** $n = 4$, **then** *one of the groups* U_i, U_{i+1} *is abelian, while the other is of class at most 2.*

Proof. If U_i is non-abelian, then by the previous lemma $U_i \cap Z(U_{[i,i+2]})$ is trivial. But by Lemma 3 and Lemma 4, $[U_{i+1}, U_{i+1}] \leq U_{i+1} \cap Z(U_{[i,i+3]})$, hence U_{i+1} is abelian. By Lemma 4, all U_i are of class at most 2. QED

Proof of the Theorem for $n = 4$.

By Lemma 6, we may assume that U_{2i} is abelian. Then (4M1) and (4M2) are immediate consequences of Lemma 6. Using Lemma 4, we have $[U_{2i+1}, U_{2i+1}] = [U_{2i}, V_{2i+2}] \leq [U_{2i}, U_{2i+2}] = V_{2i+1}$, hence (4M3). By Lemma 4, $[V_{2i-1}, U_{2i+1}] = [U_{2i}, U_{2i}]$ is trivial (since U_{2i} is abelian), hence (4M4). Condition (4M5) follows immediately from the definition of V_{2i+1}. We now show condition (4M6).

Let $u \in U_{2i}$ and $v \in V_{2i-3}$ be arbitrary and put $[v, u] = u_{2i-2} u_{2i-1}$. We have to show that $u_{2i-1} \in V_{2i-1}$. By Lemma 2 we have $u_{2i-1} = v^{\mu(u)}$. If $v = [u_{2i-4}', u_{2i-2}']$, then $u_{2i-1} = [(u_{2i-4}')^{\mu(u)}, (u_{2i-2}')^{\mu(u)}] \in [U_{2i}, U_{2i-2}]$ by Lemma 5.2.6 on page 176.

This completes the proof for $n = 4$. Note that if all root groups are abelian and if V_{2i+1} is trivial for all $i \in \mathbb{Z}$, then Γ belongs to the root system of type C_2. This happens for example if $V_{2i} = U_{2i}$ and U_{2i+1} is abelian, for all $i \in \mathbb{Z}$.

We now continue with some lemmas essentially leading to the case $n = 6$. By the foregoing we may assume $n \in \{6, 8\}$, but these restrictions are not needed in the next two results below, and so give extra information on the case $n = 4$.

Lemma 7. *Suppose $n \geq 4$. Let $u \in U_{i+1}$ and $v \in U_{i+n-1}^*$ be arbitrary and suppose $[u, v] = u_{i+2} \ldots u_{i+n-2}$. Then we have*

$$u_{i+n-2} = [\omega^{-1}(v), u]^{\mu(v)}.$$

If u_{i+n-2} is trivial and $n > 4$, then $u_{i+n-3} = u^{\mu(v)}$.

Proof. Put $\omega(v) = v''$, $\omega^{-1}(v) = v'$ and $\mu(v) = \mu(v') = m$. Since $m = v'vv''$, we have

$$
\begin{aligned}
u^{mv''^{-1}} &= u^{v'v} \\
&= (u^{v'})^v \\
&= (u[u, v'])^v \\
&= u^v[u, v']^v \\
&= u[u, v][u, v']^v \in U_{i+1}U_{[i+2, i+n-2]}U_{[i, i+n-2]}.
\end{aligned}
$$

The left-hand side equals $(u^m)^{v''^{-1}}$ and hence this is inside $U_{[i, i+n-3]}$. The right-hand side is the product of two elements in $U_{[i, i+n-2]}$, hence by the consequence of Lemma 1, the product of the factors in U_{i+n-2} of u^v and $[u, v']^v$ must be trivial. Using Lemma 2, we obtain

$$u_{i+n-2}[u, v']^m = 1,$$

hence $u_{i+n-2} = [v', u]^m$, as required. If $[v', u]$ is trivial, then we compare the factors in U_{i+n-3}. For the left-hand side, this is clearly u^m, while for the right-hand side, this is u_{i+n-3} by definition, unless $n = 4$ (in which case the factor in $U_{i+n-3} = U_{i+1}$ on the right-hand side is equal to u).   QED

We forget the notation of V_i (which was introduced to prove the theorem for $n = 4$) above and define a new V_i as the largest subgroup of U_i which centralizes both U_{i-2} and U_{i+2}.

Lemma 8. *We have $[U_i, U_i] \leq V_i$.*

Proof. This follows from $[u_i, u_i']^{u_{i+2}} = [u_i^{u_{i+2}}, u_i'^{u_{i+2}}] = [u_i[u_i, u_{i+2}], u_i'[u_i', u_{i+2}]] = [u_i, u_i']$ since $[U_i, U_{i+2}] \leq U_{i+1}$ centralizes $U_{[i, i+2]}$. Similarly for an element u_{i-2}.   QED

Note that Lemma 8 also follows from the so-called *three-subgroup lemma*. This lemma states that, if A, B, C are three subgroups of a group, then all of the groups

$[A, [B, C]]$, $[B, [C, A]]$ and $[C, [A, B]]$ are trivial whenever at least two of them are. If we put $A = B = U_i$ and $C = U_{i\pm2}$, then Lemma 8 follows.

The following lemma starts the case $n = 6$.

Lemma 9. **If** $n = 6$, **then** *we have* $[V_i, U_{i+4}] \leq U_{i+1}U_{i+2}$ *and* $[V_i, V_{i+4}] \leq U_{i+2}$.

Proof. Put $v \in V_i$, $u \in U_{i+4}$, then by Lemma 7, the factor of $[v, u]$ in U_{i+3} equals $[\omega^{-1}(u), v]^{\mu(u)}$. But $\omega^{-1}(u) \in U_{i-2}$ and so by definition of V_i, the commutator $[\omega^{-1}(u), v]$ is trivial. The second assertion follows from the first and a "flip" thereof. QED

Let us denote the set of all $\mu(u_i)$, $u_i \in U_i$, by M_i and the group generated by all elements of the sets M_i, $i \in \mathbb{Z}$, by M. Also, we denote by T the subgroup of M fixing pointwise the apartment Σ.

Lemma 10. **If** $n = 6$ *and* $[u_{i-2}, u_{i+2}] \in U_i$, **then** $[u_{i-2}, u_{i+2}] \in u_{i-2}^{M_{i+2}} \cap u_{i+2}^{M_{i-2}}$. *Moreover, if one non-trivial member of* $\{u_{i-2}, u_{i+2}, [u_{i-2}, u_{i+2}]\}$ *belongs to* V_j *for suitable* $j \in \mathbb{Z}$, *then they all do (for other suitable values of* j *of course).*

Proof. The first assertion follows immediately from the second assertion of Lemma 7 (using a "flip" as well). Suppose now $u_{i-2} \in V_{i-2}$. By the first part of the lemma, we can write $[u_{i-2}, u_{i+2}] = u_{i-2}^m$ with $m \in M_{i+2}$. We have $[u_{i-2}^m, U_{i-2}] = [u_{i-2}, U_i]^m$ and the latter is trivial by assumption.
Also, $[u_{i-2}^m, U_{i+2}] = [u_{i-2}, U_{i-4}]^m$, again trivial by assumption.
Hence $[u_{i-2}, u_{i+2}] \in V_i$. By the first part of the lemma again, we can write $u_{i+2} = [u_{i-2}, u_{i+2}]^{m'}$ with $m' \in M_{i-2}$. As above, one shows $u_{i+2} \in V_{i+2}$. The statement is now clear. QED

Lemma 11. **If** $n = 6$, **then** $[V_{i-2}, V_{i+2}] = V_i$ *and hence* V_i *is abelian.*

Proof. By Lemma 9, $[V_{i-2}, V_{i+2}] \leq U_i$ and so by Lemma 10, $[V_{i-2}, V_{i+2}] \leq V_i$. Given $v_{i+2} \in V_{i+2}$ and $v_i \in V_i$, it follows from the second assertion of Lemma 7 that $[v_i^{\mu(v_{i+2})^{-1}}, v_{i+2}] = v_i$, hence $[V_{i-2}, V_{i+2}] = V_i$. Since V_i commutes with both V_{i-2} and V_{i+2} by definition, it commutes with their commutator, but that is V_i itself, as we have just shown. QED

Lemma 12. **If** $n = 6$, **then** $\omega(V_i) = V_{i+6}$.

Proof. Take $u \in V_i$ and $v \in V_{i-4}$ arbitrary and put $\mu(u) = u'uu''$, hence $\omega(u) = u''$. We have

$$
\begin{aligned}
v^{\mu(u)} &= v^{u'uu''} \\
&= v^{uu''} \\
&= (v[v, u])^{u''} \\
&= v(v^{\mu(u)})^{u''}
\end{aligned}
$$

(using the above lemmas). This implies that $v = [(v^{\mu(u)})^{-1}, u'']$, with $u'' \in U_{i-6}$ and $(v^{\mu(u)})^{-1} \in V_{i-2}$. By Lemma 10, $u'' \in V_{i-6}$ and so $\omega(V_i) \subseteq V_{i+6}$. The result follows by symmetry.   QED

The next lemma will enable us to draw conclusions for every element of V_i given a certain property for at least one non-trivial element of V_i, because all elements of V_i are conjugate. Moreover, once V_i is non-trivial, it must coincide with U_i:

$\boxed{\textbf{Lemma 13.}}$ *Let* $n = 6$. **If** V_i *is non-trivial,* **then** $V_i = U_i$. *Moreover, all non-trivial elements of* V_i *are conjugate to one another by elements of* T. *Finally, all root groups* U_j *are abelian.*

Proof. By a "flip" of Lemma 9, $[U_{i-2}, V_{i+2}] \le U_i U_{i+1}$. By Lemma 12, $V_{i+2} = \omega^{-1}(V_{i-4})$. By Lemma 7, the factor in U_{i+1} of $[u_{i-2}, v_{i+2}] = [u_{i-2}, \omega^{-1}(v_{i-4})]$ (obvious notation) is conjugate to $[\omega(\omega^{-1}(v_{i-4})), u_{i-2}] = [v_{i-4}, u_{i-2}]$ which is trivial by definition of V_{i-4}. Hence

$$[U_{i-2}, V_{i+2}] \le U_i,$$

implying by the second assertion of Lemma 10 that $U_{i-2} = V_{i-2}$ whenever V_{i+2} is non-trivial.

By Lemma 10 we know that for arbitrary non-trivial $v_{i\pm2} \in V_{i\pm2}$, the commutator $[v_{i-2}, v_{i+2}]$ is conjugate to both v_{i-2} and v_{i+2} by elements of M. Hence v_{i-2} is conjugate to v_{i+2}. So all non-trivial elements of V_{i-2} are conjugate to all non-trivial elements of V_{i+2} by elements of M and hence all non-trivial elements of V_{i-2} are mutually conjugate by elements of T.

If V_j is non-trivial, then $U_j = V_j$ and so by Lemma 11, U_j is abelian. If V_j is trivial, then by Lemma 8, $[U_j, U_j]$ is also trivial.   QED

The key property to prove for $n = 6$ is undoubtly the commutation relation $[U_i, U_{i+3}] = \{1\}$. The following result will be very useful to that end. We formulate it for the general case, because we will need it later for $n = 8$ as well.

$\boxed{\textbf{Lemma 14.}}$ *Let* $n = 2n'$ *be even and suppose that* $u_i \ne 1$ *commutes with* u_j, $i < j < i + n - 1$.

(i) *If* $j > i + n'$, *then* $u_j = 1$.

(ii) *If* $j = i + n'$, *then* u_j *also commutes with* $\mu(u_i)$ *and with* $\omega(u_i)$.

(iii) *If* $j < i + n'$, *then*

$$[u_j, \omega(u_i)] \in U_{j+1} \ldots U_{n+2i-j-1} u_j^{\mu(u_i)}.$$

Proof. Put $m = \mu(u_i) = u_i\omega(u_i)\omega^2(u_i)$ (see Lemma 5.2.6 on page 176). We have

$$(u_j^m)^{(\omega^2(u_i))^{-1}} = (u_j^{u_i})^{\omega(u_i)}. \tag{5.2}$$

Since $u_j^m \in U_{n+2i-j}$, the left-hand side of Equation (5.2) is in $U_{[i+1,n+2i-j-1]}u_j^m$. Since $u_j^{u_i} = u_j$ (by assumption), the right-hand side of Equation (5.2) lives in $u_j U_{[j+1,n+i-1]}$. If $j > i+n'$, then the cosets $U_{[i+1,n+2i-j-1]}u_j^m$ and $u_j U_{[j+1,n+i-1]}$ can only have the identity element in common (because they live in subgroups only meeting in the identity). Hence in this case $u_j = u_j^m = 1$, showing (*i*). If $j = i+n'$, then the cosets must meet inside $U_j = U_{i+n'}$ and hence they meet **if and only if** $u_j^m = u_j$, i.e., $[u_j, \mu(u_i)] = 1$. In that case, both sides of Equation (5.2) are equal to u_j and so $u_j^{\omega(u_i)} = u_j$, proving (*ii*). Finally, if $j < i+n'$, then $u_j^{\omega(u_i)}$ belongs to the intersection of the two cosets mentioned above, and that is $u_j U_{[j+1,n+2i-j-1]}u_j^m$. The result follows. QED

Remark 1. The first assertion above also follows immediately from the first paragraph of the proof of Lemma 5 in the proof of Theorem 5.3.3 on page 181. Similarly, one shows that, if $u \in U_{[j,i+n-1]}$ and $j > i+n'$, then u and u_i commute **if and only if** $u = 1$.

We now arrive at the crux of the proof for the case $n = 6$. Enjoy Tits' beautiful argument.

Lemma 15. *Let* $n = 6$. **If** *there exists* $j \in \mathbb{Z}$ *such that* V_j *is non-trivial,* **then** $[U_i, U_{i+3}]$ *is trivial for all* $i \in \mathbb{Z}$.

Proof. Suppose V_j is non-trivial. By Lemma 13, $V_j = U_j$. Assume that $[U_j, U_{j+3}]$ is non-trivial and let $u \in U_{j+3}$ be such that it does not centralize U_j. We have $U_j^u \leq U_{[j,j+2]}$. On the other hand, we know by Lemma 11 that $U_j = [U_{j-2}, U_{j+2}]$, hence $U_j^u = [U_{j-2}^u, U_{j+2}^u] \leq [U_{[j-2,j+2]}, U_{j+2}]$. Since $U_{j+2} = V_{j+2}$ (by a "shift" of our assumption), since $U_{[j-2,j+2]} = U_{[j-2,j+1]}U_{j+2}$ and since U_{j+2} is abelian, we have that $[U_{[j-2,j+2]}, U_{j+2}] \leq U_{[j-2,j+1]}$ (use Lemma 1). Hence $U_j^u \leq U_{[j-2,j+1]} \cap U_{[j,j+2]} = U_{[j,j+1]}$. This immediately implies that $[u, U_j] \subseteq U_{j+1}$. We will now show that in fact $[u, U_j] \subseteq V_{j+1}$.

We know that $U_{[j-2,j+2]}$ is centralized by $U_j = V_j$, hence also by $[u, U_j]$ because it is normalized by $u \in U_{j+3}$. Since $[u, U_j] \subseteq U_{j+1}$, and since $[u, U_j]$ centralizes U_{j-2}, it also centralizes some element m of M_{j-2} by Lemma 14. Hence $[u, U_j]$, which is non-trivial by assumption, also centralizes $U_{[j-2,j+2]}^m = U_{[j,j+4]}$; in particular, it centralizes both U_{j-1} and U_{j+3}. Consequently $[u, U_j] \subseteq V_{j+1}$.

So there exists a non-trivial element v of V_{j+1} commuting with both U_{j-2} and U_{j+4}. Since $U_{j-2}^T = U_{j-2}$, and since all elements of $V_{j+1} = U_{j+1}$ are conjugate by elements of T, we conclude that $[U_{j+1}, U_{j-2}]$ is trivial. The lemma now follows by "shifting" and "flipping". QED

We need one more lemma before we can finish the case $n = 6$.

Lemma 16. If $n = 6$, **then** *we can choose indices such that* $V_0 = U_0$.

Proof. If V_1 is non-trivial, then we add one integer to every index. Suppose V_1 is trivial. Since U_1 is abelian (see Lemma 13), it is contained in the centre of $U_{[0,2]}$. Since U_3 normalizes $U_{[0,2]}$, the commutator $[U_1, U_3]$ centralizes $U_{[0,2]}$. By symmetry (or a "flip"), it also centralizes $U_{[2,4]}$; in particular, it centralizes both U_0 and U_4. Hence $[U_1, U_3] \leq V_2$. If we show that $[U_1, U_3]$ is non-trivial, then we are done by Lemma 13. But if $[U_1, U_3]$ is trivial, then so is $[U_{-1}, U_1]$, and hence $U_1 = V_1$ by definition, contradicting our hypothesis.   QED

Proof of the Theorem for $n = 6$.

By Lemma 16 we may suppose that $V_{2i} = U_{2i}$. By definition, this means that $[U_{2i}, U_{2i+2}]$ is trivial. Together with Lemma 15, this shows condition (6M2). Condition (6M1) follows from Lemma 13. Finally, Lemma 11 shows condition (6M3) and the proof of the case $n = 6$ is complete.

We now continue with the last case, $n = 8$. Our first aim is to define new subgroups V_{2i} for all $i \in \mathbb{Z}$. Therefore, we remark that from Lemmas 8 and 9 of the proof of Theorem 5.3.3 it follows that we can choose the indices such that $Z(U_{[2i+1,2i+8]})$ contains a non-trivial element u fixing all elements of Γ at distance at most 4 from x_{2i+8}. Hence $u \in U_{2i+4}$. Now u stabilizes $\Gamma_1(x_{2i+4})$ (which is pointwise fixed by U_{2i}) and since U_{2i} stabilizes $\Gamma_{\leq 4}(x_{2i+8})$ (which is pointwise fixed by u), the commutator $[u, U_{2i}]$ fixes $\Gamma_{\leq 4}(x_{2i+8}) \cup \Gamma_1(x_{2i+4})$. Since only the identity can do this, we have $u \in Z(U_{[2i,2i+8]}) \cap U_{2i+4}$. So we define

$$V_{2i} = Z(U_{[2i-4,2i+4]}) \cap U_{2i},$$

and by the discussion above, this is non-trivial. We also define the set

$$V_{2i}^o = \omega^{-1}(V_{2i-8}^*) \cup \{1\}.$$

Remark 2. Note that by the proof of Lemma 10 (see page 183) of the proof of Theorem 5.3.3, no non-trivial element of U_{2i+1} centralizes $U_{[2i-3,2i+5]}$ since such an element would have to fix the set $\Gamma_{\leq 4}(x_{2i+5})$ pointwise (reverse the above argument or use Lemma 7 of the proof of Weiss' theorem). We will use this at the end of the proof of Lemma 19 below.

The next lemma essentially shows that V_{2i} and V_{2i}^o meet in only the identity.

Lemma 17. If $n = 8$, then $v \in V_{2i}$ implies that $\omega(v) \notin V_{2i+8}$.

Proof. Suppose that $\omega(v) \in V_{2i+8}$. Let $u \in U_{2i+3}^*$ be arbitrary. Since $[v, u] = 1$, it follows from Lemma 14(*iii*) that $[u, \omega(v)] \in U_{2i+4} u^{\mu(v)}$. Since u normalizes $U_{[2i+4,2i+10]}$ and the latter is centralized by $\omega(v)$ by assumption, the commutator $[u, \omega(v)]$ centralizes $U_{[2i+4,2i+10]}$ and, in particular, it centralizes U_{2i+10}. By Remark 1, this can only happen if $[u, \omega(v)] = 1$, which implies that $u^{\mu(v)} = 1$ and hence $u = 1$. The lemma is proved.   QED

Lemma 18. *Suppose $i < j < i + n'$ (with $2n' = n$, n even). Let $Y_j \leq U_j$ and $Y_{n+2i-j} \leq U_{n+2i-j}$ be two subgroups with the property $Y_j^{M_i} = Y_{n+2i-j}$. Let $X \leq Y = Y_j U_{[j+1,n+2i-j-1]} Y_{n+2i-j}$ be normalized by U_i. The smallest subgroup $\hat{X}$ containing X and normalized by both U_i and U_{i+n} is contained in Y. In particular, if $X = Y_j$, then $\hat{X} = Y_j X^\dagger Y_{n+2i-j}$ for some $X^\dagger \leq U_{[j+1,n+2i-j-1]}$.*

Proof. Let $u \in U_{i+n}$. We show that $X^u \leq Y$. Let $m = \mu(u) = u'uu''$ with $u' = \omega^{-1}(u)$ and $u'' = \omega(u)$. Notice that $u', u'' \in U_i$. Then we have

$$X^u = X^{u'u} = X^{mu''^{-1}} \leq Y^{mu''^{-1}} \leq Y^{u''^{-1}} \leq U_{[i+1,n+2i-j-1]} Y_{n+2i-j}.$$

Similarly, one has also (by symmetry) $X^u \leq Y_j U_{[j+1,i+n-1]}$. Taking intersections yields $X^u \leq Y$. Now the group X_1 generated by X and $X^{U_{i+n}}$ is clearly normalized by U_{i+n} and it is contained in Y. Similarly (by a symmetric argument), the group X_2 generated by X_1 and $X_1^{U_i}$ is normalized by U_i and it is also contained in Y. Continuing in this way, we clearly have $\hat{X} = \bigcup_k X_k \leq Y$.

The second assertion readily follows since the group generated by U_i and U_{i+n} contains M_i. QED

The next lemma will show, amongst other things, that the groups U_{2i+1} are abelian.

Lemma 19. *If $n = 8$, then the following statements hold:*

(i) $[U_{2i+1}, V_{2i+6}] \leq V_{2i+4}$.

(ii) *For $v \in V_{2i+6}^*$, the mapping $u \mapsto [u, v]$ of U_{2i+1} into V_{2i+4} is an injective group homomorphism.*

(iii) *For $u \in U_{2i+1}^*$ and $v \in V_{2i+6}$, we have $[u, v] = (v^{-\mu(u)})^{-1}$.*

(iv) *The groups U_{2i+1} are abelian.*

(v) *The restriction $\mu | U_{2i+1}^*$ of μ to U_{2i+1}^* is injective.*

(vi) *For $u \in U_{2i+1}^*$, we have $\omega(u) = u^{\mu(u)}$, $\omega^2(u) = u$ and hence $\mu(u) = uu^{\mu(u)}u$.*

Proof. Since U_{2i+1} centralizes V_{2i+4}, and since $V_{2i+4}^{M_{2i+1}} = V_{2i+6}$, we can define a group $W_{2i+5} \leq U_{2i+5}$ such that $V_{2i+4} W_{2i+5} V_{2i+6}$ is the smallest subgroup of G containing V_{2i+4} and normalized by U_{2i+1} and U_{2i+9}, by Lemma 18. We define furthermore the group X_{2i+1} or Y_{2i+1} as the set of elements of U_{2i+1} centralizing W_{2i+5} or W_{2i-3} respectively. We will derive some properties of the groups W_{2i+1}, X_{2i+1} and Y_{2i+1} which will prove our assertions in case W_{2i+1} is trivial (and thus $X_{2i+1} = Y_{2i+1} = U_{2i+1}$). Only at the end we will show that this actually is the case.

(a)   If $u \in X_{2i+1}$, then by definition it centralizes W_{2i+5}. By Lemma 14(ii), $\mu(u)$ also centralizes W_{2i+5}, and hence $X_{2i+1}^{\mu(u)} \leq Y_{2i+9}$. By symmetry, we have equality. Also, again by Lemma 14(ii), $\omega(u)$ centralizes W_{2i+5} and hence $\omega(X_{2i+1}^*) \subseteq Y_{2i+9}^*$ and $\omega^2(X_{2i+1}^*) \subseteq X_{2i+1}^*$.

(b)   Since $W_{2i+5} \leq U_{2i+5}$ and the latter maps W_{2i+1} under conjugation into the product $V_{2i}W_{2i+1}V_{2i+2}$, we have $[W_{2i+1}, W_{2i+5}] \subseteq V_{2i}W_{2i+1}V_{2i+2}$. By symmetry, we also have the inclusion $[W_{2i+1}, W_{2i+5}] \subseteq V_{2i+4}W_{2i+5}V_{2i+6}$. Hence we obtain the relation $[W_{2i+1}, W_{2i+5}] = \{1\}$ and so $W_{2i+1} \leq X_{2i+1}$.

(c)   Let $u \in X_{2i+1}$ and $v \in V_{2i+6}$ be arbitrary. By definition of W_{2i+5}, we have that $v^u \in V_{2i+4}W_{2i+5}V_{2i+6}$, hence $[v, u] \in V_{2i+4}W_{2i+5}$. Define $\psi(v, u) \in V_{2i+4}$ as the factor of $[v, u]$ in U_{2i+4}. If $v' \in V_{2i+6}$, then the factor of $[v, u]^{v'}$ in U_{2i+4} is exactly equal to $\psi(v, u)$ since v' centralizes $V_{2i+4}W_{2i+5}$. The same remark holds good for $[v, u]^{u'}$ with $u' \in X_{2i+1}$ (use the definition of X_{2i+1} to see that u' centralizes W_{2i+5}). Noting that $[vv', u] = [v, u]^{v'}[v', u]$, we now easily see that ψ is multiplicative in its first argument, and similarly is multiplicative in its second argument. Also, by the "flip" of Lemma 14(iii), we obtain for $u \neq 1$,

$$\psi(v, u) = (v^{-1})^{\mu(u)} = v^{-\mu(u)}$$

(the inverse of v is there because of the "flip"). So if, moreover, $v \neq 1$, then $\psi(v, u) \neq 1$. Hence the group homomorphism $u \mapsto \psi(v, u)$ is injective (it has trivial kernel). But this implies in turn that $\mu(u) \neq \mu(u')$ if $u \neq u'$, $u, u' \in X_{2i+1}^*$. Hence the restriction $\mu|X_{2i+1}^*$ is injective. The same is true by symmetry for $\mu|Y_{2i+9}$ of course. Hence, if $u \in X_{2i+1}$, then since $\omega(u) \in Y_{2i+9}$ and $u^{\mu(u)} \in Y_{2i+9}$ (by (a)), Corollary 5.2.7 implies $\omega(u) = u^{\mu(u)}$ and since $\omega^2(u) \in X_{2i+1}$, the same Corollary 5.2.7 implies $\omega^2(u) = u$.

(d)   All assertions now follow if W_{2i+1} is trivial as remarked at the beginning of this proof. So suppose W_{2i+1} is not trivial and let $w \in W_{2i+1}^*$. Let $u \in U_{2i+5}$ be arbitrary. We claim that $[w, u] = 1$, which will imply that $[W_{2i+1}, U_{2i+5}]$ is trivial, hence $Y_{2i+5} = U_{2i+5}$, and by a "flip", $X_{2i+1} = U_{2i+1}$.

So put $v = [w, u]$, $m = \mu(w)$ and $w' = \omega(w)$. Since $w \in W_{2i+1} \leq X_{2i+1}$, we have $m = ww'w$. Note that $[w, u] = w^{-1} \cdot w^u \in V_{2i}W_{2i+1}V_{2i+2} \cap U_{[2i+2,2i+4]} = V_{2i+2}$ (by definition of W_{2i+1} and by Lemma 5.2.3(ii)). Similarly (and since $u^m \in U_{2i+5}$) $[u^m, w^{-1}] \in V_{2i+2}$. We have:

$$v^{w'}u^{-w'} = (vu^{-1})^{w'} = (u^{-w})^{w'} = u^{-mw^{-1}} = (u^{-m})^{w^{-1}},$$

and the latter is equal to $[w^{-1}, u^m]u^{-m}$ and hence belongs to $V_{2i+2}U_{2i+5}$. So $v^{w'} \in V_{2i+2}U_{2i+5} \cdot u^{w'} \subseteq V_{2i+2} \cdot U_{[2i+5,2i+8]}$, which implies that $[v, w'] \in U_{[2i+5,2i+8]}$. By (the "flip" of) Lemma 2, and since $v \in U_{2i+2}$, $w' \in U_{2i+9}$, we obtain $v = 1$ and our claim follows.

(e)   We still have to show that W_{2i+1} is trivial. Let w be as above. We claim that w centralizes $U_{[2i-3,2i+5]}$. Indeed, V_{2i} centralizes $U_{[2i-2,2i+4]}$ (by definition).

The latter is normalized by U_{2i-3} and by U_{2i+5}. Hence it is centralized by the group generated by all conjugates of V_{2i} by elements of $U_{2i-3} \cup U_{2i+5}$, and that is by definition $V_{2i} W_{2i+1} V_{2i+2}$. In particular, W_{2i+1} centralizes $U_{[2i-3,2i+5]}$, since $[W_{2i+1}, U_{2i-3}]$ is trivial by (d) above, and $[W_{2i+1}, U_{2i+5}]$ is trivial by a "flip". This shows our claim. But this contradicts Remark 2.

Noting that (iv) is a direct consequence of (ii) (since V_{2i+4} is abelian), the proof of the lemma is complete. QED

The next lemma takes care of condition (8M4).

Lemma 20. *Let* $n = 8$, $u \in U^*_{2i+1}$, $v \in V^*_{2i+6}$, $m = \mu(v)$ *and* $v_* = v^{-\mu(u)} \in V_{2i+4}$. *The following statements hold:*

(i) $[u, \omega(v)] = v_*^{-m} \in V_{2i}$;

(ii) $[\omega(v), v_*] = (v_*^m) u^{-1} (u^m) \in V_{2i} U_{2i+1} U_{2i+3}$;

(iii) $[U_{2i-1}, U_{2i+1}]$ *and* $[U_{2i-1}, U_{2i+3}]$ *are trivial.*

Proof. Put $v' = \omega(v)$ and $v'' = \omega^2(v)^{-1}$. Then $m = vv'v''^{-1}$. By the previous lemma, we have $v_* = [v, u] = u^{-v}u$. So

$$v_*^{v'} = u^{-vv'}u^{v'}$$
$$= u^{-mv''}u^{v'}$$
$$= (u^{-m})^{v''}u^{v'}.$$

(a) By the "flip" of Lemma 14, $v_*^{v'} = v_*[v_*, \omega(v)] \in v_* U_{2i+3} U_{2i+2} U_{2i+1} v_*^{-m}$.

(b) Clearly, $u^{-m} \in U_{2i+3}$, so $(u^{-m})^{v''} \in U_{2i+5} U_{2i+4} u^{-m}$. Also, $u^{v'} \in u U_{2i} U_{2i-1}$.

(c) Properties (a) and (b) imply that

$$v_*^{v'} \in v_* U_{2i+3} U_{2i+2} U_{2i+1} v_*^{-m} \cap U_{2i+5} U_{2i+4} u^{-m} u U_{2i} U_{2i-1} = \{v_* u^{-m} u v_*^{-m}\},$$

from which (ii) readily follows. Now (i) follows from (b).

We now show (iii). Let u_{2i-1} be arbitrary. We have $[v', v_*]^{u_{2i-1}} = [v', v_*^{u_{2i-1}}]$ and by Lemma 19(i), the latter belongs to $[v', V_{2i+2} v_*] = \{[v', v_*]\}$ by definition of V_{2i+2}. Hence u_{2i-1} centralizes $[v', v_*]$. Since it also centralizes $v_*^{-m} \in V_{2i}$, it centralizes by (ii) the element $u^{-m} \cdot u$. But this means that

$$1 = [u_{2i-1}, u^{-m} \cdot u] = [u_{2i-1}, u] \cdot [u_{2i-1}, u^{-m}]^u,$$

and noting that $[u_{2i-1}, u^{-m}]^u = [u_{2i-1}, u^{-m}]$ (since U_{2i+1} centralizes $U_{[2i,2i+2]}$ by Lemma 19(iv)), we obtain

$$[u, u_{2i-1}] = [u_{2i-1}, u^{-m}].$$

Hence, since u and u_{2i-1} are arbitrary, $[U_{2i-1}, U_{2i+3}] \leq U_{2i}$. But by symmetry also $[U_{2i-1}, U_{2i+3}] \leq U_{2i+2}$, so $[U_{2i-1}, U_{2i+3}]$ is trivial. But then $[u_{2i-1}, u^{-m}] = 1$, implying $[U_{2i-1}, U_{2i+1}]$ is trivial. QED

Lemma 21. If $n = 8$, then $[V_{2i}, U_{2i+6}] \le U_{[2i+1,2i+3]}V_{2i+4}$ and $[V_{2i}, V_{2i+6}] \le V_{2i+2}U_{2i+3}V_{2i+4}$. *Moreover, if* $[v_{2i}, u_{2i+6}] \in V_{2i+2}V_{2i+4}$, **then either** v_{2i} **or** u_{2i+6} *is equal to 1 (where* $v_{2j} \in V_{2j}$, $j = i, i+3$). *It follows that, for* $u_{2i+6} \ne 1$, *the map* $\psi_{u_{2i+6}} : V_{2i} \to U_{2i+3}$ *defined by:* "$\psi_{u_{2i+6}}(v_{2i})$ *is the factor of* $[u_{2i+6}, v_{2i}]$ *in* U_{2i+3}" *is an injective group homomorphism.*

Proof. The first assertion follows directly from Lemma 14(iii). By symmetry and by taking intersections, one obtains

$$[V_{2i}, V_{2i+6}] \le U_{[2i+1,2i+3]}V_{2i+4} \cap V_{2i+2}U_{[2i+3,2i+4]} = V_{2i+2}U_{2i+3}V_{2i+4}.$$

Now put $v = v_{2i}$ and $u = u_{2i+6}$ and suppose that $v \ne 1 \ne u$ with $[v, u] = u'u'' \in V_{2i+2}V_{2i+4}$ (obvious notation). Put $v' = \omega(v)$, $v'' = \omega^2(v)^{-1}$ and $m = \mu(v) = v \cdot v' \cdot v''^{-1}$. We compute the element $u^{mv''}$:

$$
\begin{aligned}
(u^m)^{v''} &= u^{vv'} \\
&= (u[u,v])^{v'} \\
&= u^{v'} \cdot u''^{-1} \cdot u'^{-v'} \\
&= u \cdot [u, v'] \cdot u''^{-1} \cdot u'^{-v'} \in U_{2i+6} \cdot U_{2i+7} \cdot U_{2i+4} \cdot u'^{-1}U_{[2i+3,2i+6]}.
\end{aligned}
$$

Since $u^{mv''} \in U_{[2i+1,2i+2]}$, we must certainly have $[u, v'] = 1$, $u^{mv''} \in U_{2i+2}$ and hence $u^{mv''} = u'^{-1}$. This implies that

$$u'^{-1} = u^{mv''} = uu''^{-1}u'^{-v'},$$

from which it follows that $u'' \cdot u^{-1} = [v', u'] \in U_{[2i+3,2i+5]}V_{2i+6}$, hence $u \in V_{2i+6}$ and $[v', u'] = u''u^{-1} \in V_{2i+4}V_{2i+6}$. The same argument (using a "shift") implies $v' \in V_{2i+8}$, but this contradicts Lemma 17.

The last assertion is proved similarly to Lemma 19(ii), noting that V_{2i} commutes with $V_{2i+2}U_{2i+3}V_{2i+4}$. \hfill QED

Lemma 22. *Let* $n = 8$. *Suppose* $v \in V_{2i+8}^*$ *and* $u \in U_{2i+3}^*$. *Put* $m = \mu(v)$, $u' = [u, v]^{m^{-1}}$ *(which belongs to* V_{2i+2}^* *by Lemma 19(i)) and* $m' = \mu(u')$. *Then* $[v, u'] = v^{-m'}u^m u'^{-m}$. *Hence the map* ψ_v *of Lemma 21 is an isomorphism.*

Proof. Let v_{2i+2} be an arbitrary element of V_{2i+2}^*. A double application of Lemma 14(iii) together with the definition of the map ψ_v of Lemma 21 readily implies

$$[v, v_{2i+2}] = v^{-\mu(v_{2i+2})} \cdot \psi_v(v_{2i+2}) \cdot v_{2i+2}^{-m}. \tag{5.3}$$

Now put $u_* = \psi_v(u') \in U_{2i+5}^*$, $v' = \omega^{-1}(v)$, $v'' = \omega(v)^{-1}$ and $w = [v', u_*^{m^{-1}}] \in U_{[2i+1,2i+2]}$.

We claim that $w \in V_{2i+2}$. Indeed, applying Lemma 18, let X be the smallest subgroup of $V_{2i+2}U_{[2i+3,2i+5]}V_{2i+6}$ containing V_{2i+2} and normalized by both U_{2i}

and U_{2i+8}. Since V_{2i+4} centralizes $U_{[2i,2i+8]}$, the product $Y = XV_{2i+4}$ is also normalized by both U_{2i} and U_{2i+8}. Note that $u' \in Y$ (since it is in X) and hence also $[v, u'] = u'^{-1}u'^v$. Since clearly both $v^{-m'}$ (as a member of V_{2i+4}) and u'^{-m} (as a member of V_{2i+6}) belong to Y, substituting u' for v_{2i+2} in Equation (5.3) implies $u_* \in Y$, hence $u_*^{m^{-1}} \in Y$ (since m is generated by elements of U_{2i} and U_{2i+8}). We conclude that $w = (u_*^{m^{-1}})^{-v'} \cdot u_*^{m^{-1}}$ also belongs to Y. This shows our claim.

If $w \neq 1$, then we have, using some equalities we have established in the previous paragraph,

$$
\begin{aligned}
[v'', u_*^{-1}] &= u_*^{v''}u_*^{-1} \\
&= (u_*^{m^{-1}})^{mv''}u_*^{-1} \\
&= (u_*^{m^{-1}})^{v'v}u_*^{-1} \\
&= (u_*^{m^{-1}}w^{-1})^v\psi_v(u')^{-1} \\
&= u_*^{m^{-1}}[u_*^{m^{-1}}, v][v, w]w^{-1}\psi_v(u'^{-1}) \\
&= u_*^{m^{-1}}[u_*^{m^{-1}}, v](v^{-\mu(w)}\psi_v(w)w^{-m})w^{-1}\psi_v(u'^{-1}) \\
&= w^{-1} \cdot u_*^{m^{-1}} \cdot v^{-\mu(w)} \cdot \psi_v(wu'^{-1}) \cdot (w^{-m}[u_*^{m^{-1}}, v]) \\
&\in V_{2i+2} \cdot U_{2i+3} \cdot V_{2i+4} \cdot U_{2i+5} \cdot V_{2i+6}
\end{aligned}
$$

(using Lemma 19(i) to obtain $[u_*^{m^{-1}}, v] \in V_{2i+6}$). If $w = 1$, then the above equations are still true when we put $v^{-\mu(w)}$ equal to 1. Now clearly $[v'', u_*^{-1}] \in U_{[2i+1,2i+4]}$. Hence all factors in U_{2i+5} and U_{2i+6} are 1, i.e., $w^m = [u_*^{m^{-1}}, v]$ and $w = u'$ (since ψ_v is injective). This means that

$$
[u, v] = [u_*^{m^{-1}}, v].
$$

Lemma 19(ii) implies that $u = u_*^{m^{-1}}$, hence $u_* = u^m$. QED

Lemma 23. *For* $n = 8$ *we have the inclusions* $[U_{2i}, U_{2i+5}] \leq V_{2i+2}U_{[2i+3,2i+4]}$ *and* $[U_{2i+5}, U_{2i+8}] \leq V_{2i+6}$.

Proof. We again use Lemma 18. Let X be the smallest subgroup of $U_{[2i+3,2i+5]}$ such that $Y = V_{2i+2}XV_{2i+6}$ is normalized by both U_{2i} and U_{2i+8}. Now let $u \in U_{2i+5}$ be arbitrary. By the previous lemma, there exist $v \in V_{2i+8}$ and $u' \in V_{2i+2}$ such that $[v, u'] = v_{2i+4}uv_{2i+6} \in V_{2i+4}uV_{2i+6}$. But since $u' \in Y$, we also have $[v, u'] = u'^{-v}u' \in Y$. Hence $v_{2i+4}u \in V_{2i+2}X$ and so $u \in V_{2i+4}X$. Since V_{2i+4} centralizes U_{2i} and X, we have $[U_{2i}, u] \subseteq [U_{2i}, X] \leq V_{2i+2}X$, which implies $[U_{2i}, u] \subseteq V_{2i+2}U_{[2i+3,2i+4]}$. Similarly $[u, U_{2i+8}] \subseteq [X, U_{2i+8}] \leq XV_{2i+6}$, implying $[u, U_{2i+8}] \subseteq V_{2i+6}$. QED

Lemma 24. If $n = 8$, **then** *the only elements of U_{2i} centralizing a non-trivial element of U_{2i-3} belong to V_{2i}.*

Proof. Denote by u a non-trivial element of U_{2i-3} centralized by some $v \in U_{2i}$. We apply Lemma 14(iii) and obtain $[v, \omega(u)] \in U_{2i+1}v^{\mu(u)}$. But by the previous lemma, we must have $[v, \omega(u)] \in V_{2i+2}U_{2i+3}U_{2i+4}$. Hence $v^{\mu(u)} \in V_{2i+2}$ and so $v \in V_{2i}$.   QED

The relation $[U_{2i}, U_{2i}] \le V_{2i}$ is not explicitly shown in TITS [1983], but it follows there from other properties e.g. on the order of elements of V_{2i} and U_{2i}/V_{2i} (Propositions 3.5, 3.9 and Lemma 3.16 of *op. cit.*). Since we do not need these properties to prove the theorem, we prefer here a direct proof of that fact.

Lemma 25. *For $n = 8$ we have $[U_{2i}, U_{2i}] \le V_{2i}$.*

Proof. Let $u, u' \in U_{2i}$ and $w \in U_{2i+3}^{*}$ be arbitrary. We have

$$[uu', w] = [u, w]^{u'}[u', w] = [u, w] \cdot [u', w] = [u', w] \cdot [u, w] = [u', w]^{u}[u, w] = [u'u, w]$$

since $[u, w], [u', w] \in V_{2i+2}$. We deduce that $[[U_{2i}, U_{2i}], w]$ is trivial, which implies by Lemma 24 that $[U_{2i}, U_{2i}] \le V_{2i}$.   QED

Lemma 26. **If** $n = 8$, **then** $[U_{2i}, V_{2i+6}] \le V_{2i+2}U_{2i+3}V_{2i+4}U_{2i+5}$.

Proof. By (the "flip" of) Lemma 21, we already have $[U_{2i}, V_{2i+6}] \le V_{2i+2}U_{[2i+3,2i+5]}$. Now let $v_{2i+2}u_{2i+3}u_{2i+4}u_{2i+5} \in [U_{2i}, V_{2i+6}]$ (with obvious notation). Since $[U_{2i+1}, U_{2i}]$ is trivial and $[U_{2i+1}, V_{2i+6}] \le V_{2i+4}$ (by Lemma 19(i)), we have $[U_{2i+1}, [U_{2i}, V_{2i+6}]] \le V_{2i+4}$. Hence

$$[U_{2i+1}, v_{2i+2}u_{2i+3}u_{2i+4}u_{2i+5}] \subseteq V_{2i+4}.$$

Since U_{2i+1} and u_{2i+4} commute with v_{2i+2}, u_{2i+3} and u_{2i+5}, this implies that $[U_{2i+1}, u_{2i+4}] \subseteq V_{2i+4}$, hence $[U_{2i+1}, u_{2i+4}]$ is trivial. The result now follows from Lemma 24.   QED

We now prepare the proof of (8M3).

Lemma 27. *Let $n = 8$. Suppose $u \in U_{2i+2}$ and $v \in V_{2i+8}^{*}$. Then*

(i) *there exists $v' \in V_{2i+2}$ such that $[v, uv'^{-1}] \in V_{2i+4}V_{2i+6}U_{2i+7}$,*

(ii) *if $[v, u] \in V_{2i+4}V_{2i+6}U_{2i+7}$, then $u \in V_{2i+2}^{o}$.*

Proof. We first show (i). Let u' be the factor of $[v, u]$ in U_{2i+5}. By Lemma 22, there exists $v' \in V_{2i+2}$ such that the factor of $[v, v']$ in U_{2i+5} is exactly u' (indeed, put $v' = [u'^{\mu(v)^{-1}}, v]^{\mu(v)^{-1}}$). Now we note that

$$
\begin{aligned}
[v, uv'^{-1}] &= v' \cdot [v, v']^{-1} \cdot [v, u] \cdot v'^{-1} \\
&\in V_{2i+2} \cdot V_{2i+4}u'^{-1}V_{2i+6} \cdot V_{2i+4}u'V_{2i+6}U_{2i+7} \cdot V_{2i+2}
\end{aligned}
$$

(by the previous lemma). Since all factors of this product commute except that, by Lemma 19(i), $[V_{2i+2}, U_{2i+7}] \le V_{2i+4}$ (but the latter again commutes with all other factors), we easily see that the factor of $[v, uv'^{-1}]$ in U_{2i+5} is equal to $u'^{-1}u' = 1$. This shows (i).

Suppose now $[v, u] \in V_{2i+4}V_{2i+6}U_{2i+7}$ and let $u \ne 1$. Let w be the factor of $[v, u]$ in V_{2i+4}. Put $u' = \omega^{-1}(u)$, $u'' = \omega(u)$ and $m = \mu(u) = u'uu''$. Since $u' \in U_{2i+10}$ commutes with v and $u'' \in U_{2i+10}$ normalizes $W := V_{2i+6}U_{2i+7}V_{2i+8}$ (by Lemma 23), we can write

$$
\begin{aligned}
v^m \;&=\; v^{uu''} = (v^u)^{u''} = (v[v,u])^{u''} = ([v,u]v)^{u''} \\
&\in\; (wW)^{u''} \subseteq w^{u''}W = w[w,u'']W \subseteq wU_{[2i+5,2i+9]}.
\end{aligned}
$$

Since $v^m, w \in U_{2i+4}$, this implies $v^m = w$ which implies in turn $w \in w^{u''}W$, hence $[w, u''] \in W$. Now let $v' \in V_{2i+10}$ be such that $[w, u''v'^{-1}] \in U_{2i+5}V_{2i+6}V_{2i+8}$ (possible by a "shift" and a "flip" of (i)). By Lemma 21, $[w, v'^{-1}] \in W$, hence $[w, u''v'^{-1}] = [w, v'^{-1}][w, u'']^{v'^{-1}} \in W$ since v' centralizes W. So in fact

$$
[w, u''v'^{-1}] \in V_{2i+6}V_{2i+8}.
$$

By Lemma 21 again, either $u'' = v'$ or $w = 1$. But the latter is impossible since $w = v^m$ and $v \ne 1$. Hence $\omega(u) = u'' = v' \in V_{2i+10}$ and $u = \omega^{-1}(u'') \in \omega^{-1}(V_{2i+10}^*) \subseteq V_{2i+2}^o$, showing (ii). \hfill QED

Lemma 28. *For $n = 8$ we have the commutation relation $[U_{2i-2}, U_{2i+2}] = \{1\}$.*

Proof. We already know that $[U_{2i-2}, V_{2i+2}]$ is trivial. By Lemma 14(ii) we also have that $[U_{2i-2}, V_{2i+2}^o]$ is trivial. The lemma will be proved if we show that V_{2i+2} and V_{2i+2}^o generate U_{2i+2}.

So let $u \in U_{2i+2}$. Let $v \in V_{2i+8}^*$ be arbitrary and let $v' \in V_{2i+2}$ be as described in Lemma 27(i). Then $v^o = uv'^{-1} \in V_{2i+2}^o$ and hence $u = v^o \cdot v' \in V_{2i+2}^o \cdot V_{2i+2}$. The lemma is proved. \hfill QED

Lemma 29. *For $n = 8$ we have the commutation relation $[U_{2i+1}, U_{2i+7}] \le U_{2i+3}U_{2i+5}$.*

Proof. Let $u \in U_{2i+1}$ and $v \in U_{2i+7}$. By Lemma 14(iii) we have $[U_{2i+1}, U_{2i+7}] \le U_{[2i+2,2i+5]}$ (since U_{2i-1} and U_{2i+1} commute by Lemma 20(iii)), and hence also $[U_{2i+1}, U_{2i+7}] \le U_{[2i+3,2i+6]}$, implying $[U_{2i+1}, U_{2i+7}] \le U_{[2i+3,2i+5]}$ and hence also $[U_{2i+3}, U_{2i+9}] \le U_{[2i+5,2i+7]}$. Bearing this and Lemma 20(iii) in mind and putting $m = \mu(u)$ and $u'^{-1} = \omega(u) = \omega^{-1}(u)$ (see Lemma 19(vi), although this is not essential in our argument below), we have

$$
\begin{aligned}
[u,v] &= v^{-u}v \\
&= (v^{-u'})^{mu'}v \\
&= (v^{-m})^{u'}v \\
&= v^{-m}\cdot[v^{-m},u']\cdot v \\
&\in U_{2i+3}\cdot U_{[2i+5,2i+7]}\cdot U_{2i+7},
\end{aligned}
$$

from which the result follows.   QED

Lemma 30. *For $n=8$ we have $[U_{2i},U_{2i+5}]\leq V_{2i+2}U_{2i+3}V_{2i+4}$.*

Proof. By Lemma 23, we already have $[U_{2i},U_{2i+5}]\leq V_{2i+2}U_{2i+3}U_{2i+4}$. Let $u\in U_{2i}$ and $v\in U_{2i+5}$ with $[u,v]=v_{2i+2}u_{2i+3}u_{2i+4}$. Lemma 20(*iii*) implies $[U_{2i+1},[u,v]]=\{1\}$. Since $[U_{2i+1},v_{2i+2}u_{2i+3}]$ is trivial by Lemma 20(*iii*) again, this implies that $[U_{2i+1},u_{2i+4}]$ is trivial. The result now follows directly from Lemma 24.   QED

Proof of the Theorem for $n=8$.

Condition (8M1) follows from Lemma 19(*iv*).
Condition (8M2) follows from the definition of V_{2i} and Lemma 25.
Condition (8M3) follows from Lemma 28.
Condition (8M4) follows from Lemma 20(*iii*).
Condition (8M5) follows from the definition of V_{2i}.
Condition (8M6) follows from Lemma 30.
Condition (8M7) follows from Lemma 23.
Condition (8M8) follows from Lemma 19(*i*).
Condition (8M9) follows from Lemma 26.
Condition (8M10) follows from Lemma 21.
Condition (8M11) follows from Lemma 2.
Condition (8M12) follows from Lemma 29.

This completes the proof of the theorem.   □

This theorem has an interesting corollary, noticed by various authors, e.g. RONAN [1980a] or WALKER [1982], [1983]. Remember that a **central elation** in a generalized $2m$-gon Γ is a non-trivial collineation fixing $\Gamma_{\leq m}(x)$ for some point x; an **axial elation** fixes dually $\Gamma_{\leq m}(L)$ for some line L.

5.4.7 Corollary. *Every Moufang $2m$-gon contains non-trivial central or axial elations.*

Proof. Let Γ be a Moufang $2m$-gon, $m=2,3,4$. We use the same notation as before. From Theorem 5.4.6 it follows immediately that, up to renumbering the roots, there is a non-trivial subgroup $V_0\leq U_0$ such that $[V_0,U_1U_2\cdots U_m]$

is trivial (alternatively, this also follows directly from Lemma 5 of the proof of Theorem 5.3.3). Now let $u \in V_0$ be arbitrary but non-trivial. It suffices to show that u fixes every element of Γ at distance m from x_m. Let $z \in \Gamma_m(x_m)$ be arbitrary and suppose $z \, \mathbf{I} \, z_1 \, \mathbf{I} \, z_2 \, \mathbf{I} \, \cdots z_{m-1} \, \mathbf{I} \, x_m$. By symmetry we may assume that $z_{m-1} \neq x_{m+1}$. Hence there exists an element $u_m \in U_m$ mapping z_{m-1} onto x_{m-1}. Since $\delta(z^{u_m}, x_m) = m$, we know that $z_{m-2}^{u_m} \neq x_m$. Hence there exists $u_{m-1} \in U_{m-1}$ mapping $z_{m-2}^{u_m}$ onto x_{m-2}. Continuing in this way, we obtain an element u'^{-1} of $U_m U_{m-1} \cdots U_1 = U_1 \cdots U_m$ mapping z onto x_0. Since u fixes x_0 and since u commutes with u', we have $z^u = x_0^{u'u} = x_0^{uu'} = x_0^{u'} = z$.

The corollary is proved. $\qquad\qquad\qquad\qquad\qquad\qquad\qquad\qquad\qquad\qquad\qquad$ $\square$

Another consequence is that the group U_+ is soluble (see Section 4.3 for the definition).

5.4.8 Lemma. *With the previous notation, the group $U_+ = U_1 U_2 \ldots U_n$ is soluble.*

Proof. We show by induction on $i \leq n$ that $U_1 U_2 \ldots U_i$ is soluble. For $i = 1$, this follows immediatley from Theorem 5.4.6 which says that U_1 is nilpotent of class at most 2. Now suppose the result is true for $i < n$. By part (*iii*) of Lemma 5.2.3 (see page 175) we have

$$U_1 U_2 \ldots U_{i+1} / U_1 U_2 \ldots U_i \cong U_{i+1}.$$

Since U_{i+1} is soluble, as is $U_1 U_2 \ldots U_i$ by the induction hypothesis, the result follows (see Section 4.3). $\qquad\qquad\qquad\qquad\qquad\qquad\qquad\qquad\qquad\qquad$ $\square$

5.4.9 Remark. In fact one can prove, using Theorem 5.4.6, that U_+ is nilpotent. This is an easy, but tedious exercise. Let us do it for $n = 4$. Without loss of generality we may assume that U_{2i} is abelian. We use the notation of Subsection 5.4.3. Since $[U_i, U_j] \leq V_1 U_2 U_3$, $1 \leq i, j \leq 4$, we immediately have (using the elementary rules of commutators of products)

$$[U_+, U_+] = [U_1 U_2 U_3 U_4, U_1 U_2 U_3 U_4] \leq V_1 U_2 U_3.$$

Similarly,

$$[U_+, V_1 U_2 U_3] \leq U_2 V_3,$$

because $[U_i, U_2 U_3 V_4] \leq U_2 V_3$, for all $i = 1, 2, 3, 4$. Since $[U_i, U_2] \leq V_3$, $i = 1, 2, 3, 4$, we can easily calculate $[U_+, U_2 V_3] = V_3$. Finally, $[U_+, V_3]$ is trivial. So we see that U_+ is of nilpotency class at most 4.

We also want to prove the following result for Moufang quadrangles.

5.4.10 Lemma (Tits [19]).** *Let Γ be a Moufang quadrangle such that, with the above notation, the elements of the groups U_{2i} are line-elations. Then Γ is point-regular **if and only if** $[U_{2i}, U_{2i+2}]$ is trivial. If $[U_{i-1}, U_{i+1}]$ is trivial for all i, **then** all non-trivial elements of U_i are involutions.*

Proof. If Γ is point-regular, then clearly all line-elations are central elations. With above notation, we hence have that U_0 fixes $\Gamma_2(x_2)$ pointwise, while U_2 stabilizes $\Gamma_2(x_2)$. Hence every element of $[U_0, U_2]$ fixes $\Gamma_2(x_2)$. So $[U_0, U_2] \leq U_0$. But $[U_0, U_2] \leq U_1$, hence $[U_0, U_2] \leq U_0 \cap U_1 = \{1\}$.

Conversely, if $[U_0, U_2]$ is trivial, then so is $[U_0, U_1 U_2]$. Let $x \in \Gamma_2(x_2)$. If $x \mathbf{I} x_3$, then it is fixed by every element of U_0, by definition. Suppose now $x \notin \Gamma_1(x_3)$. There exists a unique element $u_2 \in U_2$ mapping x onto the line x_1, hence every element $u_0 \in U_0$ fixes x^{u_2}. Consequently $x^{u_2} = x^{u_2 u_0} = x^{u_0 u_2}$, implying $x = x^{u_0}$. Hence U_0 consists of central collineations. If y is opposite x_0 such that $\{x_2, x_{-2}\} \subseteq x_0^y$, then there exists a unique $u \in U_0$ with $y^u = x_4$. Since u fixes all points collinear with x_0, it folows that $x_0^y = x_0^{x_4}$. This implies that x_0 is a regular point and the first assertion is proved.

Now suppose $[U_{i-1}, U_{i+1}]$ is trivial for all i. The lemmas that we will refer to in the rest of this proof are those of Theorem 5.4.6. First we remark that all U_i are commutative by Lemma 4. Indeed, all V_i are trivial here. Let $u \in U_i^*$, $v \in U_{i+3}^*$; then by Lemma 2 and its "flip" we have,

$$[u, v] = v^{-\mu(u)} u^{\mu(v)}.$$

Since U_i commutes with U_{i+2}, it is easily seen that, for $u' \in U_i$, $[uu', v] = [u', v][u, v] = [u, v][u', v]$. We deduce that

$$v^{-\mu(uu')} = v^{-\mu(u)} v^{-\mu(u')}.$$

Now let $u_* \in U_i$ be such that $\mu(u) = \mu(u_*)$. Let u' be such that $u_* = uu'$. Suppose $u' \neq 1$; then the previous equality becomes $v^{-\mu(u_*)} = v^{-\mu(u_*)} v^{-\mu(u')}$, so $v^{-\mu(u')} = 1$, a contradiction. Hence μ is an injective mapping.

Now note that by Lemma 5.2.7 $\mu(v) = \mu(\omega^2(v))$ and $\mu(\omega^{-1}(v)) = \mu(\omega(v))$, hence by the injectivity of μ, we conclude that $\omega^2(v) = v$ and $\omega^{-1}(v) = \omega(v)$. Put $\omega(v) = v'$; then this means $\mu(v) = vv'v = v'vv'$. We do the following calculation, keeping in mind that $\mu(v)$ commutes with all elements of U_{i+1}, and in particular with $v^{-\mu(u)}$, and that $v' \in U_{i-1}$ commutes with both $u \in U_i$ and $v^{-\mu(u)} \in U_{i+1}$:

$$\begin{aligned}
u^{\mu(v)} &= u^{v'vv'} \\
&= u^{vv'} \\
&= (u[u, v])^{v'} \\
&= uv^{-\mu(u)} u^{\mu(v)v'} \\
&= uv^{-\mu(u)} u^{vv'vv'} \\
&= uv^{-\mu(u)} (u[u, v])^{\mu(v)} \\
&= uv^{-\mu(u)} u^{\mu(v)} v^{-\mu(u)} u^{\mu(v)\mu(v)} \\
&= (uu^{\mu(v)\mu(v)}) \cdot (v^{-\mu(u)})^2 \cdot u^{\mu(v)} \in U_i \cdot U_{i+1} \cdot U_{i+2}.
\end{aligned}$$

Hence $(v^{-\mu(u)})^2$ is trivial and so every element of U_{i+1} is an involution. The second assertion is proved. $\square$

5.5 Commutation relations and classification of Moufang polygons

In this section, we give a rough outline of how the classification of all Moufang polygons is achieved. It turns out that there is no general approach to this question and so we have to consider projective planes, generalized quadrangles, hexagons and octagons separately, after some general remarks.

Suppose first that the root system to which a Moufang polygon Γ belongs contains multiple roots. Let V_i be the root group corresponding to the long root α_i' (and suppose the roots are ordered as previously). Let N be the group generated by all $\mu(u)$, $u \in U_\alpha^*$, for all roots α. Then it is possible to choose the V_i such that they are permuted amongst themselves by the action of conjugation by N (see TITS [1994a]). In the case of generalized quadrangles for instance, this is true whenever V_i satisfies

$$[U_{i-1}, U_{i+1}] \leq V_i \leq U_i \cap Z(U_{[i-2,i+2]}).$$

Now we assume the general case, so Γ is a Moufang n-gon belonging to a root system, possibly with multiple roots. The classification of all Moufang polygons is reduced to the determination of all groups generated by subgroups with certain properties by the following result. It is due to TITS (unpublished).

5.5.1 Theorem (Tits). *Let G be a group and U_i ($i \in \mathbb{Z}$, $U_i = U_{i+2n}$) subgroups of G which generate G. Then there exists a generalized n-gon Γ on which G acts flag-transitively and an apartment $\Sigma = (x_i : i \in \mathbb{Z}, x_i = x_{i+2n})$ of Γ such that U_i acts faithfully on Γ as in* Lemma 5.2.3 *(and hence Γ is a Moufang polygon) **if and only if** the following properties hold:*

(i) *for all $i < j < i + n$, one has $[U_i, U_j] \leq U_{[i+1,j-1]} = \langle U_{i+1}, U_{i+2}, \ldots, U_{j-1} \rangle$;*

(ii) *the map $U_i \times \cdots \times U_{i+n-1} \to G$, $(u_i, \ldots, u_{i+n-1}) \mapsto u_i u_{i+1} \cdots u_{i+n-1}$ is injective for all $i \in \mathbb{Z}$;*

(iii) *for every $i \in \mathbb{Z}$ and every $u \in U_i^*$, there exist $u', u'' \in U_{i+n}$ such that $u'uu''$ conjugates U_j onto U_{n+2i-j}, $j \in \mathbb{Z}$.*

Proof. We give an outline.
If Γ is a Moufang polygon, then the assertion follows from Lemma 5.2.3 and Lemma 5.2.6. Suppose now G is a group with the above mentioned properties. Let H be the intersection of all subgroups $N_G(U_i)$, $i \in \mathbb{Z}$. Then define the points of Γ as the cosets of the group $HU_{[0,n]}$; the lines of Γ are the cosets of $HU_{[1,n+1]}$. Incidence is "containing a common coset of $HU_{[1,n]}$". It is clear how to reconstruct the apartment Σ. Also, it can easily be shown that every path of length no more than $n + 1$ in Γ can be transformed under an element of G into a path contained in Σ. Finally, one shows that distances in Σ agree with distances of elements of Σ measured in Γ. $\square$

For the classification problem, this group G is too big. In fact, we can do the following. Put $U_+ = \langle U_1, U_2, \ldots, U_n \rangle = U_{[1,n]}$. Then it is not so hard to see that the system $(U_+; U_1, \ldots, U_n)$ determines Γ completely and unambigously. This is essentially a consequence of the fact that U_+ acts sharply transitively on the set of flags opposite $\{x_n, x_{n+1}\}$ (with the notation of Definitions 5.2.1). Hence one might ask under which conditions a group U defines a Moufang polygon for which $U_+ \cong U$, thus trying to reduce the classification of Moufang polygons to the determination of all systems $(U_+; U_1, \ldots, U_n)$ satisfying certain conditions. The answer to this question is given by the next result, due to TITS (unpublished); see also TITS & WEISS [19**]. We omit the proof.

5.5.2 Theorem (Tits & Weiss [19]).** *Suppose U_+ is a group generated by nontrivial subgroups $U_1, \ldots, U_n$, $n \in \mathbb{N}$, $n > 2$. Suppose, furthermore, that U_+ satisfies the following conditions:*

  (i) *for $1 \leq i < j \leq n$, one has $[U_i, U_j] \leq U_{[i+1,j-1]} = \langle U_{i+1}, U_{i+2}, \ldots, U_{j-1} \rangle$;*

  (ii) *the map $U_1 \times \cdots \times U_n \to U_+$, $(u_1, \ldots, u_n) \mapsto u_1 u_2 \cdots u_n$ is injective (and hence bijective);*

(Prol)$_0$ *there exists an automorphism $r \in \mathrm{Aut}(U_{[1,n-1]})$ which permutes U_i and U_{n-i}, $1 \leq i \leq n-1$, and there exist maps $u \mapsto \overline{u}$ and $u \mapsto \overline{\overline{u}}$ mapping U_n^* to itself such that for all $u \in U_n^*$, the element $\overline{u} u^r \overline{\overline{u}}$ permutes U_i and U_{n-i}, $1 \leq i \leq n-1$, and also permutes U_n^r and U_n by conjugation in $\mathrm{Aut}(U_{[1,n-1]}) \geq U_n$;*

(Prol)$_{n+1}$ *there exists an automorphism $r \in \mathrm{Aut}(U_{[2,n]})$ which permutes U_i and U_{n-i+2}, $2 \leq i \leq n$, and there exist maps $u \mapsto \overline{u}$ and $u \mapsto \overline{\overline{u}}$ mapping U_0^* to itself such that for all $u \in U_0^*$, the element $\overline{u} u^r \overline{\overline{u}}$ permutes U_i and U_{n-i+2}, $2 \leq i \leq n$, and also permutes U_0^r and U_0 by conjugation in $\mathrm{Aut}(U_{[2,n]}) \geq U_0$.*

Then there exists a Moufang generalized n-gon Γ and an apartment Σ in Γ such that $U_1, U_2, \ldots U_n$ are appropriate root groups related to Σ. □

Hence it suffices to determine all systems $(U_+; U_1, \ldots, U_n)$ satisfying the conditions of the previous theorem. In order to describe such a group U_+, it suffices to give the U_i, $1 \leq i \leq n$, and the commutation relations $[u_i, u_j]$ for all $u_i \in U_i$ and $u_j \in U_j$, $1 \leq i < j \leq n$. In fact, the classification of all Moufang polygons in TITS & WEISS [19**] is achieved by classifying all possible commutation relations.

We now look at projective planes, quadrangles, hexagons and octagon separately.

We will display some explicit commutation relations. However, we will denote the groups U_i additively, hence the identity will be written as 0. Accordingly, the identity element in U_+ will be denoted by 0. In many cases, some U_is are just the additive group of a field; the identity element of the multiplicative group will be denoted by 1.

5.5.3 Projective planes

For projective planes, the classification is well known and it is in fact an old result. First, one shows that every Moufang plane is coordinatized by an *alternative division ring* (see MOUFANG [1933]), and then one classifies all such rings. They turn out to be *fields, skew fields* or *octonion division algebras*; see KLEINFELD [1951] and BRUCK & KLEINFELD [1951]. In Appendix B we offer a streamlined proof of this classification due to TITS (unpublished).

5.5.4 Commutation relations for Moufang planes

Let $\mathbb{D}$ be any octonion division algebra, field or skew field. Let U_i, $i = 1, 2, 3$, be three copies of the additive group of $\mathbb{D}$. For $x \in \mathbb{D}$, we denote by x_i, $i = 1, 2, 3$, its image in U_i. Then $U_+ = U_1 U_2 U_3$ is completely defined by the commutation relations

$$[U_1, U_2] = [U_2, U_3] = \{0\}$$

and

$$[x_1, y_3] = (xy)_2.$$

5.5.5 Generalized quadrangles

This is in fact the most difficult case. The classification consists of four major steps. This division, however, is not very strict. The scheme below is one possibility available; it is not clear yet which approach will be taken by Tits and Weiss in their forthcoming book.

1. **Preliminary remarks.**
   Let Γ be a Moufang quadrangle. We use the following convention. The long roots of a root system of type C_2 and the multiple roots of a root system of type BC_2 correspond to point-elations of Γ. Up to duality, there are the following possibilities (and each dual possibility is obtained by interchanging B and C).

   (a) **Case $(C)_2$.**
   The quadrangle Γ belongs to a root system of type C_2, but not to one of type BC_2, and Γ^D belongs neither to a root system of type C_2 nor to one of type BC_2.

   (b) **Case $(BC)_2$.**
   The quadrangle Γ belongs to a root system of type BC_2, but not to one of type C_2, and its dual belongs neither to a root system of type C_2 nor to one of type BC_2.

   (c) **Case $(B - BC)_2$.**
   The quadrangle Γ belongs to a root system of type BC_2, but not to one of type C_2. Its dual Γ^D belongs to one of type C_2, but we do not care whether it belongs to one of type BC_2.

(d) **Case** $(BC - CB)_2$.
   Neither Γ nor Γ^D belongs to a root system of type C_2, but both Γ and Γ^D belong to one of type BC_2.

(e) **Case** $(B - C)_2$.
   Both Γ and Γ^D belong to a root system of type C_2.

(f) **Case** $(C - BC)_2$.
   This case is reserved for all Γ belonging to a root system of type C_2 and to one of type BC_2, and such that Γ^D belongs neither to a root system of type C_2 nor to one of type BC_2.

It follows easily that no other cases can occur. A logical plan would now be to consider all cases separately. This is more or less what happens, but some types are handled together.

2. **Step I.**
   This step consists of classifying the Moufang quadrangles in the cases $(C)_2$ and $(C - CB)_2$. This turns out to be the class of Moufang quadrangles that TITS [1995] calls "the class of Moufang quadrangles **properly belonging to a root system of type** C_2". The classification implies that in the case $(C - CB)_2$, no Moufang quadrangle belongs to a root system of type BC_2. Now it is clear that in the cases $(C)_2$ and $(C-CB)_2$, if $[U_0, U_2]$ is trivial, then $[U_1, U_3] = U_2$, for otherwise either the corresponding generalized quadrangle Γ^D belongs to a root system of type C_2 (if $[U_1, U_3]$ is trivial), or Γ belongs to a root system of type BC_2 (if $[U_1, U_3]$ is non-trivial).

   So a Moufang quadrangle properly belongs to a root system of type C_2 **if and only if**, up to renumbering the root groups, the commutation relations satisfy $[U_{2i}, U_{2i+2}] = \{0\}$ and $[U_{2i-1}, U_{2i+1}] = U_{2i}$. It turns out that we here find only classical quadrangles.

   Since the cases $(B)_2$ and $(C)_2$ are dual to each other, there is not much point in keeping two such names. We redefine the notation as follows. We retain $(C)_2$ for the orthogonal quadrangles of that type, and $(B)_2$ for the Hermitian quadrangles of that type. More precisely, we now give a detailed description.

   Let $\mathbb{K}$ be a (skew) field and let σ be an anti-automorphism of $\mathbb{K}$ with $\sigma^2 = 1$. With the notation of Proposition 2.3.4 (see page 55), let $q : V \to \mathbb{K}/\mathbb{K}_\sigma$ (with $\mathbb{K}_\sigma = \{t^\sigma - t : t \in \mathbb{K}\}$) be a non-degenerate σ-quadratic form with associated $(\sigma, 1)$-linear form $g : V \times V \to \mathbb{K}$. Let q have standard equation with respect to the right vector space

$$V = e_{-2}\mathbb{K} \bigoplus e_{-1}\mathbb{K} \bigoplus W_0 \bigoplus e_1\mathbb{K} \bigoplus e_2\mathbb{K}$$

over $\mathbb{K}$. Put $f_0(x, y) = g(x, y) + g(y, x)^\sigma$, for $x, y \in W_0$ (see Section 2.3; we have avoided using V_0 in order not to cause confusion with the notation for the root subgroups $V_i \leq U_i$). Define again $\widehat{X} = \{(x, y) : x \in W_0, y \in -q(x)\}$ and recall that we turn $\widehat{X}$ into a group by defining

$$(x, y) \oplus (x', y') = (x + x', y + y' - f_0(x, x'))$$

(see also Subsection 3.4.7 on page 104). Then the Moufang quadrangles properly belonging to a root system of type C_2 are, up to duality, the orthogonal quadrangles which are not mixed quadrangles, and the Hermitian quadrangles with $W_0 = \{0\}$. We give the commutation relations and redefine the types $(C)_2$ and $(B)_2$.

---

* Quadrangles of type $(C)_2$ and $(C - CB)_2$.

Suppose in the above that $\sigma = 1$ and f_0 is surjective onto $\mathbb{K}$ (the latter is automatically satisfied whenever $\sigma = 1$ and the characteristic of $\mathbb{K}$ is not 2; indeed, there exists $x \in V$ such that $g(x, x) \neq 0$ for otherwise $q(x) = 0$ for all $x \in V$ and hence q cannot be non-degenerate; it follows, using the $(1, 1)$-linearity, that g, and hence f_0, is surjective. Note also that f_0 is surjective whenever it is not 0). Let U_1 and U_3 be copies of $\mathbb{K}$ and let U_2 and U_4 be copies of $\widehat{X}$. For $x \in \mathbb{K}$, we denote by x_i the image of x in U_i under the identification of $\mathbb{K}$ with U_i, $i = 1, 3$, and similarly for the elements of $\widehat{X}$. Then $U_+ = U_1 U_2 U_3 U_4$ is determined by the commutation relations

$$[U_1, U_2] = [U_2, U_3] = [U_3, U_4] = [U_1, U_3] = \{0\}$$

and

$$\begin{cases} [(x, y)_2, (x', y')_4] &= (f_0(x, x'))_3, \\ [t_1, (x, y)_4] &= (xt, -yt^2)_2 (yt)_3. \end{cases}$$

These are the commutation relations of orthogonal quadrangles which are not mixed quadrangles. If $Y_0 = \{x \in W_0 : f_0(x, x') = 0, \forall x' \in V_0\}$ has size 1, then we have a Moufang quadrangle **of type** $(C)_2$; if Y_0 has size greater than 1, then we have a Moufang quadrangle **of type** $(C - CB)_2$. Indeed, in the latter case, we can take for $V_{2i} \leq U_{2i}$ the set of all $(x, y)_{2i} \in U_{2i}$ with $x \in Y_0$. Note that this can only happen in characteristic 2. There is a subquadrangle associated with the root groups U_{2i+1} and V_{2i} and it is a mixed quadrangle. These assertions can be verified directly with explicit computations.

---

* Quadrangles of type $(B)_2$.

Suppose now that $\sigma \neq 1$ and $W_0 = \{0\}$. Let U_1 and U_3 again be copies of $\mathbb{K}$ and let U_2 and U_4 be copies of $\mathbb{K}_\sigma$. Then $U_+ = U_1 U_2 U_3 U_4$ is determined by the commutation relations

$$[U_1, U_2] = [U_2, U_3] = [U_3, U_4] = [U_2, U_4] = \{0\}$$

and

$$\begin{cases} [t_1, t_3'] &= (0, t't - t^\sigma t'^\sigma)_2, \\ [t_1, (0, y)_4] &= (0, -tyt^\sigma)_2 (yt^\sigma)_3. \end{cases}$$

These Hermitian quadrangles and their duals will now be called **of type** $(B)_2$.

These are all Moufang quadrangles properly belonging to a root system of type C_2. Note that geometrically, this condition implies that all points — or lines — are regular, but no line — or point — is. Moufang quadrangles of type $(C)_2$ and $(C - CB)_2$ correspond to orthogonal and symplectic groups (disregarding some of them in characteristic 2); those of type $(B)_2$ to the unitary groups in projective dimension 3.

3. **Step II.**
Suppose now Γ is a Moufang quadrangle in the cases $(BC)_2$, $(CB)_2$, $(B - CB)_2$ or $(C - BC)_2$ (as before, we shall redefine these types according to our definition of type $(B)_2$ and $(C)_2$). Using standard notation, the groups U_i belonging to the simple roots and the groups V_i belonging to the long roots define, up to duality, a full subquadrangle belonging to the root system of type C_2 if one takes $V_i = [U_{i-1}, U_{i+1}]$ (which is always possible). Hence we obtain Moufang subquadrangles of type $(B)_2$, $(C)_2$ or $(C - CB)_2$. But in the latter case, it is readily seen that, for Γ, we have the case $(BC - CB)_2$, a contradiction. So in order to find all Moufang quadrangles in the cases $(BC)_2$, $(CB)_2$ and $(B - CB)_2$, one must extend the groups U_+ of the Moufang quadrangles of type $(B)_2$ or $(C)_2$ of the first step. This gives the following results (remember that proofs can be found in TITS & WEISS [19**]).

* | Type $(BC)_2$. |

The extensions of quadrangles of type $(C)_2$ are precisely the **exceptional Moufang quadrangles of type** (E_i), $i = 6, 7, 8$. As for the notion of "classical", we also call the dual of an exceptional quadrangle exceptional, whereas this will not be done for a quadrangle of type (E_i), $i = 6, 7, 8$, see below. There are three cases. Without going into details about algebraic groups, we would like to give some geometric information about these exceptional examples; see also Appendix C.

- *Quadrangles of type* (E_6)
  Consider a D_5-quadrangle Γ', i.e., a generalized quadrangle arising from a quadric of Witt index 2 in nine-dimensional projective space. The point-elation root groups are isomorphic to the additive group of the underlying field $\mathbb{K}$ and the others are isomorphic to the additive group of the six-dimensional vector space $\mathbb{K}^6$ over $\mathbb{K}$. The E_6-quadrangle Γ extends Γ' in such a way that Γ still has root groups U_{2i} isomorphic to $(\mathbb{K}^6, +)$, but the other root groups

U_{2i+1} are now non-abelian extensions U of $\mathbb{K}$ such that $U/\mathbb{K}$ is isomorphic to $(\mathbb{K}^8, +)$. Suppose now the root groups are with respect to the apartment Σ. The stabilizer H of Σ contains a group $\widetilde{H}$ of type $D_3 = A_3$ (i.e., an amost simple group with socle $\mathbf{PSO}_6$). The group $\widetilde{H}$ acts by conjugation on the groups U_{2i} and $U_{2i+1}/\mathbb{K}$. The respective representations induced are the standard six-dimensional representation and the direct sum of two conjugated half-spin representations.

In this case, the field $\mathbb{K}$ is neither finite nor p-adic. But the reals $\mathbb{R}$ and number fields are possible; see also Tits [1966b]. In fact, it is necessary and sufficient that there exists an octonion divisuin algebra over $\mathbb{K}$. We denote the corresponding exceptional Moufang quadrangle by $Q(E_6, \mathbb{K}, q)$, where q is the appropriate form of E_6. For $\mathbb{K} = \mathbb{R}$, there is only one isomorphism class of such forms, and so we can denote $Q(E_6, \mathbb{R})$; see also Chapter 9.

- *Quadrangles of type (E_7)*
  Consider a D_6-quadrangle Γ', i.e., a generalized quadrangle arising from a quadric of Witt index 2 in 11-dimensional projective space. The point-elation root groups are isomorphic to the additive group of the underlying field $\mathbb{K}$ and the others are isomorphic to the additive group of the eight-dimensional vector space $\mathbb{K}^8$ over $\mathbb{K}$. The E_7-quadrangle Γ extends Γ' in such a way that Γ still has root groups U_{2i} isomorphic to $(\mathbb{K}^8, +)$, but the other root groups U_{2i+1} are now non-abelian extensions U of $\mathbb{K}$ such that $U/\mathbb{K}$ is isomorphic to $(\mathbb{K}^{16}, +)$. Suppose again the root groups are with respect to the apartment Σ. The stabilizer H of Σ contains a group $\widetilde{H}$ of type $A_1 \times D_4$ (i.e., an extension of $\mathbf{PSL}_2 \times \mathbf{PSO}_8$). The group $\widetilde{H}$ acts by conjugation on the groups U_{2i} and $U_{2i+1}/\mathbb{K}$. The respective representations induced are standard representation of $\mathbf{PSO}_8$ and the standard two-dimensional representation of $\mathbf{PSL}_2$ tensored with the half-spin representation of $\mathbf{PSO}_8$. Note that the half-spin representation and the standard representation are equivalent here (they are both eight-dimensional).

In this case, the field $\mathbb{K}$ is neither finite nor p-adic, nor isomorphic to $\mathbb{R}$ or $\mathbb{C}$. But some number fields are possible; see also Tits [1966b]. We denote the corresponding exceptional Moufang quadrangle by $Q(E_7, \mathbb{K}, q)$, where again, q is some form.

- *Quadrangles of type (E_8)*
  Consider a D_8-quadrangle Γ', i.e., a generalized quadrangle arising from a quadric of Witt index 2 in 15-dimensional projective space. Again, the point-elation root groups are isomorphic to the additive group of the underlying field $\mathbb{K}$ and the others are isomorphic to

the additive group of the 12-dimensional vector space $\mathbb{K}^{12}$ over $\mathbb{K}$. The E_8-quadrangle Γ extends Γ' in such a way that Γ still has root groups U_{2i} isomorphic to $(\mathbb{K}^{12}, +)$, but the other root groups U_{2i+1} are now non-abelian extensions U of $\mathbb{K}$ such that $U/\mathbb{K}$ is isomorphic to $(\mathbb{K}^{32}, +)$. Suppose again the root groups are with respect to the apartment Σ. The stabilizer H of Σ contains a group $\widetilde{H}$ of type D_6 (i.e., an almost simple group with socle $\mathbf{PSO}_{12}$). The group $\widetilde{H}$ acts by conjugation on the groups U_{2i} and $U_{2i+1}/\mathbb{K}$. The respective representations induced are the standard 12-dimensional representation and the half-spin representation. Again, there is some form q of E_8 associated with each such quadrangle and we denote the corresponding quadrangle by $\mathsf{Q}(E_8, \mathbb{K}, q)$.

In this case, the field $\mathbb{K}$ is neither finite nor p-adic, nor can it be isomorphic to $\mathbb{R}$ or $\mathbb{C}$, nor can it be a number field; see also TITS [1966b]. But the transcedental extension $\mathbb{Q}(t, t')$ of $\mathbb{Q}$ of transcendency degree 2 is an example over which there exist exceptional quadrangles $\mathsf{Q}(E_i, \mathbb{Q}(t, t'), q_i)$, $i = 6, 7, 8$, for some form q_i (TITS, unpublished).

It is conjectured by TITS (unpublished) that whenever a quadrangle Γ of type E_7 (E_8) exists over $\mathbb{K}$, then one of type E_6 (E_7) exists over $\mathbb{K}$ and is a subquadrangle of Γ.

The exceptional Moufang quadrangles of type (E_i), $i = 6, 7, 8$, are by definition the Moufang quadrangles **of type** $(BC)_2$ (they extend Moufang quadrangles of type $(C)_2$).

* | Quadrangles of type $(CB)_2$ and $(B - CB)_2$. |

These are extensions of the Moufang quadrangles of type $(B)_2$. All classical quadrangles related to unitary groups in projective dimension at least 4 are obtained here. We give an explicit description using the commutation relations. Let $\mathbb{K}$, σ, q, W_0, f_0 and $\widehat{X}$ be as before. Suppose that $\sigma \neq 1$ and $W_0 \neq \{0\}$. We again define U_+. Let U_1 and U_3 be copies of $\mathbb{K}$ and let U_2 and U_4 be copies of $\widehat{X}$. Then $U_+ = U_1 U_2 U_3 U_4$ is determined by the commutation relations

$$[U_1, U_2] = [U_2, U_3] = [U_3, U_4] = \{0\}$$

and

$$\begin{cases} [(x,y)_2, (x',y')_4] &=& (f_0(x,x'))_3, \\ [t_1, t'_3] &=& (0, t't - t^\sigma t'^\sigma)_2 \\ [t_1, (x,y)_4] &=& (xt^\sigma, -tyt^\sigma)_2 (yt^\sigma)_3. \end{cases}$$

We define now the quadrangles **of type** $(BC)_2$ as the Hermitian quadrangles without any regular point and without any regular line (automatically with $f_0 \not\equiv 0$), and we redefine the quadrangles **of type**

$(B-CB)_2$ as the Hermitian quadrangles in projective dimension greater than 3 with $f_0 \equiv 0$ (which do have regular points and occur only in characteristic 2).

4. **Step III.**

Here, the Moufang quadrangles in the case $(B - C)_2$ are treated. Remark that for such a quadrangle $[U_{i-1}, U_{i+1}]$ is trivial for all i. Hence all U_i are abelian. More precisely, by Lemma 5.4.10, all non-trivial elements of U_i are involutions. So any non-trivial subgroup of U_{2i} is a candidate for a V_{2i}. But not all of them actually give rise to a root system of type BC_2. For example, if Γ is a finite symplectic quadrangle over a field of characteristic 2, then no suitable V_{2i} can be found, since there are no proper full subquadrangles. If $\mathbb{K}$ is not perfect, then proper full subquadrangles of $\mathsf{W}(\mathbb{K})$ exist, but no ideal ones. Here, we have a sort of case $(B-C-CB)_2$. A typical mixed quadrangle can have full subquadrangles, and ideal subquadrangles. This gives rise to a case $(B - BC - C - CB)_2$. We gather all these cases under **type** $(B - C)_2$.

Now let Γ be a Moufang quadrangle of type $(B-C)_2$. Since Γ and Γ^D belong to a root system of type C_2, it follows from Lemma 5.4.10 that all points and lines of Γ are regular. The classification of all such Moufang quadrangles is written down in an unpublished manuscript by TITS [19**]. The result is that all such quadrangles are mixed quadrangles; see Subsection 3.4.2. For future use, we list now the commutation relations. They can directly be computed from the coordinatization in Subsection 3.4.2. So let $\mathbb{K}$ be a field of characteristic 2, $\mathbb{K}'$ a subfield of $\mathbb{K}$ containing all squares, let L be a subspace of $\mathbb{K}$ considered as a vector space over $\mathbb{K}'$ and let L' be a subspace of $\mathbb{K}'$ considered as a vector space over $\mathbb{K}^2$. Suppose that $1 \in L \cap L'$ and that L and L' generate $\mathbb{K}$ and $\mathbb{K}'$, respectively, viewed as rings. Then we identify U_1 and U_3 with L, and we identify U_2 and U_4 with L'. The commutation relations are given by

$$[U_1, U_2] = [U_2, U_3] = [U_3, U_4] = [U_1, U_3] = [U_2, U_4] = \{0\}$$

and

$$[x_1, y_4] \quad = \quad (x^2 y)_2 (xy)_3.$$

5. **Step IV.**

This step treats the Moufang quadrangles of type $(BC - CB)_2$, and they extend Moufang quadrangles of type $(C - CB)_2$ and duals of these. In fact, every such Moufang quadrangle has two classes of subquadrangles, one class of full subquadrangles (of type $(C-CB)_2$), one class of ideal subquadrangles (in the dual class of type $(B - BC)_2$). An ideal subquadrangle which has an apartment in common with a full subquadrangle intersects the latter in a mixed quadrangle.

This class of quadrangles was missing in Tits' original conjecture, and it has only recently been discovered by Richard Weiss (in February 1997); see TITS & WEISS [19**].

We present the commutation relations (and I would like to thank Richard Weiss and Jacques Tits for communicating these relations to me, and their kind permission to reproduce them here).

Let $\mathbb{K}$ be a field of characteristic 2 and let $\mathbb{L}$ be a separable quadratic extension of $\mathbb{K}$. Denote by $x \mapsto \bar{x}$ the non-trivial (involutory) field automorphism of $\mathbb{L}$ fixing $\mathbb{K}$ pointwise. Let $\mathbb{K}'$ be a subfield of $\mathbb{K}$ containing the field $\mathbb{K}^2$ of all squares of $\mathbb{K}$ and let $\mathbb{L}'$ be the subfield of $\mathbb{L}$ generated by $\mathbb{L}^2$ and $\mathbb{K}'$. We then have that $\mathbb{L}^2 \subseteq \mathbb{L}' \subseteq \mathbb{L}$ and $\mathbb{L}'$ is a separable quadratic extension of $\mathbb{K}'$ (because the map $x \mapsto \bar{x}$ restricts to an automorphism of $\mathbb{L}'$ and the fixed subfield is exactly $\mathbb{K}'$). Now let there be given two elements $\alpha \in \mathbb{K}'$ and $\beta \in \mathbb{K}$ such that, for all $u, v \in \mathbb{L}$, and all $a \in \mathbb{K}'$,

$$u\bar{u} + \alpha v\bar{v} + \beta a = 0$$

implies that $u = v = a = 0$, and, for all $x, y \in \mathbb{L}'$, and all $b \in \mathbb{K}$,

$$x\bar{x} + \beta^2 y\bar{y} + \alpha b^2 = 0$$

implies that $x = y = b = 0$. We identify U_1 and U_3 with the direct product $\mathbb{L}' \times \mathbb{L}' \times \mathbb{K}$ (additively), and U_2 and U_4 with $\mathbb{L} \times \mathbb{L} \times \mathbb{K}'$. We define the quadrangle $Q(\mathbb{K}, \mathbb{L}, \mathbb{K}', \alpha, \beta)$ as the Moufang quadrangle with commutation relations

$$[U_1, U_2] = [U_2, U_3] = [U_3, U_4] = \{0\}$$

and

$$
\begin{aligned}
[(x,y,b)_1, (x',y',b')_3] &= (0, 0, \alpha(x\bar{x}' + x'\bar{x} + \beta^2(y\bar{y}' + y'\bar{y})))_2 \\
[(u,v,a)_2, (u',v',a')_4] &= (0, 0, \beta^{-1}(u\bar{u}' + u'\bar{u} + \alpha(v\bar{v}' + v'\bar{v})))_3 \\
[(x,y,b)_1, (u,v,a)_4] &= (bu + \alpha(\bar{x}v + \beta y\bar{v}), bv + xu + \beta y\bar{u}, \\
&\quad b^2 a + a\alpha(x\bar{x} + \beta^2 y\bar{y}) \\
&\quad + \alpha(u^2 x\bar{y} + \bar{u}^2\bar{x}y + \alpha(\bar{v}^2 xy + v^2\bar{x}\bar{y})))_2 \\
&\quad \cdot(ax + \bar{u}^2 y + \alpha v^2 \bar{y}, ay + \beta^{-2}(u^2 x + \alpha v^2 \bar{x}), \\
&\quad ab + b\beta^{-1}(u\bar{u} + \alpha v\bar{v}) \\
&\quad + \alpha(\beta^{-1}(xu\bar{v} + \bar{x}\bar{u}v) + y\bar{u}\bar{v} + \bar{y}uv))_3.
\end{aligned}
$$

We leave it as an interesting exercise to the reader to recognize the full and the ideal orthogonal subquadrangles. It is rather easily seen that, by restricting U_1 and U_3 to $\{0\} \times \{0\} \times \mathbb{K}$, and U_2 and U_4 to $\{0\} \times \{0\} \times \mathbb{K}'$, one obtains the mixed quadrangle $Q(\mathbb{K}, \mathbb{K}'; \mathbb{K}, \mathbb{K}')$. Some properties of the Moufang quadrangles in the case $(BC - CB)_2$ are mentioned and/or proved in Section 7.4. We say that these quadrangles have **type** $(BC - CB)_2$. Also, we call these quadrangles the **exceptional Moufang quadrangles of type** (F_4); see Remark 5.5.21.

From this discussion we readily obtain the following (for *projective points*, we refer to Definition 1.9.4):

5.5.6 Corollary. *Every Moufang quadrangle having projective points is isomorphic to a symplectic quadrangle.* □

Later on (see Theorem 6.2.1 on page 240), we will show that every generalized quadrangle all points of which are projective is automatically isomorphic to a symplectic quadrangle.

The next lemma follows immediately from Lemma 5.4.10.

5.5.7 Lemma. *If* Γ *is a Moufang quadrangle belonging to a root system of type* C_2, **then**, *up to duality, all points are distance-2-regular.* □

An immediate consequence of the above discussion is the following:

5.5.8 Lemma. *If* Γ *is a Moufang quadrangle,* **then** Γ *is a mixed quadrangle* **if and only if** *all points and all lines of* Γ *are distance-2-regular.* □

5.5.9 Lemma. *Let* Γ *be a Moufang quadrangle. Then all root groups are commutative* **if and only if** *up to duality all points are regular, or* Γ *is of type* $(BC - CB)_2$.

Proof. For Moufang quadrangles belonging to a root system of type C_2, this is clear. Suppose now Γ is of type $(CB)_2$; then we have to show that there are no regular elements. Suppose, with above standard notation, that U_{2i} is commutative (and its elements are line-elations). Then $[U_{2i}, U_{2i+2}] = V_{2i+1}$ is non-trivial, hence by Lemma 5.4.10 the points are not regular. Suppose now that the lines are regular. Then $[U_{2i-1}, U_{2i+1}] = V_{2i}$ is trivial. Lemma 4 of the proof of Theorem 5.4.6 implies that U_{2i-1} is commutative, a contradiction. Suppose now that Γ is of type $(BC)_2$; then Γ is classical and the result follows from the fact that $f_0(x, x') = f_0(x', x)$ for all $x, x' \in W_0$ **if and only if** $f_0 \equiv 0$ (for non-trivial σ), noting that the root group U_2 (with above notation) is isomorphic to the group R_2 of Subsection 3.4.7 on page 104. Finally, if Γ is of type $(BC - CB)_2$, then all root groups are commutative by definition (in fact, all root groups have exponent 2, i.e., all elations are involutions). □

So we may describe the different classes of Moufang quadrangles as follows. Quadrangles of type $(C)_2$ are the orthogonal quadrangles having regular lines, but no regular points; quadrangles of type $(B)_2$ are the Hermitian quadrangles with standard embedding in projective 3-space (they have regular points, but no regular lines); quadrangles of type $(BC)_2$ (the exceptional quadrangles of type $(E)_i$, $i = 6, 7, 8$) and of type $(CB)_2$ (the Hermitian quadrangles with standard embedding in projective space of dimension at least 4 and $f_0 \neq 0$) have neither regular points nor regular lines (and they have both abelian and non-abelian root groups);

Type	Class	Regularity	C_2-subsystem
$(B-C)_2$	Mixed	Regular points Regular lines	$(B-C)_2$
$(C-CB)_2$	Orthogonal, $0 \not\equiv f_0$ degenerate	No regular points Regular lines	$(B-C)_2$
$(C)_2$	Orthogonal, f_0 non-degenerate	No regular points Regular lines	
$(BC-CB)_2$	Exceptional of type (F_4)	No regular points No regular lines	$(C-CB)_2$
$(BC)_2$	Exceptional of type (E_i)	No regular points No regular lines	$(C)_2$
$(B)_2$	Hermitian, $W_0 = \{0\}$	Regular points No regular lines	
$(B-CB)_2$	Hermitian, $W_0 \neq \{0\}$, $f_0 \equiv 0$	Regular points No regular lines	$(B)_2$
$(CB)_2$	Hermitian, $f_0 \not\equiv 0$	No regular points No regular lines	$(B)_2$

Table 5.1. Classification of Moufang quadrangles.

quadrangles of type $(B-CB)_2$ have regular points and arise from a σ-quadratic form in some vector space of dimension at least 5 with $\sigma \neq 1$ (and all root groups are again abelian); quadrangles of type $(BC-CB)_2$ (exceptional of type (F_4)) have neither regular points nor regular lines, but all root groups are abelian; finally, quadrangles of type $(B-C)_2$ have both regular lines and regular points and all root groups are abelian. We gather some of this information in Table 5.1. A C_2-**subsystem** is the type of a minimal full or ideal subquadrangle arising from the root subgroups V_{2i} (standard notation). Note that only types $(BC)_2$ and $(CB)_2$ have non-abelian root groups. Therefore, we have not included this information in the table.

Let me end the discussion on Moufang quadrangles by mentioning that FAULKNER [1977] classified (albeit partially) the Moufang quadrangles of type $(C)_2$ and $(B)_2$ (with, at a certain point, the additional restriction that there are no involutive root elations). He also proved some interesting properties in the general case.

5.5.10 Generalized hexagons

Here, the generalized polygon Γ belongs to a root system of type G_2. Hence all root groups are commutative. The long roots form a subsystem of type A_2. Assume that they correspond to the root groups U_i with i even. We first show that, up to duality, all points are distance-2-regular (see also RONAN [1980a]). Therefore, we suppose that the long roots correspond to line-elation root groups. Let $\Sigma = (p_1, L_2, p_3, L_4, \ldots, p_{11}, L_{12}, p_1)$ be an (ordered) apartment of Γ. Let U_{2i+1},

$i \in \{1, 2, 3\}$ be the root group related to Σ fixing all elements incident with one of $p_{2i+1}, L_{2i+2}, \ldots, p_{2i+5}$; let U_{2i}, $i \in \{0, 1, 2\}$ be the root group related to Σ fixing all elements incident with one of $L_{2i}, p_{2i+1}, \ldots, L_{2i+4}$ (subscripts to be taken modulo 12). Let x be any point collinear with p_5. If x is incident with L_4 or L_6, then it is fixed by all elements of U_4; in the other case, there is a unique element $u_6 \in U_6$ mapping the line xp_5 to L_4. Let $u_4 \in U_4$ be arbitrary. Since $[U_4, U_6]$ is trivial, we have $x^{u_6 u_4} = x^{u_4 u_6}$, which implies (since $(x^{u_6})^{u_4} = x^{u_6}$) that $x^{u_6} = x^{u_4 u_6}$, hence x is fixed by u_4. A similar argument, now using the fact that $[U_3, U_6]$ is trivial, shows that every element of U_3 fixes x. Symmetrically, U_2 also fixes x. Now let p be any point of Γ belonging to $\Gamma_4(p_3) \cap \Gamma_4(p_7) \cap \Gamma_6(p_5)$. Suppose $p \,\mathbf{I}\, L \,\mathbf{I}\, p' \,\mathbf{I}\, L' \,\mathbf{I}\, p_7$. An element $u_2 \in U_2$ maps L' to L_8; an element $u_3 \in U_3$ maps p'^{u_2} to p_9; an element $u_4 \in U_4$ maps $L^{u_2 u_3}$ to L_{10}. Since $\delta(p_3, p) = 4$, the element $u_2 u_3 u_4$ maps p to p_{11}. Since $u_2 u_3 u_4$ fixes all points collinear with p_5, we have $p_5^p = p_5^{p_{11}}$, and so p_5 is distance-2-regular. We have shown (see RONAN [1980a], who provided the first published proof, using the — at that time in the literature unproved — commutation relations induced by the root system of type G_2):

5.5.11 Lemma (Ronan [1980a]). *If Γ is a Moufang hexagon, **then**, up to duality, all points are distance-2-regular.* □

Now note that, by Section 1.9.8 (see page 40), there is a weak subhexagon $\Gamma(p_1, p_7)$, which is the double of a projective plane π. Clearly, the groups U_{2i} induce root groups in π and hence π is a Moufang projective plane. Let $u_3 \in U_3$ be non-trivial. Then, with the notation of Lemma 5.2.6, the collineation $\mu(u_3)$ induces a correlation in π and a "reflection" in Σ. Moreover, it commutes with all elements of U_6 (because $[U_3, U_6]$ and $[U_3, U_{12}]$ are trivial). Proposition B.4 implies that π is Pappian and hence U_6 is the additive group of a field.

RONAN [1980a] also proves (without using the classification of Moufang hexagons) the following result (which we will prove in Chapter 6 using the classification):

5.5.12 Proposition (Ronan [1980a]). *Every Moufang hexagon contains, up to duality, an ideal split Cayley subhexagon.* □

5.5.13 Classification of Moufang hexagons — outline

Now consider, with the usual indexing, the commutation relation $[U_1, U_6]$, where even indices refer to long roots. If $u_1 \in U_1$ and $u_6 \in U_6$, then we know by Lemma 2 of the proof of Theorem 5.4.6 on page 191 that the factor of $[u_1, u_6]$ in U_5 equals $u_1^{\mu(u_6)}$. One now shows that, identifying U_6 with a field $\mathbb{K}$, and identifying U_1 and U_5 with some set V, the mapping $\mathbb{K} \times V \to V : (u_1, u_6) \mapsto u_1^{\mu(u_6)}$ turns V into a vector space over $\mathbb{K}$. It follows that the map $K_{i,j;\ell}$ mapping a pair $(u_i, u_j) \in U_i \times U_j$, $i < j < i + 6$, to the factor of $[u_i, u_j]$ in U_ℓ, $i < \ell < j$, is a polynomial of bidegree (a, b) (i.e., in the first variable the degree is a, in the second it is b) when

$\alpha_\ell = a\alpha_i + b\alpha_j$ (where α_k is the root corresponding to the root group U_k, $k \in \mathbb{Z}$). After some calculations, one finds that the product of bidegree $(1,2)$

$$(x,y) \mapsto f(x,y) \cdot y - g(x,h(y)), \qquad\qquad x,y \in V,$$

defines the quadratic mapping $Q : V \to \mathrm{End}V$, with $Q(y)(x) = (x,y)$, of a quadratic Jordan algebra of degree 1 or 3 on V, where $f(x,y)_2 = [x_1, y_3]$, $g(x,y)$ is the factor in U_3 of $[x_1, y_5]$ and $h(x)$ is the factor in U_3 of $[x_1, 1_6]$. The problem is reduced to determining all these Jordan algebras, and one finally obtains, up to duality, the following list (see also TITS [1976a], RONAN [1989] (page 174) and, to a certain extent, FAULKNER [1977], who considers Moufang hexagons belonging to the root system of type G_2, with an additional assumption on the characteristic (namely, in one of his cases, that there are root elations not of order 3; consequently he does not find the mixed hexagons, for example); his final list of possibilities is not so transparent, in fact he does not classify the algebraic structures he obtains):

1. The case where $V = \mathbb{K}$. Here, Γ is the split Cayley hexagon over the commutative field $\mathbb{K}$.

2. The case where V is a separable cubic extension of $\mathbb{K}$; if this extension is a Galois extension, then we obtain the twisted triality hexagons; otherwise we obtain the hexagons of type $^6D_4$ (see Subsection 3.5.9 on page 116).

3. If V is a pure inseparable extension in characteristic 3 such that $V^3 \subseteq \mathbb{K}$, then we obtain the mixed hexagons; see Section 3.5.3.

4. Here, V is a skew field of dimension 9 over its centre $\mathbb{K}$ (hence this can be considered as a Jordan division algebra of degree 3 over $\mathbb{K}$).

5. Here, V is the (Jordan) algebra of fixed elements of an involution σ of the second kind in a skew field of dimension 9 over its centre $\mathbb{L}$ which is a quadratic separable extension of a field $\mathbb{K}$ ("the second kind" means that σ does not induce the identity in $\mathbb{L}$; hence σ induces a non-trivial element of $\mathrm{Gal}(\mathbb{L}/\mathbb{K})$ in $\mathbb{L}$). The dimension of V over $\mathbb{K}$ is 9.

6. Here, V is a 27-dimensional exceptional Jordan algebra over $\mathbb{K}$.

If we denote in cases 1 to 5 the multiplication in the skew field by $x * y$, then, with the above notation, we have $Q(y)(x) = (x,y) = y * x * y$. For cases 4 and 5, we remark that the dimension of a skew field $\mathbb{K}'$ over its centre $\mathbb{F}'$ is always a square d^2 (if finite), and the number d is called the **degree** of $\mathbb{K}'$. In this case, d is the dimension of $\mathbb{K}'$ over a maximal commutative subfield $\mathbb{L}'$ of $\mathbb{K}'$ containing $\mathbb{F}'$, and this also equals the dimension of $\mathbb{L}'$ over $\mathbb{F}'$.

We remark that there is a certain similarity with the classification of alternative division rings; see Appendix B. Indeed, all elements of V satisfy a third-degree equation with coefficients in $\mathbb{K}$; for alternative division rings, this is a quadratic equation. The dimension of V over $\mathbb{K}$ can be $3^0, 3^1, 3^2$ or 3^3; compare with the dimension of any subring $\mathbb{L}$ of an alternative division ring $\mathbb{D}$ over the centre $\mathbb{K}$ of $\mathbb{D}$ (and such that $\mathbb{L}$ contains $\mathbb{K}$): this can be $2^0, 2^1, 2^2$ or 2^3. The cases 2^0 and 2^1 correspond to fields; the cases 3^0 and 3^1 correspond likewise to fields (cases 1,2,3 in the above list). The case 2^2 corresponds to skew fields; the case 3^2 corresponds likewise to skew fields (cases 4 and 5 in the above list). Finally, the case 2^3 is the only proper alternative case; the case 3^3 yields also the only "proper" (exceptional) Jordan algebras.

We now comment on the Jordan algebra of the last type. Upon suitable scalar extension, such a Jordan algebra becomes isomorphic to the Jordan algebra of all Hermitian 3×3 matrices over a split Cayley algebra. The exceptional 27-dimensional Jordan algebra V is also called an **Albert division algebra**, because ALBERT [1958b], [1965] gave the first examples of this type (essentially, he proved that the above Jordan algebras of Hermitian 3×3 matrices are division algebras in certain cases, and have zero divisors in other cases, as e.g. in the finite case; the general construction, by the way, works over any field, but usually yields Jordan algebras with zero divisors). He showed for instance, that there were examples over the field $\mathbb{K}(t)$ of rational functions if there exists a skew field of degree 3 over $\mathbb{K}$; one can for example take $\mathbb{K} = \mathbb{Q}$ or any p-adic field. These Jordan algebras are called *exceptional* because they cannot be embedded into an associative algebra, equipped with the Jordan product

$$\frac{1}{2}(ab + ba)$$

(if $\mathbb{K}$ has characteristic 2 then one has to use a more complicated definition), as for instance in the other cases above. The construction of ALBERT has been greatly simplified by TITS (unpublished), see JACOBSON [1968](Chapter IX). In fact, TITS (unpublished) and MCCRIMMON [1970] have shown that every Albert division algebra, i.e., every exceptional Jordan algebra, can be obtained from one of two construction methods of TITS (unpublished); see also JACOBSON [1968](Chapter IX). The little projective group of the corresponding hexagon obtained from such a Jordan algebra is an algebraic group of type E_8; see also Appendix C for the other Moufang hexagons.

If a Moufang hexagon has projective points, then all root groups have to be parametrized by isomorphic fields. By the above enumeration, we can conclude:

5.5.14 Corollary. *A Moufang hexagon with projective points is isomorphic to a split Cayley hexagon.* □

In the same way we conclude by inspection (though a direct proof along the lines of TITS [19**] is possible):

5.5.15 Corollary. *A Moufang hexagon with distance-2-regular points and distance-2-regular lines is isomorphic to a mixed hexagon.* □

Later on (see Theorem 6.3.1 on page 241 and Theorem 6.3.2 on page 243), we will show that every generalized hexagon, every point of which is distance-2-regular, is automatically a Moufang hexagon and hence a split Cayley hexagon whenever it contains at least one projective point.

An easy consequence of the proof of Corollary 5.4.7 is the following:

5.5.16 Corollary. *All elations belonging to a root group corresponding to a long root in the root system of type G_2 in a Moufang hexagon Γ are axial elations, if one fixes the duality class by requiring that Γ contains an ideal split Cayley subhexagon.*

Proof. Indeed, using standard notation, if U_0 belongs to a long root, then $[U_0, U_{[0,3]}]$ is trivial. The result now easily follows from the proof of Corollary 5.4.7. $\qquad\square$

We conclude this section with the explicit form of the commutation relations for the split Cayley hexagon; see TITS [1976a].

5.5.17 Commutation relations for $\mathsf{H}(\mathbb{K})$

Let U_i, $i \in \{1, 2, \ldots, 6\}$, be a copy of a field $\mathbb{K}$. Again we denote the image of $x \in \mathbb{K}$ in U_i by x_i. Then $U_+ = U_1 U_2 U_3 U_4 U_5 U_6$ is completely determined by the relations (up to an even "shift" of the indices):

$$[U_1, U_2] = [U_1, U_4] = [U_2, U_3] = [U_2, U_4] = [U_2, U_5] = \{0\}$$

and

$$\begin{cases} [x_1, y_3] &= (3xy)_2, \\ [x_1, y_5] &= (-3x^2y)_2 \cdot (2xy)_3 \cdot (-3xy^2)_4, \\ [x_1, y_6] &= (-x^3y)_2 \cdot (x^2y)_3 \cdot (-x^3y^2)_4 \cdot (xy)_5, \\ [x_2, y_6] &= (xy)_4. \end{cases}$$

5.5.18 Generalized octagons

We give a brief outline of the classification of Moufang octagons. Referring to the notation of the proof of Theorem 5.4.6, one first identifies the groups U_{2i+1} and V_{2i} and denotes them by $\mathbb{K}$; then one uses the commutation relation $[V_{2i}, U_{2i+5}] \leq V_{2i+2}$ to define a product in the additive group $\mathbb{K}$ which turns $\mathbb{K}$ into a field of characteristic 2. We define an endomorphism $\sigma : x_1 \mapsto x_1^{\mu(e_1)\mu(x_1)}$, where e is the unit element of $\mathbb{K}$, using the identification of $x_1 \in U_1$ with $x \in \mathbb{K}$. Alternatively, one could use the fact that the V_{2i} form a subsystem of type C_2 corresponding to a Moufang quadrangle Γ' of type $(B - C)_2$ (a mixed quadrangle). The root groups of Γ' can be parametrized by $\mathbb{K}$ and one can construct geometrically a polarity in Γ' (by considering a root elation corresponding to another apartment); it follows that Γ' is a Suzuki quadrangle and so the Tits endomorphism σ is obtained. Then one verifies that all the commutation relations are determined by the geometric information obtained so far (there is no need to carry out the actual computations). Hence Γ must be isomorphic to a Ree–Tits octagon over $\mathbb{K}$. Details of this proof can be found in TITS [1983]; see also TITS & WEISS [19**] for the alternative approach.

Similar to Corollary 5.5.16, one can show easily (especially using the commutation relations below):

5.5.19 Corollary. *All line-elations of order 2 in a Ree–Tits octagon are central elations. No other elation is central or axial.* $\qquad\square$

5.5.20 Commutation relations for Ree–Tits octagons

Let us write down the commutation relations in a Ree–Tits octagon in a somewhat different way as in TITS [1983]. Put $U_{2i+1} = \mathbb{K}$, where $\mathbb{K}$ is a field of characteristic 2 admitting a Tits endomorphism σ. We use the same notation as in Section 3.2. In particular, we recall that $\mathbb{K}_\sigma^{(2)} = \mathbb{K} \times \mathbb{K}$ is a group with operation law $(k_0, k_1) \oplus (l_0, l_1) = (k_0 + l_0, k_1 + l_1 + l_0 k_0^\sigma)$. Furthermore, for $k = (k_0, k_1)$, we put $\mathrm{Tr}(k) = k_0^{\sigma+1} + k_1$ and $\mathrm{N}(k) = k_0^{\sigma+2} + k_0 k_1 + k_1^\sigma$; we define the multiplication $a \otimes k = a \otimes (k_0, k_1) = (ak_0, a^{\sigma+1}k_1)$ for $a \in \mathbb{K}$ and $k \in \mathbb{K}_\sigma^{(2)}$ and we write $(k_0, k_1)^\sigma$ for (k_0^σ, k_1^σ). We put $U_{2i} = \mathbb{K}_\sigma^{(2)}$. As usual, the corresponding element in U_{2i+1} of an element x of $\mathbb{K}$ is denoted by x_{2i+1}; similarly for U_{2i}. Also, the groups V_{2i} correspond to the subgroup $\mathbb{K}$ of $\mathbb{K}_\sigma^{(2)}$ obtained by considering only those elements with first coordinate equal to 0; alternatively, these elements are exactly the elements of order 2 or 1 in U_{2i}. Then we have (up to an even "shift"):

$$
\begin{aligned}
[U_1, U_2 U_3 V_4 U_5] &= \{0\}; \\
[U_2, U_3 V_4 U_6] &= \{0\}; \\
[a_1, k_4] &= (0, ak_0)_2; \\
[a_1, k_6] &= (0, ak_0)_2 \cdot (ak_0^\sigma)_3 \cdot (0, a\,\mathrm{Tr}(k))_4; \\
[a_1, b_7] &= (a^\sigma b)_3 \cdot (ab^\sigma)_5; \\
[a_1, k_8] &= (a \otimes k)_2 \cdot (a^{\sigma+1}\,\mathrm{N}(k))_3 \cdot (a^\sigma \otimes (\mathrm{Tr}(k), k_0\,\mathrm{N}(k)^\sigma))_4 \cdot \\
&\quad \cdot (a^{\sigma+1}\,\mathrm{N}(k)^\sigma)_5 \cdot (a \otimes (k_1, \mathrm{Tr}(k)\,\mathrm{N}(k)))_6 \cdot (a\,\mathrm{N}(k))_7; \\
[k_2, l_4] &= (k_0 l_0)_3; \\
[k_2, a_5] &= (0, ak_0)_4; \\
[k_2, a_7] &= (0, a\,\mathrm{Tr}(k))_4 \cdot (ak_0^\sigma)_5 \cdot (0, a^\sigma k_0)_6; \\
[k_2, l_8] &= (l_0\,\mathrm{Tr}(k))_3 \cdot ((l_0 \otimes (k_0, \mathrm{Tr}(k))^\sigma) \oplus (0, l_1\,\mathrm{N}(k)))_4 \cdot \\
&\quad \cdot (k_1 l_1)_5 \cdot ((k_0 \otimes l^\sigma) \oplus (0, k_1\,\mathrm{N}(l)))_6 \cdot (k_0\,\mathrm{Tr}(l))_7.
\end{aligned}
$$

It is straightforward to derive these relations from the fomulae (1.7.1) of TITS [1983], keeping in mind that the commutator brackets are defined there differently and that group elements operate there on the left.

5.5.21 Remark. We now see that *every Moufang polygon arises from an algebraic or classical group*, in other words, every Moufang polygon has a natural construction inside some building related to an or classical algebraic group. Let me give some examples. Except for the mixed quadrangles, the Moufang quadrangles of type $(BC - CB)_2$, and the Ree–Tits octagons, these can be found in TITS [1966b]. The non-Desarguesian Moufang planes arise as the subcomplex fixed pointwise by an involution in a building of type E_6; for the classical quadrangles and the exceptional Moufang quadrangles, see Appendix C. The mixed quadrangles live inside an appropriate symplectic quadrangle, which is a classical quadrangle and hence related to an algebraic group. The Moufang quadrangles of type $(BC - CB)_2$ can be obtained as the subcomplex fixed pointwise by an involution in certain "mixed F_4-buildings", which in their turn are contained in F_4-buildings naturally associated with algebraic groups (MÜHLHERR & VAN MALDEGHEM, [19**]); therefore I have called these quadrangles "of type (F_4)". For the Moufang hexagons, we refer again to Appendix C. The Ree–Tits octagons finally arise from polarities in F_4-buildings, as we have seen in Chapter 2.

5.6 Another result of Weiss

Weiss' theorem (Theorem 5.3.3 on page 179) rules out the existence of (G,n)-Moufang trees for $n \neq 3, 4, 6, 8$. A natural question to ask is if one can classify the remaining cases. More exactly, is every (G, n)-Moufang tree the universal cover of a Moufang polygon? Before stating the general answer to that question, we show how one can fold a tree to obtain a generalized polygon.

An **apartment** of a tree is a subgraph every vertex of which has valency 2 (i.e., every vertex is adjacent to exactly two vertices). Hence an apartment of a tree has no end-points. An n-**path** is a non-stammering path of length n, $n \in \mathbb{N}$; see also Subsection 1.2.2 on page 3.

The next result is essentially contained in DELGADO & STELLMACHER [1985]. We give the version of WEISS [1995a]. See also OTT [1991].

5.6.1 Theorem (Delgado & Stellmacher [1985]). *Let* $n \geq 3$. *Suppose* (X, E) *is a thick tree and* $\mathcal{A}$ *a family of apartments of* (X, E) *such that*

(Un) *every* $(n + 1)$-*path of* (X, E) *lies on a unique element of* $\mathcal{A}$;

(Ex) *if* $(x_0, \ldots, x_{2n})$ *and* $(y_0, \ldots, y_{2n})$ *are two* $2n$-*paths each contained in an element of* $\mathcal{A}$ *and such that* $x_i = y_i$ *for* $0 \leq i \leq n$, *but* $x_{n+1} \neq y_{n+1}$, *then there is a third element of* $\mathcal{A}$ *containing the path* $(x_{2n}, \ldots, x_{n+1}, x_n, y_{n+1}, \ldots, y_{2n})$.

Let $\sim$ *be the relation on* X *defined by setting* $x \sim y$ *whenever there is an element of* $\mathcal{A}$ *containing both* x *and* y *and* x *is at distance* $2n$ *from* y. *Let* $\approx$ *be the transitive closure of* $\sim$ *and let* Γ *be the graph with vertex set* $X/\approx$, *where two equivalence*

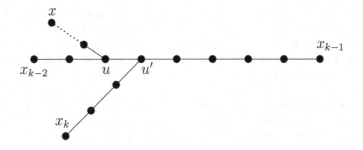

Figure 5.5.

classes are adjacent in Γ if they contain elements adjacent in (X, E). Then Γ is the incidence graph of a generalized n-gon, and the natural map ρ from X to $X/\approx$ induces a bijection from the set of vertices adjacent to some vertex x to the set of vertices adjacent to x^ρ.

Proof. We denote the distance in (X, E) by δ.

(1) Let $\{x, y\} \in E$ and suppose $x \approx x'$, $x, y, x' \in X$. We show that there exists $y' \in X$ such that $\{x', y'\} \in E$ and $y \approx y'$. Clearly, it suffices to show this for $x \sim x'$. Let $\Sigma \in \mathcal{A}$ contain x and x'. If it also contains y, then the claim follows. If it does not contain y, then the unique element $\Sigma' \in \mathcal{A}$ containing y and u, where $\delta(u, x) = \delta(u, x')$, $u \in X$, meets Σ precisely in the n-path $(x, \ldots, u)$. The vertex $y'' \in X$ belonging to $\Sigma' \setminus \Sigma$ at distance $n-1$ from u satisfies $y \sim y''$. By condition (Ex) there is a $\Sigma'' \in \mathcal{A}$ containing y'', u and x'. The vertex y' in Σ'' adjacent to x' and at distance $2n$ from y'' satisfies $y' \sim y''$. Hence $y \approx y''$ and the claim follows.

(2) Suppose $x \approx y$, $x, y \in X$. Let $x \sim x_1 \sim \cdots \sim x_k = y$ be minimal. Then we claim that $\delta(x, y) > \delta(x, x_{k-1})$. Indeed, for $k = 1$, this is obvious. Now let $k > 1$. We proceed by induction on k. Let u be the unique vertex of the $2n$-path $(x_{k-2}, \ldots, x_{k-1})$ closest to x (see Figure 5.5, where we have drawn a possible situation for $n = 4$). By the induction hypothesis $\delta(x_{k-1}, u) \geq n+1$. Hence the vertex u' at distance $n + 1$ from x_{k-1} and at distance $n - 1$ from x_{k-2} belongs to the path connecting u with x_{k-1}. Since (X, E) is a tree and $\delta(x_{k-1}, x_k) = 2n$, the only way that x_k could be closer to x than x_{k-1} is when $\delta(x_k, u') = n-1$, i.e., u' belongs to the apartment Σ which contains x_{k-1} and x_k. But condition (Un) implies that $x_{k-2} = x_k$ in that case, a contradiction. Hence the claim. Note that an almost identical argument shows that $\delta(x, y)$ is even.

Claims (1) and (2) already imply that the natural map ρ from X to $X/\approx$ induces a bijection from the set of vertices adjacent to some vertex x to the set of vertices adjacent to x^ρ. In particular, every vertex of Γ is adjacent to at least three vertices (by the thickness of the tree). Also, the last remark in (2) implies that the graph $(X, E)/\approx$ is bipartite.

(3) Suppose that there is a cycle γ in Γ of length $m < 2n$. By (1), this cycle corresponds to an m-path $(x_0, x_1, \ldots, x_m)$ with $x_0 \approx x_m$. By (2), we have $m \geq 2n$, a contradiction. Hence the girth of Γ is equal to $2n$.

(4) Let x_0 be an arbitrary vertex of (X, E). We claim that for every vertex $x \in X$, there exists a vertex $x' \in X$ such that $x \approx x'$ and $\delta(x_0, x') \leq n$. Indeed, suppose $\delta(x_0, x) > n$ and let u be the unique vertex at distance $n+1$ from x and at distance $\delta(x_0, x) - (n+1)$ from x. Let $\Sigma \in \mathcal{A}$ contain both x and u (possible by condition (Un)). The unique vertex x'' of Σ at distance $2n$ from x and at distance $n - 1$ from u satisfies $\delta(x_0, x'') < \delta(x_0, x)$ and $x \sim x''$. Induction on $\delta(x_0, x)$ completes the proof of the claim.

(5) By (4), it is clear that the diameter of Γ is less than or equal to n. Since the girth is $2n$, the diameter is exactly n.

The result follows from Lemma 1.3.6 on page 8. $\square$

Condition (Un) expresses a <u>*uniqueness*</u>; condition (Ex) an <u>*exchange*</u>.

This beautiful geometric characterization of generalized polygons by trees, together with techniques similar to those used in the proofs of Theorem 5.3.3 and Theorem 5.4.6, enabled WEISS [1995a], [1996], [1995b], [1997] to show the following result.

5.6.2 Theorem (Weiss [1995a], [1995b], [1996], [1997]). *Let (X, E) be a (G, n)-Moufang tree, for $n \geq 3$, and let G be some automorphism group of (X, E). Suppose G is generated by all n-root elations. Then (X, E) is the universal cover of the incidence graph of a Moufang generalized n-gon Γ. Also, **if** D is the group of all deck transformations, **then** the group $G/(D \cap G)$ acting on Γ is the little projective group of Γ.* $\square$

5.7   Finite Moufang polygons

5.7.1 A theorem of Fong & Seitz

Let Γ be a finite generalized polygon. If Γ satisfies the Moufang condition, then by Section 5.5, it is, up to duality, one of the following examples: a projective plane over a finite field; a symplectic quadrangle; a Hermitian quadrangle in three or four dimensions; a split Cayley hexagon; a twisted triality hexagon; a Ree–Tits octagon. The proof of this result follows from work of FONG & SEITZ [1973],

[1974] on so-called split (B, N)-pairs of rank 2. This proof is over 100 pages long and consists of purely group-theoretical arguments, apart from the very beginning, where Theorem 1.7.1 is used. A geometric approach for generalized quadrangles was started by PAYNE & THAS [1984] and finished by KANTOR [1991]. Also for generalized hexagons of order (s, t) with $s^3 \neq t$ and $t^3 \neq s$ a geometric proof is possible, using the commutation relations deduced from Theorem 5.4.6. Indeed, we have already remarked in Section 5.5.10 on page 220 that, up to duality, all points of a Moufang hexagon Γ (of order (s, t), say) are distance-2-regular. The (geometric) proof of Theorem 6.3.2 implies that Γ has an ideal split Cayley sub-hexagon Γ' of order (t, t). By Theorem 1.8.8(ii) (see page 36), this implies that either $s = t^3$ (a contradiction), or $t = s$. Hence Γ coincides with Γ' and the proof is complete. If $s = t^3$, then another geometric argument seems very likely to exist. For octagons, this is not at all so obvious.

In view of these remarks, we content ourselves with mentioning a consequence of the result of FONG & SEITZ [1973], [1974].

5.7.2 Theorem (Fong & Seitz [1973], [1974]). *Let Γ be a finite generalized polygon, let G be an automorphism group of Γ acting transitively on ordered apartments, let Σ be an apartment of Γ, let $F = \{x, L\}$ be a flag (with x a point, L a line) in Σ, let B be the pointwise stabilizer of F in G, let H be the pointwise stabilizer of Σ in G and let s_x or s_L be an element of G preserving Σ and x or L but not fixing L or x, respectively. Then Γ is isomorphic to a finite classical polygon* **if and only if** *one of the following equivalent conditions holds:*

(i) *there exists a normal nilpotent subgroup U of B such that $B = UH$;*

(ii) *there exists a normal subgroup U of B such that $B = UH$, $H \cap U = \{1\}$, $U \cap U^{s_x} \trianglelefteq U$ and $U \cap U^{s_L} \trianglelefteq U$.* $\qquad\square$

Now let Γ be a finite Moufang n-gon and let G be its little projective group. With the notation of Section 5.2 on page 174, we put $U_+ = U_1 \cdot U_2 \cdots U_n$ and $x_n = x$, $x_{n+1} = L$. From Theorem 5.2.3 on page 175 it follows easily that $U_+ \cap U_+^{s_x} = (U_1 \cdots U_n) \cap (U_0 \cdots U_{n-1}) = U_1 \cdots U_{n-1} \trianglelefteq U_+$ (indeed, U_+ fixes L; an element of $U_+^{s_x}$ fixes L **if and only if** it has no factor in U_0). Similarly $U_+ \cap U_+^{s_L} = (U_1 \cdots U_n) \cap (U_2 \cdots U_{n+1}) = U_2 \cdots U_n \trianglelefteq U_+$. Let F' be any flag opposite F. Then by Lemma 5.2.4(i), there exists a unique element of U_+ mapping F' into Σ. This implies that $B = U_+H$ and, by the uniqueness, $H \cap U_+ = \{1\}$. Hence a finite Moufang polygon satisfies condition (ii) of Theorem 5.7.2.

5.7.3 Half Moufang polygons

Let Γ be a generalized n-gon with n even. Then there are two kinds of $(n-2)$-paths (indeed, such paths can be bounded by two points, or by two lines). Correspondingly, there could be two kinds of non-trivial root elations, namely, point-elations and line-elations; If every $(n-2)$-path of one kind is a Moufang path, then we say that Γ is a **half Moufang** polygon. Of course every Moufang n-gon with n even is also a half Moufang n-gon. The converse seems to be a rather difficult problem to consider in general. For infinite polygons, the question is completely open, even for $n \geq 10$. In the finite case, however, the problem is completely solved if one uses the classification of the finite simple groups. A classification-free proof is at the moment only available for $n = 4$, using the machinery developed by PAYNE & THAS [1984]. It is not our aim to prove these results since the proofs do not reveal interesting geometric features in the cases $n = 6$ and $n = 8$. The case $n = 4$, however, would perfectly fit in *op. cit.*

5.7.4 Theorem (Thas, Payne & Van Maldeghem [1991]). *A finite half Moufang quadrangle is a Moufang quadrangle.* $\square$

5.7.5 Theorem (Buekenhout & Van Maldeghem [1994]). *Given the classification of finite simple groups, any finite half Moufang polygon is a Moufang polygon.* $\square$

Finally, we mention a little stronger version of Theorem 4.8.4 (see page 165), from the point of view of automorphism groups. The proof is certainly beyond the scope of this book!

5.7.6 Theorem (Buekenhout & Van Maldeghem [1994]). *Given the classification of finite simple groups, any finite Tits polygon with respect to the group G is a Moufang polygon with respect to G.* $\square$

5.8 Simplicity of the little projective group

In this section, we show that the little projective group of every Moufang polygon not isomorphic to $\mathsf{W}(2)$, $\mathsf{H}(2)$, $\mathsf{O}(2)$ or to the dual of one of these, is simple. We start with the main lemma. The proof we present uses the results obtained in this chapter; an alternative proof could be inspired by BOREL & TITS [1965] (2.9). Yet another approach is to look at the "rank 1 groups" generated by "opposite" root groups: usually their commutator subgroup contains all root elations. We will use this method in one case in the proof of the next lemma.

5.8.1 Lemma. *Let Γ be any Moufang polygon and let G be its little projective group. Then $G' = [G, G] = G$ except, up to duality, in the following cases, where G' has index 2 in G:*

(i) $\Gamma \cong \mathsf{W}(2)$, $G \cong \mathbf{S}_6$ and $G' \cong \mathbf{A}_6$;

(ii) $\Gamma \cong \mathsf{H}(2)$, $G \cong \mathbf{G}_2(2) \cong \mathbf{PGU}_3(3)$ and $G' \cong \mathbf{PSU}_3(3)$;

(iii) $\Gamma \cong \mathsf{O}(2)$, $G \cong {}^2\mathbf{F}_4(2)$ and $G' \cong \mathbf{T}$.

Proof. We treat the cases of projective planes, generalized quadrangles, generalized hexagons and generalized octagons separately. In all cases, we choose an apartment Σ and we number the corresponding root groups as in Section 5.5. The exceptions are well-known groups for which we do not want to go into detail here. Hence we only show the "general" part of the lemma.

Moufang planes

If $x_2 \in U_2$ is arbitrary, then $[1_1, x_3] = x_2$ according to Section 5.5.4, hence the result.

Moufang quadrangles

First we remark that, if G' contains some root group U_i, then $G' = G$. Indeed, suppose without loss of generality that G' contains U_2. By Lemma 2 of the proof of Theorem 5.4.6, there exists $x_1 \in U_1$ and $y_4 \in U_4$ such that $[x_1, y_4] \in U_2 \cdot z_3$ for any given $z_3 \in U_3$. It follows that $z_3 \in G'$ and hence by conjugation, all U_i belong to G', whence the result.

Now suppose that Γ is of type $(C)_2$, $(B)_2$ or $(C - CB)_2$. Then the indices can be chosen in such a way that $[U_1, U_3] = U_2$ (by step I of Section 5.5.5 on page 211). Hence $U_2 \subseteq G'$ and by our first remark, $G' = G$.

If Γ is a Moufang quadrangle of type $(BC)_2$, $(CB)_2$, $(B - CB)_2$ or $(BC - CB)_2$, then it contains a full or ideal subquadrangle Γ' of type $(B)_2$, $(C)_2$ or $(C - CB)_2$; see step II and step IV of Section 5.5.5. By the previous paragraph, G' contains all root elations of Γ', in particular the full root groups U_{2i}, for possibly a renumbering of the root groups. Hence the result again follows from our first remark.

Finally, let Γ be a mixed quadrangle (type $(B - C)_2$). Consider the commutation relations on page 217. We may suppose that $\mathbb{K}$ has more than two elements. Let $x \in L'$ be arbitrary but distinct from 1 and 0. Then

$$
\begin{aligned}
[1_1, 1_4] \cdot [(x+1)_1, ((x+1)^{-1})_4] &= 1_2 \cdot 1_3 \cdot (x+1)_2 \cdot 1_3 \\
&= (1 + (x+1))_2 \cdot (1+1)_3 \\
&= x_2,
\end{aligned}
$$

hence, since $1 = (x+1) + x$, we conclude that $U_2 \subseteq G'$. By our first remark again, the result follows.

Moufang hexagons

We may suppose that U_{2i} is a copy of the additive group of some field $\mathbb{K}$ and U_{2i+1} is a copy of a vector space of V over $\mathbb{K}$. If $|\mathbb{K}| = 2$, then, up to duality, we may assume $\Gamma \cong \mathsf{T}(8, 2)$. The group generated by two root groups (of point-elations) corresponding to opposite (short) roots in the root system of type G_2 is $\mathbf{PGL}_2(8)$, and the elements of order 2 of that group correspond precisely to root elations; this can easily be deduced from Proposition 4.6.9 (see page 158). Hence G' contains all point-elations, and similarly also all line-elations. Hence we may assume that $|\mathbb{K}| > 2$.

Remark that the commutation relation (6M3) implies that $U_{2i} \subseteq G'$.

Now let $r \in \mathbb{K}$ be arbitrary but fixed and distinct from 0 and 1. Let $v \in V$ be arbitrary, but distinct from 0. Put $w = \frac{1}{r^2-r}v$. With the notation of Section 5.5.13 (see page 221), we can consider for any $k \in \mathbb{K}$ the commutator

$$[v_1, k_6] \in U_2 \cdot (kh(v))_3 \cdot U_4 \cdot (kv)_5,$$

from which we deduce that the element

$$P(k, v) := (kh(v))_3 \cdot (kv)_5$$

belongs to G'. Recall that h is a polynomial form of degree 2, hence

$$P(-r, w) \cdot P(r^3, \frac{1}{r}w) = (-rh(w))_3 \cdot (-rw)_5 \cdot (rh(w))_3 \cdot (r^2 w)_5$$

$$\in U_4 \cdot ((r^2 - r)w)_5 = U_4 \cdot v_5.$$

Therefore $v_5 \in G'$ and the result follows.

Moufang octagons

We adopt the notation of Section 5.5.20, in particular the commutation relations on page 225. We immediately see that $[U_2, U_4] = U_3$ and $[U_2, U_5] = V_4$, hence $U_{2i+1} \subseteq G'$ and $V_{2i} \subseteq G'$. Furthermore, we deduce that for all $a \in \mathbb{K}$ and all $k = (k_0, k_0^{\sigma+1})$,

$$[a_1, k_8] \in (ak_0, 0)_2 \cdot V_2 \cdot U_3 \cdot V_4 \cdot U_5 \cdot (ak_0^{\sigma+1}, 0)_6 \cdot U_7.$$

By multiplying on the left and on the right with appropriate elements of V_2, U_3, V_4, U_5 and U_7 (which all belong to G'), we obtain that $Q(a, k_0) := (ak_0, 0)_2 \cdot (ak_0^{\sigma+1}, 0)_6$ belongs to G'. We may assume that $\mathbb{K}$ is not the field of two elements. Let $k \in (\mathbb{K} \setminus \{0, 1\})^\sigma$ be arbitrary and put $a^\sigma = k + 1$. One computes easily

$$Q(1, 1) \cdot Q(a^{-1}, a) = (1, 0)_2 (1, 0)_2 \cdot (1, 0)_6 (a^\sigma, 0)_6$$

$$\in V_2 \cdot (1 + a^\sigma, 0)_6 \cdot V_6 = V_2 \cdot (k, 0)_6 \cdot V_6.$$

Hence $(U_6)^\sigma \leq G'$. Next, we deduce similarly that for all $k = (k_0, 0)$ and for $l = (1, 0)$,

$$[k_2, l_8] \in U_3 \cdot U_4^\sigma \cdot U_5 \cdot k_6 \cdot U_7.$$

By multiplying on the left and on the right with appropriate elements of U_3, U_4^σ, U_5 and U_7 (which all belong to G'), we see that $k_6 \in G'$, hence $U_6 \leq G'$ and the assertion follows.

This completes the proof of the lemma. □

The next result is essentially due to TITS [1964], but we adapt it to our situation.

5.8.2 Theorem (Tits [1964]). *Let G be the little projective group of a Moufang polygon Γ. If $G = G'$, then G is a simple group.*

Proof. By Theorem 5.2.9 (see page 177), Γ is a Tits polygon with respect to G. Let $F = \{p, L\}$ be any flag of Γ and let B be the stabilizer of F in G. Let A be a proper non-trivial normal subgroup of G. Then AB is a subgroup of G containing B, hence a parabolic subgroup; see Subsection 4.7.4 on page 162. From the description of all parabolic subgroups in Theorem 4.7.5 (page 162) we infer that $AB = G_p$, $AB = G_L$, $AB = B$ or $AB = G$. In the first three cases A fixes p or L, hence it fixes all points or all lines (since G is transitive on points and lines), thus A is trivial. In the last case, we consider a group $U_+ \trianglelefteq B$ acting regularly on the set of flags opposite F (and whose existence is ensured by Lemma 5.2.4(i), page 175). It follows that $AU_+ \trianglelefteq AB = G$. Since all conjugates of U_+ generate G (after all, U_+ contains at least one root group of each type), it follows that $AU_+ = G$. But now

$$G/A = AU_+/A \cong U_+/(U_+ \cap A)$$

is soluble by Lemma 5.4.8, contradicting the fact that $(G/A)' = G'/A = G/A \neq \{1\}$. $\square$

As a corollary, we immediately obtain:

5.8.3 Corollary (Tits [1964]). *Let G be the little projective group of a Moufang polygon Γ. Then G is a simple group, except, and up to duality, in the following three cases, where $G' \trianglelefteq G$ has index 2 in G and G' is a simple group:*

(i) $\Gamma \cong \mathsf{W}(2)$, $G \cong \mathbf{S}_6$ *and* $G' \cong \mathbf{A}_6$;

(ii) $\Gamma \cong \mathsf{H}(2)$, $G \cong \mathbf{G}_2(2) \cong \mathbf{PGU}_3(3)$ *and* $G' \cong \mathbf{PSU}_3(3)$;

(iii) $\Gamma \cong \mathsf{O}(2)$, $G \cong {}^2\mathbf{F}_4(2)$ *and* $G' \cong \mathbf{T}$. $\square$

5.8.4 Corollary. *If Γ is a Moufang polygon not isomorphic to $\mathsf{W}(2)$, $\mathsf{H}(2)$ or $\mathsf{O}(2)$, then the set of point-elations (or line-elations) generates the little projective group of Γ.*

Proof. Clearly the group generated by all point-elations is a non-trivial normal subgroup of the little projective group. By the simplicity of the latter, the result follows. $\square$

If $\Gamma \cong \mathsf{W}(2)$, then Corollary 5.8.4 is still true (this follows from an easy computation in $\mathbf{S}_6$; the point-elations are the products of three disjoint transpositions, the line elations are the transpositions themselves). If $\Gamma \cong \mathsf{H}(2)$, then the line-elations generate $\mathbf{PSU}_3(3)$; the point-elations generate the whole little projective group $\mathbf{PGU}_3(3)$ (the latter is easily seen: the commutation relations imply that all line-elations are generated by the point-elations). Finally, if $\Gamma \cong \mathsf{O}(2)$, then the point-elations generate the Tits group $\mathbf{T}$ and the line-elations generate the whole little projective group, as again follows from the commutation relations.

5.9   Point-minimal and line-minimal Moufang polygons

Recall from Definition 1.8.9 (see page 37) that a **line-minimal polygon** is a generalized polygon without any proper full subpolygons, and dually, we defined a **point-minimal polygon**; a **minimal polygon** is polygon which is both point-minimal and line-minimal, i.e., it has neither proper full nor ideal subpolygons. In this section, we will identify some line-minimal, point-minimal and minimal Moufang polygons. The result for quadrangles will be used later on (see the proof of Theorem 8.5.11 in Chapter 8 on page 385).

We consider the cases $n = 3, 4, 6, 8$ separately. Of course, every projective plane is line-minimal and point-minimal by Corollary 1.8.3 (see page 34), so it suffices to look at the cases where n is even.

5.9.1 Lemma. *If Γ is a Moufang quadrangle of type $(C)_2$ or $(C - CB)_2$ (see Subsection 5.5.5), then Γ is point-minimal. If Γ is a Moufang quadrangle of type $(B)_2$, then Γ is a line-minimal quadrangle.*

Proof. We have to show that, if Γ has root groups U_i (standard notation), and Γ' is a subquadrangle containing the groups U_{2i} (this is true for every full or ideal subquadrangle by Lemma 5.2.2), with $[U_2, U_4] = U_3$ (this follows from the hypotheses and step I of Subsection 5.5.5), then all root groups U_{2i+1} are contained in Γ' (indeed, this readily implies that $\Gamma' = \Gamma$). But this is trivial.   □

5.9.2 Lemma. *Let Γ be a Moufang quadrangle of type $(CB)_2$ (see Subsection 5.5.5, step II). Then Γ is point-minimal, but not line-minimal.*

Proof. Let U_1 and U_3 be root groups of line-elations. We know by step II of Subsection 5.5.5 that the restricted Hermitian form f_0 is not identically zero. So the commutation relation $[(u, v)_1, (u', v')_3] = (f_0(u, u'))_2$ (see again Subsection 5.5.5) shows that $[U_1, U_3] = U_2$. Hence, as in the previous proof, we may conclude that Γ is point-minimal. Since Γ has proper full subquadrangles of type $(B)_2$, the lemma follows.   □

5.9.3 Lemma. *Let Γ be a Moufang quadrangle of type $(B - CB)_2$ (see again Subsection 5.5.5, step II). Let σ be the corresponding involutive anti-automorphism of the underlying skew field $\mathbb{K}$ and suppose that g is the corresponding σ-quadratic form. If the skew field generated by all elements of $\mathbb{K}$ of the form $g(x, x)$ coincides with $\mathbb{K}$, then Γ is a point-minimal quadrangle. Also, Γ is never a line-minimal quadrangle.*

Proof. We consider the notation as in Subsection 5.5.5, in particular, $f_0 \equiv 0$. We coordinatize Γ as in Subsection 3.4.7 (see page 104) and we may assume without loss of generality that a given ideal subquadrangle Γ' contains the point (1) and the hat-rack. If $k = (k_0, k_1)$, then one can check that the line $L = [k, k_1, k]$ is incident with $(1, 0, 0)$ and so belongs to Γ'. Hence the point $(0, k_1)$ belongs to Γ'. But k_1 is an arbitrary element of the form $g(x, x)$. Projecting the point $(0, a')$, with $a' = g(x, x)$, onto the line L, and then projecting the image onto $[\infty]$, we obtain that $(1 + a'k_1^{-1})$, and hence $(a'k_1^{-1})$ is a point of Γ'. Proceeding as before by substituting $(1, 0, 0)$ with any element $(a, 0, 0)$ such that (a) belongs to Γ', we see that the set of elements $x \in \mathbb{K}$ such that (x) belongs to Γ' contains the skew subfield generated by all elements of the form $g(y, y)$. If this coincides with $\mathbb{K}$, then clearly Γ is point-minimal. The last assertion follows as in the previous lemma. $\square$

5.9.4 Proposition. *Let $\Gamma = Q(d, \mathbb{K}, q)$ be an orthogonal quadrangle. Then Γ is line-minimal **if and only if** $d = 4$. Also, **if** Γ is not isomorphic to a mixed quadrangle, **then** Γ is point-minimal. In particular, every symplectic quadrangle not over a field of characteristic 2 is minimal.*

Proof. If $d > 4$, then by considering a suitable hyperplane H in $\mathbf{PG}(d, \mathbb{K})$, the intersection of H with $Q(d, \mathbb{K}, q)$ is a full subquadrangle $Q(d - 1, \mathbb{K}, q')$ of Γ. So let $d = 4$. By Proposition 3.4.13 on page 109, Γ is dual to the symplectic quadrangle $W(\mathbb{K})$. We consider a coordinatization of $W(\mathbb{K})$ as in Subsection 3.4.1 (page 100). We have to show that $W(\mathbb{K})$ is point-minimal. So we consider an ideal subquadrangle Γ'. Without loss of generality, we may assume that Γ' contains the hat-rack of the coordinatization and the additional point $(1, 0, 0)$ (Γ' is thick). The projection of $[k]$, $k \in \mathbb{K}$, onto $(1, 0, 0)$ is readily computed to be $[k, -k, k]$. Intersecting with $[0, k]$ and projecting this intersection onto $[0, 0]$, we see that Γ' contains the point $(0, -k)$, for all $k \in \mathbb{K}$. Hence Γ' coincides with Γ.

The second assertion follows immediately from Lemma 5.9.1 above, noting that every orthogonal quadrangle which is not a mixed quadrangle is of type $(C)_2$ or $(C - CB)_2$ (see Subsection 5.5.5). $\square$

5.9.5 Proposition. *Let $\Gamma = Q(\mathbb{K}, \mathbb{K}'; L, L')$ be a mixed quadrangle. Then Γ is line-minimal **if and only if** $\mathbb{K}' = L' = \mathbb{K}^2$. Dually, Γ is point-minimal **if and only if** $\mathbb{K} = L = \mathbb{K}'$. Hence Γ is a minimal quadrangle **if and only if** Γ is a symplectic quadrangle over a perfect field of characteristic 2.*

Proof. If L' strictly contains $\mathbb{K}^2$, then $Q(\mathbb{K}, \mathbb{K}^2; L, \mathbb{K}^2)$ is a proper full subquadrangle of Γ, hence Γ is not line-minimal. Now let $\mathbb{K}' = L' = \mathbb{K}^2$. Suppose that Γ' is a full subquadrangle of Γ and consider again a coordinatization of Γ such that Γ' contains the corresponding hat-rack and an additional line $[1]$. Similarly (but dually) as in the previous proof one shows that Γ' contains the line $[a^2]$, for all $a \in L$. By the same token, but substituting $[a^2]$ for $[1]$, Γ' must contain $[a^2b^2]$ for every $a, b \in L$. Combining this with the Moufang property, we see that all lines $[k]$ belong to Γ', where k belongs to the subring of $\mathbb{K}$ generated by L^2. But L generates $\mathbb{K}$ as a ring, hence L^2 generates $\mathbb{K}^2$ as a ring, and the result follows.

The second assertion is the dual of the first and follows from Proposition 3.4.4 on page 102.

The third assertion is now clear.   □

5.9.6 Proposition. *Let* $\Gamma = \mathsf{H}(3, \mathbb{K}, q, \sigma)$ *be a Hermitian quadrangle with* $\mathbb{K}$ *non-commutative. Then* Γ *is always line-minimal, but not necessarily point-minimal.*

Proof. The first assertion follows directly from Lemma 5.9.1. As for the second, it suffices to consider the quadrangle $\mathsf{Q}(7, \mathbb{R})$, which is dual to the complex quaternion Hermitian quadrangle $\mathsf{H}(3, \mathbb{H}, \mathbb{C})$ by Proposition 3.4.11 (see page 108), and then apply Proposition 5.9.4.   □

5.9.7 Example. An example of a Hermitian quadrangle $\mathsf{H}(3, \mathbb{K}, q, \sigma)$ which is point-minimal is given by the real quaternion Hermitian quadrangle $\mathsf{H}(3, \mathbb{H}, \mathbb{R})$. Indeed, using coordinates again, and supposing that (1) belongs to a subquadrangle Γ' which also contains the hat-rack, one calculates as in the previous proofs that every (a) belongs to Γ', where a can be written as $(t^\sigma - t)(u^\sigma - u)^{-1}$. Writing $\mathbb{H} = \mathbb{R} + i\mathbb{R} + j\mathbb{R} + k\mathbb{R}$ in a standard way, and letting u and t vary over $i\mathbb{R}$, $j\mathbb{R}$ and $k\mathbb{R}$, we see that, using the Moufang condition (which means that $(a + b)$ belongs to Γ' whenever (a) and (b) do), every point (a), $a \in \mathbb{H}$, belongs to Γ'. Hence Γ is point-minimal.

5.9.8 Lemma. *No Moufang quadrangle of type* $(BC - CB)_2$ *is line-mininal or point-minimal.*

Proof. This follows immediately from the fact that these quadrangles and their duals have (proper) ideal subquadrangles of type $(C - CB)_2$; see step IV of Subsection 5.5.5.   □

Finally, we consider the Moufang quadrangles of type $(BC)_2$. First, we prove a lemma.

5.9.9 Lemma. *Suppose that* Γ *is an orthogonal quadrangle over the field* $\mathbb{K}$, *and identify the group* A *of line-elations belonging to any 2-path with the vector space* $\mathbb{K}^d$ *(over* $\mathbb{K}$), *for some positive integer* $d \geq 1$. *Then the subgroup* B *of* A *induced by any full subquadrangle is some subspace (over* $\mathbb{K}$*) of* $\mathbb{K}^d$.

Proof. By the line-minimality of $\mathsf{Q}(4, \mathbb{K})$, we see that B contains a 1-space — for which we may take $\mathbb{K}$ — of A. After suitable coordinatization (identifying R_2 with A), one easily sees that, if $v_1, v_2 \in B$, then also $v_1 + k v_2 \in B$, for all $k \in \mathbb{K}$ (this follows from projecting the line $[v_2]$ onto the point $(k, v_1, 0)$ and projecting the result onto (0), which produces the line $[0, v_1 + k v_2]$). The result follows.   □

5.9.10 Proposition. *Let Γ be an exceptional Moufang quadrangle of type (E_i), $i = 6, 7, 8$. Then Γ is line-minimal, but never point-minimal.*

Proof. To show that Γ is line-minimal, we consider an exceptional Moufang quadrangle $\Gamma \cong Q(E_6, \mathbb{K}, q)$ of type (E_6). The other two cases are treated similarly. Let Γ' be a full subquadrangle of Γ. Let Σ be an apartment in Γ' and use the same notation for the root groups as in Subsection 5.5.5, step II, type $(BC)_2$, case (E_6). In particular, U_i is a group of point-elations if i is odd. Since U_1 is non-commutative, it follows immediately from Lemma 4 of the proof of Theorem 5.4.6 that $V_2 = [U_1, U_3]$ is non-trivial. Let $u_2 = [u_1, u_3]$ be a non-trivial element of V_2. By Subsection 5.5.5, we may identify the elements of U_2 with the points of a vector space $\mathbb{K}^6$. Let Γ^* be the ideal orthogonal subquadrangle of type (D_5). The quadrangle $\Gamma' \cap \Gamma^*$ is a full subquadrangle of Γ^* and hence the subgroup of U_2 induced in $\Gamma' \cap \Gamma^*$ (and thus also in Γ') is a subspace B of $\mathbb{K}^6$, which we may denote after reselecting the basis by $\mathbb{K}^\ell$, $1 \leq \ell \leq 6$. Note that $u_2 \in \mathbb{K}^\ell$, hence $u_2\mathbb{K} \in B$. Now let $\mathbf{SO}_6(\mathbb{K})$ act on U_i, $i \in \mathbb{Z}$, as in Subsection 5.5.5 and let $\theta \in \mathbf{SO}_6(\mathbb{K})$. Then $[u_1^\theta, u_3^\theta] = u_2^\theta$ belongs to B, hence the orbit of $u_2\mathbb{K}$ under $\mathbf{SO}_6(\mathbb{K})$ is contained in B. But that orbit spans $\mathbb{K}^6$ (whether or not u_2 is an isotropic vector with respect to $\mathbf{SO}_6(\mathbb{K})$), hence $B = \mathbb{K}^6$. This shows that $\Gamma' = \Gamma$. By altering the indices, we obtain the same result for $Q(E_7, \mathbb{K}', q')$ and $Q(E_8, \mathbb{K}'', q'')$.

The second assertion follows directly from the existence of orthogonal ideal subquadrangles of type (D_ℓ), $\ell = 5, 6, 8$, respectively, in Γ. $\qquad\square$

We now turn to the Moufang hexagons, where the situation is much simpler.

5.9.11 Proposition. *Let Γ be a Moufang hexagon which is not a mixed hexagon. To fix the duality class, assume that Γ has regular points. Then Γ is a line-minimal hexagon. Also, Γ is a point-minimal hexagon* **if and only if** *Γ is a split Cayley hexagon.*

Proof. By Corollary 5.5.15, Γ does not contain regular lines, hence by Subsection 5.5.10, we may assume that, with the usual notation, the commutator $[U_1, U_3]$ is non-trivial, and that U_2 centralizes U_0 and U_4. Let $u_2 = [u_1, u_3]$ be a non-trivial element of $[U_1, U_3]$ and let u be arbitrary but not trivial in U_2. By Lemma 13 of the proof of Theorem 5.4.6, there exists an automorphism θ of Γ such that $u_2^\theta = u$. Hence $[u_1^\theta, u_3^\theta] = u$. So if Γ' is a full subhexagon of Γ, then it contains U_1 and U_3, and hence U_2. Consequently $\Gamma = \Gamma'$. Now by Proposition 5.5.12, Γ contains an ideal subhexagon Γ' isomorphic to a split Cayley hexagon. So if Γ is not itself a split Cayley hexagon, then Γ' is a proper ideal subhexagon. This shows that, if Γ is point-minimal, then Γ is a split Cayley hexagon. Conversely, suppose that Γ is a split Cayley hexagon $\mathsf{H}(\mathbb{K})$. Let Γ' be an ideal subhexagon, and coordinatize Γ such that Γ' contains the hat-rack and the additional point $(1, 0, 0, 0, 0)$. Projecting this point onto $[k]$, $k \in \mathbb{K}$, we obtain $(k, -k)$ in Γ'. Projecting this point onto $[0, 0, 0, 0]$, we see that $(0, 0, 0, 0, -k)$ belongs to Γ', for every $k \in \mathbb{K}$, hence Γ' is full and consequently coincides with Γ. The proposition is proved. $\qquad\square$

5.9.12 Proposition. *Let* $\Gamma = \mathsf{H}(\mathbb{K}, \mathbb{K}')$ *be a mixed hexagon. Then* Γ *is point-minimal if and only if* $\mathbb{K} = \mathbb{K}'$, *i.e.,* Γ *is a split Cayley hexagon. Dually,* Γ *is line-minimal if and only if* $\mathbb{K}' = \mathbb{K}^3$, *i.e.,* Γ *is a dual split Cayley hexagon. In particular, if a mixed hexagon is not isomorphic or dual to a split Cayley hexagon, then it is neither a line-minimal nor a point-minimal hexagon. Also, a mixed hexagon is minimal if and only if it is a split Cayley hexagon over a perfect field (of characteristic 3).*

Proof. If Γ is a split Cayley hexagon, then as in the last part of the proof of Proposition 5.9.11 above, Γ is point-minimal. If Γ is not a split Cayley hexagon, i.e., if $\mathbb{K}' \neq \mathbb{K}$, then we have the ideal subhexagon $\mathsf{H}(\mathbb{K}', \mathbb{K}')$. This proves the first assertion. The other assertions now follow easily.   □

Finally, we consider Ree–Tits octagons. The next result is also contained in Joswig & Van Maldeghem [1995].

5.9.13 Proposition (Joswig & Van Maldeghem [1995]). *Every Ree–Tits octagon* Γ *is a minimal polygon.*

Proof. As before, Γ is point-minimal in view of the commutation relation $[k_2, l_4] = (k_0 l_0)_3$ of Subsection 5.5.20. Now suppose that Γ' is a full suboctagon of Γ. Using the notation of Subsection 5.5.20 again, we may suppose that $(k_0, k_1)_6 \in U_6$ belongs to $\mathrm{Aut}(\Gamma')$, for some $(k_0, k_1) \in \mathbb{K}_\sigma^{(2)}$. If $k_0 \neq 0$, then $[a_3, k_6] = (0, ak_0)_4$, for all $a \in \mathbb{K}$, implies that, with obvious notation, $V_{2i} = (0, \mathbb{K})_{2i} \leq U_{2i}^\dagger$, where $U_{2i}^\dagger$ is the group induced in Γ' by U_{2i} (and we will use similar notation, also for products of groups, from now on), $i \in \mathbb{Z}$. Suppose now $k_0 = 0$. Then $[a_1, k_6] = (0, ak_1)$, implying again $V_{2i} \leq U_{2i}^\dagger$. Now the commutation relation for $[a_1, (0, k_1)]$ tells us that $b = (a^\sigma k_1, 0)_4 (ak_1, 0)_6 \in (U_4 U_6)^\dagger$. Similarly computing $[1_1, (0, ak_1)]$, we see that $b' = (ak_1, 0)_4 (ak_1, 0)_6 \in (U_4 U_6)^\dagger$. Computing bb'^{-1}, it is now clear that $((a^\sigma + a)k_1, 0)_4 \in U_4^\dagger$, for all $k_1 \in \mathbb{K}$. Since there exists at least one $a \in \mathbb{K}$ such that $a \neq a^\sigma$, we conclude with the Moufang condition (which implies that $U_{2i}^\dagger$ is closed under addition) that $U_{2i}^\dagger = U_{2i}$, hence $\Gamma = \Gamma$.

This completes the proof of the proposition.   □

5.9.14 Remark. Let Σ be an apartment of a Moufang n-gon Γ. We say that Γ **is generated by the points on** Σ if the smallest full weak sub-n-gon Γ' of Γ containing all sets $\Gamma_1(L)$, with L a line of Σ, coincides with Γ. Noting that for $n = 4, 6$, this is equivalent to saying that Γ is line-minimal and Γ does not have regular lines (still assuming that Γ is a Moufang polygon!), the results of this section permit a complete classification of Moufang polygons generated by the points of an apartment Σ. Indeed, a little exercise shows that these are the Moufang planes, the dual orthogonal quadrangles which are not mixed quadrangles, the Hermitian quadrangles arising from a σ-quadratic form in a vector space of dimension 4, the exceptional Moufang quadrangles of type (E_i), $i = 6, 7, 8$, the Moufang hexagons with regular points which are not mixed hexagons (or equivalently, which do not have regular lines), and finally the dual Ree–Tits octagons.

Chapter 6

Characterizations

6.1 Introduction

The main question of this chapter is: "how do we recognize the Moufang polygons?". Of course, checking the Moufang condition is one answer, but in a lot of cases, other criteria are needed. From a geometric point of view for instance, one would like to identify Moufang polygons by certain geometric or, in the finite case, combinatorial properties. This also means that, in the case of generalized n-gons, $n \notin \{3, 4, 6, 8\}$, we would like to have geometric properties eliminating these geometries, analogous to the result of TITS [1976b], [1979] and WEISS [1979] (see Theorem 5.3.4 on page 185).

One geometric tool that is available in a generalized n-gon for every $n > 2$ is the notion of regularity. As we shall see, the presence of a certain number of regular points in a certain position will force n to be 3, 4 or 6; see for instance Subsection 1.9.7 on page 40 for the case $n = 5$. And in fact, the only regular generalized n-gons for $n > 4$ are the Moufang hexagons. If $n = 4$, then we need an extra assumption on derivation, for instance assuming that at least one point is a projective point will nail down the generalized quadrangle to a symplectic one. In the finite case, we can weaken the hypotheses a little and require span-regularity of points on an ovoid (in the cases $n = 4$ and $n = 6$). Related to these results are also some characterizations by means of *long hyperbolic lines* and *long imaginary lines*.

Another geometric property is the presence of a lot of "large" subpolygons, such as full ones. We will give results in that direction, but restricted to the finite case and only for $n = 4$ and $n = 6$.

The Moufang condition in the case of projective planes translates into the presence of a lot of little Desargues configurations. A generalization to all $n \geq 3$ exists and we shall explain it below.

For the finite Moufang polygons, there are a few of additional group-theoretical characterizations that we prove in this chapter. One of these is a characterization by generalized homologies, another is a characterization assuming a collineation group acting transitively on ordered ordinary $(n+1)$-gons, and a third one replaces the Moufang condition by the k-Moufang condition. We shall provide the original geometric proof for $n = 8$ of the latter result for $k = 3$ (by VAN MALDEGHEM (unpublished)).

Finally, we review (without their long proofs) some other characterizations, mainly group-theoretical ones, but also a geometric one for the Ree–Tits octagons.

6.2   Regularity in generalized quadrangles

There is no classification available in the literature for regular generalized quadrangles, not even in the finite case. However, if every point of a generalized quadrangle Γ is projective (see Definition 1.9.4) — and in the finite case this happens whenever every point is regular and $s = t$ for the order (s, t) of Γ — then one shows that Γ must be isomorphic to a symplectic generalized quadrangle. In the finite case, several authors have proved this; see PAYNE & THAS [1984], page 77 for more details. In the infinite case, a proof appeared in SCHROTH [1992]. We will actually prove a slightly more general result.

6.2.1 Theorem. *A generalized quadrangle Γ is isomorphic to a symplectic generalized quadrangle* $\mathsf{W}(\mathbb{K})$ *for some commutative field* $\mathbb{K}$ **if and only if** *all points of Γ are regular and at least one point is projective.*

Proof. If $\Gamma \cong \mathsf{W}(\mathbb{K})$, then all points are projective by Theorem 2.3.19 (see page 65).

Conversely, suppose now that every point is regular and at least one point is projective. We first show that every point of Γ is projective. So let p be a projective point of Γ and let x be any other point. Let y be a point opposite both x and p. It is enough to show that any two traces in $y^{\perp}$ meet in at least one point. Let X be a trace in $y^{\perp}$ and suppose $x_1, x_2 \in X$, $x_1 \neq x_2$. The point x_i, $i \in \{1, 2\}$, defines the line $\{p, x_i\}^{\perp}$ of Γ_p; notice that this is the trace p^{x_i} if x_i is not collinear with p, and the line px_i if x_i is collinear with p. The lines $\{p, x_1\}^{\perp}$ and $\{p, x_2\}^{\perp}$ meet in a unique point u of Γ_p. Since u is collinear with at least two points of X, namely x_1 and x_2, we have $X = y^u$. If X' is a second trace in $y^{\perp}$, then similarly there exists a point u' of Γ_p such that $X' = y^{u'}$. Now note that $u, u' \notin p^y$ (which justifies the notation y^u and $y^{u'}$). So the line L of Γ_p determined by u and u' meets p^y in a unique point w. There are unique points v and v' on the line wy and the trace X or X'. The point v is collinear with both u and w, so $p^v = L$ (if v is not collinear with p) or $pv = L$ (if v is collinear with p). Similarly either $p^{v'} = L$ or $pv' = L$. Consequently $v = v'$ and X and X' meet in v. Hence y is projective, and by a similar argument, x is projective.

Now we show that Γ is isomorphic to $\mathsf{W}(\mathbb{K})$ for some commutative field $\mathbb{K}$. We use a little bit of the theory of buildings here, or rather, of diagram geometry; see TITS [1974] and BUEKENHOUT [1979]. We define the following rank 3 geometry $\Omega(\Gamma)$. Its elements are points, lines and planes. The points of $\Omega(\Gamma)$ are the points of Γ; the lines of $\Omega(\Gamma)$ are all lines and all traces of Γ, and the planes are the sets $x^\perp$ with x a point of Γ. Incidence is inclusion. Clearly, $\Omega(\Gamma)$ is a thick chamber geometry; see Subsection 1.3.7 (page 8). In order to prove that $\Omega(\Gamma)$ is a projective space, it suffices to show that the geometry of elements incident with a given element e is a projective plane if x is a point or a plane, and a generalized 2-gon if x is a line (see TITS [1981]). The latter is trivially true. So let e be a plane, say $e = x^\perp$. The points and lines of $\Omega(\Gamma)$ incident with e form the projective plane Γ_x. Now let e be a point. The lines and planes of $\Omega(\Gamma)$ incident with e form the span-geometry in e, which is also a projective plane; see Subsection 1.9.6 on page 40.

So we have shown that $\Omega(\Gamma)$ is a projective space of dimension 3. The map $\rho :$ $\Omega(\Gamma) \to \Omega(\Gamma)^D$ mapping a point x to $x^\perp$, a line x^y to the line $(x^y)^\perp$ and a plane $x^\perp$ to the point x is clearly a polarity of symplectic type, hence the theorem. $\qquad\square$

One can also use the characterization of projective spaces due to VEBLEN & YOUNG [1910] in the previous proof.

6.3 Regularity in generalized hexagons

RONAN [1980a] classified all regular generalized hexagons. We will break up his result into two theorems, one similar to Theorem 6.2.1 of the previous section and another one for the general situation. The proof we give roughly follows Ronan's original proof, with some minor differences, and up to the case $t = 2$.

The proof of the following theorem is a little different from that which one could derive from *op. cit.* We avoid the technicalities about embeddable polar spaces of rank 3. Instead, we use some results of Chapter 5 on the classification of Moufang hexagons.

6.3.1 Theorem (Ronan [1980a]). *A generalized hexagon Γ is isomorphic to a split Cayley hexagon $\mathsf{H}(\mathbb{K})$ for some commutative field $\mathbb{K}$* **if and only if** *all points of Γ are distance-2-regular and at least one point is projective.*

Proof. If $\Gamma \cong \mathsf{H}(\mathbb{K})$ for some field $\mathbb{K}$, then all points are projective (they are all polar points) by Theorem 2.4.16 (see page 75).

So we may assume that Γ is a generalized hexagon in which all points are regular and at least one point is projective. Essentially the same proof as above (Theorem 6.2.1) is used to show that every point of Γ is projective. Hence by Proposition 1.9.13 on page 44, the derived geometries $\Gamma_p^\triangle$ and $\Gamma_p^\square$ are projective planes and generalized quadrangles, respectively, as p ranges over all points of Γ.

Our first aim is to construct the underlying polar space (the quadric $\mathbf{Q}(6, \mathbb{K})$); see Subsection 2.4.13 on page 73. So consider the following geometry Δ: the points of

Δ are the points of Γ, the lines of Δ are all lines and all distance-2-traces of Γ and the planes of Δ are all sets $x^\perp$ where x ranges over the set of points of Γ, and all sets $\Gamma^+(p,q)$, where p and q are opposite points of Γ. One easily verifies that Δ is a polar space of rank 3. Indeed, apart from obvious non-degeneracy conditions, we must, according to BUEKENHOUT & SHULT [1974] (see also Definitions 2.4.1 on page 66), show that

(1) for every point p and every line L of Δ, either all points on L are collinear with p (in Δ), or exactly one point of L is collinear with p;

(2) both possibilities in (1) effectively occur. Also, if a point p is collinear (in Δ) with all points of a plane π, then p is incident with π.

Condition (1) is the Buekenhout–Shult one-or-all axiom already mentioned in Chapter 2, see page 66. Condition (2) expresses that the rank is exactly equal to 3.

Condition (1) and the first part of condition (2) are readily verified noting that two points p and q are collinear in Δ **if and only if** $\delta(p,q) \leq 4$ in Γ.

We now prove the second part of (2). If p is collinear in Δ with all points of a plane $x^\perp$, then $\delta(x,p) = 4$ is impossible since any point $y \perp x$, y not incident with $\mathrm{proj}_x p$, is opposite p in Γ, a contradiction. Now suppose that p is collinear in Δ with every point of $\Gamma^+(x,y)$, for two opposite points x,y in Γ. Clearly, we only have to consider the cases $\delta(x,p) = 2$ and $\delta(x,p) = 4$. Suppose first $\delta(x,p) = 2$. Then it is easily seen that $p \notin x^y$ and that $\delta(z,p) = 6$ for every $z \in y^x \setminus y^p$. Suppose now $\delta(x,p) = 4$. Put $u := x \bowtie p$. If u is opposite y, then p is opposite v, where $\{v\} = y^x \cap y^u$, a contradiction as $v \in \Gamma^+(x,y)$. Hence $u \in x^y$. We may assume that $p \neq y \bowtie u$. Now it is clear that p is at distance 4 from each element of y^x, hence by definition of $\Gamma(x,y)$ (see page 41), the point p belongs to $\Gamma^+(x,y)$.

So Δ is a rank 3 polar space.

Now assume that Γ' is a second generalized hexagon with the following properties:

(i) The set of points of Γ' is exactly the set of points of Δ.

(ii) The set of lines of Γ' is a subset of the set of lines of Δ.

(iii) The set of lines through a point in Γ' is the set of lines through that same point in some fixed plane in Δ.

(iv) Two points are non-collinear in Δ **if and only if** they are opposite in Γ'.

Note that $\Gamma = (\mathcal{P}, \mathcal{L}, I)$ has all of these properties by construction of Δ. Assume that Γ' and Γ share at least an apartment Σ and all lines concurrent with one of three consecutively concurrent lines L_1, L_2, L_3 of that apartment. We claim that necessarily $\Gamma = \Gamma'$. Remark first that, if x and y are opposite points in

Γ', then the set x^y is precisely the set of points collinear with x in Γ' and collinear with y in Δ. Now let Σ consist of the points and lines of the circuit $(L_1, p_1, L_2, p_2, \ldots, L_6, p_6, L_1)$. Let x be a point incident with L_4, $x \neq p_3$. On L_1, there is a unique point y collinear with x in Δ (otherwise all points of L_1 are collinear with x, contradicting (iv)). All points collinear with y, except for the points on a particular line M through y, have distance 6 from x, but only one point x' on M is collinear in Δ with p_3. It follows that in both Γ and Γ' the points x and x' are collinear, so all lines through x in Γ' lie in the plane determined by L_4 and xx', hence they are the same as in Γ. A similar argument works for all points on one of the lines L_4, L_5, L_6. Of course, also the point x' above is incident with the same lines in Γ as in Γ'. Now let u be a point at distance 3 from one of the lines of Σ, say L_1. There is a unique point w on L_3 collinear with u in Δ, so w is at distance 4 from u in both Γ and Γ'. As above, this determines uniquely the line uu' with u' collinear with w and so uu' is a line of both Γ and Γ'. Hence $\Gamma_1(u) = \Gamma_1'(u)$. If finally v is a point at distance 5 from all lines of Σ, then v is collinear with at least two points v_1 and v_2 at distance 3 from some lines of Σ, hence vv_1 and vv_2 are lines of both Γ and Γ' and so again $\Gamma_1(v) = \Gamma_1'(v)$. This shows our claim.

Since Δ is a polar space of rank 3, it has the Moufang property; see TITS [1974], page 274. There are two kinds of root elations in Δ. The roots of the first kind are bounded by a sequence $(p_1, \pi_1, L_1, \pi_2, , p_2, \pi_3, L_2, \pi_4, p_1)$, where p_1, p_2 are (opposite) points, L_1, L_2 lines and $\pi_1, \pi_2, \pi_3, \pi_4$ are planes in Δ. If we take for L_1 and L_2 respectively the traces $p_1^{p_2}$ and $p_2^{p_1}$ of Γ, for π_1 and π_3 the planes $p_1^{\perp}$ and $p_2^{\perp}$, and for the planes π_2 and π_4 respectively $\Gamma^+(p_2, p_1)$ and $\Gamma^+(p_1, p_2)$, then a root elation with respect to any root with that boundary is a line-elation in Γ, since it maps Γ onto a generalized hexagon Γ' having the properties (i), (ii), (iii) and (iv) and satisfying the condition of our claim above. A similar reasoning (composing a root of the other type in Δ with a root of the first type) leads to the existence of all point-elations in Γ. Hence Γ is a Moufang hexagon. But the only Moufang hexagons with projective points are the split Cayley hexagons $\mathsf{H}(\mathbb{K})$ with $\mathbb{K}$ a field; see Corollary 5.5.14 on page 223. This completes the proof of the theorem. $\quad\square$

6.3.2 Theorem (Ronan [1980a]). *If Γ is a point-distance-2-regular generalized hexagon, then all points and lines are distance-3-regular and Γ is a Moufang hexagon. Conversely, up to duality, all points of any Moufang hexagon are regular.*

Proof. Suppose first that Γ is a point-distance-2-regular generalized hexagon of order (s, t) (s and/or t possibly infinite). Note that by Proposition 1.9.17 on page 46 all points and lines of Γ are distance-3-regular. Recall also from Subsection 1.9.16 (see page 46) that for two opposite lines L and M, we denote by $R(L, M)$ the set of all lines at distance 3 from all points at distance 3 from both L and M and we have called it the *line regulus* defined by L and M. Dually, we have defined *point reguli*.

Complementary reguli are reguli $R(L, M)$ and $R(p, q)$ such that each element of $R(L, M)$ lies at distance 3 from each element of the other regulus $R(p, q)$.

Since every point p is distance-2-regular, there is, according to Subsection 1.9.8, a unique ideal weak subhexagon $\Gamma(p, q)$ with two points per line containing two arbitrary opposite points p and q. Its point set consists of the union of the point sets of the two projective planes $\Gamma^+(p, q)$ and $\Gamma^-(p, q)$. Recall that $\Gamma^+(p, q)$ consists of all points at distance 0 or 4 from p; the lines of $\Gamma^+(p, q)$ are the traces x^y, where x is a point of $\Gamma^-(p, q)$, and y is a point of $\Gamma^+(p, q)$ opposite x. Similarly for $\Gamma^-(p, q) = \Gamma^+(q, p)$.

From now on, we fix the opposite points p and q and the corresponding ideal weak subhexagon $\Gamma(p, q)$.

 Centric planes

Let z be a point of the regulus $R(p, q)$, $p \neq z \neq q$. Define the following geometry $C(z)$. The points of $C(z)$ are the points collinear with z at distance 4 from some point of $\Gamma^+(p, q)$ not on q^p, together with the point z itself; the lines of $C(z)$ are all lines through z together with the distance-2-traces joining two points of $C(z)$ at distance 4 in Γ.

Every point x of $\Gamma^+(p, q)$ defines a unique line $L(x)$ in $C(z)$. Indeed, the points which are opposite z define a trace; those which are not opposite z are collinear with q and are at distance 3 from a unique line through z.

Now let u^v be a line of $\Gamma^+(p, q)$, with u a point of $\Gamma^-(p, q)$ and v some point of $\Gamma^+(p, q)$. If $u = q$, then all lines $L(x)$ with $x \in q^v$ are incident with z, and so q^v defines the point z of $C(z)$. If u is collinear with p, then all lines $L(x)$, with $x \in u^v$, pass through the unique point $y = u \bowtie z$ (u and z lie indeed at distance 4 from each other by the distance-3-regularity). Hence the line u^v defines the unique point y of $C(z)$. Suppose now that $u \neq q$ and u is not collinear with p. Then u is collinear with exactly one point a collinear with q. This point a lies on a unique line M (of Γ) of the regulus R' complementary to $R(p, q)$. Let M' be the unique line (in Γ) through z meeting M. Let x be a point on M' at distance 4 from at least one element b of u^v different from a. Then $\{a, b\} \subseteq u^v \cap u^x$, so $u^v = u^x$ and hence x is at distance 4 from all points of u^v. Hence u^v defines a unique point x of $C(z)$.

Conversely, if x is a point of $C(z)$, then, if $x = z$, it is defined by q^p; if it lies on some line M of R', then it is defined by the unique point of M collinear with p (in Γ); and in the other cases, the line zx meets a unique line M of R' and x is defined by the unique line of $\Gamma^+(p, q)$ containing the points a and b, where a is the point of $\Gamma^+(p, q)$ on M, and b is some other point of $\Gamma^+(p, q)$ at distance 4 from x (b exists by definition of x being a point of $C(z)$).

Suppose now that X is a line of $C(z)$. If X is a line in Γ, then $X = L(x)$, where x is the unique point collinear with q and at distance 3 from X (in Γ). If $X = z^p$, then $X = L(p)$. So suppose X is a distance-2-trace in $z^\perp$, different from z^p. Then there are two different points a, b of X which do not lie on any element of R'. These points are defined, as above, by two different lines A and B of $\Gamma^+(p, q)$. Every point of A and B lies at distance 4 from a and b, respectively. So the intersection point x in $\Gamma^+(p, q)$ of A and B lies at distance 4 from both a and b. Clearly x is not collinear with q, so since X is a distance-2-trace, it must be equal to z^x. This shows that every line of $C(z)$ not through z is obtained by a trace z^x with x a point of $\Gamma^+(p, q)$.

Hence $C(z)$ is a projective plane. Indeed, the correspondences above preserve the incidence and are bijective. We call $C(z)$ a **centric plane centred at** z. By varying q over $R(p, q)$, we can construct a centric plane $C'(z)$ centred at z containing any point x collinear with z. And by varying p and q, we can have a centric plane centred at z containing any prescribed distance-2-trace in $z^\perp$ and any point collinear with z. Note that, if two centric planes centred at z share at least one distance-2-trace and one point off that trace and different from z, then they coincide (since the intersection is also a projective plane necessarily equal to both original centric planes).

From the preceding paragraphs it also follows that the centric plane $C'(z)$ centred at z obtained from $\Gamma^-(p, q)$ coincides with $C(z)$ obtained from $\Gamma^+(p, q)$. Indeed, they have $z^p = z^q$ in common, and if x is any point of $C(z)$ not on z^p and different from z, then there is a point u of $\Gamma^+(p, q)$ at distance 4 from x and not collinear with q. The line $\mathrm{proj}_u(x)$ contains a unique point v of $\Gamma^-(p, q)$ and v lies at distance 4 from x, since $\delta(v, x) = 2$ would imply $z \in \Gamma(p, q)$. Hence we can say that $C(z)$ is obtained from $\Gamma(p, q)$.

| Centric planes attached to $\Gamma(p, q)$ |

Let $C(p)$ be any centric plane centred at p and containing p^q. There is a natural way to define a centric plane in every point x of $\Gamma(p, q)$ opposite p. Indeed, every point of $C(p)$ opposite x defines a unique distance-2-trace in $x^\perp$ and every point at distance 4 from x defines a unique closest line of Γ through x. Also, every line U of $C(p)$ which is a distance-2-trace can be written as p^u, where u is a point collinear with x. Indeed, let v be the intersection point of U and p^x in $C(p)$ and put $L = \mathrm{proj}_x(v)$. Let v' be any point of U, $v' \neq v$ and let u be the projection of v' onto L. Then clearly p^u contains v and v' and therefore $U = p^u$. By this correspondence, we obtain a centric plane $C(x)$ centred at x, dual to $C(p)$. We say that $C(x)$ **is obtained by** $C(p)$ **via projection**. So we can define a centric plane centred in any element x of $\Gamma(p, q)$ opposite p. Of course, we can continue in this way and obtain a centric plane centred at any element opposite any element opposite p, etc. The question arises if we obtain the same centric plane by projections via different elements. In fact, if $t > 2$, it is not so hard to see that the answer is positive if we can show that, whenever q' and p' are points of $\Gamma(p, q)$ such that q' is opposite both p and p', and such that the point $x = p \bowtie p'$ is not collinear with $y = q \bowtie q'$, then the centric plane $C(p')$ obtained via projection from $C(q)$ which in turn is obtained via projection from $C(p)$, coincides with the the centric plane $C'(p')$ obtained via projection from $C(q')$ which in turn is obtained via projection from $C(p)$. If $t = 2$, then this situation never occurs; on the contrary, x is always collinear with y. But in this case a similar argument to the one below can be given.

First note that all centric planes centred at a point $a \in \Gamma(p,q)$ contain the trace a^b, where b is some point in $\Gamma(p,q)$ opposite a. So it suffices to show that $C(p')$ and $C'(p')$ share a point different from p' and not lying in $\Gamma(p,q)$.

Note that p, p' and y are points of $\Gamma^+(p,q)$ and q, q', x are points of $\Gamma^-(p,q)$. So $a = p \bowtie y$, $b = q \bowtie x$, $a' = p' \bowtie y$ and $b' = q' \bowtie x$ are well defined. Now the regulus $R(ap, a'p')$ contains the lines bq and $b'q'$. The complementary regulus is $R(x,y)$. Now take any point z in $R(x,y)$, $x \neq z \neq y$. This point z is collinear with some points u, u', v, v' on, respectively, $ap, a'p', bq, b'q'$. Of course, we can choose z in such a way that u belongs to $C(p)$. But now $\delta(u,v) = 4$ implies $v \in C(q)$, $\delta(v, u') = 4$ implies $u' \in C(p')$; $\delta(u, v') = 4$ implies $v' \in C(q')$ and $\delta(v', u') = 4$ implies $u' \in C'(p')$, hence the result.

The H($\mathbb{K}$) subhexagon

We fix an arbitrary centric plane $C(p)$ centred at p and containing p^q. By the previous paragraphs, there are unique centric planes $C(x)$ centred at any point x of $\Gamma(p,q)$ and obtained by projecting $C(p)$ (possibly more than once). We define the following subgeometry Δ of Γ. It has three types of points. Type I consists of the points of $\Gamma(p,q)$. Type II consists of all points not of type I but lying in some centric plane $C(x)$, with x a point of $\Gamma(p,q)$. Type III are all points z contained in some regulus $R(x,y)$, where x, y are points of $\Gamma(p,q)$, with the additional restriction that z is collinear with a point of type II. The lines of Δ are the lines of Γ containing at least two points of Δ. We will show that Δ is a subhexagon.

First we claim that all points of Δ that are collinear with a point z of type III are themselves of type III except for one distance-2-trace z^x, where x is a point of type I and $z \in R(x,y)$ for some second point y of type I (and y opposite x). The points x and y are readily found: by definition, $z \in R(x,y)$ for some points x, y of type I and z is collinear with at least one point u of type II. Note that the only lines of $\Gamma(p,q)$ at distance 3 from z are the lines of the regulus R^* complementary to $R(x,y)$. This implies that u belongs to one of the elements of R^*. We may suppose that u belongs to $C(x \bowtie u)$. By projecting $C(x \bowtie u)$ to any point v of $\Gamma(p,q)$ collinear with y except $y \bowtie u$, we see that all points collinear with z on elements of R^* belong to at least one centric plane $C(v)$. Clearly z is not collinear with any point of type I. Hence the claim. Now let, with the same notation, v_1 and v_2 be two different specific choices for v, and suppose u_1 and u_2 are points collinear with z and with v_1 and v_2, respectively. Then by projecting $C(v_1)$ onto $C(v_2 \bowtie x)$, and projecting $C(v_2 \bowtie x)$ onto $C(y \bowtie u)$, we see that u belongs to $C(y \bowtie u)$. Hence the set of points of Δ collinear with a point a of $\Gamma(p,q)$ is exactly the set of points of the centric plane $C(a)$ (up to a itself of course). We refer to that last result by (†).

Now we claim that the points of Δ collinear with a point z of type III are exactly the points belonging to the unique centric plane $C(z)$ centred at z and defined by $\Gamma(p, q)$. If u is a point of Δ collinear with z, then clearly it belongs to $C(z)$, because if it is of type II, then this follows from the preceding paragraph, and if it is of type III, then it lies by definition at distance 4 from at least one point of $\Gamma(p, q)$, hence the result. So suppose now conversely that u is a point of $C(z)$, not of type II (if it has a type at all). Then u is at distance 4 from at least two points a, b of $\Gamma^+(p, q)$ neither collinear with x nor collinear with y (where $z \in R(x, y)$ for points x, y of $\Gamma(p, q)$). The projections of u onto a or b are lines which contain one further point a' or b' respectively of $\Gamma(p, q)$ (since the latter is an ideal subhexagon). Hence u belongs to the regulus complementary to $R(aa', bb')$. If a'' is the projection of u onto aa', then $\delta(a'', c) = 4$, where c is incident with zu and lies on an element of the regulus R^* complementary to $R(x, y)$. This shows that a'' is a point of Δ of type II (since c is) and the claim follows. From this argument it follows also that, if a is any point of $\Gamma(p, q)$ and L any line containing a point z of type III, then the points a'' and u, respectively, at distance 2 and 4 from a and at distance 3 and 1 from L belong to Δ (the general case u of type III is the preceding argument, the special case that u is of type II is trivial, and clearly u cannot be of type I). We refer below to this result as (*).

There are two types of lines in Δ: lines of type α are lines of $\Gamma(p, q)$, lines of type β are lines which do not contain any point of $\Gamma(p, q)$ and so they contain exactly one point of type II (containing two points of type II leads to a pentagon in Γ) and all other points are of type III.

Suppose z is a point and L a line in Δ with $\delta(z, L) = 5$ in Γ. Let a and b be the unique points of Γ at distance 2 and 4 from z and distance 3 and 1 from L, respectively. If we show that a and b are points of Δ, then Δ is an ideal subhexagon.

(*i*) If z has type I, then the result follows from (*) if L has type β. For type α, the result follows from $\Gamma(p, q)$ being a subhexagon.

(*ii*) If z has type II and L type α, then let u be a point of $\Gamma(p, q)$ collinear with z (then z lies in $C(u)$). We may assume that neither a nor b belongs to $\Gamma(p, q)$ (otherwise there is nothing to prove). We claim that $\delta(u, L) = 5$. Indeed, we certainly have $\delta(u, L) \geq 3$ and if $\delta(u, L) = 3$, then $u = a$ and $b = \mathrm{proj}_L(a)$ both belong to $\Gamma(p, q)$. The claim follows. Hence there is a point x on L lying in $\Gamma(p, q)$ opposite u. By projecting $C(u)$ to $C(x)$, the point z is mapped onto a distance-2-trace containing the projection b of z onto L. Hence b belongs to Δ (type II) and $a = z \bowtie b$ is by definition of type III.

(*iii*) Suppose now that z has type II and L has type β. We first claim that, if r is a point of type III and $r \in R(x, y)$ with x, y points of $\Gamma(p, q)$, then the centric plane $C(r)$ defined by $\Gamma(p, q)$ is the same as the centric plane $C'(r)$ projected from $C(x)$. Indeed, it already shares the trace r^x with $C(r)$. Let M be any line through r and let w be in $\Gamma(p, q)$ collinear with x and at distance 5 from M. Let N be any line through w different from wx and different from the line joining w and $w \bowtie r$. The line N contains one further point c of $\Gamma(p, q)$ and c is at distance 4 from a unique point v on M. Put $c' = c \bowtie v$ and let e be the unique point on cc' belonging to $\Gamma(p, q)$ and different from c. Finally, let v' be the unique point on wx at distance 4 from v. Since v is at distance 4 from c, it belongs to $C(r)$. By the distance-3-regularity, there is a point u collinear with e and at distance 4 from both $w \bowtie r$ and v'. Hence $\{w \bowtie r, c, v'\} \subseteq w^u$. Since both c and $w \bowtie r$ belong to $C(w)$, w^u is a line of $C(w)$ and hence v' is a point of $C(w)$. By (†), v' is a point of $C(x)$, and since $\delta(v, v') = 4$, the plane $C'(r)$ contains v. Hence $C(r)$ and $C'(r)$ share a distance-2-trace and a point v off that distance-2-trace and so they coincide.

Now let z' be the unique point of type II on L. If there are points x, x' of $\Gamma(p, q)$ such that $z \perp x \perp x' \perp z'$, then we have the situation of the preceding paragraph with $M = L$ and r the projection of w on L, where w is the unique point of type I different from x and collinear with z. So the point b (recall that b is the projection of z onto L) is contained in $C(r)$. Also, the line ab has type β and a is the projection of x onto ab. So by part (i) above, a is a point of Δ. If there are no such points x, x', then we can still consider the situation of the preceding paragraph, with (using the notation of that paragraph) z on cc' and L some line through r. Let z'' be the projection of z onto the line L' joining r and $w \bowtie r$. The line L' has type β, $w \bowtie r$ is on L' and has type II, and $z \perp c \perp w \perp w \bowtie r$ with $c, w \in \Gamma(p, q)$. So by our argument above, z'' is a point of Δ of type III and hence it belongs to $C(r)$. Denote by $C'(r)$ the centric plane projected from $C(c)$. It contains r^c and z'', but so does $C(r)$, hence $C(r)$ coincides with $C'(r)$. It follows that b is also a point of $C(r)$, hence b belongs to Δ. As above, a also belongs to Δ.

(iv) Now let z be a point of type III. Then the line za contains a unique point z' of type II and if $z' \neq a$, then by (ii) and (iii) the result follows. Suppose now $a = z'$. If ab is not a line of $\Gamma(p, q)$, then we can replace z by any point of $\Gamma(p, q)$ collinear with a and apply (i). If ab *is* a line of $\Gamma(p, q)$, then we may replace z by a suitable point of $\Gamma(p, q)$ at distance 4 from a and apply (i) again to obtain that b is a point of Δ.

We have shown that Δ is an ideal subhexagon. Clearly all points are distance-2-regular and the points of types I and III are projective. By Theorem 6.3.1, Δ is isomorphic to $\mathsf{H}(\mathbb{K})$ for some commutative field $\mathbb{K}$.

Note that, if $x \in R(p, q)$, $p \neq x \neq q$, then there is a unique ideal subhexagon Δ isomorphic to some $\mathsf{H}(\mathbb{K})$ containing p, q and x. Indeed, there is a unique centric plane $C(p')$ containing $x \bowtie p'$, where p' is some point of $\Gamma(p, q)$ collinear with p (and hence at distance 4 from x by the distance-3-regularity), and this defines a subhexagon Δ containing p, q, x. Every such subhexagon must contain $C(x)$, the centric plane centred at x and defined by $\Gamma(p, q)$, and hence it must contain Δ. But no split Cayley hexagon contains an ideal subhexagon isomorphic to some other split Cayley hexagon; see Corollary 2.4.17 on page 75.

Note that, by Corollary 5.5.16 (see page 224), all line-elations of Δ are axial. We will use that information to construct all (axial) line-elations in Γ.

$\boxed{\Gamma \text{ is half Moufang}}$

Let L be any line of Γ and let M, M' be two lines at distance 4 from L meeting in a point p at distance 3 from L. We show that there exists a collineation θ fixing all points at distance at most 3 from L and mapping M to M'.

First, we define the action of θ on $\Gamma_{\leq 3}(L)$ to be the trivial action.

Next, we define the action of θ on lines at distance 4 from L. Let N be such a line. First suppose that the projection x of N on L coincides with the projection of M onto L, but assume that N does not meet the line px. Then the regulus $R(M, N)$ has a unique element X meeting L, $X \neq L$. Note that the points collinear with x and on elements of $R(M, N)$ form a distance-2-trace T in $x^{\perp}$. Let r be the point in that distance-2-trace incident with N. Then there exists a unique element N' of the regulus $R(X, M')$ incident with r, since the distance-2-trace in $x^{\perp}$ determined by $R(X, M')$ coincides with T (having the points p and $X \cap L$ in common). We define $N^{\theta} = N'$. So we have defined a permutation of all lines through r, different from rx. We say that $M \mapsto M'$ **induces** $N \mapsto N'$ **at** r, or that $M \mapsto M'$ **induces this permutation at** r. We can also go back to any line through p and define its image under θ using (N, N^{θ}). This switching from p to r over and over again always gives us the same permutation since we can embed p, r and x in an ideal $\mathsf{H}(\mathbb{K})$ subhexagon, in which the axial elation with axis L and corresponding couple (M, M') uniquely determines these permutations. So we can say that **a permutation at** p **induces a permutation at** r, in symbols $p \to r$. A similar procedure can now be carried out to define N^{θ} if N meets xp, starting from any line P at distance 4 from L for which P^{θ} is already defined as above. We have to show that θ is well defined on $\Gamma_4(L) \cap \Gamma_3(x)$. For this, it suffices to show that, if r_1, r_2, r_3, r_4 are four points in $\Gamma_2(x) \cap \Gamma_3(L)$ such that r_1 and r_4 are collinear with neither r_2 nor r_3, or $r_1 = r_2$ and r_1 is not collinear with r_4, then $r_1 \to r_2 \to r_4$ yields the same as $r_1 \to r_3 \to r_4$. If $r_1 = r_2$ and r_1 is not collinear with r_4, then the result follows from the fact that we can put r_1, r_3, r_4 and x into an ideal $\mathsf{H}(\mathbb{K})$ subhexagon, which admits all axial elations. Let us refer to that special case by (‡). So suppose $r_1 \neq r_2$. If r_1 is not collinear with r_4, then by (‡), $r_1 \to r_2 \to r_4$ is the same as $r_1 \to r_4$ and this is in turn the same as $r_1 \to r_3 \to r_4$. So we may assume that r_1 and r_4 are collinear. If r_2 and r_3 are not collinear, then $r_1 \to r_2 \to r_4$ is the same as $r_1 \to r_3 \to r_2 \to r_4$ by (‡), and this is the same as $r_1 \to r_3 \to r_4$ again by (‡). If r_2 and r_3 are collinear, then we can take a point r_0 collinear with x, neither on L, nor on xr_1, nor on xr_2 (note that we can assume $t \geq 3$ since for $t = 2$ there is a unique non-trivial permutation of $t = 2$ elements) and we see that $r_1 \to r_2 \to r_4$ is the same as $r_1 \to r_0 \to r_4$, which is the same as $r_1 \to r_3 \to r_4$. So θ is well defined on $\Gamma_4(L) \cap \Gamma_3(x)$.

Now let N be a line at distance 4 from L and 5 from x. Let y be the projection of L onto N and let r be any point collinear with x, but not on L. Then r and y are opposite and so there is a unique line R through r at distance 4 from N. There is also a unique line N' through y at distance 4 from R^{θ}. We define $N^{\theta} = N'$. This is independent of the choice of r, because if r' is another such point, then there is a unique weak ideal subhexagon of order $(1, t)$ containing r and y (also containing L), and there is at least one centric plane centred at x containing r'. So r, r', L and y can be put into one ideal $\mathsf{H}(\mathbb{K})$ subhexagon again and so the result follows. This shows that we have θ well defined on $\Gamma_{\leq 4}(L)$.

Now let z be a point at distance 5 from L and let $z \, \mathbf{I} \, N \, \mathbf{I} \, y \perp x \, \mathbf{I} \, L$ (x can be viewed as an arbitrary point on L now). We define z^{θ} as the unique point of the distance-2-trace in $y^{\perp}$ containing z and x, incident with N^{θ}. In every $\mathsf{H}(\mathbb{K})$ subhexagon containing z and L, the corresponding axial elation indeed maps z to that point. If any point w is collinear with z, then w^{θ} is collinear with z^{θ} because again we can put w, z and L in the same ideal $\mathsf{H}(\mathbb{K})$ subhexagon and w and z are moved by θ according to the definition of an axial collineation with axis L in that $\mathsf{H}(\mathbb{K})$ subhexagon.

By Lemma 1.3.14 on page 14, we see that θ defines a collineation of Γ. Since θ fixes $\Gamma_{\leq 3}(L)$ pointwise, θ is an axial elation with axis L in Γ. So Γ is half Moufang. It admits all line-elations.

Construction of point-elations in Γ

We have shown that the automorphism group of Γ restricted to any ideal $\mathsf{H}(\mathbb{K})$ subhexagon contains every line-elation. Let us consider one particular ideal $\mathsf{H}(\mathbb{K})$ subhexagon Δ and suppose it is coordinatized as in Subsection 3.5.1. The following collineation is the product of three axial elations:

$$
\begin{aligned}
E[K, K', K''] \; & : \; \Delta \to \Delta \\
& : \; (a, l, a', l', a'') \; \mapsto \; (a, l - a^3 K, a' + a^2 K + K'', \\
& \qquad\qquad\qquad\qquad\quad l' + a^3 K^2 + lK + 3aa'K + K', a'' + aK) \\
& \quad\;\; [k, b, k', b', k''] \; \mapsto \; [k + K, b, k' - kK'' - KK'' + K', b', \\
& \qquad\qquad\qquad\qquad\quad k'' + K'']
\end{aligned}
$$

It is easily seen that $E[K, K', K''] = E[K, 0, 0]E[0, K', 0]E[0, 0, K'']$ preserves the incidence relation, for all $K, K', K'' \in \mathbb{K}$. Similarly, the following two maps are point-elations:

$$
\begin{aligned}
E_1(A) \; & : \; \Delta \to \Delta \\
& : \; (a, l, a', l', a'') \; \mapsto \; (a + A, l, a', l', a'') \\
& : \; [k, b, k', b', k''] \; \mapsto \; [k, b - Ak, k' + k^2 A^3 - 3kbA^2 + 3b^2 A, \\
& \qquad\qquad\qquad\qquad\quad b' + kA^2 - 2bA, k'' + kA^3 - 3bA^2 + 3b'A]
\end{aligned}
$$

$$
\begin{aligned}
E_2(A') \; & : \; \Delta \mapsto \Delta \\
& : \; (a, l, a', l', a'') \; \mapsto \; (a, l - 3aA', a' + A', l', a'') \\
& : \; [k, b, k', b', k''] \; \mapsto \; [k, b, k' - 3bA', b' + A', k'']
\end{aligned}
$$

Note that $E_2(A')$ is a $((0, 0), [0], (\infty), [\infty], (0))$-elation. From this, we calculate the following product of two axial collineations:

$$
P(A, K) = E[K, 0, 0]^{E_1(A)} E[-K, 0, 0] \; : \; \Delta \to \Delta
$$

with the following action on the points:

$$
\begin{pmatrix} a \\ l \\ a' \\ l' \\ a'' \end{pmatrix} \mapsto \begin{pmatrix} a \\ l + 3a^2AK - 3aA^2K + A^3K \\ a' - 2aAK + A^2K \\ l' - 3a'AK - 2A^3K^2 + 3aA^2K^2 \\ a'' - AK \end{pmatrix},
$$

and on the lines:

$$
\begin{pmatrix} k \\ b \\ k' \\ b' \\ k'' \end{pmatrix} \mapsto \begin{pmatrix} k \\ b - AK \\ k' - kA^3K - 3bA^2K + A^3K^2 \\ b' + A^2K \\ k'' + A^3K \end{pmatrix}.
$$

It is now straightforward to calculate

$$
P(A, K)P(1, -AK) = \\
E_2(A^2K - AK)E[0, A^3K^2 - 2A^2K^2, 0]E[0, 0, A^3K - AK].
$$

This means that

$$
E_2(A^2K - AK) = \\
P(A, K)P(1, -AK)E[0, 0, AK - A^3K]E[0, 2A^2K^2 - A^3K^2, 0]
$$

is written as a product of axial collineations all of which fix the line $[\infty]$ pointwise. Hence $E_2(A^2K - AK)$ can be extended to a collineation θ_{A^2K-AK} in Γ and it fixes all points on the line $[\infty]$ and also all lines through the points (0), (∞) and $(0, 0)$ (since Δ is an ideal subhexagon). If $t \geq 3$, we can choose $A \in \mathbb{K}$ such that $A^2 \neq A$, and hence $A^2K - AK = (A^2 - A)K$ runs through $\mathbb{K}$ as K runs through $\mathbb{K}$. Since we can embed $(0, 0)$, (∞), (0), $(0, 0, 0)$ and any other point of Γ on $[0, 0]$ in an ideal $\mathsf{H}(\mathbb{K})$ subhexagon, it remains to show that θ_{A^2K-AK} fixes every point of $[0]$.

Put $\theta = \theta_{A^2K-AK}$. Let p be any point on $[0]$ and suppose that p is not a point of Δ (if it is, then it is certainly fixed since θ restricted to Δ is a root elation) We know that θ fixes every point q of Δ collinear with (∞) (in the centric plane $C((\infty))$, θ fixes two lines pointwise). Let q be a point of Δ collinear with (∞) and having coordinates (k, b), with $k \neq 0$. Then there is a unique trace X in $(\infty)^{\perp}$ containing p and q. Since θ fixes q, $X \cup \Gamma_1((\infty))$ and $[0]$, it fixes X and hence p.

The case $t = 2$

In this case, the line-elations of $\mathsf{H}(2)$ do not generate any point-elation, hence, in order to produce a point-elation, we should consider line-elations in different copies of $\mathsf{H}(2)$. Unlike RONAN [1980a], we present an entirely geometric argument.

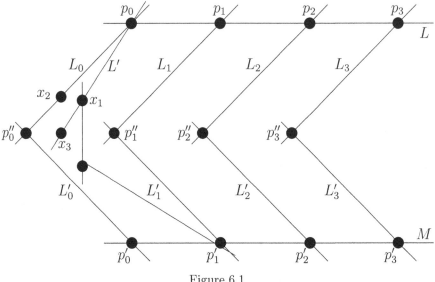

Figure 6.1.

We may assume that $\Gamma = (\mathcal{P}, \mathcal{L}, \mathbf{I})$ has order $(s, 2)$, with $s > 2$. We note that if $s < \infty$, then by Theorem $1.7.1(iv)$, $2s$ is a perfect square, hence $s \geq 8$. Since we must also deal with infinite s, we will not use the fact that, in the finite case, $s \leq 2^3 = 8$.

Let L and M be two opposite lines of Γ (see Figure 6.1). Let $\{p_i : i \in I\}$, with I some index set of size $s + 1$, be the set of points incident with L. For convenience, we may assume that, if s is finite, then $\{0, 1, 2, \dots, s\} = I$, and if s is infinite, then $\mathbb{N} \subseteq I$. Let, for each $i \in I$, the elements $L_i, L_i' \in \mathcal{L}$ and $p_i', p_i'' \in \mathcal{P}$ be (uniquely) defined as

$$p_i \, \mathbf{I} \, L_i \, \mathbf{I} \, p_i'' \, \mathbf{I} \, L_i' \, \mathbf{I} \, p_i' \, \mathbf{I} \, M.$$

We denote the unique non-trivial axial elation with axis L_i by θ_i. Also, let L' be the third line of Γ through p_0, i.e., $L_0 \neq L' \neq L$. For two non-collinear points x, y collinear with p_0, we shall denote the unique trace through x and y by $\langle x, y \rangle$.

Now pick three arbitrary elements in $I \setminus \{0\}$, which we may, without loss of generality, call $1, 2, 3$. The axial collineation θ_1 maps p_0'' to a point, say, x_1. The point x_1 is the unique point on L' belonging to the trace $(p_0)^{p_1'}$, in other words, x_1 is completely determined by p_1 and p_0'': it is the intersection of L' with $\langle p_1, p_0'' \rangle$. Similarly, θ_2 maps x_1 to a point, say, x_2, which is incident with L_0. Note that x_2 is the intersection of L_0 with $\langle p_2, x_1 \rangle$, and hence, $x_2 \neq p_0''$. In a similar way, we construct the point x_3, which is the image of x_2 under θ_3, and which is the intersection of L' and $\langle p_3, x_2 \rangle$. So $x_3 \neq x_1$. Now let p_i, $i \in I$, be the intersection of L with the trace $\langle p_0'', x_3 \rangle$. It is readily seen that $p_i \neq p_1, p_3$. Suppose it is equal to

p_2. Then the centric plane centred at p_0 containing $\langle x_1, x_2 \rangle$ and x_3 contains the points p_0, p_1, p_2, p_3 on the line L, in contradiction with the fact that such planes have order 2. So we may take $i = 4$. Note that since the element i is determined by the elements $1, 2, 3 \in I$, we may write $i = 4 = f(1, 2, 3)$ as a function of $1, 2, 3 \in I$. So we see that $\theta = \theta_1 \theta_2 \theta_3 \theta_4$ fixes p_0''. Also, it fixes every line L_i, $i \neq 1, 2, 3, 4$, and it does not fix the lines $L_1, \ldots, L_4$. Furthermore, it fixes every point on L, hence also every point on L' and on L_0. Indeed, every trace $\langle p_0'', p_i \rangle$ is fixed pointwise, $i \in I$, and the intersection of such a trace with L' gives an arbitrary point of L'. Similarly for the points on L_0. If L_0' is fixed by θ, then we put $\sigma = \theta$, otherwise we put $\sigma = \theta \theta'$, where θ' is the unique non-trivial axial elation with axis L. But now we see that σ fixes the path $(L_5, p_5, L, p_0, L_0, p_0'', L_0')$, and moreover every point on L, every point on L_0, and — trivially — every line through one of the points p_5, p_0 and p_0''. Hence σ is a $(p_5, L, p_0, L_0, p_0'')$-elation. But σ does not fix L_1, so σ is a non-trivial point-elation.

Of course we can start all over again with three other distinct axial collineations $\theta_i, \theta_j, \theta_k$, $\{i, j, k\} \not\subseteq \{1, 2, 3, 4\}$, $0, 5 \notin \{i, j, k\}$ to obtain a point-elation σ'. Since the actions of σ and σ' on the lines concurrent with L do not coincide, we have that $\sigma \neq \sigma'$. Now suppose that s is finite. Note that, if we choose $i = 1, j = 2$ and $k \neq 3$, then by the construction above, $f(1, 2, k) \neq f(1, 2, 3)$. So there is at most one $k \in I$ with $f(1, 2, k) = 5$. Similarly, for each pair $i, j \in I \setminus \{0, 5\}$, there is at most one $k \in I$ with $f(i, j, k) = 5$. This means that there are at least $(s-1)(s-2)(s-4)$ choices for the triple (i, j, k) with $f(i, j, k) \neq 5$. We conclude that we have at least $(s-1)(s-2)(s-4)/24$ distinct non-trivial $(p_5, L, \ldots, p_0'')$-elations. Since there are clearly at most $s - 1$ such, this gives the condition

$$\frac{1}{24}(s-1)(s-2)(s-4) \leq s - 1,$$

which simplifies to $(s-8)(s+2) \leq 0$. Hence $s \leq 8$. Since $s \geq 8$, we have $s = 8$ and there are exactly seven non-trival $(p_5, L, p_0, L_0, p_0'')$-elations. Since L, L_0 and L_5 are essentially arbitrary, we have shown that every 4-path is a Moufang path, hence Γ is a Moufang hexagon.

It remains to deal with the case where s is infinite. In this case, we have shown in the previous paragraph that there are infinitely many $(p_5, L, p_0, L_0, p_0'')$-elations. All these elations fix all but exactly eight lines confluent with L. Consider again σ, as above. Now there are also infinitely many $(p_0, L_0, p_0'', L_0', p_0')$-elations, hence at least one such elation σ_0 maps p_1 onto a point p_i with $i \notin \{1, 2, 3, 4\}$. Without loss of generality, we may take $i = 5$. The collineation $\sigma_1 = \sigma_0 \sigma \sigma_0^{-1}$ is clearly a $(p_1, L, p_0, L_0, p_0'')$-elation. Moreover, its action on $\Gamma(L_0')$ coincides with the action of σ on that set. And note that σ_1 fixes all but exactly eight lines confluent with L. Now consider the product $\sigma \theta_0 = \sigma_1'$ (recall that θ_0 is the unique non-trivial axial elation with axis L_0). Then σ_1' is also a $(p_1, L, p_0, L_0, p_0'')$-elation whose action on $\Gamma(L_0')$ coincides with the action of σ, and hence σ_1, on that set. So clearly $\sigma_1 = \sigma_1'$. But σ_1' obviously fixes exactly eight lines not incident with p_0 and confluent with L. This contradicts the fact that σ_1 fixes infinitely many lines confluent with L.

This concludes the case $t = 2$.

To show the converse, we remark that is enough to prove that, up to duality, all points of a Moufang hexagon are distance-2-regular. But that follows immediately from Lemma 5.5.11 on page 221.

This completes the proof of the theorem.   □

Next, we look at distance-3-regular generalized hexagons. In order to recognize these as Moufang hexagons, we need an additional assumption.

6.3.3 Definition. Let Γ be a generalized hexagon and let x, y be two points at distance 4 from each other. For every pair of points $\{z_1, z_2\}$ collinear with y and opposite x, the set $x^{z_1} \cap x^{z_2}$ is called an **intersection set** if $x^{z_1} \neq x^{z_2}$. Note that an intersection set is never empty (since $x \bowtie y \in x^{z_1} \cap x^{z_2}$).

Intersection sets are introduced by RONAN [1981]. GOVAERT [1997] generalizes and slightly modifies this concept to give a characterization of distance-i-regular points in a generalized n-gon, $2 \leq i \leq n/2$, along the lines of Remark 6.3.6 below.

It is easy to see that in a point-regular generalized hexagon all intersection sets have size 1. Conversely, it is not so difficult to prove (see Theorem 6.3.4 below) that, if all intersection sets of a distance-3-regular generalized hexagon Γ have size 1, then Γ is point-regular and hence Γ is a Moufang hexagon. But it is a little surprising that we can also recognize the line-regular hexagons via a condition on the intersection sets.

6.3.4 Theorem (Ronan [1981]). *Let Γ be a distance-3-regular generalized hexagon. If all intersection sets have size 1, then Γ is point-regular (and hence it is a Moufang hexagon). If no intersection set has size 1, then Γ is line-regular (and hence it is a Moufang hexagon).*

Proof. First suppose that in the distance-3-regular hexagon Γ all intersection sets have size 1. Let p be a point of Γ and let x, y be two non-collinear points both collinear with p. Suppose z and u are such that $\{x, y\} \subseteq p^u \cap p^z$. Let q be the projection of $x \bowtie z$ onto $\mathrm{proj}_y(u)$ and let r be the projection of $y \bowtie u$ onto $\mathrm{proj}_x(z)$. Put $z' = (x \bowtie z) \bowtie q$ and $u' = (y \bowtie u) \bowtie r$. The sets $p^z \cap p^{z'}$ and $p^u \cap p^{u'}$ are intersection sets of size greater than 1 unless $p^z = p^{z'}$ and $p^u = p^{u'}$. But since p is distance-3-regular, we have $p^{u'} = p^{z'}$. Hence $p^u = p^z$ and p is distance-2-regular. This shows the first part of the statement.

Now suppose that no intersection set has size 1. Let L be a line, let M_1 and M_2 be two lines opposite L at distance 4 from each other such that the line L' meeting both M_1 and M_2 has distance 4 from L (see Figure 6.2). Put $M = L \bowtie L'$. Suppose there exists another line $M' \in \Gamma_2(L)$, $M' \neq M$, at distance 4 from both M_1 and M_2. Let p be any point on L, $(L \cap M) \neq p \neq (L \cap M')$. We show that the projection L_1 of M_1 onto p coincides with the projection L_2 of M_2 onto p. Suppose by way

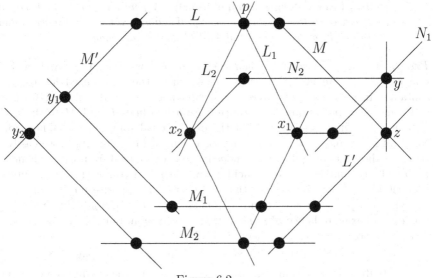

Figure 6.2.

of contradiction that $L_1 \neq L_2$. Let x_i be the projection of M_i onto L_i, $i = 1, 2$, and let z be the projection of M_1 onto M. The intersection set $z^{x_1} \cap z^{x_2}$ must contain by assumption at least one point $y \neq L \cap M$. Let y_i be the projection of M_i onto M', $i = 1, 2$. Now $z_{[3]}^{x_i} \cap z_{[3]}^{y_i}$ contains L and M_i, for $i = 1, 2$. Let N_i be the projection of x_i onto y, then $N_i \in z_{[3]}^{x_i}$, and hence since Γ is distance-3-regular, we must have $\delta(y_i, N_i) = 3$. But $\delta(N_1, N_2) = \delta(y_1, y_2) = 2$ and we obtain a circuit of length 10 in Γ, a contradiction. Hence $L_1 = L_2$ and this shows that in Γ^D all intersection sets have size 1. By the first part of the theorem, the result follows. $\square$

6.3.5 Remark. In the finite Moufang case, intersection sets have size 1, 1, 2 and $q^2 + 1$ in $\mathsf{H}(q)$, $\mathsf{T}(q^3, q)$, $\mathsf{H}(q)^D$ (q not a power of 3 in the latter) and $\mathsf{T}(q, q^3)$, respectively. The first three cases are almost immediate, the last one will be shown below after the proof of Theorem 6.3.8.

6.3.6 Remark. For the first part of the previous theorem, we did not need the full distance-3-regularity. What we used in the proof was only the fact that, whenever three different points x, y and z are at distance 3 from two different lines, then $x^y = x^z$ (and of course, the fact that all intersection sets have size 1). We will use this weaker version later on.

There is an interesting corollary to Ronan's result.

6.3.7 Corollary (Van Maldeghem, unpublished). *A generalized hexagon* Γ *is point-regular (and hence has the Moufang property)* **if and only if** *every ordinary heptagon of* Γ *is contained in at least one ideal split Cayley subhexagon.*

Proof. If Γ is point-regular, then the result follows from the proof of Theorem 6.3.2. Suppose now that every heptagon of the generalized hexagon Γ is contained in an ideal split Cayley subhexagon. Let p be opposite q (two points of Γ) and let r be a point at distance 3 from two lines L and M, with $L, M \in \Gamma_3(p) \cap \Gamma_3(q)$. It is easy to see that there is an ideal split Cayley subhexagon Γ' containing p, q and r. Since all points of Γ' are distance-3-regular, we easily see that p is distance-3-regular in Γ (indeed, r has distance 3 from every element of $\Gamma'_3(p) \cap \Gamma'_3(q)$; but the latter coincides with $\Gamma_3(p) \cap \Gamma_3(q)$). Also, if an intersection set $p^x \cap p^y$ has size greater than 1 (where p is opposite both x and y, and $\delta(x, y) = 4$, $\delta(p, x \bowtie y) = 4$), then the points p, x, y and $x \bowtie y$ lie in an ideal split Cayley subhexagon, hence $p^x = p^y$ in that subhexagon, but this is also true in Γ. The result now follows from Theorem 6.3.4. $\square$

For a stronger result in the finite case, see Theorem 6.7.1 on page 278.

For the finite extremal Moufang generalized hexagons, the condition on the intersection sets can be replaced by the extremal condition, i.e., the order (s, t) satisfies $s = t^3$ or $t = s^3$.

6.3.8 Theorem (Ronan [1980b]). *A finite extremal generalized hexagon* Γ *is classical* **if and only if** *it is distance-3-regular.*

Proof. Since distance-3-regularity is a self-dual condition, we may as well suppose that Γ has order (s, s^3). It suffices to show that every intersection set has size at least 2. So let $p^x \cap p^y$ be an intersection set (p, x, y are points of Γ, p opposite x, y, $\delta(x, y) = 4$ and $\delta(p, x \bowtie y) = 4$). Let L be the projection of $x \bowtie y$ onto p. By Subsection 1.7.6 (see page 26), there are s^3 points — and hence at least one point z — at distance 4 from both x and y and at distance 3 from L, but not collinear with $x \bowtie y$. The unique point z' of the regulus $R(z, x \bowtie y)$ collinear with p is an element of $p^x \cap p^y$ distinct from $\text{proj}_L(x)$. This shows the result. $\square$

From the above proof it follows that equally many points at distance 4 from both x and y are collinear with any point z incident with L, for $z \neq p \bowtie (x \bowtie y)$. Since in total, there are s^3 such points, $p^x \cap p^y$ has size $(s^3/s) + 1 = s^2 + 1$.

6.3.9 Theorem (Ronan [1980a]). *Suppose* Γ *is a half Moufang hexagon and all corresponding elations are axial, i.e., for every line* L *of* Γ, *the group of axial elations with axis* L *acts transitively on the set of lines at distance 4 from* L *incident with an arbitrary point of* $\Gamma_3(L)$. *Then* Γ *is point-distance-2-regular. In particular,* Γ *is a Moufang hexagon.*

Proof. Let x be any point of Γ and suppose that $|x^y \cap x^z| \geq 2$ with $y, z \in \Gamma_6(x)$. Assume, moreover, that there is a line L at distance 3 from x, y and z. Applying an arbitrary axial elation with axis L, we see that $x^y = x^z$. Assuming $\delta(y,z) = 4$, we deduce that every intersection set has size 1. Now assume $\delta(y,z) = 6$. The same argument as above shows that, if $|x^y_{[3]} \cap x^z_{[3]}| \geq 2$, then $x^y_{[3]} = x^z_{[3]}$. Hence Γ is point-distance-3-regular and the assertion follows from Theorem 6.3.4. $\square$

Finally, we mention without proof some recent characterizations of the finite Moufang hexagons.

6.3.10 Theorem (Govaert & Van Maldeghem [19]).** *Let Γ be a finite generalized hexagon of order (s,t). For every point x and any pair of points $y,z \in \Gamma_6(x)$, we put $x^{\{y,z\}} := \{u \in x^y \cap x^z : \mathrm{proj}_u y \neq \mathrm{proj}_u z\}$.*

(i) *If for all points x,y,z, with $y,z \in \Gamma_6(x)$, the conditions $0 \neq |x^{\{y,z\}}| \neq |x^y \cap x^z| \geq 2$ imply that $|x^y \cap x^z| \leq t/s + 1$, then $\Gamma \cong \mathsf{H}(s)^D$, s not divisible by 3, or $\Gamma \cong \mathsf{T}(s,s^3)$.*

(ii) *Let $s = t$. If for all points x,y,z, with $y,z \in \Gamma_6(x)$, the condition $|x^{\{y,z\}}| \geq 2$ implies that $|x^{\{y,z\}}| = |x^y \cap x^z| \geq 3$, then $\Gamma \cong \mathsf{H}(s)^D$, s not divisible by 3.*

(iii) *Let Γ be extremal. If for every pair of opposite lines L,M of Γ, and for every point x at distance 4 from a unique element y of $\Gamma_3(L) \cap \Gamma_3(M)$ and opposite every other element of $\Gamma_3(L) \cap \Gamma_3(M)$, the point $x \bowtie y$ is not opposite any member of $\Gamma_3(L) \cap \Gamma_3(M)$, then $\Gamma \cong \mathsf{T}(s,s^3)$ or t is odd and $\Gamma \cong \mathsf{T}(t^3,t)$.*

Conversely, the hexagons obtained in the statements satisfy the given respective conditions.

6.4 Regularity in generalized polygons

In this section, we will essentially prove that point-regular generalized n-gons exist only for $n = 3, 4$ and 6. In fact, stronger results can be proved. We will only focus on generalized polygons all points and/or lines of which satisfy some distance-i-regularity condition. The reader will see that the proofs do not require that and that usually a much weaker condition is enough. For instance, we have shown (see Subsection 1.9.7 on page 40) that two collinear regular points in a generalized pentagon do not exist. We will again prove that fact below (in a more general setting), but now assuming all points of the pentagon are regular. For details we refer to Van Maldeghem [1995b]. Note that we only state the results for point-regularity. All results can be dualized for line-regularity.

6.4.1 Definitions. A generalized polygon which is both point-distance-i-regular and line-distance-i-regular, for some i, is called **super-distance-i-regular**.

Let v and w be two opposite elements of some generalized n-gon Γ. Let $2 \leq i \leq n/2$. Then we call the pair (v,w) **distance-i-regular** if $|v^w_{[i]} \cap v^{w'}_{[i]}| \geq 2$ implies $v^w_{[i]} = v^{w'}_{[i]}$, for every element w' opposite v, and $|w_{[i]}^v \cap w_{[i]}^{v'}| \geq 2$ implies $w_{[i]}^v = w_{[i]}^{v'}$, for every element v' opposite w. If a pair is distance-i-regular for all suitable i, then we say that the pair is **regular**. If n is even, then all pairs of opposite points (lines) are distance-i-regular **if and only if** Γ is point-distance-i-regular (line-distance-i-regular). If n is odd, then all pairs of opposite elements are distance-i-regular **if and only if** Γ is super-distance-i-regular. These are immediate consequences of the definitions.

6.4.2 Theorem (Van Maldeghem [1995b]). *Let Γ be a generalized n-gon and suppose $2 \leq i < n/2$.* **If** *i is even and Γ is point-distance-i-regular,* **then** *i divides n.* **If** *i is odd and Γ is super-distance-i-regular,* **then** *i divides n.*

Proof. Suppose i does not divide n. Let k be the unique positive integer such that $\frac{n}{k+1} < i < \frac{n}{k}$, i.e., $k = \lfloor \frac{n}{i} \rfloor$. Consider a $(k+2)$-tuple $\gamma = (v_0, v_1, \ldots, v_{k+1})$ of elements (points and lines) of Γ, all contained in one apartment Σ and such that $\delta(v_0, v_\alpha) = i\alpha$ for $0 \leq \alpha \leq k$, and $\delta(v_0, v_{k+1}) = 2n - i(k+1)$. Let w_0 be the element of Σ opposite v_1. Then $\delta(w_0, v_{k+1}) = n - ki$. Let w_1 be any element at distance i from w_0 and at distance $n - i$ from v_1. We choose $w_1 \notin \Sigma$ (using the thickness assumption). Clearly v_0 and w_1 are opposite and so, again by the thickness, there are elements $(w_2, w_3, \ldots, w_{k+1})$ on a path γ_0 of length n from w_1 to v_0 with the property $\delta(w_0, w_\alpha) = i\alpha$, for all $\alpha \in \{0, 1, \ldots, k\}$, and $\delta(w_0, w_{k+1}) = 2n - i(k+1)$, and such that v_1 and w_2 are opposite. Now let v_1' be the unique element at distance i from v_1 and $n - 2i$ from w_1.

The distance-i-trace $(v_1)_{[i]}^{w_0}$ contains v_0, v_1' and v_2. The distance-i-trace $(v_1)_{[i]}^{w_2}$ contains v_0 and v_1', hence by assumption, it should also contain v_2, so $\delta(w_2, v_2) = n - i$. Let $\{v_2'\} = \Gamma_i(v_2) \cap \Gamma_{n-2i}(w_2)$. Note that $\text{proj}_{w_2}(v_2) \notin \gamma_0$ and $\text{proj}_{v_2}(w_2) \notin \Sigma$ since $\delta(w_1, v_2') + \delta(v_2', v_{k+1}) + \delta(v_{k+1}, w_1) \leq 2n$ (and each of the three terms is smaller than n) and $\delta(w_3, v_2') + \delta(v_2', v_0) + \delta(v_0, w_3) \leq 2n$ (and again each of the three terms is smaller than n), hence w_0, w_2 and v_2, and v_0, v_2 and w_2, respectively, are contained in an apartment by Lemma 1.5.7 on page 20. So it follows that the distance-i-trace $(v_2)_{[i]}^{w_1}$ contains v_1, v_2' and v_3. The distance-i-trace $(v_2)_{[i]}^{w_3}$ contains v_1 and v_2', hence also v_3 and so $\delta(w_3, v_3) = n - i$. Put $v_3' = \Gamma_i(v_3) \cap \Gamma_{n-2i}(w_3)$. Considering the distance-i-traces $(v_3)_{[i]}^{w_2}$ and $(v_3)_{[i]}^{w_4}$, we obtain similarly $\delta(w_4, v_4) = n - i$. Continuing in this way, we finally obtain $\delta(w_{k+1}, v_{k+1}) = n - i$, with $\text{proj}_{w_{k+1}} v_{k+1} \notin \gamma_0$ and $\text{proj}_{v_{k+1}} w_{k+1} \notin \Sigma$. So there are two different paths going from w_0 to w_{k+1} and having length $(n - i) + (n - ki)$, namely one via v_0 and one via v_{k+1}. Hence $2n - ki - i \geq n$, implying $i \leq \frac{n}{k+1}$, contradicting our assumption on i and k. This proves the theorem. $\qquad\square$

6.4.3 Definitions. Let Γ now be any weak generalized n-gon. Then we call Γ **distance-i-thick**, $1 \leq i \leq n$, if for every thick element v of Γ every element at distance i from v is also thick. In the proof of Theorem 6.4.2, we only used the distance-i-thickness of the generalized n-gon. Noting this, and noting the fact that, if Γ is a generalized n-gon, then the multiple $m\Gamma$ is a weak generalized mn-gon which is super-distance-m-regular in an almost trivial way, we can now reformulate Theorem 6.4.2 as follows.

6.4.4 Corollary (Van Maldeghem [1995b]). *Let Γ be a distance-i-thick, point-distance-i-regular (for i even) or super-distance-i-regular (for i odd) weak generalized n-gon, $1 \leq i \leq n/2$, with at least one thick point. Then i divides n. For each i dividing n, there exist point-distance-i-regular (for i even) or super-distance-i-regular (for i odd) distance-i-thick weak generalized n-gons (which are, however, not thick in general) with at least one thick point.* $\square$

The previous results restrict the possibilities for i to the divisors of n. The next result will restrict the possibilities for i to a rather small interval.

6.4.5 Theorem (Van Maldeghem [1995b]). *Let Γ be a generalized n-gon and let $2 \leq i \leq n/2$.*

(i) **If** Γ *contains a distance-i-regular pair of elements,* **then** $i = n/2$ *or* $i \leq (n+2)/4$.

(ii) **If** $i = (n+2)/4$ *and* Γ *contains a distance-i-regular pair of elements v and w,* **then** (v,w) *is also a distance-$(n/2)$-regular pair of elements.*

(iii) **If** Γ *contains two distance-i-regular elements at distance j from each other, $i \leq j < n/2$,* **then** $i \leq (n-j)/2$.

(iv) **If** $n/4 < i < n/2$, n even, and Γ *contains a distance-i-regular element v such that every element at distance $n/2$ from v is also distance-i-regular,* **then** v *is a distance-$n/2$-regular element.*

Proof. First suppose that a generalized n-gon Γ has a distance-i-regular pair (v,w), $2 \leq i < n/2$. Let Σ be an apartment containing v and w (see Figure 6.3, where we have depicted Σ as an oval, and points and lines as dots; also some distances are shown). Let v' and w' be opposite elements in Σ at some fixed distance j from v and w, respectively, $i \leq j \leq n/2$. Let v_1 and v_2 be the two elements of Σ at distance i from v and let w_1 and w_2 be the two elements of Σ at distance i from w. By the thickness assumption, there exist paths γ and γ' of length n joining v and w, and v' and w', respectively, and such that neither γ nor γ' is contained in Σ. Let w_0 and v_0 be the elements of γ at distance i from w and v, respectively. Let w_0' and v_0' be the elements of γ' at distance j from w' and v', respectively. The distance-i-traces $v_{[i]}^w$ and $w_{[i]}^v$ obviously contain v_0, v_1, v_2 and w_0, w_1, w_2, respectively. Also, the distance-i-traces $v_{[i]}^{w_0'}$ and $w_{[i]}^{v_0'}$ contain v_1, v_2 and w_1, w_2, respectively. Hence, by the distance-i-regularity of (v,w), we must have $\delta(v_0, w_0') = \delta(v_0', w_0) = n - i$. So there are paths γ_v and γ_w of length $n - i$

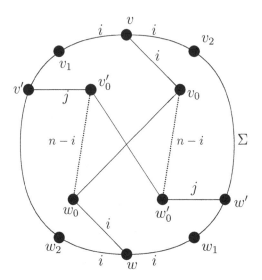

Figure 6.3.

connecting w_0' with v_0 and v_0' with w_0, respectively. The path γ_v has only w_0' in common with γ' (otherwise there is a circuit of length at most $2n$ through such common element and v) and it has only v_0 in common with the unique path joining v_0 and v. Similarly for γ_w.

First suppose that $j < n/2$. Then clearly $v_0' \neq w_0'$. This means that the closed path γ^* starting in w_0', going through v_0, through w_0, through v_0' and back to w_0' contains a circuit $\mathcal{C}$. The length of $\mathcal{C}$ is at most $(n-i)+(n-2i)+(n-i)+(n-2j)$ and this must be at least $2n$. This implies $2i \leq n - j$.

If n is odd, then we can take $j = \frac{n-1}{2}$, so in this case

$$i \leq \frac{1}{2}\left(n - \frac{n-1}{2}\right) = \frac{n+1}{4}.$$

If n is even, then we can choose $j = \frac{n-2}{2}$. In this case, we have

$$i \leq \frac{n+2}{4}.$$

This proves (i).

Now suppose that $i = (n+2)/4$ and put $j = n/2$. The path γ^* defined above has length $3n - 4i = 2n - 2$. Hence one can check that the paths γ_v and γ_w both must pass through the element x, with $\{x\} = \Gamma_{j-i}(v_0) \cap \Gamma_{j-i}(w_0)$. But this proves (ii).

Now suppose the generalized n-gon Γ has two distance-i-regular elements v, w at distance j from each other, with $2 \leq i \leq j \leq n/2$. Let Σ be any apartment containing v and w and let v' and w' be opposite v and w, respectively, inside Σ. By the thickness assumption, there exist paths γ_v and γ_w of length n connecting v with v' and w with w', respectively, and not lying in Σ. Let $v'' \in \gamma_v$ and $w'' \in \gamma_w$ be at distance j from v' and w', respectively; let v_1, v_2 and w_1, w_2 be the elements in Σ at distance i from v and w, respectively, and let $v_0 \in \gamma_v$ and $w_0 \in \gamma_w$ be at distance i from v and w, respectively. The distance-i-traces $v_{[i]}^{v'}$ and $w_{[i]}^{w'}$ contain v_0, v_1, v_2 and w_0, w_1, w_2, respectively. The distance-i-traces $v_{[i]}^{v''}$ and $w_{[i]}^{v''}$ contain v_1, v_2 and w_1, w_2, respectively. Hence $\delta(w'', v_0) = \delta(v'', w_0) = n - i$. Completely similarly to the first part above, we first assume $j < n/2$ and we obtain a circuit $\mathcal{C}$ in the path starting at w'', going through v_0, through v'', through w_0, back to w'' (this is indeed a circuit if the path going from v'' to w_0 does not pass through w'', but in this case we have a circuit through v', v'', w'' and w' of length no more than $4j$, contradicting $j < n/2$). The length of $\mathcal{C}$ is at most $(n - i) + (n - i - j) + (n - i) + (n - i - j)$. This should be at least $2n$. We have proved (iii). Secondly, we assume $j = n/2$ and $i > n/4$. Then $\mathcal{C}$ has length at most $3n - 4i < 2n$. Hence the path going from v'' to w_0 contains v'' and one sees that $\delta(v'', w'') = n/2$. Varying w, this shows (iv). $\qquad\square$

The case $n = 6$, $i = 2$ in (ii) and (iv) of the previous theorem is a special case of Proposition 1.9.17 (see page 46). The proof of (iv) provides an alternative argument for Proposition 1.9.17.

Putting $n = 8$ and $i = 3$ in the theorem above, one obtains the non-existence of a generalized octagon having a distance-3-regular pair of points (or lines). The next result says a similar thing about distance-2-regular points.

6.4.6 Theorem (Van Maldeghem [1995b]). *There does not exist any point-distance-2-regular generalized octagon.*

Proof. Let $p_0, p_1, p_2, \ldots, p_7$ be the points of an apartment in a generalized octagon Γ (where the points $p_0, \ldots, p_7, p_0$ are consecutively collinear in Γ; see Figure 6.4). Let p_0' be incident with $p_0 p_7$ but different from both p_0 and p_7 (this is possible by the thickness assumption). Construct the path

$$(p_0', p_0'p_1', p_1', p_1'p_2', p_2', p_2'p_3', p_3')$$

such that p_3' is incident with $p_3 p_4$. Let

$$(p_1, p_1 p_2'', p_2'', p_2''p_3'', p_3'', p_3''p_4'', p_4'', p_4''p_5, p_5)$$

be a path of length 8 with $p_2 \neq p_2'' \neq p_0$ (again possible by the thickness assumption). The traces $p_1^{p_5}$ and $p_1^{p_2'}$ have the points p_0 and p_2 in common, so $\delta(p_2', p_2'') = 6$ and there is a path

$$(p_2', p_2'x_1, x_1, x_1x_2, x_2, x_2p_2'', p_2'').$$

Clearly x_1 is incident with neither $p_1'p_2'$ nor $p_2'p_3'$.

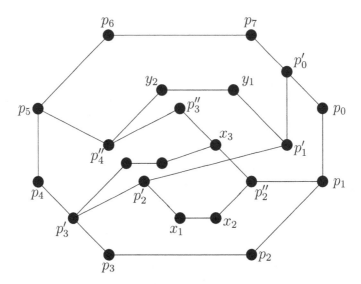

Figure 6.4.

Suppose first that x_2 is not incident with $p_2'' p_3''$. Let x_3 be the unique point on $p_2'' p_3''$ at distance 6 from p_3'. Clearly $p_2'' \neq x_3 \neq p_3''$. But the traces $(p_2'')^{p_1'}$ and $(p_2'')^{p_3'}$ share the points x_2 and p_1, hence $\delta(p_1', x_3) = 6$, so p_1' and p_3'' are opposite. The traces $p_5^{p_1}$ and $p_5^{p_1'}$ share the points p_4 and p_6, and so there is a path

$$(p_1', p_1' y_1, y_1, y_1 y_2, y_2, y_2 p_4'', p_4'').$$

Clearly y_1 is incident with neither $p_0' p_1'$ nor $p_1' p_2'$. And if y_2 were incident with $p_3'' p_4''$, then $\delta(p_1', p_3'') \leq 6$, contradicting the fact that p_1' is opposite p_3''. Since $\delta(p_2'', p_4'') + \delta(p_4'', y_1) = 8$ and $\mathrm{proj}_{p_4''}(p_2'') \neq \mathrm{proj}_{p_4''}(y_1)$, we have by Lemma 1.5.6 (page 20) that p_2'' and y_1 are opposite, but this contradicts $\{p_0', p_2', y_1\} \subseteq (p_1')^{p_5}$ and $\{p_0', p_2'\} \subseteq (p_1')^{p_2''}$. We conclude that x_2 must be incident with $p_2'' p_3''$.

So suppose x_2 is incident with $p_2'' p_3''$. By symmetry, y_2 (defined as in the previous paragraph) must be incident with $p_3'' p_4''$. But then $(p_1', y_1, y_2, p_3'', x_2, x_1, p_2', p_1')$ forms a circuit of length 14 in Γ, a contradiction.   □

6.4.7 Corollary (Van Maldeghem [1995b]). If *a generalized n-gon is point-distance-i-regular,* $2 \leq i < n/2$, *then* $i \leq (n + 2)/4$. *If, moreover,* i *is even or* Γ *is super-distance-i-regular,* **then** $n = 6$ *or* $i \leq n/4$, *and* i *divides* n.

Proof. Suppose Γ is a thick point-distance-i-regular generalized n-gon, $2 \leq i < n/2$. If n is even, then by Theorem 6.4.5(i), $i \leq \frac{n+2}{4} < \frac{n+3}{4}$. If $n \equiv 1 \bmod 4$, then we use Theorem 6.4.5(iii), putting $j = (n-1)/2$, to obtain $i \leq \frac{n+1}{4} < \frac{n+2}{4}$. If

$n \equiv 3 \bmod 4$, then we use Theorem 6.4.5(iii), putting $j = (n - 3)/2$, to obtain $i \leq \frac{n+3}{4}$. But in this case, $i \leq \frac{n+1}{4}$, because $\frac{n+3}{4}$ is not an integer. This shows the first part.

Suppose now, moreover, that i is even or Γ is super-distance-i-regular. By Theorem 6.4.2, i must divide n. If i were equal to $n/3$, then by the preceding paragraph, $n/3 \leq (n+2)/4$, hence $n \leq 6$, implying $n = 6$ or $n = 3$ (a trivial case). This shows the corollary completely. □

6.4.8 Corollary (Van Maldeghem [1995b]). *Generalized n-gons with all points regular exist only for $n \in \{3, 4, 6\}$. For $n > 4$, every such generalized n-gon is a Moufang polygon.*

Proof. The first statement follows immediately from the previous result, noting that for a point-distance-i-regular generalized n-gon with $i = 2$, n must be even, and for one with $n \geq 8$ even, there is always an integer between $(n + 2)/4$ and $n/2$. The second statement follows from Theorem 6.3.2. □

6.4.9 Corollary (Van Maldeghem, unpublished). *Suppose Γ is a half Moufang generalized $(4m + 2)$-gon, $m \geq 1$, and all corresponding elations are axial, i.e., for every line L of Γ, the group of axial elations with axis L acts transitively on the set of lines at distance $2m + 2$ from L incident with an arbitrary point of $\Gamma_{2m+1}(L)$. Then $m = 1$ and Γ is a point-distance-2-regular Moufang hexagon.*

Proof. As in the proof of Theorem 6.3.9, one shows that Γ is point-distance-$2m$-regular. Corollary 6.4.7 implies that $2m \leq m + 1$, hence $m = 1$. The result now follows from Theorem 6.3.9. □

Finally we mention without proof one further result.

6.4.10 Proposition (Van Maldeghem [1995b]). *There does not exist a super-distance-2-regular generalized 10-gon.*

6.5 Hyperbolic and imaginary lines

This section is based on VAN BON, CUYPERS & VAN MALDEGHEM [1994]. It weakens a little the conditions of Theorem 6.2.1 and Theorem 6.3.1, and it characterizes the symplectic polygons (quadrangles and hexagons) amongst all generalized $2m$-gons as those having long imaginary lines.

Our definitions differ a little from the ones in *op. cit.* in order to avoid some technicalities, but they are equivalent.

6.5.1 Definitions. Let Γ be a generalized n-gon and let x, y be two non-collinear points collinear with a given point p. The **hyperbolic line** $H(x, y)$ through x and

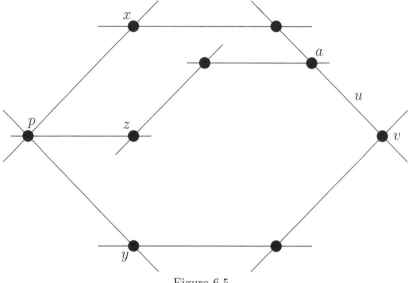

Figure 6.5.

y is defined as the intersection of all traces containing x and y. A hyperbolic line X is called **long** if the restriction to X of the projection onto any element of Γ at distance $n-1$ from every point of X is bijective whenever it is injective.

6.5.2 Lemma. *If the restriction to X of the projection of any hyperbolic line X onto some element w at distance $n-1$ from every point of X is not constant, then it is injective.*

Proof. Let $X = H(x,y)$, with x, y points of the polygon Γ, and $\delta(x,y) = 4$. Let p be some point collinear with both x and y (see Figure 6.5, where we have drawn the situation for $n = 6$). Let $z \in X$; $z \neq x_j$, then we claim that $H(x,y) = H(x,z)$. For reasons of symmetry, it is enough to show that $H(x,y) \subseteq H(x,z)$. Hence, we should show that every trace p^a containing x and z also contains y. Suppose this is not true and let p^a be a trace containing both x and z, but not y. Put $u = \text{proj}_a x$ and $v = \text{proj}_u y$. Clearly $\{x,y\} \subseteq p^v$, hence $z \in p^v$ (because $z \in H(x,y)$). This easily implies $v = a$ and hence $y \in p^a$, a contradiction. Our claim follows. One now sees that $H(x,y) = H(z,z')$ for every two distinct points z, z' in $H(x,y)$. So if the projection of $H(x,y)$ onto w is not injective, then there are two distinct points $z, z' \in H(x,y)$ having the same image w', so by the foregoing, $H(x,y) = H(z,z') \subseteq p^{w'}$. Hence the projection is constant and the lemma is proved. $\square$

The following result was essentially first noted and proved by John van Bon and Hans Cuypers and is included in VAN BON, CUYPERS & VAN MALDEGHEM [1994].

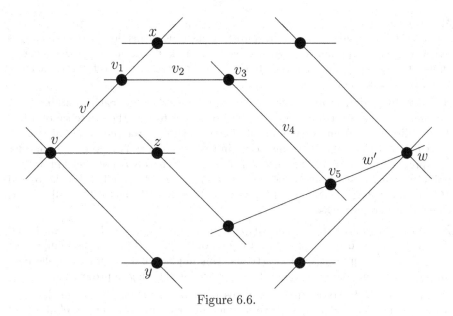

Figure 6.6.

6.5.3 Theorem (van Bon, Cuypers & Van Maldeghem [1994]). *Let* $n > 3$*. If* Γ *is a finite generalized* n*-gon of order* (s, t)*,* **then** *no hyperbolic line contains more than* $s + 1$ *points. Moreover, a hyperbolic line* X *is* **long** *if* **and only if** *it contains* $s + 1$ *points; in this case,* $s = t$*.*

If in a (not necessarily finite) generalized n*-gon all hyperbolic lines are long,* **then** *all points are projective. In particular,* n *must be even and different from 8; if* $n = 4$*, then* $\Gamma \cong \mathsf{W}(\mathbb{K})$*; if* $n = 6$*, then* $\Gamma \cong \mathsf{H}(\mathbb{K})$*, for some field* $\mathbb{K}$*.*

Hence a generalized n*-gon* Γ*,* $n > 3$ *with* $n < 10$ *if* n *is even, is Pappian if* **and only if** *all hyperbolic lines of* Γ *are long.*

Proof. We first claim that every long hyperbolic line coincides with any trace containing any two of its elements.

Let x and y be two arbitrary different points of a long hyperbolic line $H = H(x, y)$. Let v be an element at distance 2 from both x and y (see Figure 6.6, where we have put $n = 6$ to fix ideas). Let w be opposite v such that $x, y \in v^w$. By definition, H is contained in v^w. Suppose there is a point $z \in v^w$ not belonging to H. Let v' be the line vx and let w' be the projection of z onto w. Clearly v' and w' are opposite, so by the thickness assumption, there exists a path $\gamma = (v' = v_0, v_1, \ldots, v_{n-1}, v_n = w')$ with $x \neq v_1 \neq v$ and hence $v_{n-1} \neq w$. Consider the element v_{n-2}. We have $\delta(v_{n-2}, w') = 2$, $\delta(z, w') = n - 3$ and $\delta(x', w') = n - 1$, for all $x' \in H$. Hence $\delta(x', v_{n-2}) = n - 1$, for all $x' \in H$. For a similar reason, $\delta(x', v_{n-1}) = n$, hence x' is opposite v_{n-1}, for all $x' \in H$. Also similarly, one shows easily that $\delta(x, v_{n-3}) = n - 2$ and $\delta(x', v_{n-3}) = n$, for all $x' \in H$ (because

$\delta(x', v_1) + \delta(v_1, v_{n-3}) = 4 + n - 4 = n$ and $\mathrm{proj}_{v_1} x' = v' \neq v_2 = \mathrm{proj}_{v_1} v_{n-3}$; we require the latter only for $n > 4$ of course). Hence the projection of H onto v_{n-2} is not a constant mapping, so it is injective by Lemma 6.5.2, but since H is long, it should be bijective. This contradicts the fact that v_{n-1} is opposite every element of H. Hence the claim.

If Γ is finite and $H(x, y)$ is a long hyperbolic line, then it clearly contains $t + 1$ elements, by the preceding claim. In view of $n = 4, 6$ or 8, the existence of a line L at codistance 1 from every point of $H(x, y)$ such that the projection of $H(x, y)$ onto L is not constant is readily seen (indeed, take v, w as before and consider an appropriate line L at distance 3 from w, i.e., choose L not incident with a point at codistance 4 from an element of H), hence $s + 1 = t + 1$. If Γ is finite and $H(x, y)$ is not long, then consideration of a similar line L leads to $|H(x, y)| \leq s + 1$, concluding the finite case.

Now we claim that, if H is a long hyperbolic line in the generalized n-gon Γ and if x is a point at distance 2 from all points of H, then $H \cap x^v$ is non-empty, for every element v opposite x. In particular this implies that, if all hyperbolic lines all points of which lie at distance 2 from x are long, then x is projective.

So let x, v and H be as in the previous paragraph. Let $y, z \in H$ and let $w = \mathrm{proj}_v xy$, and put $u = \mathrm{proj}_w xy$. We may assume that the projections y' and z' of v onto, respectively, xy and xz are different from y and z. Hence it is clear that $\delta(x, u) = n - 2$ and $\delta(x, w) = n - 1$. Also, $\delta(y, u) = n - 2$, hence $\delta(z, u) = n$, so the projection of H onto w is bijective. Hence there is a point $a \in H$ at distance $n - 2$ from v, implying $a \in H \cap x^v$. This shows our claim.

If Γ is a generalized quadrangle or hexagon, then the remaining results follow by Theorem 6.2.1, respectively, Theorem 6.3.1. $\qquad\square$

As an immediate consequence of Theorem 6.5.3 and Theorem 1.7.1(v) (page 24), we have the following result.

6.5.4 Corollary (van Bon, Cuypers & Van Maldeghem [1994]). *No finite generalized octagon can have a long hyperbolic line.* $\qquad\square$

6.5.5 Definitions. Let Γ be a generalized $2m$-gon and let x, y be two opposite points. The set $I(x, y) = \{x, y\}^{\perp\perp\perp\perp}$ is called an **imaginary line**. An imaginary line I is called **long**, if its projection onto any (usual) line at codistance 1 from every point of I is bijective whenever it is injective. The next statement shows that only for a few generalized polygons are all imaginary lines long. Note that for $m = 2$, an imaginary line is the same thing as a hyperbolic line, and so in fact the next result tells us nothing new for generalized quadrangles.

6.5.6 Theorem (van Bon, Cuypers & Van Maldeghem [1994]). *If Γ is finite of order (s, t), **then** no imaginary line contains more than $s + 1$ points. Moreover, an imaginary line I is long **if and only if** it contains exactly $s + 1$ points; in this case, $s = t$.*

If *in a (not necessarily finite) generalized 2m-gon, all imaginary lines are long,* **then** $m = 2$ *and* $\Gamma \cong \mathsf{W}(\mathbb{K})$ *for some field* $\mathbb{K}$, *or* $m = 3$ *and* $\Gamma \cong \mathsf{H}(\mathbb{K}')$ *for some perfect field* $\mathbb{K}'$ *of characteristic* 2. *Also the converse is true, i.e., all imaginary lines of* $\mathsf{W}(\mathbb{K})$ *and* $\mathsf{H}(\mathbb{K}')$ *are long, for* $\mathbb{K}$ *any field and* $\mathbb{K}'$ *any perfect field of characteristic* 2.

Hence a generalized n-gon Γ *with n even is a (perfect) symplectic polygon (more precisely, a symplectic quadrangle or a perfect symplectic hexagon)* **if and only if** *all imaginary lines of* Γ *are long.*

Proof. Let Γ be a generalized n-gon. For $n = 4$, there is nothing to prove, so assume $n \geq 6$. It is a rather straightforward observation to show that for opposite points y and z, the imaginary line $I(y, z)$ is contained in the set of points lying at distance $m = n/2$ of any element v lying on turn at distance m from both y and z. So if v and w are two elements at distance m from both y and z, then $I(y, z) \subseteq \Gamma_m(v) \cap \Gamma_m(w)$. Hence every element of $I(y, z) \setminus \{y, z\}$ is a point opposite both y and z.

$\boxed{\text{Step 0}}$

We first show that $I(y, z) = I(x, z)$, for all $x \in I(y, z)$, $x \neq z$. First we prove that we have $y \in I(x, z)$. Suppose by way of contradiction that $y \notin I(x, z)$. Then there exists a point a not opposite both x, z, but opposite y. We may suppose that $k = \delta(z, a)$ is minimal with respect to that property. Let $\gamma = (a = z_0, z_1, \ldots z_k = z)$ be a minimal path connecting a and z. Let $\ell > 0$ be minimal with respect to the property $\delta(y, z_{2\ell}) = n$ (this is well defined since $\delta(y, z_k) = n$). By the minimality of ℓ, it is easily seen that $\delta(y, z_\ell) = n - \ell$. Let $(y = y_0, y_1, \ldots, y_{n-\ell} = z_\ell)$ be a minimal path between y and z_ℓ. The elements $z_1, \ldots z_{2\ell-1}$ and $y_{n-2\ell}, \ldots y_{n-\ell}$ are not opposite both z and y. Hence, by the definition of $I(y, z)$, all these points are not opposite x. This readily implies $\delta(x, z_\ell) \leq n - \ell$. Since both a and $y_{n-2\ell}$ are not opposite x, we have $\delta(x, z_\ell) + \ell < n$, hence, since $\delta(x, z_\ell) + \delta(z_\ell, z_{2\ell}) < n$, the point x cannot be opposite $z_{2\ell}$. But this contradicts the minimality of $\delta(z, a)$, if $z \neq z_{2\ell}$, and it contradicts the fact that x and z are opposite if $z = z_{2\ell}$.

We have shown that, if $\{y, z\}^{\perp\perp} \subseteq x^{\perp\perp}$, and $x \neq z$, then $\{x, z\}^{\perp\perp} \subseteq y^{\perp\perp}$. Hence $\{x, z\}^{\perp\perp} = \{y, z\}^{\perp\perp}$ in that case. So $I(x, z) = I(y, z)$. This now easily implies that $I(y, z) = I(x, x')$, for all $x, x' \in I(y, z)$, $x \neq x'$.

Step 0 implies that, if L is a line at codistance 1 from every point of $I(y, z)$, and the projection of $I(y, z)$ onto L is not a constant map, then it must be injective. Indeed, suppose it is not injective and let $x, x' \in I(y, z)$ have the same projection x'' onto L. For every $y' \in I(x, x') = I(y, z)$, we have that y' cannot be opposite x''. Hence $x'' = \operatorname{proj}_L y'$. So the projection of $I(y, z)$ onto L is a constant map.

$\boxed{\text{Step I}}$

If $I(y, z)$ is long, then we show that $I(y, z) = v^w_{[m]} = w^v_{[m]}$, with $v, w \in \Gamma_m(y) \cap \Gamma_m(z)$, $v \neq w$.

Suppose there is a point $x \in v_{[m]}^w$ not belonging to $I(y, z)$. Let v' be the projection of y onto v and let w' be the projection of x onto w. Clearly v' and w' are opposite, so by the thickness assumption, there exists a path $\gamma = (v' = v_0, v_1, \ldots, v_{n-1}, v_n = w')$ with $v_1 \neq v$ and $v_{n-1} \neq w$. Consider the element v_m. We have $\delta(v_m, w') = m$, $\delta(x, w') = m - 1$ and $\delta(x', w') = m + 1$, for all $x' \in I(y, z)$. Hence $\delta(x', v_m) = n - 1$, for all $x' \in I(y, z)$. For a similar reason, $\delta(x', v_{m+1}) = n$, hence x' is opposite v_{m+1}, for all $x' \in I(y, z)$. Also similarly, one shows easily that $\delta(y, v_{m-1}) = n - 2$ and $\delta(x', v_{m-1}) = n$, for all $x' \in I(y, z)$, $x' \neq y$. Hence the projection of $I(y, z)$ onto v_m is not a constant mapping, so it is injective, but since $I(y, z)$ is long, it should be bijective. This contradicts the fact that v_{m+1} is opposite every element of $I(y, z)$. Hence the result.

Step II

Let y and z again be two opposite points and let $I(y, z)$ be an imaginary line. Let x be collinear with y but opposite z. Let v be the element of Γ at distance $n/2$ from z and $n/2 - 1$ from xy. By step I, $\delta(v, a) = n/2$ for all $a \in I(y, z)$. Now let $\gamma = (x = v_0, v_1, \ldots, v_{n-1}, v_n = z)$ be any path of length n joining x with z, with $v_m \neq v$. All the elements $v_2, v_3, \ldots, v_{n-4}$ are neither opposite z nor opposite y, hence they are not opposite a, for any $a \in I(y, z)$. So let $a \in I(y, z)$, $z \neq a \neq y$. Since $\delta(a, v) = \delta(v, x) = n/2$, we have $\delta(v_2, a) = n - 2$. Suppose the path γ' joining v_2 and a contains v_j but not v_{j+1}, $2 \leq j < n/2 - 1$. Then clearly, $\delta(a, v_{2j}) = \delta(a, v_j) + \delta(v_j, v_{2j}) = n$, a contradiction. Hence γ' contains $v_{n/2-1}$. Since $\delta(x, v_{n/2-1}) = n/2 - 1$ and $\delta(x, v) = n/2$, we conclude $x_{[n/2-1]}^z = x_{[n/2-1]}^a$, for all $a \in I(y, z)$, $a \neq y$. Note that γ' does not contain $v_{n/2}$. Indeed, it certainly does not contain $v_{n/2+1}$ since a is opposite z. If it contains $v_{n/2}$, then $\text{proj}_{v_{n-1}} y$ (which is distinct from v_{n-2}) is neither opposite y, nor opposite z, but it is opposite a as $\delta(\text{proj}_{v_{n-1}} y, v_{n/2}) + \delta(v_{n/2}, a) = n$ and $\text{proj}_{v_{n/2}}(\text{proj}_{v_{n-1}} y) = v_{n/2-1} \neq \text{proj}_{v_{n/2}} a$, a contradiction.

Step III

We keep the same notation as in step II. Note that $\delta(a, v_{n/2-2}) = \delta(a, v_{n/2}) = n/2 + 2$ and $\delta(a, v_{n/2-1}) = n/2 + 1$, for all $a \in I(y, z)$, $y \neq a \neq z$. Hence $\text{proj}_{v_{n/2-1}} a \neq \text{proj}_{v_{n/2-1}} y$. Now let $a_1, a_2 \in I(y, z)$, both different from y. Since $I(y, z) = I(a_1, z)$, we can let a_1 play the role of z in the previous paragraph, and a_2 the role of a, and we obtain $\text{proj}_{v_{n/2-1}} a_1 \neq \text{proj}_{v_{n/2-1}} a_2$.

Hence we have shown that the projection ρ of $I(y, z)$ onto $v_{n/2-1}$ is injective. Now suppose that $I(y, z)$ is long. Since $\delta(v_{n/2-1}, z)$ is not equal to $n - 1$, we cannot conclude by definition that it is bijective. As a matter of fact, $v_{n/2-1}$ can be a point (if $n/2$ is odd). Nevertheless, we claim that ρ is a bijection. This will prove that, for Γ finite, $s = t$. Indeed, if $n = 6$, then $I(y, z)$ contains $s + 1$ points and $I(y, z)$ must be in bijective correspondence with the set of all lines through a point, hence

$s = t$. If $n = 8$, then $I(y, z)$ has $t+1$ points and must be in bijective correspondence with all points on a line, hence $s = t$.

Suppose by way of contradiction that ρ is not bijective and let u be an element not in the image. So u is incident with $v_{n/2-1}$ and at distance $n/2+2$ from all points of $I(y, z)$. Let v' be the projection of z onto v. Then v' and $v_{n/2-1}$ are opposite (after all, $\delta(v', z) + \delta(z, v_{n/2-1}) = n$) and there is a unique line L in Γ at distance $n/2$ from v' and distance $n/2 - 1$ from u. We have $\delta(L, z) = \delta(L, v') + \delta(v', z) = n - 1$. Let x_0 be the projection of u onto L; then $\delta(a, u) + \delta(u, x_0) = n/2 + 2 + n/2 - 2 = n$, hence x_0 is opposite every element a of $I(y, z)$. This implies that $\delta(a, L) = n - 1$ for all points a of $I(y, z)$, and since the projection of $I(y, z)$ is clearly not bijective (the image misses x_0), it must be constant. So $\delta(y, x_1) = n - 2$, where x_1 is the projection of z onto L. But x_1 is also the projection of v' onto L; indeed, $\delta(z, x_1) = \delta(z, v') + n/2 - 1$. Now $\delta(y, v') = n/2 + 1$ and so y is opposite x_1, a contradiction. This shows our claim.

Step IV

In this and the next step, we assume that *all* imaginary lines are long. Suppose z' is any other point opposite x with $\{\mathrm{proj}_v\, x, v_{n/2-1}\} \subseteq x^z_{[n/2-1]} \cap x^{z'}_{[n/2-1]}$. Suppose first also that $\delta(z', v') = n/2 - 1$. There is a unique element u' incident with $v_{n/2-1}$ at distance $n/2$ from z'. By the last claim of step III, there exists a unique $a \in I(y, z)$ such that $\delta(a, u') = n/2$. Hence $\delta(z', v) = \delta(a, v) = \delta(z', u') = \delta(a, u') = n/2$ which implies that $z' \in I(a, x)$. Hence clearly $x^a_{[i]} = x^{z'}_{[i]}$, for all $i \in \{2, \ldots, n/2\}$.

So we have shown that, whenever two distance-$(n/2 - 1)$-traces $x^z_{[n/2-1]}$ and $x^{z'}_{[n/2-1]}$ meet in at least two elements, and if **either** z and z' lie in the same imaginary line with x, **or** there exists an element v' at distance $n/2 + 1$ from x and distance $n/2 - 1$ from both z and z', then they coincide.

So now let z'' be completely arbitrary but such that $x^{z''}_{[n/2-1]}$ meets $x^z_{[n/2-1]}$ in at least two elements, say b_1 and b_2. Let u_1, u_2, u''_1 and u''_2 be the projections of z and z'' onto b_1 and b_2 (where indices and primes are self-explanatory). Put $w_1 = \mathrm{proj}_{u_1} z$. Then w_1 is opposite b_2. Let z^* be the unique point at distance $n/2$ from u''_2 and $n/2 - 1$ from w_1. By the previous paragraph, $x^z_{[n/2-1]} = x^{z^*}_{[n/2-1]}$. But now there exists a unique point $z^{**} \in I(x, z^*)$ such that the projection of z^{**} onto u''_2 coincides with the projection of z'' onto u''_2. From the previous paragraph it follows again $x^{z^*}_{[n/2-1]} = x^{z^{**}}_{[n/2-1]} = x^{z''}_{[n/2-1]}$. This shows that Γ is point-distance-$(m - 1)$-regular.

Step V

It remains to deal with the case $n = 6$, by Theorem 6.4.7. By the previous step, Γ is distance-2-regular, hence from the proof of Theorem 6.3.2 it follows that Γ contains an ideal subhexagon Γ' isomorphic to $\mathsf{H}(\mathbb{K})$ for some commutative field

$\mathbb{K}$. Embed Γ' in the quadric $\mathbf{Q}(6, \mathbb{K})$ in the standard way; see Subsection 2.4.13 on page 73. If char$(\mathbb{K}) \neq 2$, then $I(y, z)$ is the set of points on $\mathbf{Q}(6, \mathbb{K})$ conjugate (with respect to the polarity defining $\mathbf{Q}(6, \mathbb{K})$) to all points conjugate to both y and z. Clearly, this is the line yz in $\mathbf{PG}(6, \mathbb{K})$, and this line meets $\mathbf{Q}(6, \mathbb{K})$ in $\{y, z\}$. Hence $I(y, z) = \{y, z\}$ in Γ'. But this implies that $I(y, z)$ cannot be long in Γ. So char$(\mathbb{K}) = 2$. In this case, Γ' is embedded in the symplectic geometry $\mathbf{W}(5, \mathbb{K})$ in $\mathbf{PG}(5, \mathbb{K})$. Let x and y be opposite points in Γ'. Let τ be the polarity associated with $\mathbf{W}(5, \mathbb{K})$, i.e., the points, lines and planes of $\mathbf{W}(5, \mathbb{K})$ are the absolute points, lines and planes, respectively, of $\mathbf{PG}(5, \mathbb{K})$ with respect to τ. Then

$$\{x, y\}^{\perp \perp} = (x^\tau \cap y^\tau)^\tau \cap \mathcal{S}',$$

where $\mathcal{S}'$ is the point set of Γ' (embedded in $\mathbf{PG}(5, \mathbb{K})$). Clearly $(x^\tau \cap y^\tau)^\tau = xy$. Since obviously $xy \cap \mathcal{S}' = xy$ **if and only if** $\mathbb{K}$ is perfect, we infer that $\{x, y\}^{\perp \perp}$ is long **if and only if** $\mathbb{K}$ is perfect. So suppose that $\mathbb{K}$ is perfect and consider the map ρ (see step III above) in Γ'. It is bijective. But viewed as a map in Γ, it must also be bijective, showing that the imaginary lines in Γ and Γ' coincide. This immediately implies $\Gamma = \Gamma'$. If $\mathbb{K}$ is not perfect, then since Γ' has no long imaginary lines, neither will Γ have any.

Note that a direct computation with coordinates (either in $\mathsf{H}(\mathbb{K})$ or in $\mathbf{PG}(6, \mathbb{K})$) provides an alternative and more algebraic way of showing that $\mathsf{H}(\mathbb{K})$ has long imaginary lines **if and only if** $\mathbb{K}$ is perfect and has characteristic 2.

Step VI

It remains to show that in the finite case no imaginary line contains more than $s + 1$ points, where (s, t) is the order Γ. Indeed, let y, z be opposite again, and let $v \in \Gamma_{n/2-1}(y) \cap \Gamma_{n/2+1}(z)$. Let L be a line at distance $n/2$ from v such that $\text{proj}_v(y) \neq \text{proj}_v(L) \neq \text{proj}_v(z)$ (this possible by the thickness). Clearly L is at codistance 1 from every element of $I(y, z)$ and also $\text{proj}_L(y) \neq \text{proj}_L(z)$. hence the projection of $I(y, z)$ onto L is injective, and the result follows.

Since in the finite case it is clear that, if an imaginary line contains $s + 1$ points, then it is long, the proof of the theorem is complete. $\qquad \square$

6.5.7 Remark. More generally, it is shown by VAN BON, CUYPERS & VAN MAL-DEGHEM [1994] that, whenever all sets $\{x, y\}^{\perp \perp}$ are long (i.e., the projection of $\{x, y\}^{\perp \perp}$ onto an element at codistance 1 of all points of $\{x, y\}^{\perp \perp}$ is bijective whenever it is injective, or equivalently, whenever it is not constant) for all pairs of points (x, y) such that $\delta(x, y) = 2i$, for some fixed i, $2 \leq i \leq n/2$, then Γ is distance-i-regular. In particular if $i = 2$, one shows that the set $\{x, y\}^{\perp \perp}$ (with $\delta(x, y) = 4$) is long **if and only if** it coincides with any trace through x and y.

6.6 Generalized Desargues configurations

In this section, we show a generalization to all generalized polygons of a particular case of a theorem of BAER [1942] stating that in every projective plane, (p, L)-transitivity is equivalent to being (p, L)-Desarguesian. The idea of relating configurations with collineations is due to BAER [1942]. Let us get down to some definitions.

6.6.1 Definitions. Let $\Gamma = (\mathcal{P}, \mathcal{L}, I)$ be any generalized n-gon, $n > 2$. Let $\mathcal{C} = (v_1, v_2, \ldots, v_{k-1})$ be a $(k-2)$-path in Γ, $3 \leq k \leq n$. Let $\Sigma = (w_1, w_2, \ldots, w_{2n})$ and $\Sigma' = (w_1', w_2', \ldots, w_{2n}')$ be two ordered apartments in Γ. Then Σ **and** Σ' **are in perspective from** $\mathcal{C}$ if for every $i \in \{1, 2, \ldots, k-1\}$ and every $j \in \{1, 2, \ldots, 2n\}$, we have:

(i) $\delta(v_i, w_j) = \delta(v_i, w_j')$;

(ii) if v_i and w_j are not opposite, then the projection of w_j onto v_i is the same as the projection of w_j' onto v_i.

The triplet $(\mathcal{C}, \Sigma, \Sigma')$ is a **generalized k-Desargues configuration**. For $n = 3$, k must also be equal to 3 and we obtain a little Desargues configuration in a projective plane.

Now we explain what $\mathcal{C}$-Desarguesian means for the given $(k-2)$-path $\mathcal{C}$. Let $\Sigma = (w_1, w_2, \ldots, w_{2n})$ be any apartment of Γ. Recall from Lemma 1.5.9 (see page 21) that there is at least one element of Σ opposite v_2 and hence lying at distance $n-1$ from v_1, which we may, without loss of generality, assume to be w_1. We put an order on Σ by starting with w_1 (and starting with a line is allowed now). Let y be the unique element of Γ at distance $n-k+2$ from v_1 and distance $k-3$ from w_1. Let y' be any element of Γ at distance $n-k+2$ from v_1 and such that its projection on v_1 coincides with the projection of y onto v_1. If for every choice of Σ and for every choice of y', there exists an ordered apartment $\Sigma' = (w_1', w_2', \ldots, w_{2n}')$ in perspective with Σ from $\mathcal{C}$ such that $\delta(y', w_1') = k-3$, then we say that $\mathcal{C}$ **is (k-)Desarguesian**. The apartment Σ' is said to be in perspective with Σ from $\mathcal{C}$ **via** y'. If every $(k-2)$-path is Desarguesian, then we say that Γ is k-**Desarguesian**.

Let $\mathcal{C}$ be a $(k-2)$-path, $3 \leq k \leq n$, and let $\overline{\mathcal{C}}$ be a path of length k containing $\mathcal{C}$ in such a way that its extremities do not belong to $\mathcal{C}$. Using Theorem 4.4.2(v), (vi) (page 140), it is easily seen that the group of all collineations (which we will call $\mathcal{C}$-**elations**) fixing every element incident with some element of $\mathcal{C}$ acts semi-regularly on the set of apartments containing $\overline{\mathcal{C}}$. If this action is transitive, hence regular, then we say that $\mathcal{C}$ is a **Moufang path**. This definition is clearly independent of the choice of $\overline{\mathcal{C}}$. If every $(k-2)$-path is a Moufang path, then we call Γ a k-**Moufang polygon**. This generalizes the notion of Moufang polygon since the

Moufang condition is equivalent to the *n-Moufang condition* in every generalized *n*-gon.

We have the following proposition as an almost immediate consequence of these definitions.

6.6.2 Proposition. **If** a $(k-2)$-*path* C, $3 \leq k \leq n$, *is a Moufang path,* **then** *it is Desarguesian.*

Proof. It is clear that, if Σ is any apartment in Γ, and if θ is any C-elation, then Σ and Σ^θ are in perpective from C. So if y and y' are as in the definition of C being Desarguesian above, then it suffices to show that there exists a C-elation θ mapping y to y'.

By definition, $\delta(v_1, y) = n - k + 2$ and $\delta(v_1, v_{k-1}) = k - 2$. Since also $\delta(v_2, y) = n - k + 3$, we conclude that y and v_{k-1} are opposite. Similarly, y' and v_{k-1} are opposite. So if v_k is any element incident with v_{k-1}, $v_k \neq v_{k-2}$, then there exist unique apartments Σ and Σ' containing C and v_k, along with y or y' respectively. Defining $\overline{C}$ as $C \cup \{v_0, v_k\}$, where v_0 is the projection of y and y' onto v_1, we see that there exists now a C-elation θ mapping Σ to Σ' since both Σ and Σ' contain $\overline{C}$. Clearly, θ maps y to y' and the result follows. $\square$

Naturally, the question arises whether the converse of the proposition is also true. This seems to be very hard to prove in general. A positive answer is at present only obtained if $k = n$, or if $k = 3$, $n = 4$ and Γ is finite; see VAN MALDEGHEM, THAS & PAYNE [1992]. We will focus on the case $k = n$ and state the result. But we will only show it in the case $n = 6$. All arguments are, however, easily generalized to arbitrary n. The only reason for restricting to $n = 6$ is notational.

6.6.3 Theorem (Baer [1942], Thas & Van Maldeghem [1990], Van Maldeghem [1990d]). *An $(n-2)$-path in a generalized n-gon Γ is Desarguesian* **if and only if** *it is a Moufang path.*

Proof. As indicated, only the case $n = 6$ will be considered here. A detailed proof for the case $n = 4$ can be found in THAS & VAN MALDEGHEM [1990]. The general case is similar to the case $n = 6$ below.

So let Γ be a generalized hexagon and let $C = (v_1, v_2, v_3, v_4, v_5)$ be a 4-path. We coordinatize Γ in such a way that $v_5 = (\infty)$, $v_4 = [\infty]$, $v_3 = (0)$, $v_2 = [0, 0]$ and $v_1 = (0, 0, 0)$ (we may need to replace Γ by its dual to obtain this). We use the same notation as in Chapter 3. In particular, we have a (dual) hexagonal sexternary ring $(R_1, R_2, \Phi_1, \Phi_2, \Phi_3, \Phi_4)$. As for notation, we again assume that as, bs and cs come out of R_1, while ks, ls and ms come out of R_2.

We introduce two new operations, namely, we put $\Phi_2'(a, l, a', k, b, k') = l'$ **if and only if** $\delta([a, l, a', l'], [k, b, k']) = 4$; and $\Psi_3'(a, l, k, b, k', b') = a'$ **if and only if**

$\delta((a, l, a'), (k, b, k', b')) = 4$. It is easily seen that

$$(a, l, a', l', a'') \mathbf{I} [k, b, k', b', k'']$$

$$\Longleftrightarrow$$

$$\begin{cases} a'' &=& \Phi_1(a, k, b, k', b', k''), \\ l' &=& \Phi_2'(a, l, a', k, b, k'), \\ a' &=& \Phi_3'(a, l, k, b, k', b'), \\ l &=& \Phi_4(a, k, b, k', b', k''). \end{cases}$$

It is now convenient *not* to assume that $(R_1, R_2, \dots)$ is normalized, but instead to assume some other normalization condition. To that end, we suppose that there is an element $1 \in R_2$ such that the projection of the point $\mathrm{proj}_{[1,0,0,0,0]}(a)$ onto the line $[0]$ coincides with the point $(0, a)$. This means that we assume $\Phi_1(a, 1, 0, 0, 0, 0) = a$. Noting $\Phi_1(0, 1, b, 0, 0, 0) = b$, we can define an addition in R_1 by

$$a + b = \Phi_1(a, 1, b, 0, 0, 0),$$

and this has the usual property

$$0 + a = a = a + 0.$$

Let θ be a $\mathcal{C}$-elation mapping $(0, 0)$ to $(0, B)$. Obviously, $[1, 0, 0, 0, 0]$ is mapped to $[1, B, 0, 0, 0]$ and hence

$$(a, \Phi_4(a, 1, 0^4), \Phi_3(a, 1, 0^4), \Phi_2(a, 1, 0^4), \Phi_1(a, 1, 0^4))$$

is mapped to

$$(a, \Phi_4(a, 1, B, 0^3), \dots, \Phi_1(a, 1, B, 0^3)) = (a, \dots, a + B).$$

By projecting this onto the line $[0]$, we see that $(0, a)^\theta = (0, a + B)$ and by projecting the latter onto $[0, 0, 0, 0]$ and afterwards the result onto $[k]$, we see that $(k, b)^\theta = (k, b + B)$. Now it is not hard to see that

$$[k, b, k', b', k'']^\theta = [k, b + B, k', b', k''].$$

Since $[0, 0, 0, 0, l]^\theta = [0, B, 0, 0, l]$, the line $[a, l]$, which is equal to $[a, \Phi_4(a, 0^4, l)]$, is mapped onto $[a, \Phi_4(a, 0, B, 0, 0, l)]$. We write for short

$$\begin{aligned} \Upsilon_1(a, b, k) &=& \Phi_4(a, 0, b, 0, 0, k), \\ \Upsilon_2(a, b, k, c) &=& \Phi_3'(a, \Upsilon_1(a, b, k), 0, b, 0, c), \\ \Upsilon_3(a, b, k, c, l) &=& \Phi_2'(a, \Upsilon_1(a, b, k), \Upsilon_2(a, b, k, c), 0, b, l). \end{aligned}$$

Then clearly $[a, l]^\theta = [a, \Upsilon_1(a, B, l)]$ and one can check similarly that

$$(a, l, a', l', a'')^\theta = (a, \Upsilon_1(a, B, l), \Upsilon_2(a, B, l, a'), \Upsilon_3(a, B, l, a', l'), a'' + B).$$

Since θ is a collineation, $(a, l, a', l', a'') \mathbf{I} [k, b, k', b', k'']$ is equivalent to

$$(a, \Upsilon_1(a, B, l), \Upsilon_2(a, B, l, a'), \Upsilon_3(a, B, l, a', l'), a'' + B) \mathbf{I} [k, b + B, k', b', k'']. \quad (*)$$

Conversely, if $(*)$ holds whenever $(a, l, a', l', a'') \mathbf{I} [k, b, k', b', k'']$, for some $B \in R_1$, then obviously, there exists a $\mathcal{C}$-elation mapping $(0,0)$ to $(0, B)$. We will show that, assuming $\mathcal{C}$ is Desarguesian, $(*)$ holds for every $B \in R_1$ whenever $(a, l, a', l', a'') \mathbf{I} [k, b, k', b', k'']$.

We will use the following notation. If $\Sigma = (w_1, \dots, w_{12})$ is an (ordered) apartment, w_1 opposite $v_2 = [0, 0]$ and at distance 5 from $(0, 0, 0)$, and if y' is a point incident with the projection of w_1 onto $(0, 0, 0)$, then we denote an apartment in perspective with Σ from $\mathcal{C}$ via y' by $\Sigma'(w_1, y')$. In this way, we emphasize the fact that we consider w_1 to be the first element of Σ (and we may always reorder Σ such that we can have any element of Σ as first element; as an exception, we also allow lines to be the first element of Σ here).

Suppose $(a, l, a', l', a'') \mathbf{I} [k, 0^4]$, $0 \neq k \neq 1$, $a \neq 0$ and consider the apartment (defining the line M at the same time)

$$\begin{aligned} \Sigma \;=\; & ([k, 0^4], (0^5), [1, 0^4], (a'', \Phi_4(a, 1, 0^4), \Phi_3(a, 1, 0^4), \Phi_2(a, 1, 0^4), a''), \\ & M, \dots, (0, a''), \dots, (a, l, a', l', a'')). \end{aligned}$$

We construct $\Sigma' = \Sigma'(M, (0^4, B))$ (which is unique). The line L of Σ' corresponding to $[k, 0^4]$ must meet $[0^4]$ and be at distance 3 from (k, B), hence L has coordinates $[k, B, \dots, 0]$. But considering the apartment

$$\Sigma_1 = ([k, 0^4], (0^5), [0^5], (0^4), [0^3], (0^2), [0], (\infty), [k], \dots),$$

we see that $\Sigma'_1([0^3], (0^4, B))$ must contain L and the unique path connecting (k, B) with L. By definition of "being in perspective with", L must have coordinates $[k, B, 0, 0, 0]$. Similarly, $[1, 0^4]$ corresponds to $[1, B, 0^3]$ and so $(a'', \dots, a'')$ corresponds to $(a'', \dots, a'' + B)$, $(0, a'')$ to $(0, a'' + B)$ and (a, l, a', l', a'') to some point $(a, \dots, a'' + B)$. So we have

$$\Phi_1(a, k, B, 0, 0, 0) = a'' + B = \Phi_1(a, k, 0^4) + B.$$

This property, which we will refer to as property (P1), remains trivially true for $a = 0$ or $k = 0$ or $k = 1$.

Now consider a 3-path $([k, 0^4], x, M, (a, l, a', l', a''))$, for some $k, a, l, \dots$ and M, with $[k, 0^4]$ not incident with (a, l, a', l', a''). If an apartment Σ contains this path, then the element p' in $\Sigma' = \Sigma'(M, (0^4, B))$ corresponding to $p = (a, l, a', l', a'')$ is independent of Σ. Indeed, the element x' in Σ' corresponding to x is the projection onto $[k, B, 0^3]$ of the projection of x onto $[\infty]$ (so we see that whenever an element of Σ is opposite some element of $\mathcal{C}$, then the element in Σ' corresponding to the next element of Σ is uniquely determined by $\mathcal{C}$); x is opposite (0) and M is opposite $[\infty]$ or opposite $[0, 0]$, hence the claim follows from the previous remark in parentheses.

In order to know what the coordinates of p' look like, we first assume $k \neq 0$ and we choose Σ in such a way that it contains $(0, a'')$. By property (P1), Σ' contains $(0, a'' + B)$ at distance 4 from p', hence $p' = (a, \dots, a'' + B)$. Suppose now $k = 0$; then we consider the unique apartment Σ containing p, (0^5), $[1, 0^4]$, $p \bowtie (0, a'')$. The unique element of Σ collinear with $p \bowtie (0, a'')$ and different from p has also as last coordinate a'' and hence by the case $k \neq 0$ above, the corresponding element in Σ' has as last coordinate $a'' + B$ and so does p'.

So we have shown that, whenever an apartment Σ contains the 3-path

$$([k, 0^4], x, M, (a, l, a', l', a'')),$$

then the point corresponding to p in $\Sigma'(M, (0^4, B))$ has coordinates $(a, \dots, a'' + B)$. We refer to this as property (P2).

Now consider a 3-path $(L, x, M, (a, l, a', l', a''))$ with $\delta(L, (0^5)) = 3$ and $\delta(L, [0^4]) = 4$. In the same way as above, one shows that, whenever x and M are opposite some element of $\mathcal{C}$ (noting that L is always opposite $[0, 0]$), and whenever this path is contained in an apartment Σ, then the point corresponding to (a, l, a', l', a'') in $\Sigma'(L, (0^4, B))$ has coordinates $(a, \dots, a'' + B)$, where $L = [k, b, \dots]$. Call this property (P3) for further reference.

Consider a point $p = (a, l, a', l', a'')$ collinear with (0^5). Suppose first that p is not incident with $[0^5]$. Consider the apartment

$$\Sigma = (p, L, (0^5), [0^5], (0^4), [0^4, l], \dots, [a, l], (a, l, a'), [a, l, a'l']).$$

If $L = [k, \dots]$, then $\Sigma'([0^4, l], (0^4, B))$ contains correspondingly

$$(p', L', (0, 0, 0, 0, B), [0, B, 0, 0, 0], (0, B, 0, 0), [0, B, 0, 0, l])$$

(use an apartment containing $(0, 0)$ and (0^5) to obtain $[0, B, 0, 0, 0]$ and $(0, B, 0, 0)$), hence $[a, l]$ corresponds to the projection of $[0, B, 0, 0, l]$ onto $[a, l]$, which is

$$[a, \Phi_4(a, 0, B, 0, 0, l)] = [a, \Upsilon_1(a, B, l)].$$

Since p' lies at distance 3 from the latter, we have, also by property (P1), that $p' = (a, \Upsilon_1(a, B, l), \dots, a'' + B)$. Also properties (P2) and (P3) can be generalized in a similar way such that the conclusion reads $(a, \Upsilon_1(a, B, l), \dots, a'' + B)$ instead of $(a, \dots, a'' + B)$.

In fact, it is now clear that in the conclusion of properties (P1), (P2) and (P3), we can replace $(a, \dots, a'' + B)$ by

$$(a, \Upsilon_1(a, B, l), \Upsilon_2(a, B, l, a'), \Upsilon_3(a, B, l, a', l'), a'' + B).$$

Call this property (P4).

In a similar fashion, one shows that, whenever an apartment Σ contains the path

$$((k, b), [k, b, k'], (k, b, k', b'), [k, b, k', b', k'']),$$

where $L = [k, b, k', b', k'']$ is a line at distance 3 from (0^5) not meeting $[0^4]$ or $[k, 0^4]$, then $\Sigma'(L, (0^4, B))$ contains $(k, b + B)$ corresponding to (k, b). We refer to this as property (P5).

Now suppose that $p = (a, l, a', l', a'') \mathbf{I} [k, b, k', b', k''] = L$, that p is opposite (0^5), and suppose there exists $m \in R_2$ such that $m \neq k$ and all elements of the 5-path $([m, 0^4], x, M, y, N, p)$ are opposite some element of $\mathcal{C}$ (this is certainly true if there are at least four lines through any point ($t \geq 3$); note by the way that we cannot *assume* that $t \geq 3$ by the uniqueness of the finite hexagons with $t = 2$ since s could be infinite and since our proof must also be valid for general n). Consider the apartments Σ_1 containing p, L, (k, b, k', b') and $[m, 0^4]$, and Σ_2 through (k, b), (k, b, k', b') and $[m, 0^4]$. These apartments share a unique line $L_{1,2}$ meeting $[m, 0^4]$. Denote the other line of Σ_2 meeting $[m, 0^4]$ by M_2. Then it is easily checked that $\Sigma'_1(L_{1,2}, (0^4, B)) = \Sigma'_1(M, (0^4, B))$ and $\Sigma'_2(L_{1,2}, (0^4, B)) = \Sigma'_2(M_2, (0^4, B))$ and the elements p' and L' corresponding to p and L are, respectively, the same in these two apartments. But by property (P5) and the construction of L', L' has coordinates $[k, b + B, k', b', k'']$, while p' has coordinates $(a, \Upsilon_1(a, B, l), \Upsilon_2(a, B, l, a'), \Upsilon_3(a, B, l, a', l'), a'' + B)$ by property (P4). Hence, in this case,

$$[k, b + B, k', b', k''] \mathbf{I} (a, \Upsilon_1(a, B, l), \Upsilon_2(a, B, l, a'), \Upsilon_3(a, B, l, a', l'), a'' + B).$$

If $t = 2$, or if $\delta(p, (0^5)) \leq 4$, then this can be proved similarly, using different apartments. Only the case $\delta(p, [0^4]) = 3$ requires some extra work. Indeed, in this case property (P4) does not suffice and one has to show that, whenever an apartment Σ contains a 5-path beginning in $p = (a, l, a', l', a'')$ and ending with a line M at distance 3 from (0^5), and every element of that path is opposite some element of $\mathcal{C}$ (so that the corresponding path in $\Sigma' = \Sigma'(M, (0^4, B))$ is independent of Σ), then the point p' in Σ' corresponding to p has coordinates $(a, \Upsilon_1(a, B, l), \Upsilon_2(a, B, l, a'), \Upsilon_3(a, B, l, a', l'), a'' + B)$. This can be proved similarly to property (P4) itself, except that at each step, one has to use one more apartment (since the path considered is longer). With these guidelines, it should not be too hard to reconstruct all arguments in detail.

This completes the proof of the theorem.   □

As a consequence, we have the following result:

6.6.4 Theorem (Baer [1942], Thas & Van Maldeghem [1990], Van Maldeghem [1990d]). *A generalized n-gon Γ is n-Desarguesian* **if and only if** *it is a Moufang polygon.*

Proof. This follows immediately from the preceding theorem and the fact that the Moufang property is equivalent to the n-Moufang property.   □

Another corollary is related to the classical generalized quadrangles with regular lines. We first need some preliminaries.

6.6.5 Definitions. Let Γ be a generalized quadrangle and L a line of Γ. We say that L is an **axis of symmetry** if L is a regular line and if every 2-path (x, L, y), is a Moufang path. Now let Σ be an apartment of Γ and suppose that no line of Σ meets L. Then we call Σ **opposite** L. Let the points of Σ be cyclically ordered as p_1, p_2, p_3, p_4 and let L_i be the unique line through p_i meeting L, $i \in \{1, 2, 3, 4\}$. Let q_1 be any point on L_1, but not on L. Let q_2 be the projection of q_1 onto L_2, let q_3 be the projection of q_2 onto L_3, let q_4 be the projection of q_3 onto L_4 and let finally q_5 be the projection of q_4 onto L_1. If for every choice of Σ and every choice of q_1, we always have $q_5 = q_1$ and q_1, q_2, q_3 and q_4 are points of an apartment of Γ, then we call L **regular-Desarguesian**. The following result was first proved by RONAN [1980c] (although he overlooked the necessary assumption $t > 2$).

6.6.6 Corollary (Ronan [1980c]). *Suppose that Γ is a generalized quadrangle of order (s, t), s, t possibly infinite, and suppose that $t > 2$. Then a line L is regular-Desarguesian* **if and only if** *L is an axis of symmetry for Γ.*

Proof. Clearly, if L is an axis of symmetry, then it is regular-Desarguesian. Suppose now that the line L is regular-Desarguesian. We first show that L is regular. Let L' be opposite L and let M_i, $i = 1, 2, 3$, be three distinct lines meeting both L and L'. Let L'' be a line meeting M_1 and M_2, but not M_3. Let $x_1' = L' \cap M_1$, $x_1'' = L'' \cap M_1$, $x_2' = L' \cap M_2$, $x_2'' = L'' \cap M_2$ and $x_3' = L' \cap M_3$. Let x_3'' be the projection of x_2'' on M_3. Since $t > 2$, there exists a point x collinear with x_1'' and x_3'', but not incident with M_i, $i = 1, 2, 3$. The apartment with points x_1'', x_2'', x_3'', x is opposite L and since L is regular-Desarguesian, x_1', x_2', x_3', x' must be contained in an apartment, a contradiction (x' is the projection of x_3 on the line through x meeting L). Hence L is regular. But now the apartments containing p_1, p_2, p_3, p_4 and q_1, q_2, q_3, q_4, respectively, in the definition of regular-Desarguesian are in perspective from any 2-path (p, L, q), with p and q (distinct) points on L. It follows easily that every such path is Desarguesian, and hence every such path is a Moufang path by the preceding theorem. □

We end this section with a result of KISS [1990]. We omit the proof; it is similar to the case of a projective plane, proved by OSTROM [1957].

6.6.7 Theorem (Ostrom [1957], Kiss [1990]). *Every finite generalized polygon contains a generalized 3-Desargues configuration, and for each $n \geq 3$, there exists a (free) generalized n-gon containing no generalized 3-Desargues configuration.*

An infinite generalized n-gon containing no generalized 3-Desargues configuration is obtained by the free construction of Subsection 1.3.13 (on page 13) starting with an ordinary $(n + 1)$-gon.

6.7   Some combinatorial characterizations

In this section, we collect and prove some geometric characterizations of classes of finite Moufang polygons. We start with a characterization of the extremal quadrangles and hexagons as those finite polygons having a lot of full or ideal (thick) subpolygons. After this, we focus on the square finite polygons (those of order (s, s)). Lots of combinatorial characterizations of finite Moufang projective planes and generalized quadrangles are ignored here. We refer to e.g. PICKERT [1975], STEVENSON [1972] or DEMBOWSKI [1968], and PAYNE & THAS [1984].

6.7.1 Theorem (Thas [1978], De Smet & Van Maldeghem [1993a]). *Let* Γ *be a finite generalized n-gon,* $n \in \{3, 4, 6, 8\}$. **If** *every ordinary* $(n + 1)$*-gon is contained in a proper ideal subpolygon,* **then** Γ *is isomorphic to either the Hermitian quadrangle* $\mathsf{H}(3, q)$, *or the twisted triality hexagon* $\mathsf{T}(q^3, q)$.

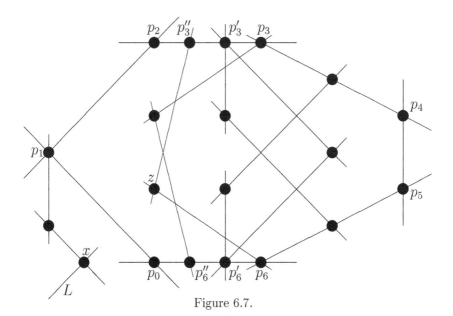

Figure 6.7.

Proof. Since no finite projective plane or generalized octagon can have a (thick and proper) ideal subpolygon (by Corollary 1.8.3 on page 34 and Theorem 1.8.8 on page 36, respectively), we already have $n = 4$ or $n = 6$. For $n = 4$, we refer to PAYNE & THAS [1984](5.3.5(ii)). So we may suppose that n equals 6. Let Ω be an ordinary heptagon in Γ with points p_i, $i = 0, 1, \ldots, 6$ such that $p_i \perp p_{i+1}$ (subscripts to be taken modulo 7); see Figure 6.7. Let Γ' be a proper ideal subhexagon containing Ω. Since $\Gamma \neq \Gamma'$, there is at least one point p_6' on the line $p_0 p_6$ not contained in Γ'. Let Ω' be the ordinary heptagon determined by the points p_0, p_1, p_2, p_3, p_6' and the line $p_3 p_4$. By assumption, there exists an ideal subhexagon Γ'' containing Ω' and by the choice of p_6', we have $\Gamma' \neq \Gamma''$. But $\Gamma' \cap \Gamma''$ is a weak ideal subhexagon since it contains the apartment Σ determined by the points p_0, p_1, p_2, p_3 and the line $p_0 p_6 = p_0 p_6'$ (compare Proposition 1.8.4). By Theorem 1.8.8(iv) on page 36, the order of Γ is (t^3, t), the order of Γ' and Γ'' is (t, t) and the order of $\Gamma' \cap \Gamma''$ is $(1, t)$, for some integer $t > 1$. Hence p_6 does not belong to Γ''. Let p_3' be the projection of p_6' onto $p_2 p_3$ and let p_5' be the projection of p_3' onto $p_5 p_6$. By assumption, there is a proper ideal subhexagon Γ''' containing $p_0, p_1, p_2, p_3', p_3' \bowtie p_5', p_5'$ and p_6.

Now let p_3'' be the projection of p_6 onto $p_2 p_3$ and put $z = p_3'' \bowtie p_6$. Clearly z does not belong to Γ'', so z does not belong to $\Gamma'' \cap \Gamma'''$. By Corollary 1.8.11(ii) (see page 37), z is at distance 3 from exactly $1 + t$ lines of Γ''. Also, by the same reference, z is at distance 3 from exactly $1+t$ lines of $\Gamma' \cap \Gamma''$ or $\Gamma'' \cap \Gamma'''$ respectively. Hence $\Gamma' \cap \Gamma'' \cap \Gamma'''$ contains $1 + t$ lines all of which are at distance 3 from z. We claim that these $1 + t$ lines are at distance 3 from p_1.

Indeed, let L be one of these lines and suppose L is at distance 5 from p_1. Let x be the projection of p_1 onto L. Then x belongs to both $\Gamma' \cap \Gamma''$ and $\Gamma'' \cap \Gamma'''$. Since $\delta(p_1, x) = 4 = \delta(p_1, p_6')$ and p_6' belongs to $\Gamma' \cap \Gamma'''$ (which has only two points per line), we must have $\delta(p_6', x) = 4$. Similarly $\delta(p_6'', x) = 4$, where p_6'' is the projection of p_3 onto $p_0 p_6$. But then we have a pentagon containing p_6', p_6'' and x, a contradiction. Hence L is at distance 3 from p_1 and our claim is proved.

But now this implies that $p_1^z = p_1^y$, for $y = p_3' \bowtie p_6'$. By varying p_6', all we need to show is that every intersection set has size 1. Remark 6.3.6 on page 255 then completes the proof of the theorem. But it is easily seen that any intersection set of size at least 2 in Γ is in fact an intersection set in a weak ideal subhexagon of order $(1, t)$, and in such a weak hexagon there are no intersection sets. The result follows. $\qquad\square$

We now consider the square finite Moufang polygons. Our first results are restatements of Theorem 6.5.3 and Theorem 6.5.6 on page 265 and page 266, respectively.

6.7.2 Theorem. *If in a finite generalized polygon* Γ *of order* (s, t) *which is not a projective plane, a hyperbolic or imaginary line has at least* $s + 1$ *points,* **then** *it contains exactly* $s + 1$ *points and* $s = t$. *If, moreover, all hyperbolic (or imaginary) lines have at least* $s + 1$ *points, then* Γ *is a Pappian (or symplectic) polygon.* $\qquad\square$

Next, we look at intersections of traces.

6.7.3 Theorem. *Let* Γ *be a finite generalized polygon of order* (s, t) *with* $s \geq t$ *which is not a projective plane. Then* Γ *is a Pappian polygon* **if and only if** *for every point* p *and every pair of points* x, y *opposite* p, *the two traces* p^x *and* p^y *meet in at least one point.*

Proof. The proof we will present treats the quadrangles, hexagons and octagons at the same time. A different proof for the case of quadrangles can be derived from PAYNE & THAS [1984], Equation (7) on page 6 and 1.3.6(ii).

Clearly the polygons $W(q)$ and $H(q)$ have the desired property.

To prove the converse, let p be any point and let x and y be both opposite p. We show that we either have $p^x = p^y$ or $|p^x \cap p^y| = 1$ and that, moreover, $s = t$. Suppose $p^x \neq p^y$ (such a pair (x, y) always exists!) and let L be a line through p such that the projection x' of x onto L differs from the projection y' of y onto L. Let M be the projection of y' onto y and let y'' be the projection of y' onto M. The restriction of the projection map is a mapping

$$\varphi : p^x \setminus \{x'\} \to \Gamma_1(M) \setminus \{y''\}$$

which is by assumption surjective. This implies that $s = t$ and φ is a bijection. Hence $|p^x \cap p^y| = 1$. Hence all points are distance-2-regular and the result follows from Corollary 1.9.5 (see page 39), Theorem 6.2.1, Theorem 6.3.1 and Theorem 6.4.6.   □

A **triad** in a generalized quadrangle is a set of three mutually non-collinear points. A **centric triad** is a triad for which there exists at least one point collinear with all points of the given triad.

The previous theorem can now be rephrased for generalized quadrangles and hexagons as follows:

6.7.4 Corollary. *Let* Γ *be a finite generalized* n-*gon of order* (s, t) *with* $s \geq t$.

(i) **(Thas [1973])** *If* $n = 4$, *then* $\Gamma \cong W(q)$ *for some prime power* q **if and only if** *every triad is centric.*

(ii) **(Thas [1980a])** *If* $n = 6$, *then* $\Gamma \cong H(q)$ *for some prime power* q **if and only if** *for any three points* x, y, z *the set* $x^\perp \cap y^{\perp\perp} \cap z^{\perp\perp}$ *is non-empty.*

Hence a finite generalized n-*gon* Γ *with* n *even and of order* (s, t), *with* $s \geq t$, *is Pappian* **if and only if** *for any three points* x, y, z *of* Γ, *the set* $x^\perp \cap y^{\perp\perp} \cap z^{\perp\perp}$ *is non-empty.*   □

The next characterization theorem deals with the finite square Moufang polygons having regular points and having an ovoid. An **ovoid** O of a generalized $2m$-gon Γ is a set of mutually opposite points such that every element x of Γ is at distance at most m from at least one point p of O. It follows that, if this distance is less than m, then p is uniquely determined by x. More about ovoids is contained in Chapter 7.

6.7.5 Theorem (Thas [1973], De Smet & Van Maldeghem [1993a]). *Let* Γ *be a finite generalized polygon of order* s *containing an ovoid* O. *Then* Γ *is point-regular (and hence it is a Pappian polygon)* **if and only if** *every point of* O *is span-regular.*

Proof. Note that Γ is either a generalized quadrangle or a generalized hexagon. If Γ is point-regular, then obviously, all points of O are span-regular. Now suppose all points of O are span-regular. If Γ is a quadrangle, the result follows from THAS [1973]; see also PAYNE & THAS [1984](5.2.5). An alternative direct proof in the same spirit as the one below for generalized hexagons may run as follows. Consider a point p of O and any dual grid $\mathcal{G}$ containing p, i.e., $\mathcal{G}$ is the set of points $p^x \cup \{p,x\}^{\perp\perp}$, for some x opposite p. Put $\mathcal{G}_1 = p^x$ and $\mathcal{G}_2 = \{p,x\}^{\perp\perp}$. Every point of $\mathcal{G}_1$ is collinear with every point of $\mathcal{G}_2$, hence there are $(s+1)^2$ lines meeting $\mathcal{G}$ in two points. The number of points on these lines is equal to $(s+1)^2(s-1)+2(s+1)$ and this is exactly the total number of points in Γ. Since every line must contain a point of O, and since $|O| = s^2 + 1$ (see Proposition 7.2.3 on page 307), there is exactly one other point q of O in $\mathcal{G}_2$. Counting the number of opposite pairs (x,y) contained in such a set $p^x \cup \{p,x\}^{\perp\perp}$ (for x varying over all points of O), we see that every pair of opposite points lies in such a set $\mathcal{G}$. By definition of $\mathcal{G}$, this implies that every point is regular, hence the result.

Now let Γ be a generalized hexagon. From Subsection 1.9.8 it follows that every pair of points of O is contained in an ideal weak subhexagon with two points per line. Counting the number of such weak subpolygons thus obtained (namely $\frac{s^3(s^3+1)}{2}$), counting the number of opposite pairs of points contained in such weak subpolygons (namely $(s^2+s+1)s^2$ in each one) and noting that no two distinct such weak subpolygons share two opposite points, we see that every pair of opposite points of Γ is contained in an ideal weak subhexagon with two points per line. So it already follows that every intersection set has size 1. Moreover, every weak subhexagon of order $(1,s)$ contains exactly two points of O.

Recall that we denote the unique ideal weak subhexagon with two points per line through two opposite points x and y by $\Gamma(x,y)$.

Now let x,y be two opposite points and let p be the unique point of O collinear with x. Let z be such that it is opposite both x and y and at distance 3 from at least two lines L and M belonging to $\Gamma_3(x) \cap \Gamma_3(y)$. We have to show that $x^y = x^z$. This is certainly true if $p = x$ or if $p \in x^y$. So suppose $p \in \Gamma_2(x) \setminus x^y$. We may assume that p is not at distance 3 from L. Since two points of $\Gamma(x,z)$ at distance 6 from x are at distance 4 from each other, there is exactly one point p_z of O in $\Gamma(x,z)$ at distance 4 from x. First suppose that p_z is collinear with a point of x^y. We may obviously assume that $\delta(p_z,L) = 3$. Put $L' := \mathrm{proj}_{p_z} x$ and remark that L' belongs to both $\Gamma(x,y)$ and $\Gamma(x,z)$. Let $a = \mathrm{proj}_M y$ and let $b = \mathrm{proj}_M z$. Let $a' = \mathrm{proj}_{L'} a$ and let $b' = \mathrm{proj}_{L'} b$. Then we may redefine y and z as, respectively, $a \bowtie a'$ and $b \bowtie b'$, and we redefine L as L'. So we have shown that we may assume that p_z is incident with L. Similarly, we may assume that the unique point p_y of O at distance 4 from x and lying in $\Gamma(x,y)$ is collinear with y. Suppose $x^z \neq x^y$. Then there is a line N through x such that the projection u of y onto N differs from the projection v of z onto N (see Figure 6.8). Let a and b be as above, i.e., $a = \mathrm{proj}_M y$ and $b = \mathrm{proj}_M z$. Note that y^x contains a, p_y, $y \bowtie p_z$ and $y \bowtie u$. Since y^z contains a and $y \bowtie p_z$, and since p_y is span-regular, y^z also contains $y \bowtie u$ by

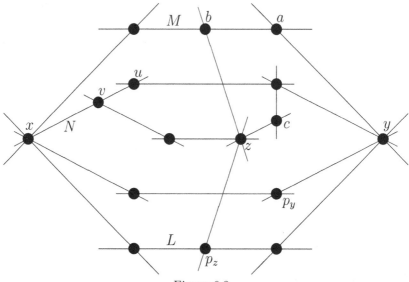

Figure 6.8.

Lemma 1.9.9 on page 41. Put $c = (y \bowtie u) \bowtie z$. Since $z^x \cap z^y$ contains b and p_z, and since $c \in z^y$, c should also be in z^x, but clearly c is opposite x, a contradiction unless $v = u$ and c is collinear with v. But this implies $x^z = x^y$.

Now suppose that p_z is not collinear with any point of $\Gamma(x, y)$. Then it is at distance 3 from a unique line xc of $\Gamma(x, y)$, with c a point of $\Gamma(x, y)$, with $\delta(p_z, c) = 4$. Put $c' = x \bowtie p_z$. Let $M' = \mathrm{proj}_y M$. Let $y' = \mathrm{proj}_{M'} c'$. By the previous argument (the conditions are indeed satisfied by replacing z with a suitable point z' of $\Gamma(x, z)$!) we have $x^z = x^{y'}$ (noting that $x^z = x^{z'}$, for all z' opposite x in $\Gamma(x, z)$). This implies that $\mathrm{proj}_L x \in x^{y'}$, hence $y = y'$. Consequently $x^y = x^z$ and the theorem follows from Remark 6.3.6 on page 255.   □

A completely similar proof applies to the following result:

6.7.6 Theorem (Payne & Thas [1984], Brouns & Van Maldeghem [19**]). *Let* Γ *be a finite generalized $2m$-gon, $m = 2, 3$, of order s and let O be the set of points at distance at most m from some given point (if $m = 2$) or line (if $m = 3$). Then* Γ *is point-regular (and hence it is a Pappian polygon)* **if and only if** *every point of* O *is span-regular.*

In fact, BROUNS & VAN MALDEGHEM [19**] prove the following more general result for finite generalized hexagons:

6.7.7 Theorem (Brouns & Van Maldeghem [19**]). *Let* Γ *be a finite generalized hexagon and let O be either an ovoid of Γ, or the set of points at distance at most*

3 *from some line in* Γ, *or the point set of a full weak subhexagon of* Γ. *Then* Γ *is point-regular (and hence it is a Moufang hexagon)* **if and only if** *every point of* O *is span-regular and every pair of opposite points in* Γ *is contained in a weak non-thick ideal subhexagon.*

We end this section by mentioning three other recent characterizations without proof.

6.7.8 Theorem (Thas & Van Maldeghem [19**b]). *A finite generalized quadrangle* Γ *of order* (s^2, s^3) *is isomorphic to* $\mathsf{H}(4, s^2)$ *if and only if any two opposite lines are contained in a proper full subquadrangle.*

6.7.9 Theorem (Van Maldeghem, unpublished, Govaert [1997]). *A finite generalized hexagon* Γ *of order* (s, t) *is isomorphic to* $\mathsf{H}(s)$, *s even, or to* $\mathsf{T}(t^3, t)$, *t even,* **if and only if** *for all lines* L, M *and every point* x *of* Γ, *the set* $\Gamma_{\leq 3}(L) \cap \Gamma_{\leq 3}(M) \cap \Gamma_{\leq 4}(x)$ *is non-empty.*

A generalization of the last result to finite octagons and infinite hexagons is contained in GOVAERT [1997].

6.7.10 Theorem (Govaert & Van Maldeghem [19**]). *Let* Γ *be a finite generalized hexagon of order* (s, t) *with s even. If for every pair of opposite lines* L, M *of* Γ, *and for every point* x *at distance 4 from a unique element* y *of* $\Gamma_3(L) \cap \Gamma_3(M)$ *and opposite every other element of* $\Gamma_3(L) \cap \Gamma_3(M)$, *the point* $x \bowtie y$ *is not opposite any member of* $\Gamma_3(L) \cap \Gamma_3(M)$, **then** Γ *is a line-distance-2-regular (and hence a Moufang) hexagon.*

6.8 Some algebraic characterizations

In this section, we state some group-theoretical characterizations of the finite Moufang polygons. In most of the cases, the proof consists of showing that the polygons under consideration satisfy the Moufang condition. All proofs need a case-by-case argument, in other words, there are different arguments for planes, quadrangles, hexagons and octagons. Usually, these proofs are long, and so we do not reproduce the full proofs here, but we choose one particular case. Since for the case of projective planes the results are classical, and since the case of generalized quadrangles needs some machinery developed by PAYNE & THAS [1984] (going beyond the scope of this book), we will, regarding proofs, mostly be concerned with the hexagons and octagons.

We remind the reader of the notions of k-*Moufang* and k-*Desarguesian* in Definitions 6.6.1 on page 271. The following lemma is an easy and direct consequence of these definitions. Afterwards we show that the notion of k-Moufang is equivalent to the notion of Moufang, whenever $k \geq 4$.

6.8.1 Lemma. *A k-Moufang polygon is a k'-Moufang polygon for $3 \leq k' \leq k$.*

$\square$

6.8.2 Theorem (Van Maldeghem & Weiss [1992]). *Let $4 \leq k \leq n$. A generalized n-gon Γ has the k-Moufang property* **if and only if** *Γ has the Moufang property.*

Proof. We may suppose that $k < n$ since for $k = n$ the result is trivially true by definition. Also, by Lemma 6.8.1, every Moufang n-gon is a k-Moufang n-gon, for every k, $3 \leq k \leq n$. So suppose from now on that Γ is a k-Moufang n-gon with $4 \leq k < n$. We show that Γ is a $(k + 1)$-Moufang polygon, from which the result follows. Let $\gamma = (v_0, v_1, \ldots, v_{k+1})$ be a $(k + 1)$-path. Let Σ and Σ' be two apartments containing γ. Since $\gamma_0 = (v_1, \ldots, v_{k+1})$ is a Moufang path by assumption, there exists a collineation θ mapping Σ onto Σ' and fixing $\Gamma_1(v_i)$ pointwise for every $i \in \{2, 3, \ldots, k\}$. Similarly, there exists a collineation θ' mapping Σ onto Σ' and fixing $\Gamma_1(v_j)$ pointwise, for all $j \in \{1, 2, \ldots, k - 1\}$. The collineation $\theta\theta'^{-1}$ fixes Σ and every set $\Gamma_1(v_l)$ pointwise, for all $l \in \{2, \ldots, k - 1\}$. Since $k - 1 \geq 3$, $\theta\theta'^{-1}$ is the identity (by Theorem 4.4.2(v), (vi); see page 140) and so $\theta = \theta'$ fixes all elements incident with v_l, $l = 1, \ldots, k$. Hence γ is a Moufang path and the theorem is proved. $\square$

The next result says that for finite generalized polygons Theorem 6.8.2 remains valid for $k = 3$.

6.8.3 Theorem (Van Maldeghem, Thas & Payne [1992], Van Maldeghem & Weiss [1992]). *A finite generalized polygon Γ has the 3-Moufang property* **if and only if** *Γ has the Moufang property.*

Proof. We show the theorem for generalized octagons, given the result for generalized quadrangles (which can be proved either by a similar method, or using the PAYNE & THAS machinery of [1984]). The proof by VAN MALDEGHEM & WEISS in [1992] is group-theoretical and we present here the original geometric proof (which is, however, a little longer). So from now on, suppose that Γ is a 3-Moufang finite generalized octagon of order (s, t) with automorphism group G.

We first introduce some notation. Throughout, we fix an apartment Σ and denote its elements by

$$p_1 \mathbf{I} L_2 \mathbf{I} p_3 \mathbf{I} L_4 \mathbf{I} \ldots \mathbf{I} L_8 \mathbf{I} p_8 \mathbf{I} L_7 \mathbf{I} p_6 \mathbf{I} \ldots \mathbf{I} L_1 \mathbf{I} p_1.$$

For a given k-path $\gamma = (v_0, v_1 \ldots, v_k)$, and for given elements w_0, w_1, we denote by $G_{(w_0, w_1)}(v_0, v_1, \ldots, v_k)$ or $G_{w_0}(v_0, v_1, \ldots, v_k)$ the group of all γ-elations fixing the elements w_0 and w_1, or w_0, respectively. Also, $G(v_0, v_1, \ldots, v_k)$ denotes the group of all γ-elations.

Now put

$$s_1 = |G_{L_7}(p_1, L_1, p_2)|, \qquad\qquad t_1 = |G_{p_6}(L_2, p_1, L_1)|,$$
$$s_2 = |G_{(L_2, L_7)}(L_1, p_2, L_3)|, \qquad t_2 = |G_{(p_3, p_6)}(p_1, L_1, p_2)|,$$
$$s_3 = |G_{(L_2, L_7)}(p_2, L_3, p_4)|, \qquad t_3 = |G_{(p_3, p_6)}(L_1, p_2, L_3)|.$$

By the Tits property (which can be easily verified), the same equalities hold for cyclic permutation of the points and lines in Σ, and also in every apartment of Γ. With this notation, we have:

Lemma 1. *The octagon Γ has the Moufang property* **if and only if** $s = s_1 = s_2 = s_3$ *and* $t = t_1 = t_2 = t_3$.

Proof. The proof of this result is similar to the proof of Theorem 6.8.2 above. In fact, one readily sees that, if $s = s_i$, $t = t_i$, $i = 1, 2, 3$, then Γ is a 4-Moufang octagon. $\hspace{4cm}$ QED

Lemma 2. *We always have* $s_2 = s$ *and* $t_2 = t$.

Proof. Note first that $G_{(p_3,p_6)}(p_1, L_1, p_2) \trianglelefteq G_{(p_3,p_6)}(p_1, L_1)$. Hence there is a natural action of the quotient group

$$H = G_{(p_3,p_6)}(p_1, L_1)/G_{(p_3,p_6)}(p_1, L_1, p_2)$$

on the set $\mathcal{S} = \Gamma_1(p_2) \setminus \{L_1, L_3\}$. Suppose an element $g \in G_{(p_3,p_6)}(p_1, L_1)$ fixes an element $L \in \mathcal{S}$. By the 3-Moufang condition, there exists a collineation $g' \in G_{(p_3,p_6)}(L_1, p_2)$ mapping L_4 to L_4^g. Hence gg'^{-1} fixes Σ, fixes every point on L_1 and fixes at least three lines through p_2. By Theorem 4.4.2(iv) (page 140), gg'^{-1} fixes a full suboctagon of order (s, t'), with $1 < t' \leq t$. By Theorem 1.8.8 (see page 36), $t = t'$ and so $g = g'$, implying that g fixes $\mathcal{S}$ elementwise. Hence the action of H on $\mathcal{S}$ is semi-regular. So $t/t_2 = |H|$ divides $t - 1 = |\mathcal{S}|$, whence $t = t_2$. Similarly (or dually) $s = s_2$, completing the proof of Lemma 2. $\hspace{2cm}$ QED

Lemma 3. *If* $s_1 \neq s$, *then* s *is a power of* 4.

Proof. As above, the group $G_{L_7}(p_1, L_1)/G_{L_7}(p_1, L_1, p_2)$ acts on the set $\mathcal{S}' = \Gamma_1(L_3) \setminus \{p_2, p_4\}$, because $G_{L_7}(p_1, L_1, p_2, L_3) = G_{L_7}(p_1, L_1, p_2)$ (remembering $s_2 = s$). If this action were semi-regular, then $s_1 = s$ as before, so we may assume that there exists $g \in G_{L_7}(p_1, L_1)$ fixing at least one but not all elements of $\mathcal{S}'$. Again we can pick a $g' \in G_{(L_2,L_7)}(L_1, p_2, L_3)$ such that gg'^{-1} fixes Σ. But gg'^{-1} also fixes all points on L_1 and at least three points on L_3. The structure of fixed elements of gg'^{-1} forms a weak suboctagon Γ' which clearly must be dual to the double of a generalized quadrangle Δ' (by Theorem 4.4.2(ii) on page 140, and the fact that it cannot be a thick generalized octagon since otherwise g fixes all elements of $\mathcal{S}'$) of order (s, t'), with $t' < s$ (otherwise g fixes all elements of $\mathcal{S}'$, contradicting our hypotheses on g and s_1). Since $s_2 = s$, Δ' is a 3-Moufang quadrangle and hence classical (by assuming the theorem for generalized quadrangles). Considering a collineation r stabilizing Σ and mapping (p_1, L_1) onto (p_2, L_3), we see that $\Gamma' \cap \Gamma'^r$ is again a weak suboctagon which is dual to the double of a generalized quadrangle $(\Delta')^*$ of order (t', t'); and this is a subquadrangle of Δ'. Hence Δ' has order (q^2, q), for some prime power q. But clearly r is a correlation of Δ^*, so q is even. This means that $s = q^2$ is a power of 4, showing the lemma. $\hspace{1cm}$ QED

Similarly, one shows that $s_3 \neq s$ implies that s is a power of 4.

Lemma 4. *We have $s_1 = s$.*

Proof. Suppose by way of contradiction that $s \neq s_1$. We keep the same notation as in the proof of the Lemma 3. In particular, we have a weak suboctagon Γ', which is dual to the double of the generalized quadrangle Δ' of order (s, t') with $t' < s$. So we can pick a point p on L_2 not in Γ'. Let $(p, L, p', L', \ldots, L_7)$ be a 7-path. Since $s_2 = s$, there is a collineation $g^* \in G_{(L_2, L_7)}(L, p', L')$ mapping p_1 to p_3. The weak octagon Γ'^r (see previous paragraph) is mapped onto a weak octagon Γ'' sharing with Γ'^r the apartment through p, L_2, p_8 and L_7 and all points of Γ'^r on L and L'. Hence Γ'' coincides with Γ'^r and g^* is a collineation of Γ'^r, which is dual to the double of a generalized quadrangle isomorphic to $\mathsf{H}(3, q^2)$, q even. But in this quadrangle, the mapping g^* induces a root elation and hence g^* restricted to Γ'^r is an involution. So $p_3^{g^*} = p_1$. But now look at the intersection of Γ' with Γ'^{g^*}. It contains Σ and all points incident with L_1 and L_4. So it induces a subquadrangle of order (s, t'') in Δ'. By Theorem 1.8.8(i) on page 36, $t'' = t'$ and hence g^* preserves Γ'. But g^* restricted to the points of L_2 is an involution, so g^* has on the set $\Gamma'_1(L_2)$ orbits of length 1 or 2 only. Since there are $t' + 1$ such points and since $t' = \sqrt{s}$ is even, there must be some fixed point on L_2 under the action of g^*. But this implies that g^* is trivial (since it is an (L, p', L')-elation). This contradiction shows $s = s_1$. The lemma is proved.   QED

Similarly $s = s_3$, and dually, one shows $t = t_1 = t_2 = t_3$. The proof is complete.   ☐

The geometric argument for generalized hexagons is much longer, because finite generalized hexagons can have ideal thick subhexagons. But roughly speaking, a similar proof applies.

6.8.4 Definition. In the definition of k-Moufang n-gons, we restricted k to $3 \leq k \leq n$. In fact, $k = 1, 2$ works as well. More precisely, let $\mathcal{C}$ be a path of length $k - 2$ (where length -1 just means that $\mathcal{C}$ is empty) in Γ, and let $\overline{\mathcal{C}}$ be a path of length k containing $\mathcal{C}$ in such a way that its extremities do not belong to $\mathcal{C}$. Let G be an automorphism group of Γ. If for all such paths $\mathcal{C}$ and $\overline{\mathcal{C}}$ the group G contains a subgroup fixing every element of Γ incident with some element of $\mathcal{C}$ and acting transitively on the set of apartments containing $\overline{\mathcal{C}}$, then we say that Γ is a k-**Moufang polygon**. It is straightforward to see that the 1-Moufang condition is equivalent to the Tits condition. Hence the Tits condition and the Moufang condition can be viewed as the extremes of the more general notion of the k-Moufang condition. We have already discussed the 1-Moufang polygons in Section 4.7. That leaves the 2-Moufang condition. Again, not so much is known in the infinite case (except for projective planes and generalized quadrangles, where the 2-Moufang condition implies the 3-Moufang condition); in the finite case a full classification is possible and a unified proof for all the cases can be given.

6.8.5 Theorem (Van Maldeghem [19**a]). *Every finite 2-Moufang polygon is a 3-Moufang polygon, and hence a Moufang polygon. Every 2-Moufang projective plane*

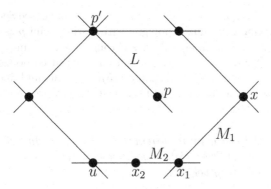

Figure 6.9.

(finite or not) is a Moufang projective plane, and every 2-Moufang generalized quadrangle (finite or not) is a 3-Moufang quadrangle.

Proof. Suppose Γ is a 2-Moufang n-gon of order (s,t), $n \geq 3$, finite if $n \geq 5$. Fix a chamber $\{p,L\}$, with p a point and L a line. Suppose first that $s = 2$. Every collineation fixing all lines through p and fixing one further point on L fixes all points on L. Hence Γ is a 3-Moufang polygon and hence a Moufang polygon by Theorem 6.8.3. Now suppose that $(s,t) = (3,3)$ and $n = 6$. Then the group H_0 fixing all points on L and all lines through p is a normal subgroup of index at most 2 of the group H fixing all points on L and fixing some chosen line M through p, $M \neq L$. If $H = H_0$, then $\{p,L\}$ is a Moufang path. Suppose now $H \neq H_0$. The group H acts transitively on 81 points (opposite p and with a fixed projection onto L). Hence $|H| = 81 \cdot r$ and $|H_0| = 81 \cdot \frac{r}{2}$. But H_0 acts semi-regularly on that set of 81 points, implying $r = 2$ and H_0 acts transitively on the set considered. This again implies that $\{p,L\}$ is a Moufang path. Hence the result follows from Theorem 6.8.3. Alternatively, the result for $(s,t) = (3,3)$ and $n = 6$ follows from Theorem 4.8.2 (which we did not prove, however). So by Theorem 1.7.13 (see page 32) and Theorem 1.7.15, we may assume that the geometry induced by Γ on the points and lines at distance $n-1$ from $\{p,L\}$ is connected. Let x and y be two points opposite p (if n is even; lines if $n = 3$) with common projection p' onto L. We show that there exists a collineation fixing all points on L and all lines through p and mapping x to y. Let $(x, M_1, x_1, M_2, x_2, \ldots, y)$ be a shortest path from x to y all elements of which are at distance $n-1$ from $\{p,L\}$. Let u be the projection of p' onto M_2. By an inductive argument, we may assume $y = u$ (see Figure 6.9, where we have put $n = 6$ for simplicity). By the 2-Moufang assumption, there exists a collineation θ of Γ fixing all lines through p and mapping x to x_1. Dually, there exists a collineation σ fixing all points on L and mapping M_1 to M_2 (and hence fixing x_1 and mapping x to u). The collineation $\theta \sigma^{-1} \theta^{-1} \sigma$ fixes all points on L and all lines through p and it maps x to u. Hence our claim. But this means that Γ is 3-Moufang, hence the theorem. $\quad\square$

Generalizing the notion of an *elation generalized quadrangle* (see Section 3.7.3), one could define an **elation generalized polygon with elation point** p as any generalized polygon admitting a group fixing all lines through p and acting regularly on the points opposite p. Dually, one defines an elation generalized polygon **with elation line**. Noting that by Lemma 5.2.4(ii) (page 175), all points and lines of a Moufang polygon are elation points and elation lines, respectively, we deduce the following from Theorem 6.8.5 above:

6.8.6 Corollary. *Let* Γ *be a finite generalized polygon or an arbitrary projective plane. Then all points of* Γ *are elation points and all lines of* Γ *are elation lines* **if and only if** Γ *is a Moufang polygon.*   □

In fact, if one takes a closer look at the proofs, one sees that it suffices to assume that the polygon has an apartment every element of which is an elation point or elation line.

It is appropriate also to mention here a result due to VAN MALDEGHEM, THAS & PAYNE [1992] about finite 3-Desarguesian quadrangles.

6.8.7 Theorem (Van Maldeghem, Thas & Payne [1992]). *A flag of a finite generalized quadrangle is Desarguesian* **if and only if** *it is a Moufang flag. Hence, a finite generalized quadrangle is* 3-*Desarguesian* **if and only if** *it is a* 3-*Moufang quadrangle* **if and only if** *it is classical.*   □

It is an open problem whether a similar result holds for finite generalized hexagons or octagons.

We can also mention here a result related to the classification of elation generalized quadrangles.

6.8.8 Theorem (Bloemen, Thas & Van Maldeghem [1996]). **If** *a finite elation generalized quadrangle has order* (p, t) *with* p *a prime number,* **then** *it is a classical quadrangle and hence isomorphic to* $\mathsf{W}(p)$, $\mathsf{Q}(4, p)$ *or* $\mathsf{Q}(5, p)$.   □

In the next theorem we classify all finite generalized n-gons admitting a group acting transitively on the ordered $(n + 1)$-gons.

6.8.9 Theorem (Thas & Van Maldeghem [1995], Van Maldeghem [1996]). *A finite generalized n-gon* Γ *admits a group* G *acting transitively on the set of all ordered* $(n + 1)$-*gons* **if and only if** Γ *is one of the following generalized polygons:*

* *the classical projective plane* $\mathbf{PG}(2, q)$, q *a power of any prime;*

* *the symplectic quadrangle* $\mathsf{W}(q)$, *the orthogonal quadrangle* $\mathsf{Q}(4, q)$ *or* $\mathsf{Q}(5, q)$ *or the Hermitian quadrangle* $\mathsf{H}(3, q)$, q *a power of any prime;*

* *the split Cayley hexagon* $H(q)$, *its dual* $H(q)^D$, *the twisted triality hexagon* $T(q^3, q)$ *or its dual* $T(q, q^3)$, *q a power of any prime.*

The group G always contains the little projective group.

Proof. For projective planes, the result follows from OSTROM & WAGNER [1959]. They state that a finite projective plane admitting a group acting 2-transitively on the set of points is classical (a proof of this is also contained in the monograph by HUGHES & PIPER [1973]). We omit the proof of this classical result here.

For generalized quadrangles, we again need the full machinery developed by PAYNE & THAS [1984]. We omit the proof. It can be found in THAS & VAN MALDEGHEM [1995]. In that paper, all possibilities for G are also determined.

For generalized octagons, one must show that there are no finite ones admitting a group acting transitively on the set of ordered nonagons. The proof of this is very similar to the case of hexagons and uses Theorem 6.4.6 on page 261 instead of Theorem 6.3.2 and Theorem 6.3.4 (which are used in the case $n = 6$).

So we assume that Γ is a finite generalized hexagon. If Γ is classical, then it has the desired properties. This follows directly from Corollary 4.5.13 (page 151).

Now let G be a group of collineations of Γ acting transitively on the set of all ordered heptagons. It is readily seen that this is equivalent to G acting transitively on the set of all *skeletons*, where we define a **skeleton** to be an apartment Σ together with a point p not belonging to Σ, but incident with a line M of Σ, and together with a line L not belonging to Σ but incident with a point q of Σ with $q\,\mathbf{I}\,M$. We will only use this notion in the present proof.

There is another notion that we would like to use in this proof. Suppose for a point p, the geometry of traces in $p^{\perp}$ satisfies the following condition:

> *every triple of pairwise non-collinear points in* $p^{\perp}$ *is contained in a unique trace* p^x, *for some point x opposite p (the point x itself may not be unique of course);*

then we call p an **anti-regular point**.

We now start the proof of the theorem.

| Sizes of the intersection sets |

We attack the problem by looking at the size of an arbitrary intersection set. Consider two points x, y at distance 4 from each other and suppose $x\,\mathbf{I}\,L_x\,\mathbf{I}\,(x \bowtie y)\,\mathbf{I}\,L_y\,\mathbf{I}\,y$. Let p be a point collinear with x but not incident with L_x. Let p_1 and p_2 be two points collinear with y, not incident with L_y and at distance 4 from p. There are two possibilities.

1. Suppose $x^{p_1} = x^{p_2}$. By the transitivity property, there is a collineation fixing $x, x \bowtie y, y$ and p_1 and mapping the line yp_2 to any desired line L through y,

$yp_1 \neq L \neq L_y$. The set x^{p_1} is preserved and p_2 is mapped onto a point p_3 on L. Obviously, $x^{p_3} = x^{p_2} = x^{p_1}$. Since L was arbitrary, every point z collinear with y, opposite x and at distance 4 from p has the property $x^z = x^{p_1}$. By the transitivity, we can now let p vary over the set of all points collinear with x but not on L_x, and hence we obtain the property that whenever z_1 and z_2 are points collinear with y and opposite x, then either $x^{z_1} = x^{z_2}$ or $x^{z_1} \cap x^{z_2} = \{x \bowtie y\}$. By transitivity, this holds for every such pair (x, y). So in this case all intersection sets of Γ have size 1.

2. Suppose $x^{p_1} \neq x^{p_2}$. Then there exists a line L through x incident with two distinct points a_1 and a_2 at distance 4 from p_1 and p_2, respectively. By the transitivity property, there is a collineation θ fixing x, y, p_1, a_1 and yp_2 and mapping a_2 to any desired point on L distinct from x and a_1. The point p_2 will be mapped onto any point p_2^θ incident with yp_2, except for y and z, where z has distance 4 from a_1. Obviously, $x^{p_1} \cap x^{p_2^\theta}$ contains p^θ and $x^{p_1} \cap x^z$ contains a_1. Since p_2^θ is essentially arbitrary, and since by transitivity also the line yp_2 is arbitrary and for the same reason p_1 as well, we conclude that whenever z_1 and z_2 are points at distance 4, collinear with y and opposite x, then $|x^{z_1} \cap x^{z_2}| \geq 2$. Reversing the roles of x and y, we also see that whenever u_1 and u_2 are two non-collinear points in $\Gamma(x)$ opposite y, then there exists a point z collinear with y and opposite x such that $\{u_1, u_2\} \subseteq x^z$.

Let us get back to the above situation involving a_1, a_2, p_1 and p_2. There also exist collineations fixing x, y, p_1, a_1, px (and hence p) and mapping a_2 to any point u of L different from x and a_1. Of course, such mappings do not preserve the line yp_2, and in fact every other choice for u gives another image of yp_2, hence $s - 1 \leq t - 1$, implying $s \leq t$.

The properties obtained in this paragraph are also valid for every choice of such a pair (x, y), by transitivity. This proves that all intersection sets of Γ have size at least 2.

From this, we derive two possibilities:

Case (i) *In either Γ or its dual, all intersection sets have size 1.*

Case (ii) *In both Γ and its dual, all intersection sets have size ≥ 2, and $s = t$.*

| Intersection sets of size 1 |

We now handle case (i). We may assume, by duality, that all intersection sets of Γ have size 1.

We consider an ordered apartment $(p_1, L_1, p_2, L_3, \ldots, p_6, L_6, \ldots, L_2, p_1)$. Let $\mathcal{S} = \Gamma_3(L_3) \cap \Gamma_3(L_4)$. Using the transitivity property as above, one shows, completely similarly to the argument in the previous paragraphs, that either $p_1^u = p_1^{p_6}$, for all $u \in \mathcal{S} \setminus \{p_1\}$, or every point on every line L through p_1, $L_1 \neq L \neq L_2$, is at distance 4 from exactly one element of $\mathcal{S}$. In the first case, we may conclude by transitivity and by Remark 6.3.6 that Γ is point-regular and hence classical. We have shown:

Lemma 1. **If** *all intersection sets have size* 1, *and* Γ *is not classical,* **then,** *with the above notation, every point on every line L through p_1 different from L_1 and L_2 is at distance 4 from exactly one element of $\mathcal{S}$.* QED

With the above notation, Lemma 1 implies that $|p_1^u \cap p_1^{p_6}| = 2$.

Lemma 2. **If** *all intersection sets have size* 1, *and* Γ *is not classical,* **then** *all points p of Γ are anti-regular.*

Proof. Let a and b be two non-collinear points collinear with p. Let x and y be two points opposite p both at distance 4 from both a and b. We must show that either $p^x = p^y$ or $p^x \cap p^y = \{a, b\}$; by Lemma 1, we already have at least one block through every triple of pairwise non-collinear points in $\Gamma(p)$. We may suppose that either $x \bowtie a$ is not collinear with $y \bowtie a$, or $x \bowtie b$ is not collinear with $y \bowtie b$, otherwise the result follows from Lemma 1. So suppose $x \bowtie b$ is not collinear with $y \bowtie b$. Let L be the line joining b and $y \bowtie b$ and let M be the line joining $a \bowtie x$. Let u and v be the points collinear with $y \bowtie b$ and $x \bowtie a$, respectively, and at distance 3 from M and L, respectively.

By the main assumption of case (i), $p^x = p^v$ and $p^y = p^u$, and by $(**)$, $p^v = p^u$, or $p^v \cap p^u = \{a, b\}$, implying either $p^x = p^y$ or $p^x \cap p^y = \{a, b\}$. This shows that p is anti-regular. QED

Now we remark that this case can never arise in the classical generalized hexagons, because Lemma 2 never holds true (on the contrary, every Moufang hexagon is distance-3-regular). So we must show that this situation can never occur. Let S be a given skeleton in Γ and let Γ' be the intersection of all subhexagons containing S. Let θ be any automorphism of Γ. If θ maps S into Γ', then θ must preserve Γ', otherwise $\Gamma'^{\theta^{-1}} \cap \Gamma'$ is a proper subpolygon of Γ'. Similarly, if θ fixes S, then it must fix every element of Γ' by Theorem 4.4.2(v) (on page 140). Hence, renaming Γ' as Γ, we may assume that G acts regularly on the set of skeletons.

Lemma 3. **If** *all intersection sets have size* 1, *and* Γ *is not classical,* **then** $t \leq s$.

Proof. Consider any point p in Γ. Fix a trace K in $p^\perp$, a point x on K and a point $y \in p^\perp$ off K, with x and y non-collinear in Γ. Remember that p is anti-regular, so every triple of points of $\Gamma_2(p)$ which are pairwise non-collinear in Γ is contained in a unique trace in $p^\perp$. Hence the number of traces through y and x meeting K in exactly two points is $t - 1$. On the other hand, there are in total s traces through x and y, at least one of which meets K exactly in x (if $K = p^u$, then there is a point w on the line joining u and $x \bowtie u$ at distance 4 from y; p^u and p^w have only x in their intersection, otherwise an ordinary j-gon, with $j < 6$, arises), so at most $s - 1$ traces through x and y meet K in a second point. This implies $t \leq s$. QED

We consider again our notation introduced in the paragraph preceding Lemma 1.

Lemma 4. If *all intersection sets have size* 1, *and* Γ *is not classical,* **then** *both* s *and* t *are prime powers.*

Proof. Let L be any line incident with p_1, $L_1 \neq L \neq L_2$. The group H_1 fixing the points p_1, p_2, p_4 and the lines L, L_1, L_2, L_3 and L_5 has order $s(s-1)$ and acts sharply doubly transitively on the set $\Gamma_1(L_2) \setminus \{p_1\}$. Hence H_1 has a unique normal regular subgroup N_1 of order s (N is the set of elements of H_1 acting without fixed points), which is elementary abelian (since H_1 acts by conjugation as an automorphism group of N_1, transitively on $N_1 \setminus \{1\}$), as is well known. So $s = \pi_1^{n_1}$ with π_1 a prime and $n_1 \in \mathbb{N}_0$. Similarly $t = \pi_2^{n_2}$, π_2 prime and $n_2 \in \mathbb{N}_0$.

QED

Lemma 5. If *all intersection sets have size* 1, *and* Γ *is not classical,* **then** $s = t$.

Proof. Let p be a point incident with L_1, $p_1 \neq p \neq p_2$. The subgroup of H_1 fixing p acts regularly on the set $\Gamma_1(L_2) \setminus \{p_1\}$, hence, with the notation of the proof of Lemma 4, this subgroup is N_1. Since p was essentially arbitrary, N_1 fixes L, L_2 and L_5, it fixes every point on L_1 and it acts regularly on $\Gamma_1(L_2) \setminus \{p_1\}$. Let N_1' be the subgroup of H_1 fixing p_3. Suppose an element $\theta \in N_1'$ fixes some point x on L_3, $p_2 \neq x \neq p_4$. Let x' be the point collinear with x and at distance 3 from L_4. By Lemma 1, $\operatorname{proj}_L x' \neq \operatorname{proj}_L p_6$, but both these points are fixed by θ. Hence θ fixes a skeleton, which implies that θ is the identity. This shows that N_1' acts regularly on the set $V = \Gamma_1(L_3) \setminus \{p_2, p_4\}$. Hence H_1 acts transitively on V and $N_1 \trianglelefteq H_1$ partitions V in orbits of equal length. But $|V| = s - 1$ is relatively prime to π_1. Since N_1 is a π_1-group, this implies that N_1 fixes all elements of L_3. Hence N_1 fixes L, L_2, L_5 and every point on L_1 and on L_3. Similarly, the dual result holds.

Note that the previous paragraph is valid for $t \leq s$. From now on we assume that $t < s$.

If we take another line M through p_1, $L_1 \neq M \neq L_2$, then we can define a regular group N_2 in the same way as N_1 was defined for L in the previous paragraphs. By the transitivity on V there exists for every $\theta \in N_1$ an element $\theta' \in N_2$ such that $(p_6^\theta)^{\theta'} = p_6$. So $\theta\theta'$ fixes Σ and all points on L_1 and on L_3, hence $\theta\theta'$ fixes a weak subhexagon of order (s, t'), implying $t = t'$ by Theorem 1.8.8(i); see page 36. Consequently $\theta' = \theta^{-1}$ and so θ fixes both L and M. Since M was essentially arbitrary, θ fixes every line through p_1. A similar argument shows that θ also fixes every line through p_4. Now consider a line X through p_2, $L_1 \neq X \neq L_3$. The group H_2 fixing Σ and X has order $s - 1$ and acts transitively on the set of points incident with L_1 and different from p_1 and p_2. Suppose some element $\varphi \in H_2$ fixes a point x on L_2, $p_1 \neq x \neq p_6$. Let w be the unique element of $\Gamma_2(x) \cap \Gamma_3(L_5)$. By Lemma 1, $\operatorname{proj}_X^w \neq \operatorname{proj}_X p_5$. Hence φ fixes a skeleton and must consequently be the identity. Hence H_2 acts regularly on the set of points of L_2 different from p_1 and p_3. Since $s > 2$ (otherwise $t = 1$), this group is non-trivial, and letting p_3 now vary over $\Gamma_1(L_2) \setminus \{p_1\}$, we obtain a group H_3 of order $s(s-1)$ acting sharply doubly transitively on $\Gamma_1(L_2) \setminus \{p_1\}$ and fixing p_1, p_2, p_4 and L_1, L_2, L_3, L_5 and X. A similar argument to that above shows that in fact N_1 is a subgroup of H_3 and hence we conclude that $(p_1, L_1, p_2, L_3, p_4)$ is a Moufang path.

By transitivity $(p_2, L_3, p_4, L_5, p_6)$ is also a Moufang path and we denote the group of all $(p_2, \ldots, p_6)$-elations by N_3. Suppose the commutator $[N_1, N_3]$ is trivial. It is easy to see that this implies that every element of N_1 fixes every line through every point of L_1 or L_3, respectively. This implies that $L_1^{L_6} = L_1^{M_6}$, where M_6 is any line opposite L_1, meeting L_5 and at distance 4 from L_2. This means that the dual Γ^D of Γ satisfies the assumption of case (i), and hence Lemma 1 holds true for Γ^D. This implies $s \le t$, a contradiction. Hence $[N_1, N_3]$ is non-trivial. But every non-trivial element θ of $[N_1, N_2]$ is an $(L_1, p_2, L_3, p_4, L_5)$-elation. By conjugating θ with the subgroup of G fixing Σ, we see that $(L_1, p_2, L_2, p_3, L_3)$ is a Moufang path. Hence Γ satisfies the Moufang condition, a contradiction as we remarked earlier. This completes the proof of Lemma 5.   QED

Now suppose $s = t$ and let p be any point of Γ. Let p' be any point at distance 4 from p and let L be the line joining p' and $p \bowtie p'$. Let x be a point collinear with p' and opposite p. By our general assumption of case (i), there are exactly s points y opposite p and collinear with p' for which $p^x = p^y$. Evidently, every other point u collinear with p' and opposite p gives rise to a different trace $p^u \ne p^x$ in $p^\perp$ and, moreover, $p^u \cap p^x = \{p \bowtie p'\}$. So p' defines exactly s traces in $p^\perp$ which meet pairwise in $p \bowtie p'$. Varying p' over L (remaining at distance 4 from p of course), we see that the points opposite p and at distance 3 from L define at most s^2 traces in $p^\perp$ We now claim that they define exactly s^2 traces. Indeed, suppose that $p^{x_1} = p^{x_2}$, with $x_1, x_2 \in \Gamma_3(L) \cap \Gamma_6(p)$, and with $\mathrm{proj}_L x_1 \ne \mathrm{proj}_L x_2$. Pick $v \in p^{x_1} \cap \Gamma_5(L)$. Let x_1' be the unique point collinear with $\mathrm{proj}_L x_1$ and at distance 3 from $\mathrm{proj}_v x_2$. By Lemma 1, we have $p^{x_2} \ne p^{x_1'}$; since all intersection sets have size 1, we have $p^{x_1'} = p^{x_1}$, a contradiction. Hence the claim.

It is now easy to see that the incidence structure $\Pi(p, L)$ with point set the set of points collinear with p but not collinear with $p \bowtie p'$, and line set the set of traces of the form p^x with x at distance 3 from L and opposite p, together with the ordinary lines through p, forms an affine plane (with the obvious incidence relation). Every point on L different from $p \bowtie p'$ symbolizes a point at infinity of $\Pi(p, L)$. Also p is a point at infinity of $\Pi(p, L)$ in an obvious way.

Lemma 6. If *all intersection sets have size 1, then Γ is classical.*

Proof. We may assume that Γ is non-classical and $s = t$. As already remarked, we may copy the first paragraph of the proof of Lemma 5. We also adopt the notation of that paragraph. In particular, we consider the group N_1 fixing L_2, L_5 and L, and fixing L_1 and L_3 pointwise. Note that this group must fix at least one other line L' through p_2, $L_1 \ne L' \ne L_3$, and one other line L'' through p_4, $L_3 \ne L'' \ne L_5$ (because N_1 is a π_1-group with π_1 prime and $t = s$ is a power of π_1). Similarly, there is a group N_4 of order $s = t$ fixing p_1, p_2, p_4 and p_6, all lines through p_2 and p_4, a point p on L_1, p' on L_3 and p'' on L_5, p, p', p'' not in Σ.

Consider the affine plane $\Pi(p_2, L_2)$. Let θ be any non-trivial element of N_1. This collineation induces in $\Pi(p_2, L_2)$ a non-trivial elation (indeed, all points of the line L_3 are fixed), hence θ has some centre. Since θ fixes the lines L', L_3 and the line at infinity of $\Pi(p_2, L_2)$, the centre must be the point p_2 at infinity, which is incident with all three fixed lines mentioned. Hence θ fixes all lines through p_2. Similarly θ induces an elation in $\Pi(p_1, L_3)$ (all points at infinity of $\Pi(p_1, L_3)$ are fixed), and hence θ has some centre, but as already three lines through the point p_1 at infinity are fixed (the line at infinity, L and L_1), p_1 must be the centre, hence θ fixes all lines through p_1 and similarly, also all lines through p_4. We conclude that θ is a $(p_1, L_1, p_2, L_3, p_4)$-elation. Conjugation with the subgroup of G fixing Σ shows that $(p_1, \ldots, p_4)$ is a Moufang path. Denote by N_3 the group of all $(p_2, L_3, \ldots, L_6, p_5)$-elations (this group has order s by transitivity). Considering again the commutator $[N_1, N_3]$ as in the last paragraph of the proof of Lemma 5, we obtain that either Γ^D has anti-regular points, in which case dually the path $(L_1, p_2, L_3, p_4, L_5)$ is a Moufang path and hence Γ is classical; or $[N_1, N_3]$ is non-trivial and Γ is classical again. This proves Lemma 6.   QED

This completes case (i).

Intersection sets of size at least 2

Here we assume that all intersection sets have size at least 2 in both Γ and Γ^D, and that $s = t$. Similarly to the previous case, it suffices to show that this situation cannot occur if G acts regularly on the set of skeletons.

Note that Γ does not contain any weak subhexagon of order $(1, s)$ or $(s, 1)$ since e.g. the former would imply that that there are intersection sets of size 1.

Consider the apartment Σ again (with previous notation). As before, we obtain a sharply doubly transitive permutation group (on the set $\Gamma_1(L_2) \setminus \{p_1\}$) and a group N_1 fixing L_2, L_5, fixing L_1 and L_3 pointwise, and fixing some lines L, L', L'' through p_1, p_2, p_4, respectively, and not contained in Σ (implied by $s = t$ being a prime power); N_1 acts regularly on $\Gamma_1(L_2) \setminus \{p_2\}$. By the transitivity property of G, any other choice of either L, L' or L'', say L_*, gives a group N_1^* which is conjugate to N_1 having the same properties. To every element θ of N_1 corresponds an element θ' of N_1^* such that $\theta\theta'$ fixes Σ elementwise. But it also fixes every point on L_1 and on L_2, hence it fixes a weak subhexagon of order (s, t'). By Theorem 1.8.8(iii) (on page 36), $t' = 1$ or $t' = s$. We have already ruled out $t' = 1$, hence $t' = s$ and so $\theta = \theta'$. We conclude that N_1 also fixes every line through p_1, every line through p_2 and every line through p_4. So $(p_1, L_1, p_2, L_3, p_4)$ is a Moufang path. Dually, every path of length 4 starting with a line is a Moufang path and hence Γ itself is a Moufang hexagon. But this is impossible since every Moufang hexagon of order (s, s) or its dual has intersection sets of size 1, see Remark 6.3.5 on page 255.

This completes the proof of the theorem.   □

The finiteness condition of the previous theorem is essential for the proof. In the infinite case, the problem remains open. One could try to carry out the free construction of Subsection 4.7.1 (see page 160) with $\mathcal{A}$ the set of ordered $(n + 1)$-gons, in order to obtain an infinite generalized n-gon with an automorphism group acting transitively on the set of ordered $(n + 1)$-gons, for all $n \geq 3$. Something in that style was considered for $n = 3$ by KEGEL & SCHLEIERMACHER [1973], although there are some problems that make it unclear whether the construction actually works (see WASSERMANN [1993]).

We refer the reader to Definitions 4.4.1 (page 140) and Definition 4.4.4 (page 143) for the notions of $\{v, v'\}$-homology, $\{v, v'\}$-transitivity and $\{v, v'\}$-quasi-transitivity, for opposite elements v and v' of a (finite) generalized polygon.

The next result provides a converse in the finite case for Proposition 4.5.10, Proposition 4.5.11 and Proposition 4.5.12 (from page 148 to page 151).

6.8.10 Theorem. *Let Γ be a finite generalized n-gon of order (s, t).*

(i) **(Baer [1942])** If $n = 3$, then Γ *is* classical **if and only if** *it is* $\{v, v'\}$-*transitive for every pair of opposite elements v and v'.*

(ii) **(Thas [1985], [1986])** If $n = 4$, then Γ *is isomorphc to one of* $\mathsf{W}(q)$, $\mathsf{Q}(4, q)$, $\mathsf{Q}(5, q)$, $\mathsf{H}(3, q^2)$ *or* $\mathsf{H}(4, q^2)$, *for some prime power* q, **if and only if** Γ *is* $\{x, y\}$-*transitive for all pairs of opposite points x and y. Moreover, Γ is isomorphic to* $\mathsf{W}(q)$, $\mathsf{Q}(4, q)$, $\mathsf{Q}(5, q)$ *or* $\mathsf{H}(3, q)$, *for some prime power* q, **if and only if** Γ *is* $\{v, v'\}$-*transitive for all pairs of opposite elements* $\{v, v'\}$.

(iii) **(Van Maldeghem [1991b])** If $n = 6$, then Γ *is* classical **if and only if** *it is* $\{v, v'\}$-*transitive for every pair of opposite elements* $\{v, v'\}$.

(iv) **(Van Maldeghem [1991a])** If $n = 8$ *and* $s \neq 2$, then Γ *is isomorphic to* $\mathsf{O}(q)$, q *an odd power of* 2, **if and only if** Γ *is* $\{x, y\}$-*transitive for every pair of opposite points* $\{x, y\}$, *and Γ is* $\{L, M\}$-*quasi-transitive for every pair of opposite lines* $\{L, M\}$. *Also, there do not exist octagons which are* $\{v, v'\}$-*transitive for all pairs of opposite elements* $\{v, v'\}$ *(even if $s = 2$).*

Proof. As usual in this section, we prove the result neither for $n = 3$ nor for $n = 4$. We remark that for $n = 3$, the result also holds in the infinite case and it is stated as an exercise in STEVENSON [1972], exercise 6, page 156. The first proof was given by BAER [1942], Lemma 3.2. For $n = 6$, it is enough to remark that every classical generalized hexagon indeed is $\{v, v'\}$-transitive for every pair of opposite elements v and v' (this follows immediately from Proposition 4.5.11, see page 150), and that the latter condition implies that Γ admits a collineation group acting transitively on the set of all ordered heptagons. The result then follows from Theorem 6.8.9. A similar remark shows that there are no generalized octagons which are $\{v, v'\}$-transitive for every pair of opposite elements v and v'. Also, the Ree–Tits octagons satisfy the given conditions by Proposition 4.5.12 on page 151.

So we may assume that $n = 8$ and that Γ is $\{x, y\}$-transitive for every pair of opposite points x, y and $\{L, M\}$-quasi-transitive for every pair of opposite lines L and M. We have to show that Γ is isomorphic to $\mathsf{O}(q)$, the Ree–Tits octagon of order (q, q^2) where q is an odd power of the prime 2. By Theorem 6.8.5, it is enough to show that Γ is a 2-Moufang octagon. We denote the full collineation group of Γ by G.

Throughout this proof, we fix an apartment Σ determined by the circuit

$$(p_1, L_1, p_2, L_3, p_4, \ldots, L_7, p_8, L_8, p_7, \ldots, L_2, p_1).$$

We also adopt the notation $G_E(D)$ for the subgroup G fixing every element of E and fixing every element incident with an element of D; E and D are sets of elements of Γ. In particular, $G(v, v')$ is the group of all $\{v, v'\}$-homologies, for opposite elements v, v'.

For v an element of Σ, we denote by $\mathcal{E}(v)$ the set of elements incident with v, but not lying in Σ, i.e., $\mathcal{E}(v) = \Gamma(v) \setminus \Sigma(v)$. This set has size $s - 1$ if v is a line, otherwise it has size $t - 1$.

Lemma 1. *The group $G(L_1, L_8)$ acts semi-regularly on $\mathcal{E}(p_1)$. Hence $s < t$.*

Proof. Otherwise there is a collineation fixing a suboctagon of order (s, t') for $1 < t' < t$, contradicting Theorem 1.8.8 (page 36). Hence, since $|G(L_1, L_8)| \geq s$, the lemma follows.   QED

Lemma 2. *The group $G(p_1, p_8)$ acts semi-regularly on $\mathcal{E}(p_3)$.*

Proof. Suppose $\sigma \in G(p_1, p_8)$ fixes some line $L \in \mathcal{E}(p_3)$. Then by Theorem 4.4.2(iii) (page 140), σ fixes the double Γ' of a generalized quadrangle of order (t, t') with $1 < t' < t$ since $t = t'$ implies that Γ' is a weak suboctagon of order $(1, t)$, contradicting $s < t$ and Proposition 1.8.7(iv) (on page 35). Choose $p \in \mathcal{E}(L_2)$ arbitrarily and let $\sigma' \in G(p_3, p_6)$ be such that $p^{\sigma \sigma'} = p$. Then $\sigma \sigma'$ fixes a suboctagon Γ'' of some order (s', t') (the same t' as above, indeed!). Of course, $1 < s' < s$. The intersection $\Gamma' \cap \Gamma''$ is a weak suboctagon of order $(1, t')$ of Γ'', hence $t' < s'$ and so $t' < s$. Since $G(L_2, L_7)$ has size at least s and since it acts semi-regularly on $\mathcal{E}(p_3)$, there exists $\sigma'' \in G(L_2, L_7)$ not preserving Γ''. But $\Gamma'' \cap \Gamma''^{\sigma''}$ is a suboctagon of Γ'' of order (s', t''), with $t'' < t'$. Hence $t'' = 1$ and $s' < t'$, a contradiction. So $G(L_1, L_8)$ acts semi-regularly on $\mathcal{E}(p_3)$.   QED

Lemma 3. *Every collineation $\theta \in G(L_1, L_8)$ that fixes an element of $\mathcal{E}(L_2)$ must fix every point on L_2.*

Proof. Suppose by way of contradiction that $\theta \in G(L_1, L_8)$ fixes at least one, but not every element of $\mathcal{E}(L_2)$. Then θ fixes the dual Γ' of the double of a generalized quadrangle of order (s, s'), $1 < s' < s$ (cf. Theorem 4.4.2(iii)). By the transitivity assumption, there exists $\theta' \in G(L_1, L_8)$ not preserving Γ. Hence the intersection $\Gamma' \cap \Gamma'^{\theta'}$ is dual to the double of a generalized quadrangle of some order (s, s''), $s'' < s'$, implying $s'' = 1$ and $s' = s$. This proves the lemma.   QED

Lemma 4. *The group $G_{L_7}(p_1, L_1)$ acts transitively on $\Gamma_1(L_2) \setminus \{p_1\}$.*

Proof. Indeed, let p_8' be any point in $\mathcal{E}(L_7)$ and let $p \in \mathcal{E}(L_1)$. For every non-trivial $\theta \in G(p_1, p_8)$, there exists a non-trivial $\theta' \in G(p_1, p_8')$ such that $p^{\theta\theta'} = p$ (by the transitivity assumption). By the semi-regularity of $G(p_1, p_8)$ (Lemma 2) on $\mathcal{E}(L_2)$, we see that $\theta\theta'$ does not fix L_2 pointwise. Obviously, the commutator $[G(L_2, L_7), \theta\theta']$ is trivial (because every element of it fixes $\Gamma_1(L_2) \cup \Gamma_1(p_1) \cup \Sigma$ elementwise), and since $G(L_2, L_7)$ acts transitively on $\mathcal{E}(L_1)$, this immediately implies that $\theta\theta'$ fixes L_1 pointwise. Conjugating $\theta\theta'$ by $G(p_1, p_8)$, the lemma follows. QED

Remark 1. Similarly, one shows that $G_{L_4, L_5}(p_1, L_2)$ acts transitively on $\Gamma_1(L_4) \setminus \{p_3\}$; that $G_{L_3, L_6}(p_1, L_2)$ acts transitively on $\Gamma_1(L_6) \setminus \{p_5\}$; and that $G_{L_8}(p_1, L_2)$ acts transitively on $\Gamma_1(L_8) \setminus \{p_7\}$.

Lemma 5. *The group $G_{p_7}(p_1, L_1)$ acts transitively on $\Gamma_1(p_7) \setminus \{L_6\}$. Symmetrically, $G_{p_6}(p_1, L_2)$ acts transitively on $\Gamma_1(p_3) \setminus \{L_2\}$.*

Proof. If p is a point of $\mathcal{E}(L_1)$, $\sigma \in G(p_1, p_8)$, $\sigma' \in G(p_1, p_8'')$ (where $p_8'' \in \Gamma_2(p_7) \cap \Gamma_6(p_2)$) and $p^{\sigma\sigma'} = p$, then $\sigma\sigma'$ does not fix any element of $\Gamma_1(p_2) \setminus \{L_1\}$ (otherwise $\sigma\sigma'$ would fix a thick suboctagon of order (s', t), a contradiction). Hence given $\sigma \in G(p_1, p_8)$, there are precisely $t - 1$ points p_8'' as above such that the corresponding elements σ' (which map p^σ back to p) map L_3 all to different lines. Indeed, if $L^{\sigma'} = L^{\sigma''}$, $\sigma'' \in G(p_1, p_8''')$ (where $p_8''' \in \Gamma_2(p_7) \cap \Gamma_6(p_2)$), and $p^{\sigma\sigma'} = p^{\sigma\sigma''} = p$, then $(p^\sigma)^{\sigma'\sigma''^{-1}} = p^\sigma$, and hence $\sigma'\sigma''^{-1}$ fixes a suboctagon of order (s', t), a contradiction if $\sigma' \neq \sigma''$. So for every line $L \in \mathcal{E}(p_2)$, there exists a collineation $\theta = \sigma\sigma' \in G_{p_7}(p_1)$ which fixes p and maps L_3 to L. But as above, the commutator $[G(p_2, p_7), \theta]$ is trivial (because every element of it fixes $\Gamma_1(p_1) \cup \Gamma_1(p_2) \cup \Sigma$ elementwise, so a weak suboctagon of order (s'', t) is fixed, implying $s'' = s$ by Proposition 1.8.7(iv)), implying as before that $\theta \in G_{p_7}(p_1, L_1)$. Since L was arbitrary, the result follows. QED

As in Lemma 5, one shows that $G_{p_4, p_5}(p_1)$ acts transitively on $\Gamma_1(p_5) \setminus \{L_4\}$. We can now prove:

Lemma 6. *The group $G(p_1)$ acts transitively on the set $\Gamma_8(p_1)$. Hence for every point p, the 0-path (p) is a Moufang path.*

Proof. Let $(x_8, M_8, x_7, M_6, x_5, M_4, x_3, L_2)$ be a 7-path. We have to show that there exists $\theta \in G(p_1)$ mapping x_8 to p_8. By Lemma 4, there exists $\theta_1 \in G(p_1)$ mapping x_3 to p_3. By Lemma 5, there exists $\theta_2 \in g(p_1)$ mapping $M_4^{\theta_1}$ to L_4. By Remark 1, there exists $\theta_3 \in G(P_1)$ mapping $x_5^{\theta_1\theta_2}$ to p_5. Continuing in this way, we finally obtain $\theta_7 \in G(p_1)$ mapping $x_8^{\theta_1\theta_2\cdots\theta_6}$ to p_8. The lemma is proved. QED

Lemma 7. *The group $G_{L_4, L_5}(L_1)$ acts transitively on $\Gamma_1(L_5) \setminus \{p_4\}$.*

Proof. Clearly $G_{L_5}(p_3, L_2) \trianglelefteq G_{L_4, p_3, L_5}$. The latter group acts transitively on $\mathcal{E}(L_1)$, hence all orbits of $G_{L_5}(p_3, L_2)$ in $\mathcal{E}(L_1)$ have equal length ℓ. So ℓ must divide $|\mathcal{E}(L_1)| = s - 1$, and it must divide $|G_{L_5}(p_3, L_2)| = s$ (the latter equality by a shift of the indices in Lemma 4). This implies $\ell = 1$, so $G_{L_5}(p_3, L_2) = G_{L_5}(p_3, L_2, L_1)$. The lemma is proved.   QED

$\boxed{\textbf{Lemma 8.}}$ *The group $G(L_1)$ acts transitively on the set $\Gamma_8(L_1)$. Hence for every line L, the 0-path (L) is a Moufang path.*

Proof. By Lemma 4, the group $G_{L_7}(p_1, L_1)$ acts transitively on $\Gamma(L_7) \setminus \{p_6\}$. With a permutation of the indices, Lemma 5 implies that $G_{p_8}(p_2, L_1)$ acts transitively on $\Gamma(p_8) \setminus \{L_7\}$. Lemma 5 itself says that $G_{p_7}(p_1, L_1)$ acts transitively on $\Gamma(p_2) \setminus \{L_1\}$. A permutation on the indices of Lemma 4 implies that $G_{L_6}(p_2, L_1)$ acts transitively on $\Gamma(L_3) \setminus \{p_2\}$. As in Lemma 5, one shows that $G_{p_5, p_4}(p_2, L_1)$ acts transitively on $\Gamma(p_4) \setminus \{L_3\}$, and symmetrically, $G_{p_3, p_6}(p_1, L_1)$ acts transitively on $\Gamma(p_6) \setminus \{L_5\}$. Together with Lemma 7, this permits us to dualize word for word the proof of Lemma 6.   QED

The theorem is proved.   □

We remark that for infinite polygons other than projective planes, no analogue to the previous theorem is known.

Finally, we mention a result due to WALKER [1982], [1983]. However, the proof is beyond the scope of this book. Moreover, we only state a simplified version of Walker's results (in our opinion the crux). Recall from Definitions 4.4.1 on page 140 that a **central elation** θ in a generalized n-gon Γ, n even, is an automorphism fixing $\Gamma_{\leq \frac{n}{2}}(p)$ for some point p, and p is then called the **centre** of θ.

6.8.11 Theorem (Walker [1982], [1983]). *Let Γ be a finite generalized hexagon or octagon. If every point of Γ is the centre of a non-trivial central elation, then Γ satisfies the Moufang property.*

6.9   A geometric characterization of the perfect Ree–Tits octagons

In this paragraph, we mention without proof the geometric characterization of the perfect Ree–Tits octagons proved in VAN MALDEGHEM [1998]. The key idea is the introduction of the so-called *block geometry* related to a point in a generalized polygon.

Let Γ be a generalized polygon and let x be a point of Γ. In a lot of cases, the set $\Gamma_1(x)$ is parametrized by a field together with a symbol (∞). From a geometric point of view, this often does not put much structure on $\Gamma_1(x)$. However, in a few cases, this set is geometrically richer. For instance, let Γ be a perfect Ree–Tits octagon; then $\Gamma_1(x)$ is parametrized by the points of a Suzuki–Tits ovoid. The idea is now to recover this ovoid in a geometrical way, and then to axiomatize the situation. To this end, we also need an axiom system for the Suzuki–Tits ovoid. We will mention part of one here, but we will postpone proofs to Chapter 7.

6.9.1 Axiomatic STi-planes

Let $\Omega = (\mathcal{P}, \mathcal{C}, \in)$ be a geometry consisting of a set $\mathcal{P}$ of points, a set $\mathcal{C}$ of subsets of $\mathcal{P}$ (the elements of $\mathcal{C}$ are called *circles*) and the set-theoretic inclusion is the incidence relation. Then Ω is called a **Suzuki–Tits inversive plane**, or briefly an **STi-plane**, if it satisfies conditions [MP1], [MP2], [CH1], [CH2], [ST] stated below (where two circles **touch** if they intersect in exactly one point).

[MP1] Any three points of Γ are contained in exactly one circle.

[MP2] For every circle $C \in \mathcal{C}$, and any two points $x, y \in \mathcal{P}$ such that $x \in C$ and $y \notin C$, there exists a unique circle $D \in \mathcal{C}$ touching C in x and containing y.

[CH1] There do not exist three circles mutually touching at distinct points.

[CH2] For every circle C and every pair of points $x, y \notin C$, either there is a unique circle containing both x and y and touching C, or all circles through x and y touch C.

[ST] The inversive plane Γ is furnished with a map $\partial : \mathcal{C} \to \mathcal{P} : C \mapsto \partial C \in C$ such that:

   [ST1] For every pair of points $x, y \in \mathcal{P}$ there is a unique circle $C \in \mathcal{C}$ containing x and y and such that $\partial C = x$.

   [ST2] For every circle C and every point $x \notin C$, there is at most one circle D containing x and ∂C and such that $\partial D \in C$.

In fact an STi-plane should be denoted by a pair (Γ, ∂). Since we will always use the same notation ∂, we will by an abuse of language call Γ an STi-plane. The map ∂ is by the way not necessarily unique, i.e., δ is not necessarily determined by Γ. For the inversive plane arising from the elliptic quadric over $\mathbf{GF}(2)$ there are essentially three distinct choices since $\mathbf{Sz}(2)$ has index 3 in $\mathbf{PSL}_2(4)$ (and the former is the full automorphism group of the corresponding STi-plane; the latter is the full automorphism group of the corresponding inversive plane).

The element ∂C, $C \in \mathcal{C}$, will be called the **corner** of C. It is exactly what is called "le nœud" in TITS [1962a]. Let a and b be two arbitrary points of an STi-plane Ω. Let C and D be two distinct circles both containing a and b. It follows from Lemma 7 of the proof of Theorem 7.6.23 (see page 334), proved there without using the extra condition [P], that for each point c, $a \neq c \neq b$, there exists a unique circle C_c containing c and touching both C and D. Moreover, C_c touches every circle containing both a and b by axiom [CH2]. Varying c, we obtain a set of circles partitioning the set $\mathcal{P} \setminus \{a, b\}$. Such a set is called **the transversal partition with extremities** a **and** b.

6.9.2 Block geometries in generalized polygons

Let x be any point of any generalized n-gon Γ, $n \geq 4$. The **block geometry** $^x\Gamma$ **in** x **of** Γ has as point set the set of lines through x. A typical block of $^x\Gamma$ is the set of lines xz such that $z \in \Gamma_2(x) \cap \Gamma_{n-2}(v_1) \cap \Gamma_{n-2}(v_2)$, where v_1 and v_2 are elements opposite x, provided that $|\Gamma_2(x) \cap \Gamma_{n-2}(v_1) \cap \Gamma_{n-2}(v_2)| \neq 1$ and $\Gamma_2(x) \cap \Gamma_{n-2}(v_1) \neq \Gamma_2(x) \cap \Gamma_{n-2}(v_2)$. The empty block is allowed.

Some geometric characterizations in this chapter can be reformulated in an elegant way using this new notion of block geometries. We give two typical examples. The block geometry in all points of a generalized quadrangle or hexagon has no blocks at all **if and only if** we are dealing with the symplectic quadrangle or the split Cayley hexagon. The block geometry in all points of a generalized hexagon contains no block or the empty block — up to duality — **if and only if** the generalized hexagon has the Moufang property. These statements follow from Theorem 6.2.1, Theorem 6.3.1 and Theorem 6.3.2.

Related to the block geometries are the following notions. Let x be a point of the generalized polygon Γ and let S be a set of points contained in some trace x^v, for some element v opposite x. Let $z \in S$. The **gate set of** S **through** z is the set of lines incident with z lying on a shortest path from x to an element v' with $S \subseteq x^{v'}$. Such a gate set is called **trivial** if it is the full set of lines incident with z. For S as above, the set $\{xz : z \in S\}$ is called **the back-up of** S **onto** x. If two traces with centre x have exactly one point in common, then we say that they **meet trivially at that point**.

We can now write down the main result of this section.

6.9.3 Theorem (Van Maldeghem [1998]). *A generalized octagon Γ is isomorphic to a perfect Ree–Tits octagon **if and only if** Γ satisfies the following axioms.*

[RT1] *The block geometry at every point of Γ satisfies conditions* [MP1], [MP2], [CH1] *and* [CH2] *(assuming the circles are the blocks).*

[RT1′] *Let x be any point of Γ and let C be any block in $^x\Gamma$. Let S be any intersection of traces in $x^\perp$ with back-up C. Then there exists a unique point z of S such that the gate set of S through z is non-trivial. Moreover, the line xz is independent of S and we denote it by ∂C. The geometry $^x\Gamma$ with mapping ∂ just defined satisfies axioms* [ST1] *and* [ST2] *(again assuming that the circles are the blocks).*

[RT2] *Let x be any point of Γ and let z_1, z_2 be two points collinear with x, with $xz_1 \neq xz_2$. Let X be any trace in $x^\perp$ containing z_1 and z_2. Let Z_1 and Z_2 be two traces in $x^\perp$ such that $X \cap Z_i = \{z_i\}$, $i = 1, 2$. Then the back-up of $Z_1 \cap Z_2$ is a member of the transversal partition of $^x\Gamma$ with extremities xz_1 and xz_2.*

[RT3] *Let x be any point of Γ and let Z be the non-trivial intersection of two distinct traces x^u and x^v, where u and v are points opposite x. Let z be the* *unique point of Z through which the gate set of Z is non-trivial and let y be any other point of Z. Then the lines at distance 5 from u (v), incident with z (y) are contained (not contained) in a circle C of $^z\Gamma$ ($^y\Gamma$) with $\partial C = xz$ ($\partial C = xy$).*

[RT4] *Let x and y be two opposite points of Γ and $L \in \Gamma_5(x) \cap \Gamma_3(y)$. Let $\{z\} = \Gamma_2(x) \cap \Gamma_3(L)$. Let $z', z'' \in x^y$ be such that the lines xz, xz', xz'' do not lie in a circle of $^x\Gamma$ with corner xz. Let $L', L'' \in \Gamma_3(x)$ be two arbitrary lines of the gate set of x^y through z' and z'', respectively. Then there exists a point y' opposite x at distance 3 from L and at distance 5 from L' and L''. Also, for every y'' opposite x, such that $x^{y''} = x^y$, $\delta(y, y'') = 6$ and $\delta(y'', L) = 3$, we have $|x^y_{[3]} \cap x^{y''}_{[3]}| > 1$.*

In the finite case, we can weaken the conditions (compare with the characterization of STi-planes given in Theorem 7.6.23, on page 334, and its finite version, Corollary 7.6.24, on page 339).

6.9.4 Theorem (Van Maldeghem [1998]). *A finite generalized octagon Γ of order (s, t) is isomorphic to a Ree–Tits octagon **if and only if** Γ satisfies axioms [RT1f], [RT1'], [RT3] and one of the conditions [RT2f] or $t = s^2$, where [RT1f] and [RT2f] are defined as follows.*

[RT1f] *The block geometry at every point of Γ satisfies conditions [MP1] and [MP2] (assuming the circles are the blocks).*

[RT2f] *If three traces with common centre meet pairwise trivially, then they all share a common point.*

6.9.5 A derived geometry for the Ree–Tits octagons

As a curiosity, we will now see how the Ree–Tits octagon $O(\mathbb{K}, \sigma)$ and the generalized quadrangle of Tits type $T_3(\mathcal{O}_{ST})$, with $\mathcal{O}_{ST}$ the Suzuki–Tits ovoid over $\mathbb{K}$, are related. Let p be any point of a perfect Ree–Tits octagon Γ. Let L be any line of Γ at distance 7 from p. The set $S_{p,L}$ of traces p^x for x opposite p and incident with L covers all points of $\Gamma_2(p)$ except for the elements of $\Gamma_1(L') \setminus \{p'\}$, where L' is the projection of L onto p and p' is the projection of L onto L'. Moreover, any two elements of $S_{p,L}$ meet in the unique point p'. We call $S_{p,L}$ a **pencil of traces** (based at p').

We now define the following geometry Γ_p, which we call the **derived geometry of Γ in p.** The points are of two types:

(i) the traces in $p^\perp$;

(ii) the points collinear with p in Γ, including p itself.

The lines are also of two types:

(a) the pencils of traces (in $p^\perp$);

(b) the lines of Γ through p.

The incidence between points of type (i) (type (ii)) and lines of type (a) (type (b)) is containment (the incidence in Γ). No point of type (i) is incident with a line of type (b). A point of type (ii) is incident with a line of type (a) if the pencil in question is based at the point in question.

For the definition of the Suzuki–Tits ovoid, we refer to Definitions 7.6.4 (page 324), see also Subsection 3.7.1 (page 122). We have the following result.

6.9.6 Theorem (Van Maldeghem [1995a]). *The geometry Γ_p as defined above, with $\Gamma \cong \mathsf{O}(\mathbb{K}, \sigma)$, is isomorphic to the generalized quadrangle of Tits type $\mathsf{T}_3(\mathcal{O}_{ST})$, with $\mathcal{O}_{ST}$ a Suzuki–Tits ovoid over $\mathbb{K}$ with corresponding Tits endomorphism σ.*

Proof. We show this result by coordinatization, unlike the original proof in VAN MALDEGHEM [1995a]. First we coordinatize Γ in such a way that $p = (\infty)$. Then we use the description of Γ displayed in 3.2 and we refer to the ith equation of the octogonal octanary ring (see 3.2 on page 119) as "Equation (i)".

Let R_1 be the field $\mathbb{K}$, and let R_2 be the Cartesian product $\mathbb{K} \times \mathbb{K}$. We will denote the elements of Γ_p with double brackets in order to clearly make the distinction with elements of Γ. We can define

$$((\infty)) = (\infty), \qquad [[\infty]] = [\infty], \qquad ((a)) = (a), \qquad [[k]] = [k],$$

for all $a \in R_1$ and all $k \in R_2$. Furthermore, we can define — not bothering about normalization — $((0, b)) = (0, b)$.

It is important to know when two points opposite p define the same trace. So let $x_i = (a_i, l_i, a_i', l_i', a_i'', l_i'', a_i''')$ be two points opposite $p = (\infty)$, $i = 1, 2$. By Equation (6), the trace p^{x_i} depends only on $a_i, a_i''', (l_i)_0$ and $(l_i'')_0$. Hence $p^{x_1} = p^{x_2}$ **if and only if** $a_1 = a_2$, $a_1''' = a_2'''$, $(l_1)_0 = (l_2)_0$ and $(l_1'')_0 = (l_2'')_0$. Hence we can identify any trace with centre p with an expression of the form

$$(a, (l_0, \bullet), \bullet, \bullet, \bullet, (l_0'', \bullet), a''').$$

It follows that the expression

$$[0, (l_0, \bullet), \bullet, \bullet, \bullet, (l_0'', \bullet)]$$

denotes a pencil of traces, all elements of which contain the point (0). Hence we may put

$$[[0, (l_0, l_0'')]] = [0, (l_0, \bullet), \bullet, \bullet, \bullet, (l_0'', \bullet)].$$

It is clear that a pencil of traces all elements of which contain the point (a), $a \in R_1$, can be represented by a set of lines which we may denote by

$$[a, (l_0, \bullet), \bullet, \bullet, \bullet, (l_0'', \bullet)].$$

Using Equations (1), (5) and (6), one calculates that a pencil of traces all elements of which contain the point (k, b), for some $k \in R_2$ and $b \in R_1$, is completely determined and uniquely defined by the set of lines of Γ having seven coordinates and sharing the first two coordinates (which must necessarily be k and b) and the first component of the third and last coordinate. Hence such a pencil can be denoted by

$$[k, b, (k_0', \bullet), \bullet, \bullet, \bullet, (k_0''', \bullet)].$$

We remark that we can use the same technique in writing down coordinates for all elements of Γ_p as we did for the generalized quadrangles in the general coordinatization theory. The reason is that there are obviously no triangles through the flag $\{((\infty)), [[\infty]]\}$, and every element of Γ_p lies in an ordinary quadrangle together with $((\infty))$ and $[[\infty]]$. One might say that Γ_p is *locally* a generalized quadrangle. Hence one easily computes that

$$((k, b)) = (k, b), \qquad [[a, (l_0, l_0'')]] = [a, (l_0, \bullet), \bullet, \bullet, \bullet, (l_0'', \bullet)],$$

$$((0, 0, a')) = (0, (0, \bullet), \bullet, \bullet, \bullet, (0, \bullet), a'),$$

$$[[0, 0, (k_0''', k_0')]] = [0, 0, (k_0', \bullet), \bullet, \bullet, \bullet, (k_0''', \bullet)].$$

Finally we obtain the coordinates of a general point and a general line:

$$[[k, b, k']] = [k, b, (k_1' + k_0^\sigma k_0', \bullet), \bullet, \bullet, \bullet, (k_0', \bullet)],$$

$$((a, l, a')) = (a, (l_0, \bullet), \bullet, \bullet, \bullet, (l_1, \bullet), a').$$

Using Equations (1), (5) and (6) again, one computes that $((a, l, a'))$ is incident in Γ_p with $[[k, b, k']]$ **if and only if**

$$\left\{ \begin{array}{rcl} b & = & a \, \mathrm{N}(k) + k_1 l_0 + k_0 l_1 + a', \\ (l_0, l_1) & = & a(k_0, k_1) + (k_0', k_1'). \end{array} \right.$$

This shows for one thing that Γ_p is a generalized quadrangle and for another that Γ_p is isomorphic to $\mathsf{T}_3(\mathcal{O}_{ST})$ by Subsection 3.7.1. $\qquad \square$

Chapter 7

Ovoids, Spreads and Self-Dual Polygons

7.1 Introduction

Ovoids and spreads are in more than one respect special configurations in generalized polygons. Roughly speaking, an ovoid in a generalized polygon is a set of mutually opposite points "of maximal size" (this will be made precise below for infinite polygons), and the dual notion is a spread. Sets of maxial size with respect to a certain property in geometries usually themselves have interesting properties. For example, they might be used to construct other geometries (as ovoids in projective spaces are used to construct generalized quadrangles; see Subsection 3.7.1 on page 120. By the way, the term "ovoid" might seem confusing here; one connection is that in a lot of cases, ovoids of $W(\mathbb{K})$, with $\mathbb{K}$ of characteristic 2, are via the standard embedding of $W(\mathbb{K})$ in $PG(3,\mathbb{K})$ ovoids of $PG(3,\mathbb{K})$; see Proposition 7.6.14 and Subsection 7.6.25 below. Another connection is that polarities of projective spaces sometimes produce ovoids in these spaces, and likewise polarities in generalized polygons sometimes produce ovoids in the polygons). Another feature of ovoids is that they sometimes have interesting automorphism groups. The Suzuki groups and the Ree groups of characteristic 2 arise in that way. We will use the ovoids to prove some properties of these groups.

There are a lot of ovoids known for many classes of (finite) generalized quadrangles. We will not review them all. We mention here that some flocks of quadratic cones define ovoids in $Q(4,q)$ (if q is odd) and hyperovals in $PG(2,q)$ (if q is even). However, we will restrict our attention to the so-called (*classical* and) *semi-classical* ovoids in the symplectic quadrangles and the split Cayley hexagons. They are particularly important because the part of their automorphism group inside the little projective group of the corresponding polygon is, up to some exceptions of small order, a simple group. But they also have interesting geometric properties. For instance in the case of quadrangles, these ovoids carry the structure of *inversive planes*, and in the case of hexagons they produce *unitals*. This additional structure on the ovoid arises naturally from the generalized polygon, as we will see (compare the original algebraic construction of, for instance, the unitals corresponding to the semi-classical ovoids in the dual split Cayley hexagons; see LÜNEBURG [1966]).

If a generalized quadrangle or hexagon has a flag $\{p, L\}$ such that both p and L are span-regular (see Subsection 1.9.6 on page 40 and Subsection 1.9.8 on page 40), and if p is a projective (see Definition 1.9.4) or polar point (see Definition 1.9.14) and L is a projective or polar line, then the polygon can be constructed by amalgamating the span-geometries in p and in L. In particular, this happens if the generalized polygon is a self-dual symplectic quadrangle or a self-dual split Cayley hexagon. We will explain this amalgamation procedure geometrically.

7.2 Generalities about polarities and ovoids

7.2.1 Definitions. A **polarity** ρ of a generalized polygon Γ is a correlation of order 2. A **self-polar** polygon is a generalized polygon admitting a polarity. An element v of Γ is said to be **absolute** (with respect to the polarity ρ) if it is incident with its image v^ρ.

For instance, in the Pappian projective plane $\mathbf{PG}(2, \mathbb{K})$ over the field $\mathbb{K}$, as described in Subsection 2.2.1 (see page 50), a polarity is given by the map $(x, y, x) \leftrightarrow [x, y, z]$. A similar polarity for the Suzuki quadrangles is described in Subsection 3.4.6 (page 103). We leave it as an exercise to the reader to find an analogous polarity in the mixed hexagons $\mathsf{H}(\mathbb{K}, \mathbb{K}^\sigma)$, where $\mathbb{K}$ is a field of characteristic 3 and σ is a Tits endomorphism in $\mathbb{K}$ (see Definition 2.5.1, page 79), by recoordinatizing $\mathsf{H}(\mathbb{K}, \mathbb{K}^\sigma)$ in the same way as the Suzuki quadrangles.

An **ovoid** $\mathcal{O}$ of a generalized n-gon Γ is a set of mutually opposite points (hence $n = 2m$ is even) such that every element v of Γ is at distance at most m from at least one element of $\mathcal{O}$.

A **spread** of a generalized polygon is the dual of an ovoid. There are other related but less important structures and we will mention some in Subsection 7.3.9 (see page 316).

We now prove some general properties of ovoids, spreads and polarities. Every result on ovoids implicitly holds dually for spreads.

7.2.2 Lemma. *Let $\mathcal{O}$ be a set of points of a generalized $2m$-gon Γ. Then $\mathcal{O}$ is an ovoid **if and only if**, for m even, every line of Γ lies at distance less than m from a unique element of $\mathcal{O}$; for m odd, every point of Γ lies at distance less than m from a unique element of $\mathcal{O}$. Also, **if** all points of $\mathcal{O}$ are mutually opposite, **then** $\mathcal{O}$ is an ovoid <u>if and only if</u>, for m even, every line of Γ lies at distance less than m from some element of $\mathcal{O}$; for m odd, every point of Γ lies at distance less than m from some element of $\mathcal{O}$.*

Proof. Suppose first that $\mathcal{O}$ is an ovoid. If m is odd, then no point lies at distance m from any other point, so every point lies at distance less than m from some element of $\mathcal{O}$. If a point p were at distance n_i less than m from $x_i \in \mathcal{O}$, $i = 1, 2$, then, by the triangle inequality, $\delta(x_1, x_2) \leq n_1 + n_2 < 2m = n$. Hence x_1 and x_2 cannot be opposite, so $x_1 = x_2$. We have shown that every point is at distance less than m from a unique element of $\mathcal{O}$. Similarly, if m is even, then every line is at distance less than m from a unique element of $\mathcal{O}$.

As in the previous paragraph, one shows that if $\mathcal{O}$ is a set of mutually opposite points, and, for m odd, every point is at distance less than m from an element of $\mathcal{O}$, and, for m even, every line is at distance less than m from an element of $\mathcal{O}$, then every point (line) is at distance less than m from a unique element of $\mathcal{O}$. So it remains to show that, if $\mathcal{O}$ is a set of points such that, for m odd (even), every point (line) lies at distance less than m from a unique element of $\mathcal{O}$, then $\mathcal{O}$ is an ovoid. Let m be odd to fix ideas (the proof for m even is similar). Then the middle element of any $(n-2)$-path γ starting with a point is itself a point at distance at most $m - 1$ from any point of γ. This implies that γ contains at most one element of $\mathcal{O}$. Consequently all elements of $\mathcal{O}$ are mutually opposite. Let v be any element of Γ. If v is a point, then by assumption, $\delta(v, x) < m$ for some $x \in \mathcal{O}$. If v is a line, then for any point v' incident with v, we have $\delta(v', x) < m$, for some $x \in \mathcal{O}$. Hence $\delta(v, x) \leq \delta(v, v') + \delta(v', x) \leq m$ and so $\mathcal{O}$ is an ovoid. $\qquad\square$

7.2.3 Proposition. *A set of mutually opposite points in a finite generalized $2m$-gon* Γ *of order* (s, t) *is an ovoid* **if and only if** *it contains*

(i) $1 + st$ *elements for* $m = 2$;

(ii) $\dfrac{(1+s)(1 + st + s^2 t^2)}{1 + s + st}$ *elements for* $m = 3$;

(iii) $1 + s^2 t^2$ *elements for* $m = 4$.

Consequently, a set of mutually opposite points in a finite generalized $2m$-gon Γ *of order* s *is an ovoid* **if and only if** *it contains* $1 + s^m$ *elements.*

Proof. The last assertion follows from the previous one by putting $s = t$. So we only have to prove (i), (ii) and (iii).

Suppose first that $\mathcal{O}$ is an ovoid. If $m = 2$, then there are $|\mathcal{O}| \cdot (1 + t)$ lines at distance 1 from some element of $\mathcal{O}$. Since every line arises in this way, we have $|\mathcal{O}| \cdot (1 + t) = (1 + t)(1 + st)$. If $m = 3$, then there are $|\mathcal{O}|$ points at distance 0 from some element of $\mathcal{O}$ and $|\mathcal{O}| \cdot (1+t)s$ points at distance 2 from some (unique) element of $\mathcal{O}$. Again, no other points are available by Lemma 7.2.2 and hence $|\mathcal{O}| \cdot (1 + s + st) = (1 + s)(1 + st + s^2 t^2)$. If $m = 4$, then there are $|\mathcal{O}| \cdot (1 + t)$ lines at distance 1 from a unique element of $\mathcal{O}$ and there are $|\mathcal{O}| \cdot (1 + t)st$ lines at distance 3 from a unique element of $\mathcal{O}$. Using Lemma 7.2.2 again, we obtain $|\mathcal{O}| \cdot (1 + t)(1 + st) = (1 + t)(1 + st)(1 + s^2 t^2)$. Hence the result.

Now suppose $\mathcal{O}$ is a set of mutually opposite points of the given size. For m even, we count the number of lines at distance less than m from some (unique) point of $\mathcal{O}$; for m odd, we count the number of points at distance less than m from some (unique) element of $\mathcal{O}$. This counting is in fact carried out in the previous paragraph and we always obtain the full number of lines (m even) or points (m odd) of Γ. Hence all lines or points lie at distance less than m from some point of $\mathcal{O}$. The result follows from Lemma 7.2.2. $\qquad\square$

The existence problem of ovoids in generalized polygons seems only to be interesting in the finite case. For instance, for infinite quadrangles of order (s,t), the method of transfinite induction (in the countable case this is just ordinary induction) guarantees the existence of ovoids and spreads if $s = t$; see CAMERON [1994] (but note that CAMERON uses in *op. cit.* another definition of ovoids, which differs in the case of generalized n-gons with $n > 4$; he considers what we will call *distance-2-ovoids* in Subsection 7.3.9). Nevertheless, one can consider ovoids satisfying some additional conditions. For instance, we will produce ovoids in infinite quadrangles which have a doubly transitive automorphism group.

7.2.4 Corollary. *No finite generalized quadrangle of order (s,s^2) has an ovoid. Also, no finite generalized hexagon of order (s,s^3) or (s^3,s) has an ovoid. Hence the classical quadrangle $\mathsf{Q}(5,q)$ has no ovoids and neither does any finite twisted triality hexagon $\mathsf{T}(q^3,q)$ or its dual $\mathsf{T}(q,q^3)$.*

Proof. The result for quadrangles can be found in PAYNE & THAS [1984](1.8.3). It amounts to counting, for a given ovoid $\mathcal{O}$ in a generalized quadrangle of order (s,s^2) and two fixed points $x, y \in \mathcal{O}$, in two ways the pairs (u, z) with $z \in \mathcal{O} \setminus \{x, y\}$ and $u \in \{x, y, z\}^{\perp}$. Taking Theorem 1.7.4$(i)$ into account, we obtain $(s^2 + 1)(s^2 - 1) = (s^3 - 1)(s + 1)$, a contradiction.

Now consider an ovoid in a finite generalized hexagon of order (s,t). If $t = s^3$ and if the number

$$\frac{(1 + s)(1 + st + s^2 t^2)}{1 + s + st}$$

is an integer, then (carrying out the division) $(1 + s)(1 + s + s^2)$ is zero (this is the remainder if $s > 2$; for $s = 2$ the above number is clearly not an integer), a contradiction. If $s = t^3$, then a similar calculation shows that the remainder $2t^3 + t^2 - 2t + 2$ must be zero. But this number is always positive for non-negative t. The corollary is proved. $\qquad\square$

The next result connects the notions of ovoid and polarity (see e.g. PAYNE [1968] or CAMERON, THAS & PAYNE [1976]).

7.2.5 Proposition. *Let ρ be a polarity of a generalized n-gon Γ. If $n = 2m$ is even,* **then** *the set of all absolute points with respect to ρ forms an ovoid of Γ.*

Proof. Let x and y be two absolute points and suppose they are not opposite each other. Since $\delta(x,y) = \delta(x^{\rho}, y^{\rho})$ and $x\,\mathbf{I}\,x^{\rho}$ and $y\,\mathbf{I}\,y^{\rho}$, we may assume upon interchanging x and y that x^{ρ} is the projection of y onto x. Indeed, otherwise

$\delta(x,y) = n$ or $\delta(x,y) = n-1$. The first possibility contradicts our assumption, the second implies that n is odd. Let p be the projection of y onto x^ρ; then $x \mathbf{I} p^\rho$. But $\delta(x,y) - 2 = \delta(p,y) = \delta(p^\rho, y^\rho) = \delta(p^\rho, x^\rho) + \delta(x^\rho, y^\rho) = 2 + \delta(x,y)$, a contradiction. Hence all absolute points are mutually opposite.

Let v be any element of Γ. Since n is even, $\delta(v, v^\rho) < n$. Now the projection v_1 of v onto v^ρ must be mapped onto the projection v_1' of v^ρ onto v; the projection of v_1 onto v_1' must be mapped onto the projection of v_1' onto v_1, etc. Continuing like this, remembering that $\delta(v, v^\rho)$ is odd, we obtain an element w, $\delta(v,w) + \delta(w, v^\rho) = \delta(v, v^\rho) < n$, incident with w^ρ. Hence either w or w^ρ is an absolute point. Suppose that w is a point. Since $\delta(v,w) = \delta(v^\rho, w^\rho)$ and $\delta(v,w) + \delta(v^\rho, w^\rho) = \delta(v, v^\rho) \pm 1$, we conclude that $\delta(v,w) \le m$. The proposition is proved. $\qquad\square$

7.2.6 Definitions. Of course the set of absolute lines of a polarity of a generalized $2m$-gon forms a spread $\mathcal{S}$. Let $\mathcal{O}$ be the set of absolute points. Then $(\mathcal{O}, \mathcal{S})$ is a pair with the following properties: $\mathcal{O}$ is an ovoid, $\mathcal{S}$ is a spread and every element of $\mathcal{O}$ (or $\mathcal{S}$) is incident with a unique element of $\mathcal{S}$ (or $\mathcal{O}$). We call such a pair an **ovoid–spread pairing**. For generalized quadrangles, every pair $(\mathcal{O}', \mathcal{S}')$ where $\mathcal{O}'$ is an ovoid and $\mathcal{S}'$ is a spread is obviously an ovoid–spread pairing. On the other hand, the only ovoid–spread pairings in finite generalized hexagons known to me arise from polarities (an ovoid–spread pairing $(\mathcal{O}, \mathcal{S})$ is said to **arise from a polarity** ρ if the set of absolute points (lines), of ρ is precisely $\mathcal{O}$ ($\mathcal{S}$)).

The next result is due to PAYNE [1968] for quadrangles, and to OTT [1981] for hexagons. The proof we offer follows the method of Payne, which only uses some standard facts about eigenvalues and eigenvectors; see also VAN MALDEGHEM [19**c]. Apparently we are the first to note that Payne's proof also applies to hexagons; Ott uses Hecke algebras, which needs more background. We remark that a similar approach for hexagons is in CAMERON, THAS & PAYNE [1976].

7.2.7 Proposition. *If there is a polarity in a finite generalized $2m$-gon Γ of order (s,t), then $s = t$ and ms is a perfect square.*

Proof. Since a polarity is a correlation, Γ must be isomorphic to its dual and hence $s = t$. In view of Theorem 1.7.1(v) (see page 24), we only have to deal with the cases $m = 2$ and $m = 3$. We do that simultaneously as far as we can go. Denote by ρ a polarity of Γ. Let N be the number of points of Γ and let $\{p_i : 1 \le i \le N\}$ be the set of points of Γ. Define the $N \times N$ matrix $M = (m_{ij})$ as follows. The entry m_{ij} is zero if p_i is not incident with p_j^ρ; it is 1 in the other case. Since ρ is a polarity, the matrix M is clearly symmetric. Hence M is diagonalizable. We compute the eigenvalues of M. Note that the sum of the elements on the diagonal of M equals $1 + s^m$, i.e., the number of absolute points with respect to ρ.

In order to determine the eigenvalues of M, we compute $A = M^2 = (a_{ij})$. It is clear that a_{ij} is equal to the number of lines incident with both p_i and p_j, $1 \le i, j \le N$.

In other words, a_{ij} equals

$$
\begin{cases}
s+1 & \text{if } i = j; \\
1 & \text{if } p_i \text{ is collinear with } p_j; \\
0 & \text{otherwise.}
\end{cases}
$$

Similarly, if $B = A^2 = (b_{ij})$, then b_{ij} is equal to

$$
\begin{cases}
2s^2 + 3s + 1 & \text{if } i = j; \\
3s + 1 & \text{if } p_i \text{ is collinear with } p_j; \\
1 + (3 - m)s & \text{if } \delta(p_i, p_j) = 4; \\
0 & \text{otherwise.}
\end{cases}
$$

If J denotes the $N \times N$ matrix all entries of which are equal to 1, then, for $m = 2$, we obtain $A^2 - 2sA = (s + 1)J$. For $m = 3$, we compute $C = A^3 = (c_{ij})$. The number c_{ij} is equal to

$$
\begin{cases}
5s^3 + 9s^2 + 5s + 1 & \text{if } i = j; \\
9s^2 + 5s + 1 & \text{if } p_i \text{ is collinear with } p_j; \\
5s + 1 & \text{if } \delta(p_i, p_j) = 4; \\
s + 1 & \text{otherwise.}
\end{cases}
$$

One computes that $A^3 - 4sA^2 + 3s^2A = (s + 1)J$. Now J has eigenvalues 0 (with multiplicity $N - 1$) and N (with multiplicity 1). To the eigenvalue N for J corresponds an eigenvalue θ for A satisfying $F_2(\theta) := \theta^2 - 2s\theta = (1+s)N$ for $m = 2$, or $F_3(\theta) := \theta^3 - 4s\theta^2 + 3s^2\theta = (1+s)N$ for $m = 3$. In both cases $\theta = (1+s)^2$ satisfies the equation and is indeed an eigenvalue of A, since it is equal to the constant row sum. Any other eigenvalue θ satisfies $F_m(\theta) = 0$. Hence the set of eigenvalues of A is $\{(1 + s)^2, 0, 2s\}$ for $m = 2$ and $\{(1 + s)^2, 0, s, 3s\}$ for $m = 3$; in both cases the multiplicity of $(1 + s)^2$ is 1. In fact, this is well known and independent of Γ admitting a polarity; it could also be deduced from the proof of Theorem 1.7.1 in Appendix A. Now we distinguish between $m = 2$ and $m = 3$.

$m = 2$. The eigenvalues of M are $1 + s$ (because this is the constant row sum; it has multiplicity 1), 0, $\sqrt{2s}$ (the latter with multiplicity, say, ℓ_+) and $-\sqrt{2s}$ (with multiplicity, say, ℓ_-). Note that ℓ_+ or ℓ_- could be 0 in which case the corresponding "eigenvalue" is "not an eigenvalue". Since the sum of all eigenvalues (taking their multiplicities into account) is equal to the sum of elements on the diagonal of M, we have

$$
1 + s + (\ell_+ - \ell_-)\sqrt{2s} = 1 + s^3,
$$

implying $\sqrt{2s} \in \mathbb{N}$ (using $s \neq 1$).

$m = 3$. The eigenvalues of M are $1+s$ (again with multiplicity 1), 0, $\sqrt{s}$ (with multiplicity, say, k_+), $-\sqrt{s}$ (with multiplicity, say, k_-), $\sqrt{3s}$ (with multiplicity, say, ℓ_+) and $-\sqrt{3s}$ (with multiplicity, say, ℓ_-). Again this notation implies that if one of the multiplicities is 0, then the corresponding "eigenvalue" does not exist. So we have the equation

$$1 + s + (k_+ - k_-)\sqrt{s} + (\ell_+ - \ell_-)\sqrt{3s} = 1 + s^3.$$

Now we consider the diagonal elements d_{ii} of the matrix $D = M^3$. If p_i is an absolute point, then d_{ii} is equal to $1 + s$ plus the number of points p_j collinear with p_i and such that $p_i \mathbf{I} p_j^\rho$. Clearly, the latter is the number of points on p_i^ρ (except p_i itself). Hence $d_{ii} = 2s + 1$. If p_i is not an absolute point, then d_{ii} is exactly the number of points p_j collinear with p_i such that $p_i \mathbf{I} p_j^\rho$. Let p_j be such a point. Consider the path $(p_i, p_i p_j, p_j, p_i^\rho)$. Since p_i is not absolute, we easily see that p_j must be absolute, hence this number is 1 and p_i is a point on an absolute line. Otherwise, this number equals 0. Noting that there are $1 + s^3$ absolute points and $s(1 + s^3)$ non-absolute points incident with some absolute line, we deduce that the sum of the entries on the diagonal of M^3 is $(1 + s^3)(1 + 3s)$. Since θ is an eigenvalue of M **if and only if** θ^3 is an eigenvalue of M^3 (and in this case the multiplicities are the same), the non-negative integers $k := k_+ - k_-$ and $\ell := \ell_+ - \ell_-$ satisfy the following system of equations:

$$\begin{cases} (1+s) + \sqrt{s} \cdot k + \sqrt{3s} \cdot \ell & = & (1 + s^3), \\ (1+s)^3 + s\sqrt{s} \cdot k + 3s\sqrt{3s} \cdot \ell & = & (1 + s^3)(1 + 3s). \end{cases}$$

Hence $k = 0$ and $\ell = \dfrac{s^3 - s}{\sqrt{3s}}$, implying $\sqrt{3s} \in \mathbb{N}$.

The proposition is proved.   □

For the sake of completeness, we mention an important result on polarities in finite projective planes. A proof can be found in HUGHES & PIPER [1973].

Recall from Subsection 3.7.1 that an **oval** in a projective plane is a set of points such that no three points are collinear and at every point of the oval there is a unique tangent line, i.e., through each point of the oval there is a unique line meeting the oval in just one point. A **unital** of order $q \in \mathbb{N}^*$ is a set of $q^3 + 1$ elements having $q^2(q^3 + 1)/2$ distinguished subsets containing $q + 1$ elements, called blocks, such that each block is determined by any two of its elements. A **secant line** of a set of points in a projective space is a line incident with at least two points of that set.

7.2.8 Theorem. *Suppose that Γ is a finite projective plane of order q and θ is a polarity of Γ. Then the number N_θ of absolute points satisfies $q+1 \le N_\theta \le q\sqrt{q}+1$. If $N_\theta = q+1$,* **then** *the set of absolute points is an oval if q is odd, and all absolute points are collinear if q is even; if $N_\theta = q\sqrt{q} + 1$,* **then** *the set of absolute points furnished with the sets of points incident with the secant lines forms a unital. Also, if q is not a square, then $N_\theta = q + 1$.*   □

7.3 Polarities, ovoids and spreads in Moufang polygons

First we deal with projective planes.

7.3.1 Proposition. *Every Moufang projective plane which is Pappian or not De-sarguesian is self-polar. A non-Pappian Desarguesian plane Γ is self-dual* **if and only if** *the corresponding coordinatizing skew field admits an anti-automorphism; it is self-polar if the corresponding coordinatizing skew field admits an involutory anti-automorphism.*

Proof. If $\mathbb{D}$ is a (non-associative) division algebra, or more generally a Cartesian number system with abelian addition, then the projective plane Γ over $\mathbb{D}$ is isomorphic to the dual of the plane Γ' coordinatized by the opposite division algebra $\mathbb{D}'$, which differs from $\mathbb{D}$ only by the new multiplication $a\square b := ba$. Indeed, the map δ with $(x,y)^\delta = [x,-y]'$, $[a,b]^\delta = (a,-b)'$, $[c]^\delta = (c)'$, $[\infty]^\delta = (\infty)'$, $(d)^\delta = [d]'$ is an isomorphism of Γ onto the dual of Γ'.

If α is an anti-automorphism of $\mathbb{D}$, then δ followed by α, applied to each coordinate (i.e., mapping $(x,y)'$ to $(x^\alpha, y^\alpha)'$, etc.) is a correlation of Γ (one can easily check this). Since every commutative field has an involutory anti-automorphism ($\alpha = 1$), as well as every non-associative alternative field (see PICKERT [1975], Chapter 6, in particular p. 170), we obtain that every Pappian projective plane and every non-Desarguesian Moufang plane is self-polar. The statement about non-Pappian Desarguesian planes follows from an easy explicit calculation and the fact that every correlation of Γ arises from a semi-linear map. $\qquad\square$

So a Desarguesian projective plane is self-dual **if and only if** the corresponding skew field $\mathbb{K}$ admits an anti-automorphism. This is not always the case; see BLANCHARD [1972], page 38. Also, a non-Pappian Desarguesian projective plane is self-polar **if and only if** $\mathbb{K}$ admits an involutory anti-automorphism. However, it seems that one does not know examples of skew fields with anti-automorphisms, but without involutory ones.

We now turn our attention to Moufang quadrangles.

7.3.2 Theorem. *A Moufang quadrangle Γ, not of type $(BC - CB)_2$, is self-dual* **if and only if** *it is isomorphic to a mixed quadrangle $Q(\mathbb{K}, \mathbb{K}'; L, L')$ for which there exists an injective endomorphism θ of $\mathbb{K}$ which maps $\mathbb{K}$ onto $\mathbb{K}'$ and which maps $\mathbb{K}'$ onto $\mathbb{K}^2$, and such that there exist elements $k \in \mathbb{K}^2$ and $k' \in \mathbb{K}'$ such that $L' = k' \cdot L^\theta$ and $L^2 = k \cdot L'^\theta$. Also, the mixed quadrangle $Q(\mathbb{K}, \mathbb{K}'; L, L')$ is self-polar* **if and only if** *there exists a Tits endomorphism θ of $\mathbb{K}$ which maps $\mathbb{K}$ onto $\mathbb{K}'$, and there exist elements $k \in \mathbb{K}^2$ and $k' \in \mathbb{K}'$, with $k'^\theta k = 1$, such that $L' = k' \cdot L^\theta$ and $L^2 = k \cdot L'^\theta$* **if and only if** *we can choose $\mathbb{K}'$, L and L' such that there exists a Tits endomorphism θ of $\mathbb{K}$ which maps $\mathbb{K}$ onto $\mathbb{K}'$ and L to L'.*

Proof. Suppose Γ is a self-dual Moufang quadrangle. Since every Moufang quadrangle has some abelian root group (it belongs to a root system of type BC_2), all root groups must be abelian. Hence by Lemma 5.5.9 of page 219, Γ has, up to duality, regular points (because we excluded type $(BC - CB)_2$). Since Γ is self-dual, all lines of Γ are also regular. So by Lemma 5.5.8 on page 219, Γ is a mixed quadrangle.

The first assertion now follows directly from Propositions 3.4.3 and 3.4.4 (on, respectively, pages 101 and 102). Note that we have swapped the symbols of k and k'.

Suppose now the map θ induces a polarity ρ. We conceive ρ as an isomorphism from $Q(\mathbb{K}, \mathbb{K}'; L, L')$ to its dual. First note that the dual of the dual of $Q(\mathbb{K}, \mathbb{K}'; L, L')$ is $Q(\mathbb{K}^2, \mathbb{K}'^2; L^2, L'^2)$ and that, according to the proof of Proposition 3.4.4 on page 102, the identity map corresponds to squaring each coordinate. From the proof of Proposition 3.4.3 on page 101 we infer

$$(a, l, a')^\rho = (X^{-\theta}a^\theta, X^{-2\theta}X'^{-\theta}l^\theta, X^{-\theta}X'^{-\theta}a'^\theta),$$

for all $a, a' \in L$ and all $l \in L'$, and with $k' = X^{-\theta}$ and $k = X'^{-\theta}$. Similarly

$$[m, b, m']^\rho = [X'^{-\theta}m^\theta, X^{-\theta}X'^{-\theta}b^\theta, X^{-2\theta}X'^{-\theta}m'^\theta],$$

for all $b \in L$ and all $m, m' \in L'$. Note that these formulae hold in $Q(\mathbb{K}, \mathbb{K}'; L, L')$. In order to calculate an image under θ in $Q(\mathbb{K}', \mathbb{K}^2; L', L^2)$, we have to do the substitution $X \mapsto X'$ and $X' \mapsto X^2$ in the above formulae. A straightforward calculation now shows that $((a, l, a')^\rho)^\rho = (a^2, l^2, a'^2)$ **if and only if** $\theta^2 = 2$ and $X^{-2}X'^{-\theta} = 1$. Substituting the values for k and k', the result of the first part of the second assertion follows.

To prove the second part of the last assertion, we remark that we can recoordinatize the symplectic quadrangle $W(\mathbb{K})$ such that the image of the point (1) under the polarity is exactly the line $[1]$. This recoordinatization defines an isomorphism from $Q(\mathbb{K}, \mathbb{K}'; L, L')$ to some $Q(\mathbb{K}, \mathbb{K}'_1; L_1, L'_1)$. This immediately implies, in view of the proof of Proposition 3.4.3, that $k = 1$ and hence also $k' = 1$ and the result follows from the previous part. $\square$

7.3.3 Remark. Applying the renaming of the coordinates as in the proof of Proposition 3.4.4 to the polarity in the above proof (which maps Γ to its dual), we easily obtain that the image of a point (a, l, a') under the given polarity with corresponding injective endomorphism $\theta : \mathbb{K} \to \mathbb{K}'$ and elements $k \in \mathbb{K}^2$, $k' \in \mathbb{K}'$, is the line $[k'a^\theta, k'k^{1/2}l^{\theta^{-1}}, k'ka'^\theta]$ of Γ, and the image of the line $[m, b, m']$ is the point $(k^{1/2}m^{\theta^{-1}}, k'kb^\theta, k'k^{1/2}m'^{\theta^{-1}})$ of Γ. We will need this explicit description of the polarity in the proof of Theorem 7.6.2 below.

We treat the Moufang quadrangles of type $(BC - CB)_2$ in a separate section; see Section 7.4 below.

7.3.4 Theorem. *A Moufang hexagon* Γ *is self-dual* **if and only if** *it is isomorphic to a mixed hexagon* $\mathsf{H}(\mathbb{K}, \mathbb{K}')$ *for which there exists an injective endomorphism* θ *of* $\mathbb{K}$ *which maps* $\mathbb{K}$ *onto* $\mathbb{K}'$ *and which maps* $\mathbb{K}'$ *onto* $\mathbb{K}^3$. *Also, the mixed hexagon* $\mathsf{H}(\mathbb{K}, \mathbb{K}')$ *is self polar* **if and only if** *there exists a Tits endomorphism* θ *of* $\mathbb{K}$ *which maps* $\mathbb{K}$ *onto* $\mathbb{K}'$.

Proof. Since every Moufang hexagon contains, up to duality, distance-2-regular points, all lines of a self-dual Moufang hexagon must also be distance-2-regular. By Corollary 5.5.15 on page 223, Γ is a mixed hexagon. The rest of the proof is similar to (but easier than) the case of quadrangles above. $\qquad\square$

7.3.5 Corollary. *A Pappian generalized* $2m$-*gon is self-dual* **if and only if** *it is defined over a perfect field of characteristic* m; *it is self polar* **if and only if** *it is defined over a perfect field of characteristic* m *admitting a Tits automorphism.*

Proof. If the Pappian $2m$-gon Γ in question is defined over the field $\mathbb{K}$, then (since either $\Gamma \cong \mathsf{Q}(\mathbb{K}, \mathbb{K}; \mathbb{K}, \mathbb{K})$, or $\Gamma \cong \mathsf{H}(\mathbb{K}, \mathbb{K})$) it follows from Theorem 7.3.2 and Theorem 7.3.4 that Γ is self-dual if and only if the characteristic of $\mathbb{K}$ is equal to m and there exists a monomorphism from $\mathbb{K}$ onto $\mathbb{K}$ which maps $\mathbb{K}$ to $\mathbb{K}^m$. Clearly the latter is equivalent to saying that $\mathbb{K} = \mathbb{K}^m$. The second assertion follows directly from the two preceding theorems. $\qquad\square$

By Theorem 1.7.1(v) on page 24 and Theorem 5.3.3 on page 179, we have:

7.3.6 Theorem. *No finite generalized octagon can be self-dual. No Moufang generalized octagon can be self-dual.* $\qquad\square$

Note, however, that, starting from an ordinary 9-gon, viewed as a partial 8-gon, and applying the free construction of Subsection 1.3.13 (page 13), one obtains a self-polar (and in particular self-dual) generalized octagon. Hence in this case, we obtain as a byproduct a very simple proof of the fact that this free octagon is certainly not a Moufang octagon!

Now we turn our attention to ovoids and spreads in Moufang polygons. Usually we will consider ovoids in a generalized polygon Γ^D rather than spreads in the original Γ. We give an exposition and occasionally a proof. We concentrate on the finite case, but include some relevant facts about infinite ovoids, too.

In the following, Γ is a generalized $2m$-gon, $m \geq 2$.

7.3.7 Ovoids in generalized octagons

If $m = 4$, then the only finite octagons known to exist are the Moufang octagons. Nothing is known about them concerning existence of ovoids and spreads. Only in the smallest case, for $\Gamma \cong \mathsf{O}(2)$, do we know that there is no ovoid, by a computer result of COOLSAET (unpublished).

7.3.8 Ovoids in generalized hexagons

Here $m = 3$. By Corollary 7.2.4, the finite twisted triality hexagons admit neither ovoids nor spreads. Hence let Γ be the split Cayley hexagon over the finite field $\mathbf{GF}(q)$. If the characteristic of $\mathbf{GF}(q)$ is 2, or if $q = 5$ or $q = 7$, then Γ has no ovoid by THAS [1981], respectively by O'KEEFE & THAS [1995]. All other cases with the characteristic of $\mathbf{GF}(q)$ not equal to 3 are open. If $q = 3^h$, then there are lots of classes of ovoids known; see Subsection 7.7.23 below.

So suppose now that Γ is the dual of the split Cayley hexagon over the finite field $\mathbf{GF}(q)$. Equivalently, we consider spreads in $\mathsf{H}(q)$. A geometric construction of a spread may run as follows (see VAN MALDEGHEM [19**b]). Let $\Gamma = \mathsf{H}(q)$ be embedded in $\Gamma' = \mathsf{H}(q^2)$ (in a natural way, i.e., with respect to some coordinatization of $\mathsf{H}(q^2)$, or with respect to the coordinates in $\mathbf{PG}(6, q^2)$ of the points of a standard embedding of $\mathsf{H}(q^2)$, the subhexagon $\mathsf{H}(q)$ arises from $\mathsf{H}(q^2)$ by restricting coordinates to $\mathbf{GF}(q)$). Let σ be the involution in $\mathsf{H}(q^2)$ defined by applying the map $x \mapsto x^q$ to every coordinate of any element of $\mathsf{H}(q^2)$. Clearly the substructure of fixed elements of $\mathsf{H}(q^2)$ is $\mathsf{H}(q)$. Now let L and M be two opposite lines of $\mathsf{H}(q)$, and let $p \, \mathrm{I} \, L$, with p a point of $\mathsf{H}(q^2)$ not belonging to $\mathsf{H}(q)$. Put $p' = \mathrm{proj}_M p^\sigma$. It is clear that the weak subhexagon $\Gamma'(p, p')$ is preserved by σ, and that $(\Gamma'^+(p, p'))^\sigma = \Gamma'^-(p, p')$. We know that $\Gamma'(p, p')$ is the double of a Desarguesian projective plane $\pi_{p,p'} \cong \mathbf{PG}(2, q^2)$ with point set $\Gamma'^+(p, p')$ and line set $\Gamma'^-(p, p')$ (see Subsection 5.5.10 on page 220). Consequently, σ induces a polarity in $\pi_{p,p'}$. Clearly the points p and $p'^\sigma = \mathrm{proj}_M p$ are absolute points in $\pi_{p,p'}$. By the distance-3-regularity of $\mathsf{H}(q^2)$, there are exactly $q + 1$ elements of $\Gamma'_3(p \bowtie p'^\sigma) \cap \Gamma'_3(p^\sigma \bowtie p')$ which belong to Γ (and hence are fixed by σ). This provides exactly $q + 1$ absolute points in $\pi_{p,p'}$ incident with $p \bowtie p'^\sigma$ (which is indeed a line of $\pi_{p,p'}$). So σ induces a unitary polarity in $\pi_{p,p'}$ and we have a set $\mathcal{S}$ of $q^3 + 1$ fixed flags, which can be seen as a set of $q^3 + 1$ fixed lines in $\mathsf{H}(q)$. As any two elements of $\mathcal{S}$ are easily seen to be opposite in $\Gamma'(p, p')$, and hence also in Γ, Proposition 7.2.3 implies that $\mathcal{S}$ is a spread of $\mathsf{H}(q)$. Viewed as a set of flags in $\pi_{p,p'} \cong \mathbf{PG}(2, q^2)$, it constitutes a Hermitian curve $\mathcal{H}$ union a dual one. A line of $\pi_{p,p'}$ meeting $\mathcal{H}$ in $q + 1$ points corresponds to a point of $\Gamma(p, p')$ at distance 3 from $q + 1$ elements of $\mathcal{S}$, hence these $q + 1$ elements form a regulus in $\mathsf{H}(q)$. We call $\mathcal{S}$ a **Hermitian** or **classical spread** of $\mathsf{H}(q)$. Note that all this implies that $\mathcal{S}$ is the intersection of the line set of $\mathsf{H}(q)$ with the line set of $\Gamma'(p, p')$. Remark also that, with a little adaptation of the arguments, this construction also works in the infinite case for fields admitting a quadratic extension.

THAS [1980b] constructs $\mathcal{S}$ as follows. Let $\Gamma = \mathsf{H}(q)$ be a standard embedding in the quadric $\mathbf{Q}(6, q)$ of $\mathbf{PG}(6, q)$. Let H be a hyperplane of $\mathbf{PG}(6, q)$ meeting $\mathbf{Q}(6, q)$ in an elliptic quadric $\mathbf{Q}^-(5, q)$. Note that the latter is the classical generalized quadrangle $\mathsf{Q}(5, q)$. Then the lines of H on Γ form a spread of both Γ and $\mathsf{Q}(5, q)$.

This spread can also be constructed group-theoretically by taking an orbit of length $1 + q^3$ of the set of lines of Γ under the action of a subgroup of the automorphism group of Γ isomorphic to the unitary group $\mathbf{PSU}_3(q)$; see CAMERON, THAS & PAYNE [1976].

A Hermitian spread in $\mathsf{H}(q)$ will be denoted by $U_H(q)$, where "U" stands for *unital* (see below) and "H" for *Hermitian*. If q is a power of 3, then the same notation is used for the corresponding ovoid (the image of a Hermitian spread under a correlation). If q is not a power of 3, then there is presently only one other class of spreads of $\mathsf{H}(q)$ known; it requires $q \equiv 1$ mod 3 (see Subsection 7.7.23 below).

We have seen in Corollary 7.3.5 that the hexagon $\Gamma = \mathsf{H}(q)$ is self-polar if and only if $\mathbf{GF}(q)$ has characteristic 3 and admits a Tits automorphism, hence if and only if $q = 3^{2h+1}$ for some $h \in \mathbb{N}$. The "only if" part of this equivalence also follows directly from Proposition 7.2.7. Hence if $q = 3^{2h+1}$, then Γ contains an ovoid $U_R(q)$, where "R" stands for *Ree* (see below), arising from a polarity. No ovoid $U_R(q)$ can be isomorphic in Γ with an ovoid $U_H(q)$. Indeed, let $H_R(q)^D$ denote a spread of Γ^D corresponding to the ovoid $U_R(q)$ in Γ and suppose that all lines of $U_R(q)^D$ are contained in a hyperplane H of $\mathbf{PG}(6, q)$ meeting $\mathbf{Q}(6, q)$ (in which we embed Γ^D in a natural way) in an elliptic quadric $\mathbf{Q}^-(5, q)$. Applying the polarity, we obtain an ovoid $\mathcal{O}$ of Γ^D all points of which are contained in H, hence in $\mathbf{Q}^-(5, q)$. If two points of $\mathcal{O}$ were collinear in $\mathbf{Q}^-(5, q)$, then they would lie at distance at most 4 in Γ^D, a contradiction. Hence by Proposition 7.2.3, the set $\mathcal{O}$ forms an ovoid in the generalized quadrangle $\mathbf{Q}(5, q)$ of order (q, q^2). This contradicts Corollary 7.2.4. Alternatively, and without using Thas' construction of $U_H(q)$, it follows in fact directly from Proposition 7.7.11 below that $U_R(q)$ does not contain any point-regulus. Since $U_H(q)$ does, the result follows. Ovoids and spreads arising from polarities in Moufang hexagons will be called **semi-classical**.

So $\mathsf{H}(3^{2h+1})$ has at least two non-isomorphic ovoids. All other known examples are derived from these two; see Subsection 7.7.23 below. The case $h = 0$ is rather special: the two ovoids are non-isomorphic in the hexagon (i.e., under Aut $\mathsf{H}(3)$), but they are isomorphic as sets of points in $\mathbf{Q}(6, 3)$ (with automorphism group $\mathbf{Sp}(6, 2)$).

7.3.9 Digression.   For Γ a generalized $2m$-gon, let us define a **distance-j-ovoid**, $1 \le j \le m$, as a set of points $\mathcal{O}$ in Γ such that any two points of $\mathcal{O}$ are at distance at least $2j$ and every element of Γ is at distance at most j from at least one element of $\mathcal{O}$. Then a distance-m-ovoid is simply an ovoid. For $j = 1$, we necessarily have that $\mathcal{O}$ must coincide with the point set of Γ. So we may assume that $j \ge 2$. This gives something new only if $m > 2$. A distance-2-ovoid is a set of non-collinear points such that every line is incident with exactly one point of that set. By CAMERON [1994], every infinite generalized $2m$-gon of order (s, t) with $s = t$ contains a distance-2-ovoid. In the finite case, nothing is known yet if $m > 2$ and $1 < j < m$ (apart from straightforward divisibility conditions), except if $m = 3$, $j = 2$ and Γ is isomorphic to $\mathsf{H}(2)$. Let, in this case, Γ' be a full weak subhexagon of order $(1, 2)$ of Γ. Then the set of points not belonging to Γ', but incident with a line of Γ', is easily checked to be a distance-2-ovoid.

A result similar to Proposition 7.2.3 can be proved for distance-j-ovoids, but by lack of examples other than ovoids in quadrangles and the sporadic example just mentioned, it is not very useful for the rest of this book.

7.3.10 Ovoids in generalized quadrangles

Finally, let $m = 2$. We mention some general methods to construct ovoids. The second assertion of (ii) of the next proposition can be found in KRAMER & VAN MALDEGHEM [19**].

7.3.11 Proposition.

(i) *Let Γ be an orthogonal or Hermitian Moufang quadrangle embedded naturally in some projective space $\mathbf{PG}(d, \mathbb{K})$. Every hyperplane of $\mathbf{PG}(d, \mathbb{K})$ not containing lines of Γ meets Γ in the point set of an ovoid of Γ.*

(ii) *Let Γ be a full subquadrangle of a generalized quadrangle Γ' and let p be a point of Γ' outside Γ. Then the set of points of Γ collinear with p in Γ' constitutes an ovoid $\mathcal{O}$ of Γ. If, moreover, Γ' is a Moufang quadrangle, then then the stabilizer G of $\mathcal{O}$ in the little projective group of Γ' acts doubly transitively on $\mathcal{O}$.*

Proof. The proof of (i) is straightforward: every line of Γ meets the hyperplane in exactly one point. As for (ii), we remark that each line of Γ is incident with a unique point collinear with p; hence $\mathcal{O}$ is an ovoid of Γ. Let q be any element of $\mathcal{O}$. We show that G_q acts transitively on $\mathcal{O} \setminus \{q\}$. Let q_1 and q_2 be two points of $\mathcal{O}$ distinct from q. Then both points q_1 and q_2 are opposite q in Γ'. Moreover, they have the same projection p onto the line pq. By Lemma 5.2.4(ii) (see page 175), there exists a unique collineation θ of Γ' fixing $\Gamma_1'(q) \cup \Gamma_1'(pq)$ elementwise and mapping q_1 to q_2. We now show that θ stabilizes $\mathcal{O}$. Therefore, it clearly is enough to prove that θ stabilizes Γ. Consider the subquadrangle Γ^θ. We have $\Gamma_1^\theta(q) = \Gamma_1(q)$ and $\Gamma_1^\theta(L) = \Gamma_1(L) = \Gamma_1'(L)$, for every line L of Γ' incident with q. Moreover, Γ and Γ^θ share an ordinary quadrangle through q and q_2. Hence by Corollary 1.8.5 we conclude that Γ and Γ^θ coincide. The proposition is proved. $\qquad\square$

7.3.12 Examples. In the finite case, we can apply Proposition 7.3.11(i) to $\mathsf{Q}(4, q)$ and $\mathsf{H}(3, q^2)$ to obtain ovoids of these quadrangles. Hence their duals $\mathsf{W}(q)$ and $\mathsf{Q}(5, q)$ admit spreads. It is also easily seen that by applying Proposition 7.3.11(ii) to $\mathsf{Q}(4, q) \subseteq \mathsf{Q}(5, q)$ and to $\mathsf{H}(3, q^2) \subseteq \mathsf{H}(4, q^2)$, we obtain the same ovoids (after all, the set of points on a quadric or a Hermitian variety collinear to a certain point is contained in a hyperplane). So usually, the two methods give the same ovoids. However, if the big quadrangle Γ' of case (ii) is neither an orthogonal nor a Hermitian one, then we cannot apply Proposition 7.3.11(i). This occurs in many non-classical quadrangles, in particular when Γ' is an exceptional Moufang quadrangle, a Moufang quadrangle of type $(BC)_2$. Hence the Moufang quadrangles of type (E_6), (E_7) and (E_8) induce 2-transitive spreads in the orthogonal quadrangles of type (D_5), (D_6) and (D_8), respectively (indeed, these quadrangles are ideal subquadrangles of the exceptional Moufang quadrangles; hence their duals are full subquadrangles of the dual Moufang quadrangles of type (E_i), $i = 6, 7, 8$). A similar phenomenom occurs in the Moufang quadrangles of type $(BC - CB)_2$. Here, 2-transitive spreads of orthogonal quadrangles of type $(C - CB)_2$ arise.

Ovoids and spreads obtained by one of the construction methods of Proposition 7.3.11 will be called **classical ovoids** and **spreads**. Ovoids and spreads obtained from a polarity in a Moufang quadrangle will be called **semi-classical ovoids** and **spreads**. The latter arise in $\mathsf{W}(q)$ for q even (with an odd exponent of 2), or more generally, in every Suzuki quadrangle, for example.

To conclude this brief discussion of ovoids and spreads in mainly finite classical quadrangles, we mention that $\mathsf{H}(4, q^2)$ has no ovoids and $\mathsf{H}(4, 4)$ has no spreads (the latter is an unpublished computer result by BROUWER), but the existence problem of spreads in $\mathsf{H}(4, q^2)$ has not yet been settled in general. Of course, by Corollary 7.2.4, the quadrangle $\mathsf{Q}(5, q)$ has no ovoids. Also $\mathsf{W}(q)$, for q odd, has no ovoids. These results are proved in PAYNE & THAS [1984].

In the next section we treat some aspects of duality of the Moufang quadrangles of type $(BC - CB)_2$. Then, we will consider in more detail the semi-classical ovoids.

7.4 Moufang quadrangles of type $(BC - CB)_2$

In this section, we just state some properties of the Moufang quadrangles of type $(BC - CB)_2$ and indicate how they can be proved. Nothing on this subject has been published yet. We use the notation of Subsection 5.5.5, step IV on page 218 throughout.

7.4.1 Proposition. *The Moufang quadrangle* $\mathsf{Q}(\mathbb{K}_1, \mathbb{L}_1, \mathbb{K}_1', \alpha_1, \beta_1)$ *of type* $(BC - CB)_2$ *is isomorphic to the quadrangle* $\mathsf{Q}(\mathbb{K}_2, \mathbb{L}_2, \mathbb{K}_2', \alpha_2, \beta_2)$ *of that type, if there exists an isomorphism* $\theta : \mathbb{L}_1 \to \mathbb{L}_2$ *which maps* $\mathbb{K}_1$ *and* $\mathbb{K}_1'$ *onto* $\mathbb{K}_2$ *and* $\mathbb{K}_2'$, *respectively, and which maps* α_1 *and* β_1 *into* $\alpha_2 \cdot \mathbb{K}_2'^2$ *and* $\beta_2 \cdot \mathbb{K}_2^2$, *respectively.*

Proof. Clearly, it suffices to show that, if $k \in \mathbb{K}_2$ and $k' \in \mathbb{K}_2'$, then $\Gamma = \mathsf{Q}(\mathbb{K}_2, \mathbb{L}_2, \mathbb{K}_2', \alpha_2, \beta_2)$ is isomorphic to $\Gamma' = \mathsf{Q}(\mathbb{K}_2, \mathbb{L}_2, \mathbb{K}_2', k'^2 \alpha_2, k^2 \beta_2)$. This is achieved by making the following substitution in the root groups of Γ:

$$\begin{cases} (u, v, a) & \mapsto & (ku, kk'v, a), & u, v \in \mathbb{L}_2, a \in \mathbb{K}_2', \\ (x, y, b) & \mapsto & (k'x, k'k^{-2}y, b), & x, y \in \mathbb{L}_2', b \in \mathbb{K}_2. \end{cases}$$

The result follows easily. $\qquad\square$

7.4.2 Proposition. *The dual of the quadrangle* $\Gamma = Q(\mathbb{K}, \mathbb{L}, \mathbb{K}', \alpha, \beta)$ *is isomorphic to the quadrangle* $\Gamma' = Q(\mathbb{K}', \mathbb{L}', \mathbb{K}^2, \beta^2, \alpha)$, *where* $\mathbb{L}'$ *is the field generated by* $\mathbb{L}^2$ *and* $\mathbb{K}'$.

Proof. As the previous proof, this is an exercise in computing. First, one rewrites the commutation relation for Γ with $\mathbb{L}^2$ and $\mathbb{L}'$ using the isomorphism $u \mapsto u^2$ between $\mathbb{L}$ and $\mathbb{L}^2$ and keeping the elements of $\mathbb{L}'$ fixed (so, with standard notation, (u, v, a) is replaced by (u^2, v^2, a) and (x, y, b) by (x, y, b^2)). Then, one makes the substitution

$$\begin{cases} (u, v, a) & \mapsto & (\beta^2 u, \beta^2 v, a), & u, v \in \mathbb{L}^2, a \in \mathbb{K}', \\ (x, y, b) & \mapsto & (\alpha^{-1} x, \alpha^{-1} y, b) & x, y \in \mathbb{L}', b \in \mathbb{K}^2. \end{cases}$$

One now obtains the commutation relations of Γ'. $\qquad\qquad\qquad\square$

7.4.3 Corollary. *The Moufang quadrangle* $Q(\mathbb{K}, \mathbb{L}, \mathbb{K}', \alpha, \beta)$ *is self-dual if there exists an isomorphism* $\theta : \mathbb{L} \to \mathbb{L}'$ *which maps* $\mathbb{L}'$, $\mathbb{K}$ *and* $\mathbb{K}'$ *onto* $\mathbb{L}^2$, $\mathbb{K}'$ *and* $\mathbb{K}^2$, *respectively, and which maps* α *and* β *into* $\beta^2 \cdot \mathbb{K}^4$ *and* $\alpha \cdot \mathbb{K}'^2$, *respectively.*

Proof. This follows immediately from the previous propositions. $\qquad\square$

7.4.4 Proposition. *The Moufang quadrangle* $Q(\mathbb{K}, \mathbb{L}, \mathbb{K}', \alpha, \beta)$ *is self-polar if there exists a Tits endomorphism* $\theta : \mathbb{L} \to \mathbb{L}'$ *which maps* $\mathbb{K}$ *onto* $\mathbb{K}'$ *and which maps* β *into* $\alpha \cdot \mathbb{K}'^2$.

Proof. This is again proved in exactly the same way as before. In fact, if the condition of the proposition is satisfied, then there is a completely symmetric form of the commutation relations, as for the Suzuki quadrangles follows from the symmetric coordinatization in Subsection 3.4.6 (page 103). $\qquad\qquad\square$

My conjecture is that all the conditions mentioned in the previous results are also sufficient. This should follow from investigating the automorphism groups of the Moufang quadrangles of type $(BC - CB)_2$.

7.5 Polarities, conics, hyperovals and unitals in Pappian planes

In this section, we mention some classical results on Pappian planes, mainly because the notions related to polarities in these planes are needed in various other parts of this book. For proofs, we refer to HUGHES AND PIPER [1973].

7.5.1 Classification of polarities in $\mathbf{PG}(2, \mathbb{K})$

Let $\mathbb{K}$ be a field. We have seen in Definitions 7.2.1 above that $\mathbf{PG}(2, \mathbb{K})$ is always self-polar. In fact, it is an easy exercise in elementary linear algebra and projective geometry to show that for every polarity ρ of $\mathbf{PG}(2, \mathbb{K})$ (using the first and second description of Subsection 2.2.1), the coordinates can be chosen in such a way that ρ maps the point (x, y, z) onto the line $[ax^\theta, by^\theta, cz^\theta]$, with $a, b, c \in \mathbb{K}^\times$, $a^\theta = a$, $b^\theta = b$, $c^\theta = c$ and θ an automorphism of $\mathbb{K}$ with $\theta^2 = 1$. There are three possibilities now (compare Subsection 2.4.18 on page 76).

1. **Orthogonal polarities**. These are polarities with $\theta = 1$ and with the characteristic of $\mathbb{K}$ different from 2. The set of absolute points forms an (absolutely irreducible) **conic** and can also be obtained by the 1-quadratic form $q : V \to V : (x, y, z) \mapsto (ax^2 + by^2 + cz^2)$, where V is a 3-dimensional vector space over $\mathbb{K}$ (the corresponding bilinear form g maps $((x, y, z), (x', y', z'))$ onto $axx' + byy' + czz'$), by considering $q^{-1}(0)$. Note that this conic can be empty.

2. **Pseudo-polarities**. Here, the characteristic of $\mathbb{K}$ is 2 and θ is the identity. The set of absolute points is either empty, or consists of a unique point, or of all points on a unique line.

3. **Unitary** or **Hermitian polarities**. In this case $\theta \neq 1$. The set of absolute points is a **Hermitian curve**, possibly empty. It can also be obtained by the θ-quadratic form

$$q : V \to V : (x, y, z) \mapsto r(axx^\theta + byy^\theta + czz^\theta) + \mathbb{K}_\theta,$$

where V is as above, and $r \in \mathbb{K}$ is such that $r + r^\theta \neq 0$ (the corresponding $(1, \theta)$-linear form g maps $((x, y, z), (x', y', z'))$ onto $r(axx'^\theta + byy'^\theta + czz'^\theta)$), by considering $q^{-1}(0)$.

7.5.2 Conics and hyperovals

Extending the notion of a conic as defined above, we say that a **conic** in $\mathbf{PG}(2, \mathbb{K})$ is the set of points (x, y, z) which satisfies a homogeneous quadratic equation $ax^2 + a'y^2 + a''z^2 + byz + b'xz + b''xy = 0$, with $a, a', a'', b, b', b'' \in \mathbb{K}$ and not all them 0. The conic is said to be **absolutely irreducible** if the quadratic equation is irreducible over every algebraic extension of $\mathbb{K}$. Any line L of $\mathbf{PG}(2, \mathbb{K})$ meets a conic in either 0, 1 or 2 points. A **tangent line** meets the conic in a unique point. If the characteristic of $\mathbb{K}$ is 2, then all tangent lines of an absolutely irreducible conic with quadratic equation as above contain the point (b, b', b''), which is called the **nucleus** of the conic. A **hyperoval** of $\mathbf{PG}(2, \mathbb{K})$ is a non-empty set of points such that every line meets this set in either 0 or 2 points. If $\mathbb{K}$ is quadratically closed, then a non-empty conic together with its nucleus forms a hyperoval.

In the finite case, i.e., when $\mathbb{K} = \mathbf{GF}(q)$, we show below that a conic is never empty and hence contains $q + 1$ points. Hence every absolutely irreducible conic in $\mathbf{PG}(2, q)$ with q even defines a hyperoval. An easy example of an oval (see Subsection 3.7.1 on page 120) in $\mathbf{PG}(2, q)$, q even and $q > 4$, which is not a conic is given by adding the nucleus to an absolutely irreducible conic and deleting one arbitrary point of the conic. If $q = 4$, then every oval is an absolutely irreducible conic and hence, since the stabilizer in $\mathbf{PGL}_3(q)$ of an absolutely irreducible conic is isomorphic to $\mathbf{PGL}_2(q)$, the stabilizer of any hyperoval in $\mathbf{PG}(2,4)$ is a 4-transitive group acting on six points. Consequently, this is the alternating group $\mathbf{A}_6$. In fact, inside $\operatorname{Aut} \mathbf{PG}(2,4)$, the stabilizer is the symmetric group $\mathbf{S}_6$. Since clearly $|\mathbf{S}_6| = |\mathbf{PSp}_4(2)| = |\operatorname{Aut} \mathsf{W}(2)|$ by Proposition 4.6.2 (see page 154), we deduce from the constructions of $\mathsf{W}(2)$ in Subsection 1.4.2 on page 15 (and in view of the uniqueness of the quadrangle of order $(2,2)$; see Theorem 1.7.9 on page 27), that $\mathbf{S}_6 \cong \mathbf{PSp}_4(2)$.

Recall that no hyperovals live in $\mathbf{PG}(2,q)$ for q odd, and that every oval in $\mathbf{PG}(2,q)$, q odd, is a conic (by SEGRE, see e.g. [1955]).

7.5.3 Hermitian curves and unitals

We defined a Hermitian curve in $\mathbf{PG}(2,\mathbb{K})$ above. In the finite case, a Hermitian curve is never empty. Indeed, this follows readily from the standard equation above, which for $\mathbb{K} = \mathbf{GF}(q^2)$ becomes $axx^q + byy^q + czz^q = 0$, with $a,b,c \in \mathbf{GF}(q)$. This equation always has solutions over $\mathbf{GF}(q)$, as no finite (reducible or not) conic is empty ($x^q = x$ for $x \in \mathbf{GF}(q)$). Now, a Hermitian curve in $\mathbf{PG}(2,q^2)$ has $q^3 + 1$ points, and hence, from Theorem 7.2.8, it follows that every Hermitian curve produces a unital. We call this unital the **Hermitian unital** and denote it by $U_H(q)$. This notation is justified by the fact that the construction of the Hermitian spread $U_H(q)$ in $\mathsf{H}(q)$ (see Section 7.3.8) makes it apparent that this Hermitian spread can be viewed as a Hermitian unital (in the plane $\pi_{p,p'}$, with the notation of Section 7.3.8).

7.5.4 Finite conics

As noted above, it is easily seen that a non-empty absolutely irreducible conic in $\mathbf{PG}(2,q)$ contains exactly $q + 1$ points (look at a pencil of lines through a point of the conic). Clearly, an absolutely reducible conic is non-empty (it contains the intersection of its components). We present two different but rather beautiful geometric arguments of the fact that every absolutely irreducible conic in $\mathbf{PG}(2,q)$ is non-empty (see DICKSON [1901]).

The first one is due to TITS (unpublished). Suppose $\mathcal{C}$ is an empty conic in $\mathbf{PG}(2,q)$. Let p_1, p_2 be two conjugate points in the projective plane over a quadratic extension of $\mathbf{GF}(q)$. Consider the bundle $\mathcal{B}$ of conics defined by $\mathcal{C}$ and the line p_1p_2 (counted twice); denote the latter by $\mathcal{C}_0$. The bundle $\mathcal{B}$ contains $q + 1$ elements, and every point of the plane is contained in a unique element of the bundle. The bundle $\mathcal{B}$ contains exactly one other conic which consists of two lines: the conic $\mathcal{C}_1$ which is the union of the tangent lines to $\mathcal{C}$ in the respective points p_1 and p_2. These lines are conjugate and hence $\mathcal{C}_1$ contains one point of $\mathbf{PG}(2,q)$. Clearly, every other element of $\mathcal{B}$ must contain either $q + 1$ points of $\mathbf{PG}(2,q)$ or none (for example $\mathcal{C}$). But $\mathcal{C}_0$ contains $q + 1$ points and $\mathcal{C}_1$ exactly one point. Since in total we must have $q^2 + q + 1$ points, there are $q - 1$ conics in the bundle other than $\mathcal{C}_0$ and $\mathcal{C}_1$ containing $q+1$ points. Since $(q-1)+2 = q+1$, the conic $\mathcal{C}$ must be among them and so $\mathcal{C}$ is non-empty. That concludes our first proof.

For our second proof, we consider the conic $\mathcal{C}$ over the quadratic extension $\mathbf{GF}(q^2)$ of $\mathbf{GF}(q)$. Let $\mathcal{C}'$ denote that extended conic in $\mathbf{PG}(2, q^2)$. Clearly, $\mathcal{C}'$ is non-empty and hence contains $q^2 + 1$ points in $\mathbf{PG}(2, q^2)$ (it does not contain a line because every line in $\mathbf{PG}(2, q^2)$ contains a point of $\mathbf{PG}(2, q)$). Every line L of $\mathbf{PG}(2, q)$ meets $\mathcal{C}'$ in two conjugate points x and x' of $\mathcal{C}'$ (conjugate with respect to the involution $a \mapsto a^q$ in $\mathbf{GF}(q^2)$), and different lines give rise to different points. But since we have $q^2 + q + 1$ lines in $\mathbf{PG}(2, q)$, the conic $\mathcal{C}'$ should contain at least $2(q^2 + q + 1)$ points, a contradiction. So we have shown that *every conic in a finite classical projective plane is non-empty.*

7.5.5 Remark. The above mentioned polarities can be easily generalized to $\mathbf{PG}(d, \mathbb{K})$. If d is odd, we have in addition a **symplectic polarity**, similar to one in $\mathbf{PG}(3, \mathbb{K})$; see Subsection 2.3.17 on page 63. Let us just mention that in the finite case, the set of absolute points of a unitary polarity is a non-empty **Hermitian variety** with standard equation $\sum_{i=0}^{d} x_i^{1+q} = 0$.

7.6   Suzuki quadrangles and Suzuki–Tits ovoids

Throughout this section, we fix a field $\mathbb{K}$ of characteristic 2 admitting a Tits endomorphism θ (see Definition 2.5.1, page 79). We denote the image of $\mathbb{K}$ under θ by $\mathbb{K}'$ and we fix the isomorphism class Λ of $\mathbb{K}'$ within $\mathbb{K}$ as well. If $\mathbb{K}''$ is another subfield of $\mathbb{K}$, then according to Proposition 3.4.3 (page 101), the mixed quadrangles $Q(\mathbb{K}, \mathbb{K}'; \mathbb{K}, \mathbb{K}')$ and $Q(\mathbb{K}, \mathbb{K}''; \mathbb{K}, \mathbb{K}'')$ are isomorphic **if and only if** $\mathbb{K}''$ belongs to Λ. As discussed earlier, not every field of characteristic 2 qualifies; for instance, if $\mathbb{K}$ is finite, then $|\mathbb{K}|$ must be a power of 2 with an odd exponent. On the other hand, θ may not be unique with respect to the property $(x^\theta)^\theta = x^2$ for all $x \in \mathbb{K}$ and not even for $\mathbb{K}^\theta = \mathbb{K}'$. Indeed, if we consider the field $\mathbb{K}$ of Laurent series over $\mathbf{GF}(2)$ with variable t where the exponents of t are elements of $\mathbb{Z} + \sqrt{2}\mathbb{Z}$, then the endomorphism

$$\theta_\epsilon : \mathbb{K} \to \mathbb{K} : f(x) \mapsto f(x^{\epsilon\sqrt{2}}),$$

for $\epsilon \in \{+, -\}$, is a Tits endomorphism and clearly $\mathbb{K}^{\theta_+} = \mathbb{K}^{\theta_-}$. We will frequently use the following easy result.

7.6.1 Lemma. *Every finite field of order 2^{2e+1}, for some non-negative integer e, admits a unique Tits endomorphism, which is always an automorphism.* $\square$

We will restrict ourselves to the mixed Suzuki quadrangles $Q(\mathbb{K}, \mathbb{K}'; \mathbb{K}, \mathbb{K}')$, $\mathbb{K}' \in \Lambda$. This is not too narrow a restriction since every other self-polar mixed quadrangle is a subquadrangle of a Suzuki quadrangle by Theorem 7.3.2. Note also that the Suzuki quadrangle $Q(\mathbb{K}, \mathbb{K}'; \mathbb{K}, \mathbb{K}')$ is a full subquadrangle of the symplectic quadrangle $W(\mathbb{K}) = Q(\mathbb{K}, \mathbb{K}; \mathbb{K}, \mathbb{K})$. We will frequently refer to that symplectic quadrangle. We denote the corresponding symplectic polarity in $\mathbf{PG}(3, \mathbb{K})$ defining $W(\mathbb{K})$ by τ. Throughout this section, we denote $\Gamma = Q(\mathbb{K}, \mathbb{K}'; \mathbb{K}, \mathbb{K}')$.

Our first result is due to TITS [1962a]. Recall that the *hat-rack* of a coordinatization of a generalized n-gon is the ordered ordinary n-gon $((\infty), [\infty], (0), [0, 0], \ldots, (0, 0)$, $[0], (\infty))$. The **dual** of this hat-rack is by definition the hat-rack in the dual generalized n-gon obtained from the original hat-rack by swapping parentheses and brackets.

Two ovoids of a polygon Γ' are called **isomorphic** if there exists a collineation of Γ' mapping the one to the other. Similarly for ovoid–spread pairings.

7.6.2 Theorem (Tits [1962a]). *The set of conjugacy classes of polarities (conjugacy with respect to* Aut Γ*) of the Suzuki quadrangle* $\Gamma := Q(\mathbb{K}, \mathbb{K}'; \mathbb{K}, \mathbb{K}')$ *is in a natural bijective correspondence with the set of conjugacy classes (with respect to the automorphism group of* $\mathbb{K}$*) of Tits endomorphisms* θ *of* $\mathbb{K}$ *such that* $\mathbb{K}' = \mathbb{K}^\theta$*. Also, this set is in a natural bijective correspondence with the set of isomorphism classes (with respect to* Aut Γ *again) of ovoid–spread pairings of* Γ *arising from polarities.*

Proof. By the last part of Theorem 7.3.2, every polarity ρ of Γ defines, for a given coordinatization, a Tits endomorphism $\theta(\rho)$ of $\mathbb{K}$ such that Γ is isomorphic to $Q(\mathbb{K}, \mathbb{K}^{\theta(\rho)}; \mathbb{K}, \mathbb{K}^{\theta(\rho)})$ and the point (1) is mapped onto the line $[1]$ (and, of course, the corresponding hat-rack is mapped to its dual). Let $\Lambda(\rho)$ be the conjugacy class of $\theta(\rho)$ with respect to Aut$(\mathbb{K})$. Then $\Lambda(\rho)$ is well defined since every recoordinatization within the restrictions of the above mentioned proof (i.e., such that the points (∞), (0), (1), $(0, 0)$ and $(0, 0, 0)$ are mapped onto the lines $[\infty]$, $[0]$, $[1]$, $[0, 0]$ and $[0, 0, 0]$, respectively) amounts to an automorphism of Γ, and hence gives rise to an automorphism of $\mathbb{K}$ conjugating the Tits endomorphism. Now suppose that $\sigma \in $ Aut Γ and consider the polarity $\sigma^{-1}\rho\sigma$. Considering new coordinates $(\ldots)'$ for points and $[\ldots]'$ for lines defined by $(\ldots)' = (\ldots)^\sigma$ and $[\ldots]' = [\ldots]^\sigma$ (where $(\ldots)$ and $[\ldots]$ represent the old coordinates), we see that $\theta(\rho) = \theta(\sigma^{-1}\rho\sigma)$ for $\sigma^{-1}\rho\sigma$ considered in the new coordinate system. Hence every polarity uniquely defines a conjugacy class of Tits endomorphisms of $\mathbb{K}$. Since a Tits endomorphism θ completely determines the polarity, once a coordinate system for $\Gamma = Q(\mathbb{K}, \mathbb{K}^\theta; \mathbb{K}, \mathbb{K}^\theta)$ is given, the first part of the theorem follows. To prove the second assertion, we remark that every polarity defines a unique ovoid–spread pairing by Proposition 7.2.5 and Definition 7.2.6. Clearly, conjugate polarities define isomorphic such pairings. We now show that isomorphic ovoid–spread pairings $(\mathcal{O}_1, \mathcal{S}_1)$ and $(\mathcal{O}_2, \mathcal{S}_2)$ are defined by conjugate polarities ρ_1 and ρ_2, respectively. By conjugating we may assume that $\mathcal{O}_1 = \mathcal{O}_2$ and $\mathcal{S}_1 = \mathcal{S}_2$. We fix a coordinatization and a hat-rack which is mapped by ρ_i, $i = 1, 2$, onto its dual and such that $(a)^{\rho_1} = [a^{\theta_1}]$, for some Tits endomorphism θ_1. It follows from the proof of Theorem 7.3.2 that there exist $k' \in \mathbb{K}$ and a Tits endomorphism θ_2 such that ρ_2

maps (a) onto $(k'a^{\theta_2})$. From Remark 7.3.3 it follows that, if $k \in \mathbb{K}^2$ is such that $k'^{\theta_2}k = 1$, then a point (a, l, a') is in $\mathcal{O}_2$ **if and only if** (a, l, a') is incident with its image $[k'a^{\theta_1}, k'k^{1/2}l^{\theta_1^{-1}}, k'ka'^{\theta_1}]$. Using the quadratic quaternary ring, one easily computes that the set of absolute points of ρ_2 is given by

$$\{(\infty)\} \cup \{(a, l, k'a^{\theta_2+1} + k'k^{1/2}l^{\theta_2^{-1}}) : a \in \mathbb{K}, l \in \mathbb{K}'\}.$$

The set of absolute points for ρ_1 is obtained by setting $k = k' = 1$ and replacing θ_2 by θ_1. Hence $\mathcal{O}_1$ is the set

$$\{(\infty)\} \cup \{(a, l, a^{\theta_1+1} + l^{\theta_1^{-1}}) : a \in \mathbb{K}, l \in \mathbb{K}'\}.$$

So it follows that, for all $a \in \mathbb{K}$ and all $l \in \mathbb{K}'$, the following identity holds:

$$a^{\theta_1+1} + l^{\theta_1^{-1}} = k'a^{\theta_2+1} + k'k^{1/2}l^{\theta_2^{-1}}.$$

Putting $l = 0$ and $a = 1$, or $a = 0$ and $l = 1$, we see that $k = k' = 1$. Putting $l = 0$ and varying a over $\mathbb{K}$, we see that θ_1 and θ_2 must coincide.

The theorem is proved. □

7.6.3 Corollary. *Two polarities of a Suzuki quadrangle having the same absolute elements coincide.*

Proof. This follows from the last part of the proof of Theorem 7.6.2 □

In fact, Tits [1962a] shows that already the ovoid determines the polarity when $\mathbb{K}$ is perfect. In order to prove that isomorphic ovoid–spread pairings define conjugate polarities in the above proof, we only used the fact that the two corresponding ovoids are isomorphic and that an isomorphism maps two lines of one spread onto the two corresponding lines of the other spread (in our proof $[\infty]$ and $[0, 0, 0]$). Later on we will generalize Tits' result to the non-perfect case and show that two polarities of a Suzuki quadrangle with the same set of absolute points must necessarily be the same; see Corollary 7.6.11 below.

7.6.4 Definitions. Now we fix an arbitrary polarity ρ and a corresponding Tits endomorphism θ. We denote, as usual, $\mathbb{K}^\theta = \mathbb{K}'$. Denote by $\mathbf{Sp}_4(\mathbb{K}, \mathbb{K}')$ the little projective group of the quadrangle $Q(\mathbb{K}, \mathbb{K}'; \mathbb{K}, \mathbb{K}')$. By Corollary 5.8.3 (see page 233) this group is simple, provided $\mathbb{K} \not\cong \mathbf{GF}(2)$. Let $\mathcal{O}$ be the set of absolute points for ρ. We call $\mathcal{O}$ a **Suzuki–Tits ovoid**. We denote by $\mathbf{Sz}(\mathbb{K}, \theta)$ the subgroup of $\mathbf{Sp}_4(\mathbb{K}, \mathbb{K}')$ stabilizing $\mathcal{O} \cup \mathcal{O}^\rho$ and call it a **Suzuki group (over $\mathbb{K}$, with respect to θ)**. If $\mathbb{K} = \mathbf{GF}(q)$ is finite, then $q = 2^{2e+1}$, the Tits endomorphism is unique (by Lemma 7.6.1) and we adopt the notation $\mathbf{Sz}(q)$. Note that, since $\mathbf{Sp}_4(\mathbb{K}, \mathbb{K}')$ can be identified with a unique subgroup of $\mathbf{PSL}_4(\mathbb{K})$ (indeed, first embed $\mathbf{Sp}_4(\mathbb{K}, \mathbb{K}')$ in $\mathbf{Sp}_4(\mathbb{K}, \mathbb{K})$, then apply Proposition 4.6.1 on page 152), we can view $\mathbf{Sz}(\mathbb{K}, \theta)$ as a subgroup of $\mathbf{PSL}_4(\mathbb{K})$.

For $\sigma \in \mathbf{Sz}(\mathbb{K}, \theta)$, the polarity $\sigma^{-1}\rho\sigma$ has the same set of absolute elements as ρ, hence by Corollary 7.6.3, σ centralizes ρ. If conversely some element $\sigma \in \mathbf{Sp}_4(\mathbb{K}, \mathbb{K}')$ centralizes ρ, then it permutes the flags fixed by ρ, hence it belongs to $\mathbf{Sz}(\mathbb{K}, \theta)$. So we have shown that $\mathbf{Sz}(\mathbb{K}, \theta)$ is the intersection with $\mathbf{Sp}_4(\mathbb{K}, \mathbb{K}')$ of the centralizer of ρ inside the full correlation group of $\mathsf{W}(\mathbb{K}, \mathbb{K}'; \mathbb{K}, \mathbb{K}')$. Our aim is to indicate how one can show that $\mathbf{Sz}(\mathbb{K}, \theta)$ is a simple group (if $\mathbb{K} \ncong \mathbf{GF}(2)$) acting 2-transitively on $\mathcal{O}$. We also want to describe the structure of the stabilizer in $\mathbf{Sz}(\mathbb{K}, \theta)$ of a point of $\mathcal{O}$.

7.6.5 Unipotent elements of $\mathbf{Sz}(\mathbb{K}, \theta)$

Again we consider a fixed coordinatization for $\Gamma = \mathsf{Q}(\mathbb{K}, \mathbb{K}'; \mathbb{K}, \mathbb{K}')$ such that the flags $\{(\infty), [\infty]\}$ and $\{(0,0,0), [0,0,0]\}$ are absolute flags for ρ and $(1)^\rho = [1]$. Let U_+ be the group generated by all root elations which fix (∞) and $[\infty]$. By Lemma 5.2.4 on page 175, U_+ acts regularly on the flags of Γ opposite $\{(\infty), [\infty]\}$. Also, U_+ is a normal subgroup of the stabilizer in $\operatorname{Aut}\Gamma$ of the flag $\{(\infty), [\infty]\}$. Put $U := U_+ \cap \mathbf{Sz}(\mathbb{K}, \theta)$. We call the elements of U, and of every conjugate of U in $\mathbf{Sz}(\mathbb{K}, \theta)$, the **unipotent elements** of $\mathbf{Sz}(\mathbb{K}, \theta)$. Also, U acts semi-regularly on $\mathcal{O} \setminus \{(\infty)\}$ and U is a normal subgroup of $\mathbf{Sz}(\mathbb{K}, \theta)_{(\infty)}$. To show the claim that U acts transitively on $\mathcal{O} \setminus \{(\infty)\}$, we exhibit a unipotent element of $\mathbf{Sz}(\mathbb{K}, \theta)$ fixing (∞) and mapping $(0,0,0)$ to $(A, A^{\theta+2} + A'^\theta, A')$, for arbitrary $A, A' \in \mathbb{K}$. Therefore, consider the following mapping $\sigma = \sigma(A, A')$ (which is the product of two root elations obtained by putting $A = 0$ or $A' = 0$):

$$\sigma : \begin{cases} (a, l, a') & \mapsto & (a + A, l + A^{\theta+2} + A'^\theta + a^2 A^\theta, a' + A' + aA^\theta), \\ [k, b, k'] & \mapsto & [k + A^\theta, b + A^{\theta+1} + A' + kA, k' + A'^\theta + kA^2]. \end{cases}$$

Clearly, $(0,0,0)$ is mapped to $(A, A^{\theta+2} + A'^\theta, A')$ and if (a, l, a') belongs to $\mathcal{O}$ (which means $l = a^{\theta+2} + a'^\theta$) then one easily verifies that

$$l + A^{\theta+2} + A'^\theta + a^2 A^\theta = (a + A)^{\theta+2} + (a' + A' + aA^\theta)^\theta,$$

hence $(a, l, a')^\sigma \in \mathcal{O}$. Hence σ preserves $\mathcal{O}$. Similarly, σ preserves $\mathcal{O}^\rho$. Hence our claim. Note that $\sigma(A, A')$ is also well defined in $\mathsf{W}(\mathbb{K})$.

Since U acts transitively on $\mathcal{O} \setminus \{(\infty)\}$, we have:

7.6.6 Theorem (Tits [1962a]). *The group $\mathbf{Sz}(\mathbb{K}, \theta)$ acts doubly transitively on the set of absolute points of ρ.*

Proof. In the coordinatization of $\mathsf{Q}(\mathbb{K}, \mathbb{K}'; \mathbb{K}, \mathbb{K}')$, we may have chosen (∞) arbitrarily in the set of absolute points for ρ. The result now follows from the discussion above. $\qquad\square$

In fact, the original construction of the Suzuki groups by SUZUKI [1960] uses a doubly transitive action on a (finite) set, but there this set is not recognized as an ovoid.

7.6.7 Structure of a point stabilizer in $\mathbf{Sz}(\mathbb{K}, \theta)$

The previous arguments imply that every unipotent element of $\mathbf{Sz}(\mathbb{K}, \theta)$ fixing (∞) is determined by the image of $(0, 0, 0)$. So we may denote such an element by $((a, a'))$ if $(a, a^{\theta+2} + a'^{\theta}, a')$ is that image (double parentheses to distinguish from points with two coordinates in the Suzuki quadrangle). Defining the multiplication $((a, a')) \cdot ((b, b'))$ as the composition of $((a, a'))$ and $((b, b'))$, one verifies with a straightforward calculation (in fact only putting $A = b$ and $A' = b'$ in the above fomulae) that

$$((a, a')) \cdot ((b, b')) = ((a + b, a' + b' + ab^{\theta})),$$

which is exactly the product used to define the group $\mathbb{K}_2^{(\theta)}$ (and this is a root group in the Ree–Tits octagons; see Subsection 3.2 on page 119).

Now we consider the stabilizer of two elements of $\mathcal{O}$, namely, of (∞) and $(0, 0, 0)$. Since we are dealing with $\mathbf{Sp}_4(\mathbb{K}, \mathbb{K}')$, there are only products of generalized homologies to consider. It is easily seen that the following mapping h_X, which is the product of two generalized homologies, preserves $\mathcal{O}$ and $\mathcal{O}^\rho$ and fixes (∞) and $(0, 0, 0)$, for all $X \in \mathbb{K}$:

$$h_X : \begin{cases} (a, l, a') & \mapsto & (Xa, X^{\theta+2}l, X^{\theta+1}a'), \\ [k, b, k'] & \mapsto & [X^\theta k, X^{\theta+1}b, X^{\theta+2}k']. \end{cases}$$

Using the correspondence between the coordinates of the generalized quadrangle and those of $\mathbf{PG}(3, \mathbb{K})$ (see Table 3.1 on page 101), it is an elementary exercise to write down the matrix of this map in $\mathbf{PG}(3, \mathbb{K})$ and to calculate its determinant. One obtains $X^{2\theta+4}$, which must be a fourth power, hence $X \in \mathbb{K}^\theta \setminus \{0\}$.

Note that h_X for $X \in \mathbb{K} \setminus \mathbb{K}^\theta$ comes from a unique element of $\mathbf{PGL}_4(\mathbb{K})$ and also belongs to the stabilizer of $\mathcal{O}$ inside $\mathbf{PGL}_4(\mathbb{K})$. We will denote that stabilizer by $\mathbf{GSz}(\mathbb{K}, \theta)$ and call it the **general Suzuki group (over $\mathbb{K}$, with respect to θ)**. In the perfect case we have $\mathbf{Sz}(\mathbb{K}, \theta) = \mathbf{GSz}(\mathbb{K}, \theta)$, so there is no need for special notation for finite $\mathbb{K}$.

The map h_X defines a multiplication $X \times ((a, a'))$ by conjugation; we obtain $X \times ((a, a')) = ((Xa, X^{\theta+1}a'))$, which is the multiplication $\otimes$ introduced in Subsection 3.2 to describe the Ree–Tits octagons. Hence we can write every element of $P := \mathbf{Sz}(\mathbb{K}, \theta)_{(\infty)}$ as a triple $((a, a'; X)) := h_X \cdot ((a, a'))$ and we have the operation

$$((a, a'; X)) \cdot ((b, b'; Y)) = ((Ya + b, Y^{\theta+1}a' + b' + Yab^\theta; XY)).$$

One now computes for all $X, x \in \mathbb{K} \setminus \{0, 1\}$ the commutation relations

$$\left[((0, 0; X)), \left(\left(0, \frac{x}{X^{\theta+1} + 1}; 1\right)\right)\right] = ((0, x; 1));$$

and

$$[((0, 0; x + 1)), ((1, 0; 1))] = ((x, 0; 1)),$$

hence, since clearly $[P, P] \subseteq U$ (by noting that $[((a, a'; X)), ((b, b'; Y))]$ is equal to some element $((\ldots, \ldots; X^{-1}Y^{-1}XY))$), we have $[P, P] = U$, given $\mathbb{K} \ncong \mathbf{GF}(2)$.

Going over to matrices in $\mathbf{PG}(3, \mathbb{K})$, a tedious computation shows that the unipotent elements of $\mathbf{Sz}(\mathbb{K}, \theta)$ generate $\mathbf{Sz}(\mathbb{K}, \theta)$. We will not perform this calculation here, but the reader can reconstruct it using the connection of the coordinates of Γ with those in $\mathbf{PG}(3, \mathbb{K})$; see Table 3.1 of Subsection 3.4.1 on page 101.

7.6.8 Remark. It is also clear that the set of elements $\{((a, a'; X)) : X \neq 1\}$ generates the subgroup of $\mathbf{Sz}(\mathbb{K}, \theta)$ fixing (∞). All these elements are the restriction to a Suzuki–Tits ovoid of products of generalized homologies in the corresponding Suzuki quadrangle.

It is now easy to show (rewriting the proof of Theorem 5.8.2 on page 233 suitably, and, in doing so, in fact using a lemma of IWASAWA [1951]):

7.6.9 Theorem (Susuki [1960], Tits [1960], [1962a]). *The Suzuki groups* $\mathbf{Sz}(\mathbb{K}, \theta)$ *are simple, provided* $|\mathbb{K}| > 2$. *If* $\mathbb{K} \cong \mathbf{GF}(2)$, *then* $\mathbf{Sz}(2)$ *is a (Frobenius) group of order 20.* $\qquad\square$

We do not need to define Frobenius groups in general to understand the action of $\mathbf{Sz}(2)$ on the corresponding ovoid (which contains five points). It is equivalent to the action of the maps $x \mapsto ax + b$, $a \in \mathbf{GF}(5)^{\times}$, $b \in \mathbf{GF}(5)$, on $\mathbf{GF}(5)$. In other words, $\mathbf{Sz}(2)$ can be seen as the stabilizer in $\mathbf{PGL}_2(5)$ of a point of $\mathbf{PG}(1, 5)$; the action on the corresponding ovoid is equivalent to the action of that stabilizer on the remaining points of $\mathbf{PG}(1, 5)$. As we will see below, $\mathbf{Sz}(2)$ is not the full automorphism group of the ovoid inside $\mathsf{W}(2)$ (i.e., in $\mathrm{Aut}\,\mathsf{W}(2)$).

7.6.10 Theorem. **If** $|\mathbb{K}| > 2$, **then** *every automorphism* σ *of the little projective group of* Γ *stabilizing the Suzuki–Tits ovoid* $\mathcal{O}$ *also stabilizes* $\mathcal{O}^{\rho}$.

Proof. By Theorem 7.6.6, we may assume that σ fixes, after suitable coordinatization, the points (∞) and $(0, 0, 0)$. Hence σ fixes the hyperbolic line containing (0) and $(0, 0)$. In $\mathbf{PG}(3, \mathbb{K})$, this means that σ has matrix

$$
\begin{pmatrix}
e & 0 & 0 & 0 \\
0 & 1 & 0 & 0 \\
0 & 0 & a & b \\
0 & 0 & c & d
\end{pmatrix},
$$

for some $a, b, c, d, e \in \mathbb{K}$. An arbitrary point of $\mathcal{O}$ can be written as $p_{(x,y)} = (x^{\theta+2} + xy + y^{\theta}, 1, y, x)$, see above. Clearly $p^{\theta}_{(x,y)} = p_{(cy+dx, ay+bx)}$. Comparing first coordinates, we obtain the identity

$$
\begin{aligned}
c^{\theta+2}y^{\theta+2} &+ c^{\theta}d^2 y^{\theta}x^2 + c^2 d^{\theta}y^2 x^{\theta} + d^{\theta+2}x^{\theta+2} + \\
acy^2 &+ (ad + bc)yx + bdx^2 + a^{\theta}y^{\theta} + b^{\theta}x^{\theta} \\
&= e(x^{\theta+2} + xy + y^{\theta}), \tag{7.1}
\end{aligned}
$$

for all $x, y \in \mathbb{K}$. Putting $x = 0$, we obtain

$$c^{\theta+2}y^{\theta+2} + acy^2 + a^\theta y^\theta \;\; = \;\; ey^\theta, \tag{7.2}$$

for all $y \in \mathbb{K}$. Putting $y = 1$, we see that $a^\theta = e + c^{\theta+2} + ac$. Substituting this back in Equation (7.2), we obtain, assuming $c \neq 0$,

$$c^{\theta+1}(y^{\theta+2} + y^\theta) + a(y^2 + y^\theta) \;\; = \;\; 0, \tag{7.3}$$

for all $y \in \mathbb{K}$. Substituting $y + 1$ for y in Equation (7.3) and adding the result to Equation (7.3), we obtain $c^{\theta+1}(y^\theta + y^2) = 0$. It is easily seen that $y^\theta + y^2 \neq 0$ whenever $y \notin \{0, 1\}$. Hence, since $|\mathbb{K}| > 2$, $c = 0$, a contradiction. Hence $c = 0$, and substituting this in Equation (7.2), we see that $a^\theta = e$. Now we put $y = 0$ in Equation (7.1) (taking into account that $c = 0$ and $a^\theta = e$) and substitute the expression thus obtained back in Equation (7.1). This gives us $adxy = a^\theta xy$, for all $x, y \in \mathbb{K}$. Hence $ad = a^\theta$. Consequently $a^\theta d^\theta = a^2$, from which we deduce that $ad^{\theta+1} = a^2$. We have shown that $a = d^{\theta+1}$ and hence $e = d^{\theta+2}$. Now again put $y = 0$ in Equation (7.1) to obtain

$$d^{\theta+2}x^{\theta+2} + bdx^2 + b^\theta x^\theta \;\; = \;\; d^{\theta+2}x^{\theta+2}, \tag{7.4}$$

for all $x \in \mathbb{K}$. Putting $x = 1$, we see that $bd = b^\theta$, hence, if $b \neq 0$, $x^2 = x^\theta$, for all $x \in \mathbb{K}$, a contradiction. We have shown that $b = 0$. But we now see that $\sigma = h_d = ((0, 0; d))$, with the above notation. The theorem is proved. $\qquad\square$

7.6.11 Corollary. *Every polarity in Γ is determined by the set of its absolute points.*

Proof. Suppose two polarities ρ and ρ' have the same set $\mathcal{O}$ of absolute points. If they also have the same set of absolute lines, then they coincide by Corollary 7.6.3. By the previous theorem, the Suzuki groups corresponding to ρ_1 and ρ_2 coincide. Since such a group fixes exactly one line through every absolute point (as can easily be derived from e.g. the explicit description of $\sigma(A, A')$ above), the result follows. $\qquad\square$

7.6.12 The finite case

In the finite case, it follows from the discussion above that the order of $\mathbf{Sz}(q)$ is equal to

$$(q^2 + 1) \cdot q^2 \cdot (q - 1),$$

and $\mathbf{Sz}(q)$ acts doubly transitive on the set of $q^2 + 1$ points of a Suzuki–Tits ovoid. If $q > 2$, then by Theorem 7.6.10, the automorphism groups (inside $\mathbf{PGL}(4, \mathbb{K})$) of $\mathcal{O}$ and $\mathcal{O} \cup \mathcal{O}^\rho$ coincide. Now, since the Tits endomorphism is unique in the finite case, we deduce from Theorem 7.6.2 that all Suzuki–Tits ovoids are isomorphic. Hence, using Proposition 4.6.2 (see page 154), we can calculate the number of Suzuki–Tits ovoids in $\mathsf{W}(q)$, $q > 2$, and we obtain

$$\frac{q^4 \cdot (q^4 - 1) \cdot (q^2 - 1)}{(q^2 + 1) \cdot q^2 \cdot (q - 1)} = q^2 \cdot (q^2 - 1) \cdot (q + 1).$$

If $q = 2$, then a direct counting shows that there are six ovoids. In this case, every Suzuki–Tits ovoid is an elliptic quadric in $\mathbf{PG}(3,2)$; this is an easy exercise. The number $q^2(q^2 - 1)(q + 1) = 36$ is now equal to the ovoid–spread pairings arising from polarities. Since there are only six ovoids and six spreads in $\mathsf{W}(2)$, we obtain the curious property that every ovoid–spread pairing $(\mathcal{O}, \mathcal{S})$ in $\mathsf{W}(2)$ arises from a polarity. Also, $|\operatorname{Aut}\mathcal{O}| = 6 \cdot 20 = 120$, hence $\operatorname{Aut}\mathcal{O} \cong \mathbf{S}_5 \cong \mathbf{P\Gamma L}_2(4)$.

We now turn to a geometric investigation of the Suzuki–Tits ovoids.

7.6.13 Suzuki–Tits ovoids in $\mathbf{PG}(3, \mathbb{K})$

We use the notation of the previous paragraphs. Using Table 3.1 of Subsection 3.4.1 on page 101, we can calculate the coordinates of the points of the Suzuki–Tits ovoid $\mathcal{O}$ in $\mathbf{PG}(3, \mathbb{K})$. We obtain

$$\mathcal{O} = \{(1,0,0,0)\} \cup \{(a^{\theta+2} + aa' + a'^{\theta}, 1, a', a) : a, a' \in \mathbb{K}\}.$$

This set of points of $\mathbf{PG}(3, \mathbb{K})$, which we also denote by $\mathcal{O}$, has the following properties.

7.6.14 Proposition (Tits [1962a]). *Any line L of $\mathbf{PG}(3, \mathbb{K})$ meets $\mathcal{O}$ in 0, 1 or 2 points. The set of lines through some point p of $\mathcal{O}$ meeting $\mathcal{O}$ in exactly one point (which must be p) is a pencil of lines in some plane through p. Hence $\mathcal{O}$ is an ovoid of $\mathbf{PG}(3, \mathbb{K})$.*

Proof. We offer an entirely geometric proof.
Let L be a line of $\mathbf{PG}(3, \mathbb{K})$ meeting $\mathcal{O}$ in at least two points x and y. Then any further point of $\mathcal{O}$ on L must lie in $\{x, y\}^{\perp\perp}$. So let $z \in \{x, y\}^{\perp\perp} \cap \mathcal{O}$, $x \neq z \neq y$. Let u be the point of x^ρ (where ρ is the polarity defining $\mathcal{O}$) collinear with y. Then u is also collinear with z. Let v be the point on y^ρ collinear with x; then similarly v is collinear with z. It is easily seen that $u^\rho = vx$, hence z^ρ meets vx. But as z is collinear with v and z^ρ is incident with z, this implies that $z^\rho = vz$. Similarly $z^\rho = zu$, a contradiction. Hence the first statement.

For the second statement, we assume that a line L meets the ovoid $\mathcal{O}$ in the point $x \in \mathcal{O}$. If L lies in the plane conjugate to x with respect to τ (recall that τ is the symplectic polarity defining $\mathsf{W}(\mathbb{K}) \supseteq \mathsf{Q}(\mathbb{K}, \mathbb{K}^\theta; \mathbb{K}, \mathbb{K}^\theta) = \Gamma$), then L either is a line of Γ or meets Γ in x. Hence L meets $\mathcal{O}$ exactly in x by the fact that $\mathcal{O}$ is an ovoid in Γ. If L does not lie in that plane, it represents some trace T through x in the symplectic quadrangle $\mathsf{W}(\mathbb{K})$. Let M_1 and M_2 be two lines of Γ through x. Let y_i be the projection onto M_i, $i = 1, 2$, of any point of T distinct from x. Since Γ is a full subquadrangle of $\mathsf{W}(\mathbb{K})$, the points y_1 and y_2 belong to Γ and so $y_1^{y_2}$ (considered as a trace in Γ) is a subset of the line L. Without loss of generality, we may assume that $M_1 = x^\rho$ and $M_2 = y_1^\rho$. Now y_2^ρ is a line through y_1 and meets the trace $y_1^{y_2}$ in some unique point y collinear with y_2. So y^ρ contains y_2 and meets y_2^ρ. Since the only line with these properties is $y_2 y$, the point y is absolute and hence belongs to $\mathcal{O}$. The proposition is completely proved. $\square$

A plane of $\mathbf{PG}(3, \mathbb{K})$ meeting $\mathcal{O}$ in exactly one point p will be called a **tangent plane (at** p**)**. A line meeting $\mathcal{O}$ in exactly one point p will be similarly called a **tangent line (at** p**)**. The previous proposition says that all tangent lines of $\mathcal{O}$ lie in one plane, the tangent plane.

There is an immediate algebraic corollary.

7.6.15 Corollary. *For any Tits endomorphism* θ *in a field* $\mathbb{K}$ *of characteristic* 2, *the equation*

$$x^{\theta+2} + xy + y^{\theta} = 0$$

in the unknowns x *and* y *has the unique solution* $x = y = 0$.

Proof. In $\mathbf{PG}(3, \mathbb{K})$, the plane conjugate to $(x_0, x_1, x_2, x_3) = (0, 1, 0, 0)$ with respect to τ is the plane with equation $X_0 = 0$. This is a tangent plane of $\mathcal{O}$ at $(0, 1, 0, 0)$ if and only if

$$a^{\theta+2} + aa' + a'^{\theta} \neq 0$$

for all $(a, a') \in \mathbb{K} \times \mathbb{K}$, $(a, a') \neq (0, 0)$. The result follows from the previous proposition. $\qquad\square$

Recall from Subsection 3.7.1 (see page 120) that a set of points in $\mathbf{PG}(3, \mathbb{K})$ having the properties of $\mathcal{O}$ mentioned in Proposition 7.6.14 is called an **ovoid** of $\mathbf{PG}(3, \mathbb{K})$. Hence every Suzuki–Tits ovoid $\mathcal{O}$ of a Suzuki quadrangle defines an ovoid $\mathcal{O}$ of the corresponding projective 3-space.

Next we look at plane sections of $\mathcal{O}$. We call a plane section **non-trivial** if it contains at least two points. Remember that an **oval** of a projective plane Γ' is a set of points $\mathcal{O}'$ such that every line of Γ' meets $\mathcal{O}'$ in 0, 1 or 2 points and every point lies on a unique tangent line (i.e., a line meeting $\mathcal{O}'$ in exactly one point); see Subsection 3.7.1. The following result is a special case of a more general result by TITS [1962a].

7.6.16 Proposition (Tits [1962a]). *Any two non-trivial plane sections of* $\mathcal{O}$ *which share exactly one point are projectively equivalent. Any two planes* z_1^{τ} *and* z_2^{τ}, *where* z_1 *and* z_2 *are points of* Γ, *meet* $\mathcal{O}$ *in projectively equivalent non-trivial plane sections (in particular, if* $\mathbb{K}$ *is perfect, then all plane sections are projectively equivalent). Also, every non-trivial plane section is an oval and all tangents of that oval are incident with one common point.*

Proof. Let π_1 and π_2 be two planes of $\mathbf{PG}(3, \mathbb{K})$ meeting $\mathcal{O}$ non-trivially and such that

$$|(\pi_1 \cap \mathcal{O}) \cap (\pi_2 \cap \mathcal{O})| = 1.$$

We may without loss of generality assume that $(\infty) \in \pi_1 \cap \pi_2 \cap \mathcal{O}$. The plane π_i, $i = 1, 2$, contains all points collinear with some point $z_i = \pi_i^{\tau}$ of $\mathsf{W}(\mathbb{K})$ (and z_i is collinear with (∞) and any other point of $\pi_i \cap \mathcal{O}$). If z_1 and z_2 are not collinear, then $z_1^\perp \cap z_2^\perp$ (in $\mathsf{W}(\mathbb{K})$) is a trace contained in a line L of $\mathbf{PG}(3,\mathbb{K})$ which must meet the ovoid at a second point by (the proof of) Proposition 7.6.14, a contradiction. So z_1 and z_2 are collinear, and since they are also collinear with (∞), they must lie on a line $[k]$, $k \in \mathbb{K} \cup \{\infty\}$. It is now an elementary exercise to find $A' \in \mathbb{K}$ such that $\sigma(0, A')$ maps z_1 to z_2. So we have found an element of $\mathbf{PSL}(4,\mathbb{K})$ (in fact also of $\mathbf{Sz}(\mathbb{K}, \theta)$) mapping $\pi_1 \cap \mathcal{O}$ bijectively onto $\pi_2 \cap \mathcal{O}$.

Now let $\pi_1 = z_1^\tau$ and $\pi_2 = z_2^\tau$, with z_1 and z_2 points of Γ, be two arbitrary planes of $\mathbf{PG}(3,\mathbb{K})$ meeting $\mathcal{O}$ non-trivially. By the 2-transitive action of $\mathbf{Sz}(\mathbb{K}, \theta)$, we may assume that both planes π_1 and π_2 contain two fixed elements x and y of $\mathcal{O}$. If we can find a non-trivial plane section $\pi \cap \mathcal{O}$ meeting $\pi_i \cap \mathcal{O}$ at exactly one point for $i = 1, 2$, then our first assertion follows from the previous paragraph. Note that z_1 is not collinear with z_2, otherwise both x and y lie on $z_1 z_2$, a contradiction. Now consider any point $z \in z_1^\perp \cap z_2^\perp$, $x \neq z \neq y$ and z in Γ. The plane $\pi = z^\tau$ meets our requirements (by the first paragraph of this proof and by the fact that each line of Γ through z contains an element of $\mathcal{O}$, hence π meets $\mathcal{O}$ in at least three points) because z is collinear with both z_1 and z_2. Hence the first two assertions.

Now let $\mathcal{C}$ be a plane section of $\mathcal{O}$, with $\mathcal{C} = \pi \cap \mathcal{O}$. Since every line in π is also a line in $\mathbf{PG}(3,\mathbb{K})$, every line of π meets $\mathcal{C}$ at 0, 1 or 2 points. A tangent line of $\mathcal{C}$ is also a tangent line of $\mathcal{O}$ in $\mathbf{PG}(3,\mathbb{K})$, hence at every point p, there is a tangent line of $\mathcal{C}$, namely the intersection of π with the tangent plane of $\mathcal{O}$ at p. Every other line in π through p does not lie in the tangent plane of $\mathcal{O}$ at p and hence meets $\mathcal{O}$, and therefore also $\mathcal{C}$, at a second point. This shows the third assertion of the proposition.

To prove the last assertion, let x be a point of $\mathsf{W}(\mathbb{K})$, not on $\mathcal{O}$, and let π be the plane x^τ. Suppose π meets $\mathcal{O}$ non-trivially. For every point $z \in \pi \cap \mathcal{O}$, the line xz is tangent to $\pi \cap \mathcal{O}$ and it contains x. The proposition is proved. $\qquad\square$

The intersection of all tangents of a non-trivial plane section of $\mathcal{O}$ will be called the **nucleus** of the oval. The existence of a nucleus is a general feature for ovals in finite projective planes of even order, as noted by QVIST [1952]; see also HUGHES & PIPER [1973], Lemma 12.10. The union of an oval and its nucleus (in a finite plane) is a hyperoval (for the definition, see Subsection 7.5.2).

In the finite case, the Suzuki–Tits ovoids in $\mathbf{PG}(3, 2^{2e+1})$, $e > 1$, are the only ovoids known to exist which are not projectively equivalent to an elliptic quadric. For $e = 1$, any Suzuki–Tits ovoid is projectively equivalent to the elliptic quadric in $\mathbf{PG}(3,2)$, and also in $\mathsf{W}(2)$. Note that the number of points of a Suzuki–Tits ovoid over $\mathbf{GF}(q)$ is $q^2 + 1$, as follows from Proposition 7.2.3.

7.6.17 Definitions. From now on we will call a non-trivial plane section of $\mathcal{O}$ a **circle (of $\mathcal{O}$)**. The proof of Proposition 7.6.16 implies that the set of nuclei of all circles of $\mathcal{O}$ is the set of points of $\mathsf{W}(\mathbb{K})$ collinear in $\mathsf{W}(\mathbb{K})$ with at least one point of $\mathcal{O}$. Furthermore, two circles meet at exactly one point **if and only if** their nuclei z and z' are collinear in $\mathsf{W}(\mathbb{K})$ and the line zz' is incident with a point of $\mathcal{O}$. Also, two distinct circles have distinct nuclei.

We will say that two circles **touch (at a point** p**)** if they share exactly one point p. We can now prove the following basic properties, which are proved in the case that $\mathbb{K}$ is perfect in VAN MALDEGHEM [1995a] (cf. VAN MALDEGHEM [1997]).

7.6.18 Proposition. *Let $\mathcal{O}$ be any Suzuki–Tits ovoid. Then $\mathcal{O}$ has the following properties:*

[MP1] *Three points of $\mathcal{O}$ are contained in exactly one circle.*

[MP2] *For every circle $\mathcal{C}$, and every two points $x, y \in \mathcal{O}$ such that $x \in \mathcal{C}$ and $y \notin \mathcal{C}$, there exists a unique circle $\mathcal{C}'$ touching $\mathcal{C}$ in x and containing y.*

[CH1] *There do not exist three circles mutually touching each other at distinct points.*

Proof.

[MP1] This is obvious since a plane in **PG**$(3, \mathbb{K})$ is determined by three non-collinear points, and no three points of $\mathcal{O}$ are collinear by Proposition 7.6.14.

[MP2] Denote the nucleus of $\mathcal{C}$ by z. It's easy to see that any circle $\mathcal{C}'$ touches $\mathcal{C}$ at x **if and only if** the nucleus z' of $\mathcal{C}'$ is incident with xz. Since we are looking for such $\mathcal{C}'$ also containing y, we see that z' must be collinear with y. The projection of y onto xz satisfies these conditions and is uniquely determined (since y is not collinear with x in $\mathsf{W}(\mathbb{K})$).

[CH1] This is also obvious: the three respective nuclei would have to be collinear in $\mathsf{W}(\mathbb{K})$ without being incident with one common line, a contradiction.

This completes the proof of the proposition.   □

From now on we assume that the field $\mathbb{K}$ is perfect. In that case, $\mathcal{O}$ has some more nice properties which eventually lead to an abstract characterization of such ovoids. Moreover, this characterization permits us to reconstruct the corresponding Suzuki quadrangle, which is a symplectic one.

A Suzuki–Tits ovoid with perfect underlying field will be called a **perfect Suzuki– Tits ovoid**. Note that by Subsection 7.6.17 two circles of a perfect Suzuki–Tits ovoid touch **if and only if** their respective nuclei are collinear in $\mathsf{W}(\mathbb{K})$. Now we show:

7.6.19 Proposition (Van Maldeghem [1997]). *Let $\mathcal{O}$ be any perfect Suzuki–Tits ovoid. Then $\mathcal{O}$ has the following properties:*

[CH2] *For every circle $\mathcal{C}$ and every pair of points $x, y \notin \mathcal{C}$, either there is a unique circle containing both x and y and touching $\mathcal{C}$, or all circles through x and y touch $\mathcal{C}$.*

[P] *For any two triples $\{\mathcal{C}_i, \mathcal{D}_i, \mathcal{E}_i\}$, $i = 1, 2$, of pairwise disjoint circles, we have that $\mathcal{E}_1$ touches $\mathcal{E}_2$ whenever both $\mathcal{C}_i$ and $\mathcal{D}_i$ touch $\mathcal{C}_j, \mathcal{D}_j$ and $\mathcal{E}_j$, $\{i, j\} = \{1, 2\}$.*

Proof.

[CH2] Denote the nucleus of C by z. We seek circles C' with nucleus z' touching C and containing x and y. These conditions immediately imply $z' \in x^\perp \cap y^\perp \cap z^\perp$. The three planes x^τ, y^τ, z^τ either meet in a unique point z', or they meet in a line, in which case every $z' \in x^\perp \cap y^\perp$ satisfies our requirements and so every circle through x and y touches C.

[P] Going over to the nuclei, the assertion just expresses the regularity in these points.

The proposition is proved. $\qquad\qquad\qquad\qquad\qquad\qquad\qquad\qquad\qquad\qquad\square$

7.6.20 Definition. Let C be a circle of a perfect Suzuki–Tits ovoid $\mathcal{O}$. Let x be its nucleus. Since x belongs to Γ, there is a unique line L of $\mathcal{O}^\rho$ incident with x. This line L is incident with a unique point ∂C of $\mathcal{O}$, which obviously belongs to C. We call this point, which we have denoted by ∂C, the **corner of** C (cf. Subsection 6.9.1 on page 299).

The corner of a circle serves as a *double condition*. Indeed, the condition on a circle C in order to contain a certain point x is that the nucleus of C is collinear in Γ with x; there are $|\mathbb{K}|^2 + |\mathbb{K}|$ such circles. However, the condition on a circle C in order to contain a certain point x as its corner is that the nucleus of C lies on x^ρ; there are only $|\mathbb{K}|$ such circles.

The same idea is expressed in the following proposition.

7.6.21 Proposition (Van Maldeghem [1997]). *Let $\mathcal{O}$ be any perfect Suzuki–Tits ovoid and let ∂ be the map which maps any circle to its corner. Then $\mathcal{O}$ has the following properties:*

[ST1] *For every pair of points x, y of $\mathcal{O}$, there is a unique circle C containing x and y and such that $\partial C = x$.*

[ST2] *For every circle C and every point $x \in \mathcal{O} \setminus C$, there is at most one circle C' containing x and ∂C and such that $\partial C' \in C$.*

Proof.

[ST1] Necessary and sufficient for the existence of C is that its nucleus z is incident with x^ρ and collinear with y. Clearly, there is exactly one such point z.

[ST2] The nucleus z' of a circle C' satisfying the given conditions must belong to $x^\perp \cap (\partial C)^\perp$ and, moreover, be incident with a line y^ρ, $y \in C$. Hence z'^ρ is incident with some $y \in C$ and meets $(\partial C)^\rho$, necessarily in the nucleus z of C. Hence the line joining ∂C and z' is mapped onto z by ρ. It follows that z' must lie on z^ρ. But at the same time z' must be found in the trace $x^\perp \cap (\partial C)^\perp$. Hence there is at most one such z'.

This completes the proof of the proposition. $\qquad\qquad\qquad\qquad\qquad\square$

In fact we will show below in the course of the proof of Theorem 7.6.23 (Lemma 2) that for every circle $\mathcal{C}$ and every point $x \in \mathcal{O} \setminus \mathcal{C}$, there is *exactly* one circle $\mathcal{C}'$ containing x and $\partial\mathcal{C}$ and such that $\partial\mathcal{C}' \in \mathcal{C}$. This can also be proved directly pushing the argument in the above proof a little further.

7.6.22 Remark. We have kept the same names of the respective properties as in the original paper (see VAN MALDEGHEM [1997]). The letters "MP" mean "Möbius plane" and the axioms [MP1] and [MP2] are the axioms for a Möbius plane. The letters "CH" mean "characteristic 2" and the axioms [CH1] and [CH2] express that every *internal affine plane* of the Möbius plane has characteristic 2. The letters "ST" mean "Suzuki–Tits" and the axioms [ST1] and [ST2] express the characteristic properties of the Suzuki–Tits ovoids related to the corners. Finally, the letter "P" stands for "perfect" and the axiom [P], together with all others, characterizes the perfect Suzuki–Tits ovoids as we will show below (actually, this is misleading since [P] is not the only axiom that expresses perfectness).

We are now ready to prove a geometric characterization of all perfect Suzuki–Tits ovoids.

7.6.23 Theorem (Van Maldeghem [1997]). *Let $\mathcal{P}$ be a set and $\mathcal{S}$ a distinguished set of subsets of $\mathcal{P}$ all containing at least three elements. Suppose there is a map $\partial : \mathcal{S} \to \mathcal{P}$ such that $\partial\mathcal{C} \in \mathcal{C}$, for all $\mathcal{C} \in \mathcal{S}$.* **If** *we call the elements of $\mathcal{S}$ circles and* **if** *we call two circles touching if they share exactly one point,* **then** *$(\mathcal{P}, \mathcal{S}, \partial)$ satisfies conditions [MP1], [MP2], [CH1], [CH2], [P], [ST1] and [ST2] (where $\mathcal{O}$ plays the role of $\mathcal{P}$) if and only if $\mathcal{P}$ can be embedded in a projective space $\mathbf{PG}(3, \mathbb{K})$, for some perfect field $\mathbb{K}$ of characteristic 2 admitting a Tits automorphism θ, such that $\mathcal{P}$ is the set of absolute points of a polarity of a certain symplectic quadrangle $\mathsf{W}(\mathbb{K})$ in $\mathbf{PG}(3, \mathbb{K})$ and the set of circles of $\mathcal{P}$ is equal to the set of plane sections of $\mathcal{P}$ in $\mathbf{PG}(3, \mathbb{K})$.*

Proof. We start with some easy consequences of the conditions. We call $\partial\mathcal{C}$ in this abstract context also the corner of $\mathcal{C}$, for any circle $\mathcal{C}$.

Lemma 1. *Let $C, D, E \in \mathcal{S}$ all contain the point x. If C touches D at x and D touches E at x, then either C touches E at x or $C = E$.*

Proof. Indeed, if C met E at a second point $y \neq x$, then there would exist two circles C and E containing y and touching D at x. The result follows from [MP2].

QED

Lemma 2. *For every circle C and every point $x \notin C$, there is a unique circle D containing x and ∂C and such that $\partial D \in C$.*

Proof. Let $y \in C$ be arbitrary but such that $y \neq \partial C$. Let E be the unique circle, whose existence is guaranteed by [ST1], containing ∂C such that $\partial E = y$. If $x \in E$, then the result follows from [ST2]. So suppose that $x \notin E$. Let D be the unique circle containing x and touching E at ∂C (see [MP2]). Note that $y \in E \cap C$ and E touches D at ∂C. Therefore, by [MP2], C does not touch D at ∂C. Thus C and D share another point z. Let F be the circle with corner z containing ∂C. Since $z \in C$, it follows from [ST2] that E and F touch at ∂C (noting $y \notin F$ because this would imply $F = C$ by [MP1]). But $z \in F \cap D$ and both D and F touch E at ∂C. This implies by Lemma 1 that $F = D$ and hence $z = \partial D \in C$. \hfill QED

Lemma 3. *If a circle C touches D at ∂D, then $\partial C = \partial D$.*

Proof. Choose $x \in C$, $x \neq \partial D$, and let C' be the circle with corner ∂D and containing x. By [ST1], C' touches D at ∂D and so [MP2] implies that $C' = C$. \hfill QED

Lemma 4. *If three circles containing a point x touch another circle which does not contain x, then these three circles have two points in common.*

Proof. Let C, D, E touch F with $x \in C \cap D \cap E$ and $x \notin F$. By [CH1], C and D meet at a second point y, $y \neq x$, and C and E also meet at a second point z, $z \neq x$. We must show $y = z$. Suppose by way of contradiction that $y \neq z$. Then $z \notin D$. By [MP2], there exists a circle G touching D at x and containing z. By [CH2], G touches F. Since $x \notin F$, this contradicts [CH1]. \hfill QED

Lemma 5. *If three circles touch two disjoint circles, then they either all have two points in common, or they are pairwise disjoint.*

Proof. Suppose C, D and E all touch the disjoint circles F and G, and suppose that C, D, E are not pairwise disjoint. By [CH1], we may assume that C and D meet at two distinct points x and y. Note that $x \in F$ would imply (by [MP2]) that $C = D$. Suppose E touches F in z. By [CH2], the circle H containing x, y and z touches both F and G. From [MP2] and Lemma 1 it follows readily that either H coincides with E, or H touches E. But the latter violates [CH1]. Whence the result. \hfill QED

Lemma 6. *There do not exist four distinct circles C_i, $i = 1, 2, 3, 4$, such that C_i touches C_{i+1} and C_i meets C_{i+2} at two distinct points, for all subscripts modulo 4.*

Proof. Let $C_1 \cap C_3 = \{x_1, x_3\}$ and $C_2 \cap C_4 = \{x_2, x_4\}$. By Lemma 1, $x_i \neq x_j$ for all $i \in \{1, 3\}$ and all $j \in \{2, 4\}$. Hence there is a unique circle D containing x_1, x_2 and x_3. By [CH2], D touches both C_2 and C_4. But this contradicts [MP2] since both C_2 and C_4 contain x_4 and touch D at x_3. \hfill QED

Lemma 7. *If C and D are two circles, then every point $x \notin C \cup D$ is contained in at least one circle D_x touching both C and D. If C and D meet at two points, then D_x is unique and the claim holds for every $x \notin C \cap D$.*

Proof. If C and D touch, then the result follows from [MP2] and Lemma 1. So we may assume that C and D do not touch.

Let $x \in \mathcal{P} \setminus (C \cup D)$. Let C_1 and C_2 be two circles touching C and containing x. By [CH1], C_1 and C_2 meet at another point $x_C \neq x$. By [CH2], all circles through x and x_C touch C. Similarly, there is a point x_D such that all circles through x and x_D touch D. Hence a circle through x, x_C and x_D touches both C and D. If C and D meet, then $x_C \neq x_D$ by Lemma 6. Moreover every circle through x touching C must contain x_C by Lemma 4 and similarly it must contain x_D. This proves the uniqueness in the case where C and D are non-disjoint. If $x \in C \setminus D$, then by considering a third circle C' containing $C \cap D$, $C \neq C' \neq D$, the result follows from the first part of this lemma (replacing C by C') and [CH2].   QED

Lemma 8. *For every point x outside any circle C, there exists a unique circle D with corner x touching C.*

Proof. There is at most one such circle by [CH1] and [ST1]. Let D_1 and D_2 be two circles containing x and touching C (these exist using [MP2] and the fact that C contains at least three elements). By [CH1], the circles D_1 and D_2 meet at a second point $y \neq x$. Let D be the circle with corner x containing y; then the result follows from [CH2].   QED

We can now reconstruct the symplectic quadrangle $\mathsf{W}(\mathbb{K})$. For that reason, we define the following geometry $\Gamma = (\mathcal{P}_\square, \mathcal{L}_\square, I)$. The justification for calling this geometry Γ, a symbol reserved for the symplectic (Suzuki) quadrangle $\mathsf{W}(\mathbb{K})$ in this section, is that we will prove that Γ is isomorphic to a symplectic quadrangle $\mathsf{W}(\mathbb{K})$ over some perfect field of characteristic 2 admitting a Tits automorphism.

Both the point set $\mathcal{P}_\square$ and the line set $\mathcal{L}_\square$ of Γ are the union of $\mathcal{P}$ and $\mathcal{S}$. Hence if $x \in \mathcal{P}$, then x can be viewed as a point or as a line of Γ. To make that difference, we write x_p or x_ℓ if we view x as an element of $\mathcal{P}_\square$ or $\mathcal{L}_\square$, respectively. Similarly for circles. We now define incidence in Γ. A point x_p, $x \in \mathcal{P}$, is incident with a line y_ℓ, $y \in \mathcal{P}$, **if and only if** $x = y$. A point x_p (line x_ℓ), $x \in \mathcal{P}$, is incident with a line C_ℓ (point C_p), $C \in \mathcal{S}$, **if and only if** $\partial C = x$. A point C_p, $C \in \mathcal{S}$, is incident with a line D_ℓ, $D \in \mathcal{S}$, **if and only if** $\partial C \in D$, $\partial D \in C$ and $\partial C \neq \partial D$.

Lemma 9. *The geometry Γ as defined above is a generalized quadrangle.*

Proof. First we claim that, if $C, D \in \mathcal{S}$, then C_p and D_p are collinear in Γ **if and only if** C touches D. If $C_p \, \mathbf{I} \, x_\ell \, \mathbf{I} \, D_p$ with $x \in \mathcal{P}$, then the claim follows from [ST1]. Suppose now $C_p \, \mathbf{I} \, E_\ell \, \mathbf{I} \, D_p$ with $E \in \mathcal{S}$. Then $\partial E \in C \cap D$ and since $D \neq C$, we have $\partial D \neq \partial C$. Clearly, also $\partial D \neq \partial E \neq \partial C$. Since $\partial C, \partial D \in E$, the result follows from [ST2]. Conversely, suppose that C and D are touching circles. If they touch in ∂C, then Lemma 3 implies $\partial C = \partial D$ and $C_p \, \mathbf{I} \, (\partial C)_\ell \, \mathbf{I} \, D_p$. So we may assume that $\{x\} = C \cap D$ with $\partial C \neq x \neq \partial D$. Let E be the circle containing ∂D, and with $\partial E = x$. By Lemma 2, there exists a circle F containing ∂C and x, and with $\partial F \in E$. Hence $F \neq E$. But Lemma 2 implies that F touches D at x. Hence, by Lemma 1, F touches C at x, contradicting $\partial C \in F \cap C$. We now have $D_p \, \mathbf{I} \, E_\ell \, \mathbf{I} \, D_p$.

Next, we show that Γ is a generalized quadrangle. So we must show that, whenever X is a point of Γ and L is a line of Γ not incident with X, then there exists a unique point–line pair (Y, M) such that $X \mathbf{I} M \mathbf{I} Y \mathbf{I} L$. There are essentially three cases to distinguish.

1. $X = x_p$ and $L = y_\ell$ with $x, y \in \mathcal{P}$. Clearly, $x \neq y$ and neither Y nor M can be elements of $\mathcal{P}$. So M must be equal to C_ℓ with $C \in \mathcal{S}$ and $\partial C = x$. Similarly $Y = D_p$ with $D \in \mathcal{S}$ and $\partial D = y$. Since C_ℓ and D_p are incident in Γ we must have that $x \in D$ and $y \in C$. But that defines C and D uniquely by [ST1].

2. $X = x_p$ and $L = C_\ell$ with $x \in \mathcal{P}$ and $C \in \mathcal{S}$. By assumption $\partial C \neq x$. Suppose first that $x \in C$. Let $D \in \mathcal{S}$ be such that $\partial D = x$ and $\partial C \in D$ (D is uniquely defined by [ST1]). Then $x_p \mathbf{I} x_\ell \mathbf{I} D_p \mathbf{I} C_\ell$. If E is any circle such that $E_\ell \mathbf{I} x_p$, then $\partial E = x$. If E_ℓ is concurrent with C_ℓ, then as for collinear points in Γ, C and E touch. But Lemma 3 implies that $\partial C = \partial E = x$, contradictory to our assumptions. So we have shown the uniqueness of Y and M if $x \in C$.

   Suppose now $x \notin C$. If F is a circle such that $F_p \mathbf{I} C_\ell$ and F_p is collinear with x_p in Γ, then as above, $\partial F \in C$ and $x, \partial C \in F$. By Lemma 2, F exists and is unique. If E is the unique circle with corner x and containing ∂F, then we have $x_p \mathbf{I} E_\ell \mathbf{I} F_p \mathbf{I} C_\ell$ and this chain is unique.

   The case where X is a circle and L is a point of Γ is completely similar.

3. $X = C_p$ and $L = D_\ell$ with $C, D \in \mathcal{S}$. First suppose that $\partial C = \partial D$. Clearly, $X \mathbf{I} (\partial C)_\ell \mathbf{I} (\partial D)_p \mathbf{I} L$. Suppose E and F are circles with $C_p \mathbf{I} E_\ell \mathbf{I} F_p \mathbf{I} D_\ell$. Then E contains ∂C, ∂F and ∂E. Since $\partial E \neq \partial F$ (by definition of incidence in Γ), we also have $E \neq F$. But also F contains $\partial C = \partial D$, ∂E and ∂F, contradicting the fact that, by definition of incidence in Γ, $\partial C \neq \partial F$ and $\partial E \neq \partial C$.

   Suppose that $\partial D \neq \partial C \in D$. Since X and L are not incident we have $\partial D \notin C$. Clearly, $X \mathbf{I} (\partial C)_\ell \mathbf{I} E_p \mathbf{I} L$, where $E \in \mathcal{S}$ with $\partial E = \partial C$ and $\partial D \in E$. Clearly, this chain is unique with the property that it contains an element of $\mathcal{P}$. Suppose now $C_p \mathbf{I} F_\ell \mathbf{I} G_p \mathbf{I} D_\ell$ with $F, G \in \mathcal{S}$. Then G contains ∂D, D contains ∂G and G touches C at, say, x. By [MP1], $x \notin D$ (because otherwise either $G = D$, a contradiction, or $G = E$ and $x = \partial C = \partial G$, and we cannot find any F_ℓ). The circle E (see above) touches C by [ST1], and it touches G by [ST2] because both G and E have their corner on D and $\partial D \in G \cap E$. Hence by [CH1] either $C \cap E \cap G$ is non-empty, or the set $\{C, E, G\}$ has two elements. In either case, one obtains $E = G$. This concludes the case $\partial C \in D$. The case $\partial D \in C$ is proved similarly.

   So we may assume that $\partial C \notin D$ and $\partial D \notin C$. Let C_1 and C_2 be two circles containing ∂D and touching C (these exist by [MP2] since a circle contains at least three points). By [CH1] C_1 and C_2 meet at a further point $x \neq \partial D$. By [CH2] all circles through ∂D and x touch C, hence so does the unique

circle E with $\{\partial D, x\} \subseteq E$ and $\partial E \in D$ (which exists by Lemma 2). Note
that E is unique with respect to the properties that $\partial E \in D$, $\partial D \in E$ and
E touches C. Indeed, if E' is any other circle meeting these conditions, then
E and E' touch by [ST2]. But by [CH1], they intersect C at the same point,
hence $E = E'$ by [MP2]. Consequently $E_p \mathbf{I} D_\ell$ and E_p is collinear with C_p.
By the uniqueness of E, we have shown condition (i) of Lemma 1.4.1 (see
page 15).

From (ST1), one deduces easily the existence of at least two circles and consequently of a circle C and a point $x \notin C$. This implies the existence of an antiflag
in Γ. In fact, it is now easy to see that for every point x, there exists a circle C with
$x \notin C$. Since $|C| \geq 3$, this implies that there are at least two circles with corner
x. Hence x_p is a thick point of Γ. But as every circle C contains at least three
points, every point C_p is thick. Dually, every line is thick and Γ is a generalized
quadrangle by Lemma 1.4.1.   QED

Lemma 10. *All points and lines of Γ are regular.*

Proof. It suffices to show that every pair of non-collinear points $\{X, Y\}$ of Γ is
regular (see Definitions 6.4.1 on page 257). So we prove that whenever U, V, W are
collinear with X and Y, and Z is collinear with U and V, then Z is collinear with
W. A similar proof will imply that all lines are regular.

1. Suppose first that $X = x_p$ and $Y = y_p$ with $x, y \in \mathcal{P}$. It is easily seen
   that U, V and W must be circles of Γ containing x and y (but not as their
   corners). So we put $U = C_\ell$, $V = D_\ell$ and $W = E_\ell$ with $C, D, E \in \mathcal{S}$. Note
   that $C \cap D \cap E = \{x, y\}$, which implies that, if Z is collinear with U and
   V, then $Z = F_p$ with $F \in \mathcal{S}$. Also we know (see above) that F touches both
   C and D. Now [CH2] implies that F also touches E, hence Z and W are
   collinear.

2. Suppose now that $X = x_p$ and $Y = G_p$ with $x \in \mathcal{P}$ and $G \in \mathcal{S}$. Again U, V
   and W must be circles and we again put $U = C_\ell$, $V = D_\ell$ and $W = E_\ell$,
   with $C, D, E \in \mathcal{S}$ and $x \in C \cap D \cap E$. We know that G touches C, D and E,
   hence by Lemma 4, C, D and E meet at a further point y. We are back at
   the situation of the preceding paragraph and so the result follows.

3. By switching the roles of X, Y, Z and U, V, W in the preceding paragraphs, we
   may now assume that they are all circles of Γ. So we put $X = H_p$, $Y = G_p$,
   $Z = F_p$, $U = C_\ell$, $V = D_\ell$ and $W = E_\ell$, with $C, D, E, F, G, H \in \mathcal{S}$. Now C
   touches H; H touches D; D touches G and G touches C, while neither C
   and D nor G and H touch each other. Hence by Lemma 6, we may assume
   that G and H are disjoint. By Lemma 5, either C, D and E share two points
   x and y and hence E touches F as before, or C, D, E are pairwise disjoint
   and the result follows directly from [P].

Hence we have shown that all points (and dually lines) of Γ are regular.   QED

Lemma 11. *All points x_p of Γ are projective (see* Definition 1.9.4*), hence Γ is isomorphic to a symplectic quadrangle* $\mathsf{W}(\mathbb{K})$ *over some field* $\mathbb{K}$.

Proof. We show that for arbitrary points Y, Z opposite x_p in Γ, the set $\{x_p, Y, Z\}^{\perp}$ is non-empty.

1. If Y and Z are elements of $\mathcal{P}$, say $Y = y_p$ and $Z = z_p$, then the point C_p of Γ, with C the unique circle containing x, y and z, is collinear with x_p, Y and Z in Γ.

2. If Y is an element of $\mathcal{P}$, say $Y = y_p$, and if Z is a circle, say $Z = C_p$, then by [CH2], there is at least one circle F containing both x and y and touching C (note that indeed $x, y \notin C$ otherwise x_p or Y is collinear with Z in Γ). The point F_p of Γ is collinear with x_p, with Y and with Z.

3. Let $Y = D_p$ and $Z = C_p$, $C, D \in \mathcal{S}$. We have by assumption $x \notin C \cup D$ and C and D do not touch each other. By Lemma 7, there is at least one circle F containing x and touching both C and D. The point F_p of Γ is again collinear with all three x_p, Y, Z.

So we have shown that Γ contains at least one projective point. Hence by Theorem 6.2.1 (see page 240), $\Gamma \cong \mathsf{W}(\mathbb{K})$, for some field $\mathbb{K}$. QED

Since dually, Γ contains at least one regular (even projective) line, $\mathbb{K}$ has characteristic 2 by Proposition 3.4.8(*iii*) (page 106). Note that the map $x_p \mapsto x_\ell$ and $C_p \mapsto C_\ell$, $x \in \mathcal{P}$ and $C \in \mathcal{S}$, induces a polarity in the symplectic quadrangle $\Gamma \cong \mathsf{W}(\mathbb{K})$. Hence by Corollary 7.3.5, $\mathbb{K}$ is a perfect field. By the definition of incidence in Γ, the set of absolute points of this polarity is exactly the set of points of Γ. An arbitrary plane section is a set of points collinear with some non-absolute point X of the quadrangle. Hence $X = C_p$ for some $C \in \mathcal{S}$. But the set of absolute points collinear with C_p is exactly C viewed as the set of elements of $\mathcal{P}$. The theorem now follows from Theorem 7.6.2. □

In the finite case, a few conditions become superfluous. Without proof, we mention:

7.6.24 Corollary (Van Maldeghem [1997]). *Let $\mathcal{P}$ be a finite set with $|\mathcal{P}|$ odd and $\mathcal{S}$ a distinguished set of subsets of $\mathcal{P}$ all containing at least three elements. Suppose there is a map $\partial : \mathcal{S} \to \mathcal{P}$ such that $\partial C \in C$, for all $C \in \mathcal{S}$. If we call the elements of $\mathcal{S}$ circles and if we call two circles touching if they share exactly one point,* **then** *$(\mathcal{P}, \mathcal{S}, \partial)$ satisfies conditions* [MP1], [MP2], [ST1] *and* [ST2] *(where $\mathcal{O}$ plays the role of $\mathcal{P}$) if and only if $\mathcal{P}$ can be embedded in a projective space* $\mathbf{PG}(3, 2^{2e+1})$, *for some $e \in \mathbb{N}$, such that $\mathcal{P}$ is the set of absolute points of a polarity of a certain symplectic quadrangle* $\mathsf{W}(2^{2e+1})$ *in* $\mathbf{PG}(3, 2^{2e+1})$ *and the set of circles of $\mathcal{P}$ is equal to the set of non-trivial plane sections of $\mathcal{P}$ in* $\mathbf{PG}(3, 2^{2e+1})$. □

A geometry $(\mathcal{P}, \mathcal{S})$ satisfying [MP1] and [MP2] is called in the literature an **inversive plane** or a **Möbius plane**, see also Subsection 6.9.1 on page 299. Theorem 7.6.23

and Corollary 7.6.24 can be formulated in terms of inversive planes. But we leave that to the reader (note also that every ovoid of $\mathbf{PG}(3,q)$ can be given the structure of an inversive plane by considering plane sections). There is also an alternative form of condition [P] which gives those perfect Suzuki–Tits ovoids for which the group $\mathbf{Sz}(\mathbb{K},\theta)$ has just two orbits in the set of lines of $\mathbf{PG}(3,\mathbb{K})$ not belonging to $\mathsf{W}(\mathbb{K})$. For details, the reader is referred to VAN MALDEGHEM [1997].

Finally, we note that the *transversal partition* defined by two points x and y (see Subsection 6.9.1 on page 299) in a Suzuki–Tits ovoid $\mathcal{O}$ is the set of non-trivial plane sections with planes containing the line $(xy)^\tau$, or equivalently, it is the set of circles with nucleus in $\{x,y\}^{\perp\perp} \setminus \{x,y\}$.

7.6.25 Finite classical and semi-classical ovoids

In this subsection we review, without proof, some other results about finite classical and semi-classical ovoids and their mutual relations. First we note that, by Theorem 7.6.2, all semi-classical ovoids are projectively equivalent. Also recall that a classical ovoid in $\mathsf{Q}(4,q)$ is just a hyperplane section in $\mathbf{PG}(4,q)$ containing no lines. For q even, $\mathsf{Q}(4,q)$ is isomorphic to $\mathsf{W}(q)$ and the image of a classical ovoid under any isomorphism is called a **classical ovoid** in $\mathsf{W}(q)$. The classical and semi-classical ovoids are the only ovoids known to exist in $\mathsf{W}(q)$ and hence also in $\mathbf{PG}(3,q)$, q even. For q odd, BARLOTTI [1955] shows that the only ovoids in $\mathbf{PG}(3,q)$ are the elliptic quadrics. For q even, THAS [1972a] shows that *every* ovoid of $\mathsf{W}(q)$, q even, is an ovoid of $\mathbf{PG}(3,q)$, and, conversely, it follows readily from SEGRE [1959] that for every ovoid of $\mathbf{PG}(3,q)$ there exists a symplectic polarity such that this ovoid is also an ovoid in the corresponding symplectic quadrangle. It is conjectured that the only ovoids in $\mathsf{W}(q)$ (or equivalently, $\mathbf{PG}(3,q)$), q even, are the classical and semi-classical ones, but as yet a proof is only known for small values of even q, in particular for $2 \leq q \leq 64$. Let us also mention that the stabilizer in $\mathbf{Sp}_4(q)$ of a classical ovoid in $\mathsf{W}(q)$ is the group $\mathbf{PSL}_2(q^2)$, acting doubly transitively on the ovoid as its natural action on the projective line $\mathbf{PG}(1,q^2)$.

There are a lot of characterizations of the classical and semi-classical ovoids in $\mathbf{PG}(3,q)$ and $\mathsf{W}(q)$ in the literature, but we restrict ourselves to mentioning only one, due to TITS [1960]; see TITS [1966a] for another characterization along the same lines.

7.6.26 Theorem (Tits [1960]). *Let $\mathcal{O}$ be an ovoid of $\mathsf{W}(q)$, q even. Then $\mathcal{O}$ is a classical or semi-classical ovoid* **if and only if** *there exists a point p of $\mathcal{O}$ such that the stabilizer in $\mathbf{Sp}_4(q)$ (viewed as collineation group of $\mathsf{W}(q)$) of $\mathcal{O} \setminus \{p\}$ contains a normal regular subgroup and at least one cyclic subgroup of order $q - 1$.* $\quad\square$

Another interesting issue is the intersection pattern of ovoids in $\mathsf{W}(q)$. This has been investigated by BAGCHI & SASTRY [1989] and we mention their results in the next theorem. An elementary proof of their principal result was given by DE SMET [1994], who also considered the intersections of the set of flags corresponding to a semi-classical ovoid (the flags fixed under the corresponding polarity).

7.6.27 Theorem (Bagchi & Sastry [1989], De Smet [1994]).

(*i*) *Two distinct classical ovoids of* $\mathsf{W}(q)$ *have either* 1 *or* $q+1$ *points in common.*

(*ii*) *Two distinct semi-classical ovoids of* $\mathsf{W}(q)$ *have either* 1, $q+1$, $q+\sqrt{2q}+1$ *or* $q-\sqrt{2q}+1$ *points in common.*

(*iii*) *A classical ovoid and a semi-classical ovoid in* $\mathsf{W}(q)$ *have either* $q+\sqrt{2q}+1$, *or* $q-\sqrt{2q}+1$ *points in common. In either case, there is a cyclic group preserving the two ovoids and acting regularly on the intersection.*

(*iv*) *Two distinct semi-classical ovoids of* $\mathsf{W}(q)$ *have* 0, 1 *or* 2 *flags in common.*

$\square$

We will later on prove (*iv*) of the previous theorem for arbitrary isomorphic Suzuki–Tits ovoids in any Suzuki quadrangle in an entirely geometric fashion (unlike the proof of DE SMET [1994]); see Corollary 8.2.4 on page 365. Also notice that in (*i*) of the previous theorem, q is an arbitrary prime power, but in (*ii*), (*iii*) and (*iv*), $q = 2^{2e+1}$, $e \in \mathbb{N}$.

The previous theorem allows one to describe all maximal subgroups of $\mathbf{Sz}(q)$ in a geometric fashion. Indeed, it follows rather easily from SUZUKI [1962] that, if M is a maximal subgroup of $\mathbf{Sz}(q)$, then either M is the stabilizer of a point of the corresponding Suzuki–Tits ovoid, or M is the stabilizer of a circle of the Suzuki–Tits ovoid (viewed as an inversive plane), or M is the stabilizer of a Suzuki–Tits sub-ovoid (over a maximal subfield), or M is the stabilizer of the intersection of a Suzuki–Tits ovoid with a classical ovoid (and isomorphic to $(q \pm \sqrt{(2q)} + 1) : 4$).

7.7 Ree hexagons and Ree–Tits ovoids

Throughout this section we fix a field $\mathbb{K}$ of characteristic 3 admitting a Tits endomorphism θ (see Definition 2.5.1). We also fix θ itself and denote $\mathbb{K}' := \mathbb{K}^{\theta}$. It follows from Theorem 7.3.4 on page 314 that the mixed hexagon $\mathsf{H}(\mathbb{K}, \mathbb{K}')$ is self-polar. If we denote by Λ the isomorphism class of $\mathbb{K}'$ within $\mathbb{K}$, then it follows from Proposition 3.5.4 on page 113 that $\mathsf{H}(\mathbb{K}, \mathbb{K}'')$, for $\mathbb{K}''$ some subfield of $\mathbb{K}$, is isomorphic to $\mathsf{H}(\mathbb{K}, \mathbb{K}')$ **if and only if** $\mathbb{K}'' \in \Lambda$.

In general, as for infinite fields of characteristic 2 (see section 7.6, page 322), the Tits endomorphism is not unique. But for finite fields, we can again state:

7.7.1 Lemma. *Every finite field of order* 3^{2e+1}, *for some non-negative integer* e, *admits a unique Tits endomorphism, which is always an automorphism.* $\square$

By the definition of the mixed hexagons, $\mathsf{H}(\mathbb{K}, \mathbb{K}')$ is a full subhexagon of a split Cayley hexagon $\mathsf{H}(\mathbb{K})$. We will frequently refer to $\mathsf{H}(\mathbb{K})$, as well as to the quadric $\mathsf{Q}(6, \mathbb{K})$ on which $\mathsf{H}(\mathbb{K})$, and hence also $\mathsf{H}(\mathbb{K}, \mathbb{K}')$, lives. We call the mixed hexagon $\mathsf{H}(\mathbb{K}, \mathbb{K}')$ a **Ree hexagon**, because of its close relation with the Ree groups in characteristic 3 (see below), although it was TITS [1960] who first defined Ree groups over non-perfect fields.

The literature on proper mixed hexagons (i.e., mixed hexagons which are not split Cayley hexagons) is almost non-existent. Some properties are hidden in TITS [1960], where the mixed hexagons are implicitly present. We prove some more properties in this section, although most of the proofs are completely similar to those of the corresponding results for Suzuki quadrangles. For instance, the following result is the analogue of Theorem 7.6.2.

7.7.2 Theorem. *The set of conjugacy classes of polarities (conjugacy with respect to* Aut Γ*) of the Ree hexagon* Γ := H($\mathbb{K}, \mathbb{K}'$) *is in a natural bijective correspondence with the set of conjugacy classes (with respect to the automorphism group of* $\mathbb{K}$*) of Tits endomorphisms* θ *of* $\mathbb{K}$*. Also, this set is in a natural bijective correspondence with the set of isomorphism classes (with respect to* Aut Γ *again) of ovoid–spread pairings of* Γ *arising from polarities.*   □

7.7.3 Corollary. *Two polarities of a Ree hexagon having the same absolute elements coincide.*   □

7.7.4 Definitions. Now we fix an arbitrary polarity ρ of H($\mathbb{K}, \mathbb{K}'$) having as corresponding Tits endomorphism θ, $\mathbb{K}^\theta = \mathbb{K}'$. We denote by $\mathbf{G}_2(\mathbb{K}, \mathbb{K}')$ the little projective group of H($\mathbb{K}, \mathbb{K}'$). By Corollary 5.8.3 (page 233) this group is a simple group. Let $\mathcal{U}$ be the set of absolute points for ρ. We call $\mathcal{U}$ a **Ree–Tits ovoid**. We denote by $\mathbf{Ree}(\mathbb{K}, \theta)$ the subgroup of $\mathbf{G}_2(\mathbb{K}, \mathbb{K}')$ stabilizing $\mathcal{U} \cup \mathcal{U}^\rho$ and call it a **Ree group (over $\mathbb{K}$, with respect to θ)**. For a finite field $\mathbb{K} \cong \mathbf{GF}(q)$, $q = 3^{2e+1}$, the Tits endomorphism is unique and hence redundant in the notation; so in this case we adopt the notation $\mathbf{Ree}(q)$. As for the Suzuki groups, one shows that $\mathbf{Ree}(\mathbb{K}, \theta)$ is the intersection with $\mathbf{G}_2(\mathbb{K}, \mathbb{K}')$ of the centralizer of ρ inside the full correlation group of H($\mathbb{K}, \mathbb{K}'$). Here, however, the group $\mathbf{Ree}(\mathbb{K}, \theta)$ is not necessarily a simple group. But we will indicate how one can show that a certain subgroup of it (which will turn out to be the commutator subgoup) is simple. We will also see that, in the finite case, $\mathbf{Ree}(q)$ is always simple, unless $\mathbb{K} \cong \mathbf{GF}(3)$. In any case, $\mathbf{Ree}(\mathbb{K}, \theta)$ acts 2-transitively on $\mathcal{U}$. We shall also describe the point stabilizer in $\mathbf{Ree}(\mathbb{K}, \theta)$ and show how one can make $\mathcal{U}$ into a *unital*.

Note that, via Corollary 4.6.6, we can again view $\mathbf{Ree}(\mathbb{K}, \theta)$ unambiguously as a subgroup of $\mathbf{PSL}_7(\mathbb{K})$.

7.7.5 Unipotent elements of $\mathbf{Ree}(\mathbb{K}, \theta)$

Put Γ = H($\mathbb{K}, \mathbb{K}'$). We can always coordinatize H($\mathbb{K}, \mathbb{K}'$) in such a way that the polarity ρ maps the corresponding hat-rack to its dual and (1) to [1]. Similarly to the case of Suzuki quadrangles, one shows that ρ takes the form

$$(a, l, a', l', a'')^\rho = [a^\theta, l^{\theta^{-1}}, a'^\theta, l'^{\theta^{-1}}, a''^\theta];$$

$$[k, b, k', b', k'']^\rho = (k^{\theta^{-1}}, b^\theta, k'^{\theta^{-1}}, b'^\theta, k''^{\theta^{-1}}),$$

for all $a, a', a'', b, b' \in \mathbb{K}$ and all $k, k', k'', l, l' \in \mathbb{K}'$ (where θ is as earlier if the recoordinatization was carried out using no field automorphism, which can always be done). It is now an elementary exercise to calculate that the point (a, l, a', l', a'') is absolute for ρ **if and only if**

$$\begin{cases} l & = & a''^{\theta} - a^{\theta+3}, \\ l' & = & a^{2\theta+3} + a'^{\theta} + a^{\theta} a''^{\theta}. \end{cases}$$

Hence the set $\mathcal{U}$ of absolute points is equal to

$$\{(\infty)\} \cup \{(a, a''^{\theta} - a^{\theta+3}, a', a^{2\theta+3} + a'^{\theta} + a^{\theta} a''^{\theta}, a'') : a, a', a'' \in \mathbb{K}\}.$$

Let $U_+ \leq \mathbf{G}_2(\mathbb{K}, \mathbb{K}')$ be the collineation group generated by all root elations which fix (∞) and $[\infty]$. Again, by Lemma 5.2.4 (see page 175) U_+ acts regularly on the flags of Γ opposite $\{(\infty), [\infty]\}$. Also, U_+ is a normal subgroup of the stabilizer in $\mathbf{G}(\mathbb{K}, \mathbb{K}')$ of the flag $\{(\infty), [\infty]\}$. Let $U := U_+ \cap \mathbf{Ree}(\mathbb{K}, \theta)$. The elements of U, and of every conjugate of U inside $\mathbf{Ree}(\mathbb{K}, \theta)$, are called the **unipotent elements** of $\mathbf{Ree}(\mathbb{K}, \theta)$. We also have that U acts semi-regularly on $\mathcal{U} \setminus \{(\infty)\}$ and $U \trianglelefteq \mathbf{Ree}(\mathbb{K}, \theta)_{(\infty)}$. We show that U acts transitively on $\mathcal{U} \setminus \{(\infty)\}$ by exhibiting an explicitly defined element $\sigma(A, A', A'')$ of U which maps $(0,0,0,0,0)$ to $(A, A''^{\theta} - A^{\theta+3}, A', A^{2\theta+3} + A'^{\theta} + A^{\theta} A''^{\theta}, A'')$. We define $\sigma := \sigma(A, A', A'')$ on the points and lines with five coordinates as follows. It maps (a, l, a', l', a'') to

$$\begin{aligned} (\ & a + A, \\ & l + A''^{\theta} - A^{\theta} a^3 - A^{\theta} A^3, \\ & a' + A' - A'' a + A^{\theta} a^2, \\ & l' + A'^{\theta} + A^{2\theta} a^3 + A^{\theta} l + A^{2\theta} A^3 + A^{\theta} A''^{\theta}, \\ & a'' + A'' + A^{\theta} a \ ) \end{aligned}$$

and $[k, b, k', b', k'']$ to

$$\begin{aligned} [\ & k + A^{\theta}, \\ & b + A'' - Ak - AA^{\theta}, \\ & k' + A'^{\theta} - A''^{\theta} k + A^3 k^2, \\ & b' + A' + A^2 k + Ab + A^2 A^{\theta} + AA'', \\ & k'' + A''^{\theta} + A^3 k \ ]. \end{aligned}$$

One can check by an elementary calculation that σ actually preserves $\mathcal{U}$.

Note that $\sigma(A, A', A'')$ is also defined in $\mathsf{H}(\mathbb{K})$.

Since U acts transitively on $\mathcal{U} \setminus \{(\infty)\}$, we obtain as in Theorem 7.6.6 (and with a similar remark concerning the original construction of the Ree groups in characteristic 3 by REE [1961]):

7.7.6 Theorem (Tits [1962a]). *The group* $\mathbf{Ree}(\mathbb{K}, \theta)$ *acts doubly transitively on the set of absolute points of* ρ. $\qquad\square$

7.7.7 Structure of a point stabilizer in $\mathbf{Ree}(\mathbb{K}, \theta)$

It follows from the discussion above that every element of U is uniquely determined by the image of $(0, 0, 0, 0, 0)$. We denote such an element by $((a, a'', a' - aa''))$, if $(a, \ldots, a', \ldots, a'')$ is that image. The reason for this unusual identification is that we want to recover exactly the same operations as defined originally by TITS [1960]. And, indeed, after a rather tedious calculation, one finds

$$((x, x', x'')) \cdot ((y, y', y'')) = ((x + y, x' + y' + xy^\theta, x'' + y'' + xy' - x'y - xy^{\theta+1})).$$

If we stabilize two points of $\mathcal{U}$, say (∞) and $(0, 0, 0, 0, 0)$, then we stabilize the hat-rack of the coordinatization and hence only elements of a torus are possible; by Proposition 4.6.7, these are generated by generalized homologies. It is easily seen that the mappings h_X, defined by

$$h_X : \begin{cases} (a, l, a', l', a'') & \mapsto & (Xa, X^{\theta+3}l, X^{\theta+2}a', X^{2\theta+3}l', X^{\theta+1}a''), \\ [k, b, k', b', k''] & \mapsto & [X^\theta k, X^{\theta+1}b, X^{2\theta+3}k', X^{\theta+2}b', X^{\theta+3}k''] \end{cases}$$

are the only elements of $\mathbf{G}_2(\mathbb{K}, \mathbb{K}')$ fixing the hat-rack of the coordinatization and stabilizing $\mathcal{U}$.

Each map h_X, $X \in \mathbb{K}$, belongs to $\mathbf{Ree}(\mathbb{K}, \theta)$ (unlike the situation for Suzuki groups where we had a restriction on X; see Subsection 7.6.7 on page 326).

Now the map h_X defines by conjugation an operation

$$X \times ((x, x', x'')) = h_X^{-1}((x, x', x''))h_X$$

and one calculates

$$X \times ((x, x', x'')) = ((Xx, X^{\theta+1}x', X^{\theta+2}x'')).$$

Again, this is identical to TITS' multiplication [1960].

The stabilizer $\mathbf{Ree}(\mathbb{K}, \theta)_{(\infty)}$ can now be described as the group containing the elements $h_X \cdot ((x, x', x'')) =: ((x, x', x''; X))$, $x, x', x'' \in \mathbb{K}$, $X \in \mathbb{K}^\times$, with the multiplication

$$((x, x', x''; X)) \cdot ((y, y', y''; Y)) = h_X h_Y \cdot \big(Y \times ((x, x', x'')) \big) \cdot ((y, y', y'')) =$$

$$((Yx + y, Y^{\theta+1}x' + y' + Yxy^\theta, Y^{\theta+2}x'' + y'' + Yxy' - Y^{\theta+1}x'y - Yxy^{\theta+1}; XY)).$$

As for Theorem 7.6.9, one can prove with a few elementary calculations:

7.7.8 Theorem (Ree [1961], Tits [1960]). *The subgroup of the Ree group $\mathbf{Ree}(\mathbb{K}, \theta)$ generated by the unipotent elements is simple, provided $|\mathbb{K}| > 3$. If the multiplicative group of $\mathbb{K}$ is generated by all squares and by the element -1, then $\mathbf{Ree}(\mathbb{K}, \theta)$ itself is simple. For $\mathbb{K} \cong \mathbf{GF}(3)$, the group $\mathbf{Ree}(3)$ has a normal simple derived group of index 3, which is isomorphic to $\mathbf{PSL}_2(8)$, hence $\mathbf{Ree}(3) \cong \mathbf{P\Gamma L}_2(8)$.* $\qquad\square$

We will comment on the proof of this theorem in Subsection 7.7.19. As for Suzuki–Tits ovoids, one shows:

7.7.9 Theorem. *Every automorphism σ of the little projective group of Γ stabilizing the Ree–Tits ovoid $\mathcal{U}$ also stabilizes $\mathcal{U}^\rho$.* $\qquad\square$

7.7.10 The finite case

In the finite case, one easily deduces from previous arguments that the order of $\mathbf{Ree}(q)$ is equal to
$$|\,\mathbf{Ree}(q)| = (q^3 + 1) \cdot q^3 \cdot (q - 1),$$
and it acts doubly transitively on the set of $q^3 + 1$ points of a Ree–Tits ovoid. Also, as for the Suzuki–Tits ovoids, we can deduce from Proposition 4.6.7 on page 157 that the number of Ree–Tits ovoids in $\mathsf{H}(q)$ is equal to

$$\frac{q^6 \cdot (q^6 - 1) \cdot (q^2 - 1)}{(q^3 + 1) \cdot q^3 \cdot (q - 1)} = q^3 \cdot (q^3 - 1) \cdot (q + 1).$$

This formula also holds for $q = 3$, in which case the Ree–Tits ovoid is not isomorphic to the Hermitian one inside $\mathsf{H}(3)$, but it is inside $\mathbf{Q}(6,3)$, where the automorphism group is isomorphic to the symplectic group $\mathbf{Sp}_6(2)$ (as already mentioned in the last paragraph of Subsection 7.3.8).

We now turn to some geometric properties of the Ree–Tits ovoids. In general, the Ree–Tits ovoids do not have many properties similar to their analogues, the Suzuki–Tits ovoids. In fact, it turns out that very different geometric properties hold, which, in turn, do not have analogues in the theory of the Suzuki–Tits ovoids. Note also that most of the results we will present have algebraic proofs which are sometimes shorter than the geometric approach that we choose (in keeping with the general spirit of this book).

Despite the remarks just made, we start with an analogue to Proposition 7.6.14. We keep the same notation as in the previous subsection. In particular, $\mathcal{U}$ is a Ree–Tits ovoid with respect to a polarity ρ in the mixed Ree hexagon $\Gamma = \mathsf{H}(\mathbb{K}, \mathbb{K}')$ corresponding to some conjugacy class of Tits endomorphisms θ for which $\mathbb{K}^\theta = \mathbb{K}'$. Our first result though can be stated in full generality.

7.7.11 Proposition. *Let Γ^* be any self-polar generalized hexagon. Let $\mathcal{U}^*$ be the set of absolute points for a certain polarity ρ^* of Γ^*. Suppose Γ^* contains an ideal non-thick weak subhexagon Γ^{**} containing a point p of $\mathcal{U}^*$. Then Γ^{**} contains a unique point q of $\mathcal{U}^*$ opposite p.*

Proof. Let $L = p^{\rho^*}$. Let x be the unique point of Γ^{**} on L with $x \neq p$. Let y be the unique point of Γ^{**} distinct from p and incident with x^{ρ^*}. Let z be the unique point of Γ^{**} distinct from x and incident with y^{ρ^*}. Let u be the unique point of Γ^*

at distance 4 from z and incident with z^{ρ^*}. Since z^{ρ^*} is incident with y, it belongs to Γ^{**}, hence u also belongs to Γ^{**}. It follows easily that the element $q = u \bowtie z$ of Γ^{**} belongs to $\mathcal{U}^*$. Clearly, q is unique because all elements of $\mathcal{U}^*$ are mutually opposite and every point of Γ^{**} is at distance 4 from either p or q. $\square$

Since in $\Gamma = \mathsf{H}(\mathbb{K}, \mathbb{K}')$ all points are span-regular (for the definition, see Subsection 1.9.8 on page 40), by Proposition 3.5.6 (see page 114), we obtain:

7.7.12 Corollary. *Every ideal non-thick weak subhexagon of Γ which contains at least one point of $\mathcal{U}$ contains exactly two points of $\mathcal{U}$.* $\square$

7.7.13 Remark. In view of the proofs of Propositions 7.6.14 and 7.7.11, it is clear that one can show in general that, whenever an ideal non-thick weak sub-$2n$-gon of a given generalized $2n$-gon Γ' contains an absolute point of a certain polarity of Γ', then it contains exactly two such absolute points.

7.7.14 Remark. Consider a fixed point p of $\mathcal{U}$. Then by Corollary 7.7.12, there is a natural bijective correspondence between the points of the span-geometry $\Gamma_p^{\square}$ which are not collinear with the point $p^{\perp}$ of $\Gamma_p^{\square}$ and the points of $\mathcal{U} \setminus \{p\}$. If $\mathbb{K}$ is perfect, then $\Gamma_p^{\square}$ is isomorphic to the generalized quadrangle $\mathsf{W}(\mathbb{K})$ and hence we see that the point-stabilizer of the Ree group $\mathbf{Ree}(\mathbb{K}, \theta)$ acts regularly on the set of points of $\mathsf{W}(\mathbb{K})$ opposite a given point.

The next result is algebraic and is the analogue of Corollary 7.6.15.

7.7.15 Proposition. *For any Tits endomorphism θ in a field $\mathbb{K}$ of characteristic 3, the equation*

$$x^{2\theta+4} + x^{\theta+1}y - y^{\theta+1} + xz^{\theta} - z^2 + xyz = 0$$

in the unknowns x, y and z has the unique solution $x = y = z = 0$.

Proof. The set of points $\Gamma_2((0^5))$ (where $(0^5) = (0,0,0,0,0) \in \mathcal{U}$ with respect to a suitable coordinatization, i.e., one for which the hat-rack is mapped onto its dual by ρ and also $(1)^{\rho} = [1]$) does not contain points of $\mathcal{U}$. Using the correspondence between the coordinates of Γ and the coordinates in $\mathbf{PG}(6, \mathbb{K})$ (via $\mathsf{Q}(6, \mathbb{K})$; see Subsection 3.5.1 on page 110), we see that, with the notation of Subsection 3.5.1, a point of $\mathsf{H}(\mathbb{K})$ with coordinates $(x_0, x_1, \ldots, x_6)$ in $\mathbf{PG}(6, \mathbb{K})$, is collinear with $(0,0,0,0,0)$ in $\mathsf{H}(\mathbb{K})$ **if and only if** $x_0 = 0$. Using the coordinate transformations from the hexagon to the projective space $\mathbf{PG}(6, \mathbb{K})$, we obtain that the image of the point $(0,0,0,0,0)$ under the map $((x, y, z))$ defined in Subsection 7.7.7 has $x_0 = 0$ in $\mathbf{PG}(6, \mathbb{K})$ **if and only if** x, y, z is a solution of the given equation. $\square$

We will now turn $\mathcal{U}$ into a non-trivial linear space (a **linear space** is a point–line geometry in which every two points lie on a unique line; it is **non-trivial** if all lines contain more than two points. For a recent survey of linear spaces, see

DELANDTSHEER [1995]). The geometric approach that we take is due to DE SMET & VAN MALDEGHEM [1996].

Consider two arbitrary points p, q of $\mathcal{U}$. Let L be the projection of q^ρ onto p and M the projection of p^ρ onto q. We call the regulus $R(L, M)$ (cf. Subsection 1.9.16 on page 46) the $\mathcal{U}$-**line-regulus defined by the pair** $\{p, q\}$.

7.7.16 Lemma. *Each line of the $\mathcal{U}$-line-regulus defined by any pair of points of $\mathcal{U}$ contains a point of $\mathcal{U}$.*

Proof. We use the notation preceding the statement of the lemma. Let N be any line of $R(L, M)$. Then N^ρ belongs to the point-regulus $R(L^\rho, M^\rho)$. By the distance-3-regularity of the points of Γ, we have $\delta(N, N^\rho) = 3$. Hence the projection of N^ρ onto N is an absolute point for ρ. $\qquad\square$

Let p and q be two distinct points of $\mathcal{U}$. The set of points of $\mathcal{U}$ incident with an element of the $\mathcal{U}$-line-regulus defined by the pair $\{p, q\}$ will be denoted by $\mathcal{B}(p, q)$ and called a **block (of $\mathcal{U}$)**.

7.7.17 Lemma. *With the above notation, we have $\mathcal{B}(p, q) = \mathcal{B}(x, y)$ for all $x, y \in \mathcal{B}(p, q)$, $x \neq y$.*

Proof. Clearly, we may assume that $y = q$. So let $x \in \mathcal{B}(p, q)$. Suppose that x lies on the line $N \in R(L, M)$, where L and M are the projections of q^ρ and p^ρ, respectively. From the proof of Lemma 7.7.16 it follows that $\delta(L, x^\rho) = 4$, in other words, the projection of x^ρ onto p equals L. Clearly, the projection of p^ρ onto x equals N, hence the $\mathcal{U}$-line-regulus defined by the pair $\{p, x\}$ is equal to $R(L, N) = R(L, M)$, which completes the proof of the lemma. $\qquad\square$

Recall from page 311 that a finite linear space is called a **unital** if there is a positive integer r such that there are $r + 1$ points on each line and there are $r^3 + 1$ points in total. If we denote the set of all blocks of $\mathcal{U}$ by $\mathcal{B}$, then we have practically shown the first assertion of the following theorem in the previous paragraph:

7.7.18 Theorem (De Smet & Van Maldeghem [1996]).

(i) *The space $(\mathcal{U}, \mathcal{B}, \in)$ is a non-trivial linear space. If $\mathcal{U}$ is finite, then it is a unital.*

(ii) *The set of fixed points in $\mathcal{U}$ of an involution belonging to the automorphism group* **Ree**$(\mathbb{K}, \theta)$ *of $\mathcal{U}$ is either empty or the set of points on a block. All blocks arise in this way.*

Proof. For the second assertion of (i), it suffices to remark that, if $|\mathbb{K}| = q$ is finite, then $|\mathcal{U}| = q^3 + 1$ by Proposition 7.2.3, and each block contains $|R(L, M)| = q + 1$ points. We now prove (ii).

It is enough to show the assertion for *some* block; by conjugation, all blocks will then arise in this way.

Suppose σ is an involution of $\mathbf{Ree}(\mathbb{K}, \theta)$ fixing at least one point. After a suitable coordinatization (as previously), we may suppose that σ fixes (∞). Hence σ can be written as $((x, x', x''; X))$, $x, x', x'' \in \mathbb{K}$, $X \in \mathbb{K}^\times$. Expressing that σ has order 2, we obtain

$$\begin{cases} 0 & = Xx + x, \\ 0 & = X^{\theta+1}x' + x' + Xx^{\theta+1}, \\ 0 & = X^{\theta+2}x'' + x'' + Xxx' - X^{\theta+1}x'x - Xx^{\theta+2}, \\ 1 & = X^2, \end{cases}$$

$((0,0,0;1))$ being the identity of the group. If $X = 1$, then it readily follows that $x = x' = x'' = 0$, a contradiction. Hence $X = -1$ and $x' = -x^{\theta+1}$. Conjugating $\sigma = ((x, -x^{\theta+1}, x''; -1))$ by $((x, -x^{\theta+1}, x'' - x^{\theta+2}; 1))$, we obtain $((0,0,0;-1))$, which is nothing other than the map h_{-1} of Subsection 7.7.7. It is now easy to calculate the fixed points for h_{-1} and one finds $(0, a''^\theta, 0, a'')$, $a'' \in \mathbb{K}$. These are precisely the points of $\mathcal{U}$ on the lines $[0, l, 0, 0]$, which constitute the line-regulus $R([0], [0,0,0,0])$. And the latter is the $\mathcal{U}$-line-regulus defined by the pair $\{(\infty), (0,0,0,0,0)\}$. This completes the proof of the theorem.   □

Property (ii) of the previous theorem was used to define the blocks in Ree–Tits unitals by LÜNEBURG [1966]. The Ree–Tits ovoids themselves were discovered by TITS [1960], where, in fact, the unitals are also implicitly present in the description of the point stabilizer.

7.7.19 Simple Ree groups

We now comment on the proof of Theorem 7.7.8. Using the explicit form of $\mathbf{Ree}(\mathbb{K}, \theta)_{(\infty)}$, one deduces readily that the stabilizer of an ordered pair of points of the Ree–Tits unital (e.g. the points (∞) and $(0,0,0,0)$) is isomorphic to the multiplicative group of $\mathbb{K}$. However, restricted to its action on the block B containing these two points, we have a group isomorphic to $\mathbb{K}^\times / \langle -1 \rangle$, the element -1 of $\mathbb{K}^\times$ corresponding to the involution fixing B pointwise. Consider now the subgroup N of $\mathbf{Ree}(\theta)$ generated by all unipotent elements. Clearly, N acts doubly transitively on $\mathcal{O}$ and $N_{(\infty)}$ is solvable. Also, by a direct computation $[N, N] = N$ (given $|\mathbb{K}| > 3$) since $[N_{(\infty)}, N_{(\infty)}]$ contains all unipotent elements of $N_{(\infty)}$. Hence N is simple. Looking at the stabilizer H in N of a block B of $\mathcal{O}$ (we may view B as a $\mathbf{PG}(1, \mathbb{K})$), we see that N induces in B a group containing $\mathbf{PSL}_2(\mathbb{K})$. Also, it is easy to exhibit an involution r in H which does have fixed points in $\mathcal{O}$; consequently r fixes some block B' pointwise. Switching the roles of B and B', we now see that H contains $\mathbf{PSL}_2(\mathbb{K})$ and the map $x \mapsto -x$. In the finite case -1 is not a square in $GF(3^{2e+1})$, hence H is isomorphic to $\mathbf{PGL}_2(3^{2e+1})$. Consequently, the finite Ree groups are simple. In the general case, $\mathbf{Ree}(\mathbb{K}, \theta)/N \cong \mathbb{K}^\times/M$, where M is a subgroup of the multiplicative group $\mathbb{K}^\times$ containing all squares and -1. I refer the reader to TITS [1960] for more information.

7.7.20 Finite classical and semi-classical unitals

In the next subsections we mention some other results concerning the finite classical and semi-classical ovoids in $\mathsf{H}(q)$. We use the notation of Subsection 7.3.8 and so denote a Hermitian ovoid in $\mathsf{H}(q)$, q any prime power, by $U_H(q)$ — and we call it a **classical unital** (we justify the word *unital* in a second) — and a Ree–Tits unital in $\mathsf{H}(q)$, $q = 3^{2e+1}$, by $U_R(q)$ — and call it a **semi-classical unital**. To make $U_H(q)$ into a unital, we remark that from the geometric construction of $U_H(q)$ it follows that (though dualizing), whenever two points x and y belong to the unital, then all points of the point-regulus $R(x,y)$ also belong to $U_H(q)$. By definition, these reguli are the blocks of the classical unital. It is easily seen that we indeed obtain a unital this way, and that it is isomorphic to the Hermitian unital defined in Subsection 7.5.3.

A first noteworthy result is a partial analogue to Theorem 7.6.27. It is due to DE SMET & VAN MALDEGHEM [1996], [1995]; see also DE SMET [1994]. In (*iii*) of the next theorem, we view a semi-classical unital as a set of flags in the obvious way.

7.7.21 Theorem (De Smet & Van Maldeghem [1996], [1995]).

(*i*) *Two distinct classical unitals in* $\mathsf{H}(q)$ *have either* 1 *or* $q+1$ *points in common. In the latter case, the intersection is a block in both unitals.*

(*ii*) *A classical unital and a semi-classical unital in* $\mathsf{H}(q)$ *have either* $q + \sqrt{3q}+1$, $q - \sqrt{3q}+1$ *or* $q+1$ *points in common. In the first two cases, there is a cyclic group preserving the two unitals and acting regularly on the intersection.*

(*iii*) *Two distinct semi-classical unitals of* $\mathsf{H}(q)$ *have* 0, 1, 2 *or* $q + 1$ *flags in common (the points of the latter do* <u>not</u> *form a block). Also, there are exactly* q^2 *semi-classical unitals meeting a given unital in precisely* $q + 1$ *flags.* $\square$

Of course, in (*ii*) and (*iii*), we necessarily have $q = 3^{2e+1}$. $e \in \mathbb{N}$.

The previous theorem allows one to describe geometrically the maximal subgroups of $\mathbf{Ree}(q)$ as found by KLEIDMAN [1988]. Indeed, if $M \leq \mathbf{Ree}(q)$ is a maximal subgroup, then either M stabilizes a point of the corresponding Ree–Tits unital, or M stabilizes a block of the unital, or M stabilizes a subunital (defined over a maximal subfield of $\mathbf{GF}(q)$), or M stabilizes the intersection of a classical unital with a semi-classical one. In the latter case, M is isomorphic to $(q \pm \sqrt{3q} + 1) : 6$ or $(2^2 \times D_{\frac{q+1}{2}}) : 3$ (using the notation of CONWAY et al. [1985]).

The last case of Theorem 7.7.21 is particularly interesting since it allows the construction of a twisted field projective plane using the Ree–Tits unitals. A *twisted field projective plane* is a projective plane coordinatized by a certain *twisted field*. Such planes were discovered by ALBERT [1952], [1958a], [1960]; see also DEMBOWSKI [1968]. We state the result without proof, and we again view a semi-classical unital as a set of flags.

7.7.22 Theorem (De Smet & Van Maldeghem [1995]). *Let F be a fixed flag of $\mathsf{H}(q)$, $q = 3^{2e+1}$. Then "intersecting in 1 or $q + 1$ flags" is an equivalence relation in the set of all semi-classical unitals containing F. Every equivalence class contains $2q^3$ elements. Let us fix one such equivalence class $\mathcal{R}$. Let $\mathcal{S}$ be the set of q^2 semi-classical unitals meeting a fixed unital $U_R(q)$ of $\mathcal{R}$ in $q + 1$ flags. Then there exist exactly $q - 1$ other elements of $\mathcal{R}$ meeting all members of $\mathcal{S}$ also in $q + 1$ flags. Hence we obtain a set of q semi-classical unitals in $\mathcal{R}$, which we call a sleeper. The set $\mathcal{R}$ is partitioned into $2q^2$ sleepers. The graph $\Delta(F)$ with as set of vertices the sleepers contained in $\mathcal{R}$ and such that two vertices are joined by an edge if every unital of the corresponding sleeper meets every unital of the other sleeper in $q+1$ flags, is the incidence graph of a bi-affine plane, i.e., an affine plane with one parallel class of lines removed. The corresponding projective plane has a ternary field defined by the operation*

$$T(m, a, b) = -(m^\phi) \cdot (a^\phi)^\theta - (a^\phi) \cdot (m^\phi)^\theta + b,$$

where $a^\phi = b$ **if and only if** *$b + b^\theta = a$. This plane is non-Desarguesian* **if and only if** *$q \neq 3$.*   □

7.7.23 Some further results in the finite case

THAS [1981] shows that in the finite case, every ovoid of $\mathsf{H}(q)$ is also an **ovoid** of $\mathbf{Q}(6, q)$, i.e., a set of points of $\mathbf{Q}(6, q)$ such that every plane of $\mathbf{Q}(6, q)$ is incident with exactly one point of that set, and, conversely, every ovoid of $\mathbf{Q}(6, q)$ is an ovoid of every split Cayley hexagon with a standard embedding in $\mathbf{Q}(6, q)$. In particular, every Ree–Tits ovoid is an ovoid of $\mathbf{Q}(6, q)$. This is no longer true in the infinite case. For example, in the non-perfect case there are planes of $\mathbf{Q}(6, \mathbb{K})$ meeting $\mathsf{H}(\mathbb{K}, \mathbb{K}')$ in exactly one — non-absolute — point!

There are also some other ovoids and spreads known in $\mathsf{H}(q)$. We survey them briefly.

1. Let $q = 3^e$, and consider any ovoid $\mathcal{U}$ of $\mathsf{H}(q)$. We may then consider the image $\mathcal{U}^\sigma$ of $\mathcal{U}$ under an automorphism σ of $\mathbf{Q}(6, q)$ which does not preserve $\mathsf{H}(q)$, and interpret the set $\mathcal{U}^\sigma$ again in $\mathsf{H}(q)$. We obtain a new ovoid $\mathcal{U}^\sigma$ in $\mathsf{H}(q)$. Then we can apply a correlation to obtain a spread $\mathcal{S}$ of $\mathsf{H}(q)$. One special case is worth mentioning. By BLOEMEN, THAS & VAN MALDEGHEM [19**], it is possible to start with a Hermitian ovoid $U_H(q)$ and to choose σ such that the spread $\mathcal{S}$ has a unique line L for which the following property holds: whenever $M \in \mathcal{S}$, then all elements of the regulus $R(L, M)$ belong to $\mathcal{S}$. We say that $\mathcal{S}$ is **locally Hermitian in** L. If we consider a point x on L and the set of $1 + q^2$ distance-2-traces of $\mathsf{H}(q)$ through p which meet exactly $1 + q$ lines of $\mathcal{S}$ (these $1 + q$ lines form a regulus), then this set of $1+q^2$ lines constitutes an ovoid in the generalized quadrangle $\mathsf{H}(q)_x^\square$. Ovoids thus arising are isomorphic to those of THAS & PAYNE [1994]; see again BLOEMEN, THAS & VAN MALDEGHEM [19**].

2. One can easily calculate (see BLOEMEN, THAS & VAN MALDEGHEM [19**])
   that, using coordinates, the set

   $$\{[\infty]\} \cup \{[\gamma b', -\gamma k'', k', b', k'']|k', b', k'' \in \mathbf{GF}(q)\},$$

   for any non-square γ, is a classical spread in $\mathsf{H}(q)$. A little distortion now
   yields new spreads for $q \equiv 1 \bmod 3$, namely, the set

   $$\mathcal{S}_{[9]} = \{[\infty]\} \cup \{[9\gamma b', -\gamma k'', k', b', k'']|k', b', k'' \in \mathbf{GF}(q)\}$$

   is a spread of $\mathsf{H}(q)$, not isomorphic to a previous mentioned one, see *op. cit.*,
   where it is also shown that $\mathcal{S}_{[9]}$ is locally Hermitian in $[\infty]$.

BLOEMEN, THAS & VAN MALDEGHEM [19**] have also shown that a spread $\mathcal{S}$ of
$\mathsf{H}(q)$ is classical **if and only if** it is locally Hermitian in at least two of its elements.

Let us also mention that HÖLZ [1981] defines additional blocks in $U_H(q)$, q odd, to
obtain a $2 - (q^3 + 1, q + 1, q + 2)$-design such that the derived designs become isomorphic to the quadrangles of Payne type obtained from $\mathsf{W}(q)$; see Subsection 3.7.2 on
page 123. For the Ree–Tits unitals, a $2 - (q^3 + 1, q + 1, q + 2)$-design may be defined
as follows. For any line of $\mathsf{H}(q)$ not incident with a point of the Ree–Tits unital,
an additional block is defined as the set of points of the unital at distance 3 from
that line. This design, however, has no generalized quadrangle as derivation. It was
constructed algebraically by ASSMUS & KEY [1989]; the geometric description we
have given here is due to DE SMET & VAN MALDEGHEM [1993b].

7.7.24 Remark. One question that we have not considered in this chapter concerns the full automorphism group of the geometric structures associated to the
polarities. More exactly: what is the full automorphism group of a Suzuki–Tits
ovoid, furnished with plane sections; what is the automorphism group of a Ree–
Tits ovoid, furnished with the blocks of the corresponding unital? The answer to
both questions is: the automorphism group is the stabilizer of the ovoid in the full
automorphism group of the corresponding mixed n-gon, $n = 4, 6$, given that the
base field contains more than two elements. This can be shown using results of
TITS (unpublished, course at Collège de France, 1997) on rank 1 groups.

7.8 Amalgamations

In this section, we want to give complementary information with respect to
the chapter "Generalized quadrangles as amalgamations of projective planes" in
PAYNE & THAS [1984]. In particular, we will describe the amalgamation procedure completely geometrically. As for amalgamating generalized quadrangles to
obtain a generalized hexagon, we restrict ourselves to the geometric part as well
since no new generalized hexagons have been discovered yet by this method and
so the algebraic conditions (which would be necessary to prove the existence of

a new hexagon) are not really important for us; for completeness' sake, though, we mention these conditions without proof. The results presented here are based on VAN MALDEGHEM & BLOEMEN [1993], except that we also include here the infinite case.

Generalized quadrangles as amalgamations of projective planes

Consider a generalized quadrangle Γ and suppose that Γ has a projective point p and a projective line L incident with p. A quadratic quaternary ring for Γ, when taking $(\infty) = p$ and $[\infty] = L$, basically consists of two ternary operations, each giving rise to a projective plane, namely, Γ_p^Δ and Γ_L^Δ. So algebraically Γ is completely determined by these two planes. Hence geometrically it should be possible to describe Γ in terms of these planes. That is exactly the content of what follows.

7.8.1 Definitions. Let $\Gamma^{(1)}$ and $\Gamma^{(2)}$ be two projective planes. Let $\{p, L^*\}$ and $\{p^*, L\}$ be distinguished flags in $\Gamma^{(1)}$ and $\Gamma^{(2)}$, respectively. Suppose there is a bijection θ from the set of lines of $\Gamma^{(1)}$ through p to the set of lines of $\Gamma^{(2)}$ through p^* and a bijection σ from the set of points of $\Gamma^{(2)}$ on L to the set of points of $\Gamma^{(1)}$ on L^*. Assume also that $(L^*)^\theta = L$ and $(p^*)^\sigma = p$. Given such a quadruple $(\Gamma^{(1)}, \Gamma^{(2)}; \theta, \sigma)$ (where the elements p, L, p^*, L^* are determined by θ and σ), we define the following geometry $\Gamma = (\mathcal{P}, \mathcal{L}, I)$. The elements of $\mathcal{P}$ are of two types:

(P1) the points of $\Gamma^{(1)}$;

(P2) the pairs (M_2, M_1), where M_2 is a line in $\Gamma^{(2)}$ not incident with p^*, where M_1 is a line of $\Gamma^{(1)}$ not incident with p, and where $(M_2 \cap L)^\sigma = M_1 \cap L^*$.

Similarly, the lines are also of two types:

(L1) the lines of $\Gamma^{(2)}$;

(L2) the pairs (x_1, x_2), where x_1 is a point in $\Gamma^{(1)}$ not incident with L^*, where x_2 is a point of $\Gamma^{(2)}$ not incident with L^*, and where $(x_1 p)^\theta = x_2 p^*$.

Incidence is defined as follows. The point a of Γ is incident with the line D of Γ if one of the following holds:

(I1) $a = x$ is of type (P1), $D = M$ is of type (L1), and either x is on L^* and $x^{\sigma^{-1}}$ is incident with M, or M passes through p^* and $M^{\theta^{-1}}$ is incident with x, or both conditions hold (in which case $(x, M) = (p, L)$);

(I2) $a = x$ is of type (P1), $D = (x_1, x_2)$ is of type (L2), and $x = x_1$;

(I3) $a = (M_2, M_1)$ is of type (P2), $D = M$ is of type (L1), and $M = M_2$;

(I4) $a = (M_2, M_1)$ is of type (P2), $D = (x_1, x_2)$ is of type (L2), and x_i is incident with M_i in $\Gamma^{(i)}$, $i = 1, 2$.

We denote the geometry Γ also by $\Gamma^{(1)} {}_\theta \uplus_\sigma \Gamma^{(2)}$. If it is a generalized quadrangle, then we say that $(\Gamma^{(1)}, \Gamma^{(2)}; \theta, \sigma)$ is **admissible** and we call $\Gamma^{(1)} {}_\theta \uplus_\sigma \Gamma^{(2)}$ an **amalgon**. The point p is called the **hinge point**; the line L the **hinge line** (of the amalgon).

Two quadruples $(\Gamma^{(1)}, \Gamma^{(2)}; \theta, \sigma)$ and $(\Gamma^{(3)}, \Gamma^{(4)}; \theta', \sigma')$ as above are called **equivalent** if there exist isomorphisms $g_i : \Gamma^{(i)} \to \Gamma^{(i+2)}$, $i = 1, 2$, such that $\theta g_2 = g_1 \theta'$ and $\sigma g_1 = g_2 \sigma'$. It is easily seen using standard arguments that "being equivalent" induces an equivalence relation.

Now let Γ be a generalized quadrangle with a projective point p and a projective line L incident with p. We can view the set of lines of Γ incident with p as a pencil of lines in both Γ_p^Δ and Γ_L^Δ (see Remark 1.9.19 on page 47); we may define a mapping θ from the set of lines in Γ_p^Δ incident with p to the set of lines in Γ_L^Δ incident with p as the identity. Similarly for a mapping σ from the set of points of Γ_L^Δ incident with L to the set of points of Γ_p^Δ incident with L. We say that the quadruple $(\Gamma_p^\Delta, \Gamma_L^\Delta; \theta, \sigma)$ is **naturally associated with** (Γ, p, L).

We now have the following result.

7.8.2 Theorem (Payne [1972], Van Maldeghem & Bloemen [1993]).

(i) *Given a generalized quadrangle Γ with a projective point p incident with a projective line L, then the quadruple $(\Gamma_p^\Delta, \Gamma_L^\Delta; \theta, \sigma)$ naturally associated with (Γ, p, L) is admissible and the amalgon $\Gamma_p^\Delta {}_\theta \uplus \sigma \Gamma_L^\Delta$ is isomorphic to Γ.*

(ii) *Given an admissible quadruple $(\Gamma^{(1)}, \Gamma^{(2)}; \theta, \sigma)$, then the hinge point p and the hinge line L of the amalgon $\Gamma^{(1)} {}_\theta \uplus_\sigma \Gamma^{(2)} =: \Gamma$ are projective. Also, the quadruple $(\Gamma_p^\Delta, \Gamma_L^\Delta; \theta', \sigma')$ naturally associated with (Γ, p, L) is equivalent to $(\Gamma^{(1)}, \Gamma^{(2)}; \theta, \sigma)$.*

Proof. This theorem has a very straightforward proof. Assertion (i) follows from the observation that a line (M_2, M_1) of type (L2) of the amalgon $\Gamma_p^\Delta {}_\theta \uplus_\sigma \Gamma_L^\Delta$ represents the point z of Γ incident with M_2 (this is a line of Γ meeting L in, say, the point y) and collinear with every point of M_1 (since these points form a trace in $p^\perp$ in Γ and y belongs to that trace). Also the dual holds and one can easily prove that this representation defines an isomorphism from $\Gamma_p^\Delta {}_\theta \uplus_\sigma \Gamma_L^\Delta$ to Γ.

As for (ii), let us show that the point p is projective. Suppose therefore that (M_2, M_1) is a point of Γ opposite p (the type of a point opposite p is indeed (P2) since all points of type (P1) are collinear with p in Γ). Incident with (M_2, M_1) are the lines (x_1, x_2) of type (L2) with $x_1 \, \mathbf{I} \, M_1$ in $\Gamma^{(1)}$, and also the line M_2 as a line of type (L1). From the incidence relation in Γ, it now follows easily that only the points x_1 of type (P1), with $x_1 \, \mathbf{I} \, M_1$ in $\Gamma^{(1)}$, are incident with a line N of Γ incident with (M_1, M_1) (this is clear for N of type (L2); if $N = M_2$ has type (L1), then by definition $(M_2 \cap L)^\sigma = M_1 \cap L^*$ and a point y of type (P1) lies on M_2 if y belongs to $L^* = L^{\theta^{-1}}$ and $y^{\sigma^{-1}}$ belongs to M_2, hence y must belong to $(M_2 \cap L)^\sigma$, thus to $M_1 \cap L^*$). So we have shown that the traces in $p^\perp$ are precisely the lines of $\Gamma^{(1)}$ not incident with p in $\Gamma^{(1)}$. This proves that p is a regular point and that Γ_p^{Δ} is isomorphic to $\Gamma^{(1)}$ and the isomorphism g_1 is induced by the identity on the point sets. So p is projective. Dually Γ_L^{Δ} is isomorphic to $\Gamma^{(2)}$ and the isomorphism g_2 is induced by the identity on the lines (and so L is projective).

Let $(\Gamma_p^{\Delta}, \Gamma_L^{\Delta}; \theta', \sigma')$ be naturally associated with (Γ, p, L). Let x be a point of Γ on L. Since σ' is the identity and $x^{g_1} = x$, we have to prove that $x^{g_2} = x^{\sigma^{-1}}$. Now g_2 is induced by the identity on the lines of $\Gamma^{(2)}$ and a line M_2 of $\Gamma^{(2)}$, $M_2 \neq L$, is incident with a point y (of $\Gamma^{(2)}$) on L **if and only if** the point y^{σ} is incident with M_2 in Γ. Applying g_2^{-1}, we obtain that $(y^{\sigma})^{g_2^{-1}} = y$. Putting $y = x^{\sigma^{-1}}$, the result follows. Dually, one shows $M^{\theta' g_2} = M^{g_1 \theta}$, for all lines M of Γ through p. The theorem is proved. $\square$

7.8.3 Corollary. *The admissible quadruples* $(\Gamma^{(1)}, \Gamma^{(2)}; \theta, \sigma)$ *and* $(\Gamma^{(3)}, \Gamma^{(4)}; \theta', \sigma')$ *are equivalent* **if and only if** *there exists an isomorphism between the amalgons* $\Gamma^{(1)}\,_{\theta} \uplus_{\sigma} \Gamma^{(2)}$ *and* $\Gamma^{(3)}\,_{\theta'} \uplus_{\sigma'} \Gamma^{(4)}$ *which maps the hinge points to each other and the hinge lines to each other.*

Proof. This follows from standard arguments and the previous theorem. $\square$

So much for the general theory about amalgamating two projective planes. Now we consider briefly some examples in — mainly — the finite case.

7.8.4 Examples. The problem of finding admissible quadruples is basically equivalent to finding two ternary operations T_1 and T_2 on a set R which both define a projective plane and such that the operations defined by

$$\begin{cases} \Psi_1(k, a, l, a') & := & T_1(a, k, l), \\ \Psi_2(k, a, l, a') & := & T_2(k, a, a'), \end{cases}$$

form a quadratic quaternary ring. The symplectic quadrangle is clearly of that shape, given that the field over which it is defined is perfect. The example of Section 3.8.1 on page 133 is not of this type, because the ternary operation $a \cdot a \cdot k + l$ is not bijective for $l = 0$ and $k = 1$.

In the finite case, one might take isomorphic Desarguesian planes of the same order s and then PAYNE [1977] shows that s is even. Easy examples of this type are provided by the following quadratic quaternary ring $(\mathbf{GF}(2^e), \mathbf{GF}(2^e), \Psi_1, \Psi_2)$:

$$\begin{cases} \Psi_1(k, a, l, a') & = & a^{2^i} k + l, \\ \Psi_2(k, a, l, a') & = & ka + a', \end{cases}$$

where $2^i - 1$ is relatively prime to $2^e - 1$. For $e = 3$, this provides two non-isomorphic examples: the classical one (for $i = 2$) and a non-classical one (for $i = 4$). Both are self-polar. VAN MALDEGHEM & BLOEMEN [1993] show that these two examples are the only generalized quadrangles of order 8 having an incident point–line pair $\{p, L\}$ with both p and L regular.

For more information about amalgons isomorphic to generalized quadrangles, especially the algebraic side of the story, we refer the reader to PAYNE [1972], [1977] and PAYNE & THAS [1984].

Generalized hexagons as amalgamations of generalized quadrangles

Consider a generalized hexagon Γ and suppose that Γ has a polar point p and a polar line L incident with p. A hexagonal sexternary ring for Γ, when taking $(\infty) = p$ and $[\infty] = L$, basically consists of two quadratic quaternary operations, each giving rise to a generalized quadrangle, namely, $\Gamma_p^\square$ and $\Gamma_L^\square$. So algebraically Γ is completely determined by these two quadrangles. As for generalized quadrangles arising from planes, it is possible to describe Γ geometrically in terms of these quadrangles.

7.8.5 Definitions. Let $\Gamma^{(1)}$ and $\Gamma^{(2)}$ be two generalized quadrangles. Let $\{p, L^*\}$ and $\{p^*, L\}$ be distinguished flags in $\Gamma^{(1)}$ and $\Gamma^{(2)}$ respectively, and suppose that p and L are both projective in their respective quadrangle. Suppose there are bijections θ and σ from the set of lines of $\Gamma^{(1)}$ through p to the set of lines of $\Gamma^{(2)}$ through p^*, and from the set of points of $\Gamma^{(2)}$ on L to the set of points of $\Gamma^{(1)}$ on L^*, respectively. Assume also that $(L^*)^\theta = L$ and $(p^*)^\sigma = p$. Given such a quadruple $(\Gamma^{(1)}, \Gamma^{(2)}; \theta, \sigma)$ (where the elements p, L, p^*, L^* are determined by θ and σ), we define the following geometry $\Gamma = (\mathcal{P}, \mathcal{L}, I)$. The elements of $\mathcal{P}$ are of four types:

[P1] the points of $\Gamma^{(1)}$ collinear with p (this includes p);

[P2] the pairs (M_2, M_1), where $M_2 \in \Gamma_2^{(2)}(L)$ and $M_1 \in \Gamma_2^{(1)}(L^*)$ with $(M_2 \cap L)^\sigma = M_1 \cap L^* \neq p$;

[P3] the pairs (x_2, M_1), with $M_1 \in \Gamma_4^{(1)}(L^*)$ and $x_2 \in \Gamma_2^{(2)}(p^*) \cap \Gamma_1^{(2)}((\text{proj}_p M_1)^\theta)$;

[P4] the pairs (x_1, x_2), where x_1 is opposite p in $\Gamma^{(1)}$, x_2 is opposite p^* in $\Gamma^{(2)}$ and $(\text{proj}_L x_2)^\sigma = \text{proj}_{L^*} x_1$.

Dually, the lines are also of four types:

[L1] the lines of $\Gamma^{(2)}$ concurrent with L (this includes L itself);

[L2] the pairs (x_1, x_2), where $x_1 \in \Gamma_2^{(1)}(p)$ and $x_2 \in \Gamma_2^{(2)}(p^*)$, with $(x_1 p)^\theta = x_2 p^* \neq L$;

[L3] the pairs (M_1, x_2), where $x_2 \in \Gamma_4^{(2)}(p^*)$ and $M_1 \in \Gamma_2^{(1)}(L^*) \cap \Gamma_1^{(1)}((\text{proj}_L x_2)^\sigma)$;

[L4] the pairs (M_2, M_1), where M_2 is opposite L in $\Gamma^{(2)}$, M_1 is opposite L^* in $\Gamma^{(1)}$ and $(\text{proj}_p M_1)^\theta = \text{proj}_{p^*} M_2$.

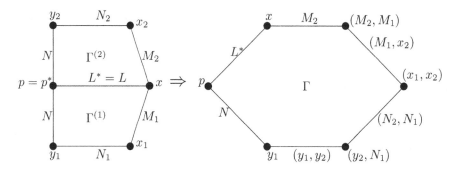

Figure 7.1.

Incidence is defined as follows. The point a of Γ is incident with the line D of Γ if one of the following holds (see also Figure 7.1, where we identify objects which correspond under θ or σ):

[I1] $a = x$ is of type [P1], $D = N$ is of type [L1], and either x is on L^* and $x^{\sigma^{-1}}$ is incident with N, or N passes through p^* and $N^{\theta^{-1}}$ is incident with x, or both conditions hold (in which case $(x, N) = (p, L)$);

[I2] $a = y$ is of type [P1], $D = (y_1, y_2)$ is of type [L2], and $y = y_1$;

[I3] $a = (M_2, M_1)$ is of type [P2], $D = M$ is of type [L1], and $M = M_2$;

[I4] $a = (M_2, M_1)$ is of type [P2], $D = (M_1', x_2)$ is of type [L3], $M_1 = M_1'$, and x_2 is incident with M_2 in $\Gamma^{(2)}$;

[I5] $D = (y_1, y_2)$ is of type [L2], $x = (y_2', N_1)$ is of type [P3], $y_2 = y_2'$, and y_1 is incident with N_1 in $\Gamma^{(1)}$;

[I6] $a = (y_2, N_1)$ is of type [P3], $D = (N_2, N_1')$ is of type [L4], $N_1 = N_1'$, and y_2 is incident with N_2 in $\Gamma^{(2)}$;

[I7] $D = (M_1, x_2)$ is of type [L3], $x = (x_1, x_2')$ is of type [P4], $x_2 = x_2'$, and x_1 is incident with M_1 in $\Gamma^{(1)}$;

[I8] $x = (x_1, x_2)$ is of type [P4], $D = (N_2, N_1)$ is of type [L4], x_1 is incident with N_1 in $\Gamma^{(1)}$, and x_2 is incident with N_2 in $\Gamma^{(2)}$.

We denote the geometry Γ also by $\Gamma^{(1)} {}_\theta \uplus_\sigma \Gamma^{(2)}$. If it is a generalized hexagon, then we say that $(\Gamma^{(1)}, \Gamma^{(2)}; \theta, \sigma)$ is **admissible** and we call $\Gamma^{(1)} {}_\theta \uplus_\sigma \Gamma^{(2)}$ an **amalgon**. The point p is called the **hinge point**; the line L the **hinge line** of the amalgon.

Two quadruples $(\Gamma^{(1)}, \Gamma^{(2)}; \theta, \sigma)$ and $(\Gamma^{(3)}, \Gamma^{(4)}; \theta', \sigma')$ as above are called **equivalent** if there exist isomorphisms $g_i : \Gamma^{(i)} \to \Gamma^{(i+2)}$, $i = 1, 2$, such that $\theta g_2 = g_1 \theta'$

and $\sigma g_1 = g_2 \sigma'$. It is again easily seen that "being equivalent" induces an equivalence relation.

Now let Γ be a generalized hexagon with a polar point p and a polar line L incident with p. We can view the set of lines of Γ incident with p as a pencil of lines in both $\Gamma_p^{\square}$ and $\Gamma_L^{\square}$ (see Remark 1.9.19 on page 47); we may define a mapping θ from the set of lines in $\Gamma_p^{\square}$ incident with p to the set of lines in $\Gamma_L^{\square}$ incident with p as the identity. Similarly for a mapping σ from the set of points of $\Gamma_L^{\square}$ incident with L to the set of points of $\Gamma_p^{\square}$ incident with L. We say that the quadruple $(\Gamma_p^{\square}, \Gamma_L^{\square}; \theta, \sigma)$ is **naturally associated with** (Γ, p, L).

Inspired by (but differently from) the notion of an anti-regular point in the proof of Theorem 6.8.9 (see page 289), we call the ordered pair (x, L) (with $\{x, L\}$ a flag) of a generalized polygon **anti-regular** if any three non-collinear points in $x^{\perp}$ one of which is incident with L are contained in a unique trace x^y, for some point y opposite x. If also y is always unique, then we say that (x, L) is a **sharply anti-regular pair**. Dually, one defines a **(sharply) anti-regular pair** (L, x).

Similar to Theorem 7.8.2, we have the following result.

7.8.6 Theorem (Van Maldeghem & Bloemen [1993]).

(i) *Given a generalized hexagon Γ with a polar point p incident with a polar line L, then the quadruple $(\Gamma_p^{\triangle}, \Gamma_L^{\triangle}; \theta, \sigma)$ naturally associated with (Γ, p, L) is admissible and the amalgon $\Gamma_p^{\square} {}_\theta \uplus_\sigma \Gamma_L^{\square}$ is isomorphic to Γ.*

(ii) *Given an admissible quadruple $(\Gamma^{(1)}, \Gamma^{(2)}; \theta, \sigma)$, then the hinge point p and the hinge line L of the amalgon $\Gamma^{(1)} {}_\theta \uplus_\sigma \Gamma^{(2)} =: \Gamma$ are polar elements. Furthermore, the ordered pairs (L^θ, p) and (p^σ, L) are sharply anti-regular pairs in $\Gamma^{(1)}$ and $\Gamma^{(2)}$, respectively; in particular the elements L^θ and p^σ are not regular in $\Gamma^{(1)}$ or $\Gamma^{(2)}$, respectively. If Γ is finite and classical of order (s, t), then $s = t$ is odd. Also, the identity on $p^{\perp}$ in Γ generates an isomorphism from $\Gamma_p^{\triangle}$ to $(\Gamma^{(1)})_p^{\triangle}$ and, dually, the identity map on $L^{\perp}$ generates an isomorphism from $\Gamma_L^{\triangle}$ to $(\Gamma^{(2)})_L^{\triangle}$. Finally, the quadruple $(\Gamma_p^{\triangle}, \Gamma_L^{\triangle}; \theta', \sigma')$ naturally associated with (Γ, p, L) is equivalent to $(\Gamma^{(1)}, \Gamma^{(2)}; \theta, \sigma)$.*

Proof. The proof of most of this theorem is straightforward and in fact similar to the proof of Theorem 7.8.2.

Let us, concerning (ii), only explicitly show that (p^σ, L) is a sharply anti-regular pair in $\Gamma^{(2)}$, and that $s = t$ is odd if Γ is classical and finite of order (s, t).

Consider three pairwise non-collinear points u_0, u_1 and u_2 in $\Gamma^{(2)}$, collinear with $p^* = p^{\sigma^{-1}}$, $u_0 \, \mathbf{I} \, L$. Suppose $u_i \, \mathbf{I} \, M_i^\theta$ (in $\Gamma^{(2)}$), $i = 1, 2$. For $i = 1, 2$, choose points $v_i \, \mathbf{I} \, M_i$ in $\Gamma^{(1)}$, $v_i \neq p$, such that v_1, v_2 and u_0^σ belong to the same trace in $p^{\perp}$. Finally, choose any line M in $\Gamma^{(1)}$ incident with u_1, but not with p.

We claim that, provided the given quadruple is admissible, the distance in the amalgon of the point (u_1, M) of type [P3] to the line (v_2, u_2) of type [L2] is bigger than 4. Indeed, there is a path from p to (u_1, M) of length 4, and to (v_2, u_2) of length 3; these paths are disjoint, up to p, and hence the claim. So there exists a unique point (z_1, z_2) of the amalgon, necessarily of type (P4), collinear with (u_1, M) and at distance 3 from (v_2, u_2). Now z_1 must be collinear with v_2 and incident with M, hence collinear with u_0^σ. Therefore, z_2 is collinear with u_0. But $\delta((z_1, z_2), (u_1, M)) = 2$ implies that z_2 is collinear with u_1 and similarly $\delta((z_1, z_2), (v_2, u_2)) = 3$ implies that z_2 is collinear with u_2 in the amalgon. The uniqueness of (z_1, z_2) implies the uniqueness of z_2 in $\Gamma^{(2)}$ with the property that $z_2 \in \{u_0, u_1, u_2\}^\perp$. Hence (p^σ, L) is a sharply anti-regular pair.

If $\Gamma^{(2)}$ is finite and classical, then, since $p^{\sigma^{-1}}$ clearly cannot be a regular point, and since $s = t$ if (s, t) is the order of $\Gamma^{(2)}$ (by the simple fact that the amalgon must have an order, necessarily $(s, s) = (t, t)$), we see that $\Gamma^{(2)}$ must be isomorphic to $Q(4, s)$, for s odd. $\square$

The following corollary is proved in a standard way, so we omit the proof.

7.8.7 Corollary. *The admissible quadruples* $(\Gamma^{(1)}, \Gamma^{(2)}; \theta, \sigma)$ *and* $(\Gamma^{(3)}, \Gamma^{(4)}; \theta', \sigma')$, *with* $\Gamma^{(i)}$ *generalized quadrangles*, $i = 1, 2, 3, 4$, *are equivalent* **if and only if** *there exists an isomorphism between the amalgons* $\Gamma^{(1)} \,_\theta \uplus_\sigma \Gamma^{(2)}$ *and* $\Gamma^{(3)} \,_{\theta'} \uplus_{\sigma'} \Gamma^{(4)}$ *which maps the hinge points to each other and the hinge lines to each other.* $\square$

7.8.8 Non-classical amalgons?

No finite amalgamations of quadrangles are known to give rise to non-classical hexagons. In fact, VAN MALDEGHEM & BLOEMEN [1993] show that amalgamating the classical generalized quadrangles of order $q \leq 9$, one can only obtain a classical hexagon isomorphic to either $H(3)$ or $H(9)$. For further algebraic results, we refer the reader to *op. cit.*, and to DE SMET [1994], who characterizes the classical hexagons of finite order 3^e by a geometric and an algebraic condition on the amalgamation.

It is possible to describe the amalgamation procedures explained in this section in a completely algebraic way, thus obtaining the conditions for a special class of sexternary rings to be hexagonal. Especially in the finite case, these conditions are not so difficult to check. But since no new examples have arisen from this approach, we simply write them down and refer the reader to VAN MALDEGHEM & BLOEMEN [1993] (see also BLOEMEN [1995]) for a proof.

7.8.9 Theorem (Van Maldeghem & Bloemen [1993]). *Let* $\{p, L^*\}$ *and* $\{p^*, L\}$ *be distinguished flags in, respectively, the quadrangles* $W(q)$ *and* $Q(4, q)$. *Given the standard embedding of* $W(q)$ *in* $\mathbf{PG}(3, q)$ *and the standard embedding of* $Q(4, q)$ *in* $\mathbf{PG}(4, q)$, *we can identify the set of lines* S_p (S_{p^*}) *of* $W(q)$ $(Q(4, q))$ *through* p (p^*) *distinct from* L^* (L) *with* $\mathbf{GF}(q)$; *similarly for the set of points* S_L (S_{L^*}) *on* L (L^*) *distinct from* p^* (p). *Let* θ (σ) *be a bijection from* S_p *to* S_{p^*} *(from* S_L *to*

S_{L^*}) *and identify* θ *and* σ *with the permutation they induce this way in* **GF**(q). *Then the quadruple* $(\mathsf{W}(q), \mathsf{Q}(4, q); \theta, \sigma)$ *is admissible* **if and only if**, *under the given conditions, the following equalities never occur simultaneously in* **GF**(q) *(where* (2) *and* (5) *must hold for* $(\phi, \psi) = (\theta^{-1}, \sigma)$ *and the others for* $(\phi, \psi) = (\theta^{-1}, \sigma)$ *and* $(\phi, \psi) = (\sigma^{-1}, \theta)$):

(1) Condition: $a \neq A \neq b \neq a$ and $K \neq k \neq L \neq K$;

$$\begin{cases} (K - k)^2(A - a) = (L - k)^2(A - b), \\ (K^\phi - k^\phi)(A^\psi - a^\psi) = (L^\phi - k^\phi)(A^\psi - b^\psi). \end{cases}$$

(2) Condition: $a \neq A \neq b \neq a$ and $K \neq k \neq L \neq K$;

$$\begin{cases} (K - k)^2(A - a) = (L - k)^2(A - b), \\ (K^\phi - k^\phi)(A^\psi - a^\psi)^2 = (L^\phi - k^\phi)(A^\psi - b^\psi)^2. \end{cases}$$

(3) Condition: $a \neq A \neq b \neq a$ and $K \neq k \neq L \neq K$;

$$\begin{cases} (K - k)^2(A - a) = (L - k)^2(A - b), \\ (L^\phi - K^\phi)(A^\psi - a^\psi)^2 = (L^\phi - k^\phi)(b^\psi - a^\psi)^2. \end{cases}$$

(4) Condition: $a \neq A \neq B \neq b \neq a \neq B \neq A \neq b$ and $k \neq K \neq P \neq L \neq k$;

$$\begin{cases} 0 = (K - k)^2(A - a) + (P - k)^2(B - A) + (L - k)^2(b - B), \\ 0 = (K^\phi - k^\phi)(A^\psi - a^\psi) + (P^\phi - k^\phi)(B^\psi - A^\psi) + \\ \qquad\qquad (L^\phi - k^\phi)(b^\psi - B^\psi), \\ 0 = (K^\phi - k^\phi)((A^\psi)^2 - (a^\psi)^2) + (P^\phi - k^\phi)((B^\psi)^2 - (A^\psi)^2) + \\ \qquad\qquad (L^\phi - k^\phi)((b^\psi)^2 - (B^\psi)^2). \end{cases}$$

(5) Condition: $B \neq X \neq A \neq a \neq B \neq Y \neq X \neq A \neq Y$ and $k \neq l \neq P \neq k \neq K \neq P \neq L \neq l$;

$$\begin{cases} (P - K)^2(X - A) + (P - k)^2(A - a) = \\ \qquad\qquad (P - L)^2(Y - B) + (P - l)^2(B - a), \\ (P^\phi - K^\phi)(X^\psi - A^\psi) + (P^\phi - k^\phi)(A^\psi - a^\psi) = \\ \qquad\qquad (P^\phi - L^\phi)(Y^\psi - B^\psi) + (P^\phi - k^\phi)(B^\psi - a^\psi), \\ (P - K)(X - A) + (P - k)(A - a) = \\ \qquad\qquad (P - L)(Y - B) + (P - l)(B - a), \\ (P^\phi - K^\phi)((X^\psi)^2 - (A^\psi)^2) + (P^\phi - k^\phi)((A^\psi)^2 - (a^\psi)^2) = \\ \qquad\qquad (P^\phi - L^\phi)((Y^\psi)^2 - (B^\psi)^2) + (P^\phi - k^\phi)((B^\psi)^2 - (a^\psi)^2). \end{cases}$$

Moreover, **if** *all these conditions are satisfied,* **then** *the automorphism group of the amalgon* Γ *obtained is transitive on the set* $\Gamma_6(p) \cap \Gamma_4(p')$ ($\Gamma_6(L) \cap \Gamma_4(L')$), *where* $p'\,\mathbf{I}\,L$, $p' \neq p$ ($L'\,\mathbf{I}\,p$, $L' \neq L$), *for every such* p' (L'). $\qquad\square$

Chapter 8

Projectivities and Projective Embeddings

8.1 Introduction

In this chapter, we aim to prove some of the main achievements in the theory of generalized polygons. First, we want to show what the little projective group and the groups of projectivities of some Moufang polygons (in particular, all finite classical polygons) look like; the latter generalizes a result of KNARR [1988]. Secondly, we want to classify all embeddings of a generalized quadrangle in a finite-dimensional projective space. In particular, such a quadrangle is a classical Moufang quadrangle. This is due to DIENST [1980a], [1980b], who extended a result of BUEKENHOUT & LEFÈVRE-PERCSY [1974] on projectively embedded finite generalized quadrangles to the infinite case. We take a somewhat different approach, avoiding the theory of *semi-quadratic sets*. We also mention some related results about embeddings of octagons, and about weak and lax embeddings of quadrangles and hexagons. These are due to THAS & VAN MALDEGHEM [1996], [19**a], [19**b], [19**c], STEINBACH [1996] and STEINBACH & VAN MALDEGHEM [19**a], [19**b]. The proofs where hexagons are involved use some very technical results about certain point sets in finite projective spaces, and this is beyond the scope of this book. Where finite generalized quadrangles are involved, one again uses the results of PAYNE & THAS [1984]. Since this area is still developing quickly, and since there seems to be a lot still to be done, we occasionally give a rough outline of a proof, sometimes a geometric proof of some parts.

In order to prepare the proof of Knarr's result (see above), we review the regularity properties of the Moufang polygons. Since we have to deal with the Ree–Tits octagon, we prove some more geometric properties of these geometries. Also, we show that for the symplectic quadrangles, for the mixed quadrangles of shape $Q(\mathbb{K}, \mathbb{K}'; \mathbb{K}, \mathbb{K}')$, for the split Cayley and mixed hexagons and for the finite twisted triality hexagons, the group induced by $\mathbf{PSL}_d(\mathbb{K})$ in the standard embedding in $\mathbf{PG}(d-1, \mathbb{K})$ is generated by all root elations of the polygon in question. For the finite Hermitian quadrangles, we have to add an extra condition; see below.

8.2 Some more properties of the Ree–Tits octagons

8.2.1 Definitions. Let $\mathbb{K}$ be a field of characteristic 2 admitting a Tits endomorphism σ and let $\mathsf{O}(\mathbb{K}, \sigma)$ be the corresponding Ree–Tits octagon. We have seen that $\mathsf{O}(\mathbb{K}, \sigma)$ contains full weak suboctagons which are all anti-isomorphic to the double of a Suzuki quadrangle over $\mathbb{K}$; see Theorem 3.6.3 on page 119. Let Γ be one such weak suboctagon, anti-isomorphic to the double of the Suzuki quadrangle $\mathsf{W}(\mathbb{K}, \sigma)$. Now notice that the points of Γ correspond to flags of $\mathsf{W}(\mathbb{K}, \sigma)$. Let us fix any Suzuki–Tits ovoid $\mathcal{O}$ on $\mathsf{W}(\mathbb{K}, \sigma)$ which corresponds to the Tits endomorphism σ, as in Theorem 7.6.2, see page 323. It is a straightforward exercise to check that the corresponding polarity is conjugate to the polarity given by swapping the parentheses and brackets in the coordinatization of Subsection 3.4.6 on page 103. We call every such polarity a **natural polarity** of $\mathsf{W}(\mathbb{K}, \sigma)$.

Now, we also know that the set of lines through any point of $\mathsf{O}(\mathbb{K}, \sigma)$ can be given the structure of a Suzuki–Tits ovoid with corresponding Tits endomorphism σ; see Section 6.9. The question naturally arising here is whether one of these ovoids is geometrically connected with $\mathcal{O}$. One way of finding out is by looking at the stabilizer G of $\mathcal{O}$ in the full automorphism group of $\mathsf{O}(\mathbb{K}, \sigma)$. It turns out that G fixes exactly one point of $\mathsf{O}(\mathbb{K}, \sigma)$. That point is the unique point in $\mathsf{O}(\mathbb{K}, \sigma)$ which lies at distance 4 from all points of $\mathcal{O}$. Moreover, it is the centre of a central elation of order 2 in $\mathsf{O}(\mathbb{K}, \sigma)$ which preserves Γ and induces a polarity in $\mathsf{W}(\mathbb{K}, \sigma)$ defining the Suzuki–Tits ovoid corresponding to $\mathcal{O}$. We call that point the **nucleus** of $\mathcal{O}$. We summarize and prove this in the following theorem.

8.2.2 Theorem (Joswig & Van Maldeghem [1995]). *Let $\mathcal{O}$ be a set of points of the Ree–Tits octagon $\mathsf{O}(\mathbb{K}, \sigma)$ contained in a full weak suboctagon $\Gamma \cong (2\,\mathsf{W}(\mathbb{K}, \sigma))^D$ and such that $\mathcal{O}$ corresponds to the set of flags of $\mathsf{W}(\mathbb{K}, \sigma)$ fixed by a natural polarity ρ. Then there exists a point $x_{\mathcal{O}}$ of $\mathsf{O}(\mathbb{K}, \sigma)$ uniquely determined by each one of the following three properties:*

(i) the point $x_{\mathcal{O}}$ is at distance 4 from each element of $\mathcal{O}$;

(ii) the point $x_{\mathcal{O}}$ is fixed by every collineation of $\mathsf{O}(\mathbb{K}, \sigma)$ stabilizing $\mathcal{O}$;

(iii) there exists an involutory central root elation (corresponding to a long root in the root system) with centre $x_{\mathcal{O}}$ stabilizing Γ and inducing the polarity ρ in $\mathsf{W}(\mathbb{K}, \sigma)$ corresponding to $\mathcal{O}$.

Proof. Without loss of generality we may take for Γ the suboctagon obtained from a coordinatization of $\mathsf{O}(\mathbb{K}, \sigma)$ by putting R_2 equal to 0; cf. the proof of Theorem 3.6.3. Also, we may take as polarity ρ the correlation induced by swapping parentheses with brackets. We obtain the following set of absolute flags (this is an easy calculation):

$$\{\{(\infty), [\infty]\}\} \cup \{\{(a, a^{\sigma+1} + a', a'), [a, a^{\sigma+1} + a', a']\} : a, a' \in \mathbb{K}\}.$$

According to the proof of Theorem 3.6.3 (page 119), the flag

$$\{(a, a^{\sigma+1} + a', a'), [a, a^{\sigma+1} + a', a']\}$$

of $\mathsf{W}(\mathbb{K}, \sigma)$ corresponds to the intersection point in Γ of the lines

$$[0, a, 0, a^{\sigma+1} + a', 0, a', 0] \text{ and } [a, 0, a^{\sigma+1} + a', 0, a', 0].$$

In view of the operations on page 119, this intersection point has coordinates $(a, 0, a^{\sigma+1} + a', 0, a', 0, a)$. Of course, the point (∞) also belongs to $\mathcal{O}$. Let G be the stabilizer of $\mathcal{O}$. If $x_\mathcal{O}$ exists, then it belongs to the distance-4-trace $(0^7)^{(\infty)}$, where with our usual convention (0^i) denotes an i-tuple of 0s, $i \in \mathbb{N}$. It is easily seen that this set consists of the point $(0, 0, 0)$ and the points $(k, 0, 0, 0)$, with $k \in R_2$ (remembering that $R_2 = \mathbb{K}_\sigma^{(2)}$). Considering $a \neq 0$ in a point of $\mathcal{O}$ having coordinates as above, we see that $(0, 0, 0)$ cannot be $x_\mathcal{O}$. On the other hand, the point $(k, 0, 0, 0)$ is at distance 4 from the point $(a, 0, a^{\sigma+1} + a', 0, a', 0, a)$ **if and only if** there exist $k'', k''' \in R_2$ and $b'' \in R_1 = \mathbb{K}$ such that the line $[k, 0, 0, 0, k'', b'', k''']$ is incident with the latter point. Considering Ψ_6, we deduce $0 = a + a \, \mathrm{N}(k)$, for all $a \in \mathbb{K}$. Hence $\mathrm{N}(k) = 1$. Considering Ψ_5, we see that $\mathrm{Tr}(k) = 0$. This implies $k_0^{\sigma+2} = 1$, where $k = (k_0, k_1)$. Applying $\sigma - 1$, it follows that $k_0^\sigma = 1$, hence $k_0 = 1 = k_1$. From the octogonal octanary ring we infer that $[(1, 1), 0, 0, 0, (a, a')^\sigma, a^{\sigma+1} + a', (a, a^{\sigma+1})]$ is incident with $(a, 0, a^{\sigma+1} + a', 0, a', 0, a)$, hence $x_\mathcal{O}$ is the point $((1, 1), 0, 0, 0)$. We have shown (i).

Since $x_\mathcal{O}$ is unique with respect to a geometric (metric) property, it will be fixed by all collineations of $\mathsf{O}(\mathbb{K}, \sigma)$ preserving $\mathcal{O}$.

Now consider the generalized homology h_A of $\mathsf{O}(\mathbb{K}, \sigma)$ mapping the point with coordinates $(a, l, a', l', a'', l'', a''')$ to

$$(Aa, A \otimes l, A^{\sigma+1}a', A^\sigma \otimes l', A^{\sigma+1}a'', A \otimes l'', Aa'''),$$

for arbitrary $A \in \mathbb{K}^\times$; see the proof of Proposition 4.5.12 (page 151). Clearly $\mathcal{O}$ is stabilized by h_A, for all $A \in \mathbb{K}^\times$. From the above expression of h_A, we immediately deduce that no point with an odd number of coordinates is fixed by all h_A, and no line with an even number of coordinates, except if all coordinates are 0. By the proof of Proposition 4.5.12, the collineation h_A maps the point (k, b, k', b', k'', b'') onto

$$(k, Ab, A \otimes k', A^{\sigma+1}b', A^\sigma \otimes k'', A^{\sigma+1}b'').$$

Hence, if some point is fixed by all these generalized homologies, then it must have the form $(k, 0^5)$, $(k, 0, 0, 0)$ or $(k, 0)$. Now, the points (∞) and (0^7) were arbitrarily chosen here, so we can consider generalized homologies fixing any two other points of $\mathcal{O}$. Furthermore, these generalized homologies induce generalized homologies in $\mathsf{W}(\mathbb{K}, \sigma)$. It is not difficult to see that, in fact, we have the stabilizer in $\mathbf{Sz}(\mathbb{K}, \theta)$ of any pair of points. We conclude with Remark 7.6.8 on page 327 that $\mathbf{Sz}(\mathbb{K}, \sigma)$ is induced by $\mathrm{Aut}(\mathsf{O}(\mathbb{K}, \sigma))$ on $\mathcal{O}$ in its natural action. In particular, there is a transitive group on $\mathcal{O}$, which implies immediately that no point at distance at most 4 from one of the elements of $\mathcal{O}$ can be fixed by G. Hence neither $(k,0)$, nor $(k,0^5)$ can be fixed by G since the first lies at distance 2 from (∞) (for all $k \in R_1$) and the last at distance 2 from (0^7) (for all $k \in R_2$). Similarly the points (0) and (0^5) are ruled out. That leaves the points $(k,0,0,0)$ and $(0,0,0)$. But these exactly form the distance-4-trace $(0^7)^{(\infty)}$. Varying the two points defining this distance-4-trace over the set $\mathcal{O}$, we see that any point fixed by G must necessarily lie at distance 4 from all points of $\mathcal{O}$. We have proved (ii).

To prove (iii), we first remark that a central elation fixes every point at distance at most 4 from the centre, hence if some central elation fixes all elements of $\mathcal{O}$, its centre must be $x_{\mathcal{O}}$. It remains to show that there is a central elation with centre $x_{\mathcal{O}}$ satisfying the properties mentioned in (iii). Therefore, it suffices to prove that the γ-elation mapping $[\infty]$ to $[0]$, where $\gamma = ([(1,1)],((1,1),0),\dots,[(1,1),0^6])$, preserves Γ, i.e., maps $[0]$ back to $[\infty]$, or in yet other words, is of order 2.

Consider a γ'-elation α which maps $[0]$ to $[k]$, where $\gamma' = ([\infty],(0),[0,0],\dots,[0^6])$. Looking at the addition in $R_2 = \mathbb{K}_{\sigma}^{(2)}$, we see that α has order 2 **if and only if** $k = (0,k_1)$, hence geometrically this is true **if and only if** the circle in the block geometry $^{(\infty)}\mathsf{O}(\mathbb{K},\sigma)$ containing $[0]$, $[\infty]$ and $[k]$ has $[\infty]$ as corner (for definitions, see Section 6.9). So we have to show that the corner of the circle containing $[0]$, $[\infty]$ and $[(1,1)]$ has the latter as corner.

So consider the Suzuki quadrangle $\mathsf{W}(\mathbb{K},\sigma)$ with the usual coordinatization inherited from $\mathsf{W}(\mathbb{K})$. The Suzuki–Tits ovoid has the points (∞) and $(a,a^{\sigma+2}+a',a')$, for all $a,a' \in \mathbb{K}$. The circle through $(0,0,0)$, (∞) and $(1,0,1)$ is the set of points collinear with the point $(1,0)$. So we should prove that this point is on the absolute line through $(1,0,1)$. But that is precisely the line $[1,0,1]$ and so the theorem is proved. $\square$

Projecting $\mathcal{O}$ onto its nucleus $x_{\mathcal{O}}$, we see that the Suzuki–Tits ovoid is mapped injectively into the set of lines through $x_{\mathcal{O}}$. This mapping is bijective **if and only if** $\mathbb{K}$ is perfect; indeed, it is readily seen that, if the set of lines through $x_{\mathcal{O}}$ is parametrized by $R_2 \cup \{\infty\}$, then the image of the projection is parametrized by $R_2^{\sigma} \cup \{\infty\}$.

We now turn our attention to regularity in $\mathsf{O}(\mathbb{K},\sigma)$. We already know by virtue of Theorem 6.4.5 on page 259 and Theorem 6.4.6 on page 261 that $\mathsf{O}(\mathbb{K},\sigma)$ has no distance-i-regular point or line for $i = 2,3$.

8.2.3 Theorem (Joswig & Van Maldeghem [1995]). *Every Ree–Tits octagon is point-distance-4-regular and line-distance-4-regular.*

Proof. Let $\Gamma = \mathsf{O}(\mathbb{K},\sigma)$ be a Ree–Tits octagon. First we prove that Γ is line-distance-4-regular. Consider two opposite lines L_1 and L_2 in Γ. We claim that these are contained in a unique weak full suboctagon Γ' which is isomorphic to the dual of the double of the Suzuki quadrangle $W(\mathbb{K}, \sigma)$. Indeed, by Theorem 3.6.3 (page 119), there are such weak suboctagons. The Moufang property (Theorem 4.5.8 on page 148) implies that the automorphism group of Γ acts transitively on opposite pairs of lines. Hence there is at least one such weak suboctagon Γ' containing L_1 and L_2. But by Corollary 1.8.6 (see page 35), it is unique and our claim follows.

The distance-4-trace $\Gamma_4(L_1) \cap \Gamma_4(L_2)$ is readily seen to correspond to a distance-2-trace in $W(\mathbb{K}, \sigma)$. The line-distance-4-regularity of Γ now follows directly from the regularity of points and lines in $W(\mathbb{K}, \sigma)$; see Corollary 3.4.5 on page 103.

Now we show that Γ is point-distance-4-regular. We assume that Γ is coordinatized as on page 119. Considering two arbitrary opposite points, we may as well take (∞) and (0^7). The set $\Gamma_4((\infty)) \cap \Gamma_4((0^7))$ consists of the points with coordinates (0^3) and $(k, 0^3)$, $k \in R_2$. Any element different from (∞) and at distance 4 from both (0^3) and (0^4) has coordinates $(0^3, l', 0^3)$, $l' \in R_2$. The point-distance-4-regularity will follow if we show that $\delta((0^3, l', 0^3), (k, 0^3)) = 4$, for all $k, l' \in R_2$. From the explicit form of the octagonal octanary ring one can check that

$$(k, 0^3) \, \mathbf{I} \, [k, 0^3, l'] \, \mathbf{I} \, (k, 0^3, l', 0) \, \mathbf{I} \, [k, 0^3, l', 0^2] \, \mathbf{I} \, (0^3, l', 0^3),$$

which proves the result. $\qquad\square$

As an application we show a generalization of Theorem 7.6.27(iv).

8.2.4 Corollary. *Let* $W(\mathbb{K}, \sigma)$ *be any Suzuki quadrangle and let* ρ_1 *and* ρ_2 *be two natural polarities of* $W(\mathbb{K}, \sigma)$. *Let* $\mathcal{F}_1$ *and* $\mathcal{F}_2$ *be sets of flags fixed by* ρ_1 *and* ρ_2, *respectively. If* $\mathcal{F}_1 \neq \mathcal{F}_2$, *then* $|\mathcal{F}_1 \cap \mathcal{F}_2| \leq 2$.

Proof. We embed the dual of the double of $W(\mathbb{K}, \sigma)$, say Γ', in the Ree–Tits octagon $O(\mathbb{K}, \sigma)$. Then we may view $\mathcal{F}_1$ and $\mathcal{F}_2$ as sets of points in $O(\mathbb{K}, \sigma)$. According to Theorem 8.2.2, there are points x_1 and x_2 at distance 4 from all points of, respectively, $\mathcal{F}_1$ and $\mathcal{F}_2$ (see Figure 8.1). Suppose now that $|\mathcal{F}_1 \cap \mathcal{F}_2| \geq 3$ and let p_1, p_2, p_3 be three elements of $\mathcal{F}_1 \cap \mathcal{F}_2$. Denote by y_1 and y_2 the two points of Γ' at distance 4 from both p_1 and p_2 in $O(\mathbb{K}, \sigma)$. These points are well defined since Γ' is a weak polygon with two lines per point. So we have

$$\{p_1, p_2\} \subseteq (y_1)^{x_1}_{[4]} \cap (x_2)^{x_1}_{[4]},$$

and consequently, by Theorem 8.2.3, $(y_1)^{x_1}_{[4]} = (x_2)^{x_1}_{[4]}$, which implies that $p_3 \in (y_1)^{x_1}_{[4]}$. Since p_3 and p_i, $i = 1, 2$, are opposite in $O(\mathbb{K}, \sigma)$, Lemma 1.5.6 on page 20 implies that $\mathrm{proj}_{y_1} p_3 \neq \mathrm{proj}_{y_1} p_i$, $i = 1, 2$. Since $\mathrm{proj}_{y_1} p_3$ clearly belongs to Γ', this contradicts the fact that the points of Γ' are not thick. Hence $x_1 = x_2$ and so $\mathcal{F}_1 = \mathcal{F}_2$. $\qquad\square$

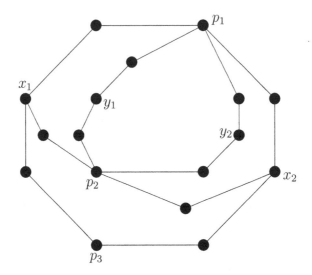

Figure 8.1.

We may now bring together all results concerning regularity of finite classical polygons. Up to duality, we consider $W(q), H(3, q^2), H(4, q^2),\ H(q), T(q^3, q)$ and $O(q)$, with q a power of a prime.

8.2.5 Theorem.

(i) *The quadrangles* $W(q)$ *and* $H(3, q^2)$ *are point-distance-2-regular. The quadrangle* $W(q)$ *is line-distance-2-regular* **if and only if** *q is a power of 2. The quadrangles* $H(3, q^2)$ *and* $H(4, q^2)$ *are never line-distance-2-regular, and the latter is never point-distance-2-regular.*

(ii) *The hexagons* $H(q)$ *and* $T(q^3, q)$ *are point-regular (i.e., point-distance-2-regular and point-distance-3-regular) and line-distance-3-regular. The hexagon* $H(q)$ *is also line-distance-2 regular (and hence line-regular)* **if and only if** *q is a power of 3. The hexagon* $T(q^3, q)$ *is never line-distance-2-regular.*

(iii) *The octagon* $O(q)$ *is neither point-distance-2-regular, nor point-distance-3-regular, nor line-distance-2-regular, nor line-distance-3-regular. But it is both point-distance-4-regular and line-distance-4-regular.*

Proof. These assertions follow from Corollary 3.4.5, Proposition 3.4.8, Proposition 3.4.13 (for (i)); Corollary 3.5.6, Corollary 3.5.7, Corollary 3.5.11 (for (ii)) and Theorem 6.4.5, Theorem 6.4.6, Theorem 8.2.3 (for (iii)).   □

8.3 The little projective groups of some Moufang polygons

In this section we denote by $G(\Gamma)$ the little projective group of a Moufang polygon Γ.

If Γ is a Moufang polygon which has a standard embedding in some projective space $\mathbf{PG}(d, \mathbb{K})$, then we have already seen that the root elations always correspond to elements of $\mathbf{PSL}_{d+1}(\mathbb{K})$. Hence the little projective group $G(\Gamma)$ of Γ is a subgroup of $\mathbf{PSL}_{d+1}(\mathbb{K})$. Since we know that $G(\Gamma)$ acts transitively on the set of all apartments of Γ, we have a complete description of $G(\Gamma)$ whenever we can decide which stabilizers of an apartment belong to $G(\Gamma)$. This usually boils down to considering generalized homologies. We want to tackle that question in a geometric way, by using the polygon rather than the corresponding algebraic forms and matrices. The following lemma is very useful to that end.

8.3.1 Lemma. *Let Γ be a Moufang generalized n-gon with n even, say $n = 2m$. Let Σ be any apartment of Γ and let L and M be two confluent lines of Σ. Also, let u and w be the two mutually opposite elements of Σ at distance m from $L \cap M$, and suppose $\delta(L, u) = m - 1$. Furthermore, let ϑ be any permutation of $\Gamma(L) \setminus \{L \cap M, \mathrm{proj}_L u\}$. If there exists a generalized homology θ of Γ fixing Σ, fixing all points incident with M and inducing ϑ on L, **then** there exists an element $g \in G(\Gamma)$ fixing Σ and inducing ϑ on L. If the elements u and w are, moreover, distance-m-regular, **then** g can be chosen to be a $\{u, w\}$-homology.*

Proof. Let x be the point of Σ opposite $y := L \cap M$. Let N be any line through x not in Σ. Since x, L and M are fixed by θ, the permutation ϑ is equal to the projectivity $[L; N; M; N^\theta; L]$. Now let γ_0, γ_1 and γ_2 be the $(n-2)$-paths the extremities of which are incident with x and y, respectively, and which contain u, N and N^θ, respectively. Denote by θ_1 (θ_2) the unique γ_1-elation (γ_2-elation) mapping L to M (M to L). It is clear that the restriction of θ_1 to $\Gamma(L)$ is equal to the projectivity $[L; N; M]$. Similarly, the restriction of θ_2 to $\Gamma(M)$ is equal to the projectivity $[M; N^\theta; L]$. Hence $\theta_1\theta_2$ fixes γ_0 and induces ϑ on L. By composing with a suitable γ_0-elation θ_0, the collineation $g := \theta_1\theta_2\theta_0$ fixes Σ and induces ϑ on L. If u and w are distance-m-regular, then the middle elements u_1 and u_2 of, respectively, γ_1 and γ_2 are at distance m from every element of $\Gamma_m(u) \cap \Gamma_m(w)$. It follows that the mapping $[u; u_1; w; u_2; u]$, which describes the action of $\theta_1\theta_2$ on $\Gamma(u)$, is the identity.

The lemma is proved. $\qquad\square$

The preceding lemma already enables us to determine the little projective group of all symplectic and Suzuki quadrangles, all split Cayley and mixed hexagons and all perfect Ree–Tits octagons.

8.3.2 Theorem.

(*i*) *Let* Γ *be a symplectic quadrangle, a split Cayley hexagon or a mixed hexagon. Let* Γ *have a standard embedding in* $\mathbf{PG}(d, \mathbb{K})$, *for some field* $\mathbb{K}$ *(for the symplectic hexagons, we may assume* $d = 5$ *or* $d = 6$*). Then the little projective group of* Γ *is the subgroup of* $\mathbf{PSL}_{d+1}(\mathbb{K})$ *stabilizing* Γ.

(*ii*) *The little projective group of a mixed quadrangle* $\Gamma = Q(\mathbb{K}, \mathbb{K}'; \mathbb{K}, \mathbb{K}')$, $\mathbb{K}^2 \leq \mathbb{K}' \leq \mathbb{K}$, *is obtained by intersecting the subgroups of* $\mathbf{PSL}_4(\mathbb{K})$ *and* $\mathbf{PSL}_4(\mathbb{K}')$ *stabilizing* Γ, *respectively* Γ^D.

(*iii*) *The little projective group of a perfect Ree–Tits octagon contains all homologies.*

Proof. First let Γ be a symplectic quadrangle $\mathsf{W}(\mathbb{K})$. We coordinatize Γ as in Subsection 3.4.1 (see page 100). From the proof of Proposition 4.6.1 (page 152), we learn that $G(\Gamma)$ is a subgroup of $\mathbf{PSL}_4(\mathbb{K})$, and that the stabilizer of the hat-rack of the coordinatization inside $\mathbf{PSL}_4(\mathbb{K}) \cap \mathrm{Aut}(\Gamma)$ takes the general form:

$$h_{A,K^2} \quad : \quad \begin{array}{ccc} (a, l, a') & \mapsto & (Aa, A^2 K^2 l, A K^2 a'), \\ [k, b, k'] & \mapsto & [K^2 k, A K^2 b, A^2 K^2 k'], \end{array}$$

for some $A, K \in \mathbb{K}^\times$. Substituting K for K^2 in the above we obtain a general element $h_{A,K}$ of the group generated by all generalized homologies stabilizing the hat-rack of the coordinatization. Note that from the shape of $h_{A,K}$, it follows that a $\{u, w\}$-homology is determined by the action on $\Gamma(v)$ for any $v \mathbf{I} u$. If u and w are lines, then a $\{u, w\}$-homology is also determined by its action on $\Gamma(v')$, for any $v' \in \Gamma_2(u) \cap \Gamma_2(w)$. Considering $h_{1,K}$, we infer from Lemma 8.3.1 above that h_{K^{-1}, K^2} belongs to $G(\Gamma)$. As $K \neq 0$ is arbitrary, we see that all $\{(0), (0, 0)\}$-homologies are contained in $G(\Gamma)$. By conjugation, also all $\{(\infty), (0, 0, 0)\}$-homologies $h_{A,1}$ are contained in $G(\Gamma)$. Composing h_{K^{-1}, K^2} and $h_{AK,1}$, we obtain that h_{A,K^2} belongs to $G(\Gamma)$.

Before we go on, we show (*ii*). So let Γ be a mixed quadrangle $Q(\mathbb{K}, \mathbb{K}'; \mathbb{K}, \mathbb{K}')$. Rereading the previous arguments, we see that all homologies $h_{K,1}$ are in $G(\Gamma)$, $K \in \mathbb{K}'$. That means that in the dual $\Gamma^D = Q(\mathbb{K}', \mathbb{K}^2; \mathbb{K}', \mathbb{K}^2)$, we have all homologies h_{1,K^2}, $K \in \mathbb{K}'$. Similarly, in Γ, we have all homologies h_{1,A^2}, $A \in \mathbb{K}$. The result follows.

Now let Γ be a split Cayley hexagon. Again we use coordinatization (see Subsection 3.5.1 on page 110). According to Proposition 4.6.6 on page 157, and Remark 4.6.8, we have to show that all generalized homologies are generated by root elations. A general element generated by homologies which fix the hat-rack looks like:

$$h_{A,K} \quad : \quad \begin{array}{ccc} (a, l, a', l', a'') & \mapsto & (Aa, A^3 K l, A^2 K a', A^3 K^2 l', A K a''), \\ [k, b, k', b', k''] & \mapsto & [K k, A K b, A^3 K^2 k', A^2 K b', A^3 K k''], \end{array}$$

with $A, K \in \mathbb{K} \setminus \{1\}$. Considering $h_{1,K}$, we see, as for the symplectic quadrangles, that the conjugates of all homologies $h_{1,K}$, $K \in \mathbb{K}$, belong to $G(\Gamma)$. One of these conjugates is $h_{A,A^{-1}}$. But $h_{1,AK}h_{A,A^{-1}} = h_{A,K}$ and the result follows.

If Γ is a mixed hexagon $\mathsf{H}(\mathbb{K}, \mathbb{K}')$, then we still have $h_{1,K}$ in $G(\Gamma)$, but only for $K \in \mathbb{K}' \setminus \{0\}$ (but that is still as general as possible). Dually, we have $h_{A,1} \in G(\Gamma)$, for all $A \in \mathbb{K} \setminus \{0\}$. Hence $h_{A,K} = h_{A,1}h_{1,K}$ is in $G(\Gamma)$ for all pairs $(A, K) \in \mathbb{K}^{\times} \times \mathbb{K}'^{\times}$. This shows the assertion for hexagons.

Now we show (iii). But that follows similarly, noting that all elements of a perfect Ree–Tits octagon $\mathsf{O}(\mathbb{K}, \sigma)$ are distance-4-regular, also remarking that both σ and $\sigma + 2$ are bijections, and using the following general form of an element generated by homologies as above (after coordinatization):

$$h_{A,B} : (a, l, a', l', a'', l'', a''') \mapsto (Aa, (AB) \otimes l, (AB^{\sigma})^{\sigma+1}a', (A^{\sigma}B^{\sigma+1}) \otimes l',$$
$$(AB^2)^{\sigma+1}a'', (AB^{\sigma+1}) \otimes l'', AB^{\sigma+2}a'''),$$
$$[k, b, k', b', k'', b'', k'''] \mapsto [B \otimes k, AB^{\sigma+2}b, (AB^{\sigma+1}) \otimes k', (AB^2)^{\sigma+1}b',$$
$$(A^{\sigma}B^{\sigma+1}) \otimes k'', (AB^{\sigma})^{\sigma+1}b'', (AB) \otimes k'''],$$

with $A, B \in \mathbb{K} \setminus \{0\}$. $\qquad\square$

We now concentrate on the finite Moufang polygons which are not covered by Theorem 8.3.2. For the next result, we cannot do without "forms".

8.3.3 Theorem. *Let Γ be the finite Hermitian quadrangle $\mathsf{H}(d, q^2)$ with standard embedding in $\mathbf{PG}(d, q^2)$, $d = 3, 4$.*

(i) **If $d = 4$, then** *the little projective group of Γ is induced by $\mathbf{PSL}_5(q^2)$ on Γ.*

(ii) **If $d = 3$, then** *the little projective group of Γ is a subgroup of index $(q+1, 2)$ of the group induced by $\mathbf{PSL}_4(q^2)$ on Γ.*

Proof. We prove the result for $d = 3$, which requires the longest proof anyway. So let Γ be the quadrangle $\mathsf{H}(3, q^2)$. Using the standard form (Proposition 2.3.4 on page 55) for a corresponding σ-quadratic form, we may assume that the points and the lines of Γ are those of the Hermitian variety in $\mathbf{PG}(3, q^2)$ with equation

$$X_{-2}X_2^q + X_{-2}^q X_2 + X_{-1}X_1^q + X_{-1}^q X_1 = 0.$$

We have seen that all collineations of Γ arise from semi-linear maps in the underlying vector space of $\mathbf{PG}(3, q^2)$; see Proposition 4.6.3 (page 155). Now we consider the following subset G of the projective group corresponding to these maps. Put

$$H = \begin{pmatrix} 0 & 0 & 0 & 1 \\ 0 & 0 & 1 & 0 \\ 0 & 1 & 0 & 0 \\ 1 & 0 & 0 & 0 \end{pmatrix}.$$

It is clear that an element of $\mathbf{PGL}_4(q^2)$ with matrix M preserves Γ (and hence induces a collineation in Γ) **if and only if** $M^t H = k \cdot H M^{-q}$, where k is some non-zero element of $\mathbf{GF}(q^2)$, M^t denotes the transpose of M, and M^{-q} is obtained from M by replacing every element by its qth power and then inverting the matrix. If we identify an element of $\mathbf{PGL}_4(q^2)$ with a corresponding matrix (defined up to a scalar non-zero multiple), then we put $M \in G$ **if and only if** there is a $k \in \mathbf{GF}(q^2)$ such that $kM \in \mathbf{SL}_4(q^2)$ and $(kM)^t H = H(kM)^{-q}$. Clearly G is a subgroup of $\mathbf{PSL}_4(q^2)$. Also, it is easy to check that every root elation of Γ belongs to G. Let us do that for one kind of root elations, say, line-elations. The root elation g, determined by the following action on the coordinates of some elements of Γ,

$$g \;:\; \begin{aligned} (a, l, a') &\mapsto (a, l + L, a') \\ [k, b, k'] &\mapsto [k, b, k' + L], \end{aligned}$$

for arbitrary $L \in R_2$ (where R_2 is the set $\{x^q - x : x \in \mathbf{GF}(q^2)\}$; see Subsection 3.4.7 starting on page 104), is easily seen to have the matrix

$$M = \begin{pmatrix} 1 & 0 & 0 & L \\ 0 & 1 & 0 & 0 \\ 0 & 0 & 1 & 0 \\ 0 & 0 & 0 & 1 \end{pmatrix}.$$

It is readily verified now that $M \in \mathbf{SL}_4(q^2)$ and $M^t H = H M^{-q}$ (noting that $L^q = -L$).

Hence $G(\Gamma) \leq G$. Let us fix an apartment Σ in Γ, and we may take without loss of generality the hat-rack of the coordinatization. We now determine the order of the subgroup G_Σ of G fixing Σ elementwise. By Proposition 4.6.3, G_Σ is a subgroup of the group G_0 generated by all generalized homologies which preserve Σ. Since $\mathbf{PGL}_4(q^2)$ induces on each point row of Γ (which is a point row of $\mathbf{PG}(3, q^2)$) and on each line pencil of Γ (which is a subset of a plane line pencil in $\mathbf{PG}(3, q^2)$) a sharply 3-transitive group, every element $h_{A,K}$ of G_Σ is determined by the image $[Kk]$ of the line $[k]$ (for some fixed $k \in R_2$ and some $K \in \mathbf{GF}(q)$) and the image (A) of the point (1). One can check that the following map determines such an element $h_{A,K}$ (see also the proof of Proposition 4.6.3 on page 155):

$$h_{A,K} \;:\; \begin{aligned} (a, l, a') &\mapsto (Aa, A^{q+1} Kl, AKa'), \\ [k, b, k'] &\mapsto [Kk, AKb, A^{q+1} Kk']. \end{aligned}$$

The corresponding matrix is a diagonal matrix $M_{A,K} = \mathrm{diag}(A^{q+1} K, A^q, A^q K, 1)$ (with obvious notation) and an elementary calculation shows that $M_{A,K}$ belongs to G **if and only if** there exists $\ell \in \mathbf{GF}(q^2)$ such that $\ell^4 = A^{3q+1} K^2$ and $\ell^{q+1} = A^{q+1} K$. By Lemma 8.3.1, the transitivity of the group $\{h_{A,1} : A \in \mathbf{GF}(q^2)^\times\}$ on the set $\Gamma_1((0)) \setminus \Sigma_1((0))$ (which can easily be read off the above explicit form of $h_{A,1}$, noting that every element of $\mathbf{GF}(q)$ can be written as A^{q+1} for some $A \in \mathbf{GF}(q^2)$) implies that the elementwise stabilizer of Σ in the little projective group $G(\Gamma)$, and hence also the one in G, acts transitively on the set of lines through (∞) distinct from $[\infty]$ and $[0]$. Hence to determine $|G_\Sigma|$, we may assume that $K = 1$. So we are looking for elements $A \in \mathbf{GF}(q^2)^\times$ such that there exists $\ell \in \mathbf{GF}(q^2)^\times$ with $\ell^4 = A^{3q+1}$ and $\ell^{q+1} = A^{q+1}$. Writing $\ell = A\theta$, with $\theta^{1+q} = 1$, we see that we are looking for the number of elements A satisfying $A^{3q-3} = \theta^4$. Since θ is itself a $(q-1)$th power, it suffices to look in the multiplicative group of $q + 1$ $(q+1)$th roots of unity. Obviously, the number of such As is exactly equal to $\frac{q^2-1}{(q+1,4)}$.

By Proposition 4.6.3, we already know that G has index $(q + 1, 2)$ in $\mathrm{Aut}(\Gamma) \cap \mathbf{PSL}_4(q^2)$. The assertion will follow if we prove that also $G(\Gamma)_\Sigma$ has order $\frac{q^2-1}{(q+1,4)}$.

Let us get back for a moment to Lemma 8.3.1 and its proof. If we use the same notation as in that proof, then from the last line of that proof, we infer that the action of the element $g \in G(\Gamma)$ on $\Gamma(u)$ is given by the projectivity $[u; u_1; w; u_1^\theta; u]$. Now we let $u = [0,0,0]$, $u_1 = [0, B, 0]$ and $w = [\infty]$. Then the projectivity $\rho = [u; u_1; w]$ is given by the map $\rho : (a, 0, 0) \mapsto (a^q B^{1-q})$. Consider, with the above notation, the generalized homology $\theta = h_{A,1}$. Then $u_2 = u_1^g = [0, BA, 0]$. The element g is now completely determined and according to the calculations we have just performed, we can write $g = h_{A^{1-q}, A^{1+q}}$. Similarly, by switching the roles of $[\infty]$ and $[0]$, we have that $g' = h_{B^2, B^{-1-q}}$ belongs to G. Again, we see that $G(\Gamma)_\Sigma$ acts transitively on the set of lines through (∞) not in Σ, so it suffices to show that the group $G^* = \{h_{A^{3-q}\theta^2, 1} : A \in \mathbf{GF}(q^2)^\times, \theta^{1+q} = 1\}$ has order $\frac{q^2-1}{(q+1,4)}$ (where we obtained a general element of G^* by composing $h_{A^{1-q}, A^{1+q}}$ and $h_{(A\theta)^2, A^{-1-q}}$). The elements A^{3-q} generate in $\mathbf{GF}(q^2)^\times$ a group of order $\frac{q^2-1}{(q-3,8)}$. If $(q - 3, 8) \in \{4, 8\}$ (which is equivalent to $(q + 1, 4) = 4$), then θ^2, with $\theta^{1+q} = 1$, generates a group of order $\frac{q^2-1}{2q-2}$, and this group contains all fourth powers of $\mathbf{GF}(q^2)$, but not all squares, since $2q - 2$ is divisible by 4, but not by 8. And if $(q - 3, 8) \in \{1, 2\}$ (and this is equivalent to $(q + 1, 4) \in \{1, 2\}$, and we have $(q + 1, 4) = (q - 3, 8)$), then, since θ^2 is always a square, the elements A^{q-3} generate a group of order $\frac{q^2-1}{(q+1,4)}$.

Hence the order of $G(\Gamma)_\Sigma$ is always equal to $\frac{q^2-1}{(q+1,4)}$. So $G = G(\Gamma)$.

The theorem is proved. Note that a general element of $G(\Gamma)_\Sigma$ may be written as gg', with the above notation. $\qquad\square$

The result $G = G(\Gamma)$ is still true in the infinite case for some fields. For instance, over the complex numbers, the same sort of argument works. Of course, in this case, we will always have that, for $h_{A,K} \in G$, or equivalently $h_{A,K} \in G(\Gamma)$, the real number K is always positive, and there is no restriction on A.

The use of Lemma 8.3.1 as in the previous proof also shows the following classical result, which we include for completeness.

8.3.4 Theorem. *Let Γ be a projective plane $\mathbf{PG}(2, \mathbb{K})$ defined over the field $\mathbb{K}$. Then the little projective group of Γ is equal to $\mathbf{PSL}_3(\mathbb{K})$.* $\qquad\square$

We now turn to the finite twisted triality hexagons.

8.3.5 Theorem. *Let Γ be the finite twisted triality hexagon $\mathsf{T}(q^3, q)$ with standard embedding in $\mathbf{PG}(7, q^3)$. Then the little projective group of Γ is induced by $\mathbf{PSL}_8(q^3)$, or equivalently, by $\mathbf{PGL}_8(q^3)$, on Γ.*

Proof. It is easy to see that every root elation in an ideal subhexagon Γ' (isomorphic to $\mathsf{H}(q)$) of Γ extends to a root elation in Γ. Indeed, this follows from the transitivity of the root elations on the appropriate sets. So $G(\Gamma)$ contains all generalized homologies of Γ'. We assume that Γ' contains the hat-rack (denote the corresponding apartment by Σ) of the coordinatization and is obtained by restricting coordinates to $\mathbf{GF}(q)$, for some coordinatization of Γ as in Subsection 3.5.8 (page 114). The stabilizer $\mathrm{Aut}(\Gamma)_\Sigma \cap \mathbf{PGL}_8(q^3)$ of Σ in the automorphism group of Γ inherited from $\mathbf{PGL}_8(q^3)$ is generated by generalized homologies, by Proposition 4.6.9 on page 158. Any element is completely determined by the images of $[1]$ and (1). A generic element looks like

$$h_{A,K} : \begin{array}{l} (a, l, a', l', a'') \\ [k, b, k', b', k''] \end{array} \begin{array}{l} \mapsto \\ \mapsto \end{array} \begin{array}{l} (Aa, \mathrm{N}(A)Kl, (\mathrm{N}(A)/A)Ka', \mathrm{N}(A)K^2l', AKa''), \\ [Kk, AKb, \mathrm{N}(A)K^2k', (\mathrm{N}(A)/A)Kb', \mathrm{N}(A)Kk'']. \end{array}$$

The matrix of the corresponding element in $\mathbf{PGL}_8(q^3)$ can be found in Appendix D (substituting σ by q there), and we see that its determinant equals $A^8 K^8 (K^{q+q^2})^4$, which is an eighth power in $\mathbf{GF}(q^3)$. So we must show that every generalized homology is generated by root elements. By our previous remark, we already have that $h_{A,K} \in G(\Gamma)$ for $A, K \in \mathbf{GF}(q)$. Now consider the group G^* generated by all γ-elations, where γ is a path of length 4 with end-points incident with, respectively, $[0, 0]$ and $[0, 0, 0]$. It is clear that this group fixes the line regulus $R([0, 0], [0, 0, 0])$ elementwise (since every generator does). It is also clear that the action of G^* on the point row $\Gamma([0, 0])$ contains $\mathbf{PSL}_2(q^3)$. Hence the stabilizer in this group of Σ is a set of generalized $\{(\infty), (0^5)\}$-homologies $\{h_{A,1} : A \in \mathcal{A}\}$ with a set $\mathcal{A}$ such that $\{A^{q+q^2} : A \in \mathcal{A}\}$ is equal to the set of all squares in $\mathbf{GF}(q^3)$. This group G_Σ^* has at most index 2 in the group $G^{**} = \{h_{A,1} : A \in \mathbf{GF}(q^3)\}$. Suppose it has exactly index 2. This means that for each $A \in \mathbf{GF}(q^3)$, either A belongs to $\mathcal{A}$ or $-A$ does (and hence q is odd). But as already remarked, $G^{**} \cap G(\Gamma)$ contains all elements $h_{A,1}$ with $A \in \mathbf{GF}(q)$. Since $q \geq 2$, we have $\{1, -1\} \subseteq \mathcal{A}$, a contradiction. So G_Σ^* is equal to G^{**}. The theorem is proved. $\qquad\square$

From the above results we infer that the little projective group $G(\Gamma)$ of a finite classical polygon has index at most 2 in the group $G^*(\Gamma)$ of collineations of Γ induced by $\mathbf{PSL}_{d+1}(q)$, for suitable d and q (i.e., for quadrangles take symplectic or Hermitian representation; for hexagons take $d = 6$ or $d = 7$). Hence it follows that, together with Corollary 5.8.3 on page 233, the group $G^*(\Gamma)'$ (the commutator subgroup of $G^*(\Gamma)$) is always a simple group.

8.3.6 Remark. Theorem 8.3.5 is not true in the infinite case. Indeed, let $\mathbb{F}$ be any field and put $\mathbb{L} = \mathbb{F}((x, y, z))$ (field of Laurent series in three variables). Let σ be the field automorphism induced by the mapping $x \mapsto y \mapsto z \mapsto x$. If $\mathbb{K} = \mathbb{L}^{(\sigma)}$, then we consider $\mathsf{T}(\mathbb{L}, \mathbb{K}, \sigma)$. From Appendix D it follows that the determinant of the matrix belonging to a generic element of the torus in $\mathrm{Aut}\, \mathsf{T}(\mathbb{L}, \mathbb{K}, \sigma) \cap \mathbf{PGL}_8(\mathbb{L})$ is equal to (with the notation of Appendix D) $(AK)^8 (A^{\sigma + \sigma^2})^4$. If we take $A = x$, $K = 1$, then this determinant becomes equal to $x^8 y^4 z^4$, which is clearly not an eighth power in $\mathbb{L}$. Hence the little projective group of $\mathsf{T}(\mathbb{L}, \mathbb{K}, \sigma)$ cannot be induced by $\mathbf{PGL}_8(\mathbb{L})$.

8.4 Groups of projectivities of some Moufang polygons

Let Γ be a Moufang polygon and $G(\Gamma)$ its little projective group. It will turn out that the group $\Pi_+(L)$ of even projectivities (i.e., the subgroup of $\Pi(L)$ consisting of all elements that can be written as a product of an even number of perspectivities) of a line L of Γ is permutation equivalent to the action of the stabilizer $G(\Gamma)_L$ of L in $G(\Gamma)$ on $\Gamma(L)$. So it is worthwhile to determine these stabilizers. We can do this for all polygons considered in the previous section.

We will use the following notation. For any field $\mathbb{K}$ of characteristic 2 and any subfield $\mathbb{K}' \leq \mathbb{K}$ with $\mathbb{K}^2 \subseteq \mathbb{K}'$, we denote by $\mathbf{PSL}_2^{\mathbb{K}'}(\mathbb{K})$ the 2-transitive subgroup of $\mathbf{PGL}_2(\mathbb{K})$ whose stabilizer of (0) and (∞) on the projective line $\mathbf{PG}_1(\mathbb{K})$ is given by the group $\{x \mapsto ax : 0 \neq a \in \mathbb{K}'\}$. It can be written as $\mathbf{PSL}_2(\mathbb{K}) : (\mathbb{K}'^\times / (\mathbb{K}^\times)^2)$ with the notation of CONWAY et al. [1985].

Also, we again use the notation $U_H(q)$ for the Hermitian unital. Recall from Subsection 7.5.3 (page 321) that this unital may be constructed as follows: the points are the points of $\mathbf{PG}(2, q^2)$ which are isotropic with respect to a Hermitian polarity; the blocks are sets of points on a secant line. We denote by $\mathbf{PGU}_3(q)$ the automorphism group of $U_H(q)$ inherited from $\mathbf{PGL}_3(q^2)$.

Also, let us, for a given perfect field $\mathbb{K}$ of characteristic 2 admitting a Tits automorphism σ, denote by $\mathcal{O}_{ST}(\mathbb{K}, \sigma)$ the corresponding Suzuki–Tits ovoid; see Section 7.6 starting on page 322.

8.4.1 Theorem. *Let Γ be a generalized polygon contained in one of the following classes: mixed quadrangles of type $\mathsf{Q}(\mathbb{K}, \mathbb{K}'; \mathbb{K}, \mathbb{K}')$, Pappian polygons, finite classical quadrangles, mixed hexagons, finite classical hexagons, perfect Ree–Tits octagons. Let G (G^*) be the stabilizer in $G(\Gamma)$ of a point row (a line pencil) of Γ, modulo the kernel of the action on that point row (line pencil). Then G and G^* are permutation equivalent to the permutation representations (G, X) and (G^*, X^*), respectively, listed in* Table 8.1 *(where the polygons in question are listed only up to duality).*

Proof. Since all point rows and line pencils in question have the natural structure of the respective sets in the permutation representations of Table 8.1, and since the Moufang condition implies that all the permutation representations are 2-transitive

The polygon Γ	*The representation* (G, X)	*The representation* (G^*, X^*)
$\mathbf{PG}(2, \mathbb{K})$	$(\mathbf{PGL}_2(\mathbb{K}), \mathbf{PG}(1, \mathbb{K}))$	$(\mathbf{PGL}_2(\mathbb{K}), \mathbf{PG}(1, \mathbb{K}))$
$\mathsf{Q}(\mathbb{K}, \mathbb{K}'; \mathbb{K}, \mathbb{K}')$	$(\mathbf{PSL}_2^{\mathbb{K}'}(\mathbb{K}), \mathbf{PG}(1, \mathbb{K}))$	$(\mathbf{PSL}_2^{\mathbb{K}^2}(\mathbb{K}'), \mathbf{PG}(1, \mathbb{K}'))$
$\mathsf{W}(\mathbb{K})$	$(\mathbf{PGL}_2(\mathbb{K}), \mathbf{PG}(1, \mathbb{K}))$	$(\mathbf{PSL}_2(\mathbb{K}), \mathbf{PG}(1, \mathbb{K}))$
$\mathsf{H}(3, q^2)$	$(\mathbf{PSL}_2(q^2), \mathbf{PG}(1, q^2))$	$(\mathbf{PGL}_2(q), \mathbf{PG}(1, q))$
$\mathsf{H}(4, q^2)$	$(\mathbf{PGL}_2(q^2), \mathbf{PG}(1, q^2))$	$(\mathbf{PGU}_3(q), U_H(q))$
$\mathsf{H}(\mathbb{K}, \mathbb{K}'; \mathbb{K}, \mathbb{K}')$	$(\mathbf{PGL}_2(\mathbb{K}), \mathbf{PG}(1, \mathbb{K}))$	$(\mathbf{PGL}_2(\mathbb{K}'), \mathbf{PG}(1, \mathbb{K}'))$
$\mathsf{H}(\mathbb{K})$	$(\mathbf{PGL}_2(\mathbb{K}), \mathbf{PG}(1, \mathbb{K}))$	$(\mathbf{PGL}_2(\mathbb{K}), \mathbf{PG}(1, \mathbb{K}))$
$\mathsf{T}(q^3, q)$	$(\mathbf{PGL}_2(q^3), \mathbf{PG}(1, q^3))$	$(\mathbf{PGL}_2(q), \mathbf{PG}(1, q))$
$\mathsf{O}(\mathbb{K}, \sigma)$	$(\mathbf{PGL}_2(\mathbb{K}), \mathbf{PG}(1, \mathbb{K}))$	$(\mathbf{Sz}(\mathbb{K}, \sigma), \mathcal{O}_{ST}(\mathbb{K}, \sigma))$

Table 8.1. Representations of stabilizers of point rows and line pencils in some Moufang polygons.

(indeed, consider the root groups associated to two "opposite" $(n - 2)$-paths), we only need to verify that the stabilizer of two elements is the desired one.

Let L be a line of Γ, let x_1, x_2 be two points incident with L, and let Σ be any apartment containing x_1, x_2 and L. We claim that the action of the stabilizer in $G(\Gamma)$ of the path (x_1, L, x_2) on the set $\Gamma(L) \setminus \{x_1, x_2\}$ is same as the action of the stabilizer in $G(\Gamma)$ of Σ on the set $\Gamma(L) \setminus \{x_1, x_2\}$.

Indeed, if an element $g \in G(\Gamma)$ stabilizes (x_1, L, x_2), then by Lemma 5.2.4(ii) (see page 175), there exists $g' \in G(\Gamma)$ fixing all elements of $\Gamma(L)$ and mapping $\Sigma^{g^{-1}}$ back onto Σ. The claim follows.

So the stabilizers we are looking for are given by the action of $G(\Gamma)_\Sigma$. We have determined that action very explicitly in the previous section. Let us give one typical example. Suppose $\Gamma = \mathsf{H}(3, q^2)$. A general element of $G(\Gamma)_\Sigma$ can be written as (see the proof of Theorem 8.3.3 above) $h_{A^{1-q}, A^{1+q}} \cdot h_{B^2, B^{-1-q}}$:

$$
\begin{aligned}
(a, l, a') &\mapsto (A^{1-q} B^2 a, B^{2q+2}(AB^{-1})^{1+q} l, A^2 B^{1-q} a'), \\
[k, b, k'] &\mapsto [(AB^{-1})^{1+q} k, A^2 B^{1-q} b, B^{2q+2}(AB^{-1})^{1+q} k'].
\end{aligned}
$$

It is clear that the stabilizer of the path $((\infty), [\infty], (0))$ acts on $\Gamma([\infty]) \setminus \{(\infty), (0)\}$ as the multiplicative group of squares of $\mathbf{GF}(q^2)$ acts on $\mathbf{GF}(q^2)^\times$ (by multiplication). Also, since the map $\mathbf{GF}(q^2) \rightarrow \mathbf{GF}(q) : x \mapsto x^{1+q}$ is surjective, we have that the stabilizer of the path $([\infty], (\infty), [0])$ acts on $\Gamma((\infty)) \setminus \{[\infty], [0]\}$ as the multiplicative group of $\mathbf{GF}(q)$ acts on itself (by multiplication). The result follows.

$\square$

8.4.2 Definitions. Let Γ be a generalized polygon. As we have seen in Section 1.5, the isomorphism class of the group $\Pi(L)$ of projectivities of a point row $\Gamma(L)$ into itself is independent of the line L. In fact, the equivalence class of the permutation representation $(\Pi(L), \Gamma(L))$ is independent of L. We denote that (abstract) permutation representation by $(\Pi(\Gamma), X(\Gamma))$ and we call $\Pi(\Gamma)$ the **general projectivity group** of Γ.

A projectivity which is the product of an even number of perspectivities will be called an **even projectivity**. Now fix a line L. Clearly the subgroup of $\Pi(L)$ consisting of all even projectivities of $\Gamma(L)$ into itself has index at most 2 in $\Pi(L)$. We denote that subgroup by $\Pi_+(L)$, and since again the permutation representations $(\Pi_+(L'), \Gamma(L'))$ are all equivalent for all lines L', we can denote the corresponding abstract representation by $(\Pi_+(\Gamma), X(\Gamma))$ (considering $\Pi_+(\Gamma)$ as a subgroup of $\Pi(\Gamma)$) and we call $\Pi_+(\Gamma)$ the **special projectivity group of** Γ. Replacing L by a point, we obtain the definitions of the **general dual projectivity group** $\Pi^*(\Gamma)$ **of** Γ with corresponding permutation representation $(\Pi^*(\Gamma), X^*(\Gamma))$, and of the **special dual projectivity group** $\Pi_+^*(\Gamma)$ **of** Γ with corresponding permutation representation $(\Pi_+^*(\Gamma), X^*(\Gamma))$.

Note that it is pointless to consider groups of projectivities whose elements are products of a multiple of ℓ perspectivities, for $\ell > 2$, since the identity can always be written as the product of two perspectivities.

Note also that, since $\Pi_+(\Gamma)$ has index at most 2 in $\Pi(\Gamma)$, we have that $\Pi_+(\Gamma) = \Pi(\Gamma)$ **if and only if** there exists a projectivity $[L; M; N; L]$ which belongs to $\Pi_+(L)$ (where L, M, N are mutually opposite lines of Γ). Also, we have that $\Pi_+(\Gamma) \neq \Pi(\Gamma)$ **if and only if** there exists a projectivity $[L; M; N; L]$ which does not belong to $\Pi_+(L)$ (where L, M, N are as above). In the latter case, $\Pi(L)$ is completely determined by $\Pi_+(L)$ and $[L; M; N; L]$.

Our aim is now to determine the general and special (dual) projectivity groups of the Moufang polygons for which we discussed the little projective group in the previous section. But as already remarked earlier, the method can be applied for every Moufang polygon, except that not much can be said for the other classes without specializing to certain (classes of) fields. In fact, we will follow roughly the ideas of KNARR [1988], who identified all general and special projectivity groups of the finite Moufang polygons. We simplify his proof at some minor points, for example, by using the distance-4-regularity of the Ree–Tits octagons, and by remarking that the existence of an axial elation in a generalized n-gon with n even implies that $\Pi_+(\Gamma) = \Pi(\Gamma)$. Moreover, we get rid of some counting arguments.

First we determine $(\Pi_+(\Gamma), X(\Gamma))$.

8.4.3 Lemma (Knarr [1988]). *Let Γ be any Moufang polygon and let L be any line of Γ. Then $(\Pi_+(\Gamma), X(\Gamma))$ is permutation equivalent to the action of the stabilizer in $G(\Gamma)$ (the little projective group of Γ) of L on $\Gamma(L)$.*

Proof. First we claim that every even projectivity $\theta : \Gamma(L) \to \Gamma(M)$, for L, M lines of Γ, is induced by an element g of $G(\Gamma)$, i.e., the restriction of g to $\Gamma(L)$ is precisely θ. It suffices to show this for a projectivity which is the product of two perpectivities. So let $\theta = [L; w; M]$ be an even projectivity, with w opposite both L and M. By Lemma 5.2.4(ii) (see page 175), there is an element $g \in G(\Gamma)$ fixing $\Gamma(w)$ and mapping L to M. For $x \, \mathbf{I} \, L$, the element x^g is precisely the projection of $\mathrm{proj}_w x$ onto M, and our claim is proved.

Now we claim that whenever $g \in G(\Gamma)$ maps a line L to some line M, then the restriction of g to $\Gamma(L)$ is an even projectivity of $\Gamma(L)$ into $\Gamma(M)$. Since $G(\Gamma)$ is generated by elations, it suffices to prove this for g an elation. In this case, g fixes some line N pointwise. Let w be some element of Γ opposite both L and N (w exists by Lemma 1.5.8 on page 20). It is clear that the restriction of g to $\Gamma(L)$ is given by the projectivity $[L; w; N; w^g; M]$. The claim follows. This completes the proof of the lemma. $\qquad\square$

8.4.4 Corollary. *Let Γ be a generalized polygon contained in one of the following classes: mixed quadrangles of type $Q(\mathbb{K}, \mathbb{K}'; \mathbb{K}, \mathbb{K}')$, Pappian polygons, finite classical quadrangles, mixed hexagons, finite classical hexagons, perfect Ree–Tits octagons. The permutation representations $(\Pi_+(\Gamma), X(\Gamma))$ and $(\Pi_+^*(\Gamma), X^*(\Gamma))$ are equivalent to the permutation representations listed in* Table 8.1 *above.* $\qquad\square$

We now determine the general (dual) projectivity groups of the above mentioned polygons. First we remark that for n odd, the special projectivity group is always the same as the general one since a projectivity which is the product of an odd number of perspectivities always maps a line pencil to a point row and vice versa. So we may restrict to the case n even, i.e., the quadrangles, hexagons and octagons. We have the following lemma.

8.4.5 Lemma (Knarr [1988]). *Let Γ be a generalized n-gon with $n = 2m$ even. If there exist three mutually opposite lines L, M, N such that $L_{[m]}^M = L_{[m]}^N$, then $\Pi(\Gamma) = \Pi_+(\Gamma)$.*

Proof. This follows from the fact that $[L; M; N; L]$ is the identity in this case. Indeed, for $x \, \mathbf{I} \, L$, the middle element w of the path $(x, \ldots, \mathrm{proj}_M x = y)$ of length $n - 2$ is at distance m from N. It is now easily seen that $\mathrm{proj}_N x = \mathrm{proj}_N y$ and the result follows. $\qquad\square$

There are two particular situations in which we can apply the previous lemma.

8.4.6 Lemma. *The general projectivity group of a generalized n-gon* Γ*,* $n = 2m$*, coincides with the special projectivity group whenever* Γ *is line-distance-m-regular, or whenever* Γ *admits an axial elation. In particular, if* Γ *is a Moufang polygon, then* $\Pi(\Gamma) = \Pi_+(\Gamma)$ *whenever* $\Pi^*(\Gamma) \neq \Pi^*_+(\Gamma)$ *(and dually).*

Proof. Clearly, if Γ is a line-distance-m-regular generalized $2m$-gon, then the condition of Lemma 8.4.5 above is satisfied. Also, if there exists an axial elation g with axis L, and if M is any line opposite L, then L, M and $N = M^g$ satisfy the assumption of Lemma 8.4.5. The last assertion follows from Corollary 5.4.7 (see page 206). $\qquad\square$

Using Corollary 3.4.5 (page 103), Proposition 3.4.8(*ii*) (page 106), Theorem 6.3.2 (page 243) and Theorem 8.2.3 (page 364), we deduce:

8.4.7 Corollary. *Let* Γ *be a mixed or orthogonal quadrangle, a Moufang hexagon or a Moufang octagon. Then* $\Pi(\Gamma) = \Pi_+(\Gamma)$*.* $\qquad\square$

We now come to the main result of this section. For the involutory automorphism $\sigma : x \mapsto x^q$ in $\mathbf{GF}(q^2)$, we denote by $\mathbf{PGL}_2^{(q)}(q^2)$ and $\mathbf{PSL}_2^{(q)}(q^2)$ the extension respectively of $\mathbf{PGL}_2(q^2)$ and $\mathbf{PSL}_2(q^2)$ in $\mathrm{Aut}(\mathbf{PGL}_2(q^2))$ by the group of order 2 generated by the semi-linear mapping with identity matrix and with corresponding field automorphism σ.

8.4.8 Theorem. *Let* Γ *be a generalized polygon contained in one of the following classes: mixed quadrangles of type* $Q(\mathbb{K}, \mathbb{K}'; \mathbb{K}, \mathbb{K}')$*, Pappian polygons, finite classical quadrangles, mixed hexagons, finite classical hexagons, perfect Ree–Tits octagons. Then the permutation representations* $(\Pi(\Gamma), X(\Gamma))$ *and* $(\Pi^*(\Gamma), X^*(\Gamma))$ *are equivalent to the permutation representations listed in* Table 8.2 *(where the polygons in question are again listed only up to duality).*

Proof. As remarked above, we should only consider $\Pi(\Gamma)$ for symplectic quadrangles and the Hermitian quadrangles $\mathsf{H}(3, q^2)$ and $\mathsf{H}(4, q^2)$, and $\Pi^*(\Gamma)$ for the Hermitian quadrangle $\mathsf{H}(4, q^2)$ (since $\mathsf{H}(3, q^2)$ is the dual of the orthogonal quadrangle $Q(5, q)$; see Proposition 3.4.9 on page 107). In order to solve these cases, it will be necessary and sufficient to perform two calculations, one of which we have already done.

First consider $\Gamma = \mathsf{W}(\mathbb{K})$, for some commutative field $\mathbb{K}$. We assume that Γ is coordinatized as in Subsection 3.4.1. We calculate the projectivity $[[0]; [0, 0]; [1, 0]; [0]]$. We have by an elementary computation:

$$(0, b) \, \mathbf{I} \, [0, b, 0] \, \mathbf{I} \, (0, 0, b) \, \mathbf{I} \, [-2b, b, 0] \, \mathbf{I} \, (1, 0, -b) \, \mathbf{I} \, [0, -b, 2b] \, \mathbf{I} \, (0, -b).$$

We infer that $\Pi(\Gamma) = \Pi_+(\Gamma) = \mathbf{PGL}_2(\mathbb{K})$ since the mapping $(b) \mapsto (-b)$, $(\infty) \mapsto (\infty)$ in $\mathbf{PG}(1, \mathbb{K})$ is an element of $\mathbf{PGL}_2(\mathbb{K})$.

The polygon Γ	The representation $(\Pi(\Gamma), X(\Gamma))$	The representation $(\Pi^*(\Gamma), X^*(\Gamma))$
$\mathbf{PG}(2, \mathbb{K})$	$(\mathbf{PGL}_2(\mathbb{K}), \mathbf{PG}(1, \mathbb{K}))$	$(\mathbf{PGL}_2(\mathbb{K}), \mathbf{PG}(1, \mathbb{K}))$
$\mathsf{Q}(\mathbb{K}, \mathbb{K}'; \mathbb{K}, \mathbb{K}')$	$(\mathbf{PSL}_2^{\mathbb{K}'}(\mathbb{K}), \mathbf{PG}(1, \mathbb{K}))$	$(\mathbf{PSL}_2^{\mathbb{K}^2}(\mathbb{K}'), \mathbf{PG}(1, \mathbb{K}'))$
$\mathsf{W}(\mathbb{K})$	$(\mathbf{PGL}_2(\mathbb{K}), \mathbf{PG}(1, \mathbb{K}))$	$(\mathbf{PSL}_2(\mathbb{K}), \mathbf{PG}(1, \mathbb{K}))$
$\mathsf{H}(3, q^2)$	$(\mathbf{PSL}_2^{(q)}(q^2), \mathbf{PG}(1, q^2))$	$(\mathbf{PGL}_2(q), \mathbf{PG}(1, q))$
$\mathsf{H}(4, q^2)$	$(\mathbf{PGL}_2^{(q)}(q^2), \mathbf{PG}(1, q^2))$	$(\mathbf{PGU}_3(q), U_H(q))$
$\mathsf{H}(\mathbb{K}, \mathbb{K}'; \mathbb{K}, \mathbb{K}')$	$(\mathbf{PGL}_2(\mathbb{K}), \mathbf{PG}(1, \mathbb{K}))$	$(\mathbf{PGL}_2(\mathbb{K}'), \mathbf{PG}(1, \mathbb{K}'))$
$\mathsf{H}(\mathbb{K})$	$(\mathbf{PGL}_2(\mathbb{K}), \mathbf{PG}(1, \mathbb{K}))$	$(\mathbf{PGL}_2(\mathbb{K}), \mathbf{PG}(1, \mathbb{K}))$
$\mathsf{T}(q^3, q)$	$(\mathbf{PGL}_2(q^3), \mathbf{PG}(1, q^3))$	$(\mathbf{PGL}_2(q), \mathbf{PG}(1, q))$
$\mathsf{O}(\mathbb{K}, \sigma)$	$(\mathbf{PGL}_2(\mathbb{K}), \mathbf{PG}(1, \mathbb{K}))$	$(\mathbf{Sz}(\mathbb{K}, \sigma), \mathcal{O}_{ST}(\mathbb{K}, \sigma))$

Table 8.2. Representations of general (dual) projectivity groups of some Moufang polygons.

Now consider $\Gamma = \mathsf{H}(3, q^2)$. From the proof of Theorem 8.3.3, we infer that the projectivity $[[\infty]; [0, 0, 0]; [0, 1, 0]; [\infty]]$ takes (b) to (b^q). Hence it follows that $\Pi(\Gamma) = \mathbf{PSL}^{(q)}(q^2)$ with the standard permutation representation. Since $\mathsf{H}(3, q^2)$ is a full subquadrangle of $\mathsf{H}(4, q^2)$, the above projectivity is also a projectivity in $\mathsf{H}(4, q^2)$, hence the result for $\Pi(\mathsf{H}(4, q^2))$.

By the last assertion of Lemma 8.4.6, we conclude $\Pi^*(\Gamma) = \Pi^*_+(\Gamma)$, for Γ isomorphic to $\mathsf{H}(d, q^2)$, $d \in \{3, 4\}$. $\qquad\square$

All projectivity groups of the Ree–Tits octagon $\mathsf{O}(\mathbb{K}, \sigma)$ ($\mathbb{K}$ not necessarily perfect) have been determined by JOSWIG & VAN MALDEGHEM [1995]. The result is that the general and special projectivity groups are extensions of $\mathbf{PSL}(2, \mathbb{K})$ by the multiplicative subgroup of $\mathbb{K}$ generated by the norms $\mathrm{N}(k)$, $k \in \mathbb{K}_\sigma^{(2)}$; the general and special dual projectivity groups are isomorphic to the general Suzuki group $\mathbf{GSz}(\mathbb{K}, \sigma)$ (with natural action on the Suzuki–Tits ovoid).

8.5 Projective embeddings of generalized quadrangles

The following definitions could be given for arbitrary geometries. But since we will not state results other than for generalized polygons, we might as well restrict ourselves to the polygons for the definitions.

8.5.1 Definitions. Let $\Gamma = (\mathcal{P}, \mathcal{L}, I)$ be a generalized polygon and $\mathbf{PG}(d, \mathbb{K})$ a d-dimensional projective space over some skew field $\mathbb{K}$ with $d \geq 3$. Let θ be an injective mapping from $\mathcal{P} \cup \mathcal{L}$ to the set of points and lines of $\mathbf{PG}(d, \mathbb{K})$ which maps points to points, lines to lines, preserves incidence and non-incidence, and such that the set $\mathcal{P}^\theta$ of points generates $\mathbf{PG}(d, \mathbb{K})$. Then we call the configuration in $\mathbf{PG}(d, \mathbb{K})$ defined by the image of $\mathcal{P} \cup \mathcal{L}$ under θ a **lax embedding of** Γ **(in** $\mathbf{PG}(d, \mathbb{K})$**)**. It is easily seen that for generalized quadrangles, the condition of preserving non-incidence follows from the other conditions. Now, a lax embedding is called a **weak embedding** if condition (W) is satisfied:

(W) For every point x of Γ, the set $(x^{\perp\perp})^\theta$ of points of $\mathbf{PG}(d, \mathbb{K})$ does not generate $\mathbf{PG}(d, \mathbb{K})$ linearly.

Let there be no confusion about *weak embeddings* and *weak quadrangles*. The distinction should always be clear from the context.

Also, a lax embedding is called a **flat embedding** if the following condition (R) is satisfied:

(R) For every point x of Γ, the set $(x^\perp)^\theta$ of points of $\mathbf{PG}(d, \mathbb{K})$ generates a plane in $\mathbf{PG}(d, \mathbb{K})$.

On the other hand, a lax embedding is called a **full embedding** if condition (F) is satisfied:

(F) For every line L of Γ, the restriction of θ to $\Gamma_1(L)$ is surjective on the set of points of $\mathbf{PG}(d, \mathbb{K})$ incident with L^θ.

Again, let there be no confusion about *full embeddings* and *full subpolygons*.

A lax embedding which is both flat and weak will be called an **ideal embedding** (or sometimes a **regular** embedding).

It is possible to classify all full ideal embeddings of finite generalized polygons (see Theorem 8.6.1 below; in particular, there are no ideal embeddings of finite octagons). For the other types of embeddings, additional assumptions must be made, e.g. on the dimension d of the projective space, on the order of the generalized n-gon Γ, or on n. Not surprisingly, the most results can be proved for $n = 4$ (also including the infinite case), a reasonable number of results are known for

$n = 6$, and only one — the above mentioned non-existence — result is known for generalized octagons.

A lot of proofs in this area use results which need additional background beyond the scope of this book. Therefore, we will not prove everything here. We will, however, provide partial proofs which, in my opinion, give the reader more geometric insight into the matter and help him to understand essentially why the given result is true.

In this section, we will be mainly concerned about full embeddings of generalized quadrangles, although some results are true for weak embeddings and thus will be proved as such.

Our eventual goal is to give a description of all full embeddings of quadrangles in projective spaces. We do that with a sequence of lemmas.

In the sequel, d denotes an integer with $d \geq 3$ and $\mathbb{K}$ is a given skew field. Also, we identify a polygon laxly embedded in $\mathbf{PG}(d, \mathbb{K})$ with its image in $\mathbf{PG}(d, \mathbb{K})$.

8.5.2 Lemma. *Let* Γ *be a generalized quadrangle laxly embedded in* $\mathbf{PG}(d, \mathbb{K})$ *and let* U *be a projective subspace of* $\mathbf{PG}(d, \mathbb{K})$. **If** U *contains at least one antiflag of* Γ, **then** *either* U *meets* Γ *in a weak full subquadrangle, or* U *meets* Γ *in subset of a pencil of lines.*

Proof. This follows immediately from the observation that, if a point p of Γ and a line L of Γ belong to U, then the unique line M of Γ meeting L and incident with p is also contained in U; and also, if a line L of Γ is in U, then all points of L in Γ are in U. $\qquad\square$

If, with the notation of the previous lemma, U intersects Γ in a weak full subquadrangle, then we denote that weak subquadrangle by $U \cap \Gamma$.

8.5.3 Lemma. *Every fully embedded weak quadrangle is also weakly embedded.*

Proof. Let Γ be a generalized quadrangle which is fully embedded in the projective space $\mathbf{PG}(d, \mathbb{K})$, for some integer $d \geq 3$ and a skew field $\mathbb{K}$. We prove the assertion by induction on d.

First let $d = 3$. Let p be any point of Γ. We have to show that all lines of Γ through p lie in some plane. Suppose this is not true. Then there is a set $\{L_1, L_2, L_3\}$ of non-coplanar lines of Γ incident with p. Let M be any line of Γ meeting L_3, but not incident with p. Then M meets the plane π spanned by L_1 and L_2 in a unique point x (of Γ) which is not collinear in Γ with p. Hence by Lemma 8.5.2, π meets Γ in a weak full subquadrangle, a contradiction (there cannot exist opposite lines in that subquadrangle because any two lines meet in π).

Now let $d \geq 4$ be arbitrary. Again let p be a point of Γ and suppose that the lines of Γ through p do not lie in some hyperplane. Then we can pick d lines $L_1, L_2, \ldots, L_d$ of Γ, all incident with p and such that the union of their point sets generates $\mathbf{PG}(d, \mathbb{K})$. It follows that $L_1, L_2, \ldots, L_{d-1}$ span a unique hyperplane η. Let M be any line of Γ meeting L_d, but not incident with p. Then M meets η in a point x of Γ and, by Lemma 8.5.2, the hyperplane η meets Γ in a full subquadrangle. The induction hypothesis now implies that $L_1, L_2, \ldots, L_{d-1}$ lie in a $(d-2)$-dimensional subspace of η, a contradiction. □

8.5.4 Lemma. *Let the weak quadrangle Γ be weakly embedded in $\mathbf{PG}(d, \mathbb{K})$. For every point p of Γ, the set of lines of Γ incident with p is contained in a unique hyperplane η_p of $\mathbf{PG}(d, \mathbb{K})$. So η_p contains all points of Γ collinear in Γ with p. Also, η_p does not contain any point of Γ opposite p in Γ.*

Proof. By Lemma 8.5.3, the points of Γ collinear in Γ with any point p of Γ generate a proper subspace η_p of $\mathbf{PG}(d, \mathbb{K})$. If η_p is not a hyperplane, then for any point x in Γ which is opposite p in Γ, the space generated by the points of η_p and $\{x\}$ is a proper subspace of $\mathbf{PG}(d, \mathbb{K})$. By Lemma 8.5.2, it meets Γ in a full and ideal subquadrangle which must coincide with Γ by Lemma 1.8.5 (see page 34), contradicting the fact that the point set of Γ generates $\mathbf{PG}(d, \mathbb{K})$.

Hence η_p is a hyperplane. But the same argument as above now shows that no point x of Γ opposite p in Γ can be contained in η_p. □

8.5.5 Notation. We will use η_p (as in the above proof) as the standard notation for the unique hyperplane containing $p^\perp$.

8.5.6 Definitions. Suppose the weak generalized quadrangle Γ is weakly embedded in $\mathbf{PG}(d, \mathbb{K})$. Let L be a line of $\mathbf{PG}(d, \mathbb{K})$ which is not a line of Γ. Then the number of points on L which belong to Γ is called the **degree of** L. The line L itself is called a **secant (line)**, provided its degree is at least 2. If the degree of a line L is equal to 1, then we call L a **tangent (line)**. Finally, if the degree of L is 0, then L is an **external line**.

8.5.7 Lemma. *All secant lines of a weakly embedded weak quadrangle have the same degree.*

Proof. Let the weak quadrangle $\Gamma = (\mathcal{P}, \mathcal{L}, \mathbf{I})$ be weakly embedded in the projective space $\mathbf{PG}(d, \mathbb{K})$. If $d = 3$, the result is true since it is easily seen that in this case the degree of any secant is equal to $t + 1$, where Γ has order (s, t), and every secant line is clearly the intersection of two planes η_x and η_y, $x, y \in \mathcal{P}$.

Now suppose $d > 3$ and let xy be a secant line, with x, y points of Γ. Let L and M be confluent lines of Γ incident with, respectively, x and y. Let L' and M' be two confluent lines meeting, respectively, L and M at unique points. Put $p = L \cap M$ and $p' = L' \cap M'$. Let $x', y' \in \Gamma_2(p')$ with $x' \mathbf{I} L'$ and $y' \mathbf{I} M'$. By the previous paragraph (intersecting Γ with a projective 3-space through L, L', M, M'), the degree of xy is equal to the degree of $x'y'$. Noting that L, M, L', M' are related via the perspectivity $[p; p']$, the result follows from Corollary 1.5.2. □

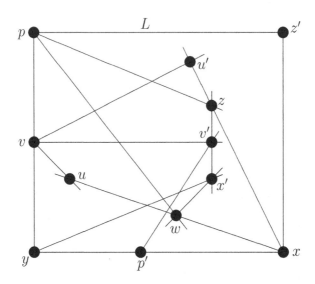

Figure 8.2.

8.5.8 Definition. We define the **degree** of a weakly embedded quadrangle as the degree of any secant line. This definition makes sense by the previous lemma.

We now embark on the classification of all generalized quadrangles which are fully embedded in some — for convenience finite-dimensional — projective space $\mathbf{PG}(d, \mathbb{K})$, with $\mathbb{K}$ any skew field. We will distinguish between the cases where the degree is 2 and larger than 2, although the statements will not be totally disjoint. We start with the case where the degree f of the embedding is larger than 2, and where $d = 3$. This is the key result for the general case $d \geq 3$. We need a proposition and a lemma first. The basic idea follows DIENST [1980b], but we have tried to be a little more explicit. Also, we have avoided the theory of semi-quadratic sets.

8.5.9 Proposition. *Let* $\Gamma = (\mathcal{P}, \mathcal{L}, \mathbf{I})$ *be a generalized quadrangle weakly embedded of degree* $f > 2$ *in* $\mathbf{PG}(3, \mathbb{K})$. *Then* Γ *is a point-regular Moufang quadrangle and the little projective group of* Γ *is induced by special projective linear transformations of* $\mathbf{PG}(3, \mathbb{K})$.

Proof. By Lemma 8.5.4, the set of points $x^{\perp}$ is contained in a unique plane, for every point x of Γ. Hence, for two opposite points x and y of Γ, the trace $x^{\perp} \cap y^{\perp}$ is a set of points incident with a secant line of $\mathbf{PG}(3, \mathbb{K})$. So it is determined by any two of its points. This shows that every point x of Γ is regular.

Now we claim that Γ is a half Moufang quadrangle, or, more exactly, Γ admits all line-elations and every line-elation is induced by an **elation** of $\mathbf{PG}(3, \mathbb{K})$ (i.e., a collineation fixing all points in a certain hyperplane H (the **axis**), and fixing all hyperplanes through a certain point (the **centre**) in H). Let p be a point of Γ and let x be any point of Γ opposite p (see Figure 8.2). Let y, z be two distinct points in p^x, and let x' be an arbitrary point in $y^\perp \cap z^\perp$ with $x' \neq p$. Then we show that there exists a central collineation with centre p (in view of the regularity of points, all line-elations must indeed be central elations) mapping xy to $x'y$, i.e., mapping x to x'. The only candidate for such a collineation is the elation θ in $\mathbf{PG}(3, \mathbb{K})$ with axis the plane η_p, centre p, and mapping x to x'. We now show that θ induces a permutation of the point set $\mathcal{P}$ of Γ. We may assume $x \neq x'$. Consider any point u in Γ which is collinear with x (in Γ). If u is collinear with p, then it is fixed under θ, and so $u^\theta \in \mathcal{P}$. If u is not collinear with p, then we may assume that $u \notin \Gamma_2(y)$ (otherwise interchange the roles of y and z). Consider the point v on py collinear with u, and consider the point w on the line xu collinear with p. Note that u is the intersection of the plane η_v with the line wx. By the regularity of p, the points x' and w are collinear in Γ. Let u' be the projection of v onto wx'. Then u' is the intersection of the plane η_v, which is fixed under θ, with the line wx', which is the image under θ of the line wx. Hence $u' = u^\theta$. By repeating this argument, we conclude with Theorem 1.7.13 (see page 32) and Theorem 1.7.15 that $\mathcal{P}^\theta = \mathcal{P}$. Considering the inverse mapping θ^{-1}, we see that θ is a permutation of $\mathcal{P}$. Since every line is determined by its points, it now follows immediately that θ induces a central collineation in Γ, which we will denote by $\theta_p(x, x')$. We have proved our claim. In fact, this claim was proved for a larger class of geometries by LEFÈVRE-PERCSY [1981b].

We now establish the point-elations in Γ. We use the same notation as in the previous paragraph. Moreover, let $v' = \mathrm{proj}_{x'z} v$ and $p' = \mathrm{proj}_{xy} v'$. We claim that there is a (p, py, y)-elation mapping x to p'. The map $\theta' = \theta_p(x, x')\theta_v(x', p')$ fixes all points on py and fixes the line xy. Moreover, it maps x to p', as required. But the restriction to the projective plane η_y of θ' is the composition of two elations with same axis py (considered as a line of $\mathbf{PG}(3, \mathbb{K})$) and distinct centres p and p' on py. Hence this restriction is again an elation with axis py and some centre y'. Since xy is fixed, it follows that $y = y'$, hence all lines of Γ through y are fixed. Suppose now that θ' maps pz to some line L. Since L must also be incident with p, we may consider the central collineation $\theta_y(z', z)$, where z' is the projection of x onto L. The elation $\theta_y(z', z)$ fixes all points on py, on xy, and all lines (of Γ) through y, and hence so does $\theta'\theta_y(z', z)$. Moreover, θ fixes pz, and a similar argument to that above shows that θ fixes all lines of Γ through p. Our claim follows. The proposition is proved. $\qquad\square$

We now wish to show that every generalized quadrangle Γ fully embedded in $\mathbf{PG}(3, \mathbb{K})$ is **embedded in a polarity**, i.e., there exists a polarity τ of $\mathbf{PG}(3, \mathbb{K})$ such that the points and lines of Γ are (some) absolute elements of τ. Therefore, we need first an algebraic result, the crux of which is due to HUA [1949], but we have adapted it to our situation.

8.5.10 Lemma. *Let σ be a permutation of the skew field $\mathbb{K}$ fixing 1 and such that*

$$(x + y)^\sigma = x^\sigma + y^\sigma, \tag{8.1}$$
$$((1 + x^{-1})^{-1})^\sigma = (1 + (x^\sigma)^{-1})^{-1}, \tag{8.2}$$

for all $x, y \in \mathbb{K}$, $x \neq 0$. Then σ is **either** *an automorphism* **or** *an anti-automorphism of $\mathbb{K}$. If, moreover, σ is involutory and*

$$(yx^\sigma x)^\sigma = x^\sigma xy^\sigma, \tag{8.3}$$

for all $x, y \in \mathbb{K}$, **then** *σ is an anti-automorphism of $\mathbb{K}$.*

Proof. We first claim that, under conditions (8.1) and (8.2), σ maps x^{-1} to the inverse of x^σ, and then we are reduced to HUA's original problem. Indeed, substitute x by $x - 1$ in (8.2). Then the left-hand side becomes, in view of (8.1),

$$((1 + (x - 1)^{-1})^{-1})^\sigma = ((x - 1)x^{-1})^\sigma$$
$$= 1 - (x^{-1})^\sigma,$$

which can be checked by multiplying $(1 + (x-1)^{-1})$ with $(x-1)x^{-1}$. The right-hand side of (8.2) becomes after substituting $x - 1$ for x,

$$(1 + ((x - 1)^\sigma)^{-1})^{-1} = (x^\sigma - 1)(x^\sigma)^{-1}$$
$$= 1 - (x^\sigma)^{-1},$$

and our claim that $(x^{-1})^\sigma = (x^\sigma)^{-1}$ is proved. Now one can verify the following identity, in the commutative case due to MENDELSOHN [1944]:

$$(x^{-1} + (y^{-1} - x)^{-1})^{-1} = x - xyx.$$

Since σ preserves addition and inverse, it preserves $x - xyx$, i.e.,

$$(xyx)^\sigma = x^\sigma y^\sigma x^\sigma. \tag{8.4}$$

With $y = 1$, we obtain $(x^2)^\sigma = (x^\sigma)^2$. Replacing x by $x + y$ in this equation, we obtain

$$(xy)^\sigma + (yx)^\sigma = x^\sigma y^\sigma + y^\sigma x^\sigma, \tag{8.5}$$

which for $x, y \neq 0$, and in view of $yx = x(y(xy)^{-1}y)x$ and (8.4), we may write as

$$(xy)^\sigma + x^\sigma y^\sigma ((xy)^\sigma)^{-1} y^\sigma x^\sigma = x^\sigma y^\sigma + y^\sigma x^\sigma.$$

We can now rewrite this as follows:

$$((xy)^\sigma - x^\sigma y^\sigma)((xy)^\sigma)^{-1}((xy)^\sigma - y^\sigma x^\sigma) = 0. \tag{8.6}$$

So this means that either $(xy)^\sigma = x^\sigma y^\sigma$, or $(xy)^\sigma = y^\sigma x^\sigma$. Now fix $x \in \mathbb{K}$. The set of all $y \in \mathbb{K}$ for which $(xy)^\sigma = x^\sigma y^\sigma$ is a subgroup S_x^+ of the additive group of $\mathbb{K}$. Similarly, the set of all $y \in \mathbb{K}$ for which we have $(xy)^\sigma = y^\sigma x^\sigma$ forms a subgroup S_x^- of $\mathbb{K}$. But by our above result, $S_x^+ \cup S_x^- = \mathbb{K}$, hence either $S_x^+ = \mathbb{K}$ or $S_x^- = \mathbb{K}$ (indeed, if $u \in S_x^+ \setminus S_x^-$ and $v \in S_x^- \setminus S_x^+$, then $u + v \in \mathbb{K} \setminus (S_x^+ \cup S_x^-)$). Now the set of all x of $\mathbb{K}$ for which $S_x^+ = \mathbb{K}$ forms a subgroup S^+ of the additive group of $\mathbb{K}$. Similarly the set of all $x \in \mathbb{K}$ for which $S_x^- = \mathbb{K}$ forms a subgroup S^- of $\mathbb{K}$. Again $S^+ \cup S^- = \mathbb{K}$, and so one of $S^+ = \mathbb{K}$ or $S^- = \mathbb{K}$ holds. This means that σ is either an automorphism or an anti-automorphism.

Suppose (8.3) holds, and that σ is an automorphism with $\sigma^2 = 1$. We want to show that in this case $\mathbb{K}$ is commutative (and hence σ is also an anti-automorphism).

Let $\mathbb{F}$ be the field of fixed elements of σ. We claim that the dimension of $\mathbb{K}$, as a left vector space over $\mathbb{F}$, is at most 2. This is clear if $\sigma = 1$; otherwise we find an element $y \in \mathbb{K} \setminus \mathbb{F}$, and we can write every element $x \in \mathbb{K}$ as

$$x = (x - (x - x^\sigma)(y - y^\sigma)^{-1}y) + (x - x^\sigma)(y - y^\sigma)^{-1}y =: a + by,$$

with $a, b \in \mathbb{F}$. This shows that $\mathbb{K} = \mathbb{F} + \mathbb{F}y$.

Replacing x by $x + 1$ in (8.3) shows that $x^\sigma + x$ belongs to the centre Z of $\mathbb{K}$, for every $x \in \mathbb{K}$, hence the image A of the $\mathbb{F}$-linear map $x \mapsto x^\sigma + x$ is contained in Z. If $A \neq \{0\}$, then A is at least one-dimensional over $\mathbb{F}$ and hence $\mathbb{K} = A + Ay' \subseteq Z + Zy'$, for some $y' \in \mathbb{K}$. This implies that $\mathbb{K}$ is commutative. So it remains to consider the case where $A = \{0\}$, i.e., we assume $x^\sigma = -x$, for all $x \in \mathbb{K}$. Putting $x = 1$, we see that $\mathbb{K}$ has characteristic 2 and $x^\sigma = x$. This implies that $\sigma = 1$ and (8.3) tells us that $x^2 \in Z$, for all $x \in \mathbb{K}$. Now suppose $x, y \in \mathbb{K}$ do not commute. Writing

$$xy = (xy)^2 y^{-2} x^{-2} yx = zyx,$$

with $1 \neq z = (xy)^2 y^{-2} x^{-2} \in Z$, we see that

$$(1 + z)yx = yx + xy = (x + y)^2 + x^2 + y^2 \in Z.$$

Hence $yx \in Z$, implying $xyx = x(yx) = (yx)x$, so $xy = yx$. We have shown that $\mathbb{K}$ is commutative.

Hence the lemma is completely proved. $\qquad\qquad\square$

We are now ready to classify all full embeddings of generalized quadrangles in $\mathbf{PG}(3, \mathbb{K})$.

8.5.11 Theorem (Dienst [1980b]). *Let Γ be a generalized quadrangle fully embedded in $\mathbf{PG}(3, \mathbb{K})$. Then one of the four following cases occurs:*

(i) *The characteristic of $\mathbb{K}$ is not equal to 2 and there exists a $(\sigma, 1)$-linear form g on the vector space V underlying $\mathbf{PG}(3, \mathbb{K})$ with σ an anti-automorphism of order 2, such that $\Gamma = Q(V, q)$, where q is the associated σ-quadratic form.*

(ii) *The characteristic of* $\mathbb{K}$ *is equal to 2 and there exist a Hermitian variety* $\mathcal{H}$
and a $(\sigma, 1)$*-linear form* g *on the vector space* V *underlying* $\mathbf{PG}(3, \mathbb{K})$ *with*
σ *an anti-automorphism of order 2 and with associated* σ*-quadratic form* q,
such that Γ *contains* $\mathsf{Q}(V, q)$ *as a full subquadrangle and such that* Γ *is itself*
contained in $\Gamma(\mathcal{H})$ *as a full subquadrangle.*

(iii) *The skew field* $\mathbb{K}$ *is commutative (and has characteristic different from 2),*
Γ *is the generalized quadrangle* $\mathsf{W}(\mathbb{K})$ *and the embedding is the standard one*
naturally arising from a symplectic polarity in $\mathbf{PG}(3, \mathbb{K})$.

(iv) *The skew field* $\mathbb{K}$ *is commutative and has characteristic 2 and* Γ *is a full*
mixed subquadrangle of a symplectic quadrangle with a standard embedding
arising from a symplectic polarity in $\mathbf{PG}(3, \mathbb{K})$.

Proof. The strategy of the proof is to construct first the anti-automorphism σ.
From this, the result will readily follow. As usual, we write scalars on the right.

So let Γ be fully embedded in $\mathbf{PG}(3, \mathbb{K})$. We consider four points p_{-2}, p_{-1}, p_1 and
p_2 of an apartment and we assume that p_i is opposite p_{-i}, $i = 1, 2$. We introduce
coordinates $p_1(1, 0, 0, 0), p_2(0, 1, 0, 0), p_{-1}(0, 0, 1, 0)$ and $p_{-2}(0, 0, 0, 1)$. We consider
an arbitrary point p, $p_1 \neq p \neq p_2$ on $p_1 p_2$ and give it coordinates $p(1, 1, 0, 0)$. The
unique point p' on $p_{-1} p_{-2}$ collinear in Γ with p is given the coordinates $p'(0, 0, 1, 1)$.
Finally we may assume that the point $p_0(0, 1, 0, 1)$ belongs to Γ, since Γ is thick,
and since the secant lines intersect Γ in at least three points. Notice that p_0 is collinear in Γ with p_1 and p_{-1}. Hence the unique point p^* of Γ collinear with p_0 and
incident with pp' lies in the plane $p_{-1} p_0 p_1$ (by Lemma 8.5.4) and therefore has coordinates $(1, 1, 1, 1)$. Similarly, the unique point p_0' on $p_0 p^*$ collinear with p_2 lies in
the plane $p_1 p_2 p_{-1}$ and so it has coordinates $(1, 0, 1, 0)$. Denote the line $p_0 p_0'$ by L_0.

It is clear that whenever we know two lines through a point, then we can calculate
the projection of that point onto any desired line of Γ by the method we applied in
the previous paragraph. Let us refer to this method as the *line–plane intersection*
method.

Now we define a permutation σ of $\mathbb{K}$ as follows. For $x \in \mathbb{K}$, we define $(0, 0, 1, x^\sigma)$
to be the point on $p_{-1} p_{-2}$ collinear with $(x, 1, 0, 0)$. Now we use Proposition 8.5.9.
We consider the $(p_{-2}, p_{-2} p_1, p_1)$-elation mapping $(0, 1, 0, 0)$ to $(y, 1, 0, 0)$, for some
$y \in \mathbb{K}$, and hence mapping $(0, 0, 1, 0)$ to $(0, 0, 1, y^\sigma)$. Since this elation comes
from a projective linear transformation of $\mathbf{PG}(3, \mathbb{K})$, it maps the point $(x, 1, 0, 0)$
onto the point $(x + y, 1, 0, 0)$, and it maps the point $(0, 0, 1, x^\sigma)$ onto the point
$(0, 0, 1, x^\sigma + y^\sigma)$. Since collinearity is preserved, we see that σ preserves the addition
in $\mathbb{K}$. Now consider the $(p_{-1}, p_{-1} p_2, p_2)$-elation θ which maps p_1 onto p. Then
θ maps $(1, x^{-1}, 0, 0)$ onto $(1, 1 + x^{-1}, 0, 0)$, hence θ maps $(x, 1, 0, 0)$ onto $((1 + x^{-1})^{-1}, 1, 0, 0)$. Applying σ, we deduce that

$$((1 + x^{-1})^{-1})^\sigma = (1 + (x^\sigma)^{-1})^{-1},$$

and so it follows from Lemma 8.5.10 that σ is an automorphism or an anti-
automorphism of $\mathbb{K}$.

By the line–plane intersection method, one calculates that the projection of $p_x = (x, 1, 0, 0)$ onto L_0 is $p'_x = (1, x^\sigma, 1, x^\sigma)$. Also, the projection of p'_x onto $p_{-2}p_{-1}$ is the point $p''_x(0, 0, x, 1)$. By definition of σ, $p''_x = (0, 0, 1, x^{-1})$ is collinear with $p'''_x = ((x^{-1})^{\sigma^{-1}}, 1, 0, 0)$, hence with $(1, x^{\sigma^{-1}}, 0, 0)$. Since the points p_{-1}, p'_x, p''_x and p'''_x are coplanar, we see that $(1, x^\sigma)$ is proportional to $((x^{-1})^{\sigma^{-1}}, 1)$, and hence σ has order at most 2.

By the same line–plane intersection method, one calculates that the projection of p'_x onto the line $p_{-2}p_1$ is the point $p^*_x(-x, 0, 0, 1)$. Similarly, the projection of p^*_x onto $p_{-1}p_2$ is the point $p^{**}_x(0, x^\sigma, 1, 0)$. Hence we conclude that the points $(x, 0, 0, 1)$ and $(0, -x^\sigma, 1, 0)$ are collinear in Γ, and consequently also the points $(0, x, 1, 0)$ and $(-x^\sigma, 0, 0, 1)$ are collinear in Γ.

Now we make the following trivial, but useful remark. Let u be any point of Γ and suppose M is a line of $\mathbf{PG}(3, \mathbb{K})$ containing u and meeting p_1p_2 in a point u_+, and meeting $p_{-1}p_{-2}$ in a point u_-. Then by Lemma 8.5.4 the unique point w_+ on p_1p_2 collinear in Γ with u is also collinear with u_- (because everything must happen in the plane uu_-w_+). Similarly the unique point w_- on $p_{-1}p_{-2}$ collinear in Γ with u is also collinear with u_+. Hence we have a way of determining the projections of a point onto the sides of the quadrangle $(p_1, p_2, p_{-1}, p_{-2})$. We refer to that method as the *opposite-line method*.

Now we want to prove that σ is an anti-automorphism. Let u be the point with coordinates $(0, 0, x, 1)$. Then by the line–plane intersection method, we calculate that it is collinear with the point $v(1, x^\sigma, 1, 0)$ on the line p'_0p_2. The same method allows one to find that the projection of p_1 onto the line uv has coordinates $(x, x^\sigma x, 0, -1)$. Hence also the point $w(yx^\sigma x, x^\sigma x, 0, -1)$, $y \in \mathbb{K}$, belongs to Γ. Note that the line $p_{-2}w$ of $\mathbf{PG}(3, \mathbb{K})$ meets the lines p_1p_2 and $p_{-1}p_{-2}$. By the opposite-line method we deduce that $(0, 0, 1, y^\sigma)$ lies in η_w. The line p_2w meets $p_{-1}p_{-2}$ in $(yx^\sigma x, 0, 0, -1)$; hence the opposite-line method implies that the point $(0, (yx^\sigma x)^\sigma, 1, 0)$ belongs to η_w. So the four points

$$(1, 0, 0, 0), (yx^\sigma x, x^\sigma x, 0, -1), (0, 0, 1, y^\sigma), (0, (yx^\sigma x)^\sigma, 1, 0)$$

are coplanar. This easily implies

$$(yx^\sigma x)^\sigma = x^\sigma xy^\sigma,$$

which allows us to conclude by Lemma 8.5.10 that σ is an anti-automorphism.

Now suppose first that σ is non-trivial. Let $u(x_1, x_2, x_{-1}, x_{-2})$ be an arbitrary point of Γ. Then we may apply the opposite-line method to the line of $\mathbf{PG}(3, \mathbb{K})$ joining $(x_1, x_2, 0, 0)$ and $(0, 0, x_{-1}, x_{-2})$ and we obtain that the points $w_1(x_{-1}^{-\sigma}, x_{-2}^{-\sigma}, 0, 0)$ and $w_2(0, 0, x_1^{-\sigma}, x_2^{-\sigma})$ belong to η_u. Similarly, the points $w_3(x_{-1}^{-\sigma}, 0, 0, -x_2^{-\sigma})$ and $w_4(0, x_{-2}^{-\sigma}, -x_1^{-\sigma}, 0)$ belong to η_u. Expressing in the coordinates that the points u, w_1, w_2, w_4 are coplanar (w_3 is then automatically coplanar with these points) we obtain the relation

$$x_1^\sigma x_{-1} - x_{-1}^\sigma x_1 - x_2^\sigma x_{-2} + x_{-2}^\sigma x_2 = 0,$$

(for all coordinates non-zero, but this relation is easily extended) which shows us that the points of Γ all lie on a Hermitian variety $\mathcal{H}$ with points and lines, but no planes (in view of the dimension). Now (i) immediately follows from Subsection 2.3.9, Proposition 5.9.4 (page 235) and Proposition 5.9.6 (page 236). Noting that the intersection of a $\mathsf{Q}(V, q)$ on $\mathcal{H}$ with Γ is a full subquadrangle of $\mathsf{Q}(V, q)$, we also see that (ii) is valid, by the same references and Remark 3.4.10 (see page 107).

So we may assume that σ is the identity and consequently that $\mathbb{K}$ is commutative. If we use that information in the calculations performed in the previous paragraph, then we readily obtain the equation of the plane η_u:

$$\eta_u \leftrightarrow x_{-1}X_1 - x_1X_{-1} - x_{-2}X_2 + x_2X_{-2} = 0,$$

(for all coefficients different from zero, but again this is easily extended) which shows that the lines of Γ are (some) lines of a symplectic polarity in $\mathbf{PG}(3, \mathbb{K})$. So (iii) follows now from Proposition 5.9.4 (on page 235) and (iv) is true by definition of mixed quadrangles.

This completes the proof of the theorem. $\square$

A similar precise classification exists for fully embedded quadrangles of degree 2 in finite-dimensional projective spaces. In fact, it is convenient here also to consider weak quadrangles. Before we state and prove the result, we need a lemma.

8.5.12 Lemma. *Let Γ be a generalized quadrangle and Γ' a weak full subquadrangle. If there is a point p of Γ' such that there is only one line L of Γ incident with p and not belonging to Γ', **then** $\Gamma \cong \mathsf{W}(2)$ and Γ' is a 3×3 grid.*

Proof. Let $x \mathbf{I} L$, $x \neq p$. Let $M \mathbf{I} x$, $M \neq L$. Finally, let $y \mathbf{I} M$, $y \neq x$, y not contained in Γ'. The set of points of Γ' collinear with y forms an ovoid $\mathcal{O}$ of Γ' which does not contain p. If $\mathcal{O}$ contains two different points x_1, x_2 opposite p, then at least one of these points, say x_1, is not incident with M, and hence the projection of yx_1 onto p is distinct from L and does not belong to Γ', a contradiction. Hence there is at most one point of $\mathcal{O}$ opposite p. The result now follows rather easily by considering lines at distance 3 from p in Γ', which all must contain points of $\mathcal{O}$. $\square$

8.5.13 Theorem. *Let Γ be a weak generalized quadrangle fully embedded in $\mathbf{PG}(d, \mathbb{K})$ of degree 2. Then $\mathbb{K}$ is commutative and Γ arises from a quadric Q of Witt index 2, i.e., the points and lines of Γ are (exactly) the points and lines of Q.*

Proof. We show the result by induction on d. First let $d = 3$. Clearly the degree of the embedding is equal to $t + 1$, where Γ has order $(|\mathbb{K}|, t)$, hence $t = 1$ and we are dealing with non-thick weak quadrangles. So Γ is a $(|\mathbb{K}| + 1) \times (|\mathbb{K}| + 1)$ grid. Since $|\mathbb{K}| > 1$, we can consider three lines L, M and N which are pairwise opposite. Without loss of generality, we can take $L = \langle(1, 0, 0, 0), (0, 1, 0, 0)\rangle$,

$M = \langle(0,0,1,0),(0,0,0,1)\rangle$ and $N = \langle(1,0,0,1),(0,1,1,0)\rangle$. In Γ, there is a line L_x through the point $p(x,1,0,0)$ meeting both M and N, $x \in \mathbb{K} \cup \{\infty\}$. Since in $\mathbf{PG}(3,\mathbb{K})$, there is only one such line, namely the intersection of the planes spanned by p and M, or by p and N, we can determine this line. After a short calculation one finds $L_x = \langle(x,1,0,0),(0,0,1,x)\rangle$, $x \in \mathbb{K} \cup \{\infty\}$ (note that $L_\infty = \langle(1,0,0,0),(0,0,0,1)\rangle$). Now, the point $w_y(0,1,y,0)$ belongs to Γ (it is incident with L_0) and so there should be a line in $\mathbf{PG}(3,\mathbb{K})$ through w_y meeting every line L_x. Now, the unique line K_y through w_y meeting L_1 and L_∞ is $\langle(0,1,y,0),(1,0,0,y)\rangle$. Observing that L_x meets K_y, one easily obtains $xy = yx$, hence $\mathbb{K}$ is commutative and Γ is a ruled quadric in $\mathbf{PG}(3,\mathbb{K})$ (of type D_2).

Now let $d > 3$ be arbitrary. Consider any two opposite lines in Γ and take any hyperplane H_1 of $\mathbf{PG}(d,\mathbb{K})$ through them. Clearly H_1 meets Γ in a weak subquadrangle, hence we can apply induction and get a quadric Q_1 in H_1 which is precisely the intersection of Γ with H_1. Consider in Q_1 a point p and let H' be the intersection of H_1 with η_p. Let v be a point of Γ collinear in Γ with p, but not contained in H'. Let H_2 be a hyperplane in $\mathbf{PG}(d,\mathbb{K})$ containing H' but distinct from η_p. Then H_2 does not contain v. It is easily seen that H_2 meets Γ in a weak subquadrangle (indeed, for any point u in $H_1 \setminus H'$ of Γ, there is at least one point w in $H_2 \setminus H'$ collinear with u in Γ, because $\eta_u \neq H_1$. This point w is opposite p, hence there are apartments of Γ in H_2.) By the induction hypothesis, the intersection of Γ with H_2 is a quadric Q_2. Note that there exists a line of Γ through v skew to H' (indeed, $\eta_v \neq \eta_p$ since otherwise η_v would contain points opposite v).

We now claim that there is a unique quadric Q containing the quadrics Q_1, Q_2 and the point v. Indeed, we can choose coordinates in $\mathbf{PG}(d,\mathbb{K})$ in such a way that H_1 has equation $X_0 = 0$ and H_2 has equation $X_d = 0$. Let $F'(X_1, X_2, \ldots, X_{d-1}) = X_0 = X_d = 0$ be an equation of $Q_1 \cap Q_2$ (determined up to a non-zero multiplicative constant). Let $F_1(X_1, X_2, \ldots, X_d) = X_0 = 0$ and $F_2(X_0, X_1, \ldots, X_{d-1}) = X_d = 0$ be an equation of, respectively, Q_1 and Q_2. We can choose the constants in such a way that

$$F_1(X_1, X_2, \ldots, X_d) = F'(X_1, X_2, \ldots, X_{d-1}) + \sum_{i=1}^{d-1} a_i X_i X_d,$$

$$F_2(X_0, X_1, \ldots, X_{d-1}) = F'(X_1, X_2, \ldots, X_{d-1}) + \sum_{j=0}^{d} b_j X_j X_0,$$

where $a_i, b_j \in \mathbb{K}$. Consider the quadric Q_c with equation

$$F'(X_1, X_2, \ldots, X_{d-1}) + \sum_{i=1}^{d-1} a_i X_i X_d + \sum_{j=0}^{d} b_j X_j X_0 + c X_0 X_d = 0.$$

Then we can determine c in such a way that $v \in Q_c$ because neither the X_0-coordinate nor the X_d-coordinate of v is equal to zero. It is now clear that we have a unique quadric $Q = Q_c$ as claimed above.

Now we claim that every point of Γ lies on Q. Indeed, first let v' be any point of Γ collinear with p, but not on the line pv. We can take v' outside Q_1 and Q_2 (i.e., v' does not belong to H', otherwise the claim is clear). Let M be a line of Γ through v not meeting H'. Then, since M contains at least three points of Q (one point in each intersection $M \cap Q_i$, $i = 1, 2$, and also v of course), M lies in Q. Let N be the unique line of Γ meeting M and containing v'. Note that N is not contained in η_p, otherwise $p \mathbf{I} N$ and so $N = pv$, implying $v' \mathbf{I} pv$. We may assume that $N \cap M$ is not contained in Q_1 (otherwise interchange the roles of Q_1 and Q_2). Consider any line M' in Q_1 meeting M. Then M, M', N generate a three-dimensional subspace U of $\mathbf{PG}(d, \mathbb{K})$ which clearly meets Γ in a weak subquadrangle Γ'. By the first part of the proof, Γ' is a ruled quadric Q'. This quadric Q' now meets Q_1 in two confluent lines M' and, say, M''. Note that M'' does not belong to H'. Indeed, since M'' and M belong to the same set of generators on Q' (as they both meet M'), N must meet M'' (since it meets M). But this implies that, if M'' belongs to H', then N belongs to η_p, a contradicton. The quadric Q' meets Q_2 in at least two points, hence either in a non-singular conic or in two lines. In the latter case Q' is completely determined by the five lines M, M', M'', and two lines in H_2, say, N', N'', which are different from M'' because M'' does not belong to H'. Since $U \cap Q$ must also be a quadric (determined by the same five lines), it must coincide with Q', hence $v' \in Q$. In the former case, the quadric Q' is uniquely determined by the three lines M, M', M'' and the irreducible conic $\mathcal{O}$. Indeed, if $|\mathbb{K}| > 3$, then there are at least two points x_1, x_2 on $\mathcal{O}$ which are not incident with M, M' or M''. The line L_i, $i = 1, 2$, in $\mathbf{PG}(d, \mathbb{K})$ containing x_i and meeting both M and M' must then belong to Q', and Q' is now determined by M', L_1 and L_2. If $|\mathbb{K}| = 3$, then the same argument still works if $M' \cap M''$ belongs to $\mathcal{O}$. If it does not, then we have one line $L_1 \neq M'$ meeting both M, M'' and $\mathcal{O}$ in different points. But if y is the intersection of $\mathcal{O}$ with M, and if y'' is the intersection of $\mathcal{O}$ with M'', then y and y'' are not collinear on Q'. Also, y is neither collinear with $M' \cap M''$, nor with $L_1 \cap M''$. Hence it is collinear with the unique point y' on M'' distinct from the three points mentioned, and Q' is determined by M', L_1 and yy'. If $|\mathbb{K}| = 2$, then with a similar argument we conclude that $\mathcal{O}, M, M'$ and M'' determine Q' uniquely. Hence $Q' \subseteq Q$ and so $v' \in Q$.

Now we show that we may assume that a point v' as in the previous paragraph really exists. Indeed, if not, then there is a unique line through p which does not belong to the full subquadrangle Q_1 of Γ. By Lemma 8.5.12, Γ has order $(2, 2)$ and Q_1 has order $(2, 1)$. But in that case, it is easily seen that $d = 4$, $|\mathbb{K}| = 2$ and Q is non-degenerate, hence it contains 15 points. But v together with the points of $Q_1 \cup Q_2$ already account for 14 of them. Since there must be a line on Q through p not in $H_1 \cup H_2$, we conclude that this must be the line pv.

Hence we can interchange the roles of v and v' now to obtain that all points of pv also belong to Q. Consequently all points of $\Gamma_2(p)$ belong to Q.

Now let u be any point of Γ belonging neither to $\Gamma_2(p)$, nor to $Q_1 \cup Q_2$. If L is a line of Γ incident with u and meeting pv, then it also meets Q_1 and Q_2, and the

POINTS	
Coordinates in $\mathsf{W}(\mathbb{K})$	Coordinates in $\mathbf{PG}(4, \mathbb{K})$
(∞)	$(1, 0, 0, 0, 0)$
(a)	$(a, 0, 1, 0, 0)$
(k, b)	$(-b, 0, k, 1, \sqrt{k_1})$
(a, l, a')	$(l - aa', 1, -a', -a, \sqrt{l_1})$

LINES	
Coordinates in $\mathsf{W}(\mathbb{K})$	Coordinates in $\mathbf{PG}(4, \mathbb{K})$
$[\infty]$	$\langle (1, 0, 0, 0, 0), (0, 0, 1, 0, 0) \rangle$
$[k]$	$\langle (1, 0, 0, 0, 0), (0, 0, k, 1, \sqrt{k_1}) \rangle$
$[a, l]$	$\langle (a, 0, 1, 0, 0), (l, 1, 0, -a, \sqrt{l_1}) \rangle$
$[k, b, k']$	$\langle (-b, 0, k, 1, \sqrt{k_1}), (k', 1, -b, 0, \sqrt{k_1'}) \rangle$

Table 8.3. An embedding of infinite degree of $\mathsf{W}(\mathbb{K})$ into $\mathbf{PG}(4, \mathbb{K})$.

three intersection points are distinct. But these three points belong to Q, hence L does, too, and consequently $u \in Q$. This proves our claim.

Finally we claim that all points of Q belong to Γ. Indeed, suppose there is a point z of Q not belonging to Γ. Then consider a line L_z of Q through z. Since z is contained neither in Q_1 nor in Q_2, we may assume that L_z does not meet H' (for if it does, then consider another point z' on L_z, $z' \neq z$ and z' not in H'; since the lines of Q through z and z', respectively, generate distinct hyperplanes, they cannot both contain H', hence there exists at least one line of Q through z or z' outside the hyperplane spanned by H' and z). Hence L_z meets Q_i in z_i, $i = 1, 2$, $z_1 \neq z_2$. Clearly z_i is a point of Γ and z_1 is opposite z_2. Hence we have a triangle on Q, so Q contains planes. In particular, Q contains a plane π through M (with the above notation). But then π meets Q_i in a line L_i, $i = 1, 2$, and L_1, L_2, M form a triangle in Γ, a contradiction. The claim follows.

The theorem is proved. □

Now let us turn to the general case for arbitrary degree. We cannot expect that Theorem 8.5.11 remains true for arbitrary dimension and characteristic 2, even for degree larger than 2. Indeed, let $\mathbb{K}$ be a non-perfect field of characteristic 2 and let $\mathbb{K} = V_1 \oplus V_2$, viewed as vector space over $\mathbb{K}^2$, where V_1 is one-dimensional and contains 1. So we may identify V_1 with $\mathbb{K}^2$. For $k \in \mathbb{K}$, we write $k = k_1 + k_2$, where $k_1 \in V_1 = \mathbb{K}^2$ and $k_2 \in V_2$. We now write down explicitly an embedding of the symplectic quadrangle $\mathsf{W}(\mathbb{K})$ in $\mathbf{PG}(4, \mathbb{K})$ which has infinite degree. Therefore, we use the coordinates introduced in Subsection 3.4.1 (see page 100). The embedding is given in Table 8.3, where we have denoted, for each element $k_1 \in \mathbb{K}^2$, by $\sqrt{k_1}$ the unique element k_1 of $\mathbb{K}$ such that $\sqrt{k_1}^2 = k_1$. Notice that the embedded quadrangle contains the quadric with equation $X_0 X_1 + X_2 X_3 = X_4^2$, which is an embedding of the dual symplectic subquadrangle $\mathsf{Q}(\mathbb{K}, \mathbb{K}^2; \mathbb{K}, \mathbb{K}^2)$ of $\mathsf{W}(\mathbb{K}) = \mathsf{Q}(\mathbb{K}, \mathbb{K}; \mathbb{K}, \mathbb{K})$. Also, it is readily checked that the degree of the embedding is infinite precisely because $V_2 \neq \{0\}$. Now, the point with coordinates $(0,0,0,0,1)$ is not incident (in $\mathbf{PG}(4, \mathbb{K})$) with any secant line. Indeed, this follows from the fact that the last coordinate in $\mathbf{PG}(4, \mathbb{K})$ of any point of $\mathsf{W}(\mathbb{K})$ is unambiguously determined by the other coordinates. So, if a polarity exists such that the corresponding set of absolute points and lines contains the set of points and lines (respectively) of the embedded $\mathsf{W}(\mathbb{K})$, then it must be *degenerate*. Let me illustrate what this means by the example under consideration. We define a polarity ρ as a certain "lifting" of the symplectic polarity ρ_1 in the hyperplane H with equation $X_4 = 0$. More explicitly: to obtain the image of a point $p \neq (0,0,0,0,1)$ of $\mathbf{PG}(4, \mathbb{K})$ under ρ, one considers the intersection p' of H and the line joining p and $(0,0,0,0,1)$; the image of p is now by definition the hyperplane generated by $(0,0,0,0,1)$ and p'^{ρ_1}. For $p = (0,0,0,0,1)$, the image under ρ is $\mathbf{PG}(4, \mathbb{K})$ itself. After all, ρ is not an ordinary polarity, and so this is allowed. We call ordinary and degenerate polarities **generalized polarities**. A general degenerate polarity is obtained by executing the above procedure, substituting the point $(0,0,0,0,1)$ by any suitable subspace (and working in $\mathbf{PG}(d, \mathbb{K})$ for any d).

Hence, the most general thing to prove could only be that all elements of a given embedded quadrangle are absolute elements of a certain generalized polarity. The case of degenerate polarities can be excluded from the beginning by putting in an additional assumption. That is exactly what we are going to do. The next theorem is due to DIENST [1980a], [1980b], but we proceed partly along the lines of JOHNSON [19**], though we again avoid the theory of semi-quadratic sets completely. Note that the theorem is also true for d infinite (by JOHNSON [19**]), but we do not insist on that.

8.5.14 Theorem (Dienst [1980a],[1980b]). *Let* $\Gamma = (\mathcal{P}, \mathcal{L}, \mathbf{I})$ *be a weak generalized quadrangle fully embedded in* $\mathbf{PG}(d, \mathbb{K})$*. Suppose that through every point of* $\mathbf{PG}(d, \mathbb{K})$ *there exists at least one secant line. Then there is a (non-degenerate) polarity* τ *such that all points and lines of* Γ *are absolute elements with respect to* τ*.*

Proof. The strategy of the proof is as follows. For each point p of Γ, the hyperplane η_p will be the image of p under the polarity τ. If x does not belong to Γ, then we will show that the set of points Γ_x of Γ such that $p \in \Gamma_x$ **if and only if** px is a tangent line, spans a unique hyperplane η_x. It then remains to show that the mapping $a \mapsto \eta_a$, for any point a in $\mathbf{PG}(d, \mathbb{K})$, defines a polarity. Note that we may assume that for thick quadrangles, $d > 3$ (see Theorem 8.5.11). If Γ is not thick and $d = 3$, then the result follows from Theorem 8.5.13. Indeed, if the characteristic of $\mathbb{K}$ is not equal to 2, then any quadric is the set of absolute points of a polarity; if the characteristic of $\mathbb{K}$ is equal to 2, then any ruled quadric is embedded in a symplectic polarity. So we may assume $d > 3$.

Suppose that the point x of $\mathbf{PG}(d, \mathbb{K})$ does not belong to Γ. By assumption there is at least one secant line L through x. So Γ_x does not coincide with Γ. We will show that Γ_x is **either** an ovoid of Γ, **or** a full weak subquadrangle Γ' of Γ such that every line of $\Gamma \setminus \Gamma'$ intersects Γ' in a (necessarily) unique point.

Let M be any line of Γ and denote by π the plane generated by x and M. If all points of Γ in π are incident with M, then $\Gamma_1(M) \subseteq \Gamma_x$. Now suppose there is some point p of Γ in π, but not on M. Let p' be the projection of p onto M. Then obviously π belongs to $\eta_{p'}$, hence $p'x$ is a tangent (otherwise $\eta_{p'}$ contains points of Γ opposite p'). All other lines through x in π intersect M and pp' in distinct points. Consequently p' is the unique point on M which belongs to Γ_x. So we have shown that, for each line M of Γ, **either** all points of M belong to Γ_x **or** exactly one point of M does. Hence, if Γ_x does not contain lines, then it is an ovoid of Γ.

Now suppose that Γ_x contains at least one line M. If p is any element of Γ_x not on M, then we have shown in the previous paragraph that all points of the line $\mathrm{proj}_p M$ of Γ belong to Γ_x, since the points p and $\mathrm{proj}_M p$ do. Hence if Γ_x contains at least one apartment, then Γ_x is a full weak subquadrangle with the required property (i.e., every line of $\Gamma \setminus \Gamma_x$ meets Γ_x in exactly one point). So suppose that Γ_x does not contain an apartment. Clearly Γ_x must contain two confluent lines M and M'. Let p be the common point of M and M'. Then every other point in Γ_x must be collinear with p (otherwise an apartment with M and M' arises). So Γ_x consists of lines through p. Since every line at distance 3 from p (in Γ) has a unique point collinear with p, we see that all lines through p must be contained in Γ_x. Hence $\Gamma_x = p^\perp$.

We claim that $\Gamma_x = p^\perp$ leads to a contradiction. First we prove a useful lemma.

Lemma 1. *Every line of* $\mathbf{PG}(d, \mathbb{K})$ *through p not contained in η_p is a secant line. Also, for every point y in η_p not belonging to Γ, the line py is a tangent.*

Proof. The second assertion is trivial. We show the first. Let M be a tangent line through p not contained in η_p. Let y be a point on M, $y \neq p$. Let L_y be a secant line through y and let y' be a point of Γ on L_y (then $y \neq y'$). If y' lies in η_p, then any point y'' of Γ on L_y is collinear with p (otherwise M meets the unique line of Γ through y'' and concurrent with py' at a point distinct from p, contradicting the fact that M is a tangent). But now py'' lies outside η_p, contradicting the definition of η_p. So y' does not lie in η_p. Hence no point of $p^\perp$ in Γ lies on a secant line through y. Consequently $p^\perp \subseteq \Gamma_y$ and so $p^\perp = \Gamma_y$. Let y'', $y \neq y'' \neq y'$ be another point of Γ on L_y. Let M' be any line of Γ through p. Let $y'z'$, $z' \mathbf{I} M'$ be a line of Γ and let z'' be the projection of y'' on $y'z'$. Noting that $y', y'' \in z'^\perp$, we see that $y \in \eta_{z''}$, hence yz'' is a tangent line and so $z'' = z' \in p^\perp$. Consequently the projection of y' onto any line N of Γ through p coincides with the projection of y'' on N. Now let y_1 be any point on $y'z'$ with $y' \neq y_1 \neq z'$. Let w be the intersection of the plane spanned by y and M' with the line $y''y_1$. The line wp is also a tangent line, because a point $u \in \Gamma$ on wp, $u \neq p$, gives rise to a secant line yu meeting M'. Interchanging the roles of y and w, we see that the projection of y'' onto any line of Γ through p must coincide with the projection of y_1 onto that line, clearly a contradiction, since that projection must also be equal to the projection of $y' \perp y_1$. The lemma is proved.   QED

Part of the argument in the proof of the lemma is borrowed from JOHNSON [19**].

So by the preceding lemma, we have that, if $\Gamma_x = p^\perp$, then x lies in η_p. Also, x lies in every $\eta_{p'}$, for $p' \perp p$ in Γ. So all points of the line px belong to every $\eta_{p'}$, $p' \in \Gamma_2(p)$. This implies that every point y of px has the property $\Gamma_y = \eta_p$. Now let p'' be any point of Γ opposite p. Then $\eta_{p''}$ meets px at some point z, contradicting $p'' \in \Gamma_z = \eta_p$. The claim is proved.

So either Γ_x is a full weak subquadrangle or Γ_x is an ovoid. After a general lemma, we deal with these two cases separately. Remember that the idea is to show that Γ_x is contained in a unique hyperplane. In fact, if the degree f of the embedding is equal to 2, then this follows from Theorem 8.5.13. So we may as well assume $f > 2$.

Lemma 2. *If pq is a secant of Γ with $p, q \in \Gamma_x$,* **then** *all points of Γ on pq lie in Γ_x.*

Proof. Consider any point p' in $p^\perp \cap q^\perp$ and let r be any point of Γ on pq. Since r clearly lies in $\eta_{p'}$ (as p and q do), r is collinear with p', hence $p' \in r^\perp$. This shows that $p^\perp \cap q^\perp \subseteq r^\perp$. We now claim that $p^\perp \cap q^\perp$ generates $\eta_p \cap \eta_q$. Indeed, clearly $p^\perp \cap q^\perp$ together with p and q generate a subspace U which, by intersecting with Γ, is easily seen to coincide with $\mathbf{PG}(d, \mathbb{K})$ (indeed, $U \cap \Gamma$ is ideal and full in Γ). Hence the dimension of the subspace U' generated by $p^\perp \cap q^\perp$ is equal to $d - 2$ (indeed, neither p nor q belongs to U', and p does not belong to the subspace generated by U' and q). But also the dimension of $\eta_p \cap \eta_q$ is equal to $n - 2$. The claim follows since $p^\perp \cap q^\perp \subseteq \eta_p \cap \eta_q$. Now since $p^\perp \cap q^\perp \subseteq r^\perp$, we have $\eta_p \cap \eta_q \subset \eta_r$. By Lemma 1, $x \in \eta_p \cap \eta_q \subseteq \eta_r$, so rx is a tangent by Lemma 1 again. Lemma 2 is proved.   QED

First we suppose that Γ_x is a full weak subquadrangle. Let p_1, p_2, p_3, p_4 be the points of an apartment in Γ_x. Denote by U the three-dimensional projective space generated by p_1, p_2, p_3, p_4. Then $U \cap \Gamma_x$ is a full and ideal (by Lemma 2) subquadrangle of $U \cap \Gamma$, hence $U \cap \Gamma$ is contained in Γ_x. Let U' be a space of maximal dimension d' containing U such that $U' \cap \Gamma$ is contained in Γ_x. Then d' is well defined and larger than or equal to 3. Suppose first that $3 \leq d' < d - 1$. In this case, there exists a point q of Γ_x not contained in U' (indeed, considering any $(d-1)$-dimensional subspace V of $\mathbf{PG}(d, \mathbb{K})$ containing U', we see that $V \cap \Gamma$ is a full subquadrangle with the property that every line of Γ not in V intersects $V \cap \Gamma$ at a point, hence $U' \cap \Gamma$, which is stricly contained in $V \cap \Gamma$, cannot possibly have that property). We may assume that q is collinear with p_1. Let M be any line of Γ through p_1 in the space U'' generated by q and U', and suppose that M is not contained in U'. We intend to prove that $M \in \Gamma_x$, so we may assume that q is not on M. The plane qM intersects U' in a line N. If N belongs to Γ, then by Lemma 2, all points of M belong to Γ_x. Suppose now that N does not belong to Γ, then N is a tangent line (as it belongs to the plane qM). Let p be any point of $U' \cap \Gamma$ opposite p_1. Then the space V'' generated by p, q, M is three-dimensional and intersects Γ in a full subquadrangle. Moreover, V'' intersects U' in a plane π containing the secant line $p_1 p$ and the tangent line N. So π intersects $V'' \cap \Gamma$ in an ovoid all of whose points belong to Γ_x. But also all points of $p_1 q$ belong to

Γ_x, hence all points of $V'' \cap \Gamma$ belong to Γ_x (indeed, if z is a point of $V'' \cap \Gamma$ not collinear with p_1, then $\text{proj}_z p_1 q$ is a line that contains two distinct elements — $\text{proj}_{p_1 q} z$ and a unique element of the ovoid — of Γ_x, so $z \in \Gamma_x$; if z is collinear with p_1, then every line in $V'' \cap \Gamma$ through z not incident with p_1 now contains $|\mathbb{K}|$ points of Γ_x and so $z \in \Gamma_x$) as in particular do points of M. We have shown that $U'' \cap \Gamma_x$ is a full and ideal subquadrangle of $U'' \cap \Gamma$, contradicting the maximality of the dimension of U'. Hence $d' = d - 1$. We claim that $U' \cap \Gamma = \Gamma_x$. Indeed, if $y \in \Gamma_x \setminus U'$, then all points of $y^\perp$ belong to Γ_x since every line of $\Gamma_1(y)$ meets U' at a point of Γ_x. Let y' be any point of Γ. If $y' \in U'$, then it belongs to Γ_x. If $y' \in \Gamma_2(y)$, then it belongs to Γ_x. By the same token as above, if y' is collinear with any point of $y^\perp \setminus U'$, then $y' \in \Gamma_x$. Now assume that $y'^\perp \cap y^\perp \subseteq U'$. Then this is not the case for any $y'' \in \Gamma_2(y') \setminus U'$, and y'' belongs to Γ_x by a previous argument. Hence y' also now belongs to Γ_x. The claim is proved.

So we may assume that Γ_x is an ovoid. We first claim that, if the $(d-2)$-dimensional subspace U intersects Γ_x in at least $d - 1$ linearly independent points, then $U \cap \mathcal{P}$ is contained in Γ_x. Indeed, let L be any line of Γ meeting U at a point of Γ_x. If p_0 is a point on L not in U, then the intersection of the d hyperplanes η_{p_i}, $i \in \{0, 1, 2, \ldots, d-1\}$, where $\{p_1, p_2, \ldots, p_{d-1}\}$ is a set of $d-1$ linearly independent points of Γ_x in U, is a space V of dimension at least 0, hence there exists a point y of $\mathbf{PG}(d, \mathbb{K})$ for which Γ_y contains all p_i, $i \in \{0, 1, \ldots, d-1\}$, and L. So Γ_y is a full weak subquadrangle and contains by the previous paragraph all points of $U \cap \mathcal{P}$. We can carry out this procedure again with a line L' not belonging to Γ_y and we obtain a space V' containing a point y' with the property that $\Gamma_{y'}$ is a full weak subquadrangle and contains $U \cap \Gamma$. Actually, it is easy to see that $\Gamma_y \cap \Gamma_{y'} = U \cap \Gamma$, since the right-hand side is an ovoid in both constituents of the left-hand side. Now suppose that x does not belong to the line yy'. Then xyy' is a plane π entirely contained in $\eta_{p_1} \cap \eta_{p_2} \cap \ldots \eta_{p_{d-1}}$. So both V and V' intersect π in a line, hence there is a point z in $V \cap V' \cap \pi$ having the property that Γ_z contains both Γ_y and $\Gamma_{y'}$, clearly a contradiction (since such a Γ_z must then coincide with $\mathcal{P}$). So x lies on yy'. But then, since for all points q of Γ in U we have $yy' \subseteq \eta_q$, the claim follows.

Note that Γ_x spans at least a hyperplane since otherwise clearly Γ_x cannot be an ovoid (indeed, consider the subspace W generated by Γ_x and any line of Γ incident with a point of Γ_x; if W is at most $(d-1)$-dimensional, then $W \cap \Gamma$ is a full non-ideal subquadrangle and through every point w of $W \cap \Gamma$ there is at least one line L_w of Γ not in W. Choosing $w \notin \Gamma_x$, we see that L_w does not meet Γ_x.) So there actually exists a $(d-2)$-dimensional subspace U such that $U \cap \Gamma \subseteq \Gamma_x$. And moreover, we may assume that there is at least one further point p_d in Γ_x, with $p_d \notin U$. Now let r be any point of Γ in the subspace U' generated by p_d and U, but r not in the plane $p_d yy'$, where y and y' are as in the previous paragraph. The line $p_d r$ meets U in some point z, and since $z \notin yy'$, we conclude as in the previous paragraph (letting z here play the role of x there) that Γ_z does not contain all points of $U \cap \mathcal{P}$. Hence there is some secant M through z in U. Putting two points of Γ on M together with p_d in a set of $(d-1)$ linearly independent points of Γ_x, we see that by the previous paragraph r also belongs to Γ_x. The points of $U' \cap \mathcal{P}$ in the plane $p_d y y'$ (if there are any) are obtained by interchanging the roles of p_d and r (and such a point r always exists; it suffices to choose $r \in \mathcal{P}$ on the secant line $p_0 p_d$, $p_0 \neq r \neq p_d$, p_0 any element of $\Gamma_x \cap U$, remembering $f > 2$). So we have shown that all points of Γ in the hyperplane U' are contained in Γ_x. Since this already constitutes an ovoid, no other points can be in Γ_x, hence Γ_x lies in a unique hyperplane.

Consequently, for every point x not in Γ, the set of points Γ_x always lies in a unique hyperplane, which we denote by η_x. We consider the map $x \mapsto \eta_x$, defined for all points of $\mathbf{PG}(d, \mathbb{K})$. We show that this defines a collineation of $\mathbf{PG}(d, \mathbb{K})$ to its dual. By the Fundamental Theorem of Projective Geometry, we only have to show that collinear points are mapped onto hyperplanes meeting in a common $(d-2)$-dimensional space, i.e., lines of $\mathbf{PG}(d, \mathbb{K})$ are mapped onto lines of the dual $\mathbf{PG}(d, \mathbb{K})^D$ of $\mathbf{PG}(d, \mathbb{K})$.

So let x, y, z be three collinear points of $\mathbf{PG}(d, \mathbb{K})$. If xy is a secant, then consider two points p, q of Γ on xy. Clearly the hyperplanes η_x, η_y and η_z all contain $p^\perp \cap q^\perp$, and the latter spans $\eta_p \cap \eta_q$. Suppose now that two secant lines are mapped onto the same $(n-2)$-dimensional subspace U of $\mathbf{PG}(d, \mathbb{K})$. All points w of Γ in U have the property that η_w contains the space V generated by the two secant lines. Considering two points v, u of Γ that generate together with U the whole space, we see that all points of Γ lie on a tangent line through any element of a certain subspace of V of *codimension* at most 2 (the intersection of V with $\eta_u \cap \eta_v$), and hence non-empty (the **codimension** of a subspace is i if its projective dimension is $d-i$). This contradicts our hypothesis. A similar argument in fact shows that $x \mapsto \eta_x$ is injective.

Now let xy be an external line, a tangent line, or a line of Γ. Consider two secants L_x and M_x through x with L_x, M_x and xy not coplanar (this is possible for, if x belongs to Γ, it suffices to consider lines not contained in η_x, and, if x does not belong to Γ, to join x with points of Γ not in Γ_x). For any point p of Γ on L_x, the hyperplane η_p does not contain the plane xyp (because this plane contains points of Γ opposite p, namely on the secant line L_x), hence there is a second secant line in xyp through p. So by the previous paragraph, xyp is mapped onto a plane of $\mathbf{PG}(d, \mathbb{K})^D$. Similarly, the plane xyM_x is mapped onto a plane of $\mathbf{PG}(d, \mathbb{K})^D$. Since $x \mapsto \eta_x$ is injective, the line xy is mapped onto a line of $\mathbf{PG}(d, \mathbb{K})^D$. So we can extend our mapping to a semi-linear map τ. Clearly τ is bijective (for otherwise there would be a point incident with no secants). It is also easy to see that for points p of Γ, we have $p^{\tau^2} = p$. So τ^2 restricted to the point set $\mathcal{P}$ of Γ is the identity. But it easily seen that $\mathcal{P}$ contains a frame (i.e., a set of $(d+2)$ points no $d+1$ of which lie in a hyperplane). Since $\mathcal{P}$ also contains lines, it follows that τ^2 is the identity, i.e., τ is a polarity.

The theorem is proved. $\square$

8.5.15 Remark. It is not always true that the polarity τ itself defines a generalized quadrangle. Indeed, the theorem also holds for degree $f = 2$, and in the case of the quadrangle $Q(5, q)$, q even, with standard embedding in $\mathbf{PG}(5, q)$, the polarity τ is a symplectic polarity, hence has absolute planes. On the other hand, the polarity τ may itself define a bigger generalized quadrangle. If Γ is the weak quadrangle $Q(3, q)$ arising from the ruled quadric $\mathbf{Q}^+(3, q)$, q even, then the corresponding polarity τ is a symplectic one defining $\mathsf{W}(q)$. We will call this embedding of $Q(3, q)$ in $\mathbf{PG}(3, q)$ its **standard embedding** (also for q odd).

It is a remarkable fact that the dimensions $d > 3$ can be handled synthetically for any degree, and that the case $d = 3$ needs coordinate calculations for every degree. An open question is whether synthetic arguments can prove (parts of) the result in dimension 3.

For a given embedded generalized quadrangle, we obtain an embedding and a corresponding (non-degenerate) polarity by projecting the quadrangle from the subspace U of points which are not incident with any secant line (it is easily seen that these points do indeed form a subspace) into a complementary subspace V (i.e., a subspace V skew to U such that U and V generate $\mathbf{PG}(d, \mathbb{K})$). The embedding thus obtained satisfies the conditions of the preceding theorem.

We now easily derive the result of BUEKENHOUT & LEFÈVRE-PERCSY [1974]

8.5.16 Theorem (Buekenhout & Lefèvre-Percsy [1974]). *Let Γ be a finite weak generalized quadrangle fully embedded in $\mathbf{PG}(d, q)$. Then we have a standard embedding, i.e., we have one of the following.*

(*i*) *$d = 3$ and Γ is isomophic to either $\mathsf{W}(q)$ or $\mathsf{H}(3, q)$. In every case, the embedding is the standard one and the elements of the quadrangle are all absolute points and lines of a symplectic or unitary polarity.*

(*ii*) *$d = 4$ and Γ is isomorphic to $\mathsf{H}(4, q)$ and the embedding is the standard one. The elements of the quadrangle are all absolute points and lines of a unitary polarity.*

(*iii*) *$d = 3, 4, 5$ and Γ is isomorphic to $Q(3, q)$, $Q(4, q)$ and $Q(5, q)$. The embedding is standard.*

Proof. If the degree f of the embedding is equal to 2, then (*iii*) follows from Theorem 8.5.13. So let $f > 2$. If the dimension d is equal to 3, then (*i*) follows from Theorem 8.5.11, noting that a finite field is perfect. If $d > 4$, then we can take subsequent subquadrangles in a $(d-1)$- and a $(d-2)$-dimensional subspace. In this way, we obtain a tower of (thick) full subquadrangles. Theorem 1.8.8(*iv*) (page 36) says that this is impossible, hence $d = 4$. Now let the order of Γ be (q, t). Intersecting with a suitable three-dimensional subspace U, we obtain a full subquadrangle $U \cap \Gamma$, hence $t > q$ by the dual of Proposition 1.8.7(*ii*) (see page 35).

If there exists a point x of $\mathbf{PG}(4, q)$ not incident with any secant line, then we may project Γ from x into a hyperplane and obtain a full embedding of Γ in $\mathbf{PG}(3, q)$. Hence $t \leq q$, a contradiction. By Theorem 8.5.14, Γ is embedded in a non-degenerate polarity. But there is at most one polarity in $\mathbf{PG}(4, q)$ which allows secant lines of degree greater than 2, and it is the unitary polarity, defined when q is a perfect square. So Γ is a full subquadrangle of $\mathsf{H}(4, q)$. The latter has order $(q, \sqrt{q^3})$, and so, if $t \neq \sqrt{q^3}$, then by Proposition 1.8.7(ii) (dualized) again, we must have $t \leq \sqrt{q} < \sqrt{t}$, a contradiction. Hence $t = \sqrt{q^3}$ and (ii) follows. The theorem is proved. $\qquad\square$

8.6 Ideal, weak and lax embeddings of polygons

In this section, we will be more sketchy about the proofs. In fact, no proof shall be given in detail. A lot of them require some additional results in finite incidence geometry. We occasionally highlight the geometric methods. Let us remark that this area has only recently been given much attention, starting with the work of THAS & VAN MALDEGHEM [1996] on ideal full embeddings of finite generalized hexagons.

Our first result says that all ideal full embeddings of generalized polygons in finite projective space are classified.

8.6.1 Theorem (Thas & Van Maldeghem [1996]). *Let Γ be a finite generalized n-gon, $n > 3$, ideally and fully embedded in $\mathbf{PG}(d, q)$. Then one of the following holds.*

(i) *$n = 4$, $d = 3$ and Γ is either the symplectic quadrangle $\mathsf{W}(q)$ or the Hermitian quadrangle $\mathsf{H}(3, q)$ and the embedding is standard;*

(ii) *$n = 6$, $d = 5, 6$ or 7 and Γ is isomorphic respectively to the symplectic hexagon $\mathsf{H}(q)$ with q even, the split Cayley hexagon $\mathsf{H}(q)$ for any prime power q, and the twisted triality hexagon $\mathsf{T}(q, \sqrt[3]{q})$. All embeddings are standard.*

In particular, there do not exist octagons ideally and fully embedded in finite projective space. $\qquad\square$

As to the proof, (i) of course follows directly from Theorem 8.5.11. To show (ii), one first proves that Γ is distance-2-regular. This amounts to showing that no plane generated by $x^\perp$ is contained in the subspace generated by $y^{\perp\perp}$, for x and y opposite points of Γ. Then the distance-2-traces are easily seen to be subsets of lines of $\mathbf{PG}(d, q)$, hence they are determined by any two of their points. For a generalized octagon, the proof here is already complete, by quoting Theorem 6.4.6 on page 261. For generalized hexagons, one shows that any subspace U of $\mathbf{PG}(d, q)$ which contains an apartment of Γ, defines an ideal subhexagon, i.e., the points of

Γ in U which are incident with at least two lines of Γ which belong to U form the point set of an ideal subhexagon. Using Theorem 1.8.7 (see page 35), one obtains an upper bound for d. Indeed, an apartment lies in five-dimensional space, an extra point defines a six-dimensional space, another point a seven-dimensional space, but then the three subsequent hexagons form a tower, and so the last one is Γ. Hence $d \leq 7$. Then essentially a case-by-case study gives the desired result. Note that we know that Γ satisfies the Moufang property by Theorem 6.3.2 (page 243), hence Γ is either $\mathsf{H}(q)$ or $\mathsf{T}(q, \sqrt[3]{q})$.

The next result classifies all generalized quadrangles weakly embedded in finite projective space.

8.6.2 Theorem (Lefèvre-Percsy [1981a], Thas & Van Maldeghem [19**a]). *Let Γ be a finite generalized quadrangle weakly embedded in $\mathbf{PG}(d, q)$. Then one of the following holds.*

(i) *There exists a subfield $\mathbf{GF}(q')$ of $\mathbf{GF}(q)$ and a projective subspace $\mathbf{PG}(d, q')$ of $\mathbf{PG}(d, q)$ (of the same dimension d) such that Γ is fully embedded in $\mathbf{PG}(d, q')$.*

(ii) *$n = 4$, $d = 4$, q is odd and $\Gamma \cong \mathsf{W}(2)$. For each odd prime power q there exists exactly one such weak embedding, up to a projective semi-linear transformation.* $\square$

Some comments are in order here. If $d = 3$, then we already know by Lemma 8.5.9 that the quadrangle is classical and has regular points. Hence $\Gamma \cong \mathsf{W}(q')$ or $\Gamma \cong \mathsf{H}(3, q')$. In the first case one considers the traces and the lines of the quadrangle to obtain a projective subspace over $\mathbf{GF}(q')$, in the latter case one does the same with a symplectic subquadrangle $\mathsf{W}(\sqrt{q'})$. With some additional work, the result follows easily. The result becomes harder to prove when the degree of the embedding is equal to 2. Then one has to use regularity of lines to produce affine subplanes and to obtain the field $\mathbf{GF}(q')$. But affine subspaces of order 2 do not define any field, and so that is where the example of (ii) arises. Let us now only briefly describe this example.

By Subsection 1.4.2 (page 15), we may identify the points of $\mathsf{W}(2)$ with the unordered pairs of elements of a set of six elements, say $\{1, 2, 3, 4, 5, 6\}$. The lines are then the sets of three disjoint pairs. Let $\mathbb{K}$ be any field (this also works for $\mathbb{K}$ a skew field, but there is no point here to this generality). Then we identify the points of $\mathsf{W}(2)$ with some points of $\mathbf{PG}(4, \mathbb{K})$ in the following way:

$$
\begin{array}{lll}
\{1,2\} \mapsto (1,0,0,0,0), & \{3,5\} \mapsto (0,1,0,0,1), & \{4,6\} \mapsto (-1,1,0,0,1), \\
\{1,4\} \mapsto (0,1,0,0,0), & \{3,6\} \mapsto (1,0,1,0,0), & \{2,5\} \mapsto (1,-1,1,0,0), \\
\{4,5\} \mapsto (0,0,1,0,0), & \{2,3\} \mapsto (0,1,0,1,0), & \{1,6\} \mapsto (0,1,-1,1,0), \\
\{5,6\} \mapsto (0,0,0,1,0), & \{1,3\} \mapsto (0,0,1,0,1), & \{2,4\} \mapsto (0,0,1,-1,1), \\
\{2,6\} \mapsto (0,0,0,0,1), & \{3,4\} \mapsto (1,0,0,1,0), & \{1,5\} \mapsto (1,0,0,1,-1).
\end{array}
$$

We have arranged it so that the three points on each row form a line (and this is readily checked!); the five lines thus obtained form a spread. The points on each of the other ten lines of the quadrangle can also be checked to be collinear in $\mathbf{PG}(4,\mathbb{K})$, e.g. $\{\{1,2\},\{3,4\},\{5,6\}\}$ clearly lies on the line with equations $X_1 = X_2 = X_4 = 0$. Also, the points collinear with a certain point x are automatically contained in a hyperplane since the three lines of $\mathsf{W}(2)$ through x can span at most a three-dimensional space. So we indeed have a weak embedding. If the characteristic of $\mathbb{K}$ is equal to 2, then we obviously have an embedding over the subfield $\mathbf{GF}(2)$. In general, this weak embedding has more nice properties, for instance the little projective group of $\mathsf{W}(2)$ extends to $\mathbf{PG}(4,\mathbb{K})$. Let us show this for one root elation. We may suppose that the characteristic of $\mathbb{K}$ is not equal to 2. Consider in $\mathbf{PG}(4,\mathbb{K})$ the involutory homology with axis the hyperplane with equation $X_1 = X_4$ and centre $(0,1,-1,1,-1)$. This has matrix

$$
\begin{pmatrix}
1 & 0 & 0 & 0 & 0 \\
0 & 0 & 0 & 0 & 1 \\
0 & 1 & 1 & 0 & -1 \\
0 & -1 & 0 & 1 & 1 \\
0 & 1 & 0 & 0 & 0
\end{pmatrix}.
$$

One can easily check that this element of $\mathbf{PSL}(5,\mathbb{K})$ preserves $\mathsf{W}(2)$ and that it represents a central elation in $\mathsf{W}(2)$ with centre $\{1,2\}$. Hence every automorphism of $\mathsf{W}(2)$ extends to an automorphism of $\mathbf{PG}(4,\mathbb{K})$ (because the little projective group is here the full automorphism group). This is not surprising in view of the fact that the symmetric group on six letters is a subgroup of $\mathbf{PSL}(5,\mathbb{K})$, for any field $\mathbb{K}$.

We conclude this section by mentioning some very recent results on lax embeddings of quadrangles and flat embeddings of hexagons. We restrict ourselves to two results, packed into one theorem. Recall that a projective linear transformation is an element of $\mathbf{PGL}$, a projective semi-linear transformation belongs to $\mathbf{P\Gamma L}$, and a special projective linear transformation is an element of $\mathbf{PSL}$.

8.6.3 Theorem (Thas & Van Maldeghem [19**b],[19**c]).

(i) *Let Γ be any finite generalized quadrangle of order (s,t) laxly embedded in $\mathbf{PG}(5,q)$. If $s \neq 2$ or q is even, then Γ is isomorphic to $\mathsf{Q}(5,s)$, weakly embedded in $\mathbf{PG}(5,q)$, and hence fully embedded in some subspace $\mathbf{PG}(5,s)$ over $\mathbf{GF}(s)$. For $s = 2$ and q odd, $\Gamma \cong \mathsf{Q}(5,2)$ and there exists, up to a projective linear transformation, a unique lax embedding which is not a weak embedding; moreover, the little projective group of $\mathsf{Q}(5,2)$ is induced by $\mathbf{PSL}_5(q)$; also, if q is a power of 3, then $\mathsf{Q}(5,2)$ is fully embedded in an affine subspace $\mathbf{AG}(5,3)$ of $\mathbf{PG}(5,q)$.*

(ii) *Let Γ be a finite generalized quadrangle of order (s,t) laxly embedded in $\mathbf{PG}(4,q)$. Then $s \leq t$.*

  – *Suppose $s = t$. If $s \neq 3$ or q is a power of 3,* **then** Γ *is weakly embedded in* $\mathbf{PG}(4, q)$ *(and hence determined by* Theorem 8.6.2 *above). If $s = 3$,* **then** $q \not\equiv 2 \bmod 3$. *For $s = 3$, $q \equiv 1 \bmod 3$, we have* $\Gamma \cong \mathsf{Q}(4, 3)$ *and there exists, up to a special projective linear transformation, a unique lax embedding which is not a weak embedding; moreover, the little projective group of* $\mathsf{Q}(4, 3)$ *is induced by* $\mathbf{PSL}_4(q)$; *also, if q is even, then* $\mathsf{Q}(4, 2)$ *is fully embedded in an affine subspace* $\mathbf{AG}(4, 4)$ *of* $\mathbf{PG}(4, q)$.

  – *Suppose $s^3 = t^2$. Then* $\Gamma \cong \mathsf{H}(4, s)$ *and it is weakly embedded in* $\mathbf{PG}(4, q)$.

(*iii*) *Let Γ be a finite generalized quadrangle of order $(s, \sqrt{s})$ laxly embedded in* $\mathbf{PG}(3, q)$. *Then* $\Gamma \cong \mathsf{H}(3, s)$ *and Γ is weakly embedded.*

(*iv*) *Let Γ be a finite generalized hexagon of order (s, t) fully embedded in the space* $\mathbf{PG}(d, s)$. *If* **either** *the embedding is flat and* either *$d = 7$,* or *$d = 6$ and $t^5 > s^3$,* or *$d = 5$ and $s = t$,* **or** *the embedding is weak, $d = 6$ and s is odd, then Γ is a classical point-regular hexagon and the embedding of Γ in* $\mathbf{PG}(d, s)$ *is a standard one.*

(*v*) *Let Γ be a finite generalized hexagon of order (s, t) ideally (laxly) embedded in* $\mathbf{PG}(d, q)$. *Then Γ is classical and point-regular and there exists a subspace* $\mathbf{PG}(d, s)$ *of* $\mathbf{PG}(d, q)$ *in which Γ is a standard embedding (in particular* $\mathbf{GF}(s)$ *is a subfield of* $\mathbf{GF}(q)$).

(*vi*) *Let Γ be a finite generalized hexagon of order (s, t) laxly embedded in the space* $\mathbf{PG}(d, q)$. *If* **either** *$d = 7$ and the embedding is flat,* **or** *$d \leq 5$ and the embedding is weak, then the embedding is ideal, and hence (v) applies.* $\qquad \square$

For the sake of completeness, we describe the exceptional lax embeddings of $\mathsf{Q}(5, 2)$ and $\mathsf{Q}(4, 3)$ turning up in (i) and (ii) of the previous theorem.

8.6.4 An exceptional lax embedding of $\mathsf{Q}(5, 2)$ in $\mathbf{PG}(5, q)$, q odd

Consider the subgroup G of $\mathbf{PSL}_6(q)$ generated by the special projective linear transformations corresponding to the the following six matrices:

$$
\begin{bmatrix}
-1 & -1 & 0 & 1 & 0 & 1 \\
0 & 1 & 0 & 0 & 0 & 0 \\
0 & -1 & -1 & 0 & 1 & 0 \\
0 & 0 & 0 & 1 & 0 & 0 \\
0 & 0 & 0 & 0 & 1 & 0 \\
0 & 0 & 0 & 0 & 0 & 1
\end{bmatrix},
\begin{bmatrix}
-1 & 0 & 1 & 0 & -1 & 1 \\
0 & 1 & 0 & 0 & 0 & 0 \\
0 & 0 & 1 & 0 & 0 & 0 \\
0 & 1 & 0 & -1 & -1 & 0 \\
0 & 0 & 0 & 0 & 1 & 0 \\
0 & 0 & 0 & 0 & 0 & 1
\end{bmatrix},
$$

$$\begin{bmatrix} 0 & 0 & 0 & -1 & 0 & 0 \\ 0 & 1 & 0 & 0 & 0 & 0 \\ 0 & 0 & 1 & 0 & 0 & 0 \\ -1 & 0 & 0 & 0 & 0 & 0 \\ 0 & 1 & 1 & 0 & -1 & 1 \\ 0 & 0 & 0 & 0 & 0 & 1 \end{bmatrix}, \begin{bmatrix} 1 & 0 & 0 & 0 & 0 & 0 \\ 0 & 1 & 0 & 0 & 0 & 0 \\ 1 & 0 & -1 & -1 & 0 & 0 \\ 0 & 0 & 0 & 1 & 0 & 0 \\ 0 & 1 & 0 & -1 & -1 & 1 \\ 0 & 0 & 0 & 0 & 0 & 1 \end{bmatrix},$$

$$\begin{bmatrix} -1 & 0 & 0 & 0 & 0 & 1 \\ 0 & 0 & 0 & 0 & -1 & 0 \\ 0 & -1 & -1 & 0 & 1 & 0 \\ 0 & 1 & 0 & -1 & -1 & 0 \\ 0 & -1 & 0 & 0 & 0 & 0 \\ 0 & 0 & 0 & 0 & 0 & 1 \end{bmatrix}, \begin{bmatrix} 1 & 1 & 0 & 0 & 0 & 0 \\ 0 & -1 & 0 & 0 & 0 & 0 \\ 0 & 1 & 1 & 0 & -1 & 0 \\ 0 & -1 & 0 & 1 & 1 & 0 \\ 0 & 0 & 0 & 0 & -1 & 0 \\ 0 & 1 & 0 & 0 & -1 & 1 \end{bmatrix}.$$

Then the orbit of the point $(1,0,0,0,0,0)$ is the point set of a lax embedding of $Q(5,2)$. The group G itself is the little projective group of $Q(5,2)$ and the above matrices represent elations. For q a power of 3, G fixes the hyperplane with equation $\sum_i X_i = 0$ and so in this case we obtain an *affine lax embedding*, which is full if $q = 3$.

8.6.5 An exceptional lax embedding of $Q(4,3)$ in $\mathbf{PG}(4,q)$, $q \equiv 1$ mod 3

Let $a \in \mathbf{GF}(q)$ be such that $a^2 - a + 1 = 0$; so a is a root of unity of order 6 (if q is odd) or 3 (if q is even). We assume that $q \equiv 1$ mod 3, in which case we have exactly two choices for a. Note that, if a satisfies $x^2 - x + 1 = 0$, then the other root is $1 - a = -a^2$. Consider the subgroup G_a of $\mathbf{PSL}_5(q)$ generated by the special projective linear transformations corresponding to the following five matrices:

$$\begin{bmatrix} a-1 & 1 & 0 & 0 & 1 \\ 0 & 1 & 0 & 0 & 0 \\ 0 & 0 & 1 & 0 & 0 \\ 0 & 0 & 1 & a-1 & 0 \\ 0 & 0 & 0 & 0 & 1 \end{bmatrix}, \begin{bmatrix} 1 & 0 & 0 & 0 & 0 \\ 0 & 1 & 0 & 0 & 0 \\ 0 & 1 & -a & 0 & 1 \\ 1 & 0 & 0 & -a & 0 \\ 0 & 0 & 0 & 0 & 1 \end{bmatrix}, \begin{bmatrix} 1 & 0 & 0 & 0 & 0 \\ 1 & -a & 0 & 0 & 0 \\ 0 & 0 & -a & 1 & 1 \\ 0 & 0 & 0 & 1 & 0 \\ 0 & 0 & 0 & 0 & 1 \end{bmatrix},$$

$$\begin{bmatrix} a-1 & 0 & 0 & 1 & 1 \\ 0 & a-1 & 1 & 0 & 0 \\ 0 & 0 & 1 & 0 & 0 \\ 0 & 0 & 0 & 1 & 0 \\ 0 & 0 & 0 & 0 & 1 \end{bmatrix}, \begin{bmatrix} -1 & 0 & 0 & 0 & 1+a \\ 0 & -1 & 0 & 0 & 1 \\ 0 & 0 & -1 & 0 & 2-a \\ 0 & 0 & 0 & 0 & 1 \\ 0 & 0 & 0 & -1 & 1 \end{bmatrix}.$$

Then the orbit of the point $(1,0,0,0,0)$ is the point set of a lax embedding of $Q(4,3)$. The group G_a itself is the little projective group of $Q(4,3)$ and the above matrices represent elations. For q even, G_a fixes the hyperplane with equation

$$aX_0 + X_1 + a^2 X_2 + X_3 + X_4 = 0$$

and so in this case we again obtain an *affine lax embedding*, which is full if $q = 4$. It can be easily checked that G_a and G_{-a^2} generate respective embeddings which are equivalent under the special projective linear transformation with diagonal matrix $\operatorname{diag}(a^4, -a^2, 1, -a^2, -a^2)$.

8.6.6 Other results

It is appropriate to mention the work of STEINBACH [1996] (and unpublished) who classifies weakly embedded generalized quadrangles and hexagons satisfying some additional hypotheses, but without restriction on the finiteness. The additional assumptions are in the style: Γ is a Moufang quadrangle and the little projective group is contained in the linear group of the projective space.

A complete classification of all weakly embedded generalized quadrangles in projective space has been carried out by STEINBACH & VAN MALDEGHEM [19**a] (degree > 2), [19**b] (degree 2).

8.6.7 Theorem (Steinbach & Van Maldeghem [19**a], [19**b]). If Γ *is a generalized quadrangle weakly embedded in a projective space* $\mathbf{PG}(d, \mathbb{K})$, *with* $\mathbb{K}$ *a skew field,* **then** Γ *is a classical or mixed quadrangle and all such weak embeddings can explicitly be described.* □

Finally, there are many more results available for laxly embedded finite generalized quadrangles, and for flatly/weakly laxly/fully embedded finite hexagons; see THAS & VAN MALDEGHEM [19**b], [19**c].

8.7 Embeddings of the slim Moufang polygons

8.7.1 Universal embeddings

Recall from Definitions 1.2.1 on page 2 that a slim polygon is a polygon of order $(2, t)$, for some t. The slim Moufang polygons are $\mathbf{PG}(2, 2)$, $\mathsf{W}(2)$, $\mathsf{Q}(4, 2)$, $\mathsf{H}(2)$, $\mathsf{H}(2)^D$, $\mathsf{T}(2, 8)$ and $\mathsf{O}(2)$. Let $\Gamma = (\mathcal{P}, \mathcal{L}, \mathbf{I})$ be one of them. A full embedding in $\mathbf{PG}(d, 2)$ for suitable d is obtained as follows. Let G be the elementary abelian 2-group generated by involutions σ_p, for $p \in \mathcal{P}$, and with the only additional relations $\sigma_x \sigma_y = \sigma_z$ if $\{x, y, z\}$ is the set of points incident with some line of Γ. We can view G as a vector space over $\mathbf{GF}(2)$ (of some dimension $d + 1$) and hence, by deleting the zero vector, we obtain a projective space $\mathbf{PG}(d, 2)$ in which Γ is fully embedded via $p \mapsto \sigma_p$. In general, d could be equal to 1, but it is rather easily seen that, if Γ admits some embedding in $\mathbf{PG}(d', 2)$, then it must arise from adding some further relations in G, and hence it can be seen as a projection of the embedding arising from G. Therefore, the latter is sometimes called the **universal embedding**. Hence, for Γ (which is a generalized polygon and thus admits embeddings), we will always have $d > 1$. In fact, it is not difficult to see that the full automorphism group of Γ is induced in $\mathbf{PG}(d, 2)$. Hence we may define the number d_i as the dimension of the projective space generated by the points of Γ at distance at most i from a point (for i even) or a line (for i odd), and we call d_i a **subdimension**.

Γ	d_2	d_3	d_4	d_5	d_6	d_7
PG$(2,2)$	2					
Q$(3,2)$	2	3				
Q$(4,2)$	3	4				
Q$(5,2)$	4	5				
$(2\,\mathbf{PG}(2,2))^D$	2	4	6	7		
H(2)	3	7	12	13		
H$(2)^D$	3	7	11	13		
T$(2,8)$	8	13	26	27		
$(2\,\mathsf{W}(2))^D$	2	4	6	10	13	15
O(2)	5	13	35	57	77	79

Table 8.4. Subdimensions of the universal embeddings of slim (weak) Moufang polygons.

These subdimensions were calculated by COOLSAET (unpublished) and they are summarized in Table 8.4. Note that $d = d_{n-1}$, if Γ is an n-gon, and $d_0 = 0$, $d_1 = 1$.

There are some connections here with the dimensions of embeddings of higher-rank buildings, and with dimensions of Lie algebras, but this would lead us too far. Note only that the universal embeddings of $\mathsf{W}(2) \cong \mathsf{Q}(4,2)$ and $\mathsf{Q}(5,2)$ are the standard orthogonal ones.

8.7.2 Coolsaet embeddings

Universal embeddings have the property that they occur in the maximal dimension possible for the given polygon (or in general, slim geometry). One can now project from a subspace of the projective space not meeting any secant line. In this way, we obtain embeddings in smaller-dimensional spaces. But usually, the symmetry is not preserved, and hence the automorphism group induced by the projective space will not be the full automorphism group of Γ. A computationally more direct way to decrease the dimension and preserve the group was found by COOLSAET (unpublished). It even turns out that the dimensions he obtains are minimal with respect to the property of preserving the automorphism group. His idea is very simple. Let Γ be a slim (finite) polygon and let $\tilde{G}$ be the elementary abelian 2-group generated by involutions σ_p, for $p \in \mathcal{P}$, this time with no additional relations. We now identify any point p of Γ with the product σ_p^o in G of all elements σ_x with x opposite p in Γ. If x, y, z are collinear points in Γ, then $\sigma_x^o \sigma_y^o \sigma_z^o$ is equal to the identity since every element opposite x is opposite exactly two elements of $\{x, y, z\}$ (it is not opposite the projection on xy). Hence, as before, we obtain an embedding of Γ in some projective subspace $\mathbf{PG}(d^o, 2)$ of $\mathbf{PG}(d', 2)$, where $d' + 1$ is the number of points of Γ and $\mathbf{PG}(d', 2)$ is the projective space emerging naturally from $\tilde{G}$ as above. By the symmetry in the construction, the automorphism group of $\mathbf{PG}(d^o, 2)$ preserving Γ is the full automorphism group of Γ. We call this embedding the **Coolsaet embedding** of Γ.

If we define d_i^o as the d_i in the previous section, then we obtain the subdimensions as in Table 8.5. We have not included the weak non-thick slim polygons because the subdimensions are the same as for the respective universal embeddings.

Γ	d_2^o	d_3^o	d_4^o	d_5^o	d_6^o	d_7^o
PG$(2,2)$	2					
W(2)	2	3				
Q$(5,2)$	4	5				
H(2)	2	3	4	5		
H$(2)^D$	3	7	11	13		
T$(2,8)$	8	13	24	25		
O(2)	4	7	13	19	24	25

Table 8.5. Subdimensions of the Coolsaet embeddings of slim Moufang polygons.

Note that the Coolsaet embeddings of W(2) and H(2) are the standard symplectic ones. The Coolsaet embedding of O(2) arises from the *standard embedding* of the smallest metasymplectic space (for the latter embedding, see COHEN [1995]).

One further remark. If in the universal embedding of a slim n-gon we have the relation $d_{n-2} = d_{n-1} - 1$, then we can project the universal embedding from the intersection of all hyperplanes generated by the points not opposite a given point, and we obtain the Coolsaet embedding. Equivalently, we can consider the dual projective space and identify a point x of the polygon with the point of this dual space corresponding to the hyperplane spanned by all points not opposite x. This also produces the Coolsaet embedding.

Chapter 9

Topological Polygons

9.1 Introduction

It is quite natural to marry geometry to topology, since the classical geometries over $\mathbb{R}$ or $\mathbb{C}$ carry special topologies, which are indispensable for characterizing these geometries (in HILBERT's foundations of geometry [1899], the necessary topological assumptions are disguised in terms of orderings; in order to include $\mathbb{H}$ and $\mathbb{O}$, KOLMOGOROFF [1932] states topological axioms: compactness, connectedness).

In this chapter we give a survey of the theory of topological generalized polygons, focusing on the case of compact connected topologies. For projective planes (i.e., generalized triangles) there exists a rich theory which is expounded in the monograph by SALZMANN, BETTEN, GRUNDHÖFER, HÄHL, LÖWEN & STROPPEL [1995]. As a topological analogue of Theorem 1.7.1 of FEIT & HIGMAN [1964], KNARR [1990] and KRAMER [1994a] have proved that compact connected n-gons of finite topological dimension exist only for $n \in \{3, 4, 6\}$; see Theorem 9.5.1 below. At present, no example of a non-classical compact connected hexagon is known, but many generalized quadrangles of this type have been constructed (this is analogous to the finite case; compare Section 3.8.2 on page 134).

Generalized polygons can also be considered in categories which are stronger than the topological category; we mention some recent results on smooth, algebraic or holomorphic polygons and on polygons of finite Morley rank; see Section 9.8 (page 423).

Usually the proofs of the results in this chapter are beyond the scope of this book, requiring a good deal of algebraic topology or of the theory of Lie groups. The reader is referred to KRAMER [1994a] or KRAMER [19**] for an excellent exposition (including proofs) of the theory of topological polygons.

9.2   Definition of topological polygons

The standard definition of a topological projective plane $\Gamma = (\mathcal{P}, \mathcal{L}, \mathbf{I})$ requires that the point set $\mathcal{P}$ and the line set $\mathcal{L}$ are endowed with non-trivial topologies such that the two geometric operations of joining distinct points and of intersecting distinct lines are continuous; compare SALZMANN et al. [1995](41.1). This definition can be extended naturally to generalized n-gons, by requiring that the unique path of length k between two elements x, y of distance $\delta(x, y) = k < n$ (compare Lemma 1.3.5(i) on page 7) should depend continuously on x, y. In fact, experience to date shows that it suffices to require this only for the special case $k = n - 1$; thus we adopt the following definition from KNARR [1990] and KRAMER [1994a], [19**].

9.2.1 Definition. A **topological** n-**gon** is a generalized n-gon $\Gamma = (\mathcal{P}, \mathcal{L}, \mathbf{I})$ with topologies on $\mathcal{P}$ and $\mathcal{L}$, which are not discrete, such that the map

$$f_{n-1} : \{(x, y) : x, y \in \mathcal{P} \cup \mathcal{L}, \ \delta(x, y) = n - 1\} \to \mathcal{P} \cup \mathcal{L}$$

described by $f_{n-1}(x, y) = \text{proj}_y x$ is continuous (with respect to the sum topology on $\mathcal{P} \cup \mathcal{L}$).

For $n = 3$, the map f_{n-1} is just the union of the two geometric operations of joining points and intersection of lines, hence a topological 3-gon is the same object as a topological projective plane. For special cases and variations of the above definition compare FORST [1981], GRUNDHÖFER & KNARR [1990], GRUNDHÖFER & VAN MALDEGHEM [1990], JÄGER [1994].

In this chapter, we view the incidence relation explicitly as a non-symmetric relation. In particular, $\mathbf{I}$ is a subset of $\mathcal{P} \times \mathcal{L}$. It will sometimes abusively be called the set of flags, although a flag is not ordered.

In the compact case, one has the following very simple characterization of topological polygons (in fact, this characterization could serve as a definition of compact topological polygons, as in BURNS & SPATZIER [1987]).

9.2.2 Lemma. *A generalized n-gon $\Gamma = (\mathcal{P}, \mathcal{L}, \mathbf{I})$ with compact Hausdorff topologies on $\mathcal{P}$ and on $\mathcal{L}$ is a topological n-gon if and only if the set $\mathbf{I}$ of all flags is closed in the product $\mathcal{P} \times \mathcal{L}$.*

Proof. In all topological polygons, the set $\mathbf{I}$ is closed in $\mathcal{P} \times \mathcal{L}$; see KRAMER [1994a](2.1.12). Conversely, if $\mathbf{I}$ is closed in $\mathcal{P} \times \mathcal{L}$, then the graph of the map f_{n-1} (and of all analogous maps f_k, $k < n$, which are defined on $\delta^{-1}(k)$) can be described in terms of $\mathbf{I}$ and is therefore closed in the product $\delta^{-1}(n-1) \times (\mathcal{P} \cup \mathcal{L})$. By compactness, this implies continuity of f_{n-1} (and of all maps f_k). See also *op. cit.*(2.5.4).   □

9.3 Examples

We describe some examples of topological projective planes, of topological quadrangles and of topological hexagons (but not octagons, in view of Theorem 9.5.1).

9.3.1 Projective planes

Let $\mathbb{K}$ be a topological skew field which is not discrete. Then the projective plane $\mathbf{PG}(2, \mathbb{K}) = (\mathcal{P}, \mathcal{L}, \mathbf{I})$ as described in Subsection 2.2.1 on page 50 is a topological projective plane; the topologies on $\mathcal{P}$ and on $\mathcal{L}$ can be obtained as quotient topologies with respect to the natural maps (homogeneous coordinates) $\mathbb{K}^3 \setminus \{0\} \to \mathcal{P}$ and $\mathbb{K}^3 \setminus \{0\} \to \mathcal{L}$. The plane $\mathbf{PG}(2, \mathbb{K})$ is compact if and only if $\mathbb{K}$ is locally compact.

Chapter 1 of SALZMANN et al. [1995] gives a detailed exposition of the planes $\mathbf{PG}(2, \mathbb{R})$, $\mathbf{PG}(2, \mathbb{C})$, of the quaternion plane $\mathbf{PG}(2, \mathbb{H})$ and also of the alternative (octonion) plane $\mathbf{PG}(2, \mathbb{O})$, which can also be turned into a topological plane (but the topology cannot be obtained as above!). These four planes are compact and connected (in the topological sense).

9.3.2 Classical generalized quadrangles

Let $\mathbb{K}$ be a topological skew field which is not discrete, and let σ be a continuous anti-automorphism of $\mathbb{K}$ of order at most 2. Every non-degenerate σ-quadratic form q on a vector space V of finite dimension over $\mathbb{K}$ (compare Subsection 2.3.1 on page 53) leads to a topological generalized quadrangle $\Gamma = \mathsf{Q}(V, q) = (\mathcal{P}, \mathcal{L}, \mathbf{I})$, provided that $\dim V \geq 5$ or $\sigma \neq 1$ (see Corollary 2.3.6 on page 57). Indeed, we endow V with the product topology, which leads to a topological projective space $\mathbf{PG}(V)$ (compare KÜHNE & LÖWEN [1992]), and $\mathcal{P}$ and $\mathcal{L}$ are topologized as subsets of the points and lines, respectively, of $\mathbf{PG}(V)$. The continuity of the map $\mathcal{P} \times \mathcal{L} \setminus \mathbf{I} \to \mathbf{I} : (p, L) \mapsto (\mathrm{proj}_L\, p, \mathrm{proj}_p\, L)$, as required in Definition 9.2.1, follows from the observation that the polarity of $\mathbf{PG}(V)$ induced by q is continuous.

In particular, the symplectic quadrangle $\mathsf{W}(\mathbb{K})$ is a topological quadrangle for every topological field $\mathbb{K}$. If we take $\mathbb{K}$ to be a locally compact skew field, then $\mathsf{Q}(V, q)$ is a compact generalized quadrangle, because $\mathcal{P}$ and $\mathcal{L}$ are closed in the compact projective space $\mathbf{PG}(V)$. If $\mathbb{K} \in \{\mathbb{R}, \mathbb{C}, \mathbb{H}\}$, then $\mathsf{Q}(V, q)$ is compact and connected; otherwise $\mathsf{Q}(V, q)$ is totally disconnected (in fact, $\mathcal{P}$ and $\mathcal{L}$ are homeomorphic to Cantor's triadic set $\{0, 1\}^{\mathbb{N}}$; see GRUNDHÖFER & VAN MALDEGHEM [1990](2.3) and KRAMER [1994a](2.5.6)).

9.3.3 The generalized quadrangles of Ferus, Karcher & Münzner

Let $P_0, P_1, \ldots, P_m$ be real symmetric matrices of size $2k \times 2k$ such that $P_i P_j + P_j P_i = 2\delta_{ij} I$ for all i, j (with I the identity matrix). Such a system of matrices corresponds to a representation of the Clifford algebra of the Euclidean space $\mathbb{R}^{m-1}$. We define

$$H(x) = \sum_{i=0}^{m} \langle P_i x, x \rangle$$

for $x \in \mathbb{R}^{2k}$, where $\langle \ldots, \ldots \rangle$ denotes the Euclidean scalar product of $\mathbb{R}^{2k}$. According to FERUS, KARCHER & MÜNZNER [1981], the set $\mathbf{I} = \{x \in \mathbb{S}_{2k-1} \mid H(x) = \frac{1}{2}\}$ is an isoparametric hypersurface in the sphere $\mathbb{S}_{2k-1}$ with four distinct principal curvatures, provided that $m > 0$ and $k > m + 1$; the corresponding focal sets are given by $\mathcal{P} = \{x \in \mathbb{S}_{2k-1} \mid H(x) = 0\}$ and $\mathcal{L} = \{x \in \mathbb{S}_{2k-1} \mid H(x) = 1\}$. The associated projections $\pi_1 : \mathbf{I} \to \mathcal{P}$ and $\pi_2 : \mathbf{I} \to \mathcal{L}$ lead to an embedding of $\mathbf{I}$ into $\mathcal{P} \times \mathcal{L}$, via $x \mapsto (\pi_1(x), \pi_2(x))$, and we may consider $\mathbf{I}$ as a subset of $\mathcal{P} \times \mathcal{L}$. As THORBERGSSON [1992] has proved, the incidence structure $(\mathcal{P}, \mathcal{L}, \mathbf{I})$ is a compact connected topological quadrangle (in fact, a smooth one; see Subsection 9.8.2). The topological parameters (in the sense of Subsection 9.5.2) of $(\mathcal{P}, \mathcal{L}, \mathbf{I})$ are $(k - m - 1, m)$.

This construction is very interesting because it yields many quadrangles which are not Moufang quadrangles; indeed, one can choose k and m such that the topological parameters are different from those of the compact connected Moufang quadrangles listed in Theorem 9.6.3 below. Moreover, some of these quadrangles have topological parameters like a Moufang quadrangle without being homeomorphic to that Moufang quadrangle; see WANG [1988](Theorem 1). By FERUS, KARCHER & MÜNZNER [1981](6.4), there exist examples of these quadrangles which admit automorphism groups (in fact, groups of isometries of $\mathbb{R}^{2k}$) which act transitively on the point spaces, but these examples are not Moufang quadrangles; see also Theorem 9.6.8 below.

A closer investigation of these generalized quadrangles, from the point of view of incidence geometry, is a promising project for further research.

9.3.4 Generalized quadrangles and circle geometries

Let O be an oval in $\mathbb{R}^2$ which is closed (in the topological sense). The plane sections of the cylinder over O are the circles of a locally compact Laguerre plane of topological dimension 2, and the Lie geometry (see SCHROTH [1995](3.2)) of this Laguerre plane is a compact connected quadrangle with topological parameters $(1, 1)$. This construction yields plenty of examples of non-classical compact connected quadrangles (because it is easy to find ovals O which are not quadrics).

In fact, there is a deeper connection between generalized quadrangles and circle geometries, as the following results of SCHROTH [1995](2.16, 3.2, 3.13) show.

A generalized quadrangle is called **anti-regular** if for every point z the following property holds: for every pair of non-collinear points x, y opposite z, the traces z^x and z^y either are disjoint or have exactly two elements in common.

For the definition of topological parameters, see Subsection 9.5.2 below.

9.3.5 Theorem (Schroth [1995]). *Let Γ be a compact generalized quadrangle with topological parameters $p = q \in \{1, 2\}$. Then Γ or its dual is anti-regular; if Γ is anti-regular,* **then** *the perp-geometry of any point of Γ is a locally compact Laguerre plane of topological dimension $2p$, and all locally compact Laguerre planes of finite topological dimension arise in this fashion.* $\qquad\qquad\square$

Similar connections exist with Möbius and Minkowski planes, if Γ admits an in-volution with certain properties; see SCHROTH [1995].

9.3.6 Compact quadrangles of Tits type

If $\mathcal{O}$ is some closed ovoid in $\mathbf{PG}(d, \mathbb{R})$, $d \geq 2$, then the corresponding generalized quadrangle $\mathsf{T}_d(\mathcal{O})$ has a natural topological structure which turns $\mathsf{T}_d(\mathcal{O})$ into a compact quadrangle. The point rows are one-dimensional over $\mathbb{R}$, but the line pencils can have arbitrary dimension over $\mathbb{R}$, since there exist closed ovoids in $\mathbf{PG}(d, \mathbb{R})$ for every $d \geq 2$. If $\mathcal{O}$ is a quadric, then we have a classical example with the same topology (compare Subsection 3.7.1 on page 121), but other examples exist in great numbers, for instance the surface with equation $\sum_{i=0}^{d} X_i^a = 0$, with a an even non-zero positive integer different from 2 (if $a = 2$, then we have a quadric).

Note that in $\mathbf{PG}(2, \mathbb{C})$ every closed oval is a conic by BUCHANAN [1979] (see also SALZMANN et al. [1995](55.13)). It is an easy consequence to derive from that fact that there are no closed ovoids in $\mathbf{PG}(d, \mathbb{C})$ for $d > 2$. Hence over $\mathbb{C}$, every compact quadrangle of Tits type is dual to the symplectic quadrangle $\mathsf{W}(\mathbb{C})$.

Finally, no quadrangle $\mathsf{T}_2^*(\mathcal{O})$ of $*$-Tits type, with $\mathcal{O}$ a conic, can be given a non-trivial topological structure; see GRUNDHÖFER, JOSWIG & STROPPEL [1994].

9.3.7 Classical generalized hexagons

Let $\mathbb{K}$ be a locally compact field which is not discrete. Then the split Cayley hexagon $\mathsf{H}(\mathbb{K})$ as described in Subsection 2.4.13 (see page 73) is embedded into the compact six-dimensional projective space $\mathbf{PG}(6, \mathbb{K})$; the point set and the line set are closed subsets of $\mathbf{PG}(6, \mathbb{K})$. In view of Lemma 9.2.2, this means that $\mathsf{H}(\mathbb{K})$ is a compact topological hexagon.

If we take $\mathbb{K} \in \{\mathbb{R}, \mathbb{C}\}$, then we get the compact connected hexagons $\mathsf{H}(\mathbb{R})$ and $\mathsf{H}(\mathbb{C})$. If we take $\mathbb{K}$ to be a local field, then we obtain a compact totally disconnected hexagon $\mathsf{H}(\mathbb{K})$.

In fact, the split Cayley hexagons over $\mathbb{R}$ and $\mathbb{C}$, together with their duals, are the only compact connected hexagons presently known (this is analogous to the situation of finite generalized hexagons, where all known examples are Moufang hexagons; see Section 3.8.2). Before embarking on an elaborate theory of compact topological hexagons, it seems desirable to construct non-classical examples.

9.4 General properties

Let $(\mathcal{P}, \mathcal{L}, \mathbf{I})$ be a topological n-gon, as in Definition 9.2.1. We mention some general properties of the topological spaces $\mathcal{P}$ and $\mathcal{L}$, and some related spaces such as the flag space $\mathbf{I}$, the point rows and the line pencils.

We start with three purely topological results, and end with a result connecting topology with (geometric) derivation (span-planes and span-quadrangles).

9.4.1 Lemma.

(*i*) $\mathcal{P}$ *and* $\mathcal{L}$ *are regular Hausdorff spaces* (T_3), *every point row and every pencil of lines is closed in* $\mathcal{P}$ *and* $\mathcal{L}$, *respectively, and the flag space* **I** *is closed in* $\mathcal{P} \times \mathcal{L}$.

(*ii*) *The projections* **I** $\to \mathcal{P}$ *and* **I** $\to \mathcal{L}$ *are locally trivial bundles, and hence open.*

(*iii*) *If some point row is path-connected, then* $\mathcal{P}, \mathcal{L},$ **I** *and every pencil of lines is path-connected, and the point rows and pencils are locally contractible.* □

For proofs (and sharper versions) see KRAMER [1994a](2.1); the regularity (in the topological sense) in (*i*) is due to JÄGER [1994].

9.4.2 Theorem (Grundhöfer, Knarr & Kramer [1995]). *If* $\mathcal{P}$ *is locally compact,* **then** $\mathcal{P}$ *and* $\mathcal{L}$ *are locally compact and have a countable basis; in particular,* $\mathcal{P}$ *and* $\mathcal{L}$ *are separable metrizable spaces.* □

9.4.3 Proposition (Kramer [1994a]). *If* $\mathcal{P}$ *is locally compact and connected,* **then** $\mathcal{P}$ *and* $\mathcal{L}$ *are compact.* □

This is proved in KRAMER [1994a](2.5.5). We remark that it is an open problem whether the conclusion of the previous proposition is true also for non-discrete locally compact point spaces $\mathcal{P}$ (which are totally disconnected), even in the special case of projective planes $(n = 3)$.

There are also some general results concerning the derived geometries. We use the notation of Section 1.9.

9.4.4 Theorem (Schroth [1992], Schroth & Van Maldeghem [1994]).

(*i*) *Let* Γ *be a compact connected polygon and suppose that* x *is a projective point; then the perp-geometry* $\Gamma_x^{\triangle}$ *in* x *is a compact connected projective plane.*

(*ii*) *Let* Γ *be a compact connected quadrangle and suppose that* x *is a projective point. Then the span-geometry* $\Gamma_x^{\triangledown}$ *in* x *is a compact connected projective plane, homeomorphically isomorphic with* $\Gamma_x^{\triangle}$.

(*iii*) *Let* Γ *be a compact connected hexagon and suppose that* x *is a polar point. Then the span-geometry* $\Gamma_x^{\square}$ *in* x *is a compact connected quadrangle, where the topologies on the point set and line set of* $\Gamma_x^{\square}$ *extend the topologies on* $\Gamma_i(x)$, *for* $i = 1, 2, 3, 4$. □

In the last case, we can weaken the condition and only require that x be a span-regular point. This will follow from the results in the next section.

9.5 The impact of algebraic topology

The deeper results on topological polygons require a substantial dose of algebraic topology (this is already true for topological projective planes). The following analogue of Theorem 1.7.1 (see page 24) is one of the highlights of the theory.

9.5.1 Theorem (Knarr [1990], Kramer [1994a]). *Let* $\Gamma = (\mathcal{P}, \mathcal{L}, \mathbf{I})$ *be a compact connected generalized n-gon such that the point space* $\mathcal{P}$ *has finite topological (covering) dimension. Then* $n \in \{3, 4, 6\}$*, i.e.,* Γ *is a projective plane, a generalized quadrangle or a generalized hexagon. Furthermore, each point row is homotopy equivalent to a sphere* $\mathbb{S}_p$ *of dimension* p*, and each line pencil is homotopy equivalent to* $\mathbb{S}_q$*, with the following restriction on* p *and* q :

(*i*) **If** $n = 3$**, then** $p = q \in \{1, 2, 4, 8\}$.

(*ii*) **If** $n = 4$ *and* $p, q > 1$**, then** $p + q$ *is odd or* $p = q \in \{2, 4\}$.

(*iii*) **If** $n = 6$**, then** $p = q \in \{1, 2, 4\}$. $\qquad\qquad\square$

It is an open problem whether the assumption that $\mathcal{P}$ has finite dimension is necessary (even for $n = 3$); all known examples of compact polygons have finite topological dimension.

The result above can be found in KRAMER [1994a](3.3.6); it was proved by KNARR [1990] under the stronger assumption that the point rows and line pencils are topological manifolds. According to KRAMER [1994a](3.1.5), these spaces are always generalized manifolds which are homotopy equivalent to a sphere. Furthermore, the double mapping cylinder $D\mathbf{I}$ of the flag space $\mathbf{I}$ is another generalized manifold and it is homotopy equivalent to a sphere of dimension $\frac{n}{2}(p + q) + 1$. The so-called Veronese embeddings of $\mathcal{P}, \mathcal{L}$ and $\mathbf{I}$ into $D\mathbf{I}$ lead to a topological problem on locally trivial bundles (see Lemma 9.4.1(*ii*)) which was studied by MÜNZNER [1980], [1981]. Scrutinizing Münzner's proof yields the structure of the cohomology rings of $\mathcal{P}, \mathcal{L}$ and $\mathbf{I}$ (see STRAUSS [1996]); these rings are listed in the appendix of GRUNDHÖFER, KNARR & KRAMER [1995] and in KRAMER [1994a](6.4.1). The restrictions on n, p and q are obtained from the structure of these cohomology rings. In the special case $n = 3$, these restrictions have been obtained by LÖWEN [1983].

We mention that is an open problem whether the case $n = 4$ or 6, $p = q = 4$ is possible; compare subsection 9.3.7 above.

9.5.2 Topological parameters

It is natural to regard the pair (p, q) associated as in Theorem 9.5.1 to a compact connected polygon Γ of finite topological dimension as a measure of the size of Γ, similar to the order (s, t) of a finite generalized polygon (see Definitions 1.2.1 on page 2). Therefore, we call (p, q) the **topological parameters** of Γ. It follows from STRAUSS [1996] or KRAMER [19**] that $p = q > 1$ holds precisely if at least one (and then all) of the spaces $\mathcal{P}, \mathcal{L}, \mathbf{I}$ have Euler characteristic not equal to zero. If $p \neq q$, then $n = 4$, and $\mathcal{P}, \mathcal{L}, \mathbf{I}$ all have Euler characteristic zero, and this fact accounts for many topological problems arising from these generalized quadrangles.

As an application of the topological parameters we have the following result, which enables one to weaken or reformulate the conditions of Theorem 9.4.4 above. At the same time, the application radius of that result is limited to n-gons for $n = 4, 6$.

9.5.3 Theorem (Schroth [1992], Schroth & Van Maldeghem [1994]).

(i) *Let Γ be a compact connected polygon with topological parameters (p, q) and suppose that x is a distance-2-regular point. Then $p \geq q$ and the perp-geometry $\Gamma_x^\triangle$ is a (compact connected) projective plane* **if and only if** *$p = q$.*

(ii) *Let Γ be a locally compact connected quadrangle with topological parameters (p, q) and suppose that x is a regular point. Then $p \geq q$ and the span-geometry $\Gamma_x^\triangledown$ is a (compact connected) projective plane* **if and only if** *$p = q$.*

(iii) *Let Γ be a compact connected hexagon and suppose that x is a distance-2-regular point. Then the perp-geometry $\Gamma_x^\triangle$ is a (compact connected) projective plane (hence x is a projective point). If, moreover, x is span-regular, then the span-geometry $\Gamma_x^\square$ is a (compact connected) quadrangle (and hence, x is a polar point).*

Note that (iii) is due to the fact that every compact connected hexagon has topological parameters (p, q) with $p = q$. We also have the following two geometrical and topological characterizations of the compact connected Pappian quadrangles and hexagons.

9.5.4 Theorem (Schroth [1992], Schroth & Van Maldeghem [1994]). *For a locally compact connected generalized quadrangle Γ with topological parameters (p, q), the following properties are equivalent:*

(i) *Γ is the symplectic quadrangle over the topological field $\mathbb{R}$ or $\mathbb{C}$;*

(ii) *Γ is the symplectic quadrangle over a topological field;*

(iii) *Γ is a symplectic quadrangle;*

(iv) *Γ is point-distance-2-regular and for every point x the perp-geometry $\Gamma_p^\triangle$ is a topological projective plane;*

(v) *Γ is point-distance-2-regular and for at least one point x the perp-geometry $\Gamma_p^\triangle$ is a projective plane;*

(*vi*) Γ *is point-distance-2-regular and for every point x the span-geometry $\Gamma_p^{\triangledown}$ is a topological projective plane;*

(*vii*) Γ *is point-distance-2-regular and for at least one point x the span-geometry $\Gamma_p^{\triangledown}$ is a topological projective plane;*

(*viii*) Γ *is point-distance-2-regular and $p = q$.*   □

9.5.5 Theorem (Schroth & Van Maldeghem [1994]). *For a compact connected generalized hexagon Γ the following properties are equivalent:*

(*i*) Γ *is the split Cayley hexagon over the topological field $\mathbb{R}$ or $\mathbb{C}$;*

(*ii*) Γ *is the split Cayley hexagon over a topological field;*

(*iii*) Γ *is a split Cayley hexagon;*

(*iv*) Γ *is point-distance-2-regular and for every point x the perp-geometry $\Gamma_p^{\triangle}$ is a topological projective plane;*

(*v*) Γ *is point-distance-2-regular and for every point x the span-geometry $\Gamma_p^{\square}$ is a topological generalized quadrangle;*

(*vi*) Γ *is point-distance-2-regular.*   □

Extending the definition of a compact quadrangle to a compact weak quadrangle in the obvious way (the double of a generalized digon is compact if both the set of points and the set of lines of the digon carry the structure of a compact topological space; if these two topologies are homotopy equivalent to the same sphere $\mathbb{S}_p$ of dimension p, then we say that the weak quadrangle has **topological parameters** $(0, p)$; dual definitions hold), we can define a **compact weak subquadrangle** $\Gamma' = (\mathcal{P}' \, \mathcal{L}', \mathbf{I}')$ of a given compact quadrangle $\Gamma = (\mathcal{P}, \mathcal{L}, \mathbf{I})$ as a weak subquadrangle which is compact under the relative topology, i.e., $\mathcal{P}'$ and $\mathcal{L}'$ are closed in $\mathcal{P}$ and $\mathcal{L}$, respectively. We can also define a **topologically point-minimal compact quadrangle** as a compact quadrangle which does not have proper ideal compact subquadrangles. Similar definitions can be given for compact hexagons. We now have the following results, which are the topological analogues of Proposition 1.8.7 (page 35) and Corollary 1.8.10 (page 37).

9.5.6 Proposition (Kramer & Van Maldeghem [19**]). *If Γ' is a compact weak full subquadrangle with topological parameters (p, q') of a compact quadrangle Γ with topological parameters (p, q), then $p + q' \leq q$. Also, if $p + q' = q$, then every line of Γ either belongs to Γ' or is incident with a unique point of Γ'.*   □

9.5.7 Corollary (Kramer & Van Maldeghem [19**]). *Up to duality, every compact polygon is topologically point-minimal.*   □

9.6 Transitivity properties

Let G be the group of all continuous collineations of a compact connected n-gon $\Gamma = (\mathcal{P}, \mathcal{L}, \mathbf{I})$ with $n > 2$. We endow G with the compact-open topology derived from the action on $\mathcal{P}$ (or on $\mathcal{L}$ or on $\mathbf{I}$, which gives the same topology). Then G is a topological group with countable basis, acting on $\mathcal{P}$ and on $\mathcal{L}$ as a topological transformation group; see ARENS [1946]. We refer to this topological group G as the **automorphism group** of Γ. The following results are basic to the study of the group G.

9.6.1 Theorem (Burns & Spatzier [1987]). *The automorphism group G of every compact connected polygon Γ is a locally compact group.* $\square$

Note that the metrizability assumption of BURNS & SPATZIER [1987] is satisfied in view of Theorem 9.4.2.

9.6.2 Theorem (Szenthe [1974]). *If the automorphism group G of a compact connected polygon $\Gamma = (\mathcal{P}, \mathcal{L}, \mathbf{I})$ is transitive on the point space $\mathcal{P}$, then G is a (real) Lie group, and $\mathcal{P}$ is a (real-analytic) manifold, hence of finite dimension (as in Theorem 9.5.1).* $\square$

This result is a consequence of the solution of Hilbert's fifth problem, which asks for a purely topological characterization of Lie groups; for our purpose, the solution given essentially by Szenthe in the above theorem, but adapted from GRUNDHÖFER, KNARR & KRAMER [1995](2.2), is most suitable.

Let Γ be a compact connected Moufang polygon, and let G be the group generated by all root collineations of Γ, i.e., G is the little projective group of Γ. By Corollary 5.8.3 (see page 233) and Theorem 9.6.2 above, G is a simple Lie group, and the BN-pair of G is the standard one; see BURNS & SPATZIER [1987](3.21, 4.8), compare also KRAMER [19**]. The simple Lie groups are known explicitly (compare HELGASON [1978](X, Table VI, page 532ff.); or ONISHCHIK & VINBERG [1990](Table 9, page 312)), and one only has to pick those of relative rank 2. This leads to the following result (see KRAMER [19**] and GRUNDHÖFER & KNARR [1990](5.2)), where we use the notation introduced in Subsection 2.3.14 (see Chapter 2 on page 62).

9.6.3 Theorem. *Let Γ be a compact connected Moufang polygon. Then up to an isomorphism (which is also a homeomorphism), we have one of the following cases:*

(a) *Γ is the projective plane over $\mathbb{R}, \mathbb{C}, \mathbb{H}$ or $\mathbb{O}$, with topological parameters $p = q \in \{1, 2, 4, 8\}$.*

(b$_1$) *Γ is isomorphic or dual to the real symplectic quadrangle $\mathsf{W}(\mathbb{R})$ or to the complex symplectic quadrangle $\mathsf{W}(\mathbb{C})$, with topological parameters $p = q = 1$ or $p = q = 2$, respectively.*

(b$_2$) Γ *is isomorphic or dual to the real orthogonal quadrangle* Q$(\ell, \mathbb{R})$, *with topological parameters* $p = 1$, $q = \ell - 3$, *where* $\ell \geq 5$.

(b$_3$) Γ *is isomorphic or dual to the complex Hermitian quadrangle* H$(\ell, \mathbb{C})$, *with topological parameters* $p = 2$, $q = 2\ell - 5$, *where* $\ell \geq 4$.

(b$_4$) Γ *is isomorphic or dual to the real quaternion Hermitian generalized quadrangle* H$(\ell, \mathbb{H}, \mathbb{R})$ *with topological parameters* $p = 4$, $q = 4\ell - 9$, *where* $\ell \geq 3$.

(b$_5$) Γ *is isomorphic or dual to the complex quaternion Hermitian quadrangle* H$(4, \mathbb{H}, \mathbb{C})$, *with topological parameters* $p = 4$, $q = 5$.

(b$_6$) Γ *is isomorphic or dual to the exceptional Moufang quadrangle* Q$(E_6, \mathbb{R})$, *with topological parameters* $\{p, q\} = \{6, 9\}$.

(c) Γ *is isomorphic or dual to the split Cayley hexagon* H$(\mathbb{R})$ *or* H$(\mathbb{C})$, *with topological parameters* $p = q = 1$ *or* $p = q = 2$, *respectively.* □

9.6.4 Discussion of anti-isomorphisms

In the above enumeration, we have made the various classes disjoint by the conditions on the dimension ℓ. All quadrangles appearing in Subsection 2.3.14 are in the above list. The ones not mentioned explicitly are:

1. the real orthogonal quadrangle Q$(4, \mathbb{R})$, which is dual to the real symplectic quadrangle W$(\mathbb{R})$ (see Proposition 3.4.13 on page 109);

2. the complex orthogonal quadrangle Q$(4, \mathbb{C})$, which is dual to the complex symplectic quadrangle W$(\mathbb{C})$ (see again Proposition 3.4.13);

3. the complex Hermitian quadrangle H$(3, \mathbb{C})$, which is dual to the real orthogonal quadrangle Q$(5, \mathbb{R})$ (see Proposition 3.4.9 on page 107);

4. the complex quaternion Hermitian quadrangle H$(3, \mathbb{H}, \mathbb{C})$, which is dual to the real orthogonal quadrangle Q$(7, \mathbb{R})$ (see Proposition 3.4.11 on page 108).

None of the generalized quadrangles and hexagons listed in Theorem 9.6.3 above is self-dual.

Perhaps also the compact totally disconnected (infinite) Moufang polygons can be classified explicitly. Using the results of TITS [1983], one observes that compact totally disconnected Moufang octagons do not exist. See GRUNDHÖFER & VAN MALDEGHEM [1990] for the connection with affine buildings of rank 3; in particular, the class of generalized polygons with valuation discussed in the next section is a subclass of the class of totally disconnected polygons and corresponds precisely to the subclass which is connected to affine buildings of rank 3.

A rather strong transitivity condition has been considered by JOSWIG [1995], who proved the following result, which is the topological analogue of Theorem 6.8.9 (see page 288).

9.6.5 Theorem (Joswig [1995]). *Let* Γ *be a compact connected n-gon. Then the automorphism group G of Γ acts transitively on the set of ordered ordinary $(n+1)$-gons in Γ precisely in the following cases:*

(i) $n = 3$ *and* Γ *is isomorphic to one of the classical projective planes over* $\mathbb{R}$, $\mathbb{C}$, $\mathbb{H}$ *or* $\mathbb{O}$;

(ii) $n = 4$ *and* Γ *is isomorphic or dual to an orthogonal quadrangle over* $\mathbb{R}$ *or* $\mathbb{C}$, *or to the real quaternion Hermitian quadrangle* $H(3, \mathbb{H}, \mathbb{R})$;

(iii) $n = 6$ *and* Γ *is isomorphic or dual to the split Cayley hexagon over* $\mathbb{R}$ *or* $\mathbb{C}$.

□

The Hermitian quadrangle in (ii) was overlooked by MICHAEL JOSWIG (this was communicated to me by LINUS KRAMER).

Transitivity on ordered $(n + 1)$-gons implies transitivity on ordered n-gons, i.e., the Tits condition; see Section 4.7 on page 159. Compact polygons satisfying the Tits condition have been considered by BURNS & SPATZIER [1987](3.12) (under the name "topologically Moufang polygons"); they proved the following result (in view of Theorem 9.4.2 above).

9.6.6 Theorem (Burns & Spatzier [1987]). *Let* Γ *be a compact connected n-gon with $n \geq 3$. If a closed subgroup G of the full automorphism group of Γ acts transitively on the set of ordered ordinary n-gons in Γ,* **then** Γ *is a connected compact Moufang polygon (all such polygons are listed in* Theorem 9.6.3 *above), and G contains the simple Lie group belonging to this Moufang polygon.*   □

This result is the analogue of Theorem 5.7.6 (page 230) on finite polygons.

In fact, the much weaker condition of flag-transitivity suffices to conclude that Γ is a Moufang polygon:

9.6.7 Theorem (Grundhöfer, Knarr & Kramer [1995], [19**]). *Let* Γ *be a compact connected polygon. If* Γ *admits a flag-transitive group of automorphisms,* **then** Γ *is a compact connected Moufang polygon.*   □

The corresponding closed flag-transitive automorphism groups can be found in GRUNDHÖFER, KNARR & KRAMER [1995], [19**].

Theorem 9.6.7 above says that the Moufang condition, the Tits condition, and flag-transitivity coincide for compact connected polygons (see Section 4.8 for the finite case). Point-transitivity is a strictly weaker condition: there exist compact connected quadrangles which are not Moufang quadrangles but admit point-transitive automorphism groups; see Subsection 9.3.3. The classification of all point-transitive compact connected polygons remains a challenging open problem. A first step in that direction is given by the following result.

9.6.8 Theorem (Kramer [1994a]). *Let Γ be a compact connected n-gon which admits a point-transitive group of automorphisms. Then Γ has finite topological dimension, hence finite topological parameters (p, q) as in* Theorem 9.5.1 *above. If $p = q$, then Γ is a Moufang polygon; in fact, Γ is isomorphic or dual to the projective plane over $\mathbb{R}, \mathbb{C}, \mathbb{H}$ or $\mathbb{O}$ or to the real or complex symplectic quadrangle, or to the real or complex split Cayley hexagon.* $\square$

For $n = 3$, the above result is essentially due to SALZMANN [1975]; see SALZMANN et al. [1995](63.8).

9.6.9 Salzmann-type results

For compact connected projective planes Γ, a large number of results of the following type have been obtained by Salzmann and his school (see SALZMANN et al. [1995]): if Γ admits a closed automorphism group G with $\dim G \geq c$, then Γ is known explicitly; here c is a bound which depends on the topological parameter $p = q \in \{1, 2, 4, 8\}$ of Γ. More precisely, if $p = 1, 2, 4, 8$ and $\dim G > 4, 8, 18, 40$, respectively, then Γ is the (Moufang) plane over $\mathbb{R}, \mathbb{C}, \mathbb{H}, \mathbb{O}$, respectively, and all planes Γ with $\dim G = 4, 8, 18, 40$, respectively, are known explicitly.

It is tempting to develop a similar theory for generalized quadrangles (and generalized hexagons, as soon as non-Moufang examples are known). First steps in this direction have been taken in STROPPEL & STROPPEL [1996], [19**]. A result of SCHROTH (unpublished) says that every compact connected quadrangle with topological parameters $(1, 1)$ which admits an automorphism group of dimension at least 6 is isomorphic or dual to the real symplectic quadrangle.

9.7 Polygons with valuation

Perhaps the analogue of a compact connected polygon in the disconnected case is a polygon with valuation, as introduced by VAN MALDEGHEM [1989b]. This class of polygons has been defined because of their strong connection with affine buildings of rank 3. In fact, it has been proved that generalized n-gons with valuation, $n \neq 6$, are equivalent with affine buildings whose spherical building at infinity is a generalized n-gon. For $n = 6$, this is a conjecture which still remains unproved. It is beyond the scope of this book to dive into the theory of affine buildings of rank 3, but we will introduce generalized polygons with valuation, state the main results, and give some examples.

9.7.1 Definition. Let $\Gamma = (\mathcal{P}, \mathcal{L}, \mathbf{I})$ be a generalized n-gon. Let Ω be some ordinary $(n + 1)$-gon in Γ. Denote respectively by $\mathcal{P}_\perp$ and $\mathcal{L}_\perp$ the set of pairs $(x, y) \in \mathcal{P}^2$ and $(L, M) \in \mathcal{L}^2$ such that $x \perp y$ and $L \perp M$. Let

$$u : \mathcal{P}_\perp \cup \mathcal{L}_\perp \to \mathbb{N} \cup \{\infty\}$$

be a map such that $u(x, y) = u(L, M) = 0$ for any two collinear points x, y in Ω and any two confluent lines L, M in Ω. Note that, unusually our habit, we write the map u at the left of the pair (x, y) (and (L, M)) instead of as an exponent. Let there be given a sequence ζ of non-negative integers

$$\zeta = (a_1, a_2, \ldots, a_{n-1}; a_{n+1}, a_{n+2}, \ldots, a_{2n-1})$$

with greatest common positive divisor equal to 1 and containing at least one non-zero positive integer. Suppose that for every line L of Γ, the restriction of u to the pairs of points incident with L is surjective onto $\mathbb{N} \cup \{\infty\}$ and is a valuation, i.e.,

(U1) $u(x, y) = \infty$ **if and only if** $x = y$;

(U2) if $u(x, y) < u(y, z)$ then $u(x, z) = u(x, y)$,

where x, y, z are incident with L, and where we regard ∞ as the largest element in $\mathbb{N} \cup \{\infty\}$. Suppose also that for every point x the restriction of u to the pairs of lines incident with x is surjective and is a valuation as above. Note that (U2) implies that u is symmetric (by putting $x = z$). Then we call (Γ, u) a **polygon with valuation (relative to Ω)** with **weight sequence** ζ if for every ordered apartment $(w_1, w_2, w_3, \ldots, w_{2n}, w_1)$, with $w_1 \in \mathcal{P}$, we have

$$\sum_{i=1}^{n-1} a_i u(w_{i-1}, w_{i+1}) = \sum_{i=n+1}^{2n-1} a_i u(w_{i-1}, w_{i+1}).$$

We have the following result.

9.7.2 Theorem (Van Maldeghem [1989b]). *Let (Γ, u) be a generalized n-gon with valuation and weight sequence ζ. Then $n \in \{3, 4, 6\}$ and, up to duality, one of the following cases occurs:*

(*i*) $n = 3$ and $\zeta = (1, 1; 1, 1)$;

(*ii*) $n = 4$ and $\zeta = (1, 1, 1; 1, 1, 1)$;

(*iii*) $n = 6$ and $\zeta = (1, 1, 2, 1, 1; 1, 1, 2, 1, 1)$.

Moreover, the dual (Γ^D, u) is also a polygon with valuation and if ζ' is the weight sequence of this dual polygon, then one of the following cases occurs:

(*i*) $n = 3$ and $\zeta' = (1, 1; 1, 1)$;

(*ii*) $n = 4$ and $\zeta' = (1, 2, 1; 1, 2, 1)$;

(*iii*) $n = 6$ and $\zeta' = (1, 3, 2, 3, 1; 1, 3, 2, 3, 1)$. $\square$

Examples of projective planes and generalized quadrangles with valuation are constructed via their coordinatizing structure. Indeed, one can translate the axioms
of a generalized polygon to the coordinatizing algebraic ring and obtain some conditions. We will not reproduce these conditions here, since they are irrelevant to
what follows, but we mention that the last examples of Section 3.7.10 on page 130
are quadrangles with valuation, and also the first example of Section 3.8.1 on
page 133. We also mention that the conditions on the planar ternary field of a
projective plane in the case of a skew field boil down to the usual valuation on
that skew field. In order to produce at least one class of examples, let us examine
how one defines a Pappian polygon with valuation from a suitable projective space
over a field with valuation.

9.7.3 Examples of Pappian polygons with valuation

Let $\mathbb{K}$ be a field with valuation, i.e., there is a surjective map $v : \mathbb{K} \to \mathbb{Z} \cup \{\infty\}$
satisfying the following properties:

(V1) $v(x) = \infty$ **if and only if** $x = 0$;

(V2) if $v(x) < v(y)$ then $v(x + y) = v(x)$;

(V3) $v(xy) = v(x) + v(y)$,

for all $x, y \in \mathbb{K}$. Examples are provided by the field $\mathbb{F}((X))$ of formal Laurent
series over the field $\mathbb{F}$ with variable X. The valuation is the order of the Laurent
series in question (see also the last example of Subsection 3.7.10 on page 130).
Or consider the field $\mathbb{Q}$ of rational numbers and a fixed prime number p. Every
non-zero element z of $\mathbb{Q}$ can be uniquely written as $z = p^i \cdot \frac{a}{b}$ with $i, a, b \in \mathbb{Z}$, $b > 0$
and neither a nor b divisible by p. The valuation of z is now by definition equal
to i. This is the p-**adic valuation**.

We now define the map u which turns $\mathbf{PG}(2, \mathbb{K})$ into a projective plane with valuation. Let $\mathbf{PG}(2, \mathbb{K})$ be defined as in Subsection 2.2.1 (page 50) using homogeneous
coordinates. Let $p = (x, y, z)$ and $p' = (x', y', z')$ be two points of $\mathbf{PG}(2, \mathbb{K})$. By
multiplying with the inverse of the element with smallest valuation, we may assume
that, without loss of generality, $x = 1$ and $v(y), v(z) \geq 0$. Also, one of x', y', z' may
be assumed to be 1 and the others have non-negative valuation. If $x' = 1$, then we
put

$$u(p, p') = \min\{v(y - y'), v(z - z')\}.$$

If $x' \neq 0$, then we may assume without loss of generality that $y' = 1$ and
$v(x), v(x') > 0$. In that case we always put $u(p, p') = 0$. Similarly, one defines
the valuation on the set of pairs of lines. It is a tedious excercise to check that
$(\mathbf{PG}(2, \mathbb{K}), u)$ is a projective plane with valuation.

A similar construction can be done with the symplectic quadrangle $\mathsf{W}(\mathbb{K})$ over the
field $\mathbb{K}$ with valuation. For the points, one looks at the coordinates in $\mathbf{PG}(3, \mathbb{K})$
and applies the same method as above, only keeping those pairs of points which are

collinear in $\mathsf{W}(\mathbb{K})$. The valuation on pairs of concurrent lines is a little trickier: one looks at the images of the lines on the Klein quadric, i.e., one takes the Plücker coordinates of the lines, hence obtains points and applies the same method as above. It is clear how to generalize this to $\mathsf{H}(\mathbb{K})$.

Finally, we note that one can take quotients of polygons with valuation to obtain *level n Hjelmslev polygons*, which are in the case of projective planes exactly the Hjelmslev planes of level n as defined in e.g. ARTMANN [1969]. In fact, the inverse limit considered in ARTMANN [1971] is precisely a projective plane with valuation. First we mention a special case of what we have been saying here in the following theorem (unpublished in the case $n = 6$; for $n = 3, 4$, the result can easily be derived from the given references, though it is not explicitly stated there).

9.7.4 Theorem (Van Maldeghem [1988a], [1990b]). *Let (Γ, u) be a generalized n-gon with valuation. Define a relation $x \sim y$, for x and y both points or both lines, as follows: $x \sim y$ if there exists a path $(x = x_0, x_1, x_2, \dots, x_{2\ell} = y)$ such that $u(x_{2i}, x_{2i+2}) > 0$, for all i, $0 \le i < \ell$. Then the relation $\sim$ is an equivalence relation in both the set of points and the set of lines. We define $\Gamma' = \Gamma / \sim$ in the obvious way, i.e., the points of Γ' are the equivalence classes with respect to the relation $\sim$, similarly for the lines, and a point of Γ' is incident with a line of Γ' if some representatives of the respective equivalence classes are incident in Γ. Then Γ' is a generalized n-gon which is an epimorphic image of Γ.* $\qquad\square$

In the examples of Pappian polygons, the epimorphic image is again a Pappian polygon, over the residue field of the field $\mathbb{K}$ with valuation. Therefore, it is natural to call the epimorphic image associated to a polygon with valuation the **residual polygon**. We can now explain how to get a compact totally disconnected polygon from a polygon with valuation. For that, we need to refine the equivalence relation defined in the previous theorem. Also, we need the notion of *completeness*.

Let L be any line of a polygon Γ with valuation μ. The valuation μ induces on $\Gamma(L)$ a metric space (Γ, d), where d is defined by $d : \Gamma(L) \times \Gamma(L) \to \mathbb{R} : (x, y) \mapsto 2^{-\mu(x,y)}$. If this metric space is a *complete metric space*, i.e., if every Cauchy sequence in $(\Gamma(L), d)$ converges, then we say that L is a **complete** line. A similar definition holds for **complete** points. One can show that the presence of one complete line implies that all lines are complete, and dually, having one complete point forces all points to be complete. If all points and lines of (Γ, μ) are complete, then we say that (Γ, μ) is **complete**, or that Γ is complete with respect to μ. The following result can be derived from GRUNDHÖFER & VAN MALDEGHEM [1990].

9.7.5 Theorem (Grundhöfer & Van Maldeghem [1990]). *Let (Γ, μ) be a generalized n-gon with valuation. Define for all $i \in \mathbb{N}$ a relation $x \sim_i y$, for x and y both points or both lines, as follows: $x \sim_i y$ if there exists a path $(x = x_0, x_1, x_2, \dots, x_{2\ell} = y)$ such that $u(x_{2j}, x_{2j+2}) > i$, for all j, $0 \le j < \ell$. Then the relation $\sim_i$ is an equivalence relation in both the set of points and the set of lines. The set of all* *equivalence classes for all i ∈ ℕ defines a class of open sets of $\mathcal{P}$ or $\mathcal{L}$ which turns Γ into a topological totally disconnected polygon. It is compact **if and only if** the residual polygon is finite and Γ is complete with respect to μ.* ☐

We note that a valuation on a generalized polygon is not unique, even up to an automorphism or duality of the polygon. Examples of this phenomenon can be found in VAN MALDEGHEM [1988b].

The following result is a consequence of recent work on affine buildings by VAN MALDEGHEM & VAN STEEN [19**].

9.7.6 Theorem (Van Maldeghem, unpublished). *Let* Γ *be a projective plane with valuation and with finite residual plane (this happens for example when* Γ *is a compact plane with valuation). If* Γ *admits an automorpism group acting transitively on the set of ordered ordinary quadrangles,* **then** Γ *is a Desarguesian projective plane.*

My conjecture is that all generalized n-gons with valuation and with finite residual polygon admitting a collineation group which acts transitively on the set of ordered ordinary $(n+1)$-gons are Moufang n-gons. A proof of this is within reach, especially for the case of generalized quadrangles.

More information about polygons with valuation, and especially projective planes and generalized quadrangles, plus a series of examples (including a lot of non-classical ones) can be found in VAN MALDEGHEM [1987a], [1987b], [1988a], [1988b], [1989a], [1989b], [1990b], [1990c].

9.8 Other categories

Here we mention some results on generalized polygons belonging to categories which are stronger than the topological category.

9.8.1 Definition. Let $\mathcal{C}$ be a reasonable category (it should have products and coproducts). One may define a $\mathcal{C}$-**polygon** to be a generalized n-gon $\Gamma = (\mathcal{P}, \mathcal{L}, \mathbf{I})$, with $n \geq 3$, satisfying the following conditions:

(i) $\mathcal{P}$ and $\mathcal{L}$ are objects of $\mathcal{C}$.

(ii) The domain $\delta^{-1}(n-1) \subseteq (\mathcal{P} \sqcup \mathcal{L})^2$ of f_{n-1} is an object of $\mathcal{C}$, and the map $f_{n-1} : \delta^{-1}(n-1) \rightarrow \mathcal{P} \sqcup \mathcal{L}$ is a morphism of $\mathcal{C}$ (here $\mathcal{P} \sqcup \mathcal{L}$ is the coproduct of $\mathcal{P}$ and $\mathcal{L}$).

If $\mathcal{C}$ is the topological category, then this boils down to Definition 9.2.1. If $\mathcal{C}$ is the category of all locally compact connected spaces (with continuous maps), then we get all compact connected polygons, in view of Proposition 9.4.3.

Let $\mathcal{C}$ be the category of all k-times continuously differentiable manifolds and maps, for some k with $2 \leq k \leq \infty$; then we say that a $\mathcal{C}$-polygon is **smooth**. The category of complex manifolds with holomorphic maps leads to **holomorphic polygons**.

9.8.2 Smooth polygons

Some general results on smooth polygons have been obtained by BÖDI & KRAMER [1995]. They prove that *every continuous homomorphism between smooth polygons is a smooth embedding*. This has the following consequences. Every topological polygon Γ admits at most one differentiable structure which makes Γ into a smooth polygon. Furthermore, the group of all continuous collineations of a smooth polygon $(\mathcal{P}, \mathcal{L}, \mathbf{I})$ is a smooth Lie transformation group on $\mathcal{P}, \mathcal{L}, \mathbf{I}$ with respect to the compact-open topology; see *op. cit.*(4.9).

We remark that THORBERGSSON (unpublished) has shown that the generalized quadrangles of FERUS, KARCHER & MÜNZNER (see Subsection 9.3.3), are in fact smooth quadrangles.

Up to now, most of the results on smooth polygons have been concerned with projective planes. OTTE [1993], [1995] proved that the classical Moufang planes over $\mathbb{R}, \mathbb{C}, \mathbb{H}, \mathbb{O}$ are the only smooth (projective) translation planes; on the other hand, he constructed many examples of non-classical smooth projective planes (by locally disturbing the classical ternary operation $ax + b$ of $\mathbb{R}, \mathbb{C}, \mathbb{H}$ or $\mathbb{O}$ in a differentiable fashion). KRAMER [1994b] shows that the point space of every smooth projective plane is homeomorphic to the point space of the corresponding classical Moufang plane of the same dimension. A systematic investigation of smooth projective planes and many results of Salzmann type (compare Subsection 9.6.9) for these planes can be found in BÖDI [1996], always with smaller bounds on the dimension of the group.

9.8.3 Holomorphic polygons

Extending results of BREITSPRECHER [1967], [1971] on projective planes (compare SALZMANN et al. [1995](Section 75)), KRAMER [1996] proved that *every holomorphic n-gon is (geometrically and biholomorphically) isomorphic or dual to a complex Pappian polygon*. His proof uses the Riemann–Roch theorem, applied to the perp-geometry and the characterization results of Theorem 9.5.4 and Theorem 9.5.5.

9.8.4 Algebraic polygons

Let $\mathbb{K}$ be an algebraically closed field. The category of all $\mathbb{K}$-varieties and $\mathbb{K}$-morphisms leads to a definition of $\mathbb{K}$-algebraic polygons, which is a slight variation of Definition 9.8.1; see KRAMER & TENT [1996](2.1). The following is the main result of *op. cit.*: *if $\mathbb{K}$ has characteristic zero, then every $\mathbb{K}$-algebraic polygon is isomorphic or dual to a Pappian polygon over $\mathbb{K}$*. In the special case of projective planes (with somewhat stronger assumptions), this result is due to STRAMBACH [1975].

One might conjecture that this result holds true over algebraically closed fields of positive characteristic; to date this question is open.

As an application, KRAMER & TENT [1996] determine all irreducible Tits systems (B, N) of rank at least 2 in G, where G is a connected algebraic group over an algebraically closed field of characteristic zero, and B is a closed subgroup. One precisely obtains the standard Tits systems belonging to the simple algebraic groups.

We conjecture that the generalized quadrangles of FERUS, KARCHER & MÜNZNER (see Subsection 9.3.3), are real algebraic polygons; this would show that there exist real algebraic quadrangles which are not Moufang quadrangles. On the other hand, it seems that the only real algebraic projective planes presently known are the four Moufang planes over $\mathbb{R}, \mathbb{C}, \mathbb{H}$ and $\mathbb{O}$.

9.8.5 Polygons of finite Morley rank

The *Morley rank* is a model-theoretic notion of dimension; it is zero precisely if the set under consideration is finite. KRAMER, TENT & VAN MALDEGHEM [19**] have proved that, *if Γ is an infinite Moufang polygon of finite Morley rank, then Γ or its dual is (definably) isomorphic to a Pappian polygon over some algebraically closed field.* In particular, Γ is an algebraic polygon.

As a consequence of this theorem, they confirm the Cherlin–Zil'ber conjecture for infinite simple groups of finite Morley rank with spherical Moufang Tits system (or (B, N)-pair): such a group is an algebraic group over an algebraically closed field.

Appendix A

An Eigenvalue Technique for Finite Generalized Polygons

In this first appendix, we show the main part of Theorem 1.7.1 on page 24. The approach taken here is the original one by FEIT & HIGMAN [1964]. There are a lot of later versions of this proof, but the original one is amongst the most elementary and direct. Hence, it is not surprising that it requires some calculations which are uninteresting for other purposes in this book. That is why I essentially will deal with only one example (the most generic one) for each type of computation.

Let me first state precisely what I will prove.

A.1 Theorem (Feit & Higman [1964]). *Let $\Gamma = (\mathcal{P}, \mathcal{L}, \mathbf{I})$ be a finite weak generalized n-gon of order (s,t). Then we have $(s,t) = (1,1)$ or $n \in \{3, 4, 6, 8, 12\}$. If Γ is thick, i.e., if $s, t \geq 2$,* **then** *we have $n \in \{3, 4, 6, 8\}$; also st is a perfect square for $n = 6$ and $2st$ is a perfect square for $n = 8$.*

Proof. The idea of the proof is to calculate the multiplicity of a certain eigenvalue of a certain matrix, and to require that this must be a positive integer. So let us start by defining the matrix that we will be dealing with.

Let Γ be a finite weak generalized polygon of order (s,t). Let A be an incidence matrix of Γ, i.e., the the rows of A are indexed by the lines of Γ and the columns by the points of Γ, and the (L,p)th entry $A(L,p)$, $L \in \mathcal{L}$, $p \in \mathcal{P}$, is equal to 1 if $p \mathbf{I} L$, otherwise it is zero. Then we consider the matrix $M = A^T A$. Clearly both the rows and columns of the matrix M are indexed by the points of Γ, and the (x,y)th entry $M(x,y)$ is 1 if x and y are collinear but distinct, it is 0 if x and y are not collinear, and it is equal to $t+1$ if $x = y$. Now we consider the matrix M^q, for some $q \in \mathbb{N}$. By the definition of the product of matrices, the (x,y)th entry of M^q is equal to

$$
\begin{aligned}
(M^q)(x,y) &= \sum M(x,z_1)M(z_1,z_2)\ldots M(z_{q-1},y) \\
&= \sum A(L_1,x)A(L_1,z_1)A(L_2,z_1)\ldots A(L_q,z_{q-1})A(L_q,y),
\end{aligned}
$$

where the sum is over all points z_i, $i = 1, 2, \ldots, q - 1$, or over all points z_i, $i = 1, 2, \ldots, q - 1$ and all lines L_j, $j = 1, 2, \ldots, q$, respectively. But clearly the term $A(L_1, x)A(L_1, z_1) \ldots A(L_q, y)$ is equal either to 1 or to 0; and the former happens **if and only if** $(x, L_1, z_1, \ldots, L_q, y)$ is a (possibly stammering) path of length $2q$ in Γ connecting x with y. Hence $(M^q)(x, y)$ is equal to the number of paths of length $2q$ between x and y.

So our first task is to count the number of such paths. The following lemma gives recursion formulae. These formulae show at the same time that $(M^q)(x, y)$ only depends on $\delta(x, y)$ and not on x, y themselves. Indeed, for $q = 1$, $M(x, y)$ only depends on $\delta(x, y)$; the proof of the formulae below will show that, if $(M^q)(x, y)$ only depends on $\delta(x, y)$, then so does $(M^{q+1})(x, y)$. Hence for $\delta(x, y) = 2r$, we may put $(M^q)(x, y) = m_r^{(q)}$.

Lemma 1. *One has $m_0^{(0)} = 1$, $m_r^{(0)} = 0$ for $r > 0$, and the following recursion formulae hold:*

(i) $m_0^{(q+1)} = (t + 1)m_0^{(q)} + s(t + 1)m_1^{(q)}$;

(ii) $m_r^{(q+1)} = m_{r-1}^{(q)} + (s + t)m_r^{(q)} + stm_{r+1}^{(q)}$, *for $0 < 2r < n - 1$;*

(iii) $m_r^{(q+1)} = m_{r-1}^{(q)} + s(s + 2)m_r^{(q)}$, *for $2r = n - 1$ (and consequently n odd);*

(iv) $m_r^{(q+1)} = (t + 1)m_{r-1}^{(q)} + s(t + 1)m_r^{(q)}$, *for $2r = n$ (and so n is even).*

Proof. Let us only show the generic case (ii). So let x and y be two points at distance $2r$ in Γ, and they are distinct, not opposite, and not at codistance 1. The formula is all right for $q + 1 = r$, because then there is just one path, and using recursion, we indeed get $m_r^{(r)} = m_0^{(0)} = 1$. So we may assume that $q + 1 > r$. Now a path γ of length $2q + 2$ from x to y may start by $(x, L, x, \ldots)$, for any line $L \mathbf{I} x$, and there are $m_r^{(q)}$ ways of travelling to y to complete this as a path of length $2q + 2$. Obviously, there are $t + 1$ choices for L, and so we already have $(t + 1)m_r^{(q)}$ paths of the required length. But γ may also start as $(x, L, z, \ldots)$, with $z \neq x$. Then there are two possibilities. **Either** $L = \mathrm{proj}_x(y)$, **or** $L \neq \mathrm{proj}_x(y)$.

First let $L = \mathrm{proj}_x(y)$. Then there are again two possibilities. **Either** $\delta(z, y) = 2r$, in which case we can walk from z to y in $m_r^{(q)}$ different ways (with a path of length $2q$), **or** $\delta(z, y) = 2r - 2$, in which case there are $m_{r-1}^{(q)}$ possibilities to continue to y along a path of length $2q$. In the former case, there are $s - 1$ choices for z (indeed, $z \neq x$ and $z \neq \mathrm{proj}_x(y)$, but $z \mathbf{I} L$), in the latter case z is unique. Hence this gives rise to $(s - 1)m_r^{(q)} + m_{r-1}^{(q)}$ paths of length $2q + 2$ from x to y.

Finally, suppose that $L \neq \mathrm{proj}_x(y)$. Since x and y are neither opposite nor at codistance 1, we then know that whenever $z \mathbf{I} L \mathbf{I} x$, $z \neq x$, we automatically have $\delta(z, 2y) = 2r + 2$; see Lemma 1.5.6 on page 20. Hence since there are t possible choices for L, and s for z incident with L, this accounts for $stm_{r+1}^{(q)}$ paths of length $2q + 2$ between x and y. Adding all numbers obtained, the formula is proved. QED

The theorem will follow from the calculation of the multiplicity of a certain eigenvalue of M. The following lemma is the basis for that argument.

Lemma 2. *Let θ be a simple root of the polynomial $f(a)$ which has the property $f(M) = 0$, and let $m(\theta)$ be the multiplicity of θ as an eigenvalue of M. Then*

$$m(\theta) = \frac{\operatorname{tr}(f_\theta(M))}{f_\theta(\theta)},$$

where $f_\theta(a) = f(a)/(a - \theta)$.

Proof. From the definition of $f_\theta(a)$, it follows that $\operatorname{tr}(f_\theta(M))$ — which of course equals the sum $\sum m(\theta_i) f_\theta(\theta_i)$ over all eigenvalues θ_i of M with multiplicity $m(\theta_i)$ — is equal to $m(\theta) f_\theta(\theta)$, since $\operatorname{tr} f(M) = 0$, and thus $f(\theta_i) = f_\theta(\theta_i) = 0$, for θ_i an eigenvalue of M distinct from θ. The lemma follows.   QED

So we have to seek a polynomial $f(x)$ such that $f(M) = 0$, an eigenvalue θ of M which is a simple root of $f(x)$, and then we must calculate the trace of the matrix $f_\theta(M)$ and the value $f_\theta(\theta)$. In order to do so, we have the following lemmas.

Lemma 3. *The matrix M has a constant row sum equal to $(s+1)(t+1)$, in other words, $MJ = (s+1)(t+1)J$, where J is the matrix all of whose entries are equal to 1.*

Proof. The row sum corresponding to the point x is the sum of the number of paths of length 2 starting in x. To any point y with $\delta(x, y) > 2$, there are no such paths; to every point $y \in \Gamma(x)$ there is a unique such path, and there are $(t+1)s$ choices for y, and finally to x itself there are $t + 1$ such paths, hence the lemma.   QED

For the next lemma we have to introduce some notation for cutting a polynomial in suitable pieces. Suppose $f(s, t)$ is a polynomial in $\sqrt{s}$ and $\sqrt{t}$. Let $u < v$ be in $\frac{1}{2}\mathbb{N}$. Then the sum of the terms $f_{ij}\sqrt{s}^i \sqrt{t}^j$ of $f(s, t)$ with $2u < i + j \leq 2v$ will be denoted by $[f(s, t)]_u^v$. It is easy to see that the map $\zeta[u, v] : f(s, t) \mapsto [f(s, t)]_u^v$ is linear and that it satisfies

$$s^i t^j \zeta[u, v](f(s, t)) = \zeta[u + i + j, v + i + j](s^i t^j f(s, t)).$$

Since we only deal with positive exponents, this implies

$$s^i t^j \zeta[-1, v](f(s, t)) = \zeta[-1, v + i + j](s^i t^j f(s, t)).$$

We can now find an explicit expression for the $m_r^{(q)}$ introduced above, at least when $r + q < n$. Let us perform the calculations for $r \neq 0$. We try to find a polynomial $p_q(s, t)$ and a number $v_{q,r}$ such that

$$m_r^{(q)} = [p_q(s, t)]_{-1}^{v_{q,r}}. \tag{A.1}$$

If such a polynomial exists, then applying the properties of $\zeta[-1, v]$, and using Lemma 1(ii), we see that

$$m_r^{(q+1)} = [p_q(s,t)]_{-1}^{v_{q,r-1}} + [(s+t)p_q(s,t)]_{-1}^{v_{q,r}+1} + [stp_q(s,t)]_{-1}^{v_{q,r+1}+2}. \quad (A.2)$$

Hence we should have $v_{q,r-1} = v_{q,r} + 1$, implying that $v_{q,r} + r = v_{q,0} = v_q$. So (A.2) becomes

$$m_r^{(q+1)} = [p_{q+1}(s,t)]_{-1}^{v_{q+1}-r} = [(s+1)(t+1)p_q(s,t)]_{-1}^{v_q+1-r}, \quad (A.3)$$

which now implies that $v_{q+1} - r = v_q + 1 - r$, so $v_q = v + q$, for some number v. Also, it follows that $p_{q+1}(s,t) = (s+1)(t+1)p_q(s,t)$, hence $p_q(s,t) = (s+1)^{q-r}(t+1)^{q-r}p_r(s,t)$. Now, one easily calculates $m_1^{(2)} = s + 2t + 1$. Since this has degree 1 in both s and t, we deduce $1 = v_{2,1} = v + 2 - 1$, so $v = 0$. Consequently

$$s + 2t + 1 = [(s+1)(t+1)p_1(s,t)]_{-1}^1,$$

which implies that $p_1(s,t) = 1 + t + f(s,t)st$, with $f(s,t)$ a polynomial in s, t. With this expression, the case $r \neq 0$ has been dealt with fully, for every $f(s,t)$. So if there is a further restriction on $f(s,t)$, it has to come from a boundary condition for $r = 0$. Now, one readily computes

$$m_0^{(2)} = (t+1)^2 + s(t+1) = st + t^2 + s + 2t + 1.$$

This should be equal to the sum of the terms of degree at most 2 of $(s+1)(t+1)p_1(s,t)$, which is, putting the constant term of $f(s,t)$ equal to f_0, equal to the sum of the terms of degree at most 2 of $(s+1)(t+1)(1+t+f_0st)$, hence we obtain the equation

$$st + t^2 + s + 2t + 1 = (2 + f_0)st + t^2 + s + 2t + 1,$$

implying $f_0 = -1$. So we can take for $p_1(s,t)$ the polynomial $(1 + t - st)$, but also $(1+t)(1-st)$, because higher-order terms do not influence the result. This makes the result look more symmetric in s and t (but it is not). So we have shown, putting all results of this paragraph together, that

$$m_r^{(q)} = [(s+1)^{q-1}(t+1)^q(1-st)]_{-1}^{q-r}. \quad (A.4)$$

Note that the power -1 appearing for $q = 0$ is fine, since in that case also $r = 0$ and then we might expand $(s+1)^{-1}$ as a power series cut off already at its second term, i.e., we substitute $1 - s$ or even 1 for $(s+1)^{-1}$.

We can now state our next lemma.

Lemma 4. *Suppose $p(a)$ is a polynomial in a, $\sqrt{s}$ and $\sqrt{t}$, homogeneous in a, s, t of degree q. Let x and y be two points of Γ with $\frac{1}{2}\delta(x,y) = r < n - q$. Then the (x,y)th entry of $p(M)$ is equal to*

$$[(s+1)^{-1}(1-st)p((s+1)(t+1))]_{-1}^{q-r}.$$

Let N be the number of points of Γ; *then the trace of* $p(M)$ *is equal to*

$$N \cdot [(s+1)^{-1}(1-st)p((s+1)(t+1))]_{-1}^{q}.$$

Proof. In the case $k = 0$ (and only in that case), a negative exponent of $(s+1)$ turns up again. As above, this is no problem.

Let $p(a) = \sum a_i(s,t)a^i$, where $a_i(s,t)$ is homogeneous in s,t of degree $q-i$. Then the (x,y)th entry of $p(M)$ is, by the foregoing and Lemma 1, equal to

$$
\begin{aligned}
\sum a_i(s,t)m_r^{(i)} &= \sum a_i(s,t)[(s+1)^{i-1}(t+1)^i(1-st)]_{-1}^{i-r} \\
&= \sum [a_i(s,t) \cdot (s+1)^{i-1}(t+1)^i(1-st)]_{-1}^{(i-r)+(q-i)} \\
&= \left[\sum a_i(s,t) \cdot (s+1)^i(t+1)^i \cdot (s+1)^{-1}(1-st)\right]_{-1}^{q-r} \\
&= [(s+1)^{-1}(1-st)p((s+1)(t+1))]_{-1}^{q-r},
\end{aligned}
$$

showing the first part of the lemma. The trace of $p(M)$ is obtained by taking N times an (x,x)th entry of $p(M)$, and the latter is obtained by putting $r = 0$ in the first part.

This completes the proof of Lemma 4.   QED

From now on we have to distinguish three different cases: $n \equiv 1 \bmod 2$, $n \equiv 0 \bmod 4$, and $n \equiv 2 \bmod 4$. The most interesting case is the second because there the information on $n = 8$ must follow, and the somewhat special case $n = 12$ is also included. Hence we will deal in detail with $n \equiv 0 \bmod 4$, only indicating the major steps in the two other cases.

First we define an appropriate polynomial f, to which we can apply Lemma 2. This requires some preparation. First, notice that for $\ell \in \mathbb{N}$, the expression $a^{-\ell} + a^\ell$ is a polynomial of degree ℓ in $a^{-1} + a$, and hence also in $a^{-1} + a + 2$. Note that the leading coefficient of that polynomial is equal to 1. Hence the expression

$$a^{-\ell} + a^{-\ell+2} + \ldots + a^{\ell-4} + a^{\ell-2} + a^\ell = \frac{1-a^{2\ell+2}}{a^\ell(1-a^2)}$$

is a monic polynomial of degree ℓ in $a^{-1} + a + 2$. Now we put $\ell = h - 1$, where $n = 4h$, and so we obtain a polynomial

$$k(a^{-1} + a + 2) = \frac{1-a^{2h}}{a^{h-1}(1-a^2)}$$

of degree $h - 1$ and with leading coefficient 1. Next we put

$$g(a) = a(a - (s+t))(st)^{h-1}k\left(\frac{((s+t)-a)^2}{st}\right).$$

Note that, since $k(a)$ has degree $h - 1$, the largest (in absolute value) negative exponent of st resulting from the polynomial k in $g(a)$ is $h - 1$, and hence this cancels out with the $(st)^{h-1}$ in the numerator. Consequently, the leading coefficient of $g(a)$ is equal to 1. Note that $g(a)$ is homogeneous in a, s, t of degree $2h$.

Now we compute each entry of $g(M)$. Let us start with the (x, y)th entries for which x and y are opposite points. Since $m_{2h}^{(q)} = 0$ for $q < 2h$, we only have to consider the coefficient of a^{2h} in $g(a)$, i.e., 1. Since $m_{2h}^{(2h)} = t + 1$, we see that the (x, y)th entry we are looking for is equal to $(t + 1)$. Next suppose that x is not opposite y and put $\delta(x, y) = 2r$. Since $r < 2h = n - 2h$, we can apply Lemma 4. A direct computation shows that

$$(s + 1)^{-1}(1 - st)g((s + 1)(t + 1)) = (t + 1)(1 - (st)^{2h}), \qquad (A.5)$$

and so the (x, y)th entry of $g(M)$ is, since $0 < 2h - r < 2h$, equal to

$$[(t + 1)(1 - (st)^{2h})]_{-1}^{2h-r} = t + 1.$$

So we have actually shown that $g(M) = (t + 1)J$, with J as above: the $N \times N$ matrix all entries of which are equal to 1. Of course, we want a polynomial $f(a)$ such that $f(M) = 0$, but Lemma 2 now shows that $MJ - (s + 1)(t + 1)J = 0$, hence $(M - (s + 1)(t + 1)I)g(M) = 0$, where I is the $N \times N$ identity matrix. So if we put $f(a) = (a - (s + 1)(t + 1))g(a)$, then we have $f(M) = 0$.

Our next task is to find a simple root of $f(a)$. Therefore, we look first at the simple roots of $k(a)$. Clearly, any $(2h)$th root of unity, except for 1 itself, is a simple root of $k(a^{-1} + a + 2)$. Hence, any $(4h)$th root of unity ξ, except for the square roots 1 and -1, is a simple root of $k(a^{-2} + a^2 + 2) = k((a^{-1} + a)^2)$. So we put

$$(\xi^{-1} + \xi)^2 = \frac{((s + t) - a)^2}{st},$$

from which one easily deduces that $a = s + t \pm \sqrt{st}(\xi^{-1} + \xi)$ is a simple root. We can choose the plus sign (the minus sign is then produced by substituting $-\xi$ for ξ). Hence

$$\theta = s + t + \sqrt{st}(\xi^{-1} + \xi)$$

is a simple root of $f(a)$ for any nth root of unity ξ, $\xi \neq 1, -1$. For reasons that will become clear later, we also assume $\xi \neq \pm\sqrt{-1}$.

Now we have found a simple root θ, we should calculate the trace of $f_\theta(M)$. As the degree of $g_\theta(a)$ in a, s, t is equal to $2h - 1$, the degree of $f_\theta(a)$ in a, s, t is equal to $2h + 1 < 4h = n$. Hence we can apply Lemma 4. Therefore, we write $f_\theta(a)$ as a sum of homogeneous polynomials as follows:

$$f_\theta(a) = -g_\theta(a) + (a - (s + t))g_\theta(a) - stg_\theta(a), \qquad (A.6)$$

which is the sum of homogeneous polynomials in a, s, t of degree $2h - 1$, $2h$ and $2h + 1$, respectively.

In view of Lemma 2, we will need the value of $(1+s)^{-1}(1-st)g_\theta((s+1)(t+1))$, so we compute this first. But clearly this is equal to

$$(1+s)^{-1}(1-st)\frac{g((s+1)(t+1))}{(s+1)(t+1)-\theta},$$

which, in view of (A.5), is equal to

$$(t+1)\frac{1-(st)^{2h}}{1-(\xi-1+\xi)\sqrt{st}+st}.$$

The denominator is a divisor (as a polynomial in $\sqrt{st}$) of **either** $1-(st)^h$ **or** $1+(st)^h$. So we have

$$(1+s)^{-1}(1-st)g_\theta((s+1)(t+1)) =$$
$$(t+1)(1\pm(st)^h)(\pm(st)^{h-1}+G(\sqrt{st})), \qquad (A.7)$$

where $G(a)$ is some polynomial of degree $h-3$. We put the polynomial in $\sqrt{s}$ and $\sqrt{t}$ defined by (A.7) equal to $p(s,t)$.

Now noting that $(s+1)(t+1)-(s+t)=1+st$, we derive from Lemma 2, using (A.6),

$$
\begin{aligned}
\mathrm{tr}(f_\theta(M)) &= N\cdot\left(-[p(s,t)]_{-1}^{2h-1}+[(1+st)p(s,t)]_{-1}^{2h}-[stp(s,t)]_{-1}^{2h+1}\right) \\
&= N\cdot\left([p(s,t)]_{2h-1}^{2h}-[stp(s,t)]_{2h}^{2h+1}\right), \\
&= N\cdot(\pm(st)^h\pm t(st)^{h-1}),
\end{aligned}
$$

in view of (A.7). Hence we obtain

$$\mathrm{tr}(f_\theta(M)) = \pm N(t+1)(st)^h. \qquad (A.8)$$

Our last aim is to compute $f_\theta(\theta)$ explicitly. Therefore, we note that, if we denote the derivative with a dash, $f_\theta(\theta)=f'(\theta)$. We readily find (remarking $f(\theta)=g(\theta)=0$)

$$g'(\theta)=2\theta(\theta-(s+t))^2(st)^{h-2}\cdot k'((\xi^{-1}+\xi)^2). \qquad (A.9)$$

Putting $k_*(a)=k(a^{-1}+a+2)=k(b)$, we see that

$$k'(a^{-1}+a+2)=\frac{dk(b)}{db}=\frac{\dfrac{dk_*(a)}{da}}{\dfrac{db}{da}}.$$

An elementary calculation shows us

$$\frac{db}{da}=\frac{a^2-1}{a^2}, \quad \text{while} \quad \left.\frac{dk_*(a)}{da}\right|_{a=\xi^2}=-\frac{2h\xi^{2h}}{1-\xi^4},$$

which is well defined since we assumed that $\xi^4 \neq 1$. Hence we obtain

$$k'((\xi^{-1} + \xi)^2) = \frac{2h\xi^{2h+4}}{(1 - \xi^4)^2} = \frac{\pm 2h}{(\xi^{-2} - \xi^2)^2},$$

since $\xi^{2h} = \pm 1$. Combining this with (A.9), we obtain

$$g'(\theta) = \frac{\pm 4h\theta(\xi^{-1} + \xi)^2(st)^{h-1}}{(\xi^{-1} + \xi)^2(\xi^{-1} - \xi)^2},$$

which, in view of

$$f'(\theta) = (\theta - (s+1)(t+1))g'(\theta),$$

gives the final result

$$f'(\theta) = \frac{\pm 4h(st)^{h-1}(st - \sqrt{st}(\xi^{-1} + \xi) + 1)(s + t + \sqrt{st}(\xi^{-1} + \xi))}{(\xi^{-1} - \xi)^2}. \qquad \text{(A.10)}$$

So dividing (A.10) by (A.8), and omitting factors which are rational numbers, such as N, st and $4h$, Lemma 2 tells us that the number

$$\eta(\xi) = \frac{(st - \sqrt{st}(\xi^{-1} + \xi) + 1)(s + t + \sqrt{st}(\xi^{-1} + \xi))}{(\xi^{-1} - \xi)^2}$$

is a rational number, for all nth roots of unity, except for the fourth roots. In fact, using a slightly different definition for $k(a^{-1} + a + 2)$ and for $g(a)$, namely

$$k(a^{-1} + a + 2) = \frac{1 - a^{2h+1}}{a^h(1 - x)},$$

$$g(a) = a(st)^h \cdot k\left(\frac{((s+t) - a)^2}{st}\right),$$

with $n = 4h+2$, we obtain that exactly the same expression $\eta(\xi)$ should be rational for $n \equiv 2 \bmod 4$. The calculations are similar to the case where n is a multiple of 4.

To get an expression which is easier to handle, we remark that, if $\eta(\xi)$ is rational, then so is $\eta(-\xi)$, and hence so is the sum

$$\eta(\xi) + \eta(-\xi) = 2st + \frac{(st + 1)(s + t) - 4st}{\xi^{-2} + \xi^2 + 2}.$$

Now note that $(st + 1)(s + t) - 4st = s(1 - t)^2 + t(1 - s)^2$, and so it is only zero if $s = t = 1$, hence only if Γ is an ordinary n-gon. So if Γ has thick elements, then $r := \xi^{-2} + \xi^2$ must be rational for all nth roots ξ of unity different from $\pm 1, \pm\sqrt{-1}$. Hence such a root ξ satisfies $\xi^4 - r\xi^2 + 1 = 0$, $r \in \mathbb{Q}$. This means that ξ^2 satisfies some quadratic equation with coefficients in $\mathbb{Q}$. We now take for ξ the primitive nth root of unity. Any polynomial over $\mathbb{Q}$ having ξ^2 as a root has degree at least $\varphi(n/2)$, where φ is the **Euler phi function** ($\varphi(m)$ is the number of positive numbers coprime to m, including 1, smaller than m). Indeed, the minimal polynomial over $\mathbb{Q}$ for ξ^2 is given by the cyclotomic polynomial $\Phi_{n/2}$ (see JACOBSON [1985], Theorem 4.17, page 272), which has degree precisely $\varphi(n/2)$. Consequently, $\varphi(n/2) \leq 2$, hence, since 1 and $(n/2) - 1$ are always relatively prime to $n/2$, $n/2$ must have a common factor with either $(n-2)/4$ or $(n-4)/4$, if these are different from 1. This readily implies $n = 4, 6, 8$ or 12 (and $\varphi(n/2)$ is, respectively, equal to $1, 2, 2, 2$).

But similarly, we must have that $\eta(\xi) - \eta(-\xi)$ is rational, and so, calculating this, we obtain that

$$\eta(\xi) - \eta(-\xi) = 2\frac{(s-1)(t-1)\sqrt{st}(\xi^{-1} + \xi)}{\xi^{-2} + \xi^2 + 2}$$

must be rational. Suppose now that Γ is thick. Then this means that, taking a primitive root for ξ again,

$$\sqrt{st}(\xi^{-1} + \xi) = \sqrt{st}\cos\left(\frac{2\pi}{n}\right)$$

must be rational. So, if $n = 6$, then $\sqrt{st}$ must be rational, hence an integer. If $n = 8$, then $\sqrt{2st}$ must be rational, hence an integer. If $n = 12$, then $\sqrt{3st}$ must be rational. But in the case $n = 12$, we can also take for ξ a primitive sixth root, and so also $\sqrt{st}$ must be rational, a contradiction. So $n \neq 12$. This proves the theorem for n even.

Now suppose that n is odd. Then we perform the same calculations, this time with

$$k(a^{-1} + a + 2) = \frac{1 - a^{2h+1}}{a^h(1-x)},$$

$$g(a) = s^h \cdot k\left(\frac{a}{s}\right),$$

where $n = 2h + 1$ (and putting $s = t$ in the expression of $f(a)$ in terms of $g(a)$). The intermediate results are here:

$$\theta = s(\xi^{-1} + \xi + 2),$$

$$\mathrm{tr}(f_\theta(M)) = -Ns^h(s+1)\frac{\xi^{-h} + \xi^h}{\xi^{-1} + \xi},$$

$$f_\theta(\theta) = \frac{ns^{h-1}(s^2 - s(\xi^{-1} + \xi) + 1)}{(\xi^{-h} - \xi^h)(\xi^{-1} - \xi)},$$

where ξ is a primitive nth root of unity. This time the number

$$\frac{(s-1)^2}{s^2 - s(\xi^{-1} + \xi) + 1}$$

must be rational. So if Γ is thick, then, as above, $\varphi(n) \leq 2$, implying $n = 3$ (since if n is odd, $1, 2, n-2$ and $n-1$ are relatively prime to n).

This completes the proof of Theorem A.1. $\qquad\qquad\qquad\qquad\qquad\square$

Appendix B

A Proof of the Theorem of Bruck and Kleinfeld

In this appendix, we give a proof of the celebrated result of BRUCK & KLEINFELD [1951] and KLEINFELD [1951] classifying all non-associative alternative division rings. This appendix is a report on two lectures held by Jacques Tits at the University of Ghent on the occasion of his 65th birthday. Originally, he presented the proof in his lectures at the Collège de France in 1977–1978, having written it down in TITS (unpublished) in the early 1960s. It puts together various existing arguments and some new ones in order to obtain a very streamlined and beautiful proof, independent of the characteristic. Though it does not fit into my general approach of omitting everything that deals with projective planes, I think this exception can be made. My contribution to the proof below was figuring out some details which the lecturer in Ghent left the audience, and also I took the trouble to squeeze in as smoothly as possible the proofs of the basic properties of an alternative division ring, such as the fact that commutivity implies associativity (the rough ideas however borrowing from the literature, mainly TITS (unpublished) and PICKERT [1975]). Let me also point out that PICKERT [1975] played an important role in establishing the proof in this appendix. The first edition of PICKERT'S work appeared shortly after the original papers of BRUCK & KLEINFELD [1951] and KLEINFELD [1951].

Let me start by stating the theorem. Since everything is independent of the characteristic, the theorem does not describe in the most elegant way the multiplication table when specialized to characteristic 2 or characteristic different from 2. But that is exactly the price we have to pay for a unified proof. To make this appendix suitable for undergraduate students, I decided to show the complete result and to include much detailed information. By an **alternative division ring**, I understand a set $\mathbb{D}$ equipped with two operation laws, an addition $a, b \mapsto a + b$ which turns $\mathbb{D}$ into a commutative group, and a multiplication $a, b \mapsto ab$, for which there is a unit element 1 and with respect to which every non-zero element x has a two-sided inverse x^{-1} such that $x^{-1}(xy) = y = (yx)x^{-1}$ for all $x, y \in \mathbb{D}$. Moreover, both distributive laws are satisfied.

B.1 Theorem (Bruck & Kleinfeld [1951], Kleinfeld [1951]). *Let $\mathbb{D}$ be an alternative division ring which is not associative. Then there exists a subfield $\mathbb{K}$ of $\mathbb{D}$ and elements e_1, e_2, e_3 of $\mathbb{D}$ such that $\mathbb{D}$ is an eight-dimensional algebra over $\mathbb{K}$ with basis*

$$1, e_1, e_2, e_2e_1, e_3, e_3e_1, e_3e_2, e_3(e_2e_1),$$

with $e_i^2 = \ell_i \in \mathbb{K}$, $i = 2, 3$, with $e_1^2 - e_1 = \ell_1 \in \mathbb{K}$ and with multiplication table (where for clarity we denote e_ie_j by e_{ij} and $e_3(e_2e_1)$ by e)

·	e_1	e_2	e_{21}	e_3
e_1	$\ell_1 + e_1$	$e_2 - e_{21}$	$-\ell_1 e_2$	$e_3 - e_{31}$
e_2	e_{21}	ℓ_2	$\ell_2 e_1$	$-e_{32}$
e_{21}	$\ell_1 e_2 + e_{21}$	$\ell_2 - \ell_2 e_1$	$-\ell_1 \ell_2$	$-e$
e_3	e_{31}	e_{32}	e	ℓ_3
e_{31}	$\ell_1 e_3 + e_{31}$	e	$\ell_1 e_{32} + e$	$\ell_3 - \ell_3 e_1$
e_{32}	$e_{32} - e$	$\ell_2 e_3$	$\ell_2 e_3 - \ell_2 e_{31}$	$-\ell_3 e_2$
e	$-\ell_1 e_{32}$	$\ell_2 e_{32}$	$-\ell_1 \ell_2 e_3$	$-\ell_3 e_{21}$

and

·	e_{31}	e_{32}	e
e_1	$-\ell_1 e_3$	e	$\ell_1 e_{32} + e$
e_2	$-e$	$-\ell_2 e_3$	$-\ell_2 e_{31}$
e_{21}	$-\ell_1 e_{32} - e$	$-\ell_2 e_3 + \ell_2 e_{31}$	$\ell_1 \ell_2 e_3$
e_3	$\ell_3 e_1$	$\ell_3 e_2$	$\ell_3 e_{21}$
e_{31}	$-\ell_1 \ell_3$	$\ell_3 e_2 - \ell_3 e_{21}$	$-\ell_1 \ell_3 e_2$
e_{32}	$-\ell_3 e_1 + \ell_3 e_{21}$	$-\ell_2 \ell_3$	$-\ell_2 \ell_3 + \ell_2 \ell_3 e_1$
e	$\ell_1 \ell_3 e_2$	$-\ell_2 \ell_3 e_1$	$\ell_1 \ell_2 \ell_3$

Let us define $X \mapsto \overline{X}$, $X \in \mathbb{D}$, as the linear map (linear over $\mathbb{K}$) determined by

$$\overline{1} = 1, \qquad \overline{e}_1 = 1 - e_1, \qquad \overline{e}_2 = -e_2, \qquad \overline{e}_{21} = -e_{21},$$
$$\overline{e}_3 = -e_3, \qquad \overline{e}_{31} = -e_{31}, \qquad \overline{e}_{32} = -e_{32}, \qquad \overline{e} = -e.$$

Then, for any element

$$X = x_0 + e_1x_1 + e_2x_2 + e_{21}x_{21} + e_3x_3 + e_{31}x_{31} + e_{32}x_{32} + ex \in \mathbb{D},$$

with $x_i, x_{jk}, x \in \mathbb{K}$, $i = 0, 1, 2, 3$, $jk = 21, 31, 32$, the norm $n(X)$, given by $n(X) = X\overline{X}$, is equal to

$$\begin{aligned} n(X) = \ & x_0^2 - \ell_1 x_1^2 + x_0 x_1 - \ell_2 x_2^2 + \ell_1 \ell_2 x_{21}^2 - \ell_2 x_2 x_{21} \\ & -\ell_3 x_3^2 + \ell_1 \ell_3 x_{31}^2 - \ell_3 x_3 x_{31} + \ell_2 \ell_3 x_{32}^2 - \ell_1 \ell_2 \ell_3 x^2 + \ell_2 \ell_3 x_{32} x. \end{aligned}$$

Conversely, for any field $\mathbb{K}$ and given elements $\ell_i \in \mathbb{K}$, the eight-dimensional algebra over $\mathbb{K}$ with basis $\{1, e_1, e_2, e_{21}, e_3, e_{31}, e_{32}, e\}$ and multiplication given by the table above defines an alternative ring $\mathbb{D}$. *It is an alternative division ring* **if and only if** *for any non-zero 8-tuple* $(x_0, x_1, x_2, x_{21}, \ldots, x) \in \mathbb{K}^8$, *the norm as given above is never equal to zero.*

Proof. We will first prove some basic properties of an alternative division ring. These are:

(ADR1) The alternative laws $z(yy) = (zy)y$, $(yy)z = y(yz)$ for all $y, z \in \mathbb{D}$, hold.

(ADR2) We also have $(xy)x = x(yx)$ for every $x, y \in \mathbb{D}$. Also, one has the so-called *Moufang identities*, i.e., for all $a, b, c, d \in \mathbb{D}$, the following identities are always valid:

$$a(b(ac)) = (aba)c \tag{B.1}$$
$$((ab)c)b = a(bcb) \tag{B.2}$$
$$(ac)(ba) = a(cb)a \tag{B.3}$$

(ADR3) The set of elements $x \in \mathbb{D}$ which commute and associate with all elements of $\mathbb{D}$ is a subfield $\mathbb{K}$ of $\mathbb{D}$.

In the course of the proof, we will also show the following important result, which can also be shown right from the start (compare PICKERT [1975]). We postpone a proof to page 442.

| **Lemma 1.** | *A commutative alternative division ring is associative.*

Let us first show (ADR1) and (B.2). To that end, we compute, for $v \neq 0$ and $y \neq v^{-1}$,

$$
\begin{aligned}
(xy)(y^{-1} - (y + v^{-1})^{-1}) &= x - (xy)(y + v^{-1})^{-1} \\
&= (x(y + v^{-1}))(y + v^{-1})^{-1} - (xy)(y + v^{-1})^{-1} \\
&= (xy + xv^{-1} - xy)(y + v^{-1})^{-1}.
\end{aligned}
$$

Hence we deduce

$$(xy)(y^{-1} - (y + v^{-1})^{-1}) = (xv^{-1})(y + v^{-1})^{-1}. \tag{B.4}$$

Specifying $x = (y + v^{-1})v$ in (B.4), we see that

$$((y + v^{-1})v)y = (y^{-1} - (y + v^{-1})^{-1})^{-1}. \tag{B.5}$$

Substituting (B.5) in (B.4), we obtain

$$xy = ((xv^{-1})(y + v^{-1})^{-1})(((y + v^{-1})v)y). \tag{B.6}$$

Now we substitute $x = (z(y + v^{-1}))v = (zy)v + z$ in (B.6), and, after cancelling out the common term zy, there remains

$$((zy)v)y = z((yv)y). \tag{B.7}$$

Putting $v = 1$, we obtain half of (ADR1) the other half of which follows by symmetry. Hence (ADR1) is proved. We now first derive $(yv)y = y(vy)$, and then (B.7) will prove (B.2).

For this, we define and introduce the standard notation for the commutator and the associator. Let $a, b, c \in \mathbb{D}$; then the **commutator** of a and b is $(a, b) = ab - ba$ and the **associator** of a, b, c is $(a, b, c) = (ab)c - a(bc)$. The commutator and associator are both additive in each of their arguments, i.e., for instance $(a + b, c, d) = (a, c, d) + (b, c, d)$, by the distributive laws. By (ADR1), we have $(a, b, c) = 0$ whenever $a = b$ or $b = c$. Computing $0 = (a + b, a + b, c)$, we easily obtain $(a, b, c) = -(b, a, c)$. Similarly $(a, b, c) = -(a, c, b)$. So the associator is alternating and hence in particular $(b, a, b) = -(a, b, b) = 0$. Hence $(yv)y = y(vy) = yvy$ and, by the previous paragraph, (B.2) is established.

We will freely make use of the facts derived in the previous paragraph. Note that, since the associator is alternating, it remains unchanged after an even permutation of its elements, and it changes sign after an odd permutation. If $(a, b) = 0$, then we say that a and b commute, or that a commutes with b. If $(a, b, c) = 0$, then we say that a, b, c associate, or that a and b associate with c, or that a associates with b and c.

Now notice that the Moufang identity (B.1) may be rewritten as $-a(b, a, c) = (a, ba, c)$, and similarly for (B.2). Also, the Moufang identity (B.3) can be written as $a(c, b, a) = (a, c, ba)$. In view of the alternating property of the associator, and the symmetric character of the identity (B.3), we see that *each* of the three Moufang identities is equivalent to *each* of the following identities:

$$a(a, b, c) = (a, ba, c) \tag{B.8}$$

$$(a, b, c)a = (a, ab, c) \tag{B.9}$$

and hence the three Moufang identities are mutually equivalent. Whence (ADR2).

To show (ADR3), we note that it is straightforward to prove that, if a and b each commute and associate with all elements of $\mathbb{D}$, then a^{-1} and ab commute with every element of $\mathbb{D}$. Also, by multiplying $(a^{-1}x)y - a^{-1}(xy)$ on the left by a, we see that, if a commutes and associates with all elements of $\mathbb{D}$, we obtain $xy - xy = 0$, hence a^{-1} associates with all elements of $\mathbb{D}$. It remains to show that, if a and b each commute and associate with all elements of $\mathbb{D}$, then ab associates with all elements of $\mathbb{D}$. So let a and b have these properties. Then

$$(ab)(xy) = a(b(xy)) = a((bx)y) = (a(bx))y = ((ab)x)y$$

(in view of the alternating property of the associator) and the result follows.

We now start the main part of the proof of the theorem.

The general aim is to show that every element of $\mathbb{D}$ satisfies a quadratic equation with coefficients in $\mathbb{K}$, or in other words, that $\mathbb{D}$ is a **quadratic algebra (over $\mathbb{K}$)**. So if $a \in \mathbb{D} \setminus \mathbb{K}$, we seek $t(a), n(a) \in \mathbb{K}$ such that $a^2 - t(a)a + n(a) = 0$. Suppose we can prove that there are elements $b, c \in \mathbb{D}$ such that $(a, b, c) \neq 0$. Taking the associator of $a^2 - t(a)a + n(a)$ with the elements b, c, we see that

$$t(a) = (a, b, c)^{-1}(a^2, b, c).$$

We will, however, soon see that $(a^2, b, c) = a(a, b, c) + (a, b, c)a$, hence, putting $u = (a, b, c)$, we should prove that $u^{-1}(au + ua)$ is independent of b, c and is contained in $\mathbb{K}$. Calculating $n(a)$, we obtain

$$n(a) = t(a)a - a^2 = (u^{-1}(au))a + a^2 - a^2 = u^{-1}(aua)$$

by the identity $0 = (u, u^{-1}, a) = (u^{-1}, a, u)$ and the Moufang identity (B.2). Now note that $n(a) = u^{-1}(u^{-1}u(aua)) = u^{-2}(u(aua)) = u^{-2}(ua)^2$. Indeed, we have $0 = (u, a, a)u = (u, a, ua)$ by (B.9), and so $0 = (u, a, ua) = (ua)^2 - u(aua)$. Similarly, we have $t(a) = u^{-2}(u(au + ua))$. Hence we obtain

$$u^2 a^2 - (u(au + ua))a + (ua)^2 = 0, \tag{B.10}$$

and so it suffices to prove that the coefficients u^2, $u(au+ua)$ and $(ua)^2$ all belong to $\mathbb{K}$ to obtain our first goal. This motivates our strategy, which is to show first that these elements commute and associate with a, b, c. So we suppose that $a, b, c \in \mathbb{D}$ are given such that $(a, b, c) \neq 0$.

Define the **superassociator** (a, b, c, d), for $a, b, c, d \in \mathbb{D}$, as

$$(a, b, c, d) \quad = \quad (ab, c, d) - b(a, c, d) - (b, c, d)a. \tag{B.11}$$

If $b = c$, then in view of (B.8), the right-hand side of (B.11) is identically zero. Hence by computing $(a, b + c, b + c, d)$ via the linearity of the superassociator, we see that the latter is alternating in the second and third argument. Similarly, $(a, b, c, a) = 0$ and obviously $(a, b, c, c) = 0$. Hence the superassociator is alternating in all its arguments. In particular, $(a, a, c, d) = -(d, a, c, a) = (d, c, a, a) = 0$. As above, we put $u = (a, b, c)$. We can now already prove part of our goal.

Lemma 2. *The element* $u^2 = (a, b, c)^2$ *associates with* a, b, *with* b, c *and with* c, a.

Proof. Observing that $(a, a, b, c) = 0$, we obtain from (B.11)

$$(a^2, b, c) \quad = \quad au + ua. \tag{B.12}$$

Now we compute in two different ways (a^2, b, cb). First, we apply (B.8) followed by (B.12) to obtain $b(au) + b(ua)$. Secondly, we apply (B.12) followed by (B.8) to obtain $a(bu) + (bu)a$. Hence we have

$$0 \quad = \quad b(ua) - a(bu) + b(ua) - (bu)a. \tag{B.13}$$

The last two terms of (B.13) are a perfect associator; in order to deal with the first two terms, we try to isolate u using $x(yu) = (xy)u - (x, y, u)$, for $\{x, y\} =$

$\{a, b\}$. Using the alternating property of the associator, we now see that (B.13) is equivalent to

$$(a, b)u \;=\; (a, b, u). \tag{B.14}$$

By the symmetric character of $\mathbb{D}$, we may write (B.14) in the opposite order (noting that u is switched to $-u$ and (x, y, z) to $-(z, y, x)$) and we obtain

$$-u(b, a) \;=\; -(-u, b, a). \tag{B.15}$$

Hence $u(u, a, b) = u(a, b)u$ by (B.14) and $(u, a, b)u = -u(a, b)u$ by (B.15). But now (B.12) implies

$$(u^2, a, b) = u(u, a, b) + (u, a, b)u = u(a, b)u - u(a, b)u = 0. \tag{B.16}$$

Hence we have shown that u^2 associates with a, b. By cyclic permutation, it also associates with b, c and c, a. This shows Lemma 2.   QED

Proof of Lemma 1.
Now is a good time to prove Lemma 1. Noting that $0 = (a, b, b)b = (a, b^2, b)$ by (B.9), we also have $(b, c, a, a^2) = 0$. Now calculating $0 = (a, a^2, b, c)$, one obtains $(a^3, b, c) = a^2u + aua + ua^2$. Hence if $\mathbb{D}$ is commutative, then $(a^3, b, c) = 3a^2u$ and consequently

$$\begin{aligned}
(a, b, c)^3 &= ((ab)c)^3 - 3((ab)c)(a(bc))(a, b, c) - (a(bc))^3 \\
&= (a^3b^3)c^3 - a^3(b^3c^3) - (a(bc))((ab)c)a^{-2}(a^3, b, c), \tag{B.17}
\end{aligned}$$

and so, if third powers associate with everything, then the right-hand side of (B.17) becomes zero and $(a, b, c) = 0$. Applying repeatedly the Moufang identities in the form (ADR2), properties (ADR1) and the identity $(a^2, a, z) = 0$, we obtain

$$\begin{aligned}
(a^3x)y = (a(axa))y = (a(ax)a)y = a((ax)(ay)) = \\
a((ax)(ya)) = a(a(xy)a) = a^3(xy),
\end{aligned}$$

hence putting $x = b$ and $y = c$, or $x = b^3$ and $y = c^3$, the proof of Lemma 1 is complete.   QED

Lemma 3. *The element $(a, b, c)^2$ commutes with a, b and c.*

Proof. If we substitute $b + u^2$ for b in (B.14), then the right-hand side remains unchanged by Lemma 2. The left-hand side becomes $(a, b)u + (a, u^2)u$. Hence $(a, u^2)u = 0$, which implies that, if $(a, u^2) \neq 0$, then $u = 0$ and so $(a, u^2) = 0$ after all. Also, by cyclic permutation, $(b, u^2) = (c, u^2) = 0$. Lemma 3 follows.   QED

Lemma 4. *The elements $u(au + ua)$ and $(ua)^2$ commute with a, b and c.*

Proof. By Lemma 3 the element $(a, ab, c)^2 = (a, b, ac)^2 = (ua)^2$ (see (B.9)) commutes with a, b and c. Also, by Lemma 3 and noting that $(a + 1, b, c) = u$, the element $(u(a + 1))^2 = (ua + u)^2$ commutes with $a + 1$, b and c, hence also with a. It follows that $u(au + ua)$, which is equal to $(ua + u)^2 - u^2 - (ua)^2$, commutes with a, b, c. Lemma 4 is proved.   QED

Lemma 5. *If $z \neq 0$ and if $x \in \mathbb{D}$ associates with $y, z \in \mathbb{D}$, then it associates with y and z^{-1}.*

Proof. We obviously have $(y, x, z, z^{-1}) = 0$ by definition. Hence, since the superassociator is alternating, we have

$$
\begin{aligned}
0 &= (z, z^{-1}, y, x) \\
&= (1, y, x) - z^{-1}(z, y, x) - (z^{-1}, y, x)z \\
&= 0 - 0 - (z^{-1}, y, x)z,
\end{aligned}
$$

and the lemma follows, noting $z \neq 0$.   QED

Lemma 6. *If $x \in \mathbb{D}$ associates with $y, z \in \mathbb{D}$ and with y and $z' \in \mathbb{D}$, and if x commutes with y, then it associates with y and zz'.*

Proof. We use again the fact that the superassociator alternates, and so we have

$$
\begin{aligned}
0 &= (y, x, z, z') + (y, z, x, z') \\
&= (yx, z, z') - x(y, z, z') - (x, z, z')y + (yz, x, z') - z(y, x, z') - (z, x, z')y.
\end{aligned}
$$

The third and the last term of the last line cancel out, and the fifth one is zero. Noting that $yx = xy$ and $(yz, x, z') = -(x, yz, z')$, we calculate

$$
\begin{aligned}
0 &= (xy, z, z') - x(y, z, z') - (x, yz, z') \\
&= ((xy)z)z' - (xy)(zz') - x((yz)z') + x(y(zz')) \\
&\qquad\qquad\qquad -(x(yz))z' + x((yz)z') \\
&= (x, y, z)z' - (x, y, zz') \\
&= -(x, y, zz').
\end{aligned}
$$

The proof of the lemma is complete.   QED

Lemma 7. *The elements $u(au + ua)$ and $(ua)^2$ associate with a, b, with b, c and with c, a.*

Proof. As in the proof of Lemma 4, it suffices to show that $(ua)^2$ associates with every pair of elements in $\{a, b, c\}$. As $(ua)^2 = (a, ab, c)^2 = (a, b, ac)^2$ as above, we deduce from Lemma 2 that it already associates with a, c and with a, b, but also with ac, b. By Lemma 5, it also associates with b, a^{-1}. Lemma 6 now implies that $(ua)^2$ associates with b and $a^{-1}(ac) = c$.   QED

Equation (B.10) becomes trivial if $u = 0$. So we have to show that for each $a \in \mathbb{D}$, we can actually choose b and c in such a way that we obtain a non-trivial quadratic equation satisfied by a and with coefficients in $\mathbb{K}$ (we say that a is **quadratic over** $\mathbb{K}$). Therefore we distinguish between two cases.

Lemma 8. *If $a \in \mathbb{D}$ does not commute with all elements of $\mathbb{D}$, then a is quadratic over $\mathbb{K}$.*

Proof. By assumption there exists $b \in \mathbb{D}$ with $(a, b) \neq 0$. Our first task is to show that there exists $c \in \mathbb{D}$ such that $(a, b, c) \neq 0$. This requires some work. First we note that, if for $x, x', y, z \in \mathbb{D}$, one has $(x', y, z) = (x', x+y, z) \neq 0$, then $(x, x') = 0$. Indeed, this follows from applying (B.14) to both (x', y, z) and $(x', x + y, z)$ and subtracting the obtained identities from each other. This immediately implies that, if an element x associates with all other elements, and an element x' does not associate with all other elements, then $(x, x') = 0$. Now suppose that a associates with all elements of $\mathbb{D}$, and let x' be such that it does not associate with all elements of $\mathbb{D}$ (x' exists by assumption); then $(a, x') = 0$ by the foregoing. If also b associates with all elements of $\mathbb{D}$, then $x' + b$ does not associate with all elements of $\mathbb{D}$, and so $(a, x' + b) = 0$, which implies that $(a, b) = 0$, contrary to our assumption. Hence b does not associate with all elements of $\mathbb{D}$, but this implies $(a, b) = 0$, a contradiction again. We conclude that a does not associate with all elements of $\mathbb{D}$. Therefore, we may pick elements a' and c such that $(a, a', c) \neq 0$.

Suppose now $(a, b, c) = 0$. Then $(a, a', c) = (a, b + a', c) \neq 0$, and by the argument in the previous paragraph, this implies $(a, b) = 0$, a contradiction. So we finally obtain $(a, b, c) \neq 0$. Now we write (B.10) in the form

$$a^2 - u^{-1}(au + ua)a + u^{-2}(ua)^2 = 0 \qquad (B.18)$$

and we put $t_{a,b,c} = u^{-1}(au + ua)$ and $n_{a,b,c} = u^{-2}(ua)^2$. So we have

$$a^2 - t_{a,y,z}a + n_{a,y,z} = 0 \qquad (B.19)$$

for all $y, z \in \mathbb{D}$ such that $(a, y, z) \neq 0$. Now let c' be such that $(a, b, c') \neq 0$. First we note that a, b, c commute with $t_{a,b,c}$ and with $n_{a,b,c}$. This follows from the general and easy fact that, if $x \in \mathbb{D}$ commutes with y and z, then it also commutes with $y^{-1}z$. Now we remark that by Lemma 5 and Lemma 6, both $t_{a,b,c}$ and $n_{a,b,c}$ associate with every pair of elements in $\{a, b, c\}$. Of course the same is true for c substituted by c'. Now we have, by subtracting (B.19) for $y = b$ and, respectively, $z = c$ and $z = c'$,

$$t_{a,b,c}a - n_{a,b,c} \quad = \quad t_{a,b,c'}a - n_{a,b,c'}. \qquad (B.20)$$

If we take the commutator of both sides with b, then we obtain

$$t_{a,b,c}(a, b) = (t_{a,b,c}a, b) = (t_{a,b,c'}a, b) = t_{a,b,c'}(a, b)$$

(we have used the fact that $(xy, z) = x(y, z)$ if x commutes and associates with y and z). Since $(a, b) \neq 0$, we conclude that $t_{a,b,c} = t_{a,b,c'}$, and hence we may write $t_{a,b,c} = t_{a,b}$. From (B.20), it immediately follows that also $n_{a,b,c} = n_{a,b,c'} = n_{a,b}$.

So $t_{a,b}$ and $n_{a,b}$ commute with all elements z such that $(a, b, z) \neq 0$. Suppose now $(a, b, z) = 0$. Choose c such that $(a, b, c) \neq 0$, then also $(a, b, c + z) \neq 0$ and so $t_{a,b}$ and $n_{a,b}$ commute with both c and $c + z$, hence they commute with z. So they commute with every element of $\mathbb{D}$.

A completely similar argument shows that both $t_{a,b}$ and $n_{a,b}$ associate with a and any other element of $\mathbb{D}$. If $b' \in \mathbb{D}$ is such that $(a, b') \neq 0$, then, similar to (B.20), we have the identity

$$t_{a,b}a - n_{a,b} = t_{a,b'}a - n_{a,b'}. \tag{B.21}$$

Again taking the commutator with b, we obtain $t_{a,b}(a, b) = t_{a,b'}(a, b)$, hence $t_{a,b} = t_{a,b'}$. As before, this implies that $t(a) = t_{a,b}$ and $n(a) = n_{a,b}$ commutes with every element of $\mathbb{D}$ and associates with every two elements b and c in $\mathbb{D}$. Hence $t(a) \in \mathbb{K}$ and $n(a) \in \mathbb{K}$. So a is quadratic over $\mathbb{K}$ and Lemma 8 is proved.

Lemma 9. If a commutes with everything, **then** a is quadratic over $\mathbb{K}$.

Proof. By Lemma 1 we know that there exist $b, c \in \mathbb{D}$ such that $(b, c) \neq 0$. Obviously also $(a + b, c) \neq 0$. So both b and $a + b$ are quadratic over $\mathbb{K}$ by Lemma 8. So we can write

$$b^2 - t(b)b + n(b) = 0, \tag{B.22}$$
$$(a + b)^2 - t(a + b)(a + b) + n(a + b) = 0. \tag{B.23}$$

Remembering that $ab = ba$, we can subtract (B.22) and (B.23) and obtain

$$a^2 - t(a + b)a + n(a + b) - n(b) = (t(a + b) - t(b))b - 2ab. \tag{B.24}$$

If a does not associate with b and c, then we have $t(b) = t_{b,c,a}$ and likewise $t(a + b) = t_{a+b,c,a}$. Now $(b, c, a) = (a + b, c, a) = u$ and we have

$$\begin{aligned} t(a + b) &= u^{-1}((a + b)u + u(a + b)) \\ &= u^{-1}(bu + ub) + u^{-1}au + a \\ &= t(b) + 2a \end{aligned}$$

since a commutes with u. Consequently the right-hand side of (B.24) is zero and a is quadratic over $\mathbb{K}$. There remains the case that a associates with b and c. Then $2a(b, c) = (2ab, c)$. Also, a^2 commutes with c, as well as $n(a + b)$ and $n(b)$. Hence taking the commutator of both sides of (B.24) with c, we obtain

$$0 = (t(a + b) - t(b))(b, c) - 2a(b, c),$$

from which it immediately follows that the right-hand side of (B.24) is equal to zero. Hence also in this case, a is quadratic over $\mathbb{K}$ and the lemma is proved. QED

So the proof is reduced to the classification of non-associative alternative quadratic algebras.

Every element x of $\mathbb{D}$ satisfies some quadratic equation

$$x^2 - t(x)x + n(x). \tag{B.25}$$

If $x \notin \mathbb{K}$, then $t(x)$ and $n(x)$ are uniquely defined. Indeed, if $t'(x), n'(x) \in \mathbb{K}$ also satisfy $x^2 - t'(x)x + n'(x) = 0$, then $(t(x) - t'(x))x = n(x) - n'(x)$, hence $x \in \mathbb{K}$, a contradiction. For $x \in \mathbb{K}$, we define $t(x) = 2x$ and $n(x) = x^2$.

We remark that $\mathbb{K} \neq \mathbf{GF}(2)$. Indeed, for $x \in \mathbb{D} \setminus \mathbb{K}$, $t(x) = 0$ implies that $n(x) = 1$ and $(x + 1)^2 = 0$, hence $x = 1$, a contradiction. So $t(x) = 1$. If $y \in \mathbb{D} \setminus \mathbb{K}$, then similarly $t(y) = 1$. Of course, $n(x) = n(y) = 1$. Hence $(x + y)^2 + (x + y) = 0$. This means that either $x + y = 1$ or $x + y = 0$. So if $x \neq y$, we have $y = x + 1$. Hence there are only four elements in $\mathbb{D}$ and consequently $\mathbb{D} \cong \mathbf{GF}(4)$ is easily seen to be associative (it consists of $0, 1$ and two mutually inverse elements).

Now suppose that $1, x, y$ are three linearly independent elements of $\mathbb{D}$ ($\mathbb{D}$ viewed as a vector space over $\mathbb{K}$). Considering the quadratic equation satisfied by $\lambda x + \mu y$, for $\lambda, \mu \in \mathbb{K} \setminus \{0\}$, we see that

$$
\begin{aligned}
\lambda\mu(xy + yx) &= -(\lambda^2 x^2 + \mu^2 y^2) + \lambda t(\lambda x + \mu y)x + \mu t(\lambda x + \mu y)y - n(\lambda x + \mu y) \\
&= \lambda^2 n(x) + \mu^2 n(y) - n(\lambda x + \mu y) + (\lambda t(\lambda x + \mu y) - \lambda^2 t(x))x \\
&\quad + (\mu t(\lambda x + \mu y) - \mu^2 t(y))y,
\end{aligned}
$$

where we have substituted x^2 by $t(x)x - n(x)$ and similarly for y^2. Hence, by dividing by $\lambda\mu$, we see that $xy + yx$ is a linear combination of $1, x, y$, say $xy + yx = ax + by + c$, with $a, b, c \in \mathbb{K}$. The elements a, b, c are given by

$$
a = \frac{1}{\mu}t(\lambda x + \mu y) - \frac{\lambda}{\mu}t(x), \tag{B.26}
$$

$$
b = \frac{1}{\lambda}t(\lambda x + \mu y) - \frac{\mu}{\lambda}t(y), \tag{B.27}
$$

$$
c = \frac{\lambda}{\mu}n(x) + \frac{\mu}{\lambda}n(y) - \frac{1}{\lambda\mu}n(\lambda x + \mu y). \tag{B.28}
$$

But a, b, c are of course independent of λ and μ. If we put $\beta(x, y) = n(x + y) - n(x) - n(y)$ (this notation should not cause confusion with the commutator (x, y) since we do not use the latter any more), then writing out (B.28) for $\lambda = \mu = 1$, and then again substituting in (B.28), we obtain $c = -\beta(x, y)$ and

$$
n(\lambda x + \mu y) = \lambda^2 n(x) + \mu^2 n(y) + \lambda\mu\beta(x, y). \tag{B.29}
$$

This remains true for $\lambda = 0$ or $\mu = 0$ by a direct computation. Eliminating $t(\lambda x + \mu y)$ in (B.26) and (B.27), we obtain

$$
\mu a - \lambda b = \mu t(y) - \lambda t(x). \tag{B.30}
$$

Writing this out for $\lambda = \mu = 1$, we can take the result together with (B.30) in a system of equations which has as unique solution (provided there exists an element in $\mathbb{K}$ different from both 0 and 1, but that is true in view of $\mathbb{K} \neq \mathbf{GF}(2)$) $a = t(y)$ and $b = t(x)$. Substituting this again in (B.26) or (B.27), we obtain

$$
t(\lambda x + \mu y) = \lambda t(x) + \mu t(y), \tag{B.31}
$$

which remains true if $\lambda = 0$ or $\mu = 0$. Finally, we have the identity

$$xy + yx \;=\; t(y)x + t(x)y - \beta(x, y), \tag{B.32}$$

which remains true if x, y and 1 are linearly dependent, as one can verify easily using (B.25). Of course, in that case a, b, c are not uniquely determined, but nevertheless, formula (B.32) is valid.

Now we define the conjugate $\bar{x}$ of x as $\bar{x} = t(x) - x$. We derive from (B.25) that $(x - t(x))x = -n(x)$, hence $\bar{x}x = n(x)$. Note that $t(t(x) - x) = t(t(x)) - t(x) = 2t(x) - t(x) = t(x)$ and similarly $n(t(x) - x) = n(t(x)) + n(x) - \beta(t(x), x)$. But (B.32) implies, putting $y = t(x)$, that $2t(x)x = 2t(x)x + t(x)t(x) - \beta(t(x), x)$, hence $n(\bar{x}) = n(x)$ and $t(\bar{x}) = t(x)$. This implies that $\bar{\bar{x}} = x$ and $n(x) = x\bar{x}$.

Lemma 10. *For $x, y \in \mathbb{D}$, we have*

(i) $n(xy) = n(x)n(y)$,

(ii) $\beta(x, y) = t(x)t(y) - t(xy)$,

(iii) $\overline{xy} = \bar{y}\bar{x}$,

(iv) $\beta(x, y) = x\bar{y} + y\bar{x}$.

Proof. We assume that xy is not in the centre $\mathbb{K}$ of $\mathbb{D}$. If it is, then the assertions are easy to verify directly.

First we have, using $\bar{z}z = n(z)$,

$$\overline{xy} \;=\; n(xy)(xy)^{-1} = n(xy)y^{-1}x^{-1} = \frac{n(xy)}{n(x)n(y)}\bar{y}\bar{x}. \tag{B.33}$$

Next, we calculate $\bar{y}\bar{x}$ using the definition, and remembering formula (B.32):

$$
\begin{aligned}
\bar{y}\bar{x} \;&=\; (t(y) - y)(t(x) - x) \\
&=\; t(x)t(y) + yx - t(x)y - t(y)x \\
&=\; t(x)t(y) - \beta(x, y) - xy \\
&=\; t(x)t(y) - \beta(x, y) - t(xy) + \overline{xy}.
\end{aligned}
$$

Plugging this into (B.33), we obtain after rearranging the terms:

$$\left(1 - \frac{n(xy)}{n(x)n(y)}\right)\overline{xy} = -\frac{n(xy)}{n(x)n(y)}(t(x)t(y) - \beta(x, y) - t(xy)),$$

and both sides must be zero since the right-hand side belongs to $\mathbb{K}$ and the left-hand side to $\mathbb{K}\overline{xy}$. This already shows (i) and (ii). But (i) and (B.33) imply (iii). To obtain (iv), we perform a straightforward calculation:

$$
\begin{aligned}
x\bar{y} + y\bar{x} \;&=\; x(t(y) - y) + y(t(x) - x) \\
&=\; t(y)x + t(x)y - (xy + yx) \\
&=\; \beta(x, y),
\end{aligned}
$$

in view of (B.32). This completes the proof of Lemma 10. QED

We now prove the fundamental lemma (compare SPRINGER [1963]). For any subset A of $\mathbb{D}$, we define $A^{\perp}$ as the set of all elements $x \in \mathbb{D}$ such that $\beta(x,y) = 0$, for all elements $y \in A$.

Lemma 11. *Let $\mathbb{B}$ be a subalgebra of $\mathbb{D}$ such that $\mathbb{B}^{\perp} \not\subseteq \mathbb{B}$ and let $e \in \mathbb{B}^{\perp} \setminus \mathbb{B}$. Then*

(i) $\mathbb{B} \cap e\mathbb{B} = \{0\}$, $\overline{\mathbb{B}} = \mathbb{B}$;

(ii) $\mathbb{A} = \mathbb{B} + e\mathbb{B}$ *is a subalgebra of $\mathbb{D}$;*

(iii) $n(a + eb) = n(a) + n(e)n(b)$, *for any $a, b \in \mathbb{B}$;*

(iv) $\mathbb{B}$ *is associative;*

(v) $\mathbb{A}$ *is associative* **if and only if** $\mathbb{B}$ *is commutative.*

Proof. Assertion (i) is clear, noting that $\bar{x} = t(x) - x \in \mathbb{K} + \mathbb{B} = \mathbb{B}$ for $x \in \mathbb{B}$. Now remark that $\mathbb{A}$ is closed under taking inverses. Indeed, if $x \in \mathbb{A} \setminus \mathbb{K}$, then $x^{-1} = n(x)^{-1}t(x) - n(x)^{-1}x \in \mathbb{A}$ (by (B.25)). We now prove (ii) by explicitly establishing the rules for multiplication in $\mathbb{A}$. First remark that $e \in \mathbb{B}^{\perp}$ implies by Lemma 10(iv) that $e + \bar{e} = \beta(e, 1) = 0$. Hence $t(e) = 0$ and $\bar{e} = -e$. So $e^2 + n(e) = 0$ and we put $e^2 = -n(e) = \ell \in \mathbb{K}$. Also, for any $x \in \mathbb{B}$, we have $0 = \beta(e, x) = e\bar{x} + x\bar{e} = e\bar{x} - xe$, hence $xe = e\bar{x}$. This means that e can be moved to the left if we conjugate every element that e jumps over. In the following computation, we freely make use of identities derived earlier, without explicitly referring to them (this includes the Moufang identities!).

$$
\begin{aligned}
(a + eb)(c + ed) &= ac + (eb)c + a(ed) + (eb)(ed) \\
&= ac + (((\bar{b}e)c)e)e^{-1} + e^{-1}(e(a(ed))) + (eb)(\bar{d}e) \\
&= ac + (\bar{b}(ece))e^{-1} + e^{-1}((eae)d) + e(b\bar{d})e \\
&= ac + (\bar{b}\bar{c})\ell e^{-1} + e^{-1}\ell(\bar{a}d) + \ell d\bar{b} \\
&= ac + e(cb) + e(\bar{a}d) + \ell d\bar{b} \\
&= (ac + \ell d\bar{b}) + e(\bar{a}d + cb).
\end{aligned}
$$

This shows (ii). Now we show (iv). Therefore, we note that, for any $x \in \mathbb{B}$, and any $y \in \mathbb{A}$, we have

$$
\mathbb{B} \ni xn(y) = x(y\bar{y}) = n(y)^{-1}x(yy^{-1}) = n(y)^{-1}(xy)y^{-1} = (xy)\bar{y},
$$

hence $(xy)\bar{y} \in \mathbb{B}$. Taking $x = \bar{b}$ and $y = \bar{a} + ec$, for arbitrary elements $a, b, c \in \mathbb{B}$, we can compute $(xy)\bar{y}$ with the above expression, and we obtain

$$
(\bar{b}(\bar{a} + ec))(a - ec) = (n(a) - \ell n(c))\bar{b} + e(a(bc) - (ab)c).
$$

So we infer that $(a, b, c) = 0$, hence $\mathbb{B}$ is associative, and as a byproduct, we see that $n(a + eb) = n(a) - \ell n(b)$, showing (iii).

Finally, if $\mathbb{B}$ is commutative, then a direct and easy computation using the explicit form of the multiplication above shows that $\mathbb{A}$ is associative. If $\mathbb{A}$ is associative, then $((ea)(e\bar{b}))(e) = (ea)((e\bar{b})e)$, for all $a, b \in \mathbb{B}$, and by an explicit computation, we obtain $e(ab) = e(ba)$, hence $\mathbb{B}$ is commutative, whence (v).   QED

We can now finish the proof of the theorem. First suppose that $t(x) = 0$, for all $x \in \mathbb{D}$. Then obviously $\mathbb{K}$ has characteristic 2, and we have

$$xy + yx = t(x)y + t(y)x + t(x)t(y) - t(xy) = 0$$

(using Lemma 10(ii)). So $xy = -yx = yx$ for all $x, y \in \mathbb{D}$, and hence $\mathbb{D}$ is commutative. Lemma 1 now implies that $\mathbb{D}$ is associative, contradicting our assumption. So we may assume that there exists $x \in \mathbb{D}$ such that $t(x) \neq 0$. If $x \notin \mathbb{K}$, then we put $e_1' = x$. If $x \in \mathbb{K}$, then consider $y \in \mathbb{D} \setminus \mathbb{K}$. If $t(y) \neq 0$, then we put $e_1' = y$. If $t(y) = 0$, then we put $e_1' = x + y$. In either case $t(e_1') \neq 0$ and $e_1' \notin \mathbb{K}$. We now set $e_1 = t(e_1)^{-1}e_1'$. Then $t(e_1) = 1$ and we have $e_1^2 = e_1 - n(e_1)$. Put $\ell_1 = -n(e_1)$. Then we compute

$$(a + e_1 b)(c + e_1 d) = (ac + \ell_1 bd) + e_1(bc + ad + bd),$$

with $a, b, c, d \in \mathbb{K}$, and

$$n(a + e_1 b) = a^2 - \ell_1 b^2 + ab.$$

The last term may seem a little strange when taking $\mathbb{K} = \mathbb{R}$, the real numbers. In characteristic 0 (as also in every odd characteristic), however, we could consider $t(e_1) = 0$, but as we are after a uniform proof, we need the condition $t(e_1) \neq 0$ (because we need it in characteristic 2).

So we have shown that $\mathbb{B} = \mathbb{K} + e_1\mathbb{K}$ is a subalgebra of $\mathbb{D}$, which is obviously commutative (this is easily seen by the explicit formula above for the product of two elements), hence it is associative (after all, it is linearly generated by the two elements 1 and e_1), and so it is a field, since it is closed under taking inverses (this is shown as in the proof of Lemma 11). Now, as $\mathbb{B}$ is associative, $\mathbb{B} \neq \mathbb{D}$. Hence there exists some element e_2' in $\mathbb{D} \setminus \mathbb{B}$. We determine $v = a + e_1 b \in \mathbb{K} + e_1\mathbb{K} = \mathbb{B}$ such that $e_2' - v \in \mathbb{B}^\perp$. For this, it suffices to require $\beta(v, 1) = \beta(e_2', 1)$ and $\beta(v, e_1) = \beta(e_2', e_1)$. After explicit calculation, one obtains (noting that $\beta(x, 1) = t(x)$, for all $x \in \mathbb{D}$)

$$\begin{cases} a + 2b &= t(e_2'), \\ -2\ell_1 a + b &= \beta(e_2', e_1). \end{cases}$$

The determinant of this system of linear equations over $\mathbb{K}$ is equal to $1 + 4\ell_1 = (e_1 + \bar{e}_1) - 4e_1\bar{e}_1 = (e_1 - \bar{e}_1)^2$, since e_1 commutes with $\bar{e}_1$. But if $e_1 = \bar{e}_1$, then $2e_1 = t(e_1) = 1$, a contradiction. So we find exactly one v meeting our requirements. Now put $e_2 = e_2' - v$, and apply Lemma 11 to obtain an associative subalgebra

$\mathbb{A} = \mathbb{B} + e_2\mathbb{B}$. Hence $\mathbb{D} \neq \mathbb{A}$. Furthermore, $\mathbb{A}$ is not commutative. Indeed, $e_1e_2 = e_2\overline{e_1} = e_2(1 - e_1) \neq e_2e_1$. Since $\mathbb{D} \neq \mathbb{A}$, we may pick $e_3' \in \mathbb{D} \setminus \mathbb{A}$. We determine $p = v + e_2w \in \mathbb{B} + e_2\mathbb{B} = \mathbb{A}$ such that $e_3' - p \in \mathbb{A}^\perp$.

First we note that $\beta(e_2x, e_2y) = n(e_2x + e_2y) - n(e_2x) - n(e_2y) = n(e_2)\beta(x, y)$. From the previous paragraph, we know that there is a unique $v \in \mathbb{B}$ such that $e_3' - v$ belongs to $\mathbb{A}^\perp$. Now put $u = e_3' - v$; then $u - e_2w \in \mathbb{A}^\perp$ is equivalent to

$$\left\{ \begin{array}{ccccc} \beta(e_2w, e_2) & = & n(e_2)\beta(w, 1) & = & \beta(u, e_2), \\ \beta(e_2w, e_2e_1) & = & n(e_2)\beta(w, e_1) & = & \beta(u, e_2e_1), \end{array} \right. \tag{B.34}$$

since both u and e_2w are already in $\mathbb{B}^\perp$. Indeed, u was determined as such and

$$\beta(e_2w, v) = e_2w\bar{v} + v\overline{e_2w} = e_2(w\bar{v} - \bar{v}w)$$

(calculating with the rules as established in the proof of Lemma 11), and this is zero since $\mathbb{B}$ is commutative. So it suffices to determine w such that (B.34) holds. But as in the previous paragraph, this has a unique solution. Hence we can put $e_3 = e_3' - p$ and we can again apply Lemma 11. We obtain a non-associative (sub)algebra $\mathbb{D}'$. If $\mathbb{D}' \neq \mathbb{D}$, then by the same method, we can find an element $e_4 \in \mathbb{D}'^\perp \setminus \mathbb{D}'$ (there is actually another way to see that this is possible. As before, one obtains a system of linear equations over $\mathbb{K}$, this time an 8×8 system. If the determinant of that system were 0, then there would be a non-zero solution of the homogeneous system, implying the existence of a non-zero vector $p + e_3q \in \mathbb{D}'$ perpendicular to all elements of $\mathbb{D}'$. But it is readily seen that $\mathbb{A}^\perp \cap \mathbb{D}' = e_3\mathbb{A}$ and $(e_3\mathbb{A})^\perp \cap \mathbb{D}' = \mathbb{A}$, hence $p = q = 0$.) Lemma 11 now implies that $\mathbb{D}'$ is associative, a contradiction. Hence $\mathbb{D} = \mathbb{D}'$. Using the multiplication explicitly given in the proof of Lemma 11, one can calculate very easily the multiplication table given in the statement of the theorem. Also the norm $n(x)$ of any element may be computed using Lemma 11 repeatedly.

The last part of the theorem follows from the fact that the given multiplication table indeed determines an alternative multiplication. Hence the alternative ring obtained in this way is a division ring **if and only if** $n(x) \neq 0$, for every non-zero x. Indeed, one can check with the given multiplication, and the derived conjugation, that every $x \in \mathbb{D}$ satisfies $x^2 - (x + \bar{x})x + \bar{x}x = 0$, with $x + \bar{x} \in \mathbb{K}$ and $n(x) = \bar{x}x \in \mathbb{K}$. The condition on the norm now just says that $n(xy) \neq 0$ for non-zero x, y, hence $xy \neq 0$ and the theorem is proved completely. $\square$

B.2   Remark. If the characteristic is different from 2, then we have already re-marked that we can choose $t(e_1) = 0$. This gives a much simpler multiplication table. If $\mathbb{K} = \mathbb{R}$, then we might even choose all $n(e_i) = 1$. In this particular case of classical octonions $\mathbb{O}$ over the reals, we may describe the multiplication geo-metrically as follows. Choose a basis $1, e_1, e_2, \ldots, e_7$ for $\mathbb{O}$ and let $e_1, e_2, \ldots, e_7$ be identified with the points of a projective plane $\mathbf{PG}(2, 2)$ of order 2. Order each line of that plane cyclically such that, if three lines are concurrent, then the points on these respective lines following the intersection point are collinear (there is essentially only one way to do this, and an example is given by defining the lines as $\{e_i, e_{i+1}, e_{i+3}\}$, subscripts modulo 7, and then an order on each line, meeting our requirements, may be defined as (e_i, e_{i+1}, e_{i+3})), and call each such ordering **good**. Then define $e_i e_i = -1$ and $e_i e_j = \epsilon e_k$, where $\{e_i, e_j, e_k\}$ is a line of $\mathbf{PG}(2, 2)$, and where ϵ is 1 or -1 depending on whether or not (e_i, e_j, e_k) defines a good cyclic ordering. We have constructed the octonion division algebra $\mathbb{O}$ over $\mathbb{R}$.

Needless to say I learned this construction also in the above mentioned lecture given by Jacques Tits. It is contained in FREUDENTHAL [1985], where it is proved, using this construction, that the octonions cannot be generalized to a 16-dimensional division algebra over the centre such that any three elements which are not contained in a skew field, generate an octonion algebra. This is done by showing that no orientation of lines in $\mathbf{PG}(2, 3)$ can be found such that the orientation in each plane is the above one.

B.3 Corollary. *Let $\mathbb{D}$ be an alternative division ring which is not associative, and let $X \in \mathbb{D}$. If X commutes with every element of $\mathbb{D}$,* **then** *it associates with every pair of elements of $\mathbb{D}$.*

Proof. Using the notation of Theorem B.1 (in particular the multiplication table given there), one easily calculates that $X = x_0 + e_1 x_1 + \cdots + e_{32} x_{32} + ex$, with $x_0, \ldots, x \in \mathbb{K}$, commutes with e_1 **if and only if** $x_2 = x_{21} = x_3 = x_{31} = x_{32} = x = 0$. One has to use the fact that $4\ell_1 \neq -1$, which is true because otherwise $(1 - 2e_1)^2 = 0$. But now $X = x_0 + e_1 x_1$ commutes with e_2 **if and only if** $x_1 = 0$. Hence $X = x_0 \in \mathbb{K}$ and the corollary is proved. □

There is an interesting application.

B.4 Proposition. *Let Γ be a Moufang projective plane defined over the alternative division ring $\mathbb{D}$. Let $\Sigma = (p_1, L_2, p_3, L_1, p_2, L_3, p_1)$ be an ordered apartment. If there is a correlation ρ which maps Σ onto $\Sigma^\rho = (L_3, p_2, L_1, p_3, L_2, p_1, L_3)$ and which centralizes the group of all (p_3, L_1)-elations,* **then** *Γ is a Pappian projective plane.*

Proof. We use the notation of Subsection 2.2.4 and put $p_3 = (\infty)$, $L_1 = [\infty]$, $p_2 = (0)$, $L_2 = [0]$, $p_1 = (0, 0)$ and $L_3 = [0, 0]$. It is readily verified that the fact that ρ commutes with the (p_3, L_1)-elation mapping $(0, 0)$ onto $(0, y)$, $y \in \mathbb{D}$, implies that $(0, y)^\rho = [0, y]$ and $[0, y]^\rho = (0, y)$. We define $X \in \mathbb{D}$ as $(1)^\rho = [X]$. Since the point (a, b), $a, b \in \mathbb{D}$, lies on the line $[1, a + b]$ (which in turn is determined by (1) and $(0, a + b)$), the line $(a, b)^\rho$ contains the intersection point of $[X]$ and $[0, a + b]$, which is $(X, a+b)$. On the other hand, (a, b) lies on $[0, b]$ and hence $(a, b)^\rho$ is incident with $(0, b)$. It follows that $(a, b)^\rho = [-aX^{-1}, b]$. Similarly $[m, k]^\rho = (Xm, k)$, for all $m, k \in \mathbb{D}$. Observing that ρ preserves incidence, we see that $ma + b = k$ implies that $(-aX^{-1})(Xm) + k = b$. Hence we have the identity $ma = (aX^{-1})(Xm)$. If $\mathbb{D}$ is associative, then $ma = am$, for all $a, m \in \mathbb{D}$, and hence $\mathbb{D}$ is a field. If $\mathbb{D}$ is non-associative, then, putting $a = X$, we see that X commutes with every element of $\mathbb{D}$, and hence by Corollary B.3, X associates with every pair of elements of $\mathbb{D}$. This implies that $ma = (aX^{-1})(Xm) = am$ and consequently $\mathbb{D}$ is commutative, hence associative. The proposition is proved. □

Appendix C

Tits Diagrams for Moufang Quadrangles

In this appendix, I will describe some properties of the Tits diagrams for the Moufang quadrangles. All these properties follow from the more general theory of Tits diagrams (called *Witt indices* by Tits himself) for buildings related to algebraic groups, but I will not consider that. In fact, not knowing the background in algebraic groups makes these properties seem very mysterious and magical, and that is exactly what I want to achieve here.

I learned everything in this appendix from TITS [1995] through his magical lectures in Paris (1993–1994 and 1994–1995). See also TITS [1966b].

I will assume that the reader knows the definition and the basic properties of a *root system* and of a *Dynkin diagram*.

In the following, the natural number n will not be the gonality of a generalized polygon, but rather the rank of a root system or of a geometry.

Introduction

Let me first introduce Tits diagrams for generalized quadrangles. I will not give the precise definition (which requires algebraic groups, i.e., groups of matrices whose entries satisfy certain polynomial equations), but I will give the geometric essence, mentioning algebraic groups as little as possible. The starting point, though, is that there are only seven classes of simple algebraic groups over an algebraically closed field $\mathbb{K}$, and these are related to the Lie algebras of the corresponding type. With each such group is associated a building (see Subsection 1.3.7, page 8) which is obtained from that group by considering certain *maximal parabolic subgroups* whose conjugates are taken to be the elements of the building. Let us first enumerate these seven classes.

1. Type A_n, $n \geq 1$. The building can be viewed as a projective space over $\mathbb{K}$ of dimension n.

2. Type B_n, $n \geq 3$. The building can be viewed as the polar space arising from a quadric in $\mathbf{PG}(2n, \mathbb{K})$. Since $\mathbb{K}$ is algebraically closed, there is essentially only one such quadric (up to isomorphism) and it has maximal Witt index. The elements of the building are the projective subspaces contained in the quadric.

3. Type C_n, $n \geq 2$. The building can be viewed as the polar space arising from a symplectic polarity in $\mathbf{PG}(2n - 1, \mathbb{K})$. The elements of the geometry are the totally isotropic subspaces. For $n = 2$, this is the symplectic quadrangle $\mathsf{W}(\mathbb{K})$, anti-isomorphic with the orthogonal quadrangle $\mathsf{Q}(4, \mathbb{K})$, which is of type B_2.

4. Type D_n, $n \geq 4$. The building is the "oriflamme complex" of a quadric of maximal Witt index in $\mathbf{PG}(2n - 1, \mathbb{K})$. This boils in fact down to nothing more than the D_n-geometry which we considered for $n = 4$ in Chapter 2. The elements of the building are the projective subspaces of dimension not equal to $n - 2$ on the quadric; those of dimension $n - 1$ are separated into two sets of generators. For $n = 3$, the building is isomorphic to one of type A_3 via the Klein correspondence.

5. Type E_n, $n = 6, 7, 8$. These are geometries related to the exceptional Lie algebras of corresponding type. There have been successful attempts to approach these buildings via point–line geometries. For a survey, see COHEN [1995].

6. Type F_4. This is also related to an exceptional Lie algebra, but we have already met the corresponding geometry under the name "metasymplectic space"; see Chapter 2.

7. Type G_2. These are the buildings related to the split Cayley hexagons $\mathsf{H}(\mathbb{K})$.

Related to each building above is the Dynkin diagram of the corresponding Lie algebra, and, with no arrow, this diagram can be viewed as the Coxeter diagram of the building, or as the Buekenhout diagram of the geometry defined by the building. I display these diagrams in Figure C.1. Each conjugacy class of maximal parabolic subgroups represents a vertex of the diagram. For instance, $\mathbf{SL}_d(\mathbb{K})$ is a group of matrices over $\mathbb{K}$ such that the determinant, which is a polynomial in the entries of the matrix, is equal to 1. The parabolic subgroups are the stabilizers of certain projective subspaces of $\mathbf{PG}(d - 1, \mathbb{K})$. The ith node of the diagram exactly corresponds to the class of parabolic subgroups whose members are the stabilizer of a projective subspace of dimension $i - 1$.

The reason for the restriction of B_n to $n \geq 3$ and of C_n to $n \geq 2$ (and not vice versa) has been suggested to me by Tits and can be explained by the fact that, in many problems, the case $B_2 = C_2$ seems to follow the general rules of the case C_n rather than those of the case B_n. By the way, we do not take these restrictions too literally. Indeed, we will freely use the notation D_3 for A_3, or B_2 for C_2 (some authors even use E_5 for D_5).

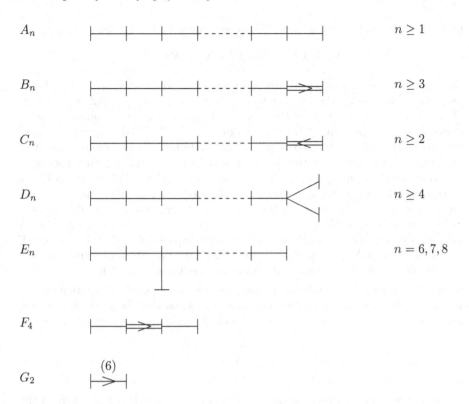

A_n $n \geq 1$

B_n $n \geq 3$

C_n $n \geq 2$

D_n $n \geq 4$

E_n $n = 6, 7, 8$

F_4

G_2 (6)

Figure C.1. Diagrams for buildings over algebraically closed fields.

Now suppose that $\mathbb{F}$ is any field, not necessarily algebraically closed, and let G be a simple algebraic group over $\mathbb{F}$. Roughly speaking, we can take the separable algebraic closure $\mathbb{K}$ of $\mathbb{F}$ and interpret the defining polynomial equations of G over $\mathbb{K}$. We obtain a simple algebraic group $\bar{G}$ over $\mathbb{K}$. The Galois group $\mathcal{G} = \mathrm{Gal}(\mathbb{K}/\mathbb{F})$ acts on $\mathbb{K}$, hence on $\bar{G}$. But it fixes $\mathbb{F}$ pointwise, hence it fixes G. Since $\mathcal{G}$ acts on $\bar{G}$, it acts on the corresponding building Ω (via the parabolic subgroups). In particular, it acts on the diagram Δ of the building corresponding to $\bar{G}$ and it induces certain orbits there. Loosely speaking, an orbit is called **isotropic** if $\mathcal{G}$ fixes at least one flag of Ω of the corresponding type. For example, if $\mathcal{G}$ induces on a diagram of type A_n a non-trivial group of order 2 (and that is possible in only one way), then a flag corresponding to the orbit $1, n$ is an incident point–hyperplane pair. The set of all such flags now forms again a building $\Omega(\mathbb{F})$. The **Tits diagram** Δ_{Tits} of $\Omega(\mathbb{F})$ is the old diagram Δ drawn in such a way that the action of $\mathcal{G}$ is made apparent (by bending the edges if there is a non-trivial action), and in addition all isotropic orbits are encircled. Let me give a very simple example. Let

$Q(6, \mathbb{R})$ be the orthogonal quadrangle corresponding to the quadratic form

$$-X_0^2 - X_1^2 + X_2^2 + X_3^2 + X_4^2 + X_5^2 + X_6^2.$$

Here, we can omit the algebraic group arguments and directly proceed with the geometry. Interpreting the bilinear form over $\mathbb{C}$, we obtain a building Ω of type B_3 (and this is independent of the number of minus signs in the quadratic form above, hence independent of the Witt index over $\mathbb{R}$). The Galois group $\mathcal{G}$ is here nothing more than complex conjugation. This acts trivially on the diagram Δ of Ω (as is also obvious from the fact that a diagram of type C_3 does not allow any non-trivial automorphism!). Hence all orbits are trivial. When is an orbit — or equivalently, a node of Δ — isotropic? When there is an element of $\mathcal{G}$ fixing an element of Ω of the corresponding type. But if some subspace of dimension ℓ is fixed under conjugation, then it is a real subspace, and hence it must lie on $Q(6, \mathbb{R})$. Consequently $\ell = 1, 2$ and exactly the subspaces of $Q(6, \mathbb{R})$ are obtained. The Tits diagram Δ_{Tits} is the C_3-diagram where the first two nodes are encircled. The building $\Omega(\mathbb{R})$ "is" precisely the Moufang quadrangle $Q(6, \mathbb{R})$.

Let me give a second example before going on. The example of a Hermitian variety is very pecular and needs separate mention. Let us consider the three-dimensional case, and the complex field $\mathbb{C}$. We consider the Hermitian variery $\mathcal{H}$ with equation

$$\sum_{i=0}^{3} X_i \bar{X}_i = 0.$$

A first observation is that $\mathbb{C}$ is already algebraically closed; hence, where is the diagram of this building (quadrangle) in Figure C.1? The answer is that $\mathbb{C}$ may well be algebraically closed, this does not matter; what matters is that the field $\mathbb{R}$, over which the polynomials that define the corresponding group are defined, is not algebraically closed. The matrix belonging to the polarity in $\mathbf{PG}(3, \mathbb{C})$ for which $\mathcal{H}$ is the set of absolute points is clearly the identity matrix. Hence an element $\theta \in \mathbf{PGL}_4(\mathbb{C})$ with corresponding matrix M preserves $\mathcal{H}$ **if and only if** $\overline{M}^T M = I$, where $\bar{\bullet}$ means the complex conjugate of $\bullet$, and $\bullet^T$ means the transpose of $\bullet$. Let G be the group of all such elements θ. The condition $\overline{M}^T M = I$ is equivalent to $\overline{M} = M^{-T}$. Hence if we extend $\mathbb{R}$ to $\mathbb{C}$, then given that the corresponding extension of G is $\mathbf{PGL}_4(\mathbb{C})$ (hence the diagram is A_3, which follows from a computation in the corresponding Lie algebra; see e.g. CARTER [1972](14.5), page 268), we see that complex conjugation (which is an element of the Galois group) maps a matrix M (where we take all the entries for simplicity and clarity again in $\mathbb{R}$) to $\overline{M} = M^{-T}$. So the subgroup H_p of matrices stabilizing a certain (real) point p of the projective space $\mathbf{PG}(3, \mathbb{C})$ is mapped under conjugation onto a subgroup $\overline{H}_p$ stabilizing a certain hyperplane (because, if a matrix acts on the points of $\mathbf{PG}(3, \mathbb{C})$, then the inverse acts on the dual of $\mathbf{PG}(3, \mathbb{C})$). Hence the Galois group always induces a non-trivial involution on the diagram.

The magic started: Tits diagrams for the exceptional Moufang quadrangles

Now the question is: how does one deduce from the Tits diagram the new root system of the geometry, in our case, C_2 or BC_2? There is a very precise answer in the general case; see TITS [1966b](2.5.2). However, in our case, we do not need the answer in full generality. Indeed, the low rank (rank 2) permits us to only look at some numbers. We proceed as follows.

First, the rank of the new diagram is equal to the number of encircled isotropic orbits. So in order to obtain Moufang quadrangles, we should only look at Tits diagrams with exactly two encircled orbits. Now, how do we know that we are going to obtain a Moufang quadrangle and not a Moufang plane, or a Moufang hexagon? The answer is given by looking at the root system belonging to Δ. Indeed, attached to every algebraic group over an algebraically closed field is a root system of the same type as the building Ω. There is an action of $\mathcal{G}$ on that root system, via the torus of $\bar{G}$, which we shall not explain in detail. The fact of the matter is that this action defines a new root system, precisely the root system belonging to the new Moufang building $\Omega(\mathbb{F})$. In our case, it is the root system belonging to a Moufang quadrangle, hence it can belong to a root system of type C_2 or BC_2. The point is that there is a method of deriving that new root system from the Tits diagram. This method works as follows. At every node of the diagram Δ, viewed as the Dynkin diagram of the root system, we write down the coefficient of the corresponding fundamental root in the unique linear combination of the fundamental roots which gives the *longest*, or *highest*, root. Then the numbers appearing at the encircled orbits must, for each orbit, add up to the numbers that appear in the diagram of a root system of type C_2 or BC_2, and in total, we must have them all, i.e., if we want a root system of type C_2, then we should have two isotropic orbits and the numbers of one of them should add up to 1, and the numbers in the other orbit should add up to 2, because the numbers of a C_2-root system are exactly 1 and 2 (in the representation of Subsection 5.4.3 on page 187, a fundamental system is given by the vectors e_0 and e_3, and the corresponding highest root is $e_2 = e_0 + 2e_3$). The numbers of a root system of type BC_2 are 2 and 2 (e_0 and e_3 are again fundamental roots, the highest root is now $f_1 = 2e_0 + 2e_3$). Hence we have to look for Tits diagrams with two encircled orbits and such that the numbers in these orbits add up to 1 and 2, or twice to 2. But which Tits diagrams are feasible? To answer this, we write down three simple rules that must be obeyed by Tits diagrams.

Rule 1. If we take the residue of any isotropic orbit (i.e., if we remove any isotropic orbit from the Tits diagram), then we must obtain a feasible Tits diagram again.

Rule 2. The isotropic orbits are scattered in the diagram symmetrically with respect to the *opposition involution*, i.e., symmetrically with respect to the unique non-trivial involution of the diagram if the diagram has type A_n, $n \geq 2$, D_{2n+1}, $n \geq 2$, or E_6.

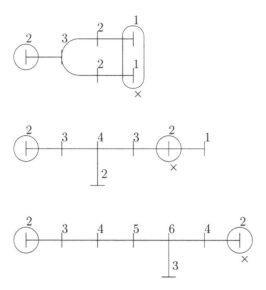

Figure C.2. The feasible Tits diagrams arising from exceptional diagrams.

Rule 3. If in a diagram of type C_n the first node is encircled, then all nodes have to be encircled.

As an example, we determine all feasible (with respect to the above rules) Tits diagrams with underlying diagram of type E_6. By Rule 2, we only have two possibilities: we can encircle two symmetrical nodes with numbers 2. But in the residue of each of them, the new Tits diagram has underlying diagram of type A_4 and only the second node is encircled, contradicting Rule 2. The second possibility is that we encircle the unique self-symmetric node (carrying the number 2) and take the two nodes carrying the number 1 together in one isotropic orbit; see Figure C.2. We can do the same exercise for the other exceptional diagrams (i.e., diagrams of type E_6, E_7, E_8, F_4, G_2) and we obtain the Tits diagrams shown in Figure C.2. It can be checked that each time we obtain the numbers 2 and 2 as sums of the numbers in the isotropic orbits. Hence all the possible Moufang quadrangles obtained are of type BC_2.

We leave it to the reader to work out some classical diagrams, i.e., the diagrams A_n, B_n, C_n and D_n. The possiblities are listed below. In order to be able to do this exercise successfully, one should have at one's disposal the coefficients of the highest root of each such diagram. For A_n, all numbers are 1; For B_n, C_n and D_n, all numbers are 2 except those of the extremities, which are always 1, except for one extremal vertex on the double edge in B_n, $n \geq 2$, and the extremal vertex not on the double edge in C_n, $n \geq 3$, both of which nevertheless carry the number 2. For completeness' sake we mention that the numbers of F_4 are (in the obvious order) $2, 3, 4, 2$, and those of G_2 are $2, 3$. For the E_n diagrams, $n = 6, 7, 8$, see Figure C.2.

Remark. There is yet another way to find out whether we obtain a generalized quadrangle by encircling two nodes. We argue at the level of Coxeter groups; see MÜHLHERR [1994]. Let w_1 and w_2 be the longest word in the Coxeter groups whose respective diagrams are the residues of the isotropic orbits in the Dynkin diagram (and ignoring the arrows); let w be the longest word in the Coxeter group whose diagram is the residue of the union of both isotropic orbits. Then the order n of $(w w_1)(w w_2)$ tells one that the rank 2 building arising from the isotropic orbits is a generalized n-gon.

Some more magic tricks

Now that we have obtained the three possible types of exceptional Moufang quadrangles of type (E_i), $i = 6, 7, 8$, we will derive from their Tits diagram (1) which isotropic orbit corresponds to the multiple root in the root system of type BC_2, (2) the dimensions of the respective root groups, and finally (3) the Tits diagram of the corresponding full (up to duality) subquadrangle belonging to the root system of type C_2 (and defined by the long roots of the root system of type BC_2; see step II of Subsection 5.5.5 on page 214). Then the reader may do the same for some classical Moufang quadrangles of type $(B - CB)_2$ and $(CB)_2$. For the exceptional Moufang quadrangles of type (F_4), see below.

Questions (1) and (2) are solved by looking in the residue of the respective isotropic orbits. If we consider such a residue, obtained by erasing one of the isotropic orbits in the Tits diagram Δ_{Tits}, then we have a new Tits diagram Δ'_{Tits} with only one isotropic orbit encircled, and with underlying Dynkin diagram Δ'. Again, we furnish Δ' with numbers as above. If the numbers of the encircled orbit add up to 2, then this orbit corresponds to the multiple root of the diagram BC_2. If on the contrary, they add up to 1, then the isotropic orbit corresponds to the simple roots of BC_2. In any exceptional case, the two cases must occur, and it is magical to see that they just do. One remark should be made here: it could happen that the diagram Δ' is disconnected. If the isotropic orbit is in one component, then we only look at that component; if the isotropic orbit is in two components, then we also take only one of these components (after all, the longest root in a reducible root system is not well defined, and so one must consider the longest root in one of the irreducible components). This happens for instance in the case of a Hermitian quadrangle in projective 4-space over a commutative field, where one of the diagrams Δ' is of type $A_1 \times A_1$. We have marked the isotropic orbit which corresponds to the multiple fundamental root with a cross ($\times$) in Figure C.2.

The dimension over $\mathbb{F}$ of the root groups can be determined in Δ'_{Tits} as follows. Suppose first we are dealing with a multiple root, and let α and 2α be the corresponding roots in the root system of type BC_2. The dimension of $U_{2\alpha}$ over $\mathbb{F}$ is the number of positive roots of a root system of type Δ' for which, when expressed as a linear combination in the fundamental roots, the coefficients of the fundamental roots corresponding to the isotropic orbit add up to 2. Similarly, the dimension of U_α over $\mathbb{F}$ is the number of positive roots of a root system of type Δ' for which the coefficients of the fundamental roots corresponding to the isotropic orbit add up to 1. The same thing holds for the root groups belonging to the simple roots. There are certain formulae for these numbers in terms of dimensions of groups over $\mathbb{K}$, but we will not go into detail here, since this geometric description is magical enough. For instance, let us prove that in all exceptional cases, the dimension of $U_{2\alpha}$ is equal to 1, hence the root groups belonging to the long roots are isomorphic to the aditive group of $\mathbb{F}$ (which conforms with the description of the exceptional Moufang quadrangles in Subsection 5.5.5, step II, in Chapter 5). There is an easy trick to see this, given the following property.

Let β be the longest root of a root system of type Δ', and let α be any fundamental root. Then the difference $\beta - \alpha$ is a root **if and only if** *in the extended diagram Δ'_{ext} the node corresponding to α is connected by an edge with the new node (hence the one corresponding to the longest root).*

Note that the extended diagram is the diagram where one takes into account all fundamental roots and also the negative of the longest root, and applying the same old rules to these roots to define the new (extended) Dynkin diagram.

Drawing these extended diagrams Δ' corresponding to the multiple roots for the exceptional cases, we see that the isotropic orbits separate the new node from the rest, hence no other fundamental root α other than one corresponding to a node in the isotropic orbit exists such that $\beta - \alpha$ is a root. This means that β is the only positive root such that the coefficients of the fundamental roots corresponding to the isotropic orbits add up to 2. See Figure C.3.

The dimensions of the root groups for the exceptional Moufang quadrangles are given in Subsection 5.5.5, step II, in Chapter 5. One can try out the magic on some small diagrams for classical Moufang quadrangles, such as the Hermitian quadrangles in projective dimension 4 over a commutative field, where one should obtain dimensions 2 and 3 for the root groups belonging to a simple root and a short root, respectively, and again 1 for those belonging to a long root.

Now we address question (3). We remark that the fundamental roots of the root system of type C_2 of a full or ideal subquadrangle can be obtained from the fundamental roots of the root system of type BC_2 by adding the negative of the longest root and deleting the fundamental root which has a multiple. Inspired by this property, we can obtain the Tits diagram of the full or ideal subquadrangle of type C_2, by adjoining the node corresponding to the negative of the longest root, and remove the isotropic orbit corresponding to the multiple root of the root system of type BC_2. The second orbit to encircle is now simply the new node. Thus we obtain a new Tits diagram of relative rank 2, i.e., there are two isotropic orbits encircled. When one does this exercise for the three types of exceptional Moufang quadrangles, than in each case one obtains a D_ℓ-quadrangle (as mentioned in Subsection 5.5.5, step II), for some ℓ. One can do the exercise for classical quadrangles

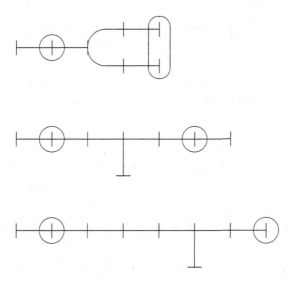

Figure C.3. Extended Tits diagrams of exceptional Moufang quadrangles.

of type BC_2, for instance, to stick with our example of $H(4, L, \sigma)$, in which case we obtain the Tits diagram of $H(3, L, \sigma)$.

Now, if one wants to test these magic tricks, than one would like to be able to recognize which Tits diagrams belong to which classical quadrangles. We now present that information in detail.

As a preceding remark, we mention that the classical quadrangle $Q(V, q)$ over the skew field $\mathbb{K}$ has no Tits diagram if V is infinite-dimensional over $\mathbb{K}$, or if $\mathbb{K}$ is infinite-dimensional over its centre, say, $\mathbb{F}$.

In what follows, we again assume standard notation and forget about the earlier meaning of $\mathbb{K}$ and $\mathbb{F}$ in this appendix. So let V be a vector space over the skew field $\mathbb{K}$ (finite-dimensional this time; see the remark made in the preceding paragraph), σ an anti-automorphism of $\mathbb{K}$ of order at most 2 and q a σ-quadratic form. More-over, we let $\mathbb{F}$ denote the centre of $\mathbb{K}$ and $\mathbb{L}$ a maximal subfield of $\mathbb{K}$ containing $\mathbb{F}$. We define d to be the dimension of $\mathbb{K}$ over $\mathbb{L}$, and that is equal to the dimension of $\mathbb{L}$ over $\mathbb{F}$; see Subsection 5.5.13. Also, ℓ shall be the rank of the Dynkin diagram underlying the Tits diagram, i.e., ℓ is the number of nodes of the Tits diagram.

The orthogonal quadrangles

These correspond to Moufang quadrangles of type $(C)_2$ and $(C - CB)_2$.
Here $\mathbb{K} = \mathbb{L} = \mathbb{F}$ is commutative and $\sigma = 1$, hence $d = 1$. We have an orthogonal quadrangle, and extended to the quadratic closure of $\mathbb{K}$, we always have either type B_ℓ or D_ℓ. The action of the Galois group in the latter case can be trivial on the diagram, or not. This gives a further distinction. For instance, in the case of $\mathbb{K} = \mathbb{R}$,

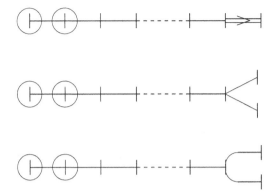

Figure C.4. Tits diagrams for orthogonal quadrangles.

complex conjugation will exactly induce the opposition map. For there are lines fixed, but no planes, hence complex conjugation must map some generator A (a maximal singular projective space) of a quadric of type D_ℓ onto another generator $\bar{A}$ with $A \cap \bar{A}$ a line. Hence A and $\bar{A}$ belong to the same set of generators **if and only if** $\dim(A) - \dim(A \cap \bar{A})$ is even. This precisely amounts to the opposition map.

Hence we have the three possibilities as shown in Figure C.4.

Hermitian quadrangles over commutative fields

Here $\sigma \neq 1$, $\mathbb{K} = \mathbb{L} = \mathbb{F}$ and we have a Hermitian form. We know that the Dynkin diagrams underlying the Tits diagrams of Hermitian forms are of type A_ℓ. Here we just want points and lines in the isotropic orbits, hence we obtain the diagrams of Figure C.5.

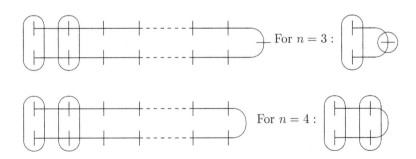

Figure C.5. The Tits diagram of a Hermitian quadrangle in the commutative case.

Hermitian quadrangles properly belonging to a root system of type C_2

These correspond to the Moufang quadrangles of type $(B)_2$.

We treat these quadrangles separately because they appear exactly as full sub-quadrangles of the other Hermitian quadrangles. First, there is the case where σ induces a non-trivial involution in the centre $\mathbb{F}$ of $\mathbb{K}$. Writing $\mathbb{K}$ as a vector space of dimension d over $\mathbb{L}$, we see that V is a $4d$-dimensional vector space over $\mathbb{L}$, where the old points have become some d-dimensional subspaces, and the old lines some $2d$-dimensional subspaces. So we have a $(4d-1)$-dimensional projective space $\mathbf{PG}(4d-1,\mathbb{L})$ with a Hermitian form (since σ does not act trivially on $\mathbb{F}$, hence neither on $\mathbb{L}$). Hence the Tits diagram is in this case a stretched version of the Hermitian commutative case above.

Now suppose that σ induces the identity in $\mathbb{F}$. There are two kinds of such involutions. Indeed, either the dimension over $\mathbb{F}$ of the subspace of fixed elements is $\frac{1}{2}d(d-1)$, or it is $\frac{1}{2}d(d+1)$; see e.g. WEIL [1960]. In the first case (which is impossible in characteristic 2) the form becomes symplectic after extension, and in the second case the form becomes orthogonal (or possibly symplectic in characteristic 2; in that case the symplectic building is isomorphic to the orthogonal building). When the form becomes orthogonal, it defines a quadric of type D_ℓ, $\ell = 2d$ (here B_ℓ is impossible since the projective dimension $4d-1$ is odd). Note that in this case the Galois group must act trivially on the Dynkin diagram because the lines of the three-dimensional quadrangle in $\mathbf{PG}(3,\mathbb{K})$, viewed as planes of V, become maximal singular subspaces of the D_ℓ quadric and they are fixed, hence the type of the generators is preserved. Let us remark that in these cases the first encircled node appears at the dth node, and *this is always a power of* 2. Added to the list of rules above, these four rules together are probably sufficient to determine all possible Tits diagrams (i.e., all diagrams for which there exist fields over which the given diagram actually occurs).

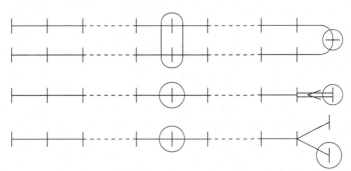

Figure C.6. The Tits diagram of a Hermitian Moufang quadrangle of type $(B)_2$.

In conclusion, the Tits diagrams of this three-dimensional case are shown in Figure C.6. The isotropic orbits all appear at the nodes d and $2d$ (counting from the left and making an arbitrary choice between the two last nodes in the D_ℓ-case).

The Hermitian Moufang quadrangles belonging to a root system of type BC_2

Here, we treat the Moufang quadrangles of type $(CB)_2$ and $(B - CB)_2$.
The story is just the same as for the three-dimensional case (see the case of
type $(B)_2$ above), except that the dimension of the vector space is bigger, and
so the Tits diagrams are stretched out. Also, there is an extra type connected
with D_ℓ, since now the Galois group can have a non-trivial orbit on the Dynkin
diagram. We simply list the diagrams; see Figure C.7. Again, the isotropic orbits
are at the places d and $2d$, and never at the rightmost orbit, except for the last
diagram. Also, in the cases B_n, C_n and D_n, the integer d is a power of 2.

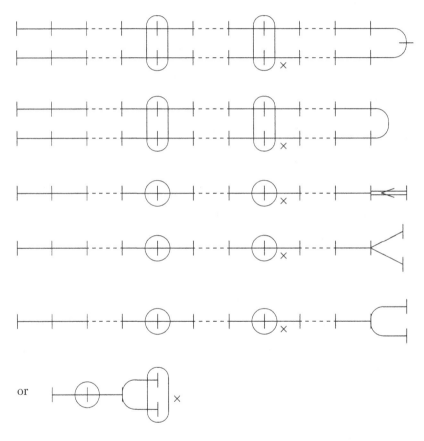

Figure C.7. The Tits diagram of a Hermitian Moufang quadrangle of type $(CB)_2$)
and $(B - CB)_2$).

Some anti-isomorphisms

One can see the sporadic isomorphisms mentioned in Proposition 3.4.9, Proposition 3.4.11 and Proposition 3.4.13 in the Tits diagrams. Indeed, the Tits diagram of the orthogonal quadrangle $Q(4, \mathbb{K}, q)$ over the field $\mathbb{K}$ is a B_2 Dynkin diagram with both nodes encircled, while a symplectic quadrangle has a C_2 Dynkin diagram with two nodes encircled as Tits diagram. Also, the Tits diagram of the Hermitian quadrangle $H(3, \mathbb{L}, q, \sigma)$, where $\mathbb{L}$ is a quadratic extension of $\mathbb{K}$ and $\sigma \in \mathrm{Gal}(\mathbb{L}/\mathbb{K})$, is a "folded" A_3 Dynkin diagram with the middle node encircled and the other two nodes in one isotropic orbit encircled, while the same thing happens for an orthogonal D_3-quadrangle $Q(5, \mathbb{K}, q')$, but with a D_3 Dynkin diagram, which is equal to an A_3 diagram. Finally, let $\mathbb{K}$ be a skew field which is a quaternion algebra over its centre $\mathbb{F}$, and let σ be a skew conjugation; see Proposition 3.4.9 on page 107. Then the Tits diagram of the Hermitian quadrangle $H(3, \mathbb{K}, q, \sigma)$ is, since σ fixes some subfield $\mathbb{L}$ over which $\mathbb{K}$ is two-dimensional, a D_4 Dynkin diagram (not folded) where the middle node is encircled as well as one of the two last nodes (see Figure C.8), while the Tits diagram of the orthogonal D_4-quadrangle $Q(7, \mathbb{F}, q')$ is a D_4 Dynkin diagram (not folded) with the first two nodes separately encircled.

Figure C.8. The Tits diagram of $H(3, \mathbb{K}, q, \sigma)$ (left) and of its dual $Q(7, \mathbb{F}, q')$.

Moufang hexagons

Basically similar things hold for Moufang hexagons, but there the magic is not so apparent since, for instance, all root group dimensions are obvious from the description of the Moufang hexagons. Also, there are no multiple roots in the root system of type G_2, and $H(\mathbb{K})$ is a full or ideal subhexagon of every Moufang hexagon. And anyway, if one applies the trick above to see the subhexagons (ignoring that there are in fact no multiple roots), then, with a little imagination, one sees the double of a Desarguesian projective plane, since all one obtains is a Tits diagram of a projective plane. This agrees with the fact that, if one deletes the short fundamental root from a fundamental system of roots of a root system of type G_2, and one adds the negative of the longest root to the fundamental system, then one obtains the fundamental system of a root system of type A_2.

For the sake of completeness, we give the Tits diagrams for the five classes of non-mixed Moufang hexagons in Figure C.9 (in the same order as the enumeration on page 222, except that we do not introduce a diagram for the mixed hexagons; see below).

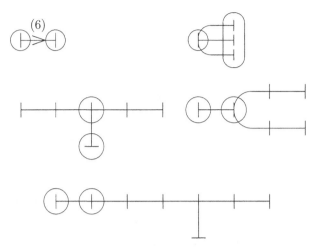

Figure C.9. Tits diagrams for Moufang hexagons.

Generalization

One could also assign a diagram to the mixed polygons, the Ree–Tits octagons, and the Moufang quadrangles of type $(BC-CB)_2$; see e.g. RONAN [1989], Appendix 2, for the mixed case and the Moufang octagons. Note, by the way, that in the case of mixed quadrangles, the diagram assigned to it by RONAN [1989] only covers the case where the root groups are fields. No diagram is introduced in *op. cit.* for the more general case of vector spaces.

But more importantly, some of the magic also holds for these diagrams, although other rules hold to produce these (for instance, Rule 3 above no longer applies; after all, in characteristic 2, types C_n and B_n are indistinguishable, hence one might think of the arrow in a Dynkin diagram as pointing in the *two* directions, using the direction that suites one best for a certain purpose). We do not go into detail, but mention one example (see MÜHLHERR & VAN MALDEGHEM [19**]). The Tits diagram for a quadrangle of type $(BC - CB)_2$ is an F_4-diagram with the two extremal nodes encircled (after all, such a quadrangle is also called "of type (F_4)"). To see the full subquadrangles of type $(C - CB)_2$, one adjoins the node symbolizing the longest root, encircles it and deletes the other extreme node (which was encircled). In this way, one obtains a B_4-diagram with the two first nodes encircled. This is indeed the Tits diagram of the full subquadrangle, which is of type $(C - CB)_2$. To obtain the full subquadrangles of this quadrangle, one adjoins the longest root to this B_4-diagram (but viewed as a C_4-diagram!), encircles the new node, deletes the encircled node which is not joined to the new node and obtains a C_2-diagram with both nodes encircled. This can be viewed as the Tits diagram for a mixed quadrangle.

Appendix D

Matrices of Root Elations
of Some Classical Polygons

In this appendix, we present the explicit form of some root elations of some Moufang polygons in both their intrinsic coordinates and those of their ambient projective space. In the latter case, this will provide a set of generating matrices for the little projective group of the polygon under consideration. We also display a general element of the torus.

In general, we will always consider a γ-elation with γ an $(n-2)$-path contained in the hat-rack and starting with either $((\infty), [\infty], \dots)$, or dually with $([\infty], (\infty), \dots)$. Also the elements of the torus are considered with respect to the hat-rack (and by *torus*, we mean the stabilizer in $\text{Aut}\,\Gamma \cap \mathbf{PGL}$ of an apartment of the generalized polygon Γ, where $\mathbf{PGL}$ is the projective linear group of the ambient projective space). Note that it suffices to give the action on points and lines with $n-1$ coordinates because the action on an element with i coordinates, $i < n-1$, is given by chopping off the $n-i-1$ last coordinates in the expression of the action on an appropriate element with $n-1$ coordinates; see also Remark 4.5.3 (page 146).

The proofs of the results in this appendix are very simple and reduce to elementary calculations, and we shall not perform them. We will always use the coordinatization as introduced in Chapter 3, and we shall freely use it without explicitly referring to it. In what follows, we will denote by K, L arbitrary elements of R_2, and by A, B arbitrary elements of R_1. We use the notation $\text{diag}(\bullet)$ for a diagonal matrix with diagonal $\bullet$.

The symplectic quadrangles

The following mapping defines a generic $([\infty], (\infty), [0])$-elation in $\mathsf{W}(\mathbb{K})$, with corresponding matrix $M_{[L]}$ in $\mathbf{PG}(3, \mathbb{K})$:

$$\begin{cases} (a, l, a') \mapsto (a, l + L, a'), \\ [k, b, k'] \mapsto [k, b, k' + L], \end{cases} \qquad M_{[L]} = \begin{pmatrix} 1 & L & 0 & 0 \\ 0 & 1 & 0 & 0 \\ 0 & 0 & 1 & 0 \\ 0 & 0 & 0 & 1 \end{pmatrix}.$$

The following mapping defines a generic $((\infty), [\infty], (0))$-elation in $\mathsf{W}(\mathbb{K})$, with corresponding matrix $M_{(B)}$ in $\mathbf{PG}(3, \mathbb{K})$:

$$\begin{cases} (a, l, a') \mapsto (a, l + 2aB, a' + B), \\ [k, b, k'] \mapsto [k, b + B, k'], \end{cases} \qquad M_{(B)} = \begin{pmatrix} 1 & 0 & 0 & -B \\ 0 & 1 & 0 & 0 \\ 0 & -B & 1 & 0 \\ 0 & 0 & 0 & 1 \end{pmatrix}.$$

The following mapping defines a generic element of the torus in $\mathsf{W}(\mathbb{K})$, with corresponding matrix M_{torus} in $\mathbf{PG}(3, \mathbb{K})$:

$$\begin{cases} (a, l, a') \mapsto (Aa, A^2 Kl, AKa'), \\ [k, b, k'] \mapsto [Kk, AKb, A^2 Kk'], \end{cases} \qquad M_{\text{torus}} = \text{diag}(A^2 K, 1, AK, A).$$

Finally, the projective linear transformation with matrix

$$M = \begin{pmatrix} 0 & 0 & 0 & 1 \\ 0 & 0 & 1 & 0 \\ 1 & 0 & 0 & 0 \\ 0 & 1 & 0 & 0 \end{pmatrix}$$

represents a collineation of $\mathsf{W}(\mathbb{K})$ which rotates the hat-rack "through 90 degrees". Hence the matrices $M, M_{(B)}, M_{[L]}$ generate the little projective group of $\mathsf{W}(\mathbb{K})$ and define the standard embedding in $\mathbf{PG}(3, \mathbb{K})$. The representations of the corresponding collineations for mixed quadrangles are easily obtained now by requiring characteristic 2 and suitably restricting the coefficients.

The classical quadrangles

Here $Q(V, q)$ is a classical quadrangle, and we again use standard notation; see Subsection 3.4.7 starting on page 104. In particular, f is the Hermitian form corresponding to the σ-quadratic form q. Furthermore, let $\mathcal{B} = \{e_i : i \in I\}$ be some basis of V_0, where I is some index set. In principle, I could be infinite, but then the matrices below must be considered as matrices in a very broad sense. But with the right interpretation, this is possible.

For a vector L_0 of V_0, we define the $|I| \times 1$ block matrix $f(L_0, \mathcal{B})$ as the matrix whose $(i, 1)$th element equals $f(L_0, e_i)$, $i \in I$. Also, we denote by $I_{|I|}$ the $|I| \times |I|$ identity matrix, while the zero matrix of any dimension will always be denoted by 0.

Put $d = 3 + |I|$. The following mapping defines a generic $([\infty], (\infty), [0])$-elation in $Q(V, q)$, with corresponding matrix $M_{[L]}$ in $\mathbf{PG}(d, \mathbb{K})$, where $L = (L_0, L_1) \in R_2$:

$$
\begin{cases}
(a, l, a') & \mapsto & (a, L \oplus l, a'), \\
[k, b, k'] & \mapsto & [k, b + f(L_0, k_0), L \oplus k'],
\end{cases}
\quad
M_{[L]} =
\begin{pmatrix}
1 & 0 & f(L_0, \mathcal{B}) & 0 & L_1 \\
0 & 1 & 0 & 0 & 0 \\
0 & 0 & I_{|I|} & 0 & L_0 \\
0 & 0 & 0 & 1 & 0 \\
0 & 0 & 0 & 0 & 1
\end{pmatrix}.
$$

The following mapping defines a generic $((\infty), [\infty], (0))$-elation in $Q(V, q)$, with corresponding matrix $M_{(B)}$ in $\mathbf{PG}(d, \mathbb{K})$:

$$
\begin{cases}
(a, l, a') & \mapsto & (a, l \oplus (0, Ba^\sigma - aB^\sigma), \\
& & a' + B), \\
[k, b, k'] & \mapsto & [k, b + B, k'],
\end{cases}
\quad
M_{(B)} =
\begin{pmatrix}
1 & -B & 0 & 0 & 0 \\
0 & 1 & 0 & 0 & 0 \\
0 & 0 & I_{|I|} & 0 & 0 \\
0 & 0 & 0 & 1 & B^\sigma \\
0 & 0 & 0 & 0 & 1
\end{pmatrix}.
$$

The following mapping defines an element of the torus in $Q(V, q)$, with corresponding matrix M_{torus} in $\mathbf{PG}(d, \mathbb{K})$ (we do not give a *generic* element of the torus here because this depends heavily on the skew field $\mathbb{K}$ and the form q):

$$
\begin{cases}
(a, l, a') & \mapsto & (Aa, A^\sigma \otimes l, Aa'), \\
[k, b, k'] & \mapsto & [k, Ab, A^\sigma \otimes k'],
\end{cases}
\quad
M_{\text{torus}} = \text{diag}(A, 1, 1, \dots, 1, A^{-\sigma}).
$$

Finally, the projective linear transformation with matrix

$$
M =
\begin{pmatrix}
0 & 1 & 0 & 0 & 0 \\
0 & 0 & 0 & 0 & 1 \\
0 & 0 & I_{|I|} & 0 & 0 \\
1 & 0 & 0 & 0 & 0 \\
0 & 0 & 0 & 1 & 0
\end{pmatrix}
$$

represents a collineation of $\mathsf{W}(\mathbb{K})$ which rotates the hat-rack "through 90 degrees". Hence the matrices $M, M_{(B)}, M_{[L]}$ generate the little projective group of $Q(V, q)$ and define the standard embedding in $\mathbf{PG}(d, \mathbb{K})$.

The symplectic hexagons

Here, $\mathbb{K}$ is a field of characteristic 2. An embedding of $\mathsf{H}(\mathbb{K})$ in $\mathbf{PG}(5, \mathbb{K})$ in this case is obtained from the standard embedding of $\mathsf{H}(\mathbb{K})$ in $\mathbf{PG}(6, \mathbb{K})$ by deleting the fourth coordinate X_3.

The following mapping defines a generic $([\infty], (\infty), [0], (0,0), [0,0,0])$-elation in $\mathsf{H}(\mathbb{K})$, with corresponding matrix $M_{[L]}$ in $\mathbf{PG}(5, \mathbb{K})$:

$$\begin{cases} (a,l,a',l',a'') & \mapsto & (a,l+L,a',l',a''), \\ [k,b,k',b',k''] & \mapsto & [k,b,k'+kL,b',k''+L], \end{cases}$$

$$M_{[L]} = \begin{pmatrix} 1 & L & 0 & 0 & 0 & 0 \\ 0 & 1 & 0 & 0 & 0 & 0 \\ 0 & 0 & 1 & 0 & 0 & 0 \\ 0 & 0 & 0 & 1 & 0 & 0 \\ 0 & 0 & 0 & L & 1 & 0 \\ 0 & 0 & 0 & 0 & 0 & 1 \end{pmatrix}.$$

The following mapping defines a generic $((\infty), [\infty], (0), [0,0], (0,0,0))$-elation in $\mathsf{H}(\mathbb{K})$, with corresponding matrix $M_{(B)}$ in $\mathbf{PG}(5, \mathbb{K})$:

$$\begin{cases} (a,l,a',l',a'') & \mapsto & (a,l+a^2 B, a', l'+aB^2 + a'B, a''+B), \\ [k,b,k',b',k''] & \mapsto & [k,b+B,k',b',k''], \end{cases}$$

$$M_{(B)} = \begin{pmatrix} 1 & 0 & 0 & 0 & B & 0 \\ 0 & 1 & 0 & B & 0 & 0 \\ 0 & 0 & 1 & 0 & 0 & 0 \\ 0 & 0 & 0 & 1 & 0 & 0 \\ 0 & 0 & 0 & 0 & 1 & 0 \\ 0 & 0 & B^2 & 0 & 0 & 1 \end{pmatrix}.$$

The following mapping defines a generic element of the torus in $\mathsf{H}(\mathbb{K})$, with corresponding matrix M_{torus} in $\mathbf{PG}(5, \mathbb{K})$:

$$\begin{cases} (a,l,a',l',a'') & \mapsto & (Aa, A^3 Kl, A^2 Ka', A^3 K^2 l', AKa''), \\ [k,b,k',b',k''] & \mapsto & [Kk, AKb, A^3 K^2 k', A^2 Kb', A^3 Kk''], \end{cases}$$

$$M_{\text{torus}} = \text{diag}(A^4 K^2, AK, A, 1, A^3 K, A^3 K^2).$$

Finally, the projective linear transformation with matrix

$$M = \begin{pmatrix} 0 & 0 & 0 & 0 & 1 & 0 \\ 0 & 0 & 0 & 0 & 0 & 1 \\ 0 & 0 & 0 & 1 & 0 & 0 \\ 0 & 1 & 0 & 0 & 0 & 0 \\ 0 & 0 & 1 & 0 & 0 & 0 \\ 1 & 0 & 0 & 0 & 0 & 0 \end{pmatrix}$$

represents a collineation of $\mathsf{H}(\mathbb{K})$ which rotates the hat-rack "through 60 degrees". Hence the matrices $M, M_{(B)}, M_{[L]}$ generate the little projective group of $\mathsf{H}(\mathbb{K})$ and define the standard embedding in $\mathbf{PG}(5, \mathbb{K})$.

The split Cayley hexagons

The following mapping defines a generic $([\infty], (\infty), [0], (0,0), [0,0,0])$-elation in $H(\mathbb{K})$, with corresponding matrix $M_{[L]}$ in $\mathbf{PG}(6, \mathbb{K})$:

$$\begin{cases} (a, l, a', l', a'') & \mapsto & (a, l + L, a', l', a''), \\ [k, b, k', b', k''] & \mapsto & [k, b, k' - kL, b', k'' + L], \end{cases}$$

$$M_{[L]} = \begin{pmatrix} 1 & -L & 0 & 0 & 0 & 0 & 0 \\ 0 & 1 & 0 & 0 & 0 & 0 & 0 \\ 0 & 0 & 1 & 0 & 0 & 0 & 0 \\ 0 & 0 & 0 & 1 & 0 & 0 & 0 \\ 0 & 0 & 0 & 0 & 1 & 0 & 0 \\ 0 & 0 & 0 & 0 & L & 1 & 0 \\ 0 & 0 & 0 & 0 & 0 & 0 & 1 \end{pmatrix}.$$

The following mapping defines a generic $((\infty), [\infty], (0), [0,0], (0,0,0))$-elation in $H(\mathbb{K})$, with corresponding matrix $M_{(B)}$ in $\mathbf{PG}(6, \mathbb{K})$:

$$\begin{cases} (a, l, a', l', a'') & \mapsto & (a, l - 3a^2 B, a' + 2aB, l' + 3aB^2 + 3a'B, a'' + B), \\ [k, b, k', b', k''] & \mapsto & [k, b + B, k', b', k''], \end{cases}$$

$$M_{(B)} = \begin{pmatrix} 1 & 0 & 0 & 0 & 0 & B & 0 \\ 0 & 1 & 0 & 0 & -B & 0 & 0 \\ 0 & 0 & 1 & 0 & 0 & 0 & 0 \\ 0 & 0 & B & 1 & 0 & 0 & 0 \\ 0 & 0 & 0 & 0 & 1 & 0 & 0 \\ 0 & 0 & 0 & 0 & 0 & 1 & 0 \\ 0 & 0 & B^2 & 2B & 0 & 0 & 1 \end{pmatrix}.$$

The following mapping defines a generic element of the torus in $H(\mathbb{K})$, with corresponding matrix M_{torus} in $\mathbf{PG}(6, \mathbb{K})$:

$$\begin{cases} (a, l, a', l', a'') & \mapsto & (Aa, A^3 Kl, A^2 Ka', A^3 K^2 l', AKa''), \\ [k, b, k', b', k''] & \mapsto & [Kk, AKb, A^3 K^2 k', A^2 Kb', A^3 Kk''], \end{cases}$$

$$M_{\text{torus}} = \text{diag}(A^4 K^2, AK, A, A^2 K, 1, A^3 K, A^3 K^2).$$

Finally, the projective linear transformation with matrix

$$M = \begin{pmatrix} 0 & 0 & 0 & 0 & 0 & 1 & 0 \\ 0 & 0 & 0 & 0 & 0 & 0 & 1 \\ 0 & 0 & 0 & 0 & 1 & 0 & 0 \\ 0 & 0 & 0 & 1 & 0 & 0 & 0 \\ 0 & 1 & 0 & 0 & 0 & 0 & 0 \\ 0 & 0 & 1 & 0 & 0 & 0 & 0 \\ 1 & 0 & 0 & 0 & 0 & 0 & 0 \end{pmatrix}$$

represents a collineation of $\mathsf{H}(\mathbb{K})$ which rotates the hat-rack "through 60 degrees". Hence the matrices $M, M_{(B)}, M_{[L]}$ generate the little projective group of $\mathsf{H}(\mathbb{K})$ and define the standard embedding in $\mathbf{PG}(6, \mathbb{K})$. The representations of the corresponding collineations for mixed hexagons are easily obtained now by requiring characteristic 3 and suitably restricting the coefficients.

The twisted triality hexagons

The following mapping defines a generic $([\infty], (\infty), [0], (0,0), [0,0,0])$-elation in $\mathsf{T}(\mathbb{L}, \mathbb{K}, \sigma)$, with corresponding matrix $M_{[L]}$ in $\mathbf{PG}(7, \mathbb{L})$:

$$\begin{cases} (a, l, a', l', a'') & \mapsto & (a, l + L, a', l', a''), \\ [k, b, k', b', k''] & \mapsto & [k, b, k' - kL, b', k'' + L], \end{cases}$$

$$M_{[L]} = \begin{pmatrix} 1 & -L & 0 & 0 & 0 & 0 & 0 & 0 \\ 0 & 1 & 0 & 0 & 0 & 0 & 0 & 0 \\ 0 & 0 & 1 & 0 & 0 & 0 & 0 & 0 \\ 0 & 0 & 0 & 1 & 0 & 0 & 0 & 0 \\ 0 & 0 & 0 & 0 & 1 & 0 & 0 & 0 \\ 0 & 0 & 0 & 0 & L & 1 & 0 & 0 \\ 0 & 0 & 0 & 0 & 0 & 0 & 1 & 0 \\ 0 & 0 & 0 & 0 & 0 & 0 & 0 & 1 \end{pmatrix}.$$

The following mapping defines a generic $((\infty), [\infty], (0), [0,0], (0,0,0))$-elation in $\mathsf{T}(\mathbb{L}, \mathbb{K}, \sigma)$, with corresponding matrix $M_{(B)}$ in $\mathbf{PG}(6, \mathbb{L})$:

$$\begin{cases} (a, l, a', l', a'') & \mapsto & (a, l - \mathrm{Tr}(Ba^{\sigma + \sigma^2}), a' + B^\sigma a^{\sigma^2} + B^{\sigma^2} a^\sigma, \\ & & l' + \mathrm{Tr}(aB^{\sigma + \sigma^2}) + \mathrm{Tr}(a'B), a'' + B), \\ [k, b, k', b', k''] & \mapsto & [k, b + B, k', b', k''], \end{cases}$$

$$M_{(B)} = \begin{pmatrix} 1 & 0 & 0 & 0 & 0 & B & 0 & 0 \\ 0 & 1 & 0 & 0 & -B & 0 & 0 & 0 \\ 0 & 0 & 1 & 0 & 0 & 0 & 0 & 0 \\ 0 & 0 & -B^\sigma & 1 & 0 & 0 & 0 & 0 \\ 0 & 0 & 0 & 0 & 1 & 0 & 0 & 0 \\ 0 & 0 & 0 & 0 & 0 & 1 & 0 & 0 \\ 0 & 0 & B^{\sigma + \sigma^2} & -B^{\sigma^2} & 0 & 0 & 1 & B^\sigma \\ 0 & 0 & B^{\sigma^2} & 0 & 0 & 0 & 0 & 1 \end{pmatrix}.$$

The following mapping defines a generic element of the torus in $\mathsf{T}(\mathbb{L}, \mathbb{K}, \sigma)$, with corresponding matrix M_{torus} in $\mathbf{PG}(7, \mathbb{L})$:

$$\begin{cases} (a, l, a', l', a'') & \mapsto & (Aa, \mathrm{N}(A)Kl, A^{\sigma + \sigma^2}Ka', \mathrm{N}(A)K^2 l', AKa''), \\ [k, b, k', b', k''] & \mapsto & [Kk, AKb, \mathrm{N}(A)K^2 k', A^{\sigma + \sigma^2}Kb', \mathrm{N}(A)Kk''], \end{cases}$$

$$M_{\text{torus}} = \text{diag}(A\,\mathrm{N}(A)K^2, AK, A, AA^\sigma K, 1, \mathrm{N}(A)K, \mathrm{N}(A)K^2, AA^{\sigma^2}K).$$

Finally, the projective linear transformation with matrix

$$M = \begin{pmatrix} 0 & 0 & 0 & 0 & 0 & 1 & 0 & 0 \\ 0 & 0 & 0 & 0 & 0 & 0 & 1 & 0 \\ 0 & 0 & 0 & 0 & 1 & 0 & 0 & 0 \\ 0 & 0 & 0 & 0 & 0 & 0 & 0 & 1 \\ 0 & 1 & 0 & 0 & 0 & 0 & 0 & 0 \\ 0 & 0 & 1 & 0 & 0 & 0 & 0 & 0 \\ 1 & 0 & 0 & 0 & 0 & 0 & 0 & 0 \\ 0 & 0 & 0 & 1 & 0 & 0 & 0 & 0 \end{pmatrix}$$

represents a collineation of $\mathsf{T}(\mathbb{L}, \mathbb{K}, \sigma)$ which rotates the hat-rack "through 60 degrees". Hence the matrices M, $M_{(B)}$, $M_{[L]}$ generate the little projective group of $\mathsf{T}(\mathbb{L}, \mathbb{K}, \sigma)$ and define the standard embedding in $\mathbf{PG}(7, \mathbb{L})$.

Appendix E

The Ten Most Famous Open Problems

In this last appendix, I gather some famous and important open problems. Organizing them a little, exactly ten major problems are obtained. I comment on each of them. Of course such a list is subjective and reflects my personal interests and taste. Nevertheless, most of the problems are recognized as important by many other specialists in the field; see KANTOR [1986a] and THAS [1995].

Problem 1. *Construction of non-classical finite generalized hexagons or octagons.*

This problem has kept a few mathematicians busy! It led KANTOR [1986a] to state the conjecture that *no non-classical finite generalized hexagon or octagon exists.* It is always worth trying to solve this problem, and most of those who have tried have discovered something interesting, such as new quadrangles (KANTOR [1980]), a new construction of the Moufang hexagons (BADER & LUNARDON [1993] and LUNARDON [1993]) or an amalgamation procedure for generalized quadrangles and hexagons (VAN MALDEGHEM & BLOEMEN [1993]). A topological analogue of the problem seems also to be very interesting: the construction of a compact connected generalized hexagon other than the two classical ones, $H(\mathbb{R})$ and $H(\mathbb{C})$. Of course, proving that no such construction is possible is also a satisfactory answer to the question . . . .

In the same spirit, it is also an open question whether there exist non-classical finite generalized quadrangles of odd order s, or of order (q^2, q^3).

Problem 2. *Construction of a finite polygon with non-classical perp-plane or non-classical span-quadrangle.*

We stay with construction problems. Here, we ask for a finite generalized quadrangle or hexagon Γ such that there exists some projective point p or some polar point x (for Γ a hexagon) with the property that $\Gamma_p^\triangle$ or $\Gamma_x^\square$ is non-classical. Infinite examples exist; see Subsection 3.8.1 on page 133. Maybe these things do not exist in the finite case. Especially for the span-quadrangles of a finite hexagon, one is tempted to conjecture that they always have to be classical. But so far, no results in that direction are known. Loosely connected with this problem, but also with the previous one, is the question whether the amalgamation procedure using two classical quadrangles, where the corresponding maps are automorphisms of $\mathbf{GF}(q)$, always yields the classical hexagons. Here the conjecture is definitely *yes*, and a proof might not be too hard.

Problem 3. *Explicit construction of a generalized pentagon.*

The third and last problem concerned with constructions of polygons is related to the "non-classical" values for n, and primarily with $n = 5$. Every construction of generalized pentagons known to me uses a kind of free construction, or universal property, but no explicit example has been produced. One could say that a construction is explicit when the generalized pentagon (or any generalized n-gon with $n \neq 3, 4, 6, 8$) can be coordinatized explicitly (again using the word "explicit" without any sharp meaning).

Problem 4. *Which orders of finite generalized polygons are feasible?*

The feasibility of orders of generalized polygons has kept some mathematicians busy, and many of them have been successful; see Section 1.7. Any further restriction on the possible orders for finite polygons would be very interesting, and one question is particularly fascinating: is it always true that for the order (s, t) of a finite polygon $s - t$ is even? In other words, do s and t always have the same parity? Also, no analogue of the theorem of BRUCK & RYSER [1949] (see Subsection 1.7.7 on page 27) is known for generalized n-gons, $n > 3$. Any result in that direction would be wonderful.

Even some special cases seem attractive: does there exist a generalized octagon of order $(3, 6)$? Or what about a generalized quadrangle of order $(6, 6)$, or a generalized hexagon of order $(3, 12)$?

In the topological case, it is still an open question whether there exists a compact connected quadrangle or hexagon with topological parameters $(4, 4)$.

Problem 5. *Are there any semi-finite polygons?*

This is an old question posed by TITS (unpublished).

Recall that a **semi-finite** polygon is a generalized polygon of order (s, t) with s finite and t infinite. A semi-finite quadrangle with $s = 2, 3, 4$ does not exist; see Subsection 1.7.8 on page 27. Any other result in this spirit would be an important step towards the solution of the problem. Especially tempting seems to be the case of a hexagon with $s = 2$. But also non-classical values for n are open here, for example, does a semi-finite n-gon exist for n even and $n > 8$?

A group-theoretic counterpart of this problem is the question whether half Moufang implies Moufang for generalized $2n$-gons. Here, the conjecture is "yes", but what about a proof? Corolarry 6.4.9 on page 263 is a very partial contribution.

Problem 6. *Prove or disprove the uniqueness of the polygons with small parameters.*

This is the last problem related to orders. The question is simply to contribute to the classification of polygons with small order; see Subsection 1.7.11 on page 32 for an overview. Particularly interesting is the (possible) uniqueness of the Ree–Tits octagon $O(2)$ of order $(2, 4)$, and the (possible) uniqueness of the split Cayley hexagon $H(3)$ of order $(3, 3)$, especially since the latter is self-dual.

Problem 7. *Classification of finite flag-transitive polygons.*

For this problem, use of the classification of finite simple groups is allowed; see also Subsection 4.8.6 on page 167. The most promising case seems here to be $n = 6, 8$. I personally believe that the case $n = 6, 8$, together with the additional hypothesis of a primitive action of the flag-transitive group on the point set of the generalized n-gon, can be solved completely.

Related to this problem is the need for an elementary proof of the fact that every finite Tits polygon is a Moufang polygon. "Elementary" here means not using the classification of finite simple groups. As a first approximation, one could try to solve this question for square polygons.

Another interesting related question is whether there is an elementary proof of the fact that every finite half Moufang hexagon or octagon is a Moufang polygon; see Theorem 5.7.5 on page 230.

Problem 8. *Do there exist point-distance-2-regular $2m$-gons, with $m \neq 2, 3, 4$?*

This question is related to the main theme of Section 6.4 (see page 257). Since 2 divides $2m$, we are not able to conclude with the results of Section 6.4 that a point-distance-2-regular generalized $2m$-gon does not exist for $m > 4$. In fact, one might first look at generalized $2m$-gons all of whose points are projective. Since $m > 4$, this is a typical infinite case.

But there is a finite version of the problem. Indeed, all point-distance-2-regular hexagons are classified, but this is not true for finite quadrangles. Two first approximations arise: (1) classify all finite point-regular quadrangles, and (2) classify all generalized quadrangles which are both point-regular and line-regular. A recent contribution to (1) is the result of THAS & VAN MALDEGHEM [19**b] stating that for $s > 2$, no finite point-regular quadrangle of order $(s, s - 2)$ exists.

Problem 9. *Ovoids and spreads in the finite polygons.*

There does not exist any known example of an ovoid or a spread in any finite generalized octagon. The task is to find examples or to prove that in certain cases there do not exist examples. Here, one may also consider distance-j-ovoids, for $j = 2, 3$. The only result available is the non-existence of ovoids in $O(2)$; see Subsection 7.3.7 on page 314.

For the finite split Cayley hexagon $H(q)$, it is not known whether it admits an ovoid if the characteristic of $\mathbf{GF}(q)$ is odd and not equal to 3. Only partial results are available (see Subsection 7.3.8, page 315). The question of existence and construction of distance-2-ovoids in $H(q)$ or in its dual has only been settled in the case $s = 2$ (see Subsection 7.3.9 on page 316), apart from some immediate divisibility conditions that one can derive.

Finally, the question whether there exist spreads in the Hermitian quadrangle $H(4, q^2)$ can also be put here (and only the non-existence of such objects in $H(4, 4)$ is known; see Subsection 7.3.12 on page 317).

Problem 10. *Find an elementary geometric construction of the (perfect) Ree–Tits octagons.*

This book has avoided the introduction of the Ree groups in characteristic 2 to construct the Ree–Tits octagons by displaying a coordinatization of the latter. It would be very interesting to have a direct geometric construction of $O(\mathbb{K}, \sigma)$, for instance as an embedding in a projective space, such that one can actually prove that the geometry one has defined is a generalized octagon. Maybe some analogue of the trilinear forms exists to describe in a symmetric way the ambient metasymplectic spaces, and that allows one to describe in a direct way the polarities that give rise to the corresponding Ree–Tits octagons.

More directly, one would like a better understanding and an elementary description of the embedding of the Ree–Tits octagons in 25-dimensional projective space.

Related to this is a question on the classification of embeddings of generalized hexagons: which assumptions does one need to characterize the embeddings of the dual split Cayley hexagons and the dual triality hexagons?

Bibliography

Abramenko P.

[1996] *Twin Buildings and Applications to S-Arithmetic Groups*, Lecture Notes in Math. **1641**, Springer, Berlin Heidelberg. (351)

Abramenko P. and H. Van Maldeghem

[19**] Connectedness of opposite-flag geometries in Moufang polygons, *preprint* (1998). (33)

Ahrens R. W. and G. Szekeres

[1969] On a combinatorial generalization of 27 lines associated with a cubic surface, *J. Austral. Math. Soc.* **10**, 485–492. (123)

Albert A. A.

[1952] On nonassociative division algebras, *Trans. Amer. Math. Soc.* **72**, 296–309. (349)

[1958a] Finite noncommutative division algebras, *Proc. Amer. Math. Soc.* **9**, 928–932. (349)

[1958b] A construction of exceptional Jordan division algebras, *Ann. Math.* **67**, 1–28. (223)

[1960] Finite division algebras and finite planes, *Proc. Sympos. Appl. Math.* **10**, 53–70. (349)

[1965] On exceptional Jordan division algebras, *Pacific J. Math.* **15**, 277–404. (223)

Arens R.

[1946] Topologies for homeomorphic groups, *Amer. J. Math.* **68**, 593–610. (416)

Artin E.

[1957] *Geometric Algebra*, Interscience, New York. (61, 62)

Artmann B.

[1969] Hjelmslev-Ebenen mit verfeinerten Nachbarschaftsrelationen, *Math. Z.* **112**, 163–180. (422)

[1971] Existenz und projektive Limiten von Hjelmslev-Ebenen n-ter Stufe, *Atti del Convegno di Geometria Combinatoria e sue Applicazioni*. Proc. Perugia 1970, Univ. Perugia, 27–41. (422)

Assmus E. F. and J. D. Key

[1989] Arcs and ovals in the Hermitian and Ree unitals, *European J. Combin.* **10**, 297–308. (351)

Bader L. and G. Lunardon

[1993]   Generalized hexagons and BLT-sets, in *Finite Geometry and Combinatorics*, Proceedings Deinze 1992 (ed. F. De Clerck *et al.*), Cambridge University Press, Cambridge, *London Math. Soc. Lecture Note Ser.* **191**, 5–16. (87, 127, 132, 133, 475)

Bader L. and S. E. Payne

[19**]   On infinite K-clan geometry, *J. Geom.*, to appear. (130)

Baer R.

[1942]   Homogeneity of projective planes, *Amer. J. Math.* **64**, 137–152. (51, 271, 272, 276, 295)

Bagchi B. and N. S. N. Sastry

[1989]   Intersection pattern of the classical ovoids in symplectic 3-space of even order, *J. Algebra* **126**, 147–160. (340)

Barlotti A.

[1955]   Un'estensione del teorema di Segre-Kustaanheimo, *Boll. Un. Mat. Ital.* (3) **10**, 498–506. (340)

Blanchard A.

[1972]   *Les Corps non Commutatifs*, Presses Univ. de France, 1972. (312)

Bloemen I.

[1995]   *Substructures and Characterizations of Finite Generalized Polygons*, Doctoral thesis, Universiteit Gent, Belgium. (358)

Bloemen I., J. A. Thas and H. Van Maldeghem

[1996]   Elation generalized quadrangles of order (p, t), p prime, are classical, *J. Statist. Plann. Inference* **56**, 49–55. (288)

[19**]   Translation ovoids of generalized quadrangles and hexagons, *Geom. Dedicata*, to appear. (350, 351)

Bödi R.

[1996]   *Smooth Stable and Projective Planes*, Thesis (Habilitationsschrift), Universität Tübingen, Germany. (424)

Bödi R. and L. Kramer

[1995]   On homomorphisms between generalized polygons, *Geom. Dedicata*, **58**, 1–14. (136, 138, 424)

Borel A. and J. Tits

[1965]   Groupes réductifs, *Inst. Hautes Études Sci. Publ. Math.* **27**, 55–151. (230)

[1973]   Homomorphismes "abstraits" de groupes algébriques, *Ann. Math.* **97**(3) (1973), 499–571. (110)

Bose R. C. and S. S. Shrikhande

[1972]   Geometric and pseudo-geometric graphs $(q^2 + 1, q + 1, 1)$, *J. Geom.* **2**, 75–94. (25)

Breitsprecher S.

[1967] Einzigkeit der reellen und der komplexen projektiven Ebene, *Math. Z.* **99**, 429–432. (424)

[1971] Projektive Ebenen, die Mannigfaltigkeiten sind, *Math. Z.* **121**, 157–174. (424)

Brouns L., K. Tent and H. Van Maldeghem

[19**] Groups of projectivities of generalized quadrangles, *preprint* (1997). (18)

Brouns L. and H. Van Maldeghem

[19**] Characterizations for classical finite hexagons, to appear in *Finite Geometry and Combinatorics*, Proceedings Deinze 1997, *Bull. Belg. Math. Soc. Simon Stevin.* (282)

Brouwer A. E.

[1991] A non-degenerate generalized quadrangle with lines of size four is finite, in *Adv. Finite Geom. and Designs*, Proceedings Third Isle of Thorn Conference on Finite Geometries and Designs, Brighton 1990 (ed. J. W. P. Hirschfeld *et al.*), Oxford University Press, Oxford, 47–49. (27, 30)

[1993] The complement of a geometric hyperplane in a generalized polygon is usually connected, in *Finite Geometry and Combinatorics*, Proceedings Deinze 1992 (ed. F. De Clerck *et al.*), Cambridge University Press, Cambridge, *London Math. Soc. Lecture Note Ser.* **191**, 53–57. (32)

Brown K. S.

[1989] *Buildings*, Springer-Verlag, New York. (8)

Bruck R. H. and E. Kleinfeld

[1951] The structure of alternative division rings, *Proc. Amer. Math. Soc.* **2**, 878–890. (173, 211, 437)

Bruck R. H. and H. J. Ryser

[1949] The nonexistence of certain finite projective planes, *Canad. J. Math.* **1**, 88–93. (27, 484)

Bruhat F. and J. Tits

[1972] Groupes réductifs sur un corps local, I. Données radicielles valuées, *Inst. Hautes Études Sci. Publ. Math.* **41**, 5–252. (52, 54, 106)

Buchanan T.

[1979] Ovale und Kegelschnitte in der komplexen projektiven Ebene, *Math.-Phys. Semesterber.* **26**, 244–260. (411)

Buekenhout F.

[1979] Diagrams for geometries and groups, *J. Combin. Theory Ser. A* **27**, 121–151. (241)

[1995] *Handbook of Incidence Geometry, Buildings and Foundations,* North-Holland, Amsterdam. (2, 121)

Buekenhout F. and C. Lefèvre-Percsy

[1974] Generalized quadrangles in projective spaces, *Arch. Math.* **25**, 540–552. (361, 397)

Buekenhout F. and E. E. Shult

[1974]   On the foundations of polar geometry, *Geom. Dedicata* **3**, 155–170. (66, 242)

Buekenhout F. and H. Van Maldeghem

[1993]   Remarks on finite generalized hexagons and octagons with a point transitive automorphism group, in *Finite Geometry and Combinatorics*, Proceedings Deinze 1992 (ed. F. De Clerck *et al.*), Cambridge University Press, Cambridge, *London Math. Soc. Lecture Note Ser.* **191**, 89–102. (167)

[1994]   Finite distance transitive generalized polygons, *Geom. Dedicata* **52**, 41–51. (165, 234, 235)

Burns K. and R. Spatzier

[1987]   On topological Tits buildings and their classification, *Inst. Hautes Études Sci. Publ. Math.* **65**, 5–34. (408, 416, 418)

Cameron P. J.

[1975]   Partial quadrangles, *Quart. J. Math. Oxford Ser.* (2) **26**, 61–73. (25)

[1981]   Orbits of permutation groups on unordered sets, II., *J. London Math. Soc.* **23**, 49–264. (27, 28)

[1994]   Ovoids in infinite incidence structures, *Arch. Math.* **62**, 189–192. (308, 316)

Cameron P. J., J. A. Thas and S. E. Payne

[1976]   Polarities of generalized hexagons and perfect codes, *Geom. Dedicata* **5**, 525–528. (308, 316)

Cartan E.

[1938]   *Leçons sur la Théorie des Spineurs*, II, "Actu. Sci. Ind." **701**, Hermann, Paris. (70)

Carter R.

[1972]   *Simple Groups of Lie Type*, John Wiley & Sons, London, Ney York, Sydney, Toronto. (456)

Cherlin, G.

[19**]   Notes on locally finite generalized quadrangles, *preprint* (1996). (27)

Cohen A. M.

[1995]   Point-line geometries related to buildings, in *Handbook of Incidence Geometry, Buildings and Foundations*, (ed. F. Buekenhout), Chapter 9, North-Holland, Amsterdam, 647–737. (158, 405, 454)

Cohen A. M. and B. N. Cooperstein

[1992]   Generalized hexagons of even order, in *A Collection of Contributions in Honour of Jack van Lint*, *Discrete Math.* **106/107**, 139–146. (171)

Cohen A. M. and J. Tits

[1985]   On generalized hexagons and a near octagon whose lines have three points, *European J. Combin.* **6**, 13–27. (12, 32)

Conway J. H., R. T. Curtis, S. P. Norton, R. A. Parker and **R. A. Wilson**

[1985] *Atlas of Finite Groups*, Clarendon Press, Oxford. (165, 167, 349, 373)

Coxeter H. S. M.

[1987] *Projective Geometry*, Springer-Verlag, second edition (1987). (12)

De Clerck F. and **H. Van Maldeghem**

[1994] On infinite flocks of quadratic cones, *Bull. Belg. Math. Soc. Simon Stevin* **1**, 399–415. (126, 130)

[1995] Some classes of rank 2 geometries, in *Handbook of Incidence Geometry, Buildings and Foundations* (ed. F. Buekenhout), Chapter 10, North-Holland, Amsterdam, 433–475. (2)

Delandtsheer A.

[1995] Dimensional linear spaces, in *Handbook of Incidence Geometry, Buildings and Foundations* (ed. F. Buekenhout), Chapter 6, North-Holland, Amsterdam, 193–294. (347)

Delgado A. and **B. Stellmacher**

[1985] Weak (B, N)-pairs of rank 2, in *Groups and Graphs: New Results and Methods*, Birkhäuser Verlag, Basel, Boston, Stuttgart, 58–244. (226)

Dembowski P.

[1968] *Finite Geometries*, Ergeb. Math. Grenzgeb. **44**, Springer-Verlag, Berlin. (12, 120, 121, 163, 278, 349)

De Smet V.

[1994] *Substructures of Finite Classical Generalized Quadrangles and Hexagons*, Doctoral thesis, Universiteit Gent, Belgium. (340, 349, 358)

De Smet V. and **H. Van Maldeghem**

[1993a] Ovoids and windows in finite generalized hexagons, in *Finite Geometry and Combinatorics*, Proceedings Deinze 1992 (ed. F. De Clerck *et al.*), Cambridge University Press, Cambridge, *London Math. Soc. Lecture Note Ser.* **191**, 131–138. (278, 280)

[1993b] The finite Moufang hexagons coordinatized, *Beiträge Algebra Geom.* **34**, 217–232. (114, 351)

[1995] A geometric construction of a class of twisted field planes by Ree–Tits unitals, *Geom. Dedicata* **58**, 127–143. (349, 350)

[1996] Intersections of the Ree ovoids and the Hermitian ovoids in the generalized hexagon related to $G_2(3^{2h+1})$, *J. Combin. Des.* **4**, 71–81. (347, 349)

Dickson L. E.

[1901] *Linear Groups with an Exposition of the Galois Field Theory*, Teubner, Leipzig (reprint Dover, New York, 1958). (321)

Dienst K. J.

[1980a] Verallgemeinerte Vierecke in Pappusschen projektiven Räumen, *Geom. Dedicata* **9**, 199–206. (361, 392)

[1980b] Verallgemeinerte Vierecke in projektiven Räumen, *Arch. Math.* (Basel) **35**, 177–186. (361, 382, 385, 392)

Dixmier, S. and F. Zara

[19**] Étude d'un quadrangle généralisé autour de deux de ses points non liés, *preprint* (1976). (30)

Faulkner J. R.

[1977] Groups with Steinberg relations and coordinatization of polygonal geometries, *Mem. Am. Math. Soc.* (10) **185**. (173, 220, 222)

Feit W. and G. Higman

[1964] The nonexistence of certain generalized polygons, *J. Algebra* **1**, 114–131. (1, 12, 24, 25, 407, 427)

Ferus D., H. Karcher and H.-F. Münzner

[1981] Cliffordalgebren und neue isoparametrische Hyperflächen, *Math. Z.* **177**, 479–502. (410)

Fong P. and G. Seitz

[1973] Groups with a BN-pair of rank 2, I, *Invent. Math.* **21**, 1–57. (173, 229)

[1974] Groups with a BN-pair of rank 2, II, *Invent. Math.* **24**, 191–239. (173, 229)

Forst M.

[1981] Topologische 4-Gone, *Mitt. Math. Sem. Giessen* **147**, 65–129. (408)

Freudenthal H.

[1985] Oktaven, Ausnahmegruppen und Octavengeometrie, *Geom. Dedicata* **19**, 7–63. (451)

Funk M. and K. Strambach

[1991] On free constructions, *Manuscripta Math.* **71**, 335–374. (14)

[1995] Free constructions, in *Handbook of Incidence Geometry, Buildings and Foundations*, (ed. F. Buekenhout), Chapter 13, North-Holland, Amsterdam, 739–780. (14)

Govaert E.

[1997] A combinatorial characterization of some finite classical generalized hexagons, *J. Combin. Theory Ser.* **80**, 339–46. (254, 283)

Govaert E. and H. Van Maldeghem

[19**] Some combinatorial and geometric characterizations of the finite dual classical generalized hexagons, *preprint* (1997). (257, 283)

Grundhöfer T.

[1988] The groups of projectivities of finite projective and affine planes, *Ars Combin.* **25**, 269–275. (18)

Grundhöfer T., M. Joswig and M. Stroppel

[1994]   Slanted symplectic quadrangles, *Geom. Dedicata* **49**, 143 – 154. (16, 18, 123, 411)

Grundhöfer T. and N. Knarr

[1990]   Topology in generalized quadrangles, *Topology Appl.* **34**, 139–152. (408, 416)

Grundhöfer T., N. Knarr and L. Kramer

[1995]   Flag-homogeneous compact connected polygons, *Geom. Dedicata* **55**, 95–114. (412, 413, 416, 418)

[19**]   Flag-homogeneous compact connected polygons II, *preprint* (1996). (418)

Grundhöfer T. and H. Van Maldeghem

[1990]   Topological polygons and affine buildings of rank three, *Atti Sem. Mat. Fis. Univ. Modena* **38**, 459–474. (408, 409, 417, 422)

Haemers W. H.

[1979]   *Eigenvalue Techniques in Designs and Graph Theory*, Doctoral thesis, Technological University of Eindhoven, The Netherlands. (25)

Haemers W. H. and C. Roos

[1981]   An inequality for generalized hexagons, *Geom. Dedicata* **10**, 219–222. (24)

Hall M.

[1943]   Projective planes, *Trans. Am. Math. Soc.* **54**, 229–277. (13)

Hanssens G. and H. Van Maldeghem

[1987]   A new look at the classical generalized quadrangles, *Ars Combin.* **24**, 199–210. (99)

[1988]   Coordinatization of generalized quadrangles, *Ann. Discrete Math.* **37**, 195–208. (96)

[1989]   Algebraic properties of quadratic quaternary rings, *Geom. Dedicata* **30**, 43–67. (96)

Helgason S.

[1978]   *Differential Geometry, Lie Groups and Symmetric Spaces*, Academic Press, New York. (416)

Higman D. G.

[1975]   Invariant relations, coherent configurations and generalized polygons, in *Combinatorics* (ed. M. Hall and J. H. Van Lint), D. Reidel, Dordrecht, 347–363. (24)

Hilbert D.

[1899]   *Grundlagen der Geometrie*, Teubner, Leipzig. (51, 407)

Hölz G.

[1981]   Construction of designs which contain a unital, *Arch. Math.* **37**, 179–183. (351)

Hua L.-K.

[1949]   On the automorphisms of a sfield, *Proc. Nat. Acad. Sci. USA* **35**, 386–389. (383)

Hughes D. R.

[1960]   On homomorphisms of projective planes, *Proc. Sympos. Appl. Math.* **10**, 45–52. (136)

Hughes D. R. and F. C. Piper

[1973]   *Projective Planes*, Springer-Verlag, Berlin. (12, 51, 52, 87, 94, 134, 289, 311, 319, 331)

Iwasawa K.

[1951]   Über die Einfachheit der speziellen projektiven Gruppen, *Proc. Imp. Acad. Tokyo* **17**, 57–59. (327)

Jacobson N.

[1968]   *Structure and Representation of Jordan Algebras*, Amer. Math. Society, Providence RI. (223)

[1985]   *Basic Algebra* I, Freeman, New York (second edition). (435)

Jäger M.

[1994]   *Topologische Gebäude*, Doctoral thesis, Universität Kiel, Germany. (408, 412)

Johnson P.

[19**]   Semi-quadratic sets and embedded polar spaces, *J. Geom.*, to appear. (392, 393)

Joswig M.

[1994]   *Translationsvierecke*,   Doctoral thesis, Eberhard-Karls-Universität Tübingen, Germany. (171)

[1995]   Generalized polygons with highly transitive collineation groups, *Geom. Dedicata* **58**, 91–100. (418)

Joswig M. and H. Van Maldeghem

[1995]   An essay on the Ree octagons, *J. Algebraic Combin.* **4**, 145–164. (117, 132, 238, 362, 364, 378)

Kantor W. M.

[1980]   Generalized quadrangles associated with $G_2(q)$, *J. Combin. Theory Ser. A* **29**, 212–219. (87, 124, 129, 130, 169, 170, 475)

[1983]   Primitive groups of odd degree and an application to finite projective planes, *J. Algebra* **106**, 14–45. (167)

[1986a] Generalized polygons, SCABs and GABs, in: *Buildings and the Geometry of Diagrams*, Proceedings Como 1984 (ed. L. Rosati), *Lecture Notes in Math.* **1181**, Springer-Verlag, Berlin, 79–158. (1, 49, 50, 73, 114, 475)

[1986b] Some generalized quadrangles with parameters (q^2, q), *Math. Z.* **192**, 45–50. (130)

[1991]   Automorphism groups of some generalized quadrangles, in *Adv. Finite Geom. and Designs*, Proceedings Third Isle of Thorn Conference on Finite Geometries and Designs, Brighton 1990 (ed. J. W. P. Hirschfeld *et al.*), Oxford University Press, Oxford, 251–256. (164, 229)

Kegel O. and A. Schleiermacher

[1973]   Amalgams and embeddings of projective planes, *Geom. Dedicata* **2**, 379–395. (161, 295)

Kiss G.

[1990] A generalization of Ostrom's theorem in generalized n-gons, *Simon Stevin* **64**, 309–317. (277)

Kleidman P. B.

[1988] The maximal subgroups of the Chevalley groups $G_2(q)$ with q odd, of the Ree groups $^2G_2(q)$, and of their automorphism groups, *J. Algebra* **117**, 30–71. (349)

Kleinfeld E.

[1951] Alternative division rings of characteristic 2, *Proc. Nat. Acad. Sci. USA* **37**, 818–820. (173, 211, 437)

Knarr N.

[1988] Projectivities of generalized polygons, *Ars Combin.* **25B**, 265–275. (17, 18, 361, 375)

[1990] The nonexistence of certain topological polygons, *Forum Math.* **2**, 603–612. (407, 408, 413)

[1992] A geometric construction of generalized quadrangles from polar spaces of rank three, *Results Math.* **21** (1992), 332–344. (127, 132)

Kolmogoroff A.

[1932] Zur Begründung der projektiven Geometrie, *Ann. Math.* **33**, 175–176. (407)

Kramer L.

[1994a] *Compact Polygons*, Doctoral thesis, Universität Tübingen, Germany. (407, 408, 409, 412, 413, 419)

[1994b] The topology of smooth projective planes, *Arch. Math.* **63**, 85–91. (424)

[1996] Holomorphic polygons, *Math. Z.* **223**, 333–341. (424)

[19**] *Compact Polygons and Isoparametric Hypersurfaces*, Birkhäuser, *in preparation.* (407, 408, 414, 416)

Kramer L. and K. Tent

[1996] Algebraic polygons, *J. Algebra* **182**, 435–447. (425)

Kramer L., K. Tent and H. Van Maldeghem

[19**] Simple groups of Finite Morley rank and Tits-buildings, *preprint.* (1996) (425)

Kramer L. and H. Van Maldeghem

[19**] Closed ovoids in compact quadrangles, *in preparation.* (317, 415)

Kühne R. and R. Löwen

[1992] Topological projective spaces, *Abh. Math. Sem. Univ. Hamburg* **62**, 1–9. (409)

Lam C. W. H.

[1991] The search for a finite projective plane of order 10, *Amer. Math. Monthly* **98**, 305–318. (27)

Lam C. W. H., G. Kolesova and L. Thiel

[1991] A computer search for finite projective planes of order 9, *Discr. Math.* **92**, 187–195. (32)

Lam C. W. H., L. Thiel and **S. Swiercz**

[1989]   The nonexistence of finite projective planes of order 10, *Canad. J. Math.* **41**, 1117–1123. (27)

Lam T. Y.

[1973]   *The Algebraic Theory of Quadratic Forms*, Academic Press, London (1978). (63)

Lefèvre-Percsy C.

[1981a] Quadrilatères généralisés faiblement plongés dans **PG**(3, q), *European J. Combin.* **2**, 249–255. (399)

[1981b] Projectivités conservant un espace polaire faiblement plongé, *Acad. Roy. Belg. Bull. Cl. Sci.* (5) **67**, 45–50. (383)

Löwe S.

[1992]   *Über die Konstruktion von verallgemeinerten Vierecken mit einem regulären Punkt*, Habilitationsschrift, Braunschweig. (132)

Löwen R.

[1983]   Topology and dimension of stable planes: On a conjecture of H. Freudenthal, *J. Reine Angew. Math.* **343**, 108–122. (413)

Lunardon G.

[1993]   Partial ovoids and generalized hexagons, in *Finite Geometry and Combinatorics*, Proceedings Deinze 1992 (ed. F. De Clerck *et al.*), Cambridge University Press, Cambridge, *London Math. Soc. Lecture Note Ser.* **191**, 233–248. (133, 475)

Lüneburg H.

[1966]   Some remarks concerning the Ree groups of type (G_2), *J. Algebra* **3**, 256–259. (306, 348)

Lunelli L. and **M. Sce**

[1958]   *k-Archi completi nei piani proiettivi desarguesiani di rango 8 e 16*, Technical Report, Centro di Calcoli Numerici, Politecnico di Milano. (310, 124)

McCrimmon K.

[1970]   The Freudenthal-Springer-Tits constructions revisited, *Trans. Amer. Math. Soc.* **148**, 192–314. (223)

Mendelsohn N. S.

[1944]   (Problem in problem section) *Amer. Math. Monthly* **51**, 171. (384)

Mortimer B.

[1975]   A geometric proof of a theorem of Hughes on homomorphisms of projective planes, *Bull. London Math. Soc.* **7**, 267–268. (136)

Moufang R.

[1933]   Alternativkörper und der Satz vom vollständigen Vierseit, *Abh. Math. Sem. Univ. Hamburg* **9**, 207–222. (173, 211)

Mühlherr B.

[1994] *Some Contributions to the Theory of Buildings Based on the Gate Property*, Doctoral thesis, Universität Tübingen, Germany. (459)

Mühlherr B. and H. Van Maldeghem

[19**] Exceptional Moufang quadrangles of type F_4, *preprint* (1997). (256, 466)

Münzner H.-F.

[1980] Isoparametrische Hyperflächen in Sphären, I., *Math. Ann.* **251**, 57–71. (413)

[1981] Isoparametrische Hyperflächen in Sphären, II., *Math. Ann.* **256**, 215–232. (413)

O'Keefe C. M. and J. A. Thas

[1995] Ovoids of the quadric $Q(2n, q)$, *European J. Combin.* **16**, 87–92. (315)

O'Meara O. T.

[1971] *Introduction to Quadratic Forms*, Springer-Verlag, Berlin (1971). (61, 61, 63)

Onishchik A. L. and E. B. Vinberg

[1990] *Lie Groups and Algebraic groups*, Springer-Verlag, Berlin. (416)

Ostrom T. G.

[1957] Transitivities in projective planes, *Canad. J. Math.* **9**, 389–399. (277)

Ostrom T. G. and A. Wagner

[1959] On projective and affine planes with transitive collineation groups, *Math. Z.* **71**, 186–199. (165, 165, 289)

Ott U.

[1981] Eine Bemerkung über Polaritäten eines verallgemeinerten Hexagons, *Geom. Dedicata* **11**, 341–345. (309)

[1991] Generalized n-gons in trees, in *Combinatorics '88*, Proceedings Ravello, 1988 (ed. A. Barlotti *et al.*,), *Res. Lecture Notes Math.*, 73–85. (226)

Otte J.

[1993] *Differenzierbare Ebenen*, Doctoral thesis, Universität Kiel, Germany. (424)

[1995] Smooth projective translation planes, *Geom. Dedicata* **58**, 203–212. (424)

Pasini A.

[1983] The nonexistence of proper epimorphisms of finite thick generalized polygons, *Geom. Dedicata* **15**, 389 – 397. (135, 136, 138)

Payne S. E.

[1968] Symmetric respresentations of nondegenerate generalized n-gons, *Proc. Amer. Math. Soc.* **19**, 1321–1326. (308, 309)

[1971] Nonisomorphic generalized quadrangles, *J. Algebra* **18**, 201–212. (123)

[1972] Generalized quadrangles as amalgamations of projective planes, *J. Algebra* **31**, 120–136. (353, 355)

[1973] A restriction on the parameters of a subquadrangle, *Bull. Amer. Math. Soc.* **79**, 747–748. (38)

[1977] Generalized quadrangles with symmetry II, *Simon Stevin* **50**, 209–245. (354)

Payne S. E. and **J. A. Thas**

[1984]   *Finite Generalized Quadrangles*, Pitman Res. Notes Math. Ser. **110**, Longman, London, Boston, Melbourne. (11, 15, 16, 25, 32, 39, 49, 50, 121, 124, 136, 171, 229, 230, 240, 278, 278, 279, 281, 283, 289, 308, 318, 351, 355, 361)

Pickert G.

[1975]   *Projektive Ebenen*, Springer-Verlag, Berlin, second edition. (278, 312, 437, 439)

Qvist B.

[1952]   Some remarks concerning curves of the second degree in a finite plane, *Ann. Acad. Sci. Fenn. Ser.* A. I. **134**, 1–27. (331)

Ree R.

[1961]   A family of simple groups associated with the Lie algebra of type (G_2), *Amer. J. Math.* **83**, 797–462 (343, 344)

Ronan M. A.

[1980a]   A geometric characterization of Moufang hexagons, *Invent. Math.* **57**, 227–262. (46, 75, 206, 220, 221, 241, 243, 251)

[1980b]   A note on the $^3D_4(q)$ generalised hexagons, *J. Combin. Theory Ser.* A **29**, 249–250. (46, 256)

[1980c]   Semiregular graph automorphisms and generalised quadrangles, *J. Combin. Theory Ser.* A **29**, 319–328. (277)

[1981]   A combinatorial characterization of the dual Moufang hexagons, *Geom. Dedicata* **11**, 61–67. (254)

[1989]   *Lectures on Buildings*, Academic Press, San Diego, *Persp. Math.* **7**. (8, 17, 81, 162, 222, 466)

Ronan M. A. and **J. Tits**

[1987]   Building buildings, *Math. Ann.* **278**, 291–306. (81)

[1994]   Twin trees I, *Invent. Math.* **116**, 463–479. (11)

Salzmann H.

[1975]   Homogene kompakte projektive Ebenen, *Pacific J. Math.* **60**, 217–234. (419)

Salzmann H., D. Betten, T. Grundhöfer, H. Hähl, R. Löwen and **M. Stroppel**

[1995]   *Compact Projective Planes*, de Gruyter, Berlin. (407, 408, 409, 411, 419, 424)

Sarli J.

[1986]   Embedding the root group geometry of $^2F_4(q)$, **in** *Groups*, Proocedings St. Andrews 1985 (ed. E. F. Robertson *et al.*), *London Math. Soc. Lecture Note Ser.* **121**, Cambridge University Press, Cambridge, 300–304. (78, 164)

[1988]   The geometry of root subgroups in Ree groups of type $^2F_4$, *Geom. Dedicata* **26**, 1–28. (164)

Scharlau W.

[1985]   *Quadratic and Hermitian Forms*, Springer-Verlag, Berlin. (61, 63)

Schellekens G. L.

[1962a] On a hexagonic structure, I., *Indag. Math.* **24**, 201–217. (72)

[1962b] On a hexagonic structure, II., *Indag. Math.* **24**, 218–234. (72)

Schroth A. E.

[1992] Characterizing symplectic quadrangles by their derivations, *Arch. Math.* **58**, 98–104. (240, 412, 414)

[1995] *Topological Circle Planes and Topological Quadangles*, Longman/Wiley, New York. (410, 411)

Schroth A. E. and H. Van Maldeghem

[1994] Half-regular and regular points in compact polygons, *Geom. Dedicata* **51**, 215–233. (412, 414, 415)

Segre B.

[1955] Ovals in a finite projective plane, *Canad. J. Math.* **7**, 414–416. (321)

[1959] On complete caps and ovaloids in three-dimensional Galois spaces of characteristic two, *Acta Arith.* **5**, 315–332. (340)

Seitz G. M.

[1973] Flag-transitive subgroups of Chevalley groups, *Ann. Math.* **97**, 27–56. (167)

Skornjakov L. A.

[1957] On homomorphisms of projective planes and T-homomorphisms of ternary rings, *Math. Sbornik* **43**, 285–294. (136, 138)

Springer T. A.

[1963] Octaven, Jordan-Algebren und Ausnahmegruppen, Lecture notes of a seminar in the "Mathematisches Institut der Universität Göttingen", Summer 1963. (448)

Steinbach A.

[1996] Classical polar spaces (sub-)weakly embedded in projective space, *Bull. Belg. Math. Soc. Simon Stevin* **3**, 477–490. (361, 403)

Steinbach A. and H. Van Maldeghem

[19**a] Generalized quadrangles weakly embedded of degree > 2 in projective space, *preprint* (1996). (361, 403)

[19**b] Generalized quadrangles weakly embedded of degree 2 in projective space, *in preparation*. (403)

Stevenson F. W.

[1972] *Projective planes*, W. H. Freeman and Company, San Francisco (1972). (32, 278, 295)

Strambach K.

[1975] Algebraische Geometrien, *Rend. Sem. Mat. Univ. Padova* **53**, 165–210. (424)

Strauss T.

[1996] *Cohomology Rings, Sphere Bundles, and Doubly Mapping Cylinders*, Master's thesis (Diplomarbeit), Universität Würzburg, Germany. (413, 414)

Stroppel B. and M. Stroppel

[1996]   The automorphism group of a compact generalized quadrangle has finite dimension, *Arch. Math.* **66**, 77–79. (419)

[19**]   Connected orbits in topological generalized quadrangles, *Results Math.*, to appear. (419)

Suzuki M.

[1960]   A new type of simple groups of finite order, *Proc. Mat. Acad. Sci. U. S. A.* **46**, 868–870. (325, 327)

[1962]   On a class of doubly transitive groups, *Ann. Math.* **75**, 105–145. (341)

Szenthe J.

[1974]   On the topological characterization of transitive Lie group actions, *Acta Sci. Math.* (Szeged) **36**, 323–344. (416)

Taylor D. E.

[1992]   *The Geometry of the Classical Groups*, Heldermann, Berlin. (162)

Thas J. A.

[1972a]   Ovoidal translation planes, *Arch. Math.* **23**, 110–112. (340)

[1972b]   4-gonal subconfigurations of a given 4-gonal configuration, *Rend. Accad. Naz. Lincei* **53**, 520–530. (35)

[1973]   On 4-gonal configurations, *Geom. Dedicata* **2**, 317–326. (280)

[1974]   A remark concerning the restriction on the parameters of a 4-gonal configuration, *Simon Stevin* **48**, 65–68. (35, 38)

[1976]   A restriction on the parameters of a subhexagon, *J. Combin. Theory Ser.* A **21**, 115–117. (35, 38)

[1978]   Combinatorial characterizations of generalized quadrangles with parameters $s = q$ and $t = q^2$, *Geom. Dedicata* **7**, 223–232. (278)

[1979]   A restriction on the parameters of a suboctagon, *J. Combin. Theory Ser.* A **27**, 385–387. (35, 38)

[1980a]   A remark on the theorem of Yanushka and Ronan characterizing the generalized hexagon $H(q)$ arising from the group $G_2(q)$, *J. Combin. Theory Ser.* A **29**, 361–362. (280)

[1980b]   Polar spaces, generalized hexagons and perfect codes, *J. Combin. Theory Ser.* A **29**, 87–93. (315)

[1981]   Ovoids and spreads of finite classical polar spaces, *Geom. Dedicata* **10**, 135–144. (315, 350)

[1985]   Characterization of generalized quadrangles by generalized homologies, *J. Combin. Theory Ser.* A **40**, 331–341. (295)

[1986]   The classification of all (x, y)-transitive generalized quadrangles, *J. Combin. Theory Ser.* A **42**, 154–157. (295)

[1987]   Generalized quadrangles and flocks of cones, *European J. Combin.* **8**, 441–452. (124, 127, 130)

[1995]   Generalized polygons, in *Handbook of Incidence Geometry, Buildings and Foundations*, (ed. F. Buekenhout), Chapter 9, North-Holland, Amsterdam, 383–431. (72, 127, 130, 132, 136, 475)

Thas J. A. and S. E. Payne

[1994]   Spreads and ovoids in finite generalized quadrangles, *Geom. Dedicata* **52**, 227–253. (350)

Thas J. A., S. E. Payne and H. Van Maldeghem

[1991]   Half Moufang implies Moufang for finite generalized quadrangles, *Invent. Math.* **105**, 153–156. (230)

Thas J. A. and H. Van Maldeghem

[1990]   Generalized Desargues configurations in generalized quadrangles, *Bull. Soc. Math. Belg.* **42**, 713–722. (272, 276)

[1995]   The classification of all finite generalized quadrangles admitting a group acting transitively on ordered pentagons, *J. London Math. Soc.* (2) **51**, 209–218. (288)

[1996]   Embedded thick finite generalized hexagons in projective space, *Proc. London Math. Soc.* (2) **54**, 566–580. (361, 398)

[19**a] Generalized quadrangles weakly embedded in finite projective space, *J. Statist. Plann. Inference*, to appear. (361, 399)

[19**b] Lax embeddings of finite generalized quadrangles, *preprint* (1997). (361, 400, 403, 477)

[19**c] Flat lax and weak lax embeddings of finite generalized hexagons, *preprint* (1997). (283, 361, 400, 403)

Thorbergsson G.

[1992]   Clifford algebras and polar planes, *Duke Math. J.* **67**, 627–632. (410)

Tits J.

[1954]   Espaces homogènes et groupes de Lie exceptionels, *Proc. Internat. Math. Congr.*, Amsterdam, Sept. 1954, vol. I, 495–496. (49)

[1955]   Sur certaines classes d'espaces homogènes de groupes de Lie, *Mém. Acad. Roy. Belg.* **29** (3), 268 pp. (49)

[1959]   Sur la trialité et certains groupes qui s'en déduisent, *Inst. Hautes Études Sci. Publ. Math.* **2**, 13–60. (1, 6, 12, 49, 67, 70, 71, 72, 74, 76)

[1960]   Les groupes simples de Suzuki et de Ree, *Séminaire Bourbaki* **13**(210), 1–18. (49, 78, 164, 327, 340, 342, 344, 348, 348)

[1961]   Géométrie polyédriques et groupes simples, *Deuxième réunion du Groupement de mathématiciens d'expression latine*, Florence, Sept. 1961, 66–88. (159)

[1962a] Ovoïdes et groupes de Suzuki, *Arch. Math.* **13**, 187–198. (121, 299, 323, 324, 325, 327, 329, 330, 343)

[1962b] Théorème de Bruhat et sous-groupes paraboliques, *C. R. Acad. Sci. Paris Ser. I Math.* **254**, 2910–2912. (85, 159)

[1962c] Ovoïdes à translations, *Rendiconti di Matematica* **21**, 37–59. (120)

[1962d] Groupes semi-simple isotropes, *Colloque sur la Théorie des Groupes Algébriques*, C. B. R. M., Bruxelles, June 1962, 137–147. (186)

[1964]   Algebraic and abstract simple groups, *Ann. Math.* **80**, 313–329. (85, 233)

[1966a] Une propriété caractéristique des ovoïdes associés aux groupes de Suzuki, *Arch. Math.* **17**, 136–153. (340)

[1966b] Classification of algebraic semisimple groups, in *Algebraic Groups and Discontinuous Groups*, Boulder 1965, *Proc. Symp. Pure Math.* **9**, 33–62. (215, 216, 226, 453, 457)

[1974] *Buildings of Spherical Type and Finite BN-pairs*, Lecture Notes in Math. **386**, Springer-Verlag, Berlin, Heidelberg, New York. (1, 4, 9, 10, 17, 49, 52, 80, 81, 81, 85, 113, 138, 152, 163, 163, 164, 241, 243)

[1976a] Classification of buildings of spherical type and Moufang polygons: a survey, in *Coll. Intern. Teorie Combin. Acc. Naz. Lincei,* Proceedings Roma 1973, *Atti dei convegni Lincei* **17**, 229–246. (22, 49, 81, 144, 147, 147, 173, 222, 224)

[1976b] Non-existence de certains polygones généralisés, I, *Invent. Math.* **36**, 275–284. (173, 185, 239)

[1977] Endliche Spiegelungsgruppen, die als Weylgruppen auftreten, *Invent. Math.* **43**, 283–295. (1, 11, 13, 161)

[1979] Non-existence de certains polygones généralisés, II, *Invent. Math.* **51**, 267–269. (173, 185, 239)

[1981] A local approach to buildings, in *The Geometric Vein. The Coxeter Festschrift* (ed. D. Chandler *et al.*), Springer-Verlag, 317–322. (80, 241)

[1983] Moufang octagons and the Ree groups of type $^2F_4$, *Amer. J. Math.* **105**, 539–594. (78, 117, 117, 119, 148, 173, 191, 204, 225, 225, 417)

[1994a] Moufang polygons, I. Root data, *Bull. Belg. Math. Soc. Simon Stevin* **1**, 455–468. (173, 186, 191, 209)

[1994b] Résumé de cours, *Annuaire du Collège de France*, **94**, 101–114. (174, 179)

[1995] Résumé de cours, *Annuaire du Collège de France*, **95**, 79–95. (52, 174, 212, 453)

[19**] Quadrangles de Moufang, I, *preprint* (1976). (101, 102, 103, 146, 173, 207, 217, 223)

Tits J. and **R. Weiss**

[19**] The classification of Moufang polygons, *in preparation.* (173, 210, 210, 214, 218, 225)

Tutte W. T.

[1966] *Connectivity in Graphs*, University of Toronto Press, Toronto. (164)

van Bon J., H. Cuypers and **H. Van Maldeghem**

[1994] Hyperbolic lines in generalized polygons, *Forum Math.* **8**, 343–362. (263, 265, 266, 266, 270)

Van Maldeghem H.

[1987a] Non-classical triangle buildings, *Geom. Dedicata* **24**, 123–206. (423)

[1987b] On locally finite alternative division rings with valuation, *J. Geom.* **30**, 42–48. (423)

[1988a] Valuations on PTRs induced by triangle buildings, *Geom. Dedicata* **26**, 29–84. (422, 423)

[1988b] *Niet-Klassieke $\tilde{C}_2$-Gebouwen*, Thesis (Habilitationsschrift), Universiteit Gent, Belgium. (423)

[1989a] Quadratic quaternary rings with valuation and affine buildings of type $\tilde{C}_2$, *Mitt. Math. Sem. Giessen* **189**, 1–159. (130, 134, 423)

[1989b] Generalized polygons with valuation, *Arch. Math.* **53**, 513–520. (419, 420, 423)

[1990a] A note on epimorphisms of projective planes, generalized quadrangles and generalized hexagons, *Soochow J. Math.* **16**, 7–10. (139)

[1990b] Generalized quadrangles with valuation, *Geom. Dedicata* **35**, 77–87. (422, 423)

[1990c] An algebraic characterization of affine buildings of type $\tilde{C}_2$, *Mitt. Mathem. Sem. Giessen* **198**, 1–42. (423)

[1990d] A configurational characterization of the Moufang generalized polygons, *European J. Combin.* **11**, 381–392. (272, 276)

[1991a] Common characterizations of the finite Moufang polygons, in *Adv. Finite Geom. and Designs*, Proceedings Third Isle of Thorn Conference on Finite Geometries and Designs, Brighton 1990 (ed. J. W. P. Hirschfeld *et al.*), Oxford University Press, Oxford, 391–400. (295)

[1991b] A characterization of the finite Moufang hexagons by generalized homologies, *Pacific J.Math.* **151**, 357–367. (295)

[1995a] The geometry of traces in Ree octagons, in *Groups of Lie Type and their Geometries*, Proceedings Como 1993 (ed. W. M. Kantor & L. Di Martino), Cambridge University Press, Cambridge, *London Math. Soc. Lect. Notes Ser.* **207**, 157–171. (302, 332)

[1995b] The nonexistence of certain regular generalized polygons, *Arch. Math.* **64**, 86–96 (257, 259, 261, 263)

[1996] A finite generalized hexagon admitting a group acting transitively on ordered heptagons is classical, *J. Combin. Theory Ser. A* **75**, 254–269. (288)

[1997] A geometric characterization of the perfect Suzuki–Tits ovoids, *J. Geom.* **58**, 192–202. (332, 332, 333, 340)

[19**a] Some consequences of a result of Brouwer, *Ars Combin.*, to appear. (286)

[1998] A geometric characterization of the perfect Ree–Tits octagons, *Proc. London Math. Soc.* (3) **76**, no. 1, 203–256. (85, 164, 298, 300, 301)

[19**b] Ovoids and spreads arising from involutions, Proceedings of the conference *Groups and Geometries*, Siena, September 1996, to appear. (315)

[19**c] A remark on polarities of generalized hexagons and partial quadrangles, *J. Combin. Theory Ser. A*, to appear. (309)

Van Maldeghem H. and I. Bloemen

[1993] Generalized hexagons as amalgamations of generalized quadrangles, *European J. Combin.* **14**, 593–604. (44, 97, 352, 353, 354, 357, 358, 475)

Van Maldeghem H., J. A. Thas and S. E. Payne

[1992] Desarguesian finite generalized quadrangles are classical or dual classical, *Des. Codes Cryptogr.* **1**, 299–305. (272, 284, 288)

Van Maldeghem H. and K. Van Steen

[19**] Characterization by automorphism groups of some rank 3 buildings, II. A half strongly-transitive locally finite triangle building is a Bruhat-Tits building, *Geom. Dedicata*, to appear. (423)

Van Maldeghem H. and R. Weiss

[1992] On finite Moufang polygons, *Israel J. Math.* **79**, 321–330. (284)

Veblen O. and **J. W. Young**

[1910]   *Projective Geometry*, Ginn, Boston. (241)

Walker M.

[1982]   On central root automorphisms of finite generalized hexagons, *J. Algebra* **78**, 303–340. (206, 298)

[1983]   On central root automorphisms of finite generalized octagons, *European J. Combin.* **4**, 65–86. (206, 298)

Wang Q.-M.

[1988]   On the topology of Clifford isoparametric hypersurfaces, *J. Diff. Geom.* **27**, 55–66. (410)

Wassermann J.

[1993]   Freie Konstruktionen von verallgemeinerten n-Ecken, Master's thesis (Diplomarbeit), Universität Tübingen. (295)

Weil A.

[1960]   *J. Indian Math. Soc.* **24**, 589–623. (463)

Weiss R.

[1979]   The nonexistence of certain Moufang polygons, *Invent. Math.* **51**, 261–266. (173, 179, 185, 239)

[1995a] Moufang trees and generalized triangles, *Osaka J. Math.* **32**, 987–1000. (178, 226, 228)

[1995b] Moufang trees and generalized hexagons, *Duke Math. J.* **79**, 219 – 233. (178, 228)

[1996]   Moufang trees and generalized quadrangles, *Forum Math.* **8**, 485–500. (178, 228)

[1997]   Moufang trees and generalized octagons, *Duke Math. J.* **88**, 449–464. (178, 228)

Yanushka A.

[1976]   Generalized hexagons of order (t, t), *Israel J. Math.* **23**, 309–324. (165)

[1981]   On order in generalized polygons, *Geom. Dedicata* **10**, 451–458. (22)

Unpublished:

   Brouwer A. E. (318)
   Cameron P. J. (30)
   Coolsaet K. (12, 25, 314, 404)
   Cuypers H. (33)
   Kantor W. M. (27, 30, 33, 131, 165, 171)
   Mühlherr B. (164)
   Schroth A. E. (419)
   Steinbach A. (403)
   Thorbergsson G. (424)
   Tits J. (27, 79, 173, 209, 211, 216, 223, 321, 351, 437, 476)
   Van Maldeghem H. (165, 167, 240, 256, 263, 283, 423)

Index